Smart Innovation, Systems and Technologies

Volume 32

About this Series

The Smart Innovation, Systems and Technologies book series encompasses the topics of knowledge, intelligence, innovation and sustainability. The aim of the series is to make available a platform for the publication of books on all aspects of single and multi-disciplinary research on these themes in order to make the latest results available in a readily-accessible form. Volumes on interdisciplinary research combining two or more of these areas is particularly sought.

The series covers systems and paradigms that employ knowledge and intelligence in a broad sense. Its scope is systems having embedded knowledge and intelligence, which may be applied to the solution of world problems in industry, the environment and the community. It also focusses on the knowledge-transfer methodologies and innovation strategies employed to make this happen effectively. The combination of intelligent systems tools and a broad range of applications introduces a need for a synergy of disciplines from science, technology, business and the humanities. The series will include conference proceedings, edited collections, monographs, handbooks, reference books, and other relevant types of book in areas of science and technology where smart systems and technologies can offer innovative solutions.

High quality content is an essential feature for all book proposals accepted for the series. It is expected that editors of all accepted volumes will ensure that contributions are subjected to an appropriate level of reviewing process and adhere to KES quality principles.

More information about this series at http://www.springer.com/series/8767

Lakhmi C. Jain · Himansu Sekhar Behera
Jyotsna Kumar Mandal
Durga Prasad Mohapatra
Editors

Computational Intelligence in Data Mining - Volume 2

Proceedings of the International Conference on CIDM, 20-21 December 2014

Springer

Editors
Lakhmi C. Jain
University of Canberra
Canberra
Australia

and

University of South Australia
Adelaide, SA
Australia

Himansu Sekhar Behera
Department of Computer Science
and Engineering
Veer Surendra Sai University
of Technology
Sambalpur, Odisha
India

Jyotsna Kumar Mandal
Department of Computer Science
and Engineering
Kalyani University
Nadia, West Bengal
India

Durga Prasad Mohapatra
Department of Computer Science
and Engineering
National Institute of Technology Rourkela
Rourkela
India

ISSN 2190-3018 ISSN 2190-3026 (electronic)
Smart Innovation, Systems and Technologies
ISBN 978-81-322-3561-3 ISBN 978-81-322-2208-8 (eBook)
DOI 10.1007/978-81-322-2208-8

Springer New Delhi Heidelberg New York Dordrecht London

Softcover reprint of the hardcover 1st edition 2015

Printed on acid-free paper

Springer (India) Pvt. Ltd. is part of Springer Science+Business Media (www.springer.com)

Preface

The First International Conference on "Computational Intelligence in Data Mining (ICCIDM-2014)" was hosted and organized jointly by the Department of Computer Science and Engineering, Information Technology and MCA, Veer Surendra Sai University of Technology, Burla, Sambalpur, Odisha, India between 20 and 21 December 2014. ICCIDM is an international interdisciplinary conference covering research and developments in the fields of Data Mining, Computational Intelligence, Soft Computing, Machine Learning, Fuzzy Logic, and a lot more. More than 550 prospective authors had submitted their research papers to the conference. ICCIDM selected 192 papers after a double blind peer review process by experienced subject expertise reviewers chosen from the country and abroad. The proceedings of ICCIDM is a nice collection of interdisciplinary papers concerned in various prolific research areas of Data Mining and Computational Intelligence. It has been an honor for us to have the chance to edit the proceedings. We have enjoyed considerably working in cooperation with the International Advisory, Program, and Technical Committees to call for papers, review papers, and finalize papers to be included in the proceedings.

This International Conference ICCIDM aims at encompassing a new breed of engineers, technologists making it a crest of global success. It will also educate the youth to move ahead for inventing something that will lead to great success. This year's program includes an exciting collection of contributions resulting from a successful call for papers. The selected papers have been divided into thematic areas including both review and research papers which highlight the current focus of Computational Intelligence Techniques in Data Mining. The conference aims at creating a forum for further discussion for an integrated information field incorporating a series of technical issues in the frontier analysis and design aspects of different alliances in the related field of Intelligent computing and others. Therefore the call for paper was on three major themes like Methods, Algorithms, and Models in Data mining and Machine learning, Advance Computing and Applications. Further, papers discussing the issues and applications related to the theme of the conference were also welcomed at ICCIDM.

The proceedings of ICCIDM have been released to mark this great day in ICCIDM which is a collection of ideas and perspectives on different issues and some new thoughts on various fields of Intelligent Computing. We hope the author's own research and opinions add value to it. First and foremost are the authors of papers, columns, and editorials whose works have made the conference a great success. We had a great time putting together this proceedings. The ICCIDM conference and proceedings are a credit to a large group of people and everyone should be there for the outcome. We extend our deep sense of gratitude to all for their warm encouragement, inspiration, and continuous support for making it possible.

Hope all of us will appreciate the good contributions made and justify our efforts.

Acknowledgments

The theme and relevance of ICCIDM has attracted more than 550 researchers/academicians around the globe, which enabled us to select good quality papers and serve to demonstrate the popularity of the ICCIDM conference for sharing ideas and research findings with truly national and international communities. Thanks to all who have contributed in producing such a comprehensive conference proceedings of ICCIDM.

The organizing committee believes and trusts that we have been true to the spirit of collegiality that members of ICCIDM value, even as maintaining an elevated standard as we have reviewed papers, provided feedback, and present a strong body of published work in this collection of proceedings. Thanks to all the members of the Organizing committee for their heartfelt support and cooperation.

It has been an honor for us to edit the proceedings. We have enjoyed considerably working in cooperation with the International Advisory, Program, and Technical Committees to call for papers, review papers, and finalize papers to be included in the proceedings.

We express our sincere thanks and obligations to the benign reviewers for sparing their valuable time and effort in reviewing the papers along with suggestions and appreciation in improvising the presentation, quality, and content of this proceedings. Without this commitment it would not be possible to have the important reviewer status assigned to papers in the proceedings. The eminence of these papers is an accolade to the authors and also to the reviewers who have guided for indispensable perfection.

We would like to gratefully acknowledge the enthusiastic guidance and continuous support of Prof. (Dr.) Lakhmi Jain, as and when it was needed as well as adjudicating on those difficult decisions in the preparation of the proceedings and impetus to our efforts to publish this proceeding.

Last but not the least, the editorial members of Springer Publishing deserve a special mention and our sincere thanks to them not only for making our dream come true in the shape of this proceedings, but also for its brilliant get-up and in-time publication in Smart, Innovation, System and Technologies, Springer.

I feel honored to express my deep sense of gratitude to all members of International Advisory Committee, Technical Committee, Program Committee, Organizing Committee, and Editorial Committee members of ICCIDM for their unconditional support and cooperation.

The ICCIDM conference and proceedings are a credit to a large group of people and everyone should be proud of the outcome.

Himansu Sekhar Behera

About the Conference

The International Conference on “Computational Intelligence in Data Mining” (ICCIDM-2014) has been established itself as one of the leading and prestigious conference which will facilitate cross-cooperation across the diverse regional research communities within India as well as with other International regional research programs and partners. Such an active dialogue and discussion among International and National research communities is required to address many new trends and challenges and applications of Computational Intelligence in the field of Science, Engineering and Technology. ICCIDM 2014 is endowed with an opportune forum and a vibrant platform for researchers, academicians, scientists, and practitioners to share their original research findings and practical development experiences on the new challenges and budding confronting issues.

The conference aims to:

- Provide an insight into current strength and weaknesses of current applications as well as research findings of both Computational Intelligence and Data Mining.
- Improve the exchange of ideas and coherence between the various Computational Intelligence Methods.
- Enhance the relevance and exploitation of data mining application areas for end-user as well as novice user application.
- Bridge research with practice that will lead to a fruitful platform for the development of Computational Intelligence in Data mining for researchers and practitioners.
- Promote novel high quality research findings and innovative solutions to the challenging problems in Intelligent Computing.
- Make a tangible contribution to some innovative findings in the field of data mining.
- Provide research recommendations for future assessment reports.

So, we hope the participants will gain new perspectives and views on current research topics from leading scientists, researchers, and academicians around the world, contribute their own ideas on important research topics like Data Mining and Computational Intelligence, as well as network and collaborate with their international counterparts.

Conference Committee

Patron
Prof. E. Saibaba Reddy
Vice Chancellor, VSSUT, Burla, Odisha, India

Convenor
Dr. H.S. Behera
Department of CSE and IT, VSSUT, Burla, Odisha, India

Co-Convenor
Dr. M.R. Kabat
Department of CSE and IT, VSSUT, Burla, Odisha, India

Organizing Secretary
Mr. Janmenjoy Nayak, DST INSPIRE Fellow, Government of India

Conference Chairs

Honorary General Chair
Prof. P.K. Dash, Ph.D., D.Sc., FNAE, SMIEEE, Director
Multi Disciplinary Research Center, S 'O'A University, India
Prof. Lakhmi C. Jain, Ph.D., M.E., B.E.(Hons), Fellow (Engineers Australia), University of Canberra, Canberra, Australia and University of South Australia, Adelaide, SA, Australia

Honorary Advisory Chair
Prof. Shankar K. Pal, Distinguished Professor
Indian Statistical Institute, Kolkata, India

General Chair
Prof. Rajib Mall, Ph.D., Professor and Head
Department of Computer Science and Engineering, IIT Kharagpur, India

Program Chairs
Dr. Sukumar Mishra, Ph.D., Professor
Department of EE, IIT Delhi, India
Dr. R.P. Panda, Ph.D., Professor
Department of ETC, VSSUT, Burla, Odisha, India
Dr. J.K. Mandal, Ph.D., Professor
Department of CSE, University of Kalyani, Kolkata, India

Finance Chair
Dr. D. Dhupal, Coordinator, TEQIP, VSSUT, Burla

Volume Editors
Prof. Lakhmi C. Jain, Ph.D., M.E., B.E.(Hons), Fellow (Engineers Australia), University of Canberra, Canberra, Australia and University of South Australia, Adelaide, SA, Australia
Prof. H.S. Behera, Reader, Department of Computer Science Engineering and Information Technology, Veer Surendra Sai University of Technology, Burla, Odisha, India
Prof. J.K. Mandal, Professor, Department of Computer Science and Engineering, University of Kalyani, Kolkata, India
Prof. D.P. Mohapatra, Associate Professor, Department of Computer Science and Engineering, NIT, Rourkela, Odisa, India

International Advisory Committee

Prof. C.R. Tripathy (VC, Sambalpur University)
Prof. B.B. Pati (VSSUT, Burla)
Prof. A.N. Nayak (VSSUT, Burla)
Prof. S. Yordanova (STU, Bulgaria)
Prof. P. Mohapatra (University of California)
Prof. S. Naik (University of Waterloo, Canada)
Prof. S. Bhattacharjee (NIT, Surat)
Prof. G. Saniel (NIT, Durgapur)
Prof. K.K. Bharadwaj (JNU, New Delhi)
Prof. Richard Le (Latrob University, Australia)
Prof. K.K. Shukla (IIT, BHU)
Prof. G.K. Nayak (IIIT, BBSR)
Prof. S. Sakhya (TU, Nepal)
Prof. A.P. Mathur (SUTD, Singapore)
Prof. P. Sanyal (WBUT, Kolkata)
Prof. Yew-Soon Ong (NTU, Singapore)
Prof. S. Mahesan (Japfna University, Srilanka)
Prof. B. Satapathy (SU, SBP)

Prof. G. Chakraborty (IPU, Japan)
Prof. T.S. Teck (NU, Singapore)
Prof. P. Mitra (P.S. University, USA)
Prof. A. Konar (Jadavpur University)
Prof. S. Das (Galgotias University)
Prof. A. Ramanan (UJ, Srilanka)
Prof. Sudipta Mohapatra, (IIT, KGP)
Prof. P. Bhattacharya (NIT, Agaratala)
Prof. N. Chaki (University of Calcutta)
Dr. J.R. Mohanty, (Registrar, VSSUT, Burla)
Prof. M.N. Favorskaya (SibSAU)
Mr. D. Minz (COF, VSSUT, Burla)
Prof. L.M. Patnaik (DIAT, Pune)
Prof. G. Panda (IIT, BHU)
Prof. S.K. Jena (NIT, RKL)
Prof. V.E. Balas (University of Arad)
Prof. R. Kotagiri (University of Melbourne)
Prof. B.B. Biswal (NIT, RKL)
Prof. Amit Das (BESU, Kolkata)
Prof. P.K. Patra (CET, BBSR)
Prof. N.G.P.C Mahalik (California)
Prof. D.K. Pratihar (IIT, KGP)
Prof. A. Ghosh (ISI, Kolkata)
Prof. P. Mitra (IIT, KGP)
Prof. P.P. Das (IIT, KGP)
Prof. M.S. Rao (JNTU, HYD)
Prof. A. Damodaram (JNTU, HYD)
Prof. M. Dash (NTU, Singapore)
Prof. I.S. Dhillon (University of Texas)
Prof. S. Biswas (IIT, Bombay)
Prof. S. Pattnayak (S'O'A, BBSR)
Prof. M. Biswal (IIT, KGP)
Prof. Tony Clark (M.S.U, UK)
Prof. Sanjib ku. Panda (NUS)
Prof. G.C. Nandy (IIIT, Allahabad)
Prof. R.C. Hansdah (IISC, Bangalore)
Prof. S.K. Basu (BHU, India)
Prof. P.K. Jana (ISM, Dhanbad)
Prof. P.P. Choudhury (ISI, Kolkata)
Prof. H. Pattnayak (KIIT, BBSR)
Prof. P. Srinivasa Rao (AU, Andhra)

International Technical Committee

Dr. Ranjan Behera, Ph.D., Assistant Professor
Department of EE, IIT, Patna, India
Dr. A.K. Barisal, Ph.D., Reader
Department of EE, VSSUT, Burla, Odisha, India
Dr. R. Mohanty, Reader
Department of CSE, VSSUT, Burla

Conference Steering Committee

Publicity Chair
Prof. A. Rath, DRIEMS, Cuttack
Prof. B. Naik, VSSUT, Burla
Mr. Sambit Bakshi, NIT, RKL

Logistic Chair
Prof. S.P. Sahoo, VSSUT, Burla
Prof. S.K. Nayak, VSSUT, Burla
Prof. D.C. Rao, VSSUT, Burla
Prof. K.K. Sahu, VSSUT, Burla

Organizing Committee
Prof. D. Mishra, VSSUT, Burla
Prof. J. Rana, VSSUT, Burla
Prof. P.K. Pradhan, VSSUT, Burla
Prof. P.C. Swain, VSSUT, Burla
Prof. P.K. Modi, VSSUT, Burla
Prof. S.K. Swain, VSSUT, Burla
Prof. P.K. Das, VSSUT, Burla
Prof. P.R. Dash, VSSUT, Burla
Prof. P.K. Kar, VSSUT, Burla
Prof. U.R. Jena, VSSUT, Burla
Prof. S.S. Das, VSSUT, Burla
Prof. Sukalyan Dash, VSSUT, Burla
Prof. D. Mishra, VSSUT, Burla
Prof. S. Aggrawal, VSSUT, Burla
Prof. R.K. Sahu, VSSUT, Burla
Prof. M. Tripathy, VSSUT, Burla
Prof. K. Sethi, VSSUT, Burla
Prof. B.B. Mangaraj, VSSUT, Burla
Prof. M.R. Pradhan, VSSUT, Burla
Prof. S.K. Sarangi, VSSUT, Burla
Prof. N. Bhoi, VSSUT, Burla
Prof. J.R. Mohanty, VSSUT, Burla

Prof. Sumanta Panda, VSSUT, Burla
Prof. A.K. Pattnaik, VSSUT, Burla
Prof. S. Panigrahi, VSSUT, Burla
Prof. S. Behera, VSSUT, Burla
Prof. M.K. Jena, VSSUT, Burla
Prof. S. Acharya, VSSUT, Burla
Prof. S. Kissan, VSSUT, Burla
Prof. S. Sathua, VSSUT, Burla
Prof. E. Oram, VSSUT, Burla
Dr. M.K. Patcl, VSSUT, Burla
Mr. N.K.S. Behera, M.Tech. Scholar
Mr. T. Das, M.Tech. Scholar
Mr. S.R. Sahu, M.Tech. Scholar
Mr. M.K. Sahoo, M.Tech. Scholar
Prof. J.V.R. Murthy, JNTU, Kakinada
Prof. G.M.V. Prasad, B.V.CIT, AP
Prof. S. Pradhan, UU, BBSR
Prof. P.M. Khilar, NIT, RKL
Prof. Murthy Sharma, BVC, AP
Prof. M. Patra, BU, Berhampur
Prof. M. Srivastava, GGU, Bilaspur
Prof. P.K. Behera, UU, BBSR
Prof. B.D. Sahu, NIT, RKL
Prof. S. Baboo, Sambalpur University
Prof. Ajit K. Nayak, S'O'A, BBSR
Prof. Debahuti Mishra, ITER, BBSR
Prof. S. Sethi, IGIT, Sarang
Prof. C.S. Panda, Sambalpur University
Prof. N. Kamila, CVRCE, BBSR
Prof. H.K. Tripathy, KIIT, BBSR
Prof. S.K. Sahana, BIT, Meshra
Prof. Lambodar Jena, GEC, BBSR
Prof. R.C. Balabantaray, IIIT, BBSR
Prof. D. Gountia, CET, BBSR
Prof. Mihir Singh, WBUT, Kolkata
Prof. A. Khaskalam, GGU, Bilaspur
Prof. Sashikala Mishra, ITER, BBSR
Prof. D.K. Behera, TAT, BBSR
Prof. Shruti Mishra, ITER, BBSR
Prof. H. Das, KIIT, BBSR
Mr. Sarat C. Nayak, Ph.D. Scholar
Mr. Pradipta K. Das, Ph.D. Scholar
Mr. G.T. Chandrasekhar, Ph.D. Scholar
Mr. P. Mohanty, Ph.D. Scholar
Mr. Sibarama Panigrahi, Ph.D. Scholar

Mr. A.K. Bhoi, Ph.D. Scholar
Mr. T.K. Samal, Ph.D. Scholar
Mr. Ch. Ashutosh Swain, MCA
Mr. Nrusingh P. Achraya, MCA
Mr. Devi P. Kanungo, M.Tech. Scholar
Mr. M.K. Sahu, M.Tech. Scholar

Contents

Editors' Biography

Prof. Lakhmi C. Jain is with the Faculty of Education, Science, Technology and Mathematics at the University of Canberra, Australia and University of South Australia, Australia. He is a Fellow of the Institution of Engineers, Australia. Professor Jain founded the Knowledge-Based Intelligent Engineering System (KES) International, a professional community for providing opportunities for publication, knowledge exchange, cooperation, and teaming. Involving around 5,000 researchers drawn from universities and companies worldwide, KES facilitates international cooperation and generates synergy in teaching and research. KES regularly provides networking opportunities for the professional community through one of the largest conferences of its kind in the area of KES. His interests focus on artificial intelligence paradigms and their applications in complex systems, security, e-education, e-healthcare, unmanned air vehicles, and intelligent agents.

Prof. Himansu Sekhar Behera is working as a Reader in the Department of Computer Science Engineering and Information Technology, Veer Surendra Sai University of Technology (VSSUT) (A Unitary Technical University, Established by Government of Odisha), Burla, Odisha. He has received M.Tech. in Computer Science and Engineering from N.I.T, Rourkela (formerly R.E.C., Rourkela) and Doctor of Philosophy in Engineering (Ph.D.) from Biju Pattnaik University of Technology (BPUT), Rourkela, Government of Odisha respectively. He has published more than 80 research papers in various international journals and conferences, edited 11 books and is acting as a member of the editorial/reviewer board of various international journals. He is proficient in the field of Computer Science Engineering and served in the capacity of program chair, tutorial chair, and acted as advisory member of committees of many national and international conferences. His research interest includes Data Mining and Intelligent Computing. He is associated with various educational and research societies like OITS, ISTE, IE, ISTD, CSI, OMS, AIAER, SMIAENG, SMCSTA, etc. He is currently guiding seven Ph.D. scholars.

Prof. Jyotsna Kumar Mandal is working as Professor in Computer Science and Engineering, University of Kalyani, India. Ex-Dean Faculty of Engineering, Technology and Management (two consecutive terms since 2008). He has 26 years of teaching and research experiences. He was Life Member of Computer Society of India since 1992 and life member of Cryptology Research Society of India, member of AIRCC, associate member of IEEE and ACM. His research interests include Network Security, Steganography, Remote Sensing and GIS Application, Image Processing, Wireless and Sensor Networks. Domain Expert of Uttar Banga Krishi Viswavidyalaya, Bidhan Chandra Krishi Viswavidyalaya for planning and integration of Public domain networks. He has been associated with national and international journals and conferences. The total number of publications to his credit is more than 320, including 110 publications in various international journals. Currently, he is working as Director, IQAC, Kalyani University.

Prof. Durga Prasad Mohapatra received his Ph.D. from Indian Institute of Technology Kharagpur and is presently serving as an Associate Professor in NIT Rourkela, Odisha. His research interests include software engineering, real-time systems, discrete mathematics, and distributed computing. He has published more than 30 research papers in these fields in various international Journals and conferences. He has received several project grants from DST and UGC, Government of India. He received the Young Scientist Award for the year 2006 from Orissa Bigyan Academy. He has also received the Prof. K. Arumugam National Award and the Maharashtra State National Award for outstanding research work in Software Engineering for the years 2009 and 2010, respectively, from the Indian Society for Technical Education (ISTE), New Delhi. He is nominated to receive the Bharat Shiksha Ratan Award for significant contribution in academics awarded by the Global Society for Health and Educational Growth, Delhi.

Multi-objective Design Optimization of Three-Phase Induction Motor Using NSGA-II Algorithm

Soumya Ranjan and Sudhansu Kumar Mishra

Abstract The modeling of electrical machine is approached as a system optimization, more than a simple machine sizing. Hence wide variety of designs are available and the task of comparing the different options can be very difficult. A number of parameters are involved in the design optimization of the induction motor and the performance relationship between the parameters also is implicit. In this paper, a multi-objective problem is considered in which three phase squirrel cage induction motor (SCIM) has been designed subject to the efficiency and power density as objectives. The former is maximized where the latter is minimized simultaneously considering various constraints. Three single objective methods such as Tabu Search (TS), Simulated Annealing (SA) and Genetic Algorithm (GA) is used for comparing the Pareto solutions. Performance comparison of techniques is done by performing different numerical experiments. The result shows that NSGA-II outperforms other three for the considered test cases.

Keywords Multi-objective optimization · Induction motors · Multi-objective evolutionary algorithms · Single objective evolutionary algorithm

1 Introduction

Three-phase induction motors have been widely used in industrial applications. Over the past decade, there have been clear areas in motor utilization that demand higher power density and increased energy efficiency. In many industrial

---

S. Ranjan (✉)
Department of Electrical and Electronics Engineering, NIST, Berhampur, India
e-mail: Soumya.biteee@gmail.com

S.K. Mishra
Department of Electrical and Electronics Engineering, Birla Institute of Technology, Mesra, Ranchi, India
e-mail: Sudhansu.nit@gmail.com

L.C. Jain et al. (eds.), *Computational Intelligence in Data Mining - Volume 2*,
Smart Innovation, Systems and Technologies 32, DOI 10.1007/978-81-322-2208-8_1

applications, motor size and inertia are critical. Motors with high power density can offer a performance advantage in applications such as paper machines. However, high-power density cannot compromise reliability and efficiency. In such multi-objective optimization (MO), it is impossible to obtain the solution with maximizing or minimizing all objectives simultaneously because of the trade off relation between the objectives. When the MO is applied to the practical design process, it is difficult to achieve an effective and robust optimal solution within an acceptable computation time. The solutions obtained are known as *Pareto-optimal* solutions or *non-dominated solutions*. The rest is called dominated solutions. There are several methods to solve MO problems and one method of them, Pareto optimal solutions are generally used for the balanced solutions between objectives.

Appelbaum proposed the method of "boundary search along active constrains" in 1987 [1]. Madescu proposed the nonlinear analytical iterative field-circuit model (AIM) in 1996 by Madescu et al. [2]. However, these techniques have many shortcomings to provide fast and accurate solution, particularly when the optimal solution to a problem has many variables and constraints. Thus, to deal with such difficulties efficient optimization strategies are required. This can be overcome by multi-objective optimization (MO) technique [3–7].

This paper aims at MO which incorporates NSGA-II algorithm for minimization of power density and maximization of efficiency of three phases SCIM using different nonlinear constrained optimization techniques [8–10]. The Pareto-optimization technique is used in order to solve the multi-objective optimization problem of electric motor drive in a parametric fashion. It results in a set of optimal solutions from which an appropriate compromise design can be chosen based on the preference of the designer. In addition to that various SOEA techniques such as Simulated Annealing (SA), Tabu Search (TS), Genetic Algorithm (GA) is applied to compare among Pareto-optimal solutions [11]. Their performance has been evaluated by the metrics such as Delta, Convergence (C) and Spacing (S) through simulation studies.

2 Multi-objective Optimization Design

The general formulation of MOPs as [12]

Maximize/Minimize

$$f(\vec{x}) = (f_1(\vec{x}), f_2(\vec{x}), \ldots, f_M(\vec{x})) \tag{1}$$

Subjected to constraints:

$$g_j(\vec{x}) \geq 0, \quad j = 1, 2, 3 \ldots, J \tag{2}$$

$$h_k(\vec{x}) = 0, \quad k = 1, 2, 3 \ldots, K \tag{3}$$

where $\vec{x}$ represents a vector of decision variables $\vec{x} = \{x_1, x_2, \ldots, x_N\}^T$.

The search space is limited by

$$x_i^L \leq x_i \leq x_i^U, \quad i = 1, 2, 3\ldots, N \tag{4}$$

x_i^L and x_i^U represent the lower and upper acceptable values respectively for the variable x_i. N represents the number of decision variables and M represents the number of objective functions. Any solution vector $\vec{u} = \{u_1, u_2, \ldots u_K\}^T$ is said to dominate over $\vec{v} = \{v_1, v_2, \ldots, v_k\}^T$ if and only if

$$\left.\begin{array}{ll} f_i(\vec{u}) \leq f_i(\vec{v}) & \forall i \in \{1, 2, \ldots, M\} \\ f_i(\vec{u}) < f_i(\vec{v}) & \exists i \in \{1, 2, \ldots, M\} \end{array}\right\} \tag{5}$$

Those solutions which are not dominated by other solutions for a given set are considered non-dominated solutions are called Pareto optimal solution.

The practical application of genetic algorithm to multi-objective optimization problem (MOP) involves various problems out of which NSGA-II [13, 14] algorithm has been implemented to find the Pareto-optimal solution between power density and efficiency.

3 Multi-objective Evolutionary Algorithm Frameworks

A majority of MOEAs in both the research and the application areas are Pareto-dominance based which are mostly the same frameworks as that of NSGA-II. In these algorithms a selection operator based on Pareto-domination and a reproduction operator is used. The operator of the MOEAs guides the population iteratively towards non-dominated regions by preserving the diversity to get the Pareto-optimal set. The evaluate operator leads to population convergence towards the efficient frontier and helps preserve the diversity of solutions along the efficient frontier. Both goals are achieved by assigning a rank and a density value to each solution. The MOEAs provide first priority to non-dominance and second priority to diversity. However, the methods by which they achieve these two fundamental goals differ. The main difference between the algorithms lies in their fitness assignment techniques. Coello et al. Classifies the constraints handling methods into five categories: (1) penalty functions (2) special representations and operators (3) repair algorithms (4) separate objective and constraints and (5) hybrid methods [15, 16].

4 Design Optimization of Induction Motor

In this paper the design of induction motor is formulated by MOEAs based on non-dominated sorting, NSGA-II which does not combine the two objectives to obtain the Pareto-optimal solution set. Here, the two objectives are taken individually and

an attempt is made to optimize both simultaneously. The main objective is to maximize efficiency (η) and minimize power density (ξ). The proposed NSGA-II is suitably oriented in such a way as to optimize the two objectives. To express both the objectives in maximization form, the first objective ξ is expressed as $-\xi$. In addition to these objectives, different practical constraints mentioned are also considered. In order to design, the problem is expressed as Maximize η and $-\xi$ simultaneously considering all constraints [17, 18].
The sizing equation of an induction machine is

$$P_R(IM) = \frac{\sqrt{2}\pi^2}{2(1+K_\phi)} K_\omega \eta \cos\phi_r B_g A \frac{f}{p} \lambda_0^2 D_0^2 L_e \tag{6}$$

In terms of efficiency (η) can be written as

$$\eta = \frac{P_R(IM)2(1+K_\phi)}{\sqrt{2}\pi^2 K_\omega \cos\phi_r B_g A \frac{f}{p} \lambda_0^2 D_0^2 L_e} \tag{7}$$

The power density of the induction machine is given by

$$\zeta(IM) = \frac{\sqrt{2}\pi^2}{2(1+K_\phi)} K_\omega \eta \cos\phi_r B_g A \frac{f}{p} \lambda_0^2 \frac{L_g}{L_t} \tag{8}$$

where $\cos\phi_r$ is the power factor which is related to the rated power $P_R(IM)$, the pole pairs p of the machine, and the converter frequency f. The design variables for induction motor are chosen as consisting of four flux densities at the teeth and yokes for the stator and rotor, one current density in stator winding and three geometric variables. Three geometric variables are the depth of stator slot, the ratio of the rotor slot bottom width of rotor tooth width and the ratio of rotor slot top radius of the rotor slot bottom radius [19, 20].

5 Performance Measure for Comparison

The final Pareto-optimal front obtained from different MOEAs techniques is compared using performance metrics such as Spacing (S), Diversity metric (Δ), Convergence metric (C) [17]. These performance metrics set the benchmark to compare the results and select the best outcomes.

6 Simulation Results

The 5 kW, 4-pole, three-phase squirrel-cage induction motor is chosen as a sample design. The rated frequency is 50 Hz and voltage is 170 V. Also, the ratio of maximum torque to nominal torque is set 2.5 as a constraint. Lower limit of

Table 1 The performance evaluation metrics

Algorithm		SA	TS	GA	NSGA-II
S	Avg.	2.67E−5	9.73E−6	5.74E−6	4.21E−6
	Std.	6.72E−6	1.44E−6	1.43E−6	1.05E−6
Δ	Avg.	8.03E−1	7.86E−1	5.96E−1	5.53E−1
	Std.	2.08E−1	1.93E−1	1.48E−1	1.47E−1

Table 2 The result obtained for C metric

	SA	TS	GA	NSGA-II
SA	–	0.3422	0.2192	0.1932
TS	0.4443	–	0.2653	
GA	0.5653	0.4922	–	0.25373
NSGA-II	0.5988	0.5911	0.3262	–

Table 3 Comparison of CPU time (s)

Algorithms	NSGA-II		GA		TS		SA	
CPU time	Mean	Var	Mean	Var	Mean	Var	Mean	Var
	150	25	680	35	720	40	690	45

efficiency is 90 % and that of power density is 0.3 kW/kg. The population size is set to be 100. The algorithms stop after 20,000 function evaluations. Initial populations are generated by uniformly randomly sampling from the feasible search space. The uniform Crossover rate is taken 0.8. The mutation rate is 0.10 where it is taken as 1/n, i.e. n is 10, the number of decision variables.

Table 1 shows the S metric and Δ metric obtained using all four algorithms. Table 1 shows that the S and Δ metric value for NSGA-II is less than other three algorithms and hence its performance is better among all.

Table 2 shows the result obtained for Convergence (C) metrics. The values 0.5988 in the fourth row, first column means almost all solutions from final populations obtained by NSGA-II dominates the solutions obtained by SA. The values 0 in the first row, first column mean that no solution of the non-dominated population obtained by TS, GA and NSGA-II is dominated by solutions from final populations obtained by SA. From the result, it clear that the performance of NSGA-II significantly outperforms the competing algorithms in the considered optimal design of induction motor.

The comparison time computed by the CPU is shown in Table 3. The mean time and the variance (var) of time for NSGA-II algorithm is less than other algorithms. The Simulation statistics generated by the four algorithms NSGA-II, GA, TS, SA respectively are illustrated from Figs. 1, 2, 3 and 4. It is shown in Fig. 5 that NSGA-II results in wide areas of convergence and is diversified.

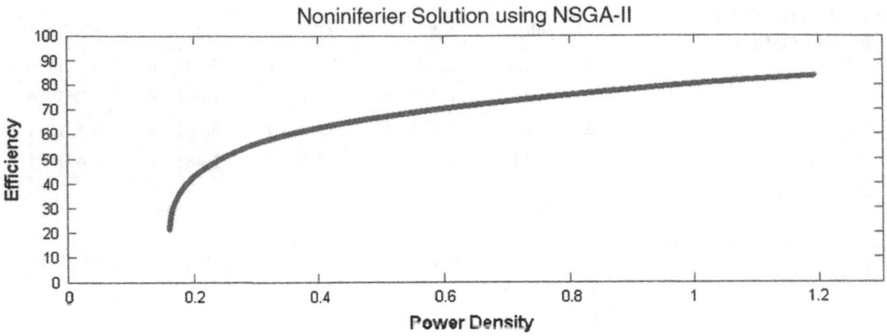

Fig. 1 Plots of Pareto fronts achieved by NSGA II

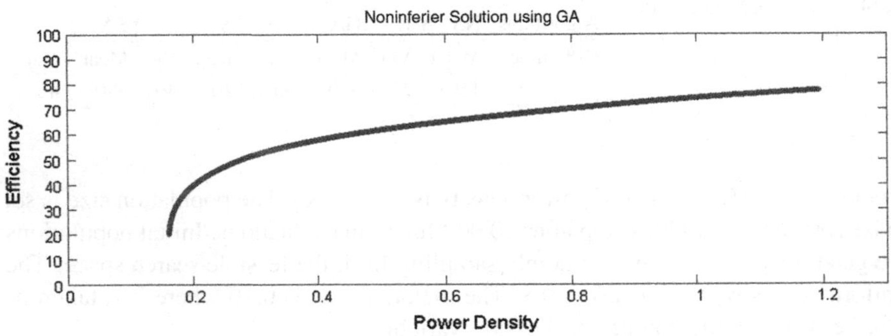

Fig. 2 Plots of Pareto fronts achieved by GA

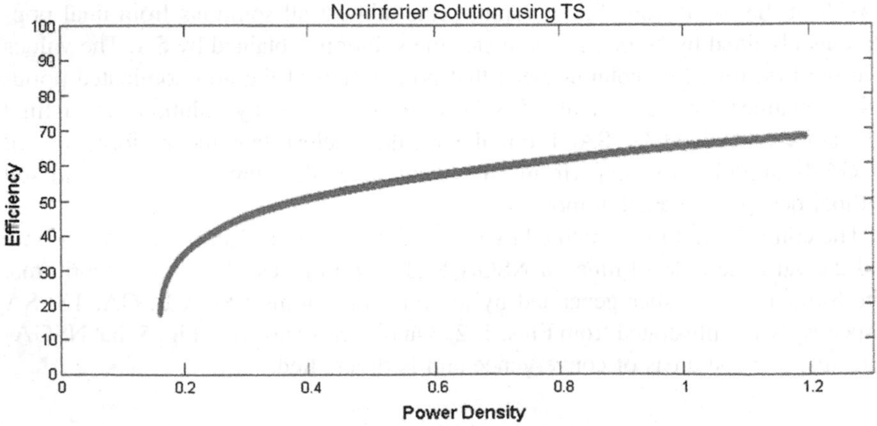

Fig. 3 Plots of Pareto fronts achieved by TS

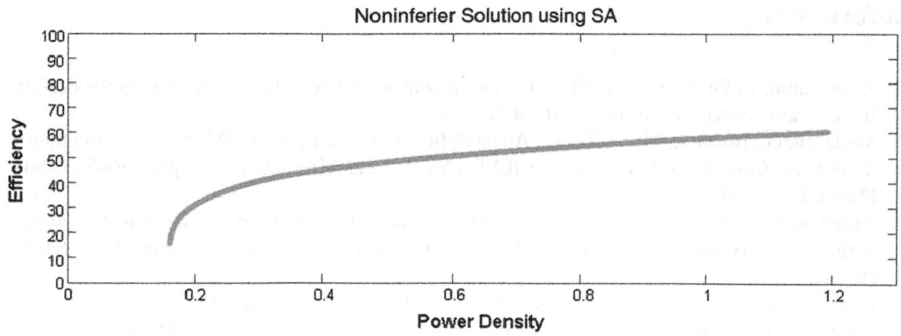

Fig. 4 Plots of Pareto fronts achieved by SA

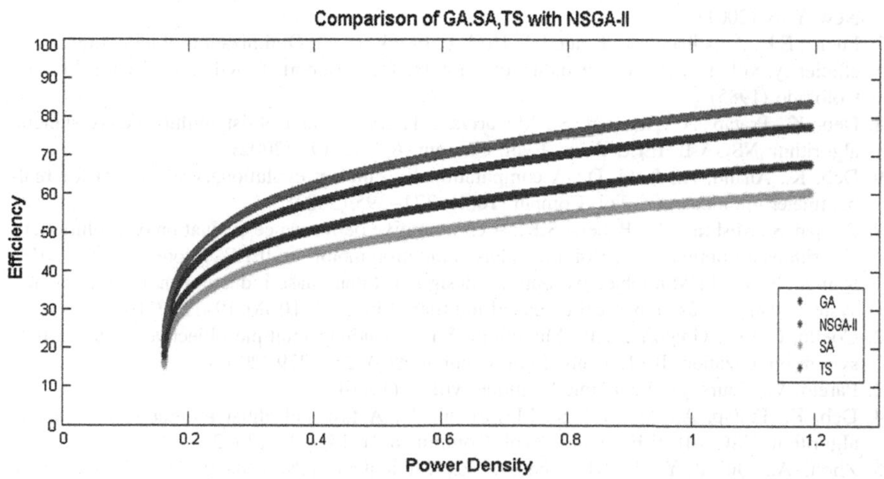

Fig. 5 Pareto front at different cardinality

7 Conclusion

In this paper, the multi-objective design optimization based on NSGA-II and size equations are applied for the three phase induction motors. In order to effectively obtain a set of Pareto optimal solutions, ranking method is applied. From the results, we can select the balanced optimal solution between the power density and efficiency. In case of optimized model, the efficiency increases at 80 % and the power density is also increased 12 kW/kg, compared to the SA, TS and GA result of the initial model. The performance metrics of NSGA-II results in best possible Pareto solutions. The proposed method can be efficiently and effectively used to multi-objectives design optimization of the machine cost and efficiency of electric machines.

References

1. Appelbaum, J., Fuchs, E.F., White, J.C.: Optimization of three-phase induction motor design. IEEE Trans. Energy Convers. **2**, 407–422 (1987)
2. Madescu, G., Bolea, I., Miller, T.J.E.: An analytical iterative model (AIM) for induction motor design. In: Conference Record of the IEEE-IAS Annual Meeting, vol. 1, pp. 556–583, San Diego, CA (1996)
3. Appelbaum, J., Khan, I.A., Fuchs, E.F.: Optimization of three-phase induction motor design with respect to efficiency. In: Proceedings of ICEM, pp. 639–642, Lousanne, Switzerland (1984)
4. Ramarathnam, R., Desai, B.G.: Optimization of poly-phase induction motor design—a nonlinear programming approach. IEEE Trans. Power Apparatus Syst. **PAS-90**, 570–578 (1971)
5. Fetih, N.H., El-Shewy, H.M.: Induction motor optimum design, including active power loss effect. IEEE Trans. Energy Convers. **1**(3), 155–160 (1986)
6. Deb, K.: Multi-Objective Optimization Using Evolutionary Algorithms, pp. 48–55. Wiley, New York (2001)
7. Fuchs, E.F., Appelbaum, J., Khan, I.A, Holl, J., Frank, U. V.: Optimization of induction motor efficiency, vol. 1. Three Phase Induction Motors, Final Report, EPRIEL-4152-ccm, Project, Colorado (1985)
8. Deb, K., Pratap, A., Agarwal, S., Meyarivan, T.: A fast and elitist multiobjective genetic algorithm: NSGA-II. IEEE Trans. Evol. Comput. **6**, 182–197 (2002)
9. Deb, K., Anand, A., Joshi, D.: A computationally efficient evolutionary algorithm for real-parameter optimization. Evol. Comput. **10**(4), 371–395 (2002)
10. Ranjan, S., Mishra, S.K., Behera, S.K.: A comparative performance evaluation of evolutionary algorithms for optimal design of three-phase induction motor. In: IEEE Xplore, pp. 1–5 (2014)
11. Ranjan, S., et al.: Multiobjective optimal design of three-phase induction motor for traction systems using an asymmetrical cascaded multilevel inverter. IJISM 19–125 (2014)
12. Coello, C.A.C., Gregorio, T.P., Maximino, S.L.: Handling multiple objectives with particle swarm optimization. IEEE Trans. Evol. Comput. **8**(3), 256–279 (2004)
13. Pareto, V., Cours, D.: Economie Politique, vol. 8 (1986)
14. Deb, K., Pratap, A., Agarwal, S., Meyarivan, T.: A fast and elitist multiobjective genetic algorithm: NSGA-II. IEEE Trans. Evol. Comput. **6**(2), 182–197 (2002)
15. Zhou, A., Qu, B.-Y., Li, H., Zhao, S.-Z., Suganthan, P.N., Zhang, Q.: Multiobjective evolutionary algorithms: a survey of the state-of-the-art. Swarm Evol. Comput. **1**(1), 32–49 (2011)
16. Montes, E.M., Coello, C.A.C.: Constraint-handling in nature-inspired numerical optimization: past, present and future. Swarm Evol. Comput. **1**(4), 173–194 (2011)
17. Huang, S., Luo, J., Leonardi, F., Lipo, T.A.: A general approach to sizing and power density equations for comparison of electrical machines. IEEE Trans. Ind. Appl. **34**(1), 92–97 (1998)
18. Kim, M.K., Lee, C.G., Jung, H.K.: Multiobjective optimal design of three-phase induction motor using improved evolution strategy. IEEE Trans. Magn. **34**(5), 2980–2983 (1998)
19. Ojo, O.: Multiobjective optimum design of electrical machines for variable speed motor drives. Tennesse Technological University, Cookeville, TN 38505

A Comparative Study of Different Feature Extraction Techniques for Offline Malayalam Character Recognition

Anitha Mary M.O. Chacko and P.M. Dhanya

Abstract Offline Handwritten Character Recognition of Malayalam scripts have gained remarkable attention in the past few years. The complicated writing style of Malayalam characters with loops and curves make the recognition process highly challenging. This paper presents a comparative study of Malayalam character recognition using 4 different feature sets—Zonal features, Projection histograms, Chain code histograms and Histogram of Oriented Gradients. The performance of these features for isolated Malayalam vowels and 5 consonants are evaluated in this study using feedforward neural networks as classifier. The final recognition results were computed using a 5 fold cross validation scheme. The best recognition accuracy of 94.23 % was obtained in this study using Histogram of Oriented Gradients features.

Keywords Offline character recognition · Feature extraction · Neural networks

1 Introduction

Offline character recognition is the process of translating handwritten text from scanned, digitized or photographed images into a machine editable format. Compared to online recognition, offline recognition is a much more challenging task due the lack of temporal and spatial information. Character recognition research has gained immense popularity because of its potential applications in the areas of postal automation, bank check processing, number plate recognition etc. Even though ambient studies have been performed in foreign languages [1], only very

A.M.M.O. Chacko (✉) · P.M. Dhanya
Department of Computer Science and Engineering, Rajagiri School of Engineering and Technology, Kochi, India
e-mail: anithamarychacko@gmail.com

P.M. Dhanya
e-mail: dhanya_pm@rajagiritech.ac.in

L.C. Jain et al. (eds.), *Computational Intelligence in Data Mining - Volume 2*,
Smart Innovation, Systems and Technologies 32, DOI 10.1007/978-81-322-2208-8_2

few works exist in the Malayalam character recognition domain. This is mainly due to its extremely large character set and complicated writing style with loops curves and holes.

Some of the major works reported in the Malayalam character recognition domain are as follows: Lajish [2] proposed the first work in Malayalam OCR using fuzzy zoning and normalized vector distances. 1D wavelet transform of vertical and horizontal projection profiles were used in [3] for the recognition of Malayalam characters. The performance of wavelet transform of projection profiles using 12 different wavelet filters were analyzed in [4]. In [5], recognition of Malayalam vowels was done using chain code histogram and image centroid. They have also proposed another method for Malayalam character recognition using Haar wavelet transform and SVM classifier [6]. Moni and Raju used Modified Quadratic Classifier and 12 directional gradient features for handwritten Malayalam character recognition [7]. Here gradient directions were computed using Sobel operators and were mapped into 12 directional codes. Recently, Jomy John proposed another approach for offline Malayalam recognition using gradient and curvature calculation and dimensionality reduction using Principal Component Analysis (PCA) [8]. A detailed survey on Malayalam character recognition is presented in [9].

A general handwritten character recognition system consists of mainly 4 phases—Preprocessing, Feature Extraction, Classification and Postprocessing. Among these, feature extraction is an important phase that determines the recognition performance of the system. To get an idea of recognition results of different feature extraction techniques in Malayalam character recognition, we have performed a comparative study using 4 different features—Zonal features, projection histograms, chain codes and Histogram of Oriented Gradients (HOG) features. The performance of these four feature sets are analyzed by using a two layer feedforward neural network as classifier.

The paper is structured as follows: Sect. 2 presents the data collection method used and the sequence of preprocessing steps done. Section 3 describes the feature extraction procedure for the four feature sets. The classifier used is introduced in Sect. 4. Section 5 presents the experimental results and discussions and finally conclusion is presented in Sect. 6.

2 Data Collection and Preprocessing

Malayalam belongs to the Dravidian family of languages which has official language status in Kerala. The complete character set of Malayalam consists of 15 vowels, 36 consonants, 5 chillu, 9 vowel signs, 3 consonant signs, 3 special characters and 57 conjunct consonants. Since a benchmarking database is not available for Malayalam, we have created a database of 260 samples for the isolated Malayalam vowels and 5 consonants ('ka','cha','tta','tha' and 'pa'). For this 13 class recognition problem, we have collected handwritten samples from 20 people

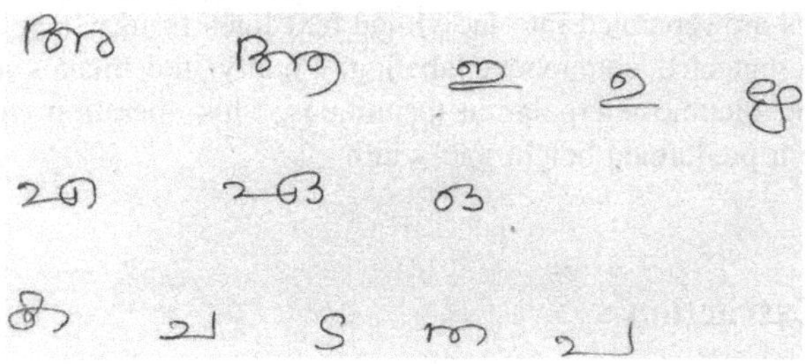

Fig. 1 Samples of handwritten Malayalam characters

belonging to different age groups and professions. Each of these 13 characters are assigned class-ids. The scanned images were then subjected to preprocessing. Figure 1 shows sample characters of the database.

2.1 Preprocessing

Preprocessing steps are carried out to reduce variations in the writing style of different people. The sequences of preprocessing steps (Fig. 2) carried out are as follows: Here, scanned images are binarized using Otsu's method of global thresholding. This method is based on finding the threshold that minimizes the intra-class variance. A large amount of noise such as salt and pepper noise may exist in the image acquired by scanning. So in order to reduce this noise to some extent, we have applied a 3 × 3 median filter. In the segmentation process, the

Fig. 2 Preprocessing steps. **a** Scanned image. **b** Binarized image. **c** Size normalized image

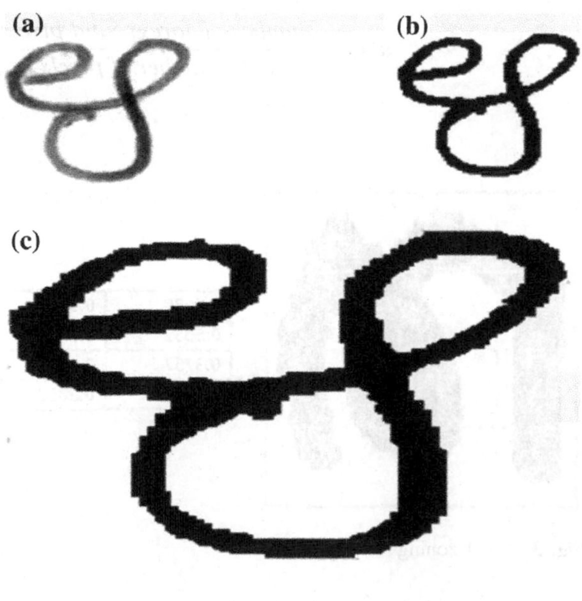

character images are separated into individual text lines from which characters are isolated using connected component labeling. Finally, the images are resized to 256 × 256 using bicubic interpolation techniques. This operation ensures that all characters have a predefined height and width.

3 Feature Extraction

The performance of an HCR system depends to a great extent on the extracted features. Over the years, many feature extraction techniques have been proposed for character recognition. A survey of feature extraction techniques is presented in [10]. In this study, we have used 4 sets of features for comparing the performance of the character recognition system: Zonal features, Projection histograms, Chain code histograms and Histogram of Oriented Gradients.

3.1 Zoning

Zoning is a popular method used in character recognition tasks. In this method, the character images are divided into zones of predefined sizes and then features are computed for each of these zones. Zoning obtains local characteristics of an image. Here, we have divided the preprocessed character images into 16 zones (4 × 4) as in and then pixel density features were computed for each of the zones (Fig. 3). The average pixel density was calculated by dividing the number of foreground pixels by the total number of pixels in each zone i.

$$d(i) = \frac{Number\ of\ foreground\ pixels\ in\ zone\ i}{Total\ number\ of\ pixels\ in\ zone\ i} \quad (1)$$

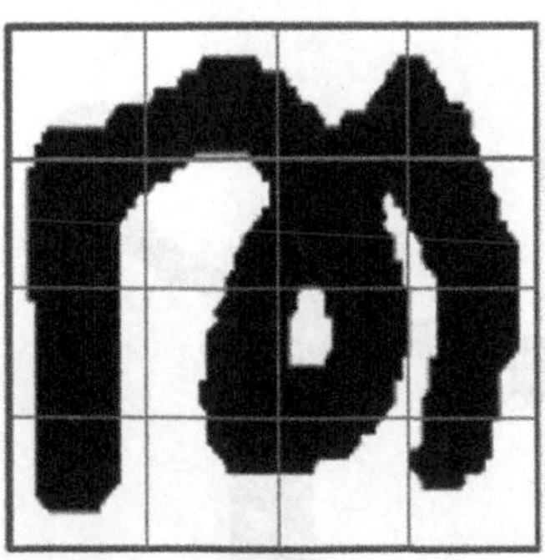

0.6638	0.3748	0.6079	0.4331
0.5652	0.5945	0.2852	0.3706
0.3757	0.4375	0.3264	0.3076
0.1465	0.2813	0.1648	0.3447

Fig. 3 4 × 4 zoning

Thus we have obtained 16 density features which are used as input to the classifier.

3.2 Projection Profile

Projection profile is an accumulation of black pixels along rows or columns of an image. The discriminating power of horizontal and vertical projection profiles make them well suitable for the recognition of a complex language like Malayalam. Projection profiles have been successfully applied for Malayalam character recognition [3, 4].

In this study, we have extracted both vertical and horizontal projection profiles by counting the pixels column wise and row wise respectively which together forms a 512 dimension feature vector (Fig. 4 shows the vertical and horizontal projection histogram for a Malayalam character 'tha').

Since, the size of the feature vector is too large, we have applied Principal Component Analysis (PCA) to reduce the dimensionality of the feature set. PCA is a technique that reduces the dimensionality of the data while retaining as much variations as possible in the original dataset. Using PCA, we have reduced the dimension of the feature vector from 512 to 260.

3.3 Chain Code Features

The chain code approach proposed by Freeman [11] is a compact way to represent the contour of an object. The chain codes are computed by moving along the boundary of the character in clockwise/anticlockwise direction and assigning each

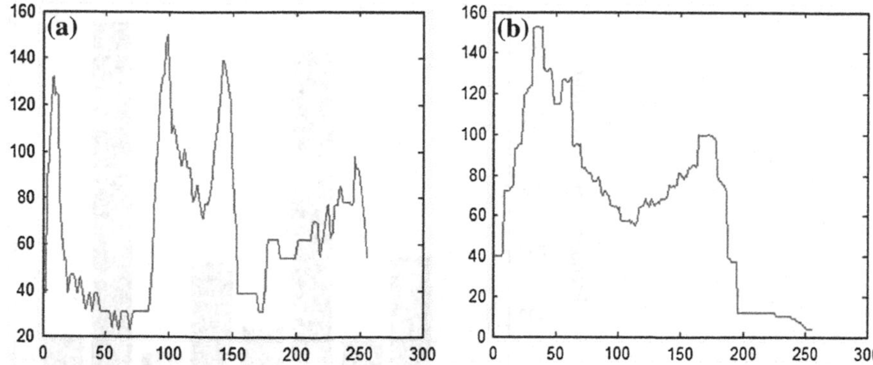

Fig. 4 Projection histogram of character 'tha'. **a** Horizontal projection histogram. **b** Vertical projection histogram

Fig. 5 8 directional chain codes

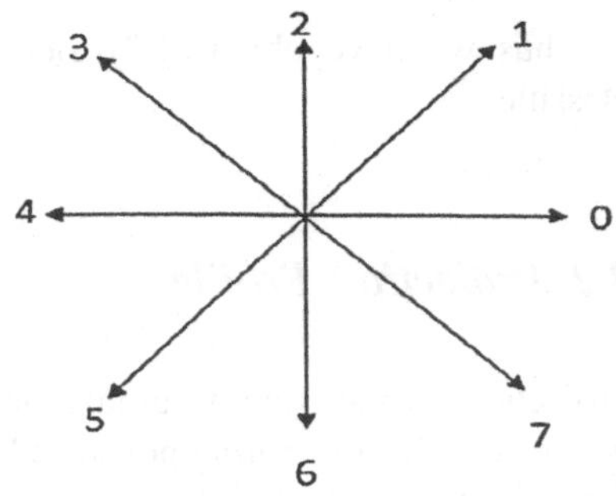

pixel along the contour a direction code (0–7) that indicates the direction of next pixel along the contour till the starting point is revisited. Here, we have used freeman chain code of eight directions (Fig. 5). Chain code and image centroid have been successfully applied for Malayalam vowel recognition in [5, 14].

Since the size of the chain code varies for different characters, we normalize it as follows: The frequency of each direction code is computed to form a chain code histogram (CCH). Figure 6 shows the chain code histogram of Malayalam character 'tha'. Image centroid is also used as an additional feature here. Thus we get a feature vector of size 10.

3.4 Histogram of Oriented Gradients

Histograms of Oriented Gradients are feature descriptors that are computed by counting the occurrences of gradient orientations in localized parts of an image. For computing these features, the image is divided into cells and histograms of gradient

Fig. 6 Chain code histogram

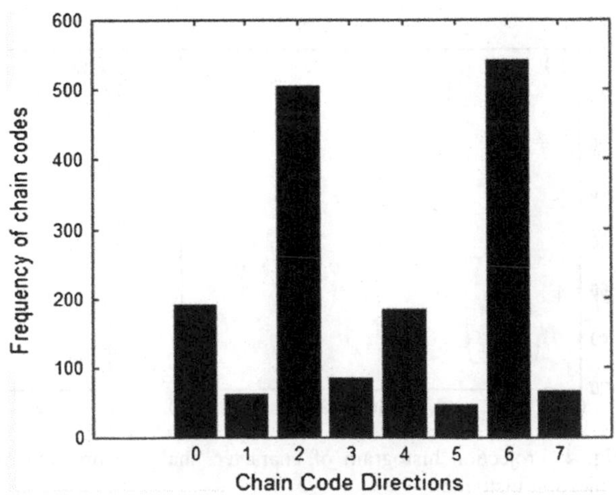

directions are formed for each of these cells. These histogram forms the descriptor. HOG features have been successfully implemented for other applications such as human detection [12], pedestrian detection [13] etc. Recently, it has also been implemented for character recognition in Hindi. However these features have not been explored for Malayalam character recognition.

In this method, the image was divided into 9 overlapping rectangular cells and for each of these cells, gradient directions were computed. Based on the gradient directions, each pixel within a cell casts a weighted vote to form an orientation based histogram channel of 9 bins. The gradient strength of each cell were normalized according to L2-norm. Thus the 9 histograms with 9 bins are concatenated to form an 81 dimensional feature vector [13] which is fed as input to the classifier.

4 Classification

Classification is the final stage of character recognition task in which character images are assigned unique labels based on the extracted features. In this study, we have used neural networks for comparing the performance of different feature sets. The principal advantage of neural networks is that they can learn automatically from examples. Here, we have used a two layer feedforward neural network consisting of a single hidden layer (Fig. 7). The input to the neural network consists of each of the feature sets that we have extracted. Thus the number of nodes in the input layer is equal to the size of the feature set that we use in each case. The output layer contains one node for each of the output classes, i.e., here we have 13 nodes in the output layer. We have used the number of hidden layer neurons to be 20 for our experiment.

5 Experimental Results and Discussions

In this section, the results of different feature sets for offline Malayalam character recognition are presented. The implementation of the system was carried out using Matlab R2013a. The results have been computed using a 5 fold cross validation

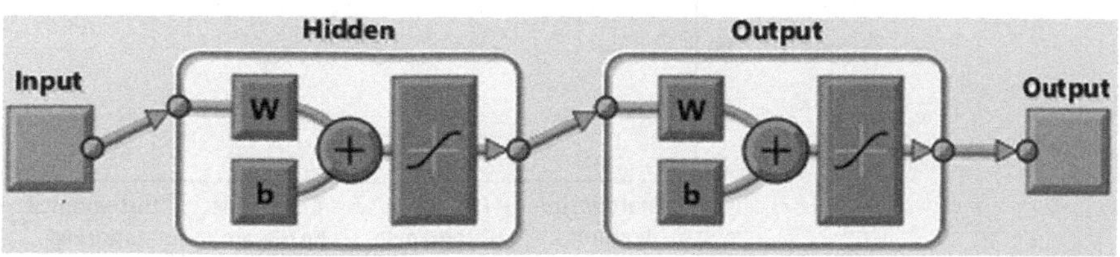

Fig. 7 Neural network model

technique. In a five-fold cross validation scheme, the entire dataset is divided into 5 subsets. During each fold, one of the subset is used for testing the classifier and the rest are used for training. The recognition rates from the test set in each fold are averaged to obtain the final accuracy of the classifier.

Table 1 summarizes the overall accuracy of the system with each of the four feature sets. A graphical representation of the recognition results is also shown (Fig. 8). From the experiment, the best recognition rate of 94.23 % was obtained for the histogram of oriented gradients features. The next highest accuracy was obtained for projection histograms with PCA. But compared to other feature sets, they need feature vectors of larger size (260). The density based features also provide good recognition accuracy with a relatively small feature vector size of 16. The chain code histogram feature gave the lowest recognition results among all the features used in this study. The recognition rate of 78.8 % was obtained using this chain code histogram feature.

From the confusion matrix obtained for each of the feature sets, we calculate precision and recall values. The plot of precision and recall values for each of the four feature sets are shown in Figs. 9 and 10 respectively. From the graphs, we noted that the HOG features achieved the highest average precision and recall values of 0.9423 and 0.9449 respectively. The lowest precision and recall values of 0.8026 and 0.7924 were shown by chain code histogram features. The average

Table 1 Recognition results of different feature sets

Feature set	Feature size	Accuracy (%)
Zonal density	16	84.6
Projection histograms + PCA	260	88.07
Chaincode histogram + centroid	10	78.8
HOG	81	94.23

Fig. 8 Recognition accuracy of different features

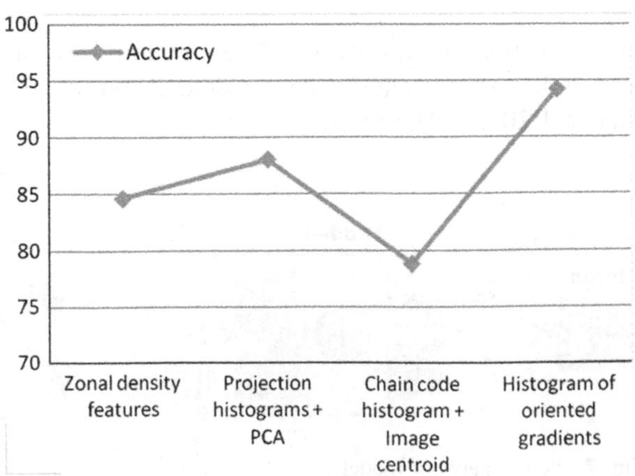

Fig. 9 Precision versus classes

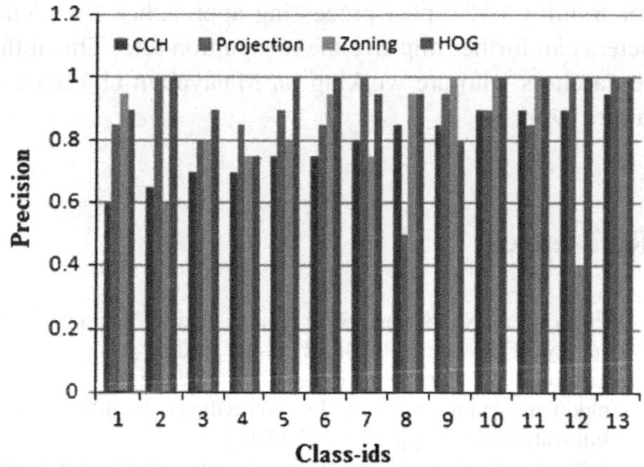

Fig. 10 Recall versus classes

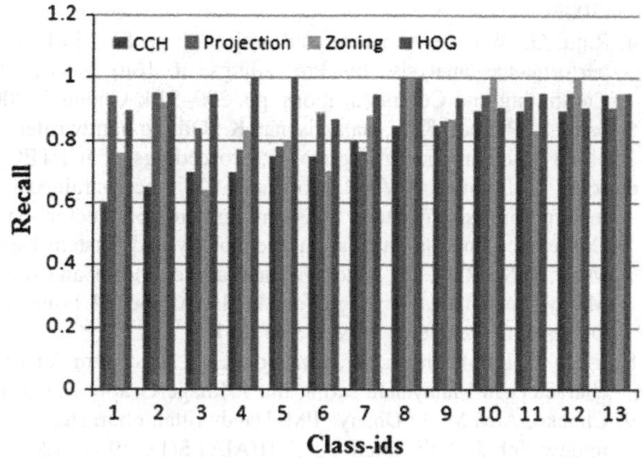

precision values for the zonal density features and projection features were 0.8423 and 0.8808 respectively. The average recall values for these features were 0.8614 and 0.8904 respectively.

6 Conclusion

The work presented in this paper analyses the performance of offline Malayalam character recognition using 4 feature sets—zone density features, projection histograms, chain code histograms and histogram of oriented gradient features. These features were classified using a two layer feedforward neural network. A five fold cross validation scheme was applied to measure the performance of the system. We have obtained the best recognition accuracy of 94.23 % using histogram of oriented gradient features. This accuracy can be further improved by using a larger dataset for training. Also post processing approaches for identifying similar shaped characters can further improve the recognition rate. The authors hope that this paper aids researchers who are working on Malayalam character recognition domain in their future works.

References

1. Plamondan, R., Srihari, S.N.: Online and offline character recognition: a comprehensive survey. IEEE Trans. PAMI **22**, 63–84 (2000)
2. Lajish, V.L.: Handwritten character recognition using perpetual fuzzy zoning and class modular neural networks. In: Proceedings of 4th International National Conference on Innovations in IT, pp. 188–192 (2007)
3. John, R., Raju, G., Guru, D.S.: 1D wavelet transform of projection profiles for isolated handwritten character recognition. In: Proceedings of ICCIMA07, pp. 481–485, Sivakasi (2007)
4. Raju, G.: Wavelet transform and projection profiles in handwritten character recognition—a performance analysis. In: Proceedings of 16th International Conference on Advanced Computing and Communications, pp. 309–314, Chennai (2008)
5. John, J., Pramod K.V., Balakrishnan K.: Offline handwritten Malayalam character recognition based on chain code histogram. In: Proceedings of ICETECT (2011)
6. John, J., Pramod, K.V., Balakrishnan, K.: Unconstrained handwritten Malayalam character recognition using wavelet transform and support vector machine classifier. In: International Conference on Communication Technology and System Design, Elsevier (2011)
7. Moni, B.S., Raju, G.: Modified quadratic classifier and directional features for handwritten Malayalam character recognition. In: IJCA Special Issue on Computational Science—New Dimensions Perspectives NCCSE (2011)
8. John, J., Balakrishnan, K., Pramod, K.V.: A system for offline recognition of handwritten characters in Malayalam script. Int. J. Image Graph. Signal Process. **4**, 53–59 (2013)
9. Chacko, A.M.M.O.: Dhanya PM, Handwritten character recognition in Malayalam scripts—a review. Int. J. Artif. Intell. Appl. (IJAIA) **5**(1), 79–89 (2014)
10. Trier, O.D., Jain, A.K., Taxt, J.: Feature extraction methods for character recognition—a survey. Pattern Recogn. **29**(4), 641–662 (1996)
11. Freeman, H.: On the encoding of arbitrary geometric configurations. IRE Trans. Electr. Comp. TC **10**(2), 260–268 (1961)
12. Dalal, N., Triggs, B.: Histograms of oriented gradients for human detection. In: IEEE Computer Society Conference on Computer Vision and Pattern Recognition, pp. 886–893 (2005)
13. Ludwig, O., Delgado, D., Goncalves, V., Nunes, U.: Trainable classifier-fusion schemes: an application to pedestrian detection. In: 12th International IEEE Conference on Intelligent Transport Systems, pp. 1–6 (2009)
14. Chacko, A.M.M.O., Dhanya, P.M.: A differential chain code histogram based approach for offline Malayalam character recognition. In: International Conference on Communication and Computing (ICC-2014), pp. 134–139 (2014)

A Context Sensitive Thresholding Technique for Automatic Image Segmentation

Anshu Singla and Swarnajyoti Patra

Abstract Recently, energy curve of an image is defined for image analysis. The energy curve has similar characteristics as that of histogram but also incorporates the spatial contextual information of the image. In this work we proposed a thresholding technique based on energy curve of the image to find out the optimum number of thresholds for image segmentation. The proposed method applies concavity analysis technique existing in the literature on the energy curve to detect all the potential thresholds. Then a threshold elimination technique based on cluster validity measure is proposed to find out the optimum number of thresholds. To assess the effectiveness of proposed method the results obtained using energy curve of the image are compared with those obtained using histogram of the image. Experimental results on four different images confirmed the effectiveness of the proposed technique.

Keywords Concavity analysis · DB index · Energy curve · Histogram · Segmentation

1 Introduction

Image segmentation plays vital role in various applications like medical imaging, object detection, fingerprint identification, text recognition etc. [1–3]. It involves segmenting an image into regions with uniform characteristics. If different objects belong to an image are sufficiently separated from each other, then the histogram of the image may have many peaks to represent different objects. The potential thresholds can be found at the valley regions of the histogram by applying

A. Singla (✉)
School of Mathematics and Computer Applications,
Thapar University, Patiala 147004, India
e-mail: asheesingla@gmail.com

S. Patra
Department of Computer Science and Engineering,
Tezpur University, Tezpur 784028, India
e-mail: swpatra@gmail.com

L.C. Jain et al. (eds.), *Computational Intelligence in Data Mining - Volume 2*,
Smart Innovation, Systems and Technologies 32, DOI 10.1007/978-81-322-2208-8_3

thresholding technique. A survey of various threshold selection techniques and their applications can be found in [4].

Thresholding techniques can be divided into bi-level and multilevel category, depending on the number of thresholds required to be detected. In bi-level thresholding, an image is segmented into two different regions depending on a threshold value selected from the histogram of the image [1, 2, 5]. The pixels with gray values greater than the threshold value are assigned into object region, and the rest are assigned into background. Multilevel thresholding segments a gray level image into several distinct regions by detecting more than one threshold [6].

Histogram of the image does not consider spatial contextual information of the image. Thus, the existing thresholding techniques based on the histogram are unable to incorporate the spatial contextual information in the threshold selection process. Recently, energy function is defined to generate the energy curve of an image by taking into an account the spatial contextual information [7]. The characteristics of this curve are similar to the histogram of an image. In this work instead of histogram, the energy curve of the image is analyzed to find out the optimal number of thresholds. To select multiple potential thresholds from the energy curve, concavity analysis technique presented in [8] is used. In order to select optimal number of thresholds from the list of potential thresholds obtained by applying concavity analysis technique, a threshold elimination technique based on cluster validity measure is proposed. The proposed threshold elimination technique remove some potential thresholds from the list to find out the optimal number of thresholds for automatic segmentation of the image.

The rest of the paper is organized as follows: Sect. 2 presents the proposed technique. Section 3 gives the detailed description of the data set and experimental results. Finally, Sect. 4 draws the conclusion of this work.

2 Proposed Method

In order to take into account the spatial contextual information of the image for threshold selection, the energy curve defined in [7] is used. The detailed steps of the proposed technique are given in the following subsections.

2.1 Energy Curve

Recently, we defined an energy function to generate the energy curve of an image [7]. This energy function computes the energy of an image at each gray value by taking into an account the spatial contextual information of the image. The characteristics of this energy curve are similar to the histogram of an image i.e., if the energy curve of an image include peaks, we can separate it into a number of modes. Each mode is expected to correspond to a region, and there exists a threshold at the valley between any two adjacent modes. Since the energy curve is generated by taking into an account the spatial contextual information of the image, it is more smooth and may contain better discriminatory capability as compared to that of histogram.

2.2 Concavity Analysis

After generating the energy curve of the image, concavity analysis technique presented in [8] is exploited to find out the list of potential thresholds from this energy curve. Let E be the energy curve of an image defined over the set of gray level $[0, L]$. For concavity analysis, we consider the subset of E say C defined over $[k_1, k_2]$ such that $C(k_1)$ and $C(k_2)$ are the first and last non-zero values of the energy curve E respectively. In order to find concavities of C, the smallest closed polygon P is constructed in the range $[k_1, k_2]$ by following the same steps as described in [8]. After that the vertical distance $V(i)$ between P and C at each gray level i, $i \in [k_1, k_2]$ is calculated as follows:

$$V(i) = P(i) - C(i) \tag{1}$$

Finally, from concavity region with respect to each edge of the polygon P, a gray value corresponding to maximum value of vertical distance $V(i)$ is selected as a potential threshold. The number of thresholds obtained by this technique may be larger than the number of objects in the image. In order to select optimal number of thresholds for segmentation the proposed threshold elimination technique is presented next.

2.3 Proposed Threshold Elimination Technique

In order to find the optimal number of thresholds for automatic segmentation, a cluster validity measure called Davies Boulding (DB) index is used [9]. It is a function of the ratio of the sum of within-object scatter to between object separations. Let $w_1, w_2, \ldots, w_k$, be the k objects defined by thresholds $t_1 < t_2 < t_3 < \cdots < t_k$. Then the DB index is defined as:

$$\begin{aligned} R_{ij} &= \frac{\sigma_i^2 + \sigma_j^2}{d_{ij}^2} \\ R_i &= \max_{j=1\ldots k; i \neq j} \{R_{ij}\} \\ DB &= \frac{1}{k}\sum_{1}^{k} R_i \end{aligned} \tag{2}$$

where σ_i^2 and σ_j^2 are the variances of object ω_i and ω_j, respectively, and d_{ij} is the distance of object centers ω_i and ω_j. Smaller the DB value, better is the segmentation as a low scatter and a high distance between object lead to small values of R_{ij}.

Let $T = \{t_1, t_2, \ldots, t_k\}$ be the set of k potential thresholds obtained by applying concavity analysis technique as described in Sect. 2.2. To find out the optimal number of objects present in the image, first, DB-index is calculated by taking into account all the k potential thresholds. In the next step, DB-index is calculated: (i) by dropping the left most trailing potential threshold t_1 from T, (ii) by dropping the right most trailing potential threshold t_k from T. If the DB-index obtained by (i) is smaller than the DB-index obtained by (ii), then drop t_1 else drop t_k from T. The process is repeated until T contain single potential threshold. As a result, different combinations of potential thresholds and their corresponding DB-index values are obtained. Finally the combination associated with smallest DB-index is selected to segment the image.

3 Experimental Results

In order to assess the effectiveness of the proposed technique four different images Man, Cameraman, Boat and Lena are considered. Figure 1e–l shows the smallest polygon obtained on histogram and energy curve of different images respectively, by applying concavity analysis technique. From these figures it is concluded that energy curve of the image is more suitable as compared to the histogram to find out

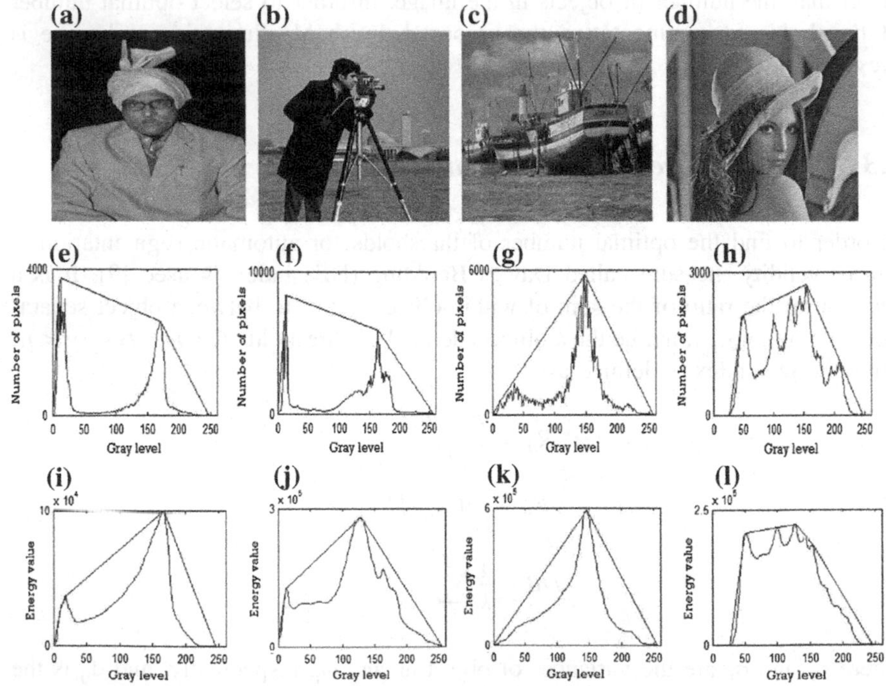

Fig. 1 Original images: **a** Man, **b** Cameraman, **c** Boat and **d** Lena. Smallest polygon that covers (**e–h**) the histogram and (**i–l**) the energy curve of the images (**a–d**)

Table 1 Combinations of potential thresholds and corresponding DB-index generated by the proposed threshold elimination technique

Image	Histogram		Energy curve	
	Thresholds	DB-index	Thresholds	DB-index
Man	5 10 13 39 172 183	0.4903	11 81 185	0.5662
	10 13 39 172 183	0.5455	11 81	0.5693
	5 10 13 39 172	0.4537	81 185	0.2046
	10 13 39 172	0.5126	**81**	**0.0291**
	5 10 13 39	0.2401	185	0.5219
	5 10 13	0.3767		
	10 13 39	0.2604		
	10 13	0.3865		
	13 39	0.2138		
	13	0.3816		
	39	**0.0397**		
Cameraman	13 26 192	0.3524	7 82 194	0.5406
	26 192	0.2461	7 82	0.6036
	13 26	0.2828	82 194	0.1566
	26	**0.0890**	**82**	**0.0613**
	13	0.2272	194	0.4101
Boat	8 115 179	0.2301	**117 184**	**0.1565**
	115 179	**0.1652**	117 184	0.1769
	8 115	0.2487		0.3401
	115	0.1672		
	179	0.3760		
Lena	37 129 168 225	0.2470	39 69 111 136 185	0.1872
	37 129 168	0.2568	36 69 111 136	0.2456
	129 168 225	**0.2160**	**69 111 136 185**	**0.1675**
	129 168	0.2187	69 111 136	0.2355
	168 225	0.2333	111 136 185	0.1950
	129	0.2323	111 136	0.2894
	168	0.2524	136 185	0.2080
			111	0.1981
			136	0.2466

The bold one is the optimal number of thresholds selected by the proposed technique

potential thresholds due to its smoothness property. Table 1 shows the different combinations of potential thresholds and their associated DB-index values obtained by using the proposed threshold elimination technique considering both the energy curve and the histogram of the images. From this table one can see that for all the considered images, the proposed technique produced better results when applied to energy curve of the images.

Fig. 2 Segmented images with respect to optimal thresholds obtained by proposed technique using **a–d** histogram, and **e–h** energy curve of the images

The optimal results obtained by the proposed technique as shown in Table 1 are highlighted in bold. For the Man and the Cameraman images, the proposed technique automatically finds out single threshold to segment the images into two regions. For the boat image it gives two thresholds to segment the image into three homogenous regions. For the Lena image, the proposed technique automatically finds three and four thresholds by analyzing histogram and energy curve of the image respectively. For qualitative analysis, Fig. 2 shows the segmented images obtained from the proposed technique considering the histogram and the energy curve of the images.

From these figures one can visualize that the proposed technique based on energy curve produced much better results.

4 Discussion and Conclusion

In this work we developed a thresholding technique that works on the context sensitive energy curve of the images to find out optimal number of thresholds for solving image segmentation problem. In our method, first concavity analysis technique existing in literature is applied on the energy curve to detect all the potential thresholds. After detecting a list of potential thresholds, a novel threshold elimination technique is used to remove some potential thresholds to find optimal number of thresholds for automatic segmentation of the image. To assess the effectiveness of proposed method four different images are considered in the experiment. From the experiment, it is observed that the proposed technique provides better results as compared to that applied on histogram of the image. This is mainly because the histogram does not consider the spatial contextual information.

References

1. Otsu, N.: A threshold selection method from gray level histograms. IEEE Trans. Syst. Man Cybern. **9**, 62–66 (1979)
2. Kapur, J.N., Sahoo, P.K., Wong, A.K.C.: A new method for gray-level picture thresholding using the entropy of the histogram. Comput. Vis. Graph. Image Process. **29**, 273–285 (1985)
3. Patra, S., Ghosh, S., Ghosh, A.: Histogram thresholding for unsupervised change detection of remote sensing images. Int. J. Remote Sens. **32**, 6071–6089 (2011)
4. Sezgin, M., Sankur, B.: Survey over image thresholding techniques and quantitative performance evaluation. J. Electron. Imaging **13**, 146–165 (2004)
5. Huang, L.K., Wang, M.J.J.: Image thresholding by minimizing the measures of fuzziness. Pattern Recogn. **28**, 41–51 (1995)
6. Huang, L.K., Wang, M.J.J.: Thresholding technique with adaptive window selection for uneven lighting image. Pattern Recogn. Lett. **26**, 801–808 (2005)
7. Patra, S., Gautam, R., Singla, A.: A novel context sensitive multilevel thresholding for image segmentations. Appl. Soft Comput. **23**, 122–127 (2014)
8. Rosenfeld, A., Torre, P.D.: Histogram concavity as an aid in threshold selection. IEEE Trans. Syst. Man Cybern. **13**, 231–235 (1983)
9. Duda, R.O., Hart, P.E., Stork, D.G.: Pattern Classif. Wiley, Singapore (2001)

Encryption for Massive Data Storage in Cloud

Veeralakshmi Ponnuramu and Latha Tamilselvan

Abstract Cloud is an evolving computing technology in which the services are provisioned dynamically through internet. The users' data are stored in the remotely located servers that are maintained by cloud providers. The major security challenge in cloud is that the users cannot have direct control over the remotely stored data. To preserve the privacy of user's data stored in the cloud, an efficient algorithm maintaining the confidentiality of data is proposed. We ensure that the data stored in the untrusted cloud server is confidential by developing a new data encryption algorithm. Unlike other encryption algorithms, our encryption algorithm needs lesser computation overhead. Encryption and decryption algorithms are developed in java and Remote Method Invocation (RMI) concepts are used for communication between client and server. Simulation environment is set up with the eucalyptus tool. This technique provides the data confidentiality with minimum computational overhead of the client.

Keywords Confidentiality · Data encryption · Cloud security · Massive storage

1 Introduction

Cloud is an on-demand, pay-by-use model for sharing a pool of computing resources like servers, CPU cycles, memory, applications, storage and services that is managed by cloud service providers. The services can be easily provisioned from the cloud providers and released with minimum endeavour by the cloud users. With cloud, IT infrastructure can be easily adjusted to accommodate the changes in demand.

V. Ponnuramu (✉) · L. Tamilselvan
B.S. Abdur Rahman University, Chennai, Tamilnadu, India
e-mail: veerphd1@gmail.com

L. Tamilselvan
e-mail: latha_tamilselvan@yahoo.com

L.C. Jain et al. (eds.), *Computational Intelligence in Data Mining - Volume 2*,
Smart Innovation, Systems and Technologies 32, DOI 10.1007/978-81-322-2208-8_4

There are various [1] issues like security, scalability, availability, resource scheduling, data migration, memory management, data privacy, data management, reliability, load balancing, access control in cloud. Cloud moves the applications, softwares and databases to the large data centers that are located anywhere in the world, where the servers can not be trustworthy. This unique feature of cloud imparts many new security challenges. The cloud offers many benefits like enhanced collaboration, limitless flexibility, portability and simpler devices. To enjoy the benefits of cloud, the users have to store their data in the encrypted format.

In symmetric algorithms, the same key can be used for both encryption and decryption. Symmetric algorithms are highly secured and can be executed in high speed. In case of asymmetric algorithms, different keys are used for encryption and decryption. Asymmetric encryption algorithms (called as public-key algorithms) need a key of 3,000 bits to produce the same level of security as that of a symmetric algorithm with a 128-bit key. Since the asymmetric encryption algorithms are slow, they cannot be used for encrypting bulk of data. In this paper, a new symmetric encryption algorithm that can be used for encrypting massive data has been proposed.

2 Related Work

2.1 Security Issues in Cloud

There are numerous security issues in cloud as the customers are not having direct control over the stored data in cloud. Jensen et al. [2] discussed the security issues arising from the usage of cloud services and by the technologies used to build the internet-connected and cross-domain collaborations. It emphases on browser security, WS-security, cloud integrity, transport layer security, and binding issues in the field of cloud.

2.2 Merkle Hash Tree (MHT)

Wang et al. [3] verified the correctness of data stored in server by allowing the third parity auditor. With the aid of Merkle hash tree it is possible for the clients to perform block-level operations on the data files by preserving the level of data correctness assurance. In this scenario, chances are there for the third party auditor to misuse the data while doing the verification operation. Lifei et al. [4] established a new mechanism to verify the correctness of computations (addition, subtraction, multiplication, division, etc.) done by the cloud provider. For that, they have used the Merkle hash tree for checking the computation correctness. The only criteria is the number of computations submitted to the server must be in the power of 2, since the Merkle hash tree has the 2^n number of leaves.

2.3 Proof of Retrievability Scheme (POR)

For verifying the data integrity some sentinel characters were embedded in the data file by Juels and Kaliski [5]. These sentinels were hidden in the data blocks. In the verification phase, the user can challenge the server by mentioning the positions of sentinels and request the provider to return the relevant sentinel values. This procedure allows the user to challenge the server for a limited number of times by knowing the positions of the sentinel values in advance. Ateniese et al. [6] proposed a new model called "Provable Data Possession" to ensure the possession of files stored on the untrusted server. They used RSA- based homomorphic tags for assessing outsourced data. Here also the user needs to pre-compute the tags and store all the tags in advance. The computation of tags requires a lot of computation overhead and storage space. The homomorphic properties were also used to check the integrity of data [7]. A storage correctness model for verifying the correctness of stored data by calculating a few number of precomputed tokens was proposed by Wang et al. [8]. For ensuring the remote integrity, the MAC and reed solomon code were used [9]. Integrity was assured by the algorithm suggested by Wang et al. [10]. But confidentiality was not discussed by them. An encryption algorithm is said to be computationally secure if it cannot be easily broken by the cryptanalyst. To protect sensitive data, a new encryption algorithm has been designed in this paper, which offers a higher level of security as compared to the present encryption algorithms.

3 Problem Definition

The important security issue in cloud computing is the integrity and confidentiality of data. Providers should ensure that the sensitive data is disclosed only to authorized users. Service providers should ensure that the applications hosted as a service through the cloud are secure. It is presumed that the cloud model comprises of n cloud servers(s1, s2,...sn) under the direct control of cloud service providers (CSP) and m cloud users (u1, u2,...un). The cloud service providers are maintaining servers across the world and the users data will be stored in the servers. The users are unknown about the location of their data. So the users must store their sensitive data in the encrypted format. There are many encryption algorithms like DES, 3DES, RC4, AES, etc. These encryption algorithms are too complex to apply for a large data file. In this article, an attempt is made to propose a novel symmetric encryption algorithm with minimum computation overhead. For data confidentiality and integrity, the client has to do the following sequence of operations.

1. Binary Sequence Generation
2. Key Generation
3. Encryption of data
4. Metadata generation for the encrypted data
5. Storing the metadata in the cloud server.

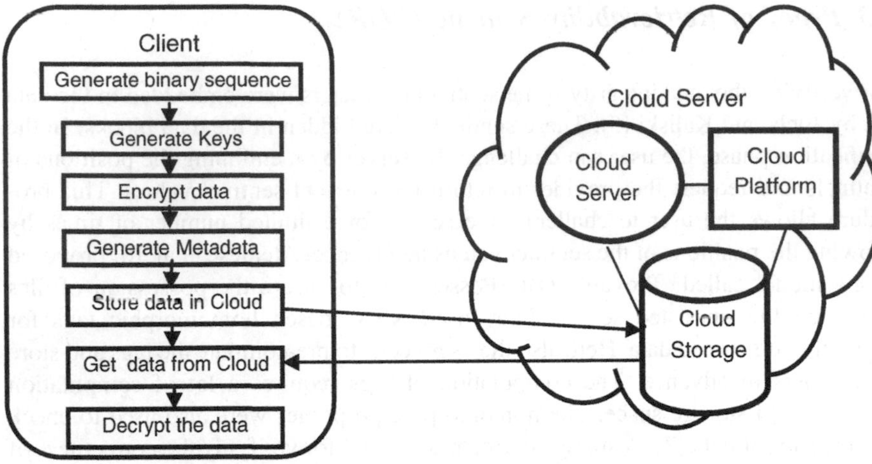

Fig. 1 System architecture

6. Integrity Verification
7. Decryption of data.

The operations metadata generation, Storing the metadata and integrity verification were conversed by Ponnuramu and Tamilselvan [11]. The remaining three operations are narrated in this paper. The operation sequence is depicted in the Fig. 1.

3.1 Binary Sequence Generation

A binary sequence is generated using the recurrence relation of the form as in the Eq. 1

$$X_{n+m} = (C_0X_n + C_1X_{n+1} + C_2X_{n+2} + \cdots C_{m-1}X_{n+m-1}) \bmod 2 \quad (1)$$

For generating the recurrence relation, the user has to choose the value m, the initial vector values like (X_1, X_2, X_3, X_4,… X_m) and the coefficient values like (C_0, C_1, C_2, C_3,…C_{m-1}). For example For the initial vector (0, 1, 0, 0, 0) ie. $X_1 = 0$; $X_2 = 1$; $X_3 = 0$; $X_4 = 0$; $X_5 = 0$ and for the coefficient (1, 0, 1, 0, 0) ie. $C_1 = 1$; $C_2 = 0$; $C_3 = 1$; $C_4 = 0$; $C_5 = 0$, the recurrence relation is like the Eq. 2.

$$X_{n+5} = X_n + X_{n+2} \quad (2)$$

The binary sequence generated for the initial vector (0, 1, 0, 0, 0) and the coefficient (1, 0, 1, 0, 0) is 0100001001011001111100011011101010000,…

3.2 Key Generation

1. The client chooses two keywords of any length (keyword1, keyword2) and one random number **randno of** any length.
2. From the keywords, generate two keys by using the following steps.
3. To generate key1

    i. Initialize the key1 as key1 = 0 and i = 0.

    ii. $$\mathrm{key1} = \mathrm{key1} + (\mathrm{ASCII}(\mathrm{keyword1}[\mathrm{i}]) * (\mathrm{i} + 1)) \tag{3}$$

    i → the index of each character in the keyword1.

    iii. Repeat the step 2 for all the characters in the keyword1.

4. Generate the key2 using the step 3 with the keyword2

3.3 Algorithm for Encryption

Divide the data file **F** into **p** data blocks **db1, db2, db3,... dbp**. Consider each **p** data blocks contains **q** bytes like **b1, b2, b3,... bq**

1. Binary sequence is generated so that the number of bits in the binary sequence in p data blocks.
2. Read the first bit in the binary sequence.
3. If the bit value is one, encrypt the block with the **key1** using the Eq. 4.

$$\mathbf{Encrypt[i,j]} = (\mathbf{key1} + \mathbf{ASCII(F[i,j])} + \mathbf{randno}^{(\mathbf{i+j})})\mathrm{mod}256 \tag{4}$$

i → refers to the block number
j → index of the character in the ith block.

4. If the bit value is zero, encrypt the block with the **key2** using the Eq. 5.

$$\mathbf{Encrypt[i,j]} = (\mathbf{key2} + \mathbf{ASCII(F[i,j])} + \mathbf{randno}^{(\mathbf{i+j})})\mathrm{mod}256 \tag{5}$$

5. Write the Encrypt[i, j] in the encdata.txt file
6. Repeat the steps 3 or 4 and 5 for all the characters in that block.
7. Read the next bit in the binary sequence.
8. Repeat the steps 3–7 for the datablocks in the text file.

3.4 Algorithm for Decryption

Generate the binary sequence from the recurrence relation initial vector and the coefficients and the keys key1, and key2 from the keywords keyword1, keyword2 respectively for decryption. Divide the encrypted datafile encdata.txt into **p** data blocks **db1, db2, db3,... dbp.** Consider each **p** data blocks contains **q** bytes like **b1, b2, b3,... bq**

1. Binary sequence is generated so that the number of bits in the binary sequence in n data blocks.
2. Read the first bit in the binary sequence.
3. If the bit value is one, encypt the block with the **key2** using the Eq. 6.

$$\mathbf{Decrypt[i,j]} = (\mathbf{encdata[i,j]} - \mathbf{key2}) - \mathbf{randno}^{(\mathbf{i+j})})\mathrm{mod}256 \tag{6}$$

   i → refers to the block number
   j → index of the character in the ith block.
4. If the bit value is zero, encrypt the block with the **key3** using the Eq. 7.

$$\mathbf{Decrypt[i,j]} = (\mathbf{encdata[i,j]} - \mathbf{key3}) - \mathbf{randno}^{(\mathbf{i+j})})\mathrm{mod}256 \tag{7}$$

   i → refers to the block number
   j → index of the character in the ith block.
5. If Decrypt[i, j] < 0 then do the Eq. 8

$$\mathbf{Decrypt[i,j]} = \mathbf{decrypt[i,j]} + \mathbf{256} \tag{8}$$

   i → refers to the block number
   j → index of the character in the ith block.
6. Write the decrypt[i, j] in the decryptdata.txt file.
7. Read the next bit in the binary sequence.
8. Repeat the steps 3–7 for the datablocks in the text file.

The information to be kept secret by the cloud user are Initial Vector values. (X_1, X_2, X_3, X_4,... X_m), Initial Coefficients. (C_0, C_1, C_2, C_3,...C_{m-1}), keyword1, keyword2, Random number randno, Number of characters in the block −q. In this, the initial vector values and the coefficients are used for generating the binary sequence. From the keyword1 and keyword2, two keys are generated. The binary sequence, the keys, randno, q are used for data encryption.

4 Simulation

A private cloud environment has been established with the open source eucalyptus cloud simulator tool. Ubuntu Enterprise Cloud (UEC) is the integrated tool consisting of ubuntu with eucalyptus. It can also host Microsoft Windows images. To install and configure an Ubuntu enterprise cloud, two Servers (Serverl and Server2) that run with 32-bit, 1 GHz server version and another machine that run as a Desktop 32-bit version (Clientl) are required. The ubuntu desktop version is installed on Client1 so that the browsers can be used to access the web interface of UEC. The experiment is conducted using a setup consisting of two servers (1 GHz, 2 GB RAM). and one desktop (1 GHz, 2 GB RAM). Encryption and decryption algorithms are implemented in java and communication between client and server is implemented with the java remote method invocation concepts.

5 Results

The encryption algorithm was implemented in Java. The code for generating the binary sequence, key generation have been successfully executed. Using the generated binary sequence and key, the plain text was encrypted. The results are illustrated with the Example1.

Example 1

Initial vector values—(0, 1, 0, 0, 0) Initial coefficients—(1, 0, 1, 0, 0).

The binary sequence generated is shown in the Fig. 2.

The keyword1 is shown in the Fig. 3.

The keyword2 is shown in the Fig. 4.

```
010010110011111000110111010100001001011001111100011011101010000100101100111
110001101110101000010010110011111000110111010100001001011001111100011011101
010000100101100111110001101110101000010010110011111000110111010100001001011
001111100011011101010000100101100111110001101110101000010010110011111000110
111010100001001011001111100011011101010000100101100111110001101110101000010
010110011111000110111010100001001011001111100011011101010000100101100111110
```

Fig. 2 Binary sequence

```
welcometoourdepartmentBSAUniversityaaaaabbbbbbbbbbbbbwwwwwwwwwwwwwwww
wwwwwrrrrrrrrrrrrrr
```

Fig. 3 Keyword1

```
Madhankumar Veeralakshmi LathaTamilselvan
```

Fig. 4 Keyword2

```
aaaaaaaaabbbbbbbbbbbccccccccccccddddddddddddeeeeeeeeeeeeffffff
import java.io.File;
import java.io.FileInputStream;
import java.io.IOException;
public class fread {
  public static void main(String[] args) {
    File file = new File("fread.txt");
            int i=1;
    if (!file.exists()) {
```

Fig. 5 Plain text

```
"!$#&%((+*-,/.103²¶µ¸·º¹¼»¾½À¿ÃÂÅÄÇÆ%$'&)(++.-
/2143658&)(+*ÓÓ18:<>Cî;1I3?½Æ??ÁÇ¿?iiÇÎÐÒÔÙ?Ñ#;%õ/8ö391<AEG%IF¼·Æ?hdÆÉÏÍÓ
Ô?ÌÆÚÈ?ÒÅ?¡ª?Õ¿ÄÎÕÉÒÐ qqssØ<(51.ê080ADð9D:5;öTÓÓèë:B.;74ðFF6H@9ùNÊÃÁ|Ì¿Ê
Î?µÙÖÐÔÐÃÈ?ÎÌÄÏ?~Ümm????¬ÒÔÐ?ÓÕÉÁ  ??ÎÈÙ?ªÐÒÎ??ÐßÑÐÀ?ÒÙÔ?? qqoÒÖß?Ö©
©Ih????ÎÊ???ÎÔÖÒ?ÔæÚ?CAùùüòPááöùøûúý/XQC3>þBGIGHBFOFK@PHA,           0ò
ôùö=G@MýJNREÛËØØ???£xt??????å×é××Ò¢ss????é|x????ÞÚóúöü=?E=        CP"HJF
??ÌÒÔÐ?ÐÍÝÀÖÑ×??? ?âss??????Áêãç×â¢æëí?ÙÚÔØáØÝ?×Ùß×£ÛÜêÇÙD;ûýþÿABNQQYI
```

Fig. 6 Cipher text

The keys generated from the keyword1 to keyword2 are key1-472974 and key2-86780. The random number used is 193 and the number of characters in a block is 20.

The part of plain text is shown in the Fig. 5.

By applying the proposed encryption algorithm, for the plain text as in the Fig. 5, we got the cipher text as like in the Fig. 6.

6 Discussions

The cryptanalysis method computes the number of occurrences of characters in the cipher text and plain text. And it also compares their frequencies. With this method, experimental analysis are conducted for different strings to measure the level of security of data. To prove the correctness of this cryptographic module, spectral analysis of the frequency of characters are used. First, as a test case we chose, a file which contained the plaintext "aaaaaaaaabbbbbbbbbbbccccccccccccddddddddddddeeeeeeeeeeeeffffff"

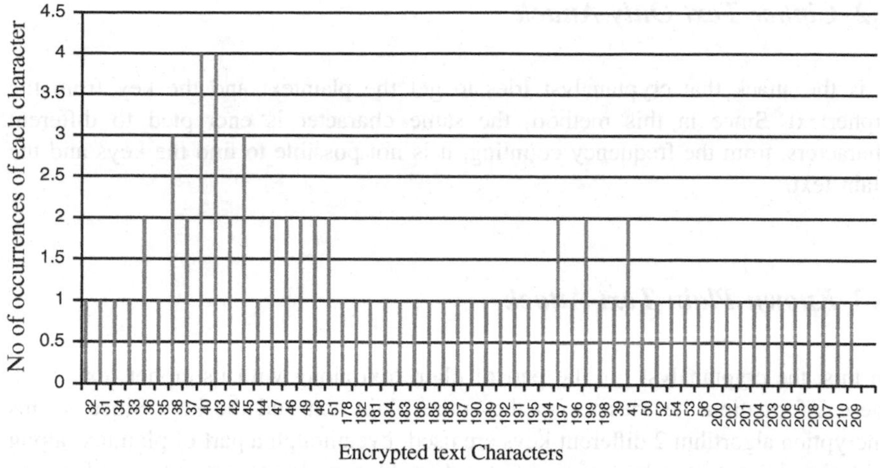

Fig. 7 Frequency of characters of encrypted data

Table 1 Time of encryption and decryption of different files

File size (KB)	Encryption time (ms)	Decryption time (ms)
1	15	20
2	25	27
3	32	35
4	35	47
5	40	50
6	44	46
7	50	48
8	51	48
9	59	54
10	64	65

and used this method to encrypt the data. Figure 7 shows the frequency of characters of the encrypted data. Table 1 displays the time required for encryption and decryption of files of different sizes.

6.1 Brute Force Attack

In this method, two keywords are used for encryption and also the keywords are of variable sizes. Along with the keywords, initial vector values. (X_1, X_2, X_3, X_4,... X_m) initial coefficients (C_0, C_1, C_2, C_3,...C_{m-1}). random number randno, number of characters in the block –q. are all the secret information maintained by the cloud user. So finding the key with the brute force attack is not possible with this system.

6.2 Cipher Text Only Attack

It is the attack that cryptanalyst tries to get the plaintext and the key from the ciphertext. Since in this method, the same character is encrypted to different characters, from the frequency counting, it is not possible to find the keys and the plain text.

6.3 Known Plain Text Attack

In this, the cryptanalyst got the part of plain text along with its cipher text. With these information, he tries to get the entire plain text and the key. Since in this encryption algortihm 2 different keys are used, eventhough a part of plaintext along with the cipher text is known to cryptanalyst, he can able to get only one key and not the second key.

6.4 Correlation Attacks

The attacks try to extract some information about the initial state from the output stream. Here, since the recurrence relation has been used to generate binary sequence, the attacker can try to get the initial seed (initial vector and initial coefficients) from the binary sequence. If the length of the initial seed is smaller, then the binary sequence will be getting repeated. Then is possible for a hacker to get initial seed. In order to avoid this the user must choose a larger initial seed.

7 Conclusion

In this paper an efficient block cipher, symmetric encryption algorithm has been proposed. This algorithm is not vulnerable to brute force attack, cipher text only attack and known plaintext attacks. Comparing to other encryption algorithms, it takes only less computation overhead and the time for encryption and decryption is very less.

The proposed scheme can be applied for massive data with minimum overhead. This algorithm can also be used for encryption of audio and video files. In future, this can be used for the encryption of data that is dynamically changing.

References

1. Pearson, S.: Taking account of privacy when designing cloud computing services. Proceedings of ICSE-Cloud'09, Vancouver (2009)
2. Jensen, M. et al.: On technical security issues in cloud computing. In: IEEE International Conference on Cloud Computing, pp. 109–116 (2009)
3. Wang, Q., Wang, Q., Li, J., Ren, K., Lou, W.: Enabling public auditability and data dynamics for storage security in cloud computing. IEEE Trans. Parallel Distrib. Syst. **22**(5), 847–858 (2011)
4. Wei, L., Zhu, H., Cao, Z., Jia, W.: SecCloud: bridging secure storage and computation in cloud. In: ICDCS'10, pp. 52–61 (2010)
5. Juels. A., Kaliski, B.S.: Pors: proofs of retrievability for large files. In: Proceedings of the 14th ACM Conference on Computer and Communications Security, pp. 584–597 (2007)
6. Ateniese, G., Burns, R., Curtmola, R., Herring, J., Kissner, L., Peterson, Z., Song, D.: Provable data possession at untrusted stores. In: Proceedings of the 14th ACM Conference on Computer and Communications Security, pp. 598–609, ACM, New York, NY, USA (2007)
7. Shacham, H., Waters, B.: Compact proofs of retrievability. In: Proceedings of 14th International Conference on Theory and Application of Cryptology and Information Security: Advances in Cryptology (ASIACRYPT '08), pp. 90–107 (2008)
8. Wang, C., Wang, Q., Ren, K., Lou, Q.: Ensuring data storage security in cloud computing. In: Proceedings of IWQoS'09, pp. 1–9 (2009)
9. Chang, E.C., Xu, J.: Remote integrity check with dishonest storage server. In: Proceedings of ESORICS'08, pp. 223–237, Springer, Berlin, Heidelberg (2008)
10. Wang, C., Wang, Q., Chow, S.S.M., Ren, K., Lou, W.: Privacy-preserving public auditing for secure cloud storage. IEEE Trans. Comput. **62**(2), 362–374 (2013)
11. Veeralakshmi, P., Tamilselvan, L.: Data integrity proof and secure computation in cloud computing. J. Comput. Sci. **8**, 1987–1995 (2012)

An Integrated Approach to Improve the Text Categorization Using Semantic Measures

K. Purna Chand and G. Narsimha

Abstract Categorization of text documents plays a vital role in information retrieval systems. Clustering the text documents which supports for effective classification and extracting semantic knowledge is a tedious task. Most of the existing methods perform the clustering based on factors like term frequency, document frequency and feature selection methods. But still accuracy of clustering is not up to mark. In this paper we proposed an integrated approach with a metric named as Term Rank Identifier (TRI). TRI measures the frequent terms and indexes them based on their frequency. For those ranked terms TRI will finds the semantics and corresponding class labels. In this paper, we proposed a Semantically Enriched Terms Clustering (SETC) Algorithm, it is integrated with TRI improves the clustering accuracy which leads to incremental text categorization. Our experimental analysis on different data sets proved that the proposed SETC performing better.

Keywords Text categorization · Clustering · Semantic knowledge · Term rank identifier · Semantically enriched terms clustering

1 Introduction

Today the world became web dependent. With the booming of the Internet, the World Wide Web contains a billion of textual documents. To extract the knowledge from high dimensional domains like text or web, our search engines are not enough smart to provide the accurate results. This factor leads the WWW to urgent need for effective clustering on high dimensional data.

K. Purna Chand (✉) · G. Narsimha
Department of CSE, JNTU College of Engineering, Kakinada, Andhra Pradesh, India
e-mail: purnachand.k@gmail.com

G. Narsimha
e-mail: narsimha06@gmail.com

L.C. Jain et al. (eds.), *Computational Intelligence in Data Mining - Volume 2*,
Smart Innovation, Systems and Technologies 32, DOI 10.1007/978-81-322-2208-8_5

Many traditional approaches are proposed and developed to analyze the high dimensional data. Text Clustering is one of the best mechanisms to identify the similarity between the documents. But most of the clustering approaches are depends upon the factors like term frequency, document frequency, feature selection and support vector machines (SVM). But there is still uncertainty while processing highly dimensional data.

This research is mainly focuses on improving the text categorization on text document clusters. The proposed TRI and SETC will boost up the text categorization by providing semantically enriched document clusters. The primary goal is to measure the most frequent terms occurring on any text document clusters with our proposed metric Term Rank Identifier (TRI). For those frequent terms the semantic relations are calculated with Wordnet Tools. The basic idea behind the frequent item selection is to reduce the high dimensionality of data. The secondary goal is to apply our proposed text clustering algorithm Semantically Enriched Terms Clustering (SETC) to cluster the documents which are measured by TRI.

2 Related Work

There exist two categories of major text clustering algorithms: Hierarchical and Partition methods. Agglomerative hierarchical clustering (AHC) algorithms initially treat each document as a cluster, uses different kinds of distance functions to compute the similarity between all pairs of clusters, and then merge the closest pair [1]. On other side Partition algorithms considers the whole database is a unique cluster. Based on a heuristic function, it selects a cluster to split. The split step is repeated until the desired number of clusters is obtained. These two categories are compared in [2].

The FTC algorithm introduced in used the shared frequent word sets between documents to measure their closeness in text clustering [3]. The FIHC algorithm proposed in [4] went further in this direction. It measures the cohesiveness of a cluster directly by using frequent word sets, such that the documents in the same cluster are expected to share more frequent word sets than those in different clusters. FIHC uses frequent word sets to construct clusters and organize them into a topic hierarchy. Since frequent word sequences can represent the document well, clustering text documents based on frequent word sequences is meaningful. The idea of using word sequences for text clustering was proposed in [5]; However, STC does not reduce the high dimension of the text documents; hence its complexity is quite high for large text databases.

The sequential aspect of word occurrences in documents should not be ignored to improve the information retrieval performance [6]. They proposed to use the maximal frequent word sequence, which is a frequent word sequence not contained in any longer frequent word sequence. So, in view of all the text clustering algorithms discussed above we proposed TRI and SETC.

Table 1 General notation of 2 × 2 contingency table

Category Term	Category 1	Category 2	Total
Term 1	a	b	a + b
Term 2	c	d	c + d
Total	a + c	b + d	a + b + c + d = n

2.1 *Traditional Text Categorization Measures*

2.1.1 χ^2 Statistics

In text mining for the information retrievals, we frequently use χ^2 Statistics in order to measure the term frequencies and term-category dependencies. It can be done by measuring the co-occurrences of the terms and listed in contingency tables (Table 1). Suppose that a corpus contains n labeled documents, and they fall into m categories. After the stop words removal and the stemming, distinct terms are extracted from the corpus.

For the χ^2 term-category dependency test, we consider two strategies one is the null hypothesis and the alternative hypothesis. The null hypothesis states that the two variables, term and category, are independent of each other. On the other hand, the alternative hypothesis states that there is some dependency between the two variables.

General formula to calculate the dependency is

$$\chi^2 = \sum_{i=1}^{k} \left| \frac{(O_i - E_i)^2}{E_i} \right| \tag{1}$$

where

O_i—the observed frequency in the ith cell of the table.
E_i—the expected frequency in the ith cell of the table

The degrees of freedom are (r − 1) (c − 1). Here r = # of rows and c = # of columns.

2.2 *Term Rank Identifier (TRI)*

In our exploration, we found that χ^2 does not fully explore all the information provided in term-category independence test. We point out where the problem is due to identifying only positive term category dependencies based upon the frequent words. In view of this, we proposed a new term-category dependency measure, denoted TRI, which identifies highly related terms based upon their frequencies and each term is assigned with ranks and is categorized by its semantics.

Table 2 Term-ranking based on their frequencies

Category / Term	C_1	C_2	C_3	Frequency	Rank
T_1	d_1	d_1, d_4	d_3	5	1
T_2	d_1, d_2	d_1, d_2		4	2
T_3	d_5	d_2		2	4
T_4	d_2, d_5	d_4		3	3

Table 3 Calculating semantically related terms

Category / Term	C_1	C_2	C_3
T_1	T_2, T_3	T_2, T_3	T_2, T_3
T_2	T_1	T_1	
T_3	T_1		T_2
T_4	T_2	T_2	
Total terms (union)	3	3	2

Example 1 For suppose a database D consists of 5 documents D = {d_1, d_2, d_3, d_4, d_5} are categorized as three categories c_1 = {d_1, d_2, d_5}, c_2 = {d_1, d_2, d_4} and C_3 = {d_3} and we observed four different terms t_1, t_2, t_3 and t_4.

The above illustrated example is represented in Table 2. If we observe closely that the term T_1 almost all occurred in all documents except in d_2, d_5. And coming to the term T_2 even its rank is 2 but it is occurred only in d_1, d_2 documents. Likewise by analyzing all the occurrences of different terms we concluded that term-category frequency is not much better in all cases. So our proposed metric Term Rank Identifier (TRI) measures the semantic relatedness (Table 3) of each term in every document.

So from Table 3 we can say that the terms T_1, T_2 and T_3 are semantically related to each and every category. Compare to c_3; c_1 and c_2 categories consists of highly related terms. So we can determine that documents of c_1 = {d_1, d_2, d_5}, c_2 = {d_1, d_2, d_4} and consisting of similar information and these documents are clustered by our proposed Semantically Enriched Terms Clustering (SETC) Algorithm.

3 Proposed Text Clustering Algorithm

3.1 Overview of Text Clustering

In many traditional text clustering algorithms, text documents are represented by using the vector space model [7]. In this model, each document d is considered as a vector in the term-space and is represented by term-frequency (TF) vector: Normally, there are several preprocessing steps, including the stop words removal and the stemming, on the documents. A widely used refinement to this model is to weight each term based on its inverse document frequency (IDF) [8] in the corpus.

For the problem of clustering text documents, there are different criterion functions available. The most commonly used is the cosine function [8]. The cosine function measures the similarity between two documents as the correlation between the document vectors representing them.

For two documents d_i and d_j, the similarity can be calculated as

$$\cos(d_i, d_j) = d_i * d_j / \|d_i\| \ \|d_j\| \tag{2}$$

where * represents the vector dot product, and $\|d_i\|$ denotes length of vector 'd_i'. The cosine value is 1 when two documents are identical and 0 if there is nothing in common between them. The larger cosine value indicates that these two documents share more terms and are more similar. The K-means algorithm is very popular for solving the problem of clustering a data set into k clusters. If the dataset contains n documents, $d_1; d_2;\ldots; d_n$, then the clustering is the optimization process of grouping them into k clusters so that the global criterion function is either minimized or maximized.

$$\sum_{j=1}^{k} \sum_{i=1}^{n} f(d_i, Cen_j) \tag{3}$$

where Cen_j represents the centroid of a cluster cj, for $j = 1;\ldots; k$, and $f(d_i,Cen_j)$ is the clustering criterion function for a document d_i, and a Centroid Cen_j. When the cosine function is used, each document is assigned to the cluster with the most similar centroid, and the global criterion function is maximized as a result.

3.2 Semantically Enriched Terms Clustering (SETC)

In the previous section we described that our proposed metric TRI identifies the semantically highly related terms. The semantic relativeness is calculated with the help of Wordnet 3.0. (Lexical Semantic Analyzer). It is used to calculate the synonyms and estimated relative frequencies of given terms.

Algorithm: The objective of the algorithm is to generate semantically highly related terms

Input: Set of different text documents and Wordnet 3.0. for Semantics.
Output: Categorized Class labels which generates taxonomies.

Step 1: Given a collection of text documents $D = \{d_1, d_2, d_3, d_4, d_5\}$. Finds the unigrams, bigrams, trigrams and multigrams for every document.

Unigram—Frequently Occurring 1 Word

Bigram—Frequently Occurring 2 Words
Trigram—Frequently Occurring 3 Words
Multigrams—Frequently Occurring 4 or more Words.

Step 2: Assign ranks to the each term based upon their relative frequencies in a single document or in clustered documents.

$$\textbf{Rank} = \textbf{Term Frequency (TF)},\ \textbf{Min\_Support} = \textbf{2}$$

Step 3: Identify the semantic relationship between the terms by using a Lexical Semantic Analyzer **Wordnet 3.0**

$$\textbf{Sem\_Rel(Terms)} = \textbf{Synonyms or Estimated Relative Frequency}$$

Step 4: Categorizing the semantically enriched terms into different categories by assigning the class labels.

Step 5: Construct taxonomies which are generated by class labels.

Primarily, we considered a single document d_1 and measured the term-category dependency and identified frequent terms and these terms are assigned with ranks based upon their frequencies in that particular document d_1. Next the semantic related ness between each terms can be measured with our metric TRI and terms are categorized according to synonymy and expected related frequencies with the help of Wordnet 3.0. Lexical Semantic Analyzer. Like that each document $d_2 \ldots d_n$ can be categorized with the help of our proposed metric TRI.

Later, our proposed Semantically Enriched Terms Clustering (SETC) Algorithm clusters all the documents into k no of clusters. Our proposed method is quite differentiated from traditional K-Means and K-Medoids partition algorithms. These algorithms do clustering as a mean of the data objects and centroid values. But compare to these traditional algorithms our proposed SETC algorithm with TRI metric is out performing and improving the accuracy of text categorization by focusing the term semantics.

4 Experimental Results

In this section, we compared our proposed metric with the existing measures like $\chi 2$ Statistics (Table 4) and observed that our metric TRI is identifying the semantically highly related terms effectively.

The performance of our integrated approach is compared with traditional and most familiar clustering algorithms like K-Means, K-Medoids and TCFS are applied

Table 4 Performance comparisons between χ2 statistics and TRI

Category	C_1		C_2		C_3	
Terms	χ^2 statistics	TRI	χ^2 statistics	TRI	χ^2 statistics	TRI
T_1	0.540	1.198	0.540	1.198	0.540	1.198
T_2	0.423	1.023	0.423	1.023	0	0
T_3	0.227	0.546	0	0	0.227	0.546
T_4	1.121	1.242	1.121	1.242	0	0

Table 5 Performance comparisons of SETC with other clustering methods

Data set	K-means	K-Medoids	TCFS with CHIR	SETC with TRI
20-News Groups	0.432	0.522	0.542	0.594
Reuters	0.562	0.584	0.608	0.806
PubMed	0.618	0.632	0.654	0.812
Wordsink	0.422	0.502	0.722	0.998

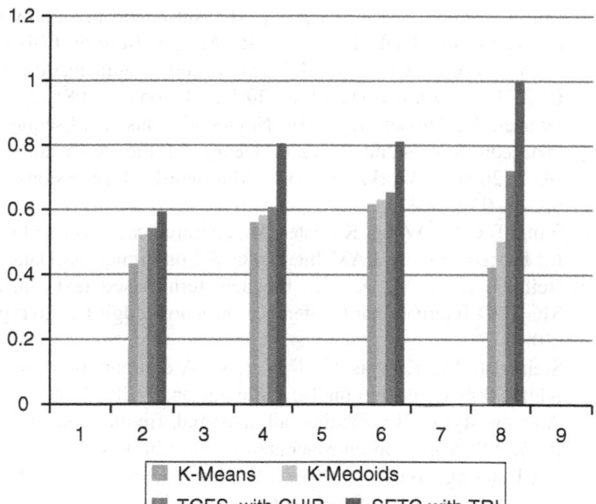

Fig. 1 Performance improvements of SETC with different clustering algorithms

on datasets like 20-News Groups, Reuters, PubMed and Wordsink, we observed that SETC (Table 5) with TRI is producing good results. The statistics are shown here.

Figure 1 represents the performance improvements of our proposed algorithm by comparing with traditional and well-known clustering algorithms.

5 Conclusion

In this paper, we introduced a new metric named as Term Rank Identifier (TRI) which calculates the highly related terms based upon their synonyms and expected relative frequencies. The comparison is made on real data sets with available measures like $\chi 2$ Statistics and GSS Coefficients; we observed that, it is performing well. And we proposed a Text Clustering algorithm named as Semantically Enriched Terms Clustering (SETC), which is integrated with TRI. Our proposed SETC algorithm is compared with other clustering and feature selection algorithms like K-Means, K-Medoids, TCFS with CHIR.The experimental results shows that our SETC is outperforming in terms of clustering accuracy on different data sets.

In Future, we enhance the text categorization and clustering capabilities by proposing additional measures which are independent of scope of the cluster. And we are planning to build ontologies automatically by introducing NLP Lexical Analyzers.

References

1. Liu, X., Song, Y., Liu, S., Wang, H.: Automatic taxonomy construction from keywords. In: Proceedings of KDD'12, pp. 12–16, August, Beijing, China (2012)
2. Li, Y., Luo, C., Chung, S.M.: Text clustering with feature selection by using statistical data. IEEE Trans. Knowl. Data Eng. **20**(5), 641–651 (2008)
3. Doucet, A., Ahonen-Myka, H.: Non-contiguous word sequences for information retrieval. In: Proceedings of 42nd Annual Meeting of the Association for Computational Linguistics (ACL-2004). Workshop on Multiword Expressions and Integrating Processing, pp. 88–95 (2004)
4. Fung, B.C.M., Wang, K., Ester, M.: Hierarchical document clustering using frequent itemsets. In: Proceedings of SIAM International Conference on Data Mining, pp. 59–70 (2003)
5. Beil, F., Ester, M., Xu, X.: Frequent term-based text clustering. In: Proceedings of ACM SIGKDD International Conference on Knowledge Discovery and Data Mining, pp. 436–442 (2002)
6. Steinbach, M., Karypis, G., Kumar, V.: A comparison of document clustering techniques. In: KDD-2000 Workshop on Text Mining, pp. 1–20 (2000)
7. Ahonen-Myka, H.: Finding all maximal frequent sequences in text. In: Proceedings of ICML-99 Workshop on Machine Learning in Text Data Analysis, pp. 11–17 (1999)
8. A Clustering Toolkit, Release 2.1.1. http://www.cs.umn.edu/karypis/cluto/
9. Beydoun, G., Garcia-Sanchez, F., Vincent-Torres, C.M., Lopez-Lorca, A.A., Martinez-Bejar, R.: Providing metrics and automatic enhancement for hierarchical taxonomies. Inf. Process. Manage. **49**(1), 67–82 (2013)
10. Pont, U., Hayegenfar, F.S., Ghiassi, N., Taheri, M., Sustr, C., Mahdavi, A.: A semantically enriched optimization environment for performance-guided building design and refurbishment. In: Proceedings of the 2nd Central European Symposium on Building Physics, pp. S. 19–26, 9–11 Sept 2013, Vienna, Austria. (2013). ISBN 978-3-85437-321-6

11. Ahonen-Myka, H.: Discovery of frequent word sequences in text. In: Proceedings of the ESF Exploratory Workshop on Pattern Detection and Discovery in Data Mining, pp. 16–19 (2002)
12. The Lemur Toolkit for Language Modeling and Information Retrieval. http://www-2.cs.cmu.edu/lemur/
13. Data Mining: Concepts and Techniques—Jiawei Han, Micheline Kamber Harcourt India, 3rd edn. Elsevier, Amsterdam (2007)

An Android-Based Mobile Eye Gaze Point Estimation System for Studying the Visual Perception in Children with Autism

J. Amudha, Hitha Nandakumar, S. Madhura, M. Parinitha Reddy and Nagabhairava Kavitha

Abstract Autism is a neural developmental disorder characterized by poor social interaction, communication impairments and repeated behaviour. Reason for this difference in behaviour can be understood by studying the difference in their sensory processing. This paper proposes a mobile application which uses visual tasks to study the visual perception in children with Autism, which can give a profound explanation to the fact why they see and perceive things differently when compared to normal children. The application records the eye movements and estimates the region of gaze of the child to understand where the child's attention focuses to during the visual tasks. This work provides an experimental proof that children with Autism are superior when compared to normal children in some visual tasks, which proves that they have higher IQ levels than their peers.

Keywords Autism spectral disorder · Mobile application · Cognitive visual tasks · Human-Computer interaction

J. Amudha (✉) · H. Nandakumar · S. Madhura · M.P. Reddy · N. Kavitha
Department of Computer Science, Amrita School of Engineering, Amrita Vishwa Vidyapeetham, Bangalore, India
e-mail: j_amudha@blr.amrita.edu

H. Nandakumar
e-mail: hitha.1511@gmail.com

S. Madhura
e-mail: 12smadhura@gmail.com

M.P. Reddy
e-mail: parinitha66@gmail.com

N. Kavitha
e-mail: kavitha4568@gmail.com

L.C. Jain et al. (eds.), *Computational Intelligence in Data Mining - Volume 2*,
Smart Innovation, Systems and Technologies 32, DOI 10.1007/978-81-322-2208-8_6

1 Introduction

Eye gaze point estimation system is a device which can accurately predict the point at which the user looks at. It can be used for different applications like behavioral analysis of children, in gaming industry for the Human-Computer interaction and it can also be used to help people with disabilities to use computer. Autism is a developmental disorder symptoms of which can be seen in the child from 2 to 3 years of age. Studies have shown that studying the visual perception in autistic children could explain about the difference in behavior exhibited by these children. Visual perception is the brain's interpretation of the visual input. They are frequently associated with developmental disabilities like avoiding contact with eyes, staring at spinning objects etc.

In spite of the behavioral defects, the children with Autism perform better in observing details in images during cognitive tasks. They have a unique way of interpreting the visual scenes. They perform faster and better in embedded figure task and odd man-out tasks when compared to normal children. This paper proposes a novel approach to study the visual behavior in children with Autism by estimating their eye gaze pattern while viewing an image using a mobile device. As the user looks at the mobile screen, the point of gaze of the user is computed and the coordinate values of the point where the user looks at is displayed. Implementation of an eye gaze tracker on a mobile device is a challenging task because the device moves continuously and tracking the eyes to estimate the gaze point thus becomes difficult. The rest of the paper is organized as follows. Section 2 gives a brief of the eye gaze point estimation algorithms relevant to the work, Sect. 3 presents the proposed system design, Sect. 4 gives the implementation details and results and Sect. 5 discusses the advantages and limitations of the system.

2 Literature Survey

Studying the visual behavior in children with Autism can give us an idea about the behavioral differences in children with Autism when compared to normal children. In [1] the author proposed a methodology to study about the gaze pattern of children with Autism using bottom-up model of Computational Visual Attention. From this work, it was concluded that the children with Autism are less affected by scene inversion while watching videos. Visual tasks were used in [2] to compare the performance of normal children and Autistic children while viewing the images.

In order to study the visual behavior of children with Autism, an eye gaze tracker system has to be designed which can give the coordinates at which is child is looking at. An efficient framework is described in [3] which can be used to detect the eye regions of the user using a mobile device. It uses Haar detector to detect the eyes continuously and hardware accelerometer to detect the head movements. The camshaft algorithm is used to track the eye movements and estimate the point of gaze of the user. Further [4] achieves eye gesture recognition in portable devices. It uses image processing algorithms, pattern recognition and computer vision to detect the eye gestures from a recorded video using a front camera. The system consists of a face detector, eye detector, eye tracker to track the movement of eyes, gaze direction estimator and finally an eye gesture recognizer which analyses the sequence of gaze directions to recognize an eye gesture. A methodology to implement a low cost eye tracker is proposed in [5]. It enables people with severe disabilities to use the computer. Fuzzy rules are used to estimate the direction of gaze of the user. A performance rate of up to 92 % was achieved using this technique. Further in [6], an application was proposed which can control the mobile phone using the eyes. Any application in the mobile phone could be opened by just looking at it and blinking.

These are some of the major works from which the idea of an eye gaze point estimation mobile application has been developed and implemented in this paper. It is a free mobile app which can be easily downloaded from the internet and used with any android mobile device to assess the behavior of the child.

3 Mobile Application

3.1 System Design

The system aims at developing an android application for finding the region of gaze of the user. In the present technology, there are only a few tools which use the idea of assessing the autistic child's gaze pattern through eye tracking. The application is made attractive in order to make it user friendly. The caretaker/parent has an equal role in handling the application. The child is asked to look at the embedded image. Simultaneously, the video of the child's eye movement is captured through the front camera of the smart phone which is stored in the media gallery of the phone at the rate of 1 frame per second. Further, frames are extracted from the video at a rate of 1 frame per second which are stored in a separate folder. To detect the gaze region, the first frame that is captured is taken first and the left or right eye is cropped carefully out of it. The pupil is touched on the cropped image of eye. This will be considered as the centre co-ordinates (x_c, y_c) as shown in Fig. 1. This co-ordinate is then used as the reference co-ordinate for the further predictions. This procedure of detecting pupil is done for each frame. The co-ordinates of pupil for these frames will be considered as (x_f, y_f) where f is the frame number.

Calibration is an important component in eye tracking experiments. Only if calibration is done for each user, the differences in the different features in each user can be taken into account. The calibration technique used in this system is simple. The user is asked to look at the centre of the screen for certain amount of time. The two variables are considered namely, m_h and m_v which are horizontal disparity and vertical disparity respectively. The variables are as shown in Fig. 2. Based on the values of m_h and m_v, the region is predicted.

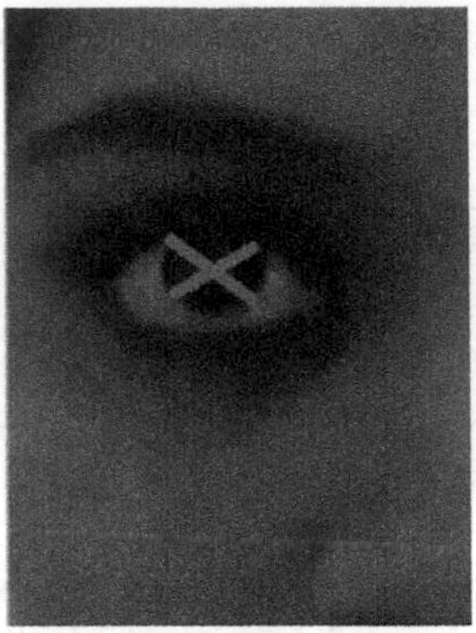

Fig. 1 The pupil is touched to retrieve the co-ordinates

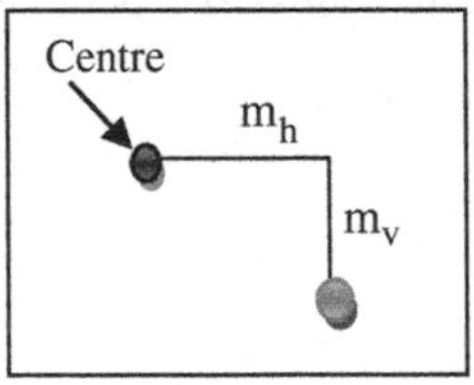

Fig. 2 The *red colour dot* indicates the position of the pupil. The m_h and m_v (*horizontal* and *vertical* disparities respectively) are as shown

$$m_h = x_f - x_c$$
$$m_v = y_f - y_c$$

Based on certain rules, the region of interest is approximately predicted. The rules are as given in Table 1.

The co-ordinates of all the pupil is stored and given as input to the decision loop to determine the region of gaze of the user. The screen of the phone is divided into 9 co-ordinates as shown in Fig. 3.

Table 1 List of rules used to design the eye gaze estimation system

Rule	m_h	m_v	Region of gaze
1	Positive	Positive	Right bottom
2	Positive	Negative	Right up
3	Negative	Positive	Left bottom
4	Negative	Negative	Left up
5	Positive	Zero	Right
6	Zero	Positive	Bottom
7	Negative	Zero	Left
8	Zero	Negative	Up
9	Zero	Zero	Centre

Left Up	Up	Right Bottom
Left	Centre	Right
Left Bottom	Down	Right Bottom

Fig. 3 The screen of the phone is divided into 9 regions

3.2 Cognitive Visual Tasks

To study the visual behavior in children, the cognitive tasks are used. Visual tasks are tasks which are used to test the attention of a person while visually scanning an environment for an object (also known as target) placed in between different objects (also known as distracters). They are used to measure the degree of attention given to stimuli. Some of the most commonly used cognitive Visual Tasks for studying the visual behavior in children are shown in Fig. 4a–o. They are used to evaluate the visual behavior of a child with our mobile application developed.

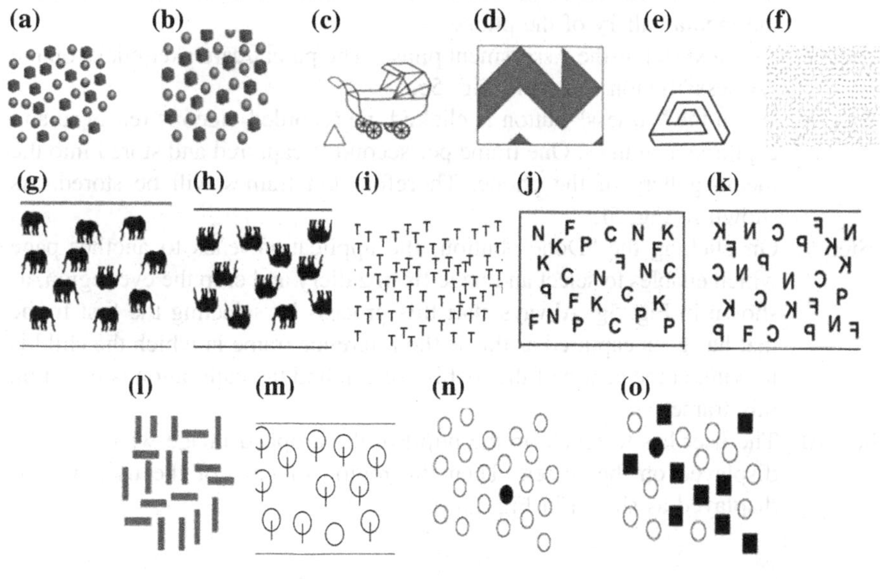

Fig. 4 Set of visual tasks. **a** Feature search task. **b** Conjunctive search task. **c** Embedded figure task. **d** Finding the hidden triangle task. **e** Complex figure task. **f** Identifying pattern task. **g** Figure identification task. **h** Inverted figure task. **i** Confusing letter pattern. **j** Letter identification task. **k** Inverted letters task. **l** Colour selection. **m–o** Odd figure identification

3.3 Implementation

The application can be downloaded easily and can be used. The working of the application is as follows.

Step 1: Download the application.
Step 2: The icon for the application will appear on the screen as shown in Fig. 5a.
Step 3: On opening the application, the home page is displayed for 5 s as shown in Fig. 5b and later it is directed to the next page.
Step 4: The start and help buttons are displayed as shown in Fig. 5c.
Step 5: The care taker/parent can click on the HELP button to understand the features and the working of the application.
Step 6: Once the caretaker/parent understands the application he/she can start using the application by clicking on the START button. The child is first asked to look at the centre of the screen. A small window appears at the left corner of the screen which records the video of the child as he looks at the mobile screen. Then an embedded image is displayed on the screen as shown in Fig. 5d and the child is asked to look at the image for 10 s. The parent/caretaker should make sure that the child looks at the image without any distraction. The mobile front camera keeps recording the eye movements of the child for 10 s. The recorded video is saved in the media gallery of the phone.
Step 7: The next step is the Assessment phase. The parent/caretaker clicks on the "Assess" button shown in Fig. 5e.
Step 8: When the "Assess" button is clicked the recorded video is retrieved and captured as frames. One frame per second is captured and stored into the media gallery of the phone. Therefore ten frames will be stored. As shown in Fig. 5f.
Step 9: On clicking the "Done" button, the application leads to another page which enables to select an image from gallery and crop the eye region, as shown in Fig. 5g. Always start this process by selecting the first frame that has been captured as this is the reference frame in which the child is looking at the centre of the mobile screen and the calibration is based on this frame.
Step 10: The user has to touch on the pupil in the cropped image which will be displayed on the screen. Then the region of gaze of the user will be displayed as shown in Fig. 5h, i.

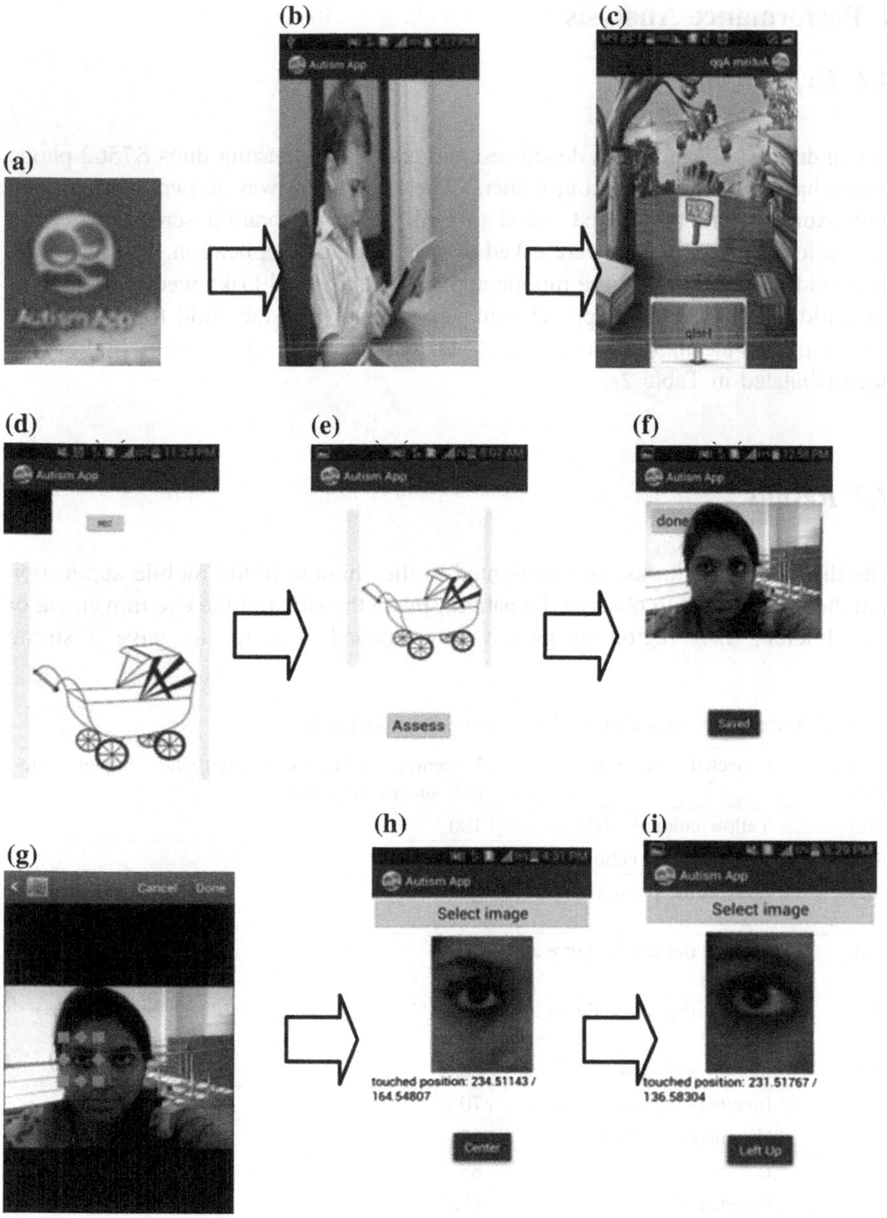

Fig. 5 Screen shot of the mobile app and the results obtained

4 Performance Analysis

4.1 Experiment

The android application was developed and tested on Samsung duos S7562 phone which has 0.3 mega pixel front camera. The application was shown to 6 children with Autism. They were first asked to look at the calibration screen. After the calibration procedure, they were asked to start using the application. The instructor also guided them in using the mobile app. Different visual tasks were presented to the children in the mobile app. The image was shown to the child for a period of 10 s and then the image was changed. The performance of Autistic children has been tabulated in Table 2.

4.2 Results

The different visual tasks were presented to the children in the mobile application and they were asked to observe the pattern. From the video of the eye movement of the children, their region of gaze was computed. The results gave a strong

Table 2 Performance of autistic children during the visual tasks

Visual task	Expected gaze location	Percentage of autistic children who completed the task successfully (%)
4(a)	Yellow colored circle	100
4(b)	Green colored cuboid	100
4(c)	Look at the pram as a whole	25
4(d)	Look at the whole figure as a whole	20
4(e)	Look at the whole figure as a whole	25
4(f)	Region with dots	66.6
4(g)	Inverted elephant	70
4(h)	Non-inverted elephant	60
4(i)	L	83.3
4(j)	Inverted F	83.3
4(k)	P	66.6
4(l)	Horizontal red color	66.6
4(m)	Circle	66.6
4(n)	Filled circle	100
4(o)	Filled circle	50

Fig. 6 The gaze pattern of autism affected child versus normal gaze pattern

validation to the fact that the children with Autism are more superior when compared to normal children during the odd-man out tasks.

The image in the Fig. 6 shows the region of interest of Autism affected child and that of the normal gaze. The patterns in green indicate the normal gaze pattern and the patterns in red indicate the Autism affected child's gaze pattern.

The gaze pattern of the Autistic child clearly indicates that he spend more time looking at the inner details of the figure (like the square and rectangle geometric shapes) rather than the overall picture. On the other hand, the normal child perceived the entire image as a whole rather than in pieces like the Autistic child. This is a strong validation to the fact that children with Autism are good in the embedded figure task when compared to normal children.

5 Conclusion and Future Work

This paper presents the development of a mobile application which can be used to analyze the gaze pattern in children with Autism by using Visual Tasks. It gives an idea about their visual perception. This difference in visual perception can give an explanation to a majority of behavioral differences seen in these children. This paper aims to serve as a bridge between technological advances and medical researches for understanding and aiding neural developmental disorders. However, every individual is distinct. The work presented here is a study based on a group of subjects only. It gives an assessment on the behavior of a major population of Autistic children. This study has been done under normal light conditions and large head movement that can occur has not been considered here. So, the study can be further extended by using mobile eye gaze tracking devices to correctly predict the coordinates of gaze of the child during large head movements.

Acknowledgments We would like to thank "Apoorva Center of Autism", Bangalore for selflessly helping us to interact with the Autistic children to gain a deeper understanding of the problem and successfully completing this work.

References

1. Shic, F., Scassellati, B., Lin, D., Chawarska K.: Measuring context: the gaze patterns of children with autism evaluated from the bottom-up. In: 6th International Conference on Development and Learning, IEEE (2007)
2. Plaisted, K., O'Riordan, M., Cohen, S.B.: Enhanced visual search for a conjunctive target in autism: a research note. J. Child Psychol. Psychiatry **39**(5), 777–783 (1998)
3. Han, S., Yang, S., Kim, J., Gerla, M.: EyeGuardian: a framework of eye tracking and blink detection for mobile device users. In: UCLA. Department of Computer Science, Los Angeles (2012)
4. Vaitukaitis, V., Bulling, A.: Eye Gesture Recognition on Portable Devices. University of Cambridge, Cambridge (2012)
5. Su, M.C., Wang, K.C., Chen, G.D.: An eye tracking system and its application in aids for people with severe disabilities. Biomed. Eng. Appl. Basis Commun. **18**(6), 319–327 (2006) (National Central University)
6. Miluzzo, E., Wang, T., Campbell, A.T.: EyePhone: Activating Mobile Phones With Your Eyes. Computer Science Department, Dartmouth College, Hanover (2010)
7. Pino, C., Kavasidis, I.: Improving Mobile Device Interaction By Eye Tracking Analysis, pp. 1199–1202. University of Catania, Catania (2012)
8. Costa, S., Lehmann, H., Soares, F.: Where is your nose?—Developing body awareness skills among children with autism using a humanoid robot. In: The 6th International Conference on Advances in Computer-Human Interaction (2013)
9. Daikin, S., Frith, U.: Vagaries of visual perception in autism. Neuron **48**, 497–507 (2005)
10. Billard, A., Robins, B., Nadel, J., Dautenhahn, K.: Building robota, a mini-humanoid robot for the rehabilitation of children with autism. RESNA Assistive Technol. J. **19**, 37–49 (2006)

FPGA Implementation of Various Image Processing Algorithms Using Xilinx System Generator

M. Balaji and S. Allin Christe

Abstract This paper makes an comparison between two architectures existing and modified for various image processing algorithms like image negative, image enhancement, contrast stretching, image thresholding, parabola transformation, boundary extraction for both grayscale and color images. The synthesis tool used is Xilinx ISE 14.7 and designed in simulink (MATLAB 2012a) workspace. The existing and modified architectures is implemented using Xilinx System Generator (XSG) and hardware software co-simulation is done using Spartan 3E FPGA board. The results and resource utilization for both the architectures is obtained the results shows that the modified architectures shows an average 39 % less resource utilization than that of the existing architecture also for boundary extraction the modified architectures produced refined results while comparing visually with the existing architecture results.

Keywords Co-simulation · Image processing · MATLAB · Simulink · Xilinx system generator

1 Introduction

Image processing is a method to convert an image into digital form and perform some operations on it, in order to get an enhanced image or to extract some useful information from it. It is a type of signal dispensation in which input is image, like video frame or photograph and output may be image or characteristics associated with that image. Usually Image Processing system includes treating images as two

M. Balaji (✉) · S.A. Christe
Department of ECE, PSG College of Technology, Peelamedu,
Coimbatore 641004, India
e-mail: jjbalaji.m@gmail.com

S.A. Christe
e-mail: sac@ece.psgtech.ac.in

L.C. Jain et al. (eds.), *Computational Intelligence in Data Mining - Volume 2*,
Smart Innovation, Systems and Technologies 32, DOI 10.1007/978-81-322-2208-8_7

dimensional signals while applying already set signal processing methods to them. It is one-among the rapidly growing technologies today, with its applications in various aspects of a business. Image Processing forms core research area within engineering and computer science disciplines too.

Image processing basically includes three steps importing the image with optical scanner or by digital photography, then analyzing and manipulating the image which includes data compression and image enhancement and spotting patterns that are not visible to human eyes like satellite photographs, finally output is the last stage in which result can be altered image or report that is based on image analysis. The purpose of image processing can be visualization, image sharpening and restoration, image retrieval, measurement of pattern, image recognition. Digital Image Processing techniques help in manipulation of the digital images by using computers. As raw data from imaging sensors from satellite platform contains deficiencies. To get over such flaws and to get originality of information, it has to undergo various phases of processing. The three general phases that all types of data have to undergo while using digital technique are pre-processing, enhancement and display, information extraction.

This paper focus on developing algorithmic models in MATLAB using Xilinx blockset for specific purpose, creating workspace in MATLAB to process image pixels in the form of multidimensional image signals for input and output images, performing hardware implementation [1] of given algorithms on FPGA.

2 Xilinx System Generator

XSG [2] is an Integrated Design Environment (IDE) for FPGAs, which uses simulink [3], as a development environment and is presented in the form of model based design.

The Xilinx System Generator for DSP is a plug-into Simulink that enables designers to develop high-performance DSP systems for Xilinx FPGAs. Designers can design and simulate a system using MATLAB, Simulink, and Xilinx library of bit/cycle-true models. The tool will then automatically generate synthesizable Hardware Description Language (HDL) code mapped to Xilinx pre-optimized algorithms. This HDL design can then be synthesized for implementation in FPGA. As a result, designers can define an abstract representation of a system-level design and easily transform this single source code into a gate-level representation. Additionally, it provides automatic generation of a HDL testbench, which enables design verification upon implementation.

3 Proposed Design

The proposed architecture for various image processing algorithms using simulink and Xilinx blocks is divided into 3 phases. In the proposed design the input image used for all the algorithms is of the size 256 × 256 but this proposed method can be used for any size of images. The image format used in this paper is png, jpg formats but the proposed method can be applied for any image formats.

- Image pre-processing
- FPGA based implementation of image processing algorithms
- Image post-processing

Generally image will be in the form of matrix but in the hardware level implementation this matrix must be an array of one dimension hence we need to pre-process [4] the image but in the case of software level simulation using simulink blocksets alone, the image can be taken as a two-dimensional arrangement such as m × n, there is no need for any image pre-processing. Also image post-processing blocks which are used to convert the image output back to floating point type is done using Simulink blocksets whose model based design in [4] is used here.

The algorithm steps that is followed for all the implemented algorithms is given below.

Step 1 Image pre-processing using simulink blocksets

- Image from file: Fetches the input image from a file.
- Resize: Resizes the input image to a defined size.
- Convert 2-D to 1-D: Converts 2-D image to a single array of pixels.
- Frame conversion: Converts entire set of elements to a single frame.
- Unbuffer: Converts frame to scalar samples output at a higher sampling rate.

Step 2 Implementation of image processing algorithms using Xilinx blocksets

- Gateway In: Used for Xilinx fixed point data type conversion.
- Hardware blocks: Here only the real processing starts. Depending on the algorithms these blocksets will change.
- Gateway Out: Convert Xilinx fixed point data type back to simulink integer.

Step 3 Image post-processing using simulink blocksets

- Data type conversion: Converts image signal to unsigned integer format.
- Buffer: Converts scalar samples to frame based output at lower sampling rate.
- Convert 1-D to 2-D: Convert 1-D image signal back to 2-D image.
- Video Viewer: To display the processed output image.

The few improved design using Xilinx blocksets for various image processing algorithms is given below.

(a) Algorithm for grayscale and color image negative, image enhancement, contrast stretching, image thresholding, parabola transformation using M-Code block
The existing architecture [4, 5] which is implemented for the image negative, image enhancement, contrast stretching, image thresholding, parabola transformation algorithm is replaced with M-code block in this modified architecture shown in Figs. 1 and 2. In proposed method why we go for M-code block is that in the existing method the above mentioned algorithms is implemented using various number of Xilinx blocksets but if we use M-code block those various number of blocksets can be replaced with a single Xilinx block which results in reduction of various factors for example resource utilization in this case.

(b) Algorithm for image boundary extraction

The existing architecture [4] which is implemented for the image thresholding algorithm is replaced with the architecture shown in Fig. 3.

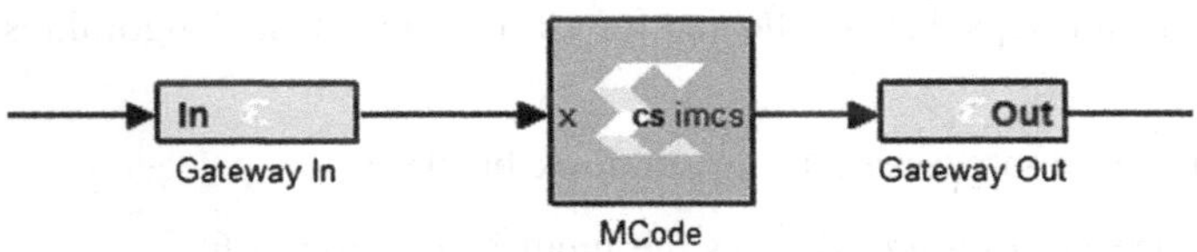

Fig. 1 Implementation for image negative, image enhancement, contrast stretching, image thresholding, parabola transformation using M-Code block for grayscale images

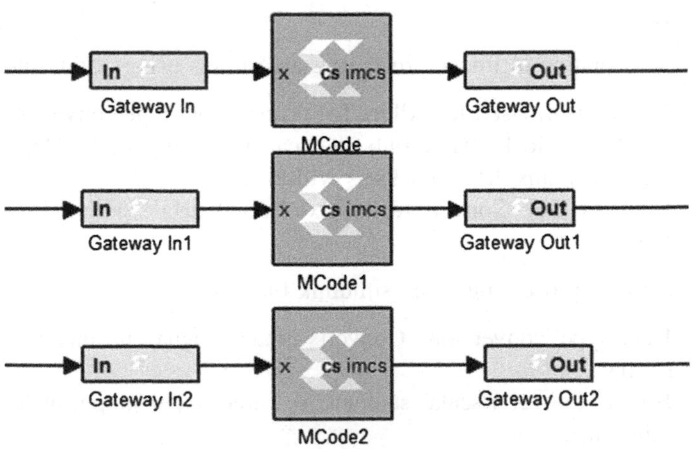

Fig. 2 Implementation for image negative, image enhancement, contrast stretching, image thresholding, parabola transformation using M-Code block for color images

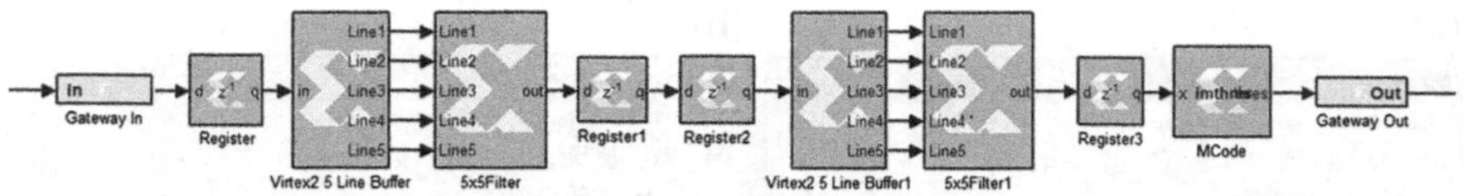

Fig. 3 Implementation for image boundary extraction

4 Results

The different image processing algorithms which are discussed above are implemented using Spartan 3E FPGA board and their corresponding hw/sw co-simulation level outputs for grayscale image which is implemented is shown in Figs. 4, 5, 6, 7 and 8.

The input and hw/sw co-simulated output images of various image processing algorithms for color image implemented using Spartan 3E FPGA board is shown in Figs. 9, 10, 11, 12 and 13.

The resources that are utilized by Spartan 3E FPGA board while implementing existing and modified architectures for various image processing algorithms for both grayscale and color images and percentage of reduction in resource utilization for proposed architecture with the existing architecture [4, 5] is given in Table 1.

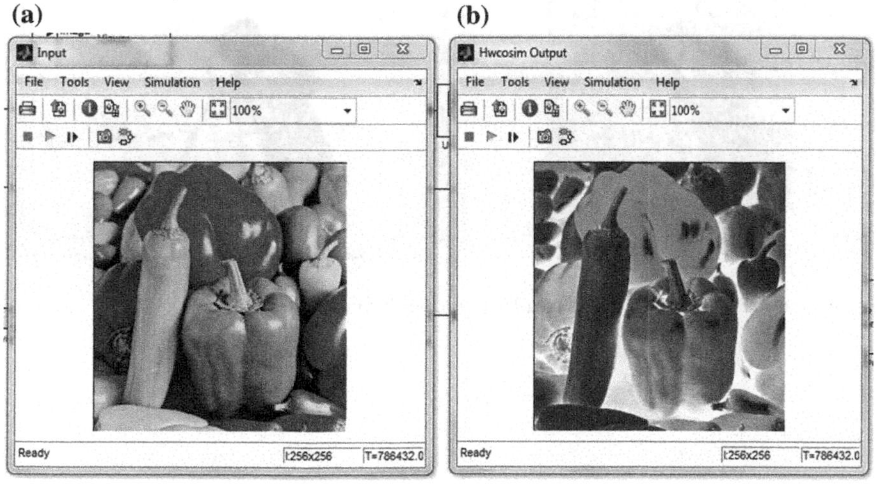

Fig. 4 Grayscale image negative. **a** Input image. **b** Output image

(a) (b)

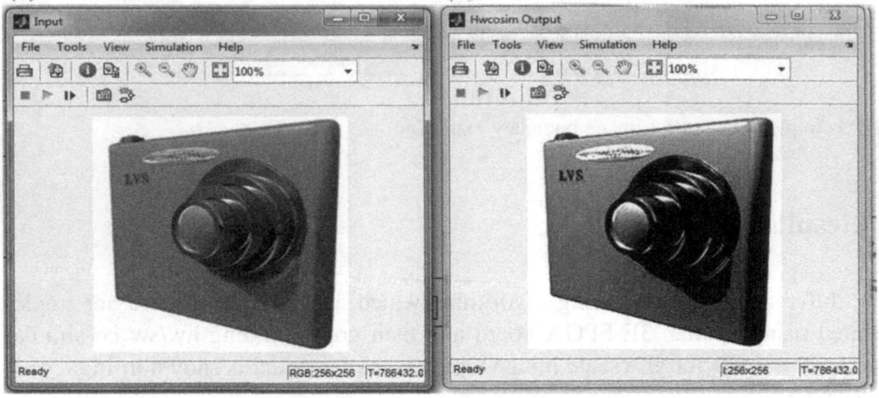

Fig. 5 Grayscale image contrast stretching. **a** Input image. **b** Output image

(a) (b)

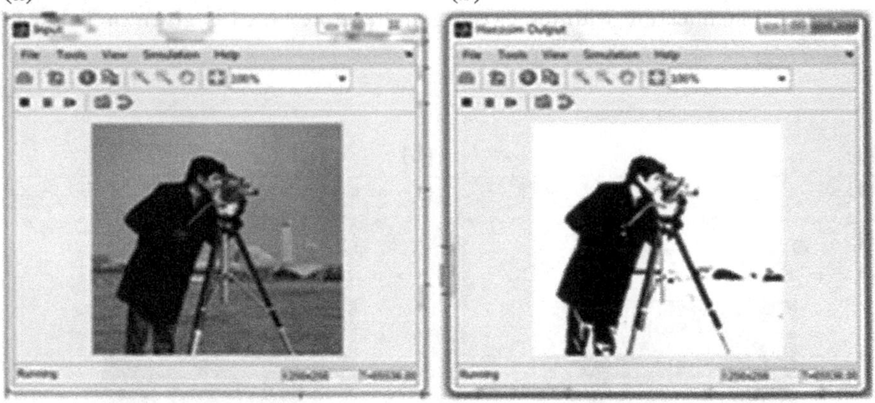

Fig. 6 Grayscale image thresholding. **a** Input image. **b** Output image

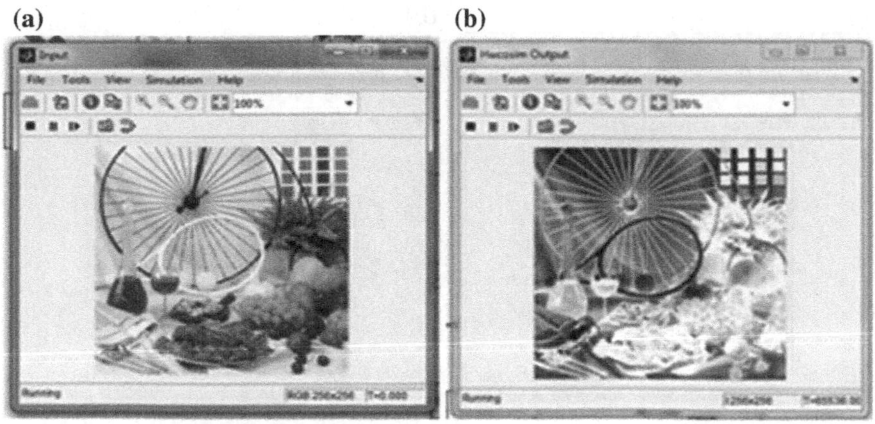

Fig. 7 Grayscale image parabola transformation. **a** Input image. **b** Output image

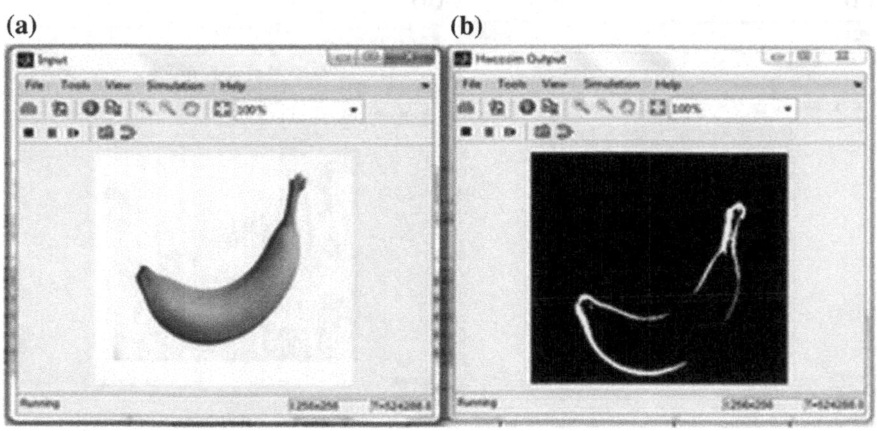

Fig. 8 Image boundary extraction. **a** Input image. **b** Output image

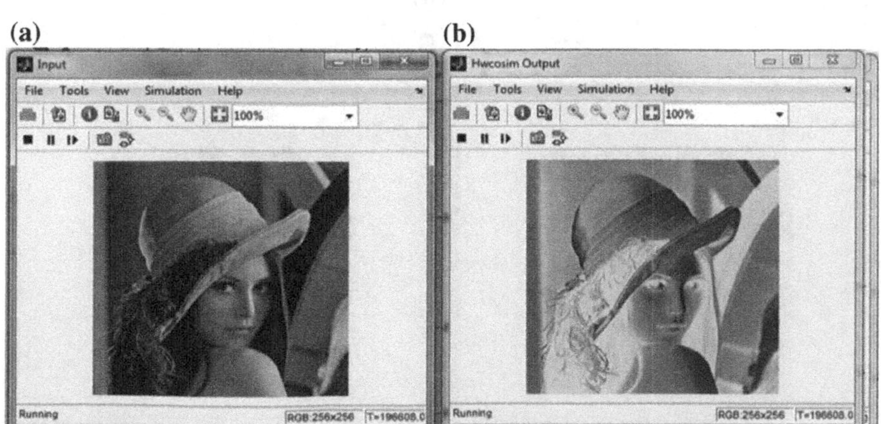

Fig. 9 Color image negative. **a** Input image. **b** Output image

(a) (b)

Fig. 10 Color image enhancement. **a** Input image. **b** Output image

(a) (b)

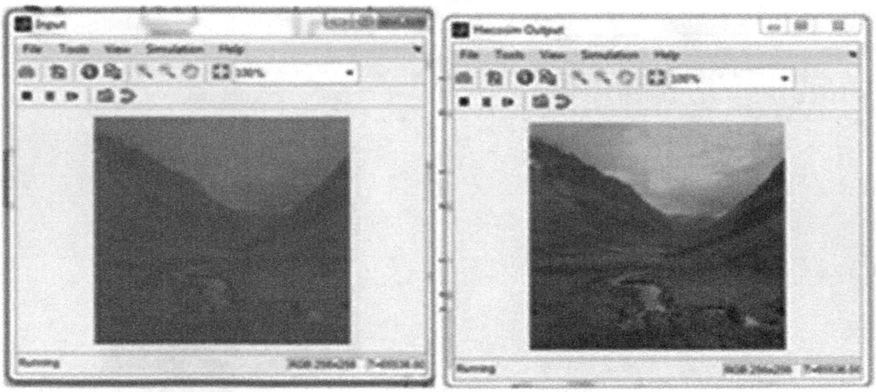

Fig. 11 Color image contrast stretching. **a** Input image. **b** Output image

(a) (b)

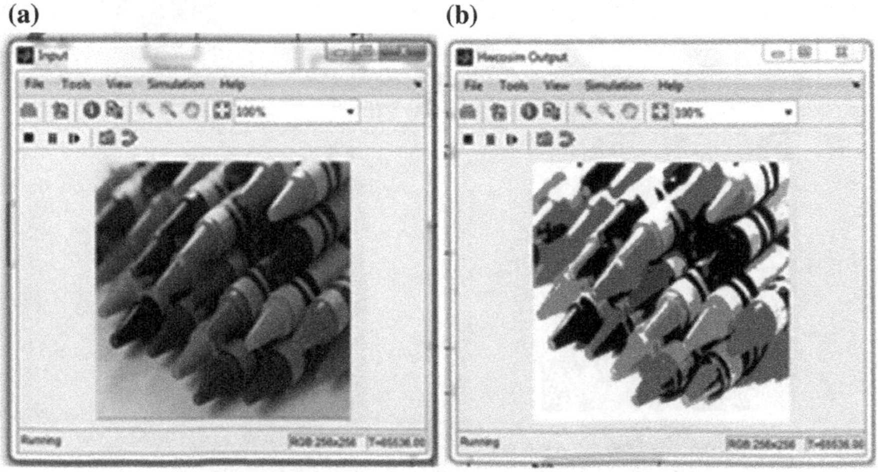

Fig. 12 Color image thresholding. **a** Input image. **b** Output image

(a) (b)

Fig. 13 Color image parabola transformation. **a** Input image. **b** Output image

Table 1 Resource utilization by Spartan 3E FPGA board for existing and modified architectures

Algorithm	Resource utilization by Spartan 3E for existing and proposed architectures								Percentage of reduction in resource utilization (%)
	Existing method				Proposed method				
	Slices	FF's	LUT's	IOB's	Slices	FF's	LUT's	IOB's	
Image negative	17	0	25	65	0	0	0	64	40.19
Color image negative	51	0	75	195	0	0	0	192	40.19
Image enhancement	16	0	2	65	7	0	2	65	10.84
Color image enhancement	48	0	12	195	21	0	6	195	12.94
Contrast stretching	85	0	82	98	28	0	36	69	49.81
Color contrast stretching	255	0	246	294	84	0	108	207	49.81
Image thresholding	6	1	12	64	6	0	11	40	31.33
Color image thresholding	18	3	36	192	18	0	33	120	31.33
Parabola transformation	908	0	1,779	195	155	0	298	108	80.53
Color parabola transformation	8,502	0	16,422	585	756	0	1,437	324	90.13
Boundary extraction	882	1,386	947	64	879	1,376	943	40	1.25

5 Conclusion

In this paper two architectures existing and modified for various image processing algorithms like image negative, image enhancement, contrast stretching, image thresholding, parabola transformation, boundary extraction for both grayscale and color images is implemented. Also resource utilization for the two implemented architectures is tabulated and compared. It can be seen from the above results that the modified methodology shows an average 39 % less resource utilization when compared with the existing methodology [4, 5]. Also when comparing visually for boundary extraction the modified methodology produces refined boundaries while comparing with the existing methodology. This modified methodology could also be extended for different image processing algorithms and also for higher end boards with proper user configuration.

References

1. Christe, S.A., Vignesh, M., Kandaswamy, A.: An efficient FPGA implementation of MRI image filtering and tumor characterization using Xilinx system generator. Int. J. VLSI Des. Commun. Syst. **2**(4) (2011)
2. Xilinx System Generator User's Guide. http://www.Xilinx.com
3. MATLAB. http://www.mathworks.com
4. Neha, P.R., Gokhale, A.V.: FPGA implementation for image processing algorithms using xilinx system generator. IOSR J. VLSI Signal Process. **2**(4), 26–36 (2013)
5. Elamaran, V., Rajkumar, G.: FPGA implementation of point processes using xilinx system generator. J. Theoret. Appl. Inf. Technol. **41**(2) (2012)

A Study of Interestingness Measures for Knowledge Discovery in Databases—A Genetic Approach

Goyal Garima and Jyoti Vashishtha

Abstract One of the vital areas of attention in the field of knowledge discovery is to analyze the interestingness measures in rule discovery and to select the best one according to the situation. There is a wide variety of interestingness measures available in data mining literature and it is difficult for user to select appropriate measure in a particular application domain. The main contribution of the paper is to compare these interestingness measures on diverse datasets by using genetic algorithm and select the best one according to the situation.

Keywords Knowledge discovery · Interestingness measures · Datasets · Genetic algorithm

1 Introduction

A lot of data is being appended in databases every day and this data may contain hidden knowledge which could help in decision making. However, it is not humanly possible to analyze this data and find the useful and interesting pieces of data for the users. Therefore, data mining algorithms/tools are used to extract valuable knowledge/patterns from this large data [1, 2]. It is essential that the discovered knowledge should be accurate, comprehensible and interesting [3]. However, data mining algorithms/tools extract patterns which are accurate and comprehensible but not necessarily interesting [4]. *Interestingness* is the most desirable property in the literature of rule mining as interesting patterns increases the quality of decision making in those rare circumstances where the existing rules

G. Garima (✉) · J. Vashishtha
Department of Computer Science and Engineering,
Guru Jambheshwar University of Science and Technology, Hisar, India
e-mail: garima.goyal1992@gmail.com

J. Vashishtha
e-mail: jyoti.vst@gmail.com

L.C. Jain et al. (eds.), *Computational Intelligence in Data Mining - Volume 2*,
Smart Innovation, Systems and Technologies 32, DOI 10.1007/978-81-322-2208-8_8

are not applicable [5]. Consequently, a variety of interestingness measures are being suggested by the data mining researchers to extract interesting patterns [6, 7]. These measures are categorized as *subjective* (takes into account both the data and the user of these data), *objective* (is based only on the raw data, no knowledge about the user is required.) and *semantics* (considers the semantics and explanations of the patterns, require user domain knowledge) measures [5].

The word 'interesting' has different meanings in different situations, in one situation a particular behavior may be interesting but it may be not interesting in other. Therefore, it is essential to devise and design techniques to analyze these interestingness measures and discover the best one applicable to the situation.

In this paper we propose a methodology based on Genetic Algorithm (GA) to analyze and select the best interesting measures for classification task. *Classification* is a form of data analysis, which extracts models (also called as classifiers) that describes important data classes. Classification is a process having two steps: First is learning—training data is analyzed by an algorithm, which builds the classifier. The second one is classification—in this test data is used to estimate the accuracy of the classification results. If the accuracy of discovered rule is considered acceptable, then the rules can be applied to the classification of new data tuples. Genetic algorithm is used because it has the capability of avoiding the convergence to local optimal solutions and it also takes care of attribute interactions while evolving rules whereas most of the other rule induction methods tend to be quite sensitive to attribute-interaction problems. These methods select one attribute at a time and evaluate a partially constructed candidate solution, rather than a full candidate rule. In case of GA, all the interactions among attributes are taken into account and fitness function evaluates the individual as a whole [8]. The paper is organized in four sections. Section 2 provides a brief call back to the earlier interestingness measures for mining interesting rules, Sect. 3 presents GA design that includes encoding scheme, fitness function, and genetic operators applied. Proposed method and experimental results are presented in Sect. 4.

2 Related Work

Recently, it has been widely acknowledged that even highly accurate knowledge might be worthless if it scores low on the qualitative parameters of comprehensibility and interestingness [9]. Therefore, various interestingness measures are used to find interesting rules. Various interestingness measures have been developed till now. The most basic are the support and confidence/precision. There are a lot others also like lift, leverage, information-gain etc.

The support and confidence [10] measures were the original interestingness measures that are proposed for association rules. As classification rules generated as a result of data mining are used for the prediction of unseen data, the most common measure that is used to evaluate the quality of classification rules is predictive accuracy [5]. Predictive Accuracy, support and confidence are the basic measures for association and classification rules but many other measures have also been proposed which have their own importance and use. Most of these are derived from these basic measures. Precision corresponds to confidence in association rule mining [11]. J-Measure [12] is used to represent the average information content of a probabilistic classification rule. This measure is used to find the best rules in context of discrete-valued attributes. Leverage [13] measures the difference of premises and consequences appearing together in the data set and what would be expected if Premises and consequences were statistically dependent. Piatetsky-Shapiro's measure [13] is used to evaluate the correlation b/w attributes in simple classification rule or between antecedent and consequent. Jaccard [14] is used for managing uncertainty in rule based system. Gini index [15] is the impurity based criteria that measures the divergences between the probability distributions of the target attribute values. Lift [16] measures how many times more often Premises and consequences occur together than expected if they were statistically independent. Certainty Factor [5] is a method for managing uncertainty in rule-based systems and expresses the degree of subjective belief that a particular item is true. Information gain [17] is the impurity based criteria that uses entropy measure as the impurity measure and it is based on pioneering work on information theory by Shannon and weaver.

All probability-based objective interestingness measures proposed for association rules can also be applied directly to classification rule estimation, as they only involve the probabilities of the consequent of a rule, the antecedent of a rule, or both. However, when these measures are used in this way, they measure the interestingness of the rule with respect to the training dataset, whereas the center of attention in classification rule mining is predictive accuracy [5].

3 Proposed GA Design

In the proposed GA, to keep the individuals simple, *Michigan-style approach* [1] is used to encode each solution in the population. Crowding GA has been implemented with the intention to discover a set of classification rules instead of a single best rule. Crowding is a generalization of pre-selection. It uses overlapping populations. In selection and reproduction a portion of the population is replaced by the newly generated individuals. The replacement is a distinct feature. Before replacement, the new offspring will execute a comparison with individuals of the population using a distance function as a measure of similarity. The population member that is most similar to the offspring gets replaced. This strategy maintains the diversity in the population and avoids premature convergence of the traditional GA to the single best solution.

The applied method will provide a systematic framework of procedures and principles to achieve the objective. This procedure is shown in Fig. 1.

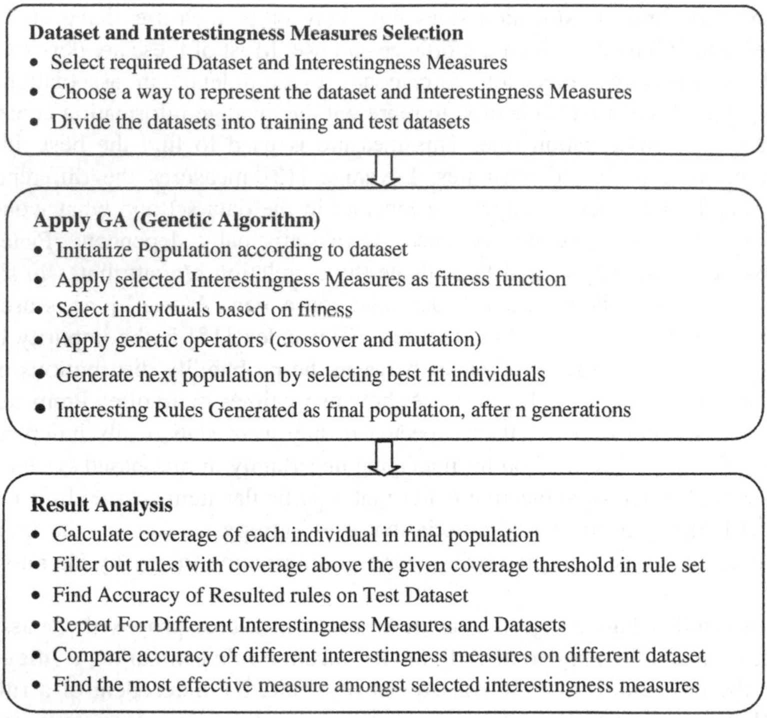

Dataset and Interestingness Measures Selection
• Select required Dataset and Interestingness Measures
• Choose a way to represent the dataset and Interestingness Measures
• Divide the datasets into training and test datasets
Apply GA (Genetic Algorithm)
• Initialize Population according to dataset
• Apply selected Interestingness Measures as fitness function
• Select individuals based on fitness
• Apply genetic operators (crossover and mutation)
• Generate next population by selecting best fit individuals
• Interesting Rules Generated as final population, after n generations
Result Analysis
• Calculate coverage of each individual in final population
• Filter out rules with coverage above the given coverage threshold in rule set
• Find Accuracy of Resulted rules on Test Dataset
• Repeat For Different Interestingness Measures and Datasets
• Compare accuracy of different interestingness measures on different dataset
• Find the most effective measure amongst selected interestingness measures

Fig. 1 Procedure for proposed design

3.1 Encoding

A binary genotypic approach is used for encoding the conditions in antecedent part of classification rule and a single bit is used for consequent part of classification rule. In a genome, the n numbers of genes or bits are used to encode an attribute with n values.

For example, let us take mushroom dataset. It contains 22 attributes, and a class attribute (i.e., the attribute which is desirable for result) which has two values edible or poisonous. Its genome is represented by 126 bits in all where 125 bits are needed for the values of 22 attributes and remaining 1 bit is used for representing class attribute i.e., if it is '0' then class is edible and otherwise if '1' then poisonous.

3.2 Initialization

The complete random initialization provides very poor initial rules and is not effective to discover optimal rule set. Therefore, the proposed algorithm uses the initial population biased towards attributes that are more relevant and informative. A better fit initial population is likely to converge to more reliable rule set with high predictive accuracy in lesser number of generations [9].

3.3 Fitness Function

Fitness functions are used to evaluate the quality of rule. The choice of the fitness function is very crucial as it leads the exploration by evolutionary algorithms towards the optimal solution. *Interestingness measures* are used as fitness functions to find out the interesting rules. There is a variety of interestingness measures have been developed till now and out of them a few famous ones are being used for evaluation. Almost all the probability based interestingness measures may be used interchangeably in both classification and association tasks [6]. The interestingness measures used as fitness functions are presented in Table 1. The traditional probability based interestingness measures are represented into the form using terms like support, precision, sensitivity, specificity, coverage, prevalence etc. Consider a rule *If P Then D*, where *P* represents the premise part or antecedent of the rule and *D* is used to represent the consequent part of the rule. Let *pcount* represents the number of records satisfying antecedent of the rule, *dcount* represents the number of records satisfying consequent of the rule and *pdcount* represents the number of records satisfying both P and D and let '*t*' be the total number of records in the dataset. The most common measure *Support* = *pdcount/t*, probability of (P · D) or the number of examples satisfying both P and D, *Precision* = *pdcount/pcount*, which is also called the confidence of the rule and is the conditional probability of (D/P) or is the number of examples satisfying P · D given P. *Coverage* = *pcount/t*, is the probability of P or the number of examples satisfying P. *Prevalence* = *dcount/t*, is the probability of D or the number of examples satisfying D. *Sensitivity* = *pdcount/dcount*, which is also called the recall of the rule and is the conditional probability of (P/D) or is the number of the tuples satisfying P · D given D. The *Specificity* = ((1 − *coverage*) * (1 − *prevalence*))/(1 − *prevalence*), is the conditional probability of (∼D/∼P).

3.4 Genetic Operators

Crossover is the process by which genetic material from one parent is combined with genetic material from the second parent for producing the potentially promising new offspring. This operation is used to avoid the solution process to converge to local optimum. Here *one-point crossover* is used.

Mutation in GAs is used to introduce diversity. It may specialize or generalize a candidate rule by inserting or removing conditional clauses in the antecedent part of the rule. As compare to crossover probability, mutation probability is kept low.

Table 1 Probability based combined objective interestingness measures

Measure	Formula	
Lift/Interest	$\frac{\text{Prob}(P \cap D)}{\text{Prob}(P)\ \text{Prob}(D)}$	$\frac{\textit{Sensitivity}}{\textit{Coverage}}$
Leverage	$\frac{\text{Prob}(P \cap D)}{\text{Prob}(P)} - \text{Prob}(P) * \text{Prob}(D)$	$\text{Precision} - \textit{Coverage} * \textit{Prevalence}$
Jaccard	$\frac{\text{Prob}(P \cap D)}{(\text{Prob}(P) + \text{Prob}(D) - \text{Prob}(P \cap D))}$	$\frac{\textit{Support}}{\textit{Coverage} + \textit{Prevalence} - \textit{Support}}$
Certainty factor	$\frac{\left(\frac{\text{Prob}(P \cap D)}{\text{Prob}(P)} - \text{Prob}(D)\right)}{(1 - \text{Prob}(D))}$	$\frac{(\text{Precision} - \textit{Prevalence})}{1 - \textit{Prevalence}}$
Piatetsky-Shapiro	$\text{Prob}(P \cap D) - \text{Prob}(P) * \text{Prob}(D)$	$\textit{Support} - \textit{Coverage} * \textit{Prevalence}$
J-measure	$\text{Prob}(P \cap D)$ $* \log\left(\frac{\text{Prob}(P \cap D)}{\text{Prob}(P) * \text{Prob}(D)}\right)$ $+ \text{Prob}(P \cap \sim D)$ $* \log\left(\frac{\text{Prob}(P \cap \sim D)}{\text{Prob}(P) * \text{Prob}(\sim D)}\right)$	$\textit{Support} * \log\left(\frac{\textit{Sensitivity}}{\textit{Coverage}}\right)$
Gini-index	$\text{Prob}(P) * \left\{\left(\frac{\text{Prob}(P \cap D)}{\text{Prob}(P)}\right)^2 + \left(\frac{\text{Prob}(P \cap \sim D)}{\text{Prob}(P)}\right)^2\right\}$ $+ \text{Prob}(\sim P) * \left\{\left(\frac{\text{Prob}(\sim P \cap D)}{\text{Prob}(\sim P)}\right)^2 + \left(\frac{\text{Prob}(\sim P \cap \sim D)}{\text{Prob}(\sim P)}\right)^2\right\}$ $- \text{Prob}(D)^2 - \text{Prob}(\sim D)^2$	$\text{Coverage} *$ $\left\{\text{Precision}^2 + \left(\frac{(1 - \text{Specificity}) * (1 - \text{Prevalence})}{\text{Coverage}}\right)^2\right\}$ $+ (1 - \text{Coverage}) *$ $\left\{\left(\frac{(1 - \text{Sensivity}) * \text{Prevalence}}{1 - \text{Coverage}}\right)^2 \left(\frac{\text{Specificity} * (1 - \text{Prevalence})}{1 - \text{Coverage}}\right)^2\right\}$ $- \textit{Prevalence}^2 - (1 - \textit{Prevalence})^2$
Information gain	$\log\left(\frac{\text{Prob}(P \cap D)}{\text{Prob}(P) * \text{Prob}(D)}\right)$	$\log\left(\frac{\textit{Sensitivity}}{\textit{Coverage}}\right)$

4 Experimental Setup and Result Analysis

The proposed GA method is implemented using MATLAB software on Intel Core i5 processor for experimental work. The performance of the suggested approach is validated on four data sets obtained from UCI Machine Learning Repository which are described in Table 2. Datasets used here are of different nature. The datasets used are divided into training (60 %) and test sets (40 %). The reason for taking training sample larger is to capture rare/interesting conditions which may be present in very few instances in the dataset. During the rule discovery, the values of various parameters used for implementation of GA, are listed in Table 3.

The interestingness measures are evaluated and analyzed on the parameters predictive accuracy (PR) and the number of rules discovered (NR) during mining process. Accuracy is the probability of total number of examples correctly classified by the rule set either positive or negative. The calculated accuracy and number of interesting rules generated on the application of various fitness functions in final population are shown in Table 4.

In Fig. 2, a graph is drawn showing the accuracy of different interesting measures for selected datasets and in Fig. 3, a graph is drawn showing the number of interesting rules for different interesting measures for selected datasets. According to results, the accuracy and number of interesting rules generated, Piatetsky-Shapiro is best for 'Mushroom' dataset with a large number of examples. The dataset 'Nursery' is larger than 'Mushroom and the measures Piatetsky-Shapiro and Jaccard give best results. The measure Jaccard is best for Tic-Tac-Toe and Car which have lesser number of examples. Therefore, the measures Jaccard and Piatetsky-Shapiro give

Table 2 Datasets description used for experimentation

Name of dataset	Number of instances	Number of attributes	Number of classes	Number of alleles with class attributes
Mushroom	5,610	22	2	126
Nursery	12,960	8	5	28
Tic-tac-toe	958	9	2	28
Car	1,728	6	4	22

Table 3 List of commonly used variables in simulation

Name of dataset	Population size	Number of generations	Probability of mutation	Probability of crossover	Confidence (greater than equal to)	Sensitivity (greater than equal to)
Mushroom	50	200	0.01	0.66	0.6	0.2
Nursery	25	50	0.01	0.66	0.75	0.06
Tic-tac-toe	50	500	0.001	0.66	0.6	0.06
Car	50	200	0.01	0.66	0.6	0.2

Table 4 Accuracy and number of interesting rules in final population for various interestingness measures on different datasets

Interestingness measure	Mushroom		Nursery		Tic-tac-toe		Car	
	Accuracy	No. of rules	Accuracy	No. of rules	Accuracy	No. of rules	Accuracy	No. of rules
Precision/Confidence	85.977	4–5	43.687	2	37.188	7–8	69.055	4–5
Leverage	80.444	9–11	42.098	4–5	29.191	12–13	67.704	10
Jaccard	95.031	3–4	51.315	3–4	84.334	5–6	71.346	3
Certainty factor	86.490	4–5	45.049	2	72.367	6	66.667	4–5
Gini index	68.214	10–12	33.816	5	28.965	12–14	53.603	10–11
J-measure	93.991	3	49.992	3	73.629	7–8	66.811	4–5
Piatetsky-Shapiro	96.234	3	51.714	2–3	72.618	6–8	68.765	3–5
Information gain	89.149	3–5	39.948	3	31.369	7–8	66.787	4–5

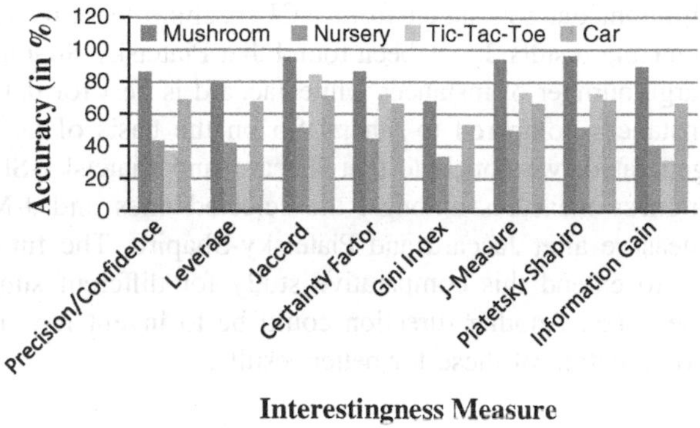

Fig. 2 Accuracy of different interestingness measures on different datasets

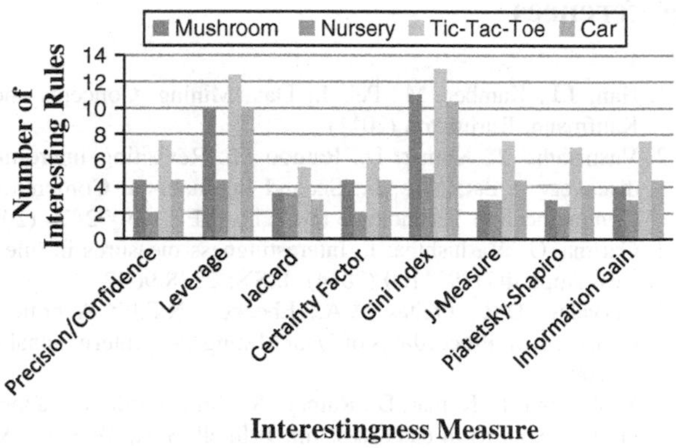

Fig. 3 Number of interesting rules for different interestingness measures on different datasets

good results and find most interesting rules and there is a very little difference in their accuracy calculations. Gini index and Leverage are generating large number of rules amongst all other measures although Leverage gives better accuracy.

J-Measure also generates the best interesting rules and its accuracy is also superior, near to the best accuracy. It can also be used for rule extraction in large as well small data sets. Among certainty factor and Information gain, both give good quality interesting rules and good accuracy however Certainty factor has better performance.

5 Conclusion

This paper has proposed an evolutionary approach to find the effectiveness of different interestingness measures and to select the appropriate interestingness measure. Eight popular interestingness measures has been studied which cover a large range of potential users' preferences. Experimental study is carried out on four different datasets taken from UCI machine learning repository and it has shown amazing results. It has been found that Piatetsky-Shapiro is best for datasets having large number of instances while Jaccard is best for datasets having less number of instances compared to others. So on the basis of accuracy and number of rules generated, we conclude that Jaccard and Piatetsky-Shapiro are the best interestingness measures amongst the selected ones and J-Measure is the second best measure after Jaccard and Piatetsky-Shapiro. The future research direction could be to extend this comparative study for different subjective measures and other measures. Another direction could be to invent new interestingness measures by combination of these for better results.

References

1. Han, J.J., Kamber, M., Pei, J.: Data Mining, Concepts and Techniques, 3rd edn. Morgan Kaufmann, Burlington (2011)
2. Vashishtha, J., Kumar, D., Ratnoo, S.: Revisiting interestingness measures for knowledge discovery in databases. In: Second International Conference on Advanced Computing and Communication Technologie (ACCT), IEEE, pp. 72–78 (2012)
3. Garima, G., Vashishtha, J.: Interestingness measures in rule mining: a valuation. Int. J. Eng. Res. Appl. **4**(7), 93–100 (2014). ISSN: 2248-9622
4. Carvalho, D.R., Freitas, A.A., Ebecken, N.F.F.: A critical review of rule surprisingness measures. In: Proceedings of Data Mining IV—International Conference on Data Mining, vol. 7 (2003)
5. Vashishtha, J., Kumar, D., Ratnoo, S.: An evolutionary approach to discover intra- and interclass exceptions in databases. Int. J. Intell. Syst. Technol. Appl. **12**, 283–300 (2013)
6. Geng, L., Hamilton, H.J.: Interestingness measures for data mining: a survey. ACM Comput. Surv. **38**(3), 1–32 (2006)
7. Triantaphyllou, E., Felici, G.: Data Mining and Knowledge Discovery Approaches Based on Rule Induction Techniques, vol. 6. Springer, Berlin (2006)
8. Freitas, A.A.: Data mining and Knowledge Discovery with Evolutionary Algorithms. Natural Computing Series. Springer, New York (2002)
9. Vashishtha, J., Kumar, D., Ratnoo, S., Kundu, K.: Mining comprehensible and interesting rules: a genetic algorithm approach. Int. J. Comput. Appl. **31**(1), 39–47 (2011) (0975–8887)
10. Agrawal, R., Imielinski, T., Swami, A.: Mining associations between sets of items in large databases. In: Proceedings of the ACM SIGMOD International Conference on Management of Data, pp. 207–216. Washington, DC. (1993)
11. Pagallo, G., Haussler, D.: Boolean feature discovery in empirical leaning. Mach. Learn. **5**(1), 71–99 (1990)
12. Smyth, P., Rodney, M.G.: Rule induction using information theory. In: Knowledge Discovery in Database, pp. 159–176. AAAI/MIT Press, Cambridge (1991)
13. Piatetsky-Shapiro, G.: Discovery, analysis, and presentation of strong rules. In: Piatetsky-Shapiro, G., Frawley, W.J. (eds.) Knowledge Discovery in Databases, pp. 229–248. MIT Press, Cambridge (1991)
14. Tan, P., Kumar, V., Srivastava, J.: Selecting the right interestingness measure for association patterns. In: Proceedings of the 8th International Conference on Knowledge Discovery and Data Mining (KDD 2002), pp. 32–41. Edmonton, Canada (2002)

15. Breiman, L., Freidman, J., Olshen, R., Stone, C.: Classification and Regression Trees. Wadsworth and Brooks, Pacific Grove (1984)
16. Brin, S., Motwani, R., Ullman, J.D., Tsur, S.: Dynamic itemset counting and implication rules for market basket data. In: Proceedings of the ACM SIGMOD International Conference on Management of Data. ACM SIGMOD, pp. 265–276 (1997)
17. Quinlan, J.R.: Induction of decision trees. Mach. Learn. **1**(1), 81–106 (1986)

Efficient Recognition of Devanagari Handwritten Text

Teja C. Kundaikar and J.A. Laxminarayana

Abstract Research in Devanagari handwritten character recognition has challenging issues due to the complex structural properties of the script. Handwritten characters are of unique style for individual persons. The process of recognition involves the preprocessing and segmentation of input document image containing Devanagari characters. It is proposed to use Kohonen Neural Network to interpret Devanagari Characters from the segmented images. The recognized Devanagari characters are stored in a file.

Keywords Handwritten text · Recognition · Kohonen neural network · Devanagari

1 Introduction

Devanagari should be given more special consideration for analysis and document retrieval due to its popularity. Even though there is a lot of ongoing research work to recognize Devanagari Characters, it is observed that recognition accuracy is not at an expected level. This work mainly focuses on improving the recognition accuracy of Devanagari characters. The main objective of this work is to recognize the Devanagari handwritten text from the scanned document. Initially, preprocessing of input image is performed to remove the noise from the input image, followed by binary conversion. The binary image is segmented to retrieve individual Devanagari characters. These characters need to be converted to binary matrix and chain code matrix, which are given as input to Kohonen Neural Network for character recognition. The recognized character is stored in file.

T.C. Kundaikar (✉) · J.A. Laxminarayana
Goa College of Engineering Farmagudi, Ponda, Goa, India
e-mail: tkundaikar10@gmail.com

J.A. Laxminarayana
e-mail: jal@gec.ac.in

L.C. Jain et al. (eds.), *Computational Intelligence in Data Mining - Volume 2*,
Smart Innovation, Systems and Technologies 32, DOI 10.1007/978-81-322-2208-8_9

2 Related Work

Handwritten character recognition research is part of character recognition which involves identifying characters written by hand. For the past years various techniques were presented to recognize English handwritten characters. The results obtained were mostly accurate. But the same techniques fall short while dealing with Devanagari scripts as local scripts usually is made up of a complex structure. There are many research works in the field of off-line handwritten Devanagari character recognition. Arora et al. [1] presented a method of Recognition of non-compound handwritten Devanagari characters using a combination of MLP and minimum edit distance. In this work, the overall global recognition accuracy of system using combined MLP is 76.67 % without confused characters.

Saraf and Rao proposed [2] 98.78 % recognition accuracy on individual character. Munjal and Sahu did study of various techniques to recognize Devanagari characters [3] and concluded that the efficiently of recognizing handwritten Devanagari characters decreases mainly due to the incorrect character segmentation of touching or broken characters.

Holambe and Thool presented [4] comparative study of different classifiers for Devanagari handwritten characters, which consists of experimental assessment of the efficiency of various classifiers in terms of accuracy in recognition. They have used one feature set and 21 different classifiers for their experiment. It gives an idea of the recognition results of different classifiers and provides new benchmark for future research. Further, they have reported the comparative study of Devanagari handwritten character recognition by using classifiers. They reported highest accuracy of 96 % using k-nearest neighbor classification.

Ramteke and Rane [5] proposed a process of segmentation that includes separating a word, a line, individual character or pseudo character images from a given script image. The large variation in handwriting style and of the script makes the task of segmentation quite difficult. Before proceeding to the process of segmentation preprocessing is need to be done. In the preprocessing step, smoothing of image using median filter, the binarization of image and the scaling are included. The process of segmentation consists of analyzing the digitalized image provided by a scanning device, so as to localize the limits of each character and to isolate them from each other. In the handwritten Devanagari script the space between the words and the characters usually varies. This produces some difficulties in the segmentation process of handwritten Devanagari script. For the segmentation of the handwritten Devanagari script into words, vertical projection profile i.e. the histogram of input image is used in which the zero valley peaks show the space between the words and characters. For the character segmentation of Devanagari, proposed method uses the vertical profile which separate the base character using clear paths between them. The segmentation accuracy for this method depends upon the proper writing i.e. non-overlapping or characters, proper space between words and characters, proper connection of characters through shirorekha. The segmentation for word gives 98 % of accuracy and for characters 97 % of accuracy.

Tathod et al. [6] presented a system to recognize handwritten Devanagari scripts. The system considers handwritten text as an input and recognizes characters using neural network approach in which a character matrix is created of every character in the training phase and the corresponding character is matched during the recognition phase. The system uses Kohonen neural network algorithm in the recognition phase to recognize the character. The system gives a step by step processing and working of neural networks in different phases. After the recognition of character is done it is replaced by the standard font to integrate information from different handwritings.

3 Existing Approaches

3.1 Preprocessing

Binarization Binarization is an important preprocessing step which converts gray image into a binary image. It finds the global threshold that minimizes the interclass variance of the resulting black and white pixels.

Skew Detection Skew the orientation of the document image to an angle determines using PCA [7].

3.2 Segmentation

Structural The structured approach of horizontal and vertical projection of histogram [13]. There are three steps of segmentation algorithm line, word and character segmentation.

3.3 Neural Network

Kohonen Neural Network The Kohonen neural network differs considerably from the feedforward back propagation neural network. The Kohonen neural network differs both in how it is trained and how it recalls a pattern. The Kohohen neural network does not use any sort of activation function. Further, the Kohonen neural network does not use any sort of a bias weight. Output from the Kohonen neural network does not consist of the output of several neurons. When a pattern is presented to a Kohonen network one of the output neurons is selected as a "winner". This "winning" neuron is the output from the Kohonen network. Often these "winning" neurons represent groups in the data that is presented to the Kohonen Neural Network.

Following are the steps to get the winning neuron [8];

1. **Normalization of Input**: To normalize the input, first calculate the "vector length" of the input data, or vector. This is done by summing the squares of the input vector.
2. **Normalization factor**: The normalization factor is the reciprocal of the square root of the vector length calculated in above step.
3. **Calculating Each Neuron's Output**: To calculate the output the input vector and neuron connection weights both are considered. First the "dot product" of the input neurons and their connection weights must be calculated. To calculate the dot product between two vectors you must multiply each of the elements in the two vectors. The Kohonen algorithm specifies that we must take the dot product of the input vector and the weights between the input neurons and the output neurons.
4. **Normalizing Output neuron**: The output is normalized by multiplying it by the normalization factor.
5. **Bipolar Mapping**: Mapping is done by adding one to each of output neuron and divide the result in half.
6. **Choosing the winner**: To choose the winning neuron the output that has the largest output value is chosen.

4 Proposed Work

It is proposed to develop a system such that the input to the system is image of the input document which can be a scanned copy. Preprocessing is performed on this input document followed by the segmentation. The subimage of the character image is mapped to a binary matrix of size 9 × 9 and given to the Kohonen neural network to recognize the respective character. This character is mapped to its corresponding standard unicode of Devanagari character and transliterated to its roman script.

At every step, skew is detected using PCA algorithm [7]. Instead of vertically projecting from top to bottom during character recognition, as done in existing work [9], a vertical histogram is projected from bottom to top to get the segmentation of the characters.

The proposed system is characterized in the form of an algorithm, which is given as follows:

Algorithm

1. The input text document is converted into an image using appropriate tools and under the controlled environment.
2. A binary image is generated for each input document image.

3. If the input characters are tilted vertically, the tilt angle is calculated and appropriate image rotation is applied in reverse direction.
4. The line segmentation and the word segmentation steps are applied on image and for each skew is detected using PCA algorithm.
5. Character is segmented such that upper modifier and lower modifier are also segmented and taken as separate character.
6. Each character image is mapped into a binary matrix of size 9 × 9.
7. Chain code of matrix is calculated using chain code algorithm and stored in an array.
8. Binary Matrix is given to the trained Kohonen Neural network for recognizing the character.
9. Chain Code Matrix is given to Kohonen Neural network for recognizing the character.
10. The Output of step 8 and 9 are compared. If they are same then label one of the character as an output character and go to step 12 else go to step 11.
11. Label the character with highest probability as an output character.
12. Write the output to a file.
13. If there are more characters to be recognized the go to step 6.

5 Implementation

The proposed algorithm for the recognition of handwritten Devanagari characters is successfully implemented for the limited set of characters. There are six classes implemented in java, which are KohonenNetwork, Network, loadimage, Sample-Data, TrainingSet and Sample. KohonenNetwork class which extends the features of Network class, implements the functions such as finding the winning neuron, normalizing the weights, calculating error and normalizing input. Network class is abstract class whose functions are implemented in KohonenNetwork class. SampleData class which assign the dataset to object of the same class. Loadimage class implements the function for preprocessing image such as binary conversion, line identification, word identification and character border identification, then this character is converted to matrix of the size 9 × 9 and this input is given to kohonen neural network whose functionality is implemented in the class KohonenNetwork which finally writes the recognized characters in a file. The dataset of 2,400 samples is used to train the Kohonen Neural Network.

For the input as shown in Fig. 1, the system computes the equivalent matrix re-presentation as shown in Fig. 2. When a group of devanagari characters are given as input as shown in Fig. 3, the system produces the matrices as shown in Fig. 4. Each matrix is given to Kohonen Neural Network, which gives the output as recognized character that is stored in file as shown in Fig. 5.

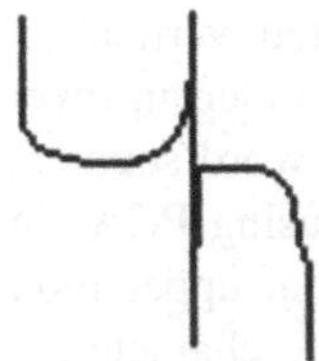

Fig. 1 Image letter ph

100001000
100001000
100011000
111111000
011101111
000001001
000001001
000001001
000001001

Fig. 2 9 × 9 Matrix of ph

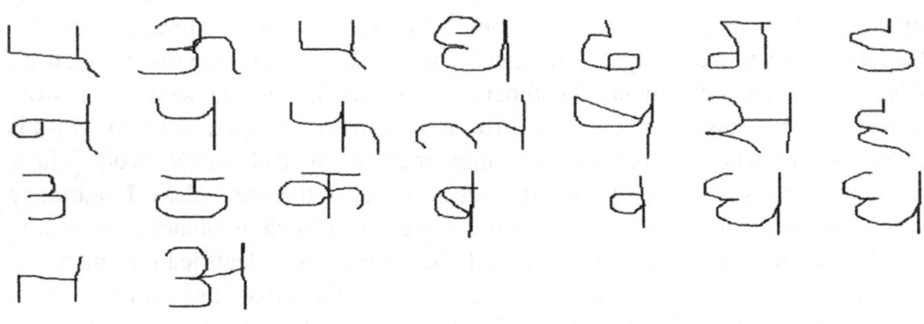

Fig. 3 Image of handwritten document

100000100 100000100 100000100 100000100 110000100 011111110 000000010 000000011 000000001	011111000 011001000 000001000 000011111 001111011 001111001 000001001 000011001 111111001	100000010 100000010 100000010 100000010 100000110 111111110 000000110 000000011 000000001	111111001 110011001 101110001 111100001 011110011 110000011 111000011 011111111 000000001	000110000 000110000 111110000 110010000 100000000 100011110 100011111 110010001 011111111	011111111 001100010 001100011 000110001 000110001 011110001 111110001 100010001 111110001	000001000 001111000 111011000 110000000 111000000 001111111 000000001 000000001 011111111
000000011 000000011 000000011 011111011 110111111 100100011 111100011 000000011 000000001	100000001 100000001 110000011 111111111 000000001 000000001 000000001 000000001 000000001	100001000 100001000 100011000 111111110 000001111 000001001 000001001 000001001 000001001	000000001 000000001 000000001 000000001 111101111 100111011 100110001 100011001 100011001	110000001 111100001 100111111 100001111 111111111 000000001 000000001 000000001 000000001	111000001 101100001 000100001 000111111 000111111 111100001 111100001 000111001 000011001	000010000 111110000 110010000 100000000 111110000 111011000 111100000 001111110 000000011
111111100 000000100 000000100 000011100 111111111 000000001 000000001 000000001 011111111	111111111 000010000 000010000 111111100 110001111 100000001 100000001 110000011 111111110	001111111 111111000 110011111 100011111 100011001 100011001 111111001 000001001 000001000	000000011 000000011 111111111 111100011 101111011 110011110 111111110 000000010 000000010	000000010 000000011 010000011 111111111 100000011 110000111 111111101 000000001 000000001	111100001 110000001 100000001 111110001 011100001 010000001 011000011 001111111 000000001	111100001 110000001 100000001 111110001 011100001 010000001 011000011 001111111 000000001
000000001 000000001 000000001 111100011 101111111 100000011 100000011 100000011 100000011	111110001 100110001 000010001 000010011 011111111 000111001 000011001 000011001 111111001					

Fig. 4 Matrix of document as shown in Fig. 3

एअएधढगङ

नपफलपयइ

आठकबवघघ

तअ

Fig. 5 Recognized text

6 Conclusion

It is observed that the methods discussed in the literature to recognize the Devanagari characters using neural networks are able to successfully recognize most of the hand written characters. However a printed image of Devanagari text given to one of the existing work was unable to segment the characters and so recognition was not successfully done. It was also observed that bold Devanagari character could not be recognized. Segmentation of tilted image could not be performed, and it was unable to recognize confused characters. The success of those methods lie in the size of database, i.e. larger the size of database used for training the neural network higher is probability of successful recognition and segmentation of the characters. To achieve the desired result by overcoming disadvantage of the existing work, we proposed the Devanagari character recognition model which improves on the segmenting the characters from the input image and chain code algorithm to overcome the recognition of confused character. The proposed methodology give overall accuracy of 71 % recognition and confused character recognition is 86 % but if the handwritten characters are badly written then accuracy is reduced to 49.99 %. The method for recognition of Devanagari characters using neural network successfully recognize most of the characters. As there are various ways in which a person can write, it does not give 100 % accuracy. However, the success of the method lies in the size of database, i.e. larger the size of database used for training the neural network higher is probability of successful recognition.

References

1. Arora, S., Bhattacharjee, D., Nasipuri, M., Basu, D.K., Kundu, M.: Recognition of non-compound handwritten devnagari characters using a combination of MLP and minimum edit distance. Int. J. Comput. Sci. Secur. (IJCSS) **4**(1) (2010)
2. Saraf, V., Rao, D.S.: Devnagari script character recognition using genetic algorithm for get better efficiency. Int. J. Soft Comput. Eng. (IJSCE) **2**(4) (2013). ISSN: 2231-2307
3. Munjal, G., Sahu, N.: Study of techniques used for devanagari handwritten character recognition. Int. Res. Eng. Sci. (IJRES) (2013). ISSN (Online) 2320-9364, ISSN (Print) 2320-9356
4. Holambe, A.N., Thool, R.C.: Comparative study of different classifiers for devanagari handwritten character. Int. J. Eng. Sci. Technol. **2**(7), 2681–2689 (2010)
5. Ramteke, A.S., Rane, M.E.: Offline handwritten devanagari script segmentation. Int. J. Sci. Technol. Res. **I** (2012)
6. Tathod, M., Shah, R., Nalamwar, S.R., Yadav, R., Shah, M.: Devnagari script recognition using kohonen neural network. IJIDCS **3**(1) (2013)
7. Shivsubramani, K., Loganathan, R., Srinivasan, C.J., Ajay, V., Soman, K.P.: Enhanced skew detection method using principal component analysis. In: IEEE International Conference on Signal and Image Processing, pp. 600–603. Macmillan India Publishers, London (2006)
8. http://www.heatonresearch.com/articles/6/page2.html
9. Goyal P., Diwakar, S., Agrawal, Devanagari, A.: Character recognition towards natural human computer interaction. In: IHCI'10 Proceedings of International Conference on Interaction Design and International Development, pp. 55–59 (2010)

Quality Assessment of Data Using Statistical and Machine Learning Methods

Prerna Singh and Bharti Suri

Abstract Data warehouses are used in organization for efficiently managing the information. The data from various heterogeneous data sources are integrated in data warehouse in order to do analysis and make decision. Data warehouse quality is very important as it is the main tool for strategic decision. Data warehouse quality is influenced by Data model quality which is further influenced by conceptual data model. In this paper, we first summarize the set of metrics for measuring the understand ability of conceptual data model for data warehouses. The statistical and machine learning methods are used to predict effect of structural metrics, on understand ability, efficiency and effectiveness of Data warehouse Multidimensional (MD) conceptual model.

Keywords Conceptual model · Data warehouse quality · Multidimensional data model · Statistical · Understand ability

1 Introduction

In order to solve the problem for data to provide correct information, organizations are adopting a data warehouse which is defined as a subject oriented, integrated, non volatile data that support decision process [1]. Data Warehouses are increasing in complexity [2] and needs a serious attention for evaluation of quality throughout its design and development [3]. Some of the serious issues such as loss of clients, financial losses and discontent among employee can be due to lack of quality in data warehouse. Information quality of the data warehouse comprise of the data

P. Singh (✉)
Jagan Institute of Management Studies, New Delhi, India
e-mail: prerna.singh@jimsindia.org

B. Suri
USICT, Dwarka, New Delhi, India
e-mail: bhartisuri@gmail.com

L.C. Jain et al. (eds.), *Computational Intelligence in Data Mining - Volume 2*,
Smart Innovation, Systems and Technologies 32, DOI 10.1007/978-81-322-2208-8_10

warehouse quality and the data presentation quality. Data Warehouse quality can be influenced by database management quality, data quality and data model quality. Data model quality can be categorized into conceptual, logical and physical [1].

According to Manuel Serrano et al. [4] Data Warehouse (DW) is used for making strategy decision. DW quality is crucial for the organization. They have used global methods for defining and obtaining correct metrics. They performed theoretical validation of the proposed metrics and had carried out family of experiments for empirical validation of metrics. The study has been replicated in this paper by performing empirical validation on the metrics defined by global methods [4]. Global Methods are used for defining and obtaining correct metrics.

In this paper, the set of structural metrics that are already defined for conceptual model for data warehouse [4] are summarized. Then empirical validation is performed by conducting family of experiments by us. The paper is structured as follows: Sect. 2 summarizes about the relevant related work. Section 3 defines the method used for defining metrics, and Sect. 4 defines the family of experiments carried out to prove empirical validation of metrics. Section 5 defines the conclusion drawn from the family of experiments.

2 Related Work

2.1 Multidimensional Modeling

A variety of multidimensional data models have been proposed by both academic and industry communities [5]. These models are represented by set of graphical notation that help in their use and reading. But none of these has been accepted as standard extension mechanism (stereotypes, tagged values and constraints) provided by the unified modeling language (UML) [6]. In [7], the conceptual model has been represented by means of class diagram in which information is organized in facts and dimensions. Facts are represented by fact classes and dimensions are represented by dimension classes. The basic notions are the dimension and the data cube. A dimension represents a business perspective under which data analysis is to be performed and is organized in a hierarchy of levels which correspond to different ways to group its elements [1]. Multi dimensional model can be represented as a data cube. A data cube represent factual data on which the analysis is focused and associate measures and coordinates defined over a set of dimension levels [1].

Data modeling is an art [8], even if the product of this activity has the prosaic name of the database scheme. The data model that allows the designer to devise schemes that are easy to understand and can be used to build a physical database with any actual software system, is called conceptual data model [5]. Conceptual data model represent concepts of the real world. It is widely recognized that there are at least two specific notations that any conceptual data model for data warehousing should include in some form: the fact (data cube) and the dimension A widespread notation used in implementation in this context is the "star schema" [2] in which fact and dimension are simply relational tables connected in a specific way. One of the uses of multidimensional model is that they can be used for documentation purposes as they are easily understood by the non-specialist. They can also be used to describe in abstract terms the content of data warehousing application already in existence. The dimensions are organized into hierarchy of levels, obtained by grouping elements of the dimension according to the analysis needs. A dimension has three main components: a set of levels, a set of level description and a hierarchy over the levels.

In this representation, levels are depicted by means of round cornered boxes and there is a direct arc between the two levels. Small diamonds depict the description of a level. A multidimensional data model is a direct reflection of the manner in which a business process is viewed [1]. The Dimensional Fact (DF) Model by Golfarelli [6, 7], the Multi-Dimensional/ER Model by Sapia et al. [9, 10], the StarER Model by Tryfona [11], the Model proposed by Husemann [12] and the Yet Another Multidimensional Model (YAM) by Abello et al. [13] are examples of multidimensional models.

3 Metrics

A metric is a way to measure the quality factor in a constant and objective manner. They are used to understand software development and maintain projects. They are used to maintain quality of the system and determine the best way which helps the user in research work. It should be defined according to organization needs. The

Table 1 Metrics description [1]

Metrics	Description
NDC(S)	Number of dimensional classes of the stars
NBC(S)	Number of the base classes of stars
NC(S)	Total number of classes of the stars NC(S) = NDC(S) + NBC (S) + 1
RBC(S)	Ratio of base classes. Number of base classes per dimensional class of stars
NAFC(S)	Number of FA attribute of the fact class of the stars
NADC(S)	Number of D and DA attribute of the dimensional classes of the stars
NABC(S)	Number of D and DA attribute of the base classes of
NA(S)	Total number of FA, D and DA attribute NA(S) = NAFC (S) + NADC (S) + NADC (S) + NABC (S)
NH(S)	Number of hierarchy relationship of the stars
DHP(S)	Maximum depth of the hierarchy relationship of the data stars
RSA(S)	Ratio of attributes of the stars. Number of attributes FA decided by the number of D and DA attributes

main goal is to assess and control the quality of the conceptual data ware house schema. This section describe the metrics proposed by Serrano et al. [4] for data warehouse multidimensional model. The proposed metrics are defined in Table 1.

4 Empirical Validation

4.1 Experimental Design

4.1.1 Variables in the Study

Dependent Variables: Understandability, Efficiency and Effectiveness are the dependent variable, which we want to predict using structural metrics(independent variables).The understand ability of the tests was measured as the time each subject took to perform the tasks of each experiment test. The experimental tasks consisted of understanding the schemas and answering the corresponding questions. Here, low value of understanding time for the schemas means it is understandable, whereas, high value of understanding time for the schemas means, it is non-understandable.

Hypothesis Formulation:

In this section, research hypothesis are presented. Null hypothesis, H_0: There is no significant effect of individual metric on schema understand ability. Alternate hypothesis, H_1: There is a significant effect of individual metric on schema understands ability.

The metrics value for all schemas is defined in Table 2. The data collected for understanding time is in Table 3.

Table 2 Metrics value for all schema

	NDC	NBC	NC	RBC	NAFC
S01	3	5	9	1.67	2
S02	4	9	14	2.25	3
S03	3	4	8	1.vel	3
S04	4	8	13	2	2
S05	3	8	12	2.67	3
S06	4	6	11	1.5	1
S07	3	5	9	1.67	5
S08	3	5	9	1.67	3
S09	3	6	10	2	3
S10	3	4	8	1.33	2
S11	3	6	10	2	3
S12	3	5	9	1.67	3

Table 3 Data collected of understanding time (s)

	S01	S02	S03	S04	S05	S06	S07	S08	S09	S10	S11	S12
Sub1	170	300	343	234	107	62	64	125	183	100	120	130
Sub2	160	200	300	200	100	70	89	100	108	107	124	120
Sub3	200	218	260	160	230	180	142	197	112	130	105	110
Sub4	230	360	262	120	180	100	191	100	120	120	100	100
Sub5	300	120	180	240	120	180	110	157	178	120	149	100
Sub6	177	135	262	175	180	120	180	240	180	178	120	166
Sub7	180	180	120	210	172	111	130	231	170	129	130	122
Sub8	157	201	171	86	134	107	–	112	78	114	89	120
Sub9	300	240	120	135	172	98	120	117	99	110	92	83
Sub10	178	200	210	118	135	90	91	125	93	100	108	83
Sub11	179	160	140	130	115	108	83	110	111	100	77	83
Sub12	180	180	120	140	95	84	119	147	120	100	105	111
Sub13	229	162	114	109	110	114	150	120	111	106	108	128
Sub14	120	180	120	110	100	120	100	90	120	60	120	90
Sub15	131	119	129	64	102	68	115	78	71	96	75	105
Sub16	188	165	200	200	100	120	110	108	128	124	100	105
Sub17	120	180	300	300	180	180	240	60	124	111	108	90
Sub18	86	86	120	240	184	122	64	65	100	120	130	120
Sub19	120	185	125	185	242	185	120	185	100	104	108	111
Sub20	140	120	240	180	300	180	300	180	107	111	105	104

4.2 Research Methodology

In this section both statistical (Logistic regression analysis) and machine learning methods (Decision Tress, Naïve Bayesian Classifier) have been used for prediction of schema understand ability.

4.2.1 Statistical Method

The logistic regression (LR) analysis is used to predict the dependent variable (Understand ability) from a set of independent variable in order to determine the percentage of variance in the dependent variable explained by independent variable [14]. There are two step wise selection methods which are forward selection and backward elimination.

The general multivariate regression formula [10] which is used as follows:-

$$\mathrm{Prob}(X1, X2, \ldots, Xn) = e^{g(x)}/1 + e^{g(x)}$$

where $g(x) = B_0 + B_1 * X_1 + B_2 * X_2 + \cdots + B_n * X_n$ 'prob' is the probability of a schema being understandable.

4.2.2 Machine Learning Methods

The machine learning methods are used for designing computer programs that improve their performance for some specific task based on past observation. The well known machine learning.

Methods are decision tress and naïve Bayesian classifier. Decision tree (DT) is a predictive machine learning methods that decides the target value (dependent variable) of a new sample based on various attribute values of the available data. The DT includes several algorithms such as Quinlan's ID3, C4.5, C5, J48 and CART [11, 13]. Here we applied J48 algorithm which is the modified version of an earlier algorithm C4.5 developed by Quinlan [15].

4.3 Analysis and Results

The machine learning methods as well as statistical methods have been applied to predict the effect of each metric on understand ability of conceptual schema. The following measures are used to evaluate the performance of each predicted understand ability model:

Sensitivity: It measures the correctness of the predicted model. It is defined as the percentage of schemas correctly predicted to be understandable.
Specificity: It also measures the correctness of the predicted model. It is defined as the percentage of schemas predicted that will not be understandable.
Accuracy: It is defined as the ratio of number of schemas that are predicted correctly to the total number of schemas.
Receiver Operating Characteristic (ROC) analysis: The performance of the outputs of the predicted models may be evaluated using ROC analysis. It is an effective method of evaluating the performance of the model predicted.

4.3.1 Descriptive Statistics

The maximum, minimum, standard deviation and average of the understand ability of the time taken by the subjects to answers the given set of questions for each schemas of the experiment is tabulated in Table 4.

4.3.2 Univariate LR Analysis Results

In univariate LR analysis, effect of each metrics on the understand ability is predicted and shown in Table 5.

Table 6 is used to predict the sensitivity, specificity, precision etc. for some metrics and model.

Table 4 Descriptive statistics of data collected

Schemas	Understanding time (in seconds)				Understandability
	Maximum (M)	Minimum (m)	Standard deviation	Average	
S01	300	86	55.58	177.2	Non-understandable
S02	360	86	63.00	189.5	Non-understandable
S05	300	95	59.65	152.6	Non-understandable
S06	180	62	40.20	119.9	Understandable
S07	300	0	65.17	125.9	Understandable

Table 5 Logistic model correlation coefficients (B)

Metric	B	SE	Sig	Odd ratio	R^2
NDC	−2.4	1.8	0.06	0.07	0.38
NBC	−0.3	2.3	0.08	2.3	0.08
NC	−0.6	3.4	1.11	2.75	0.14
RBC	−0.3	3.4	2.3	0.669	0.28

Table 6 LR analysis-sensitivity, specificity, accuracy and AUC

Metric	Sensitivity	Specificity	Accuracy	AUC
NDC	1	0.44	0.87	0.87
NBC	0.67	0.3	0.7	0.3
NC	0.8	0.33	0.6	0.7
RBC	0.9	0	0.5	0.4

4.3.3 Descision Tree (Dt) Analysis Results

The Decision Tree Analysis results are shown in Table 7.

4.4 Comparison of Prediction Methods

We have applied statistical method (LR analysis) and machine learning methods (Decision Tree) to predict understand ability of DW conceptual schema using structural metric. The result of the experiment is shown in Table 8. The method which reports high sensitivity and high specificity during prediction of understand ability of DW conceptual schema is the better method for prediction.

Table 7 DT analysis

Metric	Sensitivity	Specificity	Accuracy	AUC
NDC	1	0.44	0.87	0.87
NBC	0.67	0.3	0.7	0.3
Model	1	0.7	0.8	0.6

Table 8 Prediction method comparison

Method	Sensitivity	Specificity	Accuracy	AUC
LR analysis	0.5	0.5	0.8	0.5
J 48	0.7	0.7	0.6	0.6

4.5 Validation of Hypothesis

In this section, we validate our hypothesis formulated in Sect. 4.1.

According to results of LR analysis the individual metric has significant effect on understand ability of conceptual schema. The machine learning methods (Naive BC, DT) also confirmed the finding of regression analysis as the values of sensitivity, specificity and accuracy for some individual metric are high and is same as of LR analysis. Therefore we reject the null hypothesis and accept the alternative hypothesis (H_1).

5 Conclusion and Future Work

The quality of data warehouse is very important because when data warehouse is constructed properly, it provides organizations with a foundation that is reusable. In this paper we have used already defined metrics for data warehouse in order to control their quality. The statistical and machine learning methods have been used as research methodology to predict the understand ability of conceptual schemas of DW. We have concluded that there exists a strong correlation between the understanding time and some of the metrics. We can derive that metric NADC (Number of D and DA attribute of dimensional classes) is correlated with the Understand ability of data warehouse schemas.

References

1. Serrano, M., Trujillo, J., Calerro, C., Piattini, M.: Metrics for data warehouse conceptual model understandability. Inf. Softw. Technol. 851–890 (2007)
2. Kimball, R.: The Data Warehouse Toolkit. Wiley, New York (2011)
3. Kesh, S.: Evaluating the quality of entity relationship models. Inf. Softw. Technol. **37**, 681–689 (1995)

4. Serrano, M., Calero, C., Trujello, J.: Sergio Lujan-Mora and Mario Riattini. Empirical Validation of Metrics for Conceptual Models of Data Warehouses. In: Pearson, A., Stirna, J. (eds.) CAiSE, LNCS, vol. 3084, pp. 506–520 (2004)
5. Batini, C., Ceri S., Navathe S.: Conceptual database design: an entity relationship approach. Benjamin/Cummings
6. Jeusfeld, M., Quix, C., Jarke, M.: Design and analysis of quality information for data warehouses. In: 17th International Conference on Conceptual Modeling (ER 98), Singapore (1998)
7. Golfarelli, M., Maio, D., Rizzi S.: The dimensional fact model—a conceptual for data warehouses. Int. J. Coop. Inf. Syst. (IJCIS) **7**, 215–247 (1998)
8. Basili, V., Romach.: The tame project towards improvement oriented software environments. IEEE Trans. Soft Eng. **14**(6) 728–738 (1988)
9. Golfarelli, M., Rizzi, S.: A methodological framework for data warehouse design. In: 1st International Workshop on Data Warehousing and OLAP (Dolap 98) Maryland (USA) (1998)
10. Sapia, C.: On Modeling and Predicting Query Behavior in OLAP Systems. In: International Workshop on Design and Management of Data warehouses (DMDW '99), pp. 1–10, Heidelberg (Germany) (1999)
11. Sapia, C., Blaschka, M., Holfing, G., Dinter, B.: Extending use the E/R model for multidimensional paradigm. In: 1st International Workshop on Data Warehouse and Data mining (DWDM '98), pp. 105–116. Springer Singapore (1998)
12. Husemann, B., Lechtenborger, J., Vossen, G.: Conceptual data warehouse design. In: 2nd International Workshop on Design and Management of Data Warehouses (DMDW 2000), pp. 3–9, Stockholm (Sweden) (2000)
13. Abello, A., Samos, J., Saltor, F.: YAM2 (Yet Another Multi Dimensional Model) An Extension of UML. In: International Database Engineering and Application Symposium (IDEAS 2002), pp. 172–181. IEEE Computer Society Edmonton (Canada) (2002)
14. Caldiera, V.R.B.G., Dieter Rombach, H.: The goal question metric approach. In: Encyclopedia of Software Engineering. Wiley, New York (1994)
15. Moody, D.: Metrics for evaluating the quality of entity relationship models. In: 17th International Conference on Conceptual Modelling, pp. 213–225 (ER 98) Singapore (1998)

Design of Biorthogonal Wavelets Based on Parameterized Filter for the Analysis of X-ray Images

P.M.K. Prasad, M.N.V.S.S. Kumar and G. Sasi Bhushana Rao

Abstract The X-ray bone images are extensively used by the medical practitioners to detect the minute fractures as they are painless and economical compared to other image modalities. This paper proposes a parameterized design of biorthogonal wavelet based on the algebraical construction method. In order to assign the characters of biorthogonal wavelet, there are two kinds of parameters which are introduced in construction process. One is scale factor and another one is sign factor. In edge detection, the necessary condition of wavelet design is put forward and two wavelet filers are built. The simulation results show that the parameterized design of biorthogonal wavelet is simple and feasible. The biorthogonal wavelet zbo6.6 performs well in detecting the edges with better quality. The various performance metrics like Ratio of Edge pixels to size of image (REPS), peak signal to noise ratio (PSNR) and computation time are compared for various biorthogonal wavelets.

Keywords Biorthogonal wavelet · Symmetry · Vanishing moments · Parameterized · Support interval · Filter banks

1 Introduction

Edge detection of X-ray images delivers details about fracture in bone and plays an important role in patient diagnosis for doctors. Bone edge detection in medical images is a crucial step in image guided surgery. Human body is constructed by

P.M.K. Prasad (✉)
Department of ECE, GMR Institute of Technology, Rajam, India
e-mail: mkprasad.p@gmrit.org

M.N.V.S.S. Kumar
Department of ECE, AITAM, Tekkali, India
e-mail: muvvala_sai@yahoo.co.in

G. Sasi Bhushana Rao
Department of ECE, Andhra University College of Engineering, Visakhapatnam, India
e-mail: sasigps@gmail.com

L.C. Jain et al. (eds.), *Computational Intelligence in Data Mining - Volume 2*,
Smart Innovation, Systems and Technologies 32, DOI 10.1007/978-81-322-2208-8_11

number of bones, veins and muscles. Some of the bones are merged with muscles. Regular eye vision can't detect the fracture. Diverse sizes of bones for diverse individuals in various ages make edge detection a challenging region. For example, a child has soft bone and aged individuals might have hard bone. So, bones in medical images have various sizes and the intensity values of bone pixels are normally non uniform and noisy. Digital Imaging is now a cornerstone in the provision of diagnostic data to medical practitioners. Edge feature extraction of X-ray bone image is very useful for the medical practitioners as it provides important information for diagnosis which in turn enable them to give better treatment decisions to the patients. Presently digital images are increasingly used by medical practitioners for disease diagnosis. The images are produced by several medical equipments like MRI, CT, ultrasound and X-ray. Out of these, X-ray is one, the oldest and frequently used devices, as they are painless and economical. Edge feature extraction deals with extracting or detecting the edge of an image.

The most important information of an image is the edge. The Edges of the images have always been used as primitives in the field of image analysis, segmentation and object tracing. It is the most common approach for detecting meaningful discontinuities in the gray level [1]. When image is acquired, the factors such as the projection, mix, and noise are produced. These factors bring on image feature's blur and distortion, consequently it is very difficult to extract image feature. Moreover, due to such factors it is also difficult to detect edge [2]. Therefore an efficient technique based on wavelet transform is used for edge detection. This is because wavelet transform has the advantage of detecting edges using different scales.

The orthogonal wavelet transforms like Haar, daubechies, coiflet, symlets can be used to detect the edges of an image. But a lot of false edge information will be extracted. They are also sensitive to noise. An important property of human visual system is that people are more tolerant of symmetric errors than asymmetric ones. Therefore, it is desirable that the wavelet and scaling functions are symmetric. Unfortunately, the properties of orthogonality and symmetry conflict each other in the design of compactly supported wavelet. Owing to this analysis, it is necessary to use symmetric biorthogonal wavelets [3]. The biorthogonal wavelet is more advantages compared to o orthogonal wavelet because of more flexibility and computation time. As there are different properties of the wavelet such as orthogonality, symmetry and vanishing moments which can be varied [4], therefore the qualities of detected edge are different. Depending upon the properties of the wavelet, the quality of the edge results would be obtained [5]. So design of biorthogonal wavelet with anticipant characteristic is essential to improve the edge detection results.

There are two methods for the design of biorthogonal wavelet i.e. Multiresolution Analysis (MRA) and Lifting Wavelet transform. Though the two methods can design biorthogonal wavelet, but their realization is limited and difficult. This paper proposes a new method to design biorthogonal wavelet. This new method is parameterized filter design which is based on algebraic construction [6]. By selecting the characteristics, a serial of biorthogonal wavelets can be constructed by parameterized filter design [7–9]. The parameters can be used to adjust the designed biorthogonal wavelet.

2 Biorthogonal Wavelet Theory

The wavelet expansion system is to be orthogonal across both translations and scale gives a clean, robust, and symmetric formulation with a Parseval's theorem. It also places strong limitations on the possibilities of the system. Requiring orthogonality uses up to large number of the degree of freedom, results in complicated design equations, prevents linear phase analysis and synthesis filter banks, and prevents asymmetric analysis and synthesis systems. This develops the biorthogonal wavelet system using a nonorthogonal basis and dual basis to allow greater flexibility in achieving other goals at the expense of the energy partitioning property that Parselval's theorem states. Some researchers have considered "almost orthogonal" systems where there is some relaxation of the orthogonal constraints in order to improve other characteristics [3]. The design of orthonormal wavelets requires a step known as spectral factorization, which can make the filter lengths grow when going to coarser scales; moreover, orthogonal filters cannot be symmetric. These limitations are also encountered in classical wavelet filter design, and they can be circumvented by relaxing the orthogonality condition and considering biorthogonal wavelet. Daubechies said that the only symmetric, finite length, orthogonal filter is the haar filter [5]. While talking about the limitations of the haar wavelet, the shorter filter length sometimes fails to detect large changes in the input data. So it is necessary to to design symmetric filters of length greater than two [2]. The goal is to construct two low pass filters h and $\tilde{h}$ and their associated high pass filters g and $\tilde{g}$. Biorthogonal Wavelet is compactly supported symmetrical wavelet. The symmetry of coefficients is often desirable because it result in linear phase of transfer function. Biorthogonal wavelet has two scaling functions and two wavelet functions [10].

3 Parameterized Filter Banks Design

The design of filter banks of biorthogonal wavelet is based on low pass and perfect reconstruction condition. Then the expression of low pass decomposition filter and high pass reconstruction filter can be obtained. Multi-scale wavelet decomposition is the first step in applying the wavelet transform to the image. The high pass detail information can be obtained [5]. The edge of the image can be obtained with the high pass coefficient and reconstruction. Therefore the high pass decomposition filter plays predominant role than other filter in wavelet transform. The sign sequence of filters element gives effect to the capability of edge detection.

3.1 Parameter Filter Design Process

The Multiresolution Analysis (MRA) to design the biorthogonal wavelet is:

$$\tilde{h}, h \Rightarrow \tilde{g}, g \quad \tilde{\phi}, \phi \Rightarrow \tilde{\psi}, \psi$$

In order to increase the expected characteristics of high pass decomposition filter, the process of construction is updated.

$$\tilde{\psi} \rightarrow \phi \rightarrow \tilde{\phi} \rightarrow \psi \quad \tilde{g} \rightarrow h \rightarrow \tilde{h} \rightarrow g$$

In this case, $\tilde{g}$ and $\tilde{h}$ is the high pass and low pass decomposition filter, and is the high pass and low pass reconstruction filter. The parameterized filter design approach for biorthogonal wavelet is shown in Fig. 1. There are four steps in parameterized filter design of biorthogonal wavelet. The first step is to select the characteristics of high pass decomposition filter, such as support interval, symmetry [11] and vanishing moments [12] which are based on the application. The second step is parameterized design of high pass decomposition filter. The scale factor [k] and sign factor '±' is added in filter vector. If the support interval of filter is too long, the scale factor is [k_1, k_2, k_3 … k_n] are used. So the proportion of adjacent filter elements can be adjusted independent by scale factors. The range of scale factor determines the filter sequence in ascending or descending order form vector to side. The positive sign or the negative sign impacts the filters elements sign sequence. Third, the parameterized low pass reconstruction filter through the following equation.

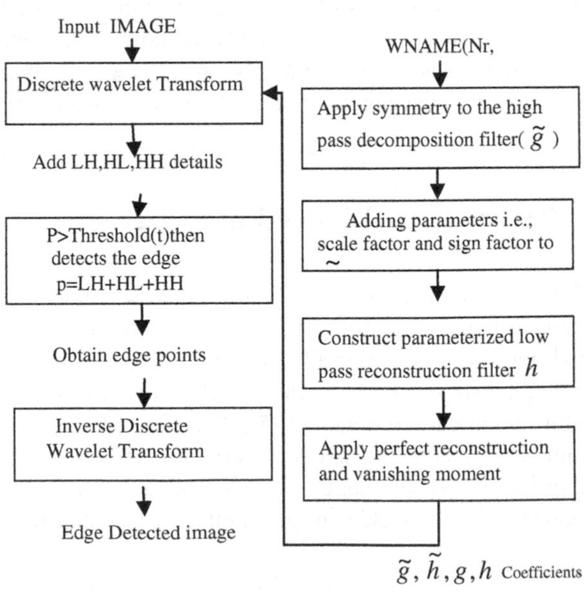

Fig. 1 Block diagram of parameterized filter design of biorthogonal wavelet

$$h_k = (-1)^{k-1}\tilde{g}_{1-k} \tag{1}$$

In the last step, using perfect reconstruction condition (PR), biorthogonal filter condition and vanishing moments, the parameterized expression of low pass decomposition filter $\tilde{h}$, high pass reconstruction filter g can be obtained.

Biorthogonal filter condition is

$$\left.\begin{aligned}\sum_k h_{2k} = \sum_k h_{2k+1} = 1/\sqrt{2}\\ \sum_k \tilde{h}_{2k} = \sum_k \tilde{h}_{2k+1} = 1/\sqrt{2}\end{aligned}\right\} \tag{2}$$

Perfect reconstruction condition is [13]

$$\begin{aligned}&\sum_k h_k \tilde{h}_{k-2n} = \delta_{0,n}\\ &\tilde{g}_n = (-1)^n h_{1-n}\\ &g_n = (-1)^n \tilde{h}_{1-n}\end{aligned} \tag{3}$$

4 Design of Biorthogonal Wavelet for Edge Detection Using Parameterized Filter

4.1 Proprieties of Biorthogonal Wavelet Used in Edge Detection

1. The high pass decomposition filter must be odd symmetry about ½ location
2. The sign of high pass decomposition filter must be monotonic
3. The rank of vanishing moments must be larger than singularity of image
4. The support interval of high pass decomposition filter should not be too long.

It is suitable at 4–6.

Parameterized algebraical construction method can be used to design biorthogonal wavelet with even support interval. The support interval of high pass decomposition filter $\tilde{g}$ is assumed as six. The expression of $\tilde{g}$ and h is

$$\begin{aligned}\tilde{g} &= \{\tilde{g}_{-2}, \tilde{g}_{-1}, \tilde{g}_0, \tilde{g}_1, \tilde{g}_2, \tilde{g}_3\}\\ h &= \{h_{-2}, h_{-1}, h_0, h_1, h_2, h_3\}\end{aligned}$$

The sign of high pass decomposition filter monotonic, the relations between high pass filter elements is set as

$$\tilde{g}_0 = -\tilde{g}_1, \quad -\tilde{g}_{-1} = -\tilde{g}_2, \quad \tilde{g}_{-2} = -\tilde{g}_3$$

and introduce two scale factors $\{k_1, k_2\}$ and $k_1 > 0$ $k_i \neq 1$ $i = 1, 2$. Defining

$$\tilde{g}_1 = k_1\tilde{g}_3 \quad \tilde{g}_2 = k_2\tilde{g}_3$$

Then, the high pass filter is expressed as

$$\tilde{g} = \{-\tilde{g}_3, -k_2\tilde{g}_3, -k_1\tilde{g}_3, k_1\tilde{g}_3, k_2\tilde{g}_3, \tilde{g}_3\}$$

From (1), the low pass reconstruction filter is

$$h = \{-\tilde{g}_3, k_2\tilde{g}_3, -k_1\tilde{g}_3, -k_1\tilde{g}_3, k_2\tilde{g}_3, -\tilde{g}_3\}$$

If defining the Support interval of low pass decomposition filter is six too. Then

$$\tilde{h} = \left\{\tilde{h}_{-2}, \tilde{h}_{-1}, \tilde{h}_0, \tilde{h}_1, \tilde{h}_2, \tilde{h}_3\right\} = \tilde{h} = \left\{\tilde{h}_3, \tilde{h}_2, \tilde{h}_1, \tilde{h}_1, \tilde{h}_2, \tilde{h}_3\right\}$$

From biorthogonal filter condition (2),

$$\left.\begin{array}{l}\tilde{h}_3 + \tilde{h}_2 + \tilde{h}_1 = \sqrt{2}/2 \\ -\tilde{g}_3 + k_2\tilde{g}_3 - k_1\tilde{g}_3 = \sqrt{2}/2\end{array}\right\} \tag{4}$$

From Perfect reconstruction condition (3),

$$2\left(-\tilde{h}_3\tilde{g}_3 + k_2\tilde{g}_3\tilde{h}_2 - k_1\tilde{g}_3\tilde{h}_1\right) = 1 \tag{5}$$

$$-\tilde{g}_3\tilde{h}_1 + k_2\tilde{g}_3\tilde{h}_1 - k_1\tilde{g}_3\tilde{h}_1 - k_1\tilde{g}_3\tilde{h}_3 = 0 \tag{6}$$

The low pass reconstruction filter be scaled then

$$\tilde{h}_2 = k_2\tilde{h}_3 \tag{7}$$

Here, the vanishing moment is chosen as three according to the support interval. The equations which got from vanishing moment condition is

$$\left.\begin{array}{l}\tilde{h}_1 = 3\tilde{h}_2 - 5\tilde{h}_3 \\ \tilde{h}_1 = 9\tilde{h}_2 - 35\tilde{h}_3\end{array}\right\} \tag{8}$$

From the system of Linear equations i.e., (4), (8), the filter coefficients are

$$\tilde{h}_1 = 0.441875, \quad \tilde{h}_2 = 0.2209375, \quad \tilde{h}_3 = 0.0441875$$

Substituting the above values in (13) and then we can get the scale factor k_2, then substitute all values in (7) and (6), then $k_1 = \frac{20}{3}$, $k_2 = 5$, $\tilde{g}_3 = -0.2651$.

The filters coefficients of biorthogonal wavelet is

$$\left.\begin{aligned} \tilde{h} &= \{0.0441875, 0.2209375, 0.441875, 0.441875, \\ &\quad 0.2209375, 0.0441875\} \\ \tilde{g} &= \left\{\frac{3\sqrt{2}}{16}, \frac{15\sqrt{2}}{16}, \frac{5\sqrt{2}}{4}, -\frac{5\sqrt{2}}{4}, -\frac{15\sqrt{2}}{16}, -\frac{3\sqrt{2}}{16}\right\} \\ h &= \left\{\frac{3\sqrt{2}}{16}, -\frac{15\sqrt{2}}{16}, \frac{5\sqrt{2}}{4}, \frac{5\sqrt{2}}{4}, -\frac{15\sqrt{2}}{16}, \frac{3\sqrt{2}}{16}\right\} \\ g &= \{0.0441875, -0.2209375, 0.441875, -0.441875, \\ &\quad 0.2209375, -0.0441875\} \end{aligned}\right\} \tag{9}$$

The biorthogonal wavelet named as 'zbo6.6', where 'zbo' is the wavelet name and '6.6' indicates the number of reconstruction filter coefficients are six, number of decomposition filter coefficients are six. If the vanishing moments is high, the support interval of high pass decomposition filter $\tilde{h}$ must be longer than 'zbo6.6'. In there, the filter $\tilde{g}$ and h kept constantly and select the support interval of filter $\tilde{h}$ as ten. The filter $\tilde{h}$ is even symmetry.

$$\begin{aligned} \tilde{h} &= \{h_{-4}, h_{-3}, h_{-2}, h_{-1}, h_0, h_1, h_2, h_3, h_4, h_5\} \\ &= \{h_5, h_4, h_3, h_2, h_1, h_1, h_2, h_3, h_4, h_5\} \end{aligned}$$

The process of construction is same as above. The filter condition and the perfect reconstruction condition is

$$\left.\begin{aligned} & -\tilde{g}_3 + k_2\tilde{g}_3 - k_1\tilde{g}_3 = \sqrt{2}/2 \\ & \tilde{h}_5 + \tilde{h}_4 + \tilde{h}_3 + \tilde{h}_2 + \tilde{h}_1 = \sqrt{2}/2 \\ & \tilde{h}_4 = k_2\tilde{h}_5 \\ & 2\left(-\tilde{h}_3\tilde{g}_3 + k_2\tilde{g}_3\tilde{h}_2 - k_1\tilde{g}_3\tilde{h}_1\right) = 1 \\ & -\tilde{g}_3\tilde{h}_2 + k_2\tilde{g}_3\tilde{h}_3 - k_1\tilde{g}_3\tilde{h}_4 - k_1\tilde{g}_3\tilde{h}_5 = 0 \\ & \tilde{g}_3\tilde{h}_1 - k_2\tilde{g}_3\tilde{h}_1 + k_1\tilde{g}_3\tilde{h}_3 - k_2\tilde{g}_3\tilde{h}_4 + \tilde{g}_3\tilde{h}_5 = 0 \end{aligned}\right\} \tag{10}$$

In this case, the equations from vanishing moments is

$$\left.\begin{aligned} \tilde{h}_1 &= 3\tilde{h}_2 - 5\tilde{h}_3 + 7\tilde{h}_4 - 9\tilde{h}_5 \\ \tilde{h}_1 &= 9\tilde{h}_2 - 35\tilde{h}_3 + 91\tilde{h}_4 - 189\tilde{h}_5 \\ \tilde{h}_1 &= 15\tilde{h}_2 - 65\tilde{h}_3 + 175\tilde{h}_4 - 369\tilde{h}_5 \\ \tilde{h}_1 &= 33\tilde{h}_2 - 275\tilde{h}_3 + 126\tilde{h}_4 - 4149\tilde{h}_5 \end{aligned}\right\} \tag{11}$$

From system of Linear equations i.e., (10), (11), the filter coefficients are

$$\left.\begin{array}{l} h_1 = 0.1547, h_2 = 0.2320, \\ h_3 = 0.2099, h_4 = 0.0939, h_5 = 0.0166 \\ k_1 = 8.6667, k_2 = 5.6667, g_3 = -0.1768 \end{array}\right\} \quad (12)$$

The filter coefficients of biorthogonal wavelet 'zbo6.10' is

$$\left.\begin{array}{l} \tilde{h} = \{0.0166, 0.0939, 0.2099, 0.2320, 0.1547, \\ \quad 0.1547, 0.2320, 0.2099, 0.0939, 0.0166\} \\ \tilde{g} = \{0.1768, 1.0017, 1.5321, -1.5321, -1.0017, -0.1768\} \\ h = \{0.1768, -1.0017, 1.5321, 1.5321, -1.0017, 0.1768\} \\ g = \{0.0166, -0.0939, 0.2099, -0.2320, 0.1547, \\ \quad -0.1547, 0.2320, -0.2099, 0.0939, -0.0166\} \end{array}\right\} \quad (13)$$

4.2 Performance Metrics

(i) Visual effects: The quality of an image is subjective and relative, depending on the observation of the user. One can only say the quality of image as good, but others may disagree.

(ii) Ratio of Edge Pixels to Size of image (REPS): The edge pixels or edge points represents the strength of the image. If the pixel value of image is greater than the threshold value, then it is considered as edge pixel. REPS can be calculated by taking the ratio of number of edge pixels to the Size of an image [14]. This result gives the information about the required percentage of pixels in order to represent the proper of edge features in the image.

$$REPS\ (\%) = \frac{No.\ of\ Edge\ Pixels}{Size\ of\ an\ image} \times 100 \quad (14)$$

(iii) Peak signal to Noise Ratio (PSNR): PSNR is one of the parameters that can be used to quantify image quality. PSNR parameter is often used as a benchmark level of similarity between reconstructed images with the original image [15]. A larger PSNR produces better image quality. PSNR equation is illustrated below

$$PSNR = 20\ \log_{10} \frac{255}{\sqrt{MSE}} \quad (15)$$

where mean square error

$$MSE = \frac{1}{mn}\sum_{y=1}^{m}\sum_{x=1}^{n}(I(x,y) - I'(x,y))^2$$

$I(x,y)$ is Original Image, $I'(x,y)$ is edge detected image.

(iv) Computation Time: It is the time taken to execute the program.

5 Results and Discussion

In this paper the biorthogonal wavelets like zbo6.6 and zbo6.10 are designed and their filter coefficients are also obtained. Figures 2 and 3 shows the waveform of high decomposition filter and low pass decomposition filter coefficients of 'zbo6.6'and zbo6.10.

The waveform shows 'zbo6.6' is the most smoothness of two wavelets. The convergence of 'zbo6.10' is faster than 'zbo6.6'. And the difference between the symmetry axis of 'zbo6.6' is largest. So in edge detection, 'zbo6.6'wavelet can enhance the edge. The high pass decomposition filter of the two wavelets is odd symmetry. And the sign of filter sequence is monotonic.

In this paper, right hand X-ray bone image with minute fracture at the neck of the fourth metacarpal bone is considered. In order to calculate the edge features of this image, the algorithm is implemented on MATLAB7.9. The biorthogonal wavelet based on parameter filter design like zbo6.6 and zbo6.10 are applied to detect the edges of bone X-ray hand image. Figure 4 shows the images processed by these methods (Tables 1 and 2).

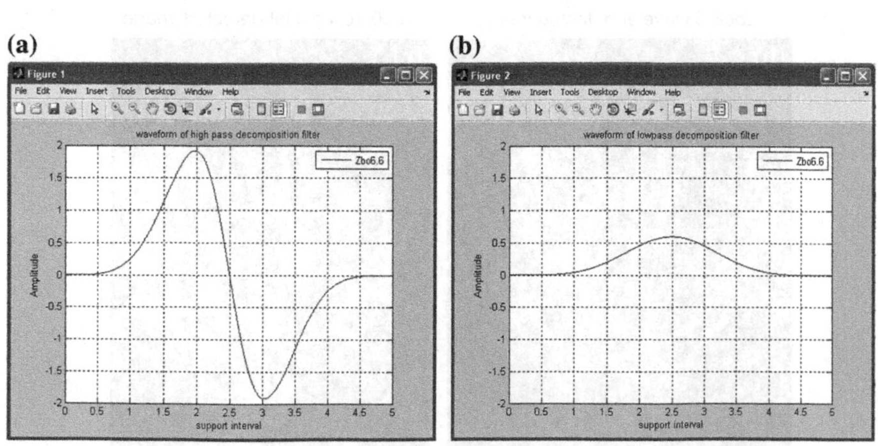

Fig. 2 Waveforms of decomposition filters zbo6.6 **a** high pass filter, **b** low pass filter

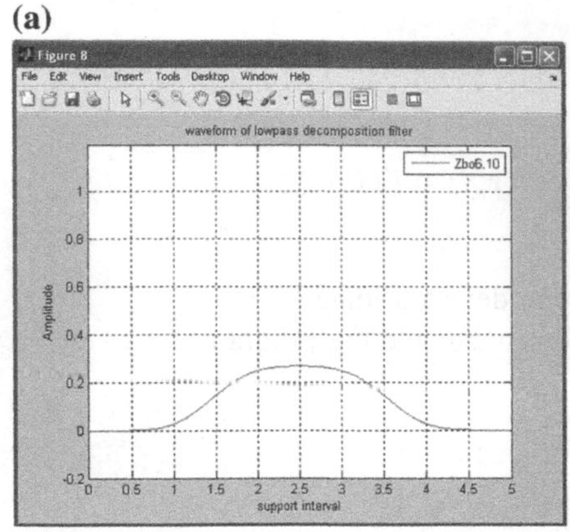

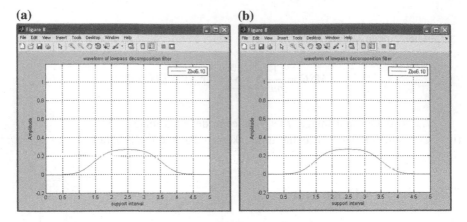

Fig. 3 Waveforms of decomposition filters zbo6.10 **a** high pass filter, **b** low pass filter

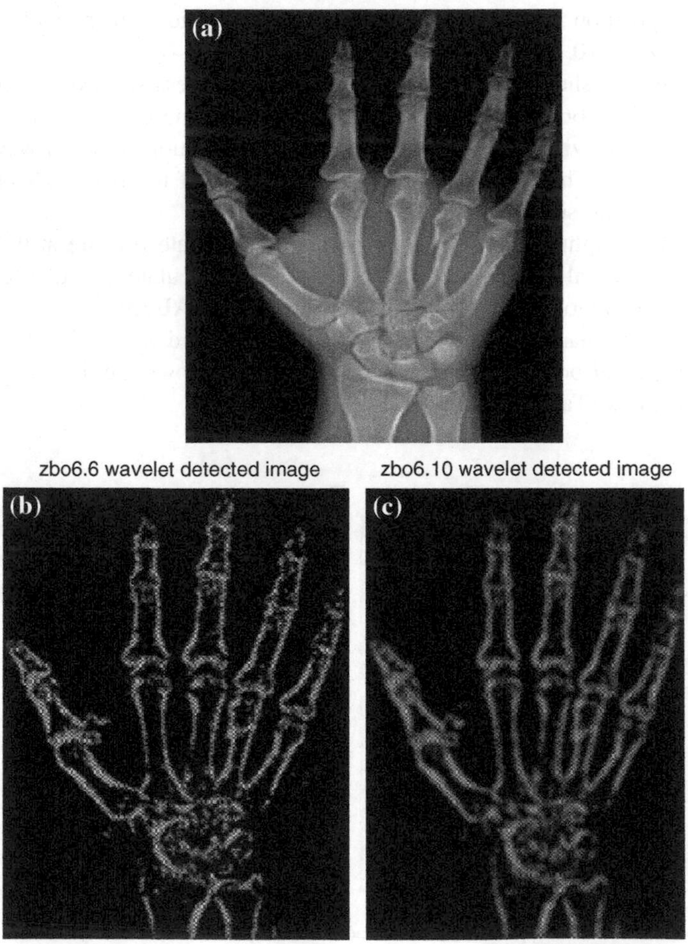

Fig. 4 **a** Original X-ray hand image (599 × 395) jpg, **b** zbo6.6 wavelet edge detected image, edge points 16,437, **c** zbo6.10 wavelet edge detected image, edge points 17,445

Table 1 Performance metrics for edge detected X-ray hand image (599 × 395) jpg for various thresholds

Wavelet type	Threshold	Edge points	PSNR (dB)	Computation time (s)	REPS (%)
Zbo6.6	0.03–0.06	6,690	11.3741	0.4375	2.83
	0.05–0.07	12,192	11.3741	0.6406	5.15
	0.09–0.1	16,437	11.3741	0.7969	6.95
Zbo6.10	0.03–0.06	7,701	11.3741	0.4375	3.25
	0.05–0.07	10,278	11.3741	0.7969	4.34
	0.09–0.1	17,445	11.3741	0.9844	7.37

Table 2 Performance metrics for edge detected X-ray hand image for 0.09–0.1 threshold

Wavelet type	Threshold	Edge points	PSNR (dB)	Computation time (s)	REPS (%)
Zbo6.6	0.09–0.1	16,437	11.3741	0.7969	6.95
Zbo6.10	0.09–0.1	17,445	11.3741	0.9844	7.37

From Fig. 4 it is clear that the edge detected image of zbo6.6 is best when compared to zbo6.10 wavelet in terms of visual perception. The main features of original image can be extracted perfectly. Although the number of edge points and REPS is less for zbo6.6 wavelet when compared to zbo6.10 but visual perception is good for zbo6.6. Though the number of edge points and REPS is high for zbo6.10 wavelet compared to zbo6.6 wavelet, but their visual quality is less due to false edge points. In this paper, the threshold is set as 0.09–0.1 as it produces more edge points. The threshold is based on selection of low frequency and high frequency range. It always produces more edge points for any image. The wavelet zbo6.6 is the best and effective in edge detection processing due to its visual quality, less computation time when compared to zbo6.10.

6 Conclusions

In this paper, parameterized filter based biorthogonal wavelet transforms are applied to the X-ray images to extract the minute fractures. In the process of analysis of edge detection, various biorthogonal wavelets are compared in terms of PSNR, Ratio of Edge pixels to size of image (REPS) and computation time. The biorthogonal wavelet zbo6.6 wavelet edge detected image gives visibly good quality edge detected image with less computation time. In process of designing wavelet, two kind parameters are introduced. One is sign, which is used to change the main characters of wavelet, such as waveform shape. Another is scale factor, which is used to adjust waveform in detail. If sign is same, we can improve the convergence of low pass filter by increasing the factor. The process of construction is simply and flexible. The zbo6.6 wavelet is faster and best due to its good visual quality when compared with zbo6.10 wavelet. The PSNR value of edge detected hand image based on zbo6.6 wavelet is 11.3741 dB which is same for zbo6.10 wavelet. The computation time of zb06.6 wavelet 0.7969 s which is less when compared to wavelet zbo6.10. The wavelet zbo6.6 gives best result when compared with zb06.10 wavelet for detecting the minute fractures of an X-ray image.

References

1. Gonzalez, R.C., Woods, R.E.: Digital Image Processing, 2nd edn. Pearson Education, Upper Saddle River (2004)
2. Cui, F.Y., Zou L.J., Song B.: Edge feature extraction based on digital image processing techniques. In: Proceedings of the IEEE International Conference on Automation and Logistics, pp. 2320–2324, Qingdao, China (2008)
3. Verma, R., Mahrishi, R., Srivastava, G.K., Siddavatam, R.: A novel image reconstruction using second generation wavelets. In: International Conference on Advances in Recent Technologies in Communication and Computing (2009). doi:10.1109/ARTCom.2009.59. ISBN 978-0-7695-3845-7
4. Soman, B.K.P., Ramachandran, B.K.P.: Insight into Wavelets from Theory to Practice, 2nd edn. PHI, New Delhi (2008)
5. Vanfleet, P.J.: Discrete Wavelet Transformations an Elementary Approach with Applications. Wiley, Hoboken
6. Han, D.B., Ron, A., Shen, Z.B.: Framelets: MRA based constructions of wavelet frames. Appl. Comput. Harmonic Anal. **14**, 146 (2003)
7. Changzhen, X.: Construction of biorthogonal two-direction refinable function and two-direction wavelet with dilation factor m. Comput. Math. Appl. **56**(7), 1845–1851 (2008)
8. Rosca, D., Antoine, J.P.: Locally supported orthogonal wavelet bases on the sphere via stereographic projection. Math. Probl. Eng. Article ID 124904 (2009)
9. Rafieea, J.: A novel technique for selecting mother wavelet function using an intelligent fault diagnosis system. Expert Syst. Appl. **36**(3), 4862–4875 (2009)
10. Khera, S., Malhotra S.: Survey on medical image de noising using various filters and wavelet transform. Int. J. Adv. Res. Comput. Sci. Softw. Eng. **4**(4), 230–234 (2014)
11. Benedetto, J.J., Li, S.: The theory of multi resolution analysis frames and applications to filter banks. Appl. Comput. Harmonic Anal. **5**(4), 389–427 (1998)
12. Chaudhury, K.N.: Construction of Hilbert transform pairs of wavelet bases and Gabor-like transforms. Signal Proc. **57**(9), 3411–3425 (2009)
13. Qiang, Z., Xiaodong, C., Hongxi, W., Ping, L.: Parametrized algebraical construction method of biorthogonal wavelet. In: 3rd International Conference on Computer Science and Information Technology (ICCSIT 2010)
14. Vidya, P., Veni, S., Narayanankutty, K.A.: Performance analysis of edge detection methods on hexagonal sampling grid. Int. J. Electron. Eng. Res. **1**(4), 313–328 (2009). ISSN 0975–6450
15. Kudale G.A., Pawar M.D.: Study and analysis of various edge detection methods for X-ray images. Int. J. Comput. Sci. Appl. (2010)
16. Lizhen, L.: Discussion of digital image edge detection method. Mapp. Aviso **3**, 40–42 (2006)

An Efficient Multi-view Based Activity Recognition System for Video Surveillance Using Random Forest

J. Arunnehru and M.K. Geetha

Abstract Vision-based human activity recognition is an emerging field and have been actively carried out in computer vision and artificial intelligence area. However, human activity recognition in a multi-view environment is a challenging problem to solve, the appearance of a human activity varies dynamically, depending on camera viewpoints. This paper presents a novel and proficient framework for multi-view activity recognition approach based on Maximum Intensity Block Code (MIBC) of successive frame difference. The experimare carried out using West Virginia University (WVU) multi-view activity dataset and the extracted MIBC features are used to train Random Forest for classification. The experimental results exhibit the accuracies and effectiveness of the proposed method for multi-view human activity recognition in order to conquer the viewpoint dependency. The main contribution of this paper is the application of Random Forests classifier to the problem of multi-view activity recognition in surveillance videos, based only on human motion.

Keywords Video surveillance · Human activity recognition · Frame difference · Motion analysis · Random forest

1 Introduction

In recent decades, the video surveillance system is an essential technology to ensure safety and security in both public and private areas like railway stations, airports, bus stands, banks, ATM, petrol/gas stations and commercial buildings due to

J. Arunnehru (✉) · M.K. Geetha
Speech and Vision Lab, Department of Computer Science and Engineering, Faculty of Engineering and Technology, Annamalai University, Annamalai Nagar, Chidambaram, Tamil Nadu, India
e-mail: arunnehru.aucse@gmail.com

M.K. Geetha
e-mail: geesiv@gmail.com

L.C. Jain et al. (eds.), *Computational Intelligence in Data Mining - Volume 2*,
Smart Innovation, Systems and Technologies 32, DOI 10.1007/978-81-322-2208-8_12

terrorist activity and other social problems. The study of human activity recognition have been extensively utilized in the area of computer vision and pattern recognition [1], whose point is to consequently segment, keep and perceive human activity progressively and maybe anticipate the progressing human activities.

Recognizing user activities mean to infer the user's current activities from available data in the form of videos, which is considered as a rich source of information. The automatic recognition of human activities opens a number of interesting application areas such as healthcare activities, gesture recognition, automatic visual surveillance, human-computer-interaction (HCI), content based retrieval and indexing and robotics using abstract information from videos [2].

Actions performed by humans are dependent on many factors, such as the view invariant postures, clothing, illumination changes, occlusion, shadows, changing backgrounds and camera movements. Different people will carry out the similar action in a different way, and even the same person will perform it another way at different times. Due to the variability in view, human body size, appearance, shape and the sophistication of human actions, the task of automatically recognizing activity is very challenging [3].

1.1 Outline of the Work

This paper accord with multi-view activity recognition, which aims to distinguish human activities from video sequences. The proposed method is evaluated using West Virginia University (WVU) multi-view activity dataset [4] considered actions such as *clapping, waving one hand, waving two hands, punching, jogging, jumping jack, kicking, picking, throwing and bowling*. Difference image is obtained by subtracting the successive frames. Motion information is obtained by the Region of Interest (ROI). The obtained ROI is separated into two blocks B1 and B2. Maximum Intensity Block Code (MIBC) is extracted as a feature from the block showing maximum motion in the ROI. The extracted feature is fed to the Random Forest (RF) classifier.

The rest of the paper is structured as follows. Section 2 reviews related work. Section 3 illustrates the feature extraction and related discussions. Section 4 explains the workflow of the proposed approach. Experimental results and the WVU multi-view activity dataset are described in Sect. 5 and Finally, Sect. 6 concludes the paper.

2 Related Work

Human action recognition in view-invariant scenes is still an open problem. Several excellent survey papers illustrated various applications related to human action analysis, representation, segmentation and recognition [1–3, 5, 6]. Motion is an important cue has been widely applied in the fields like visual surveillance and artificial intelligence [7]. In [8] present a space-time shape based on 2D human silhouettes, which have the spatial information with reference to the pose and shape descriptors which were used to recognize the human action. The temporal templates called Motion History Image (MHI) and Motion Energy Image (MEI) is used to represent the actions [9] and classification was done by SVM classifier [10]. In [11] an automatic activity recognition is carried out based on motion patterns extracted from the temporal difference image based on Region of Interest (ROI) and the experiments are carried out on the datasets like KTH and Weizmann datasets. In [12] propose a Motion Intensity Code (MIC) to recognize human actions based on motion information. The PCA and SVM were used to learn and classify action from KTH and Weizmann dataset which yields better performance. In [4] uses a Local Binary Pattern (LBP) with Three Orthogonal Planes (TOP) to represent the human movements with dynamic texture features and temporal templates is used to classify the observed movement. In [13] uses a collective approach of spatial-temporal and optical flow motion features method, which uses the improved Dynamic Time Warping (DTW) method to enhance temporal consistence of motion information, then coarse-to-fine DTW limitation on motion features pyramids to distinguish the human actions and speed up detection performance. In [14] adopted a 2D binary silhouette feature descriptor joined together with Random Forest (RF) classifier and the voting strategy is used to signify the each activity, then multi-camera fusion performance is evaluated by concatenating the feature descriptors for multiple view human activity recognition. In [15] utilize a combination of optical flow and edge features to model the each action and boosting method is adapted for efficient action recognition. In [16] proposed a method Maximum Spatio-Temporal Dissimilarity Embedding (MSTDE) based on human silhouettes sequences and recognition results obtained by similarity matching from different action classes.

3 Feature Extraction

The extraction of discriminative feature is most fundamental and important problem in activity recognition, which represents the significant information that is vital for further analysis. The subsequent sections present the depiction of the feature extraction method used in this work.

3.1 Frame Difference

To recognize the moving object across a sequence of image frames, the current image frame is subtracted either by the previous frame or successive frame of the image sequences called as temporal differencing. The difference images were obtained by applying thresholds to eliminate pixel changes due to camera noise, changes in lighting conditions, etc. This method is extremely adaptive to detect the

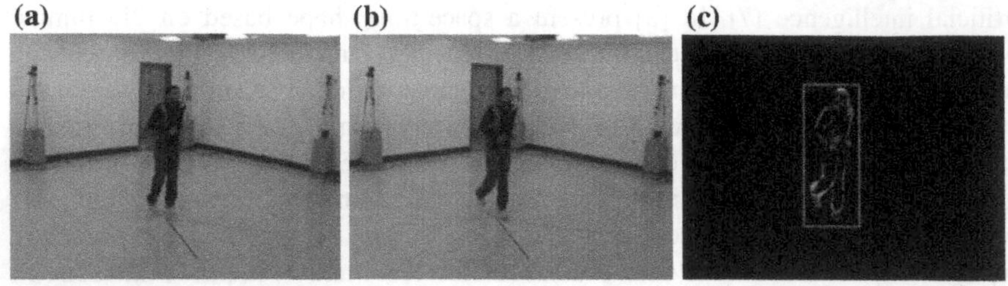

Fig. 1 **a**, **b** Two successive frames. **c** Difference image of (**a**) and (**b**) from WVU multi-view activity dataset

movable region corresponding to moving objects in dynamic scenes and superior for extracting significant feature pixels. The difference image obtained by simply subtracting the previous frame t with current frame is at time $t + 1$ on a pixel by pixel basis.

$$\begin{aligned} \mathrm{D}_s(x,y) &= |P_s(x,y) - P_{s+1}(x,y)| \\ 1 \le x \le w,&\ \ 1 \le y \le h \end{aligned} \tag{1}$$

The extracted motion pattern information is considered as the Region of Interest (ROI). Figure 1a, b illustrates the successive frames of the WVU multi-view activity dataset. The resulting difference image is shown in Fig. 1c. $D_s(x, y)$ is the difference image, $P_s(x, y)$ is the pixel intensity of (x, y) in the sth frame, h and w are the height and width of the image correspondingly. Motion information T_s or difference image is considered using

$$\mathrm{T}_s(x,y) = \begin{cases} 1, & \mathit{if}\ D_s(x,y) > t \\ 0, & \mathit{otherwise}; \end{cases} \tag{2}$$

where t is the threshold.

3.2 Maximum Intensity Block Code (MIBC)

To recognize the multi-view activity performed, motion is a significant signal normally extracted from video. In that aspect, Maximum Intensity Block Code (MIBC) is extracted from the motion information about the activity video sequences. The method for extracting the feature is explained in this section.

Figure 2 shows the motion information extracted for kicking, jogging and bowling activities from WVU multi-view activity dataset. Initially ROI is identified and considered as motion region as seen in Fig. 3a. The extracted ROIs are not fixed size. ROI separated into two blocks B1, B2 comprising of the upper part of the

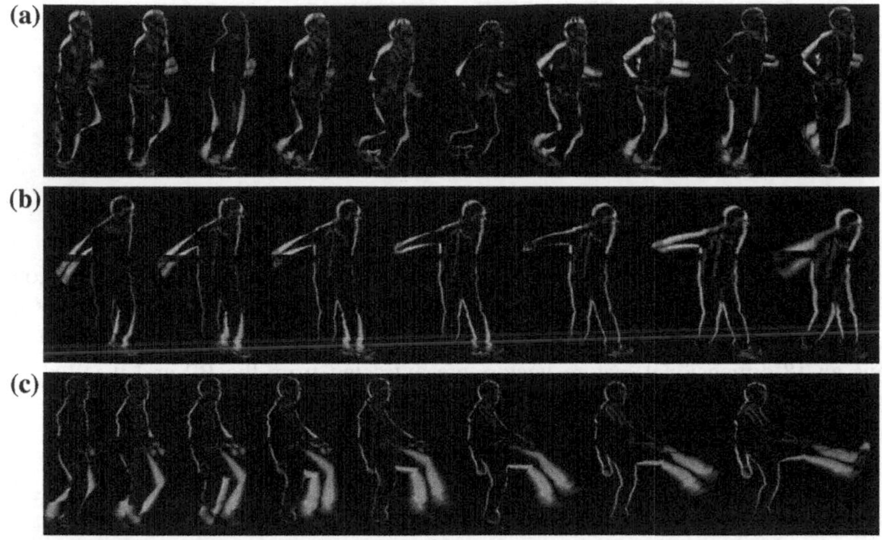

Fig. 2 Sample sequences of extracted ROIs from WVU activity dataset. **a** Kicking. **b** Jogging. **c** Bowling

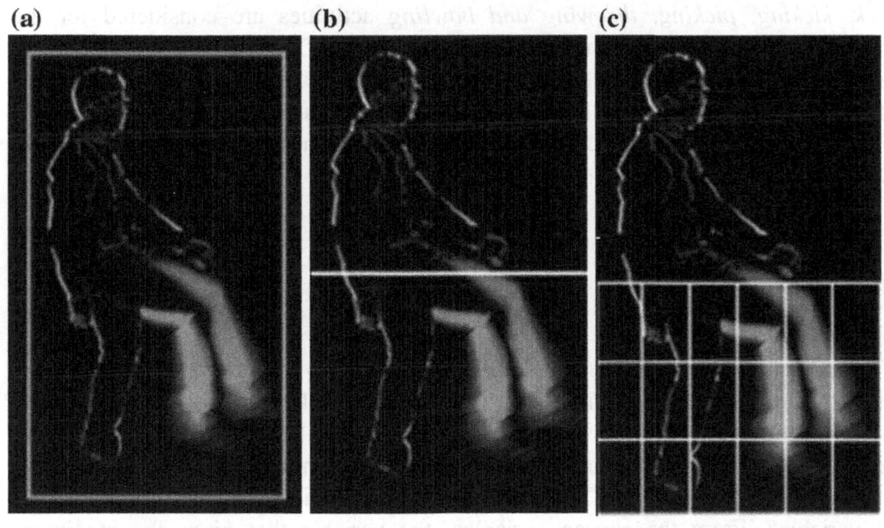

Fig. 3 **a** ROI of the motion pattern. **b** ROI divided into B1 and B2 blocks. **c** B2 is divided into 3 × 6 sub-blocks

body and lower part of the body for kicking activity captured on View 3 (V3) camera as shown in Fig. 3b. In order to reduce the amount of calculation, only the maximum motion identified block is considered for further analysis. In the Fig. 3b,

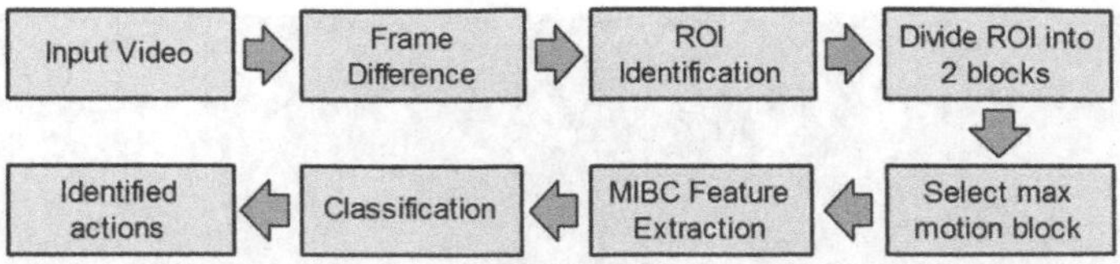

Fig. 4 Overview of the proposed approach

block B2 shows maximum motion. So, B2 alone is considered for MIBC extraction as shown in Fig. 3c. The block under consideration is equally divided into 3 × 6 sub-blocks. The average intensity of each block in 3 × 6 regions is computed and thus an 18-dimensional feature vectors are extracted from the B2 and it is fed to Random Forest (RF) classifiers for further processing.

4 Proposed Approach

The general overview of the proposed approach is illustrated in Fig. 4. This approach uses West Virginia University (WVU) multi-view activity dataset [17] such as *clapping, waving one hand, waving two hands, punching, jogging, jumping jack, kicking, picking, throwing and bowling* activities are considered for the experiments. The video frames are smoothing by a Gaussian spatial filter. It is necessary to pre-process all video frames to eliminate noise for fine features extraction and classification. ROI is extracted from the activity video sequences and MIBC features as discussed in Sect. 3.2. The obtained features are fed to the Random Forest (RF) classifier for activity recognition.

5 Random Forest

The Random Forest (RF) ensemble [18] grows with several unpruned trees (*n-tree*) on bootstrap samples of the input data. Each one of the tree is grown on a bootstrap sample (2/3 of the original input data identified as "in-bag" data) taken with substitution from the input data. Trees are divided into more than a few nodes using random subsets of variables. The default variable value is the square root of the sum of variables. From the chosen variables, the variable that gives the maximum decrease in impurity is selected to split the samples at every node. The most votes from all the trees in the ensemble determine the absolute prediction in classification. A tree has grown-up to its finer size, while not pruning till the nodes are untainted. A prediction of the response variable is formed by aggregating the prediction over all trees [19]. A divergent tree in Random Forest is a weedy classifier, for the reason that the sample of random subset is used to train a tree. The variables are chosen randomly in every node of the tree for expecting a low relationship value between the trees and over-fitting is consequently prevented and it is used to determine the robustness of a variable in a final model. The random Forest (RF) algorithm is simple to employ as only two parameters (*n-tree* and random feature samples) need to be optimized [19].

6 Experimental Results

In this section, the proposed method is evaluated using WVU multi-view activity dataset. The experiments are carried out in OpenCV 2.4 in Ubuntu 12.04 Operating System on a computer with Intel Core i7 Processor 3.40 GHz with 8 GB RAM. The obtained MIBC features are fed to Random Forest (RF) classifier using open source Machine Learning Tool WEKA [20] to develop the model for each activity and these models are used to test the performance.

6.1 WVU Multi-view Activity Dataset

The publicly-available West Virginia University multi-view activity dataset (http://csee.wvu.edu/~vkkulathumani/wvu-action.html) is used in the experiment. This dataset consists more than six subjects performing 12 different activities including *standing still, nodding head, clapping, waving one hand, waving two hands, punching, jogging, jumping jack, kicking, picking, throwing and bowling* as shown in Fig. 5. The multi-view videos have been recorded with an eight calibrated and synchronized network camera setup at a rate of 20 fps in standard definition resolution (640 × 480 pixels) for each activity in a rectangular indoor region (about 50 × 50 feet) from different viewing directions (view-angles: 0°, 45°, 90°, 135°, 180°, 225°, 270° and 315°) as illustrated in Fig. 6, resulting in a total of 960 video samples consists of approximately a 3 s of video.

6.2 Experimental Results on Random Forest

The experiment was done using an open source Machine Learning Tool WEKA on 10 different activities in eight different viewing directions of WVU multi-view activity dataset. The performance of the Random Forest (RF) classifier is measured using 10-fold cross validation model. The performance of the proposed method of each single view camera as well as for the combined multiple-view cameras. The confusion matrix for the multiple-view camera results is shown in Fig. 7a–h and the overall camera fusion (V1–V8) recognition accuracy reaches 84.3 % is shown in

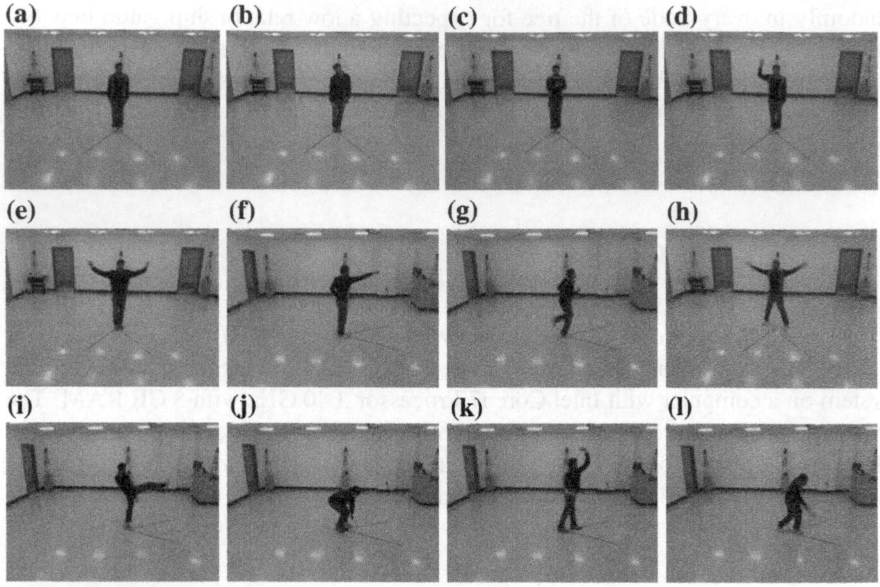

Fig. 5 Samples from WVU multi-view activity dataset **a** standing still, **b** nodding head, **c** clapping, **d** wave one hand, **e** waveing two hands, **f** punching, **g** jogging, **h** jumping jack, **i** kicking, **j** picking, **k** throwing, **i** bowling

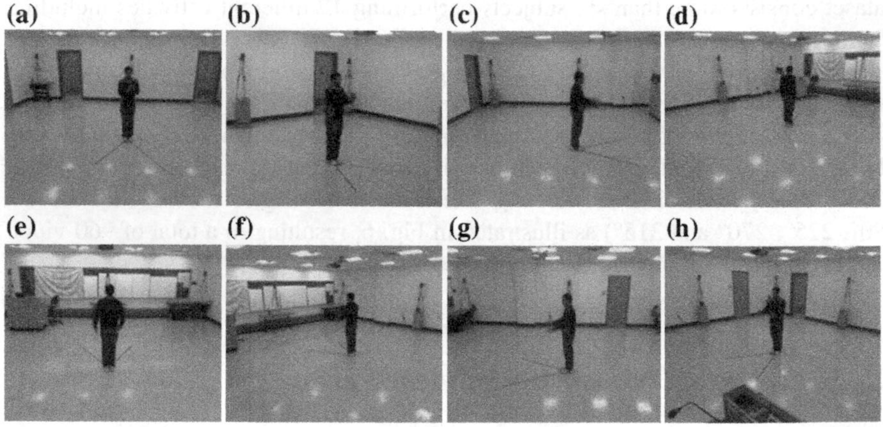

Fig. 6 Sample images of a clapping sequence captured from eight cameras positioned around a room. Each image shows a different camera view. **a** View 1. **b** View 2. **c** View 3. **d** View 4. **e** View 5. **f** View 6. **g** View 7. **h** View 8

Fig. 7i. In Fig. 7j shows the average accuracy results of each single view. As can be observed that rear and side view angles (V5, V3 and V7) shows the better recognition accuracies equal to 84.36, 83.51 and 83.71 % respectively.

(a)

	A	B	C	D	E	F	G	H	I	J
A	95.3	1.0	0.7	2.0	0.0	0.0	0.0	0.0	1.0	0.0
B	3.1	85.3	1.9	6.6	0.0	0.0	0.0	0.2	3.1	0.0
C	3.4	2.4	90.3	2.0	0.0	0.0	0.0	0.0	1.5	0.3
D	2.9	6.1	1.9	82.9	0.2	0.2	0.0	0.2	5.3	0.5
E	0.0	0.0	0.0	0.0	83.9	2.7	6.4	3.2	0.5	3.2
F	0.0	0.0	0.0	0.2	3.4	87.6	3.6	4.2	0.2	0.8
G	0.0	0.0	0.0	0.3	8.3	5.8	79.7	3.1	0.0	2.9
H	0.0	0.0	0.0	0.0	2.7	5.4	3.6	85.8	0.0	2.5
I	1.9	5.8	2.4	8.8	0.5	0.3	0.2	0.2	80.0	0.0
J	0.0	0.0	0.0	0.3	2.7	2.9	3.7	6.6	0.2	83.6

View 1

(b)

	A	B	C	D	E	F	G	H	I	J
A	93.4	2.2	1.4	1.5	0.0	0.0	0.0	0.0	1.5	0.0
B	2.2	91.2	0.8	4.7	0.0	0.0	0.0	0.0	0.7	0.3
C	2.7	1.9	94.6	0.2	0.0	0.0	0.2	0.0	0.5	0.0
D	5.6	1.9	0.5	90.2	0.0	0.0	0.0	0.0	1.7	0.2
E	0.0	0.0	0.0	0.2	88.3	4.9	2.9	2.9	0.0	0.8
F	0.0	0.0	0.0	0.5	5.3	88.1	1.4	3.2	0.2	1.4
G	0.0	0.0	0.0	0.0	5.9	5.3	83.7	3.6	0.3	1.2
H	0.0	0.0	0.2	0.0	0.3	2.2	1.5	94.7	0.0	1.0
I	1.2	3.7	0.3	2.5	0.0	0.2	0.7	0.5	90.3	0.5
J	0.0	0.2	0.0	0.0	1.5	1.2	1.5	3.7	0.3	91.5

View 2

(c)

	A	B	C	D	E	F	G	H	I	J
A	91.0	3.6	2.5	1.0	0.0	0.0	0.0	0.0	1.7	0.2
B	6.1	80.0	7.3	1.5	0.0	0.2	0.3	0.0	4.2	0.3
C	3.4	10.2	78.1	4.6	0.5	0.0	0.0	0.2	3.1	0.0
D	2.4	3.1	3.7	86.8	0.5	0.2	0.2	0.0	3.1	0.2
E	0.2	0.5	0.2	0.5	80.8	7.8	4.1	3.9	0.2	1.9
F	0.2	0.8	0.2	0.0	8.5	80.8	3.4	2.0	1.0	3.1
G	0.2	0.2	0.3	0.0	3.4	8.0	81.0	3.6	0.3	3.1
H	0.2	0.3	0.0	0.0	3.6	2.2	2.4	90.7	0.0	0.7
I	5.3	3.9	3.6	3.2	0.2	0.3	0.2	0.2	82.5	0.7
J	0.5	0.0	0.0	0.2	3.7	4.6	5.3	2.2	0.3	83.2

View 3

(d)

	A	B	C	D	E	F	G	H	I	J
A	90.2	2.7	2.4	3.9	0.0	0.0	0.0	0.0	0.8	0.0
B	5.8	83.4	5.6	2.5	0.0	0.2	0.0	0.0	2.5	0.0
C	7.1	7.6	80.3	3.7	0.0	0.0	0.0	0.0	1.2	0.0
D	4.7	7.3	3.2	82.7	0.0	0.0	0.0	0.0	2.0	0.0
E	0.0	0.0	0.0	0.3	77.6	6.3	6.1	5.8	0.2	3.7
F	0.0	0.2	0.2	0.2	8.0	79.5	3.9	4.2	0.0	3.9
G	0.0	0.0	0.0	0.2	9.3	5.9	75.6	2.9	0.0	6.1
H	0.0	0.0	0.0	0.0	2.9	2.2	2.5	89.0	0.0	3.4
I	6.3	4.6	2.4	9.0	0.0	0.2	0.0	0.3	77.1	0.2
J	0.0	0.0	0.0	0.2	5.3	5.4	11.9	5.1	0.8	71.4

View 4

(e)

	A	B	C	D	E	F	G	H	I	J
A	92.0	2.0	2.7	2.4	0.0	0.0	0.0	0.0	0.8	0.0
B	2.0	88.8	2.2	2.2	0.0	0.0	0.0	0.0	4.7	0.0
C	5.3	4.1	85.1	4.4	0.0	0.0	0.0	0.0	1.2	0.0
D	4.2	3.4	2.0	86.4	0.0	0.0	0.0	0.0	3.9	0.0
E	0.0	0.0	0.0	0.0	83.1	2.7	4.4	4.9	0.0	4.9
F	0.0	0.0	0.0	0.3	5.8	82.2	2.5	7.1	0.0	2.0
G	0.2	0.0	0.0	0.0	6.8	3.6	80.0	3.6	0.7	5.3
H	0.0	0.0	0.2	0.0	2.9	3.7	3.4	84.9	0.0	4.9
I	5.3	3.9	2.0	8.5	0.0	0.2	0.0	0.2	80.0	0.0
J	0.0	0.2	0.2	0.2	3.7	2.9	7.1	4.6	0.2	81.0

View 5

(f)

	A	B	C	D	E	F	G	H	I	J
A	87.5	3.7	2.4	3.6	0.0	0.0	0.0	0.0	2.9	0.0
B	7.5	80.8	4.7	2.0	0.0	0.0	0.0	0.0	4.9	0.0
C	4.6	5.9	85.9	1.7	0.0	0.0	0.0	0.0	1.9	0.0
D	6.4	2.5	1.2	86.9	0.0	0.0	0.2	0.0	2.5	0.2
E	0.0	0.0	0.0	0.2	80.5	3.7	5.8	8.8	0.0	1.0
F	0.0	0.2	0.0	0.2	7.5	81.4	3.6	4.2	0.2	2.9
G	0.0	0.0	0.0	0.2	7.3	4.2	76.4	7.1	0.7	4.1
H	0.0	0.2	0.5	0.0	4.6	4.2	4.1	84.7	0.2	1.5
I	5.6	4.6	1.5	5.1	0.7	0.3	0.7	0.2	81.0	0.3
J	0.0	0.3	0.0	0.7	3.6	3.1	4.1	4.7	0.2	83.4

View 6

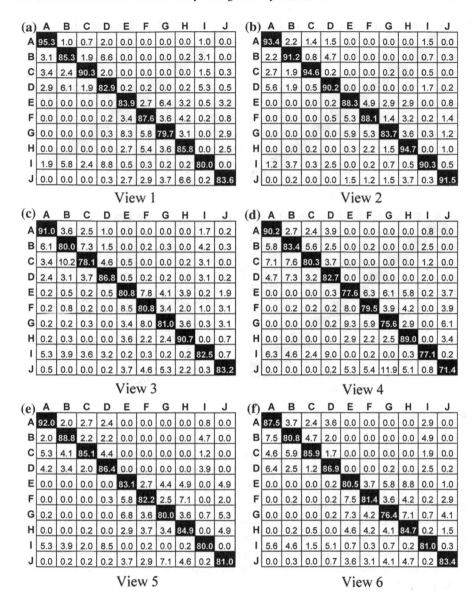

Fig. 7 **a–h** shows the confusion matrix for eight different views from WVU multi-view activity dataset using random forest (RF) classifier. **i** The confusion matrix of the multiple camera fusion results. **j** Classification accuracies (%) for multiple camera view (V1–V8) on WVU multi-view dataset, where *A* clapping, *B* waving one hand, *C* waving two hands, *D* punching, *E* jogging, *F* jumping jack, *G* kicking, *H* picking, *I* throwing, *J* bowling

(g)

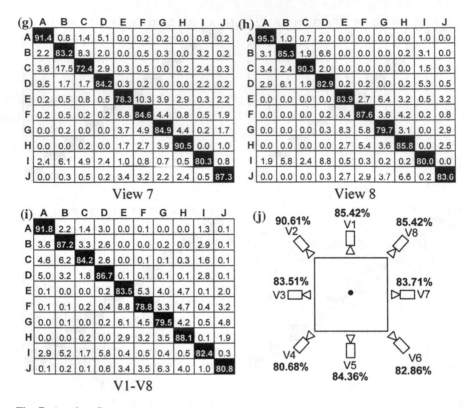

	A	B	C	D	E	F	G	H	I	J
A	91.4	0.8	1.4	5.1	0.0	0.2	0.2	0.0	0.8	0.2
B	2.2	83.2	8.3	2.0	0.0	0.5	0.3	0.0	3.2	0.2
C	3.6	17.5	72.4	2.9	0.3	0.5	0.0	0.2	2.4	0.3
D	9.5	1.7	1.7	84.2	0.3	0.2	0.0	0.0	2.2	0.2
E	0.2	0.5	0.8	0.5	78.3	10.3	3.9	2.9	0.3	2.2
F	0.2	0.5	0.2	0.2	6.8	84.6	4.4	0.8	0.5	1.9
G	0.0	0.2	0.0	0.0	3.7	4.9	84.9	4.4	0.2	1.7
H	0.0	0.0	0.2	0.0	1.7	2.7	3.9	90.5	0.0	1.0
I	2.4	6.1	4.9	2.4	1.0	0.8	0.7	0.5	80.3	0.8
J	0.0	0.3	0.5	0.2	3.4	3.2	2.2	2.4	0.5	87.3

View 7

(h)

	A	B	C	D	E	F	G	H	I	J
A	95.3	1.0	0.7	2.0	0.0	0.0	0.0	0.0	1.0	0.0
B	3.1	85.3	1.9	6.6	0.0	0.0	0.0	0.2	3.1	0.0
C	3.4	2.4	90.3	2.0	0.0	0.0	0.0	0.0	1.5	0.3
D	2.9	6.1	1.9	82.9	0.2	0.2	0.0	0.2	5.3	0.5
E	0.0	0.0	0.0	0.0	83.9	2.7	6.4	3.2	0.5	3.2
F	0.0	0.0	0.0	0.2	3.4	87.6	3.6	4.2	0.2	0.8
G	0.0	0.0	0.0	0.3	8.3	5.8	79.7	3.1	0.0	2.9
H	0.0	0.0	0.0	0.0	2.7	5.4	3.6	85.8	0.0	2.5
I	1.9	5.8	2.4	8.8	0.5	0.3	0.2	0.2	80.0	0.0
J	0.0	0.0	0.0	0.3	2.7	2.9	3.7	6.6	0.2	83.0

View 8

(i)

	A	B	C	D	E	F	G	H	I	J
A	91.8	2.2	1.4	3.0	0.0	0.1	0.0	0.0	1.3	0.1
B	3.6	87.2	3.3	2.6	0.0	0.0	0.2	0.0	2.9	0.1
C	4.6	6.2	84.2	2.6	0.0	0.1	0.1	0.3	1.6	0.1
D	5.0	3.2	1.8	86.7	0.1	0.1	0.1	0.1	2.8	0.1
E	0.1	0.0	0.0	0.2	83.5	5.3	4.0	4.7	0.1	2.0
F	0.1	0.1	0.2	0.4	8.8	78.8	3.3	4.7	0.4	3.2
G	0.0	0.1	0.0	0.2	6.1	4.5	79.5	4.2	0.5	4.8
H	0.0	0.0	0.2	0.0	2.9	3.2	3.5	88.1	0.1	1.9
I	2.9	5.2	1.7	5.8	0.4	0.5	0.4	0.5	82.4	0.3
J	0.1	0.2	0.1	0.6	3.4	3.5	6.3	4.0	1.0	80.8

V1-V8

(j)

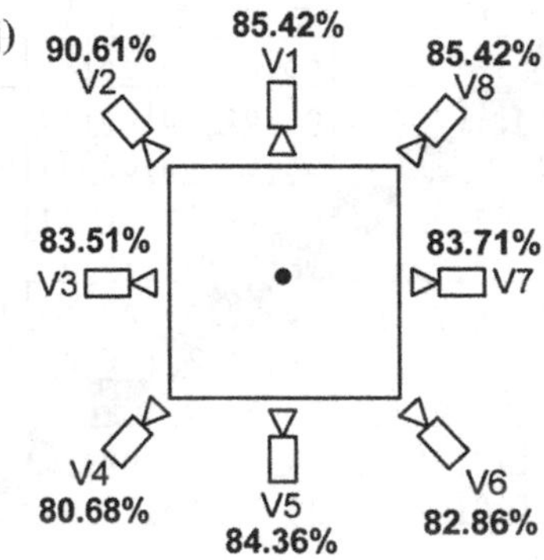

Fig. 7 (continued)

The best activity classification rates obtained from the front view and front side view angles (V1, V2 and V8) are equal to 85.42, 90.61 and 85.42 % respectively. The most of the misclassification exists in V4 and V6 due to the similarities of the postures. In Fig. 7i shows the confusion matrix for overall camera fusion (V1–V8). Find that most confusion occurs between jumping and jogging. The activity *kicking* is confused as *jogging* and *jumping*. This may occur due to the strong similarity between the action pairs. Finally, well distinguished activities such as *clapping*, *picking*, *waving one hand*, *waving two hands* and *punching* are well recognized from any view angle.

7 Conclusion and Future Work

The multi-view action recognition provides an active research fields in computer vision and pattern recognition. However, the camera viewpoints problem is complicated to solve, the appearance of an activity varies dynamically, depending on camera viewpoints. In this paper, Maximum Intensity Block Code (MIBC) method aiming at multi-view human activity recognition is presented. The proposed method was evaluated for various activities, in terms of different viewpoints. Experiments are evaluated on WVU multi-view activity dataset considering different actions. The ROI extracted from the temporal difference image are used to extract MIBC feature from maximum motion information and classification done by Random Forest classifier. The experimental results demonstrated that the proposed MIBC method performs well and achieved good recognition results for various actions. Since, there is no existing state-of-the-art available for the WVU multi-view activity dataset. Further research will deal with multi-view person interaction with complex human activities.

References

1. Vishwakarma, S., Agrawal, A.: A survey on activity recognition and behavior understanding in video surveillance. Visual Comp. **29**(10), 983–1009 (2013)
2. Weinland, D., Ronfard, R., Boyer, E.: A survey of vision-based methods for action representation, segmentation and recognition. CVIU **115**(2), 224–241 (2011)
3. Poppe, R.: A survey on vision-based human action recognition. Image Vis. Comput. **28**(6), 976–990 (2010)
4. Kellokumpu, V., Zhao, G., Pietikäinen, M.: Recognition of human actions using texture descriptors. Mach. Vis. Appl. **22**(5), 767–780 (2011)
5. Turaga, P., Chellappa, R., Subrahmanian, V.S., Udrea, O.: Machine recognition of human activities: a survey. IEEE Trans. Circ. Syst. Video Technol. **18**(11), 1473–1488 (2008)
6. Moeslund, T.B., Hilton, A., Krüger, V.: A survey of advances in vision-based human motion capture and analysis. CVIU **104**(2), 90–126 (2006)
7. Rougier, C., Meunier, J., St-Arnaud, A., Rousseau, J.: Robust video surveillance for fall detection based on human shape deformation. IEEE Trans. Circ. Syst. Video Technol. **21**(5), 611–622 (2011)
8. Blank, M., Gorelick, L., Shechtman, E., Irani, M., Basri, R.: Actions as space-time shapes. In: Tenth IEEE, International Conference on Computer Vision, vol. 2, pp. 1395–1402 (2005)
9. Bobick, A.F., Davis, J.W.: The recognition of human movement using temporal templates. IEEE Trans. Pattern Anal. Mach. Intell. **23**(3), 257–267 (2001)
10. Meng, H., Pears, N., Bailey, C.: A human action recognition system for embedded computer vision application. In: IEEE Conference on CVPR, pp. 1–6 (2007)
11. Arunnehru, J., Geetha, M.K.: Automatic activity recognition for video surveillance. Int. J. Comput. Appl. **75**(9), 1–6 (2013)
12. Arunnehru, J., Geetha, M.K.: Motion Intensity Code for Action Recognition in Video Using PCA and SVM, pp. 70–81. LNAI, Springer International Publishing (2013)
13. Zhang, W., Zhang, Y., Gao, C., Zhou, J.: Action recognition by joint spatial-temporal motion feature. J. Appl. Math. (2013)
14. Zhu, F., Shao, L., Lin, M.: Multi-view action recognition using local similarity random forests and sensor fusion. J. Pattern Recognit. Lett. **34**(1), 20–24 (2013)
15. Wang, L., Wang, Y., Jiang, T., Zhao, D., Gao, W.: Learning discriminative features for fast frame-based action recognition. J. Pattern Recogn. **46**(7), 1832–1840 (2013)
16. Cheng, J., Liu, H., Li, H.: Silhouette analysis for human action recognition based on maximum spatio-temporal dissimilarity embedding. MVAP **25**(4), 1007–1018 (2014)

17. Ramagiri, S., Kavi, R., Kulathumani, V.: Real-time multi-view human action recognition using a wireless camera network. In: ICDSC, pp. 1–6 (2011)
18. Breiman, L.: Random forests. Mach. Learn. **45**(1), 5–32 (2001)
19. Biau, G.: Analysis of a random forests model. J. Mach. Learn. Res. **98888**(1), 1063–1095 (2012)
20. Hall, M., Frank, E., Holmes, G., Pfahringer, B., Reutemann, P., Witten, I.H.: The WEKA data mining software: an update. ACM SIGKDD Explor. Newsl. **11**(1), 10–18 (2009)

Position and Orientation Control of a Mobile Robot Using Neural Networks

D. Narendra Kumar, Halini Samalla, Ch. Jaganmohana Rao, Y. Swamy Naidu, K. Alfoni Jose and B. Manmadha Kumar

Abstract In this paper, an adaptive neuro-control system with two levels is proposed for the motion control of a nonholonomic mobile robot. In the first level, a PD controller is designed to generate linear and angular velocities, necessary to track a reference trajectory. The proposed strategy is based on changing the robot control variables. Using this model, the nonholonomic constraints disappear and shows how the direct adaptive control theory can used to design robot controllers. In the second level, a neural network converts the desired velocities, provided by the first level, into a torque control. By introducing appropriate Lyapunov functions asymptotic stability of state variables and stability of system is guaranteed. The tracking performance of neural controller under disturbances is compared with PD controller. Sinusoidal trajectory and lemniscate trajectories are considered for this comparison.

Keywords Neural network · Mobile robot

D. Narendra Kumar (✉) · H. Samalla · Ch.J. Rao · Y. Swamy Naidu
Department of EEE, Sri Sivani College of Engineering, Srikakulam, Andhra Pradesh, India
e-mail: narendrakumar.dandasi@gmail.com

H. Samalla
e-mail: halini.samalla@gmail.com

K. Alfoni Jose
Department of EEE, VITS College of Engineering, Sontyam, Visakhapatnam, Andhra Pradesh, India
e-mail: alfoni.ee08@gmail.com

B. Manmadha Kumar
Department of EEE, Aditya Institute of Technology and Management, Tekkali, Andhra Pradesh, India
e-mail: boddmann@yahoo.co.in

L.C. Jain et al. (eds.), *Computational Intelligence in Data Mining - Volume 2*, Smart Innovation, Systems and Technologies 32, DOI 10.1007/978-81-322-2208-8_13

1 Introduction

Over the last few years, a number of studies were reported concerning a machine learning, and how it has been applied to help mobile robots in the navigation. Mobile robots have potential in industry and domestic applications. Navigation control of mobile robots has been studied by many authors in the last decade, since they are increasingly used in wide range of applications. The motion of mobile robot must be modeled by mathematical calculation to estimate physical environment and run in defined trajectories. At the beginning, the research effort was focused only on the kinematic model, assuming that there is perfect velocity tracking [1]. Later on, the research has been conducted to design navigation controllers, including also the dynamics of the robot [2, 3]. Taking into account the specific robot dynamics is more realistic, because the assumption "perfect velocity tracking" does not hold in practice. Furthermore, during the robot motion, the robot parameters may change due to surface friction, additional load, among others. And the environment and road conditions also always varying. Therefore, it is desirable to develop a robust navigation control, which has the following capabilities: (i) ability to successfully handle estimation errors and noise in sensor signals, (ii) "perfect" velocity tracking, and (iii) adaptation ability, in presence of time varying parameters in the dynamical model [4, 5].

Artificial neural networks are one of the most popular intelligent techniques widely applied in engineering for systems which are time variant. Their ability to learn complex input-output mapping, without detailed analytical model, approximation of nonlinear functions, and robustness for noise environment make them an ideal choice for real implementations [6, 7]. In this paper, some of the neural network techniques to stabilize the mobile robot and run it in desired trajectories are used. Neural networks are able to model of nonlinear dynamic systems without knowing their internal structure and physical description. Training data necessary to obtain neural parameters which ensures the movement of mobile robot in predefined trajectory.

Nonholonomic property is seen in many mechanical and robotic systems, particularly those using velocity inputs. Smaller control space compared with configuration space (lesser control signals than independent controlling variables) causes conditional controllability of these systems [6]. So the feasible trajectory is limited. This means that a mobile robot with parallel wheels can't move laterally. Nonholonomic constraint is a differential equation on the base of state variables, it's not integrable. Rolling but not sliding is a source of this constraint.

2 Problem Statement

The dynamics of a mobile robot is time variant and changes with disturbances. Its dynamic model is composed of two consecutive part; kinematic model and equations of linear and angular torques. By transforming dynamic error equations of kinematic model to mobile coordinates, the tracking problem changes to stabilization. In the trajectory tracking problem, the robot must reach and follow a trajectory in the Cartesian space starting from a given initial configuration. The trajectory tracking problem is simpler than the stabilization problem because there is no need to control the robot orientation: it is automatically compensated as the robot follows the trajectory, provided that the specified trajectory respects the non-holonomic constraints of the robot. Controller is designed in two consecutive parts: in the first part kinematic stabilization is done using simple PD control laws, in the second one, direct adaptive control using RBF Networks has been used for exponential stabilization of linear and angular velocities. Uncertainties in the parameters of dynamic model (mass and inertia) have been compensated using model reference adaptive control.

3 Kinematic Control

In this paper the mobile robot with differential drive is used [5] as shown in Fig. 1. The robot has two driving wheels mounted on the same axis and a free front wheel. The two driving wheels are independently driven by two actuators to achieve both the transition and orientation. The position of the mobile robot in the global frame $\{X, O, Y\}$ can be defined by the position of the mass center of the mobile robot system, denoted by C, or alternatively by position A, which is the center of mobile robot gear, and the angle between robot local frame $\{x_m, C, y_m\}$ and global frame. Kinematic equations [8] of the two wheeled mobile robot are:

$$\begin{bmatrix} \dot{x} \\ \dot{y} \\ \dot{\theta} \end{bmatrix} = \begin{bmatrix} \cos(\theta) & 0 \\ \sin(\theta) & 0 \\ 0 & 1 \end{bmatrix} \begin{bmatrix} v \\ w \end{bmatrix} \tag{1}$$

and

$$\begin{bmatrix} v \\ w \end{bmatrix} = \begin{bmatrix} r & r \\ \frac{r}{D} & -\frac{r}{D} \end{bmatrix} \begin{bmatrix} v_R \\ v_L \end{bmatrix} \tag{2}$$

3.1 Kinematic Controller Design

The robot stabilization problem can be divided into two different control problems: robot positioning control and robot orientating control. Figure 2 illustrates the robot positioning problem, where Δl is the distance between the robot and the desired reference $(x_{ref}; y_{ref})$ in the Cartesian space. The robot positioning control problem will be solved if we assure $\Delta l \rightarrow 0$. This is not trivial since the l variable does not appear in the model of Eq. 1.

Fig. 1 The representation of a nonholonomic mobile robot

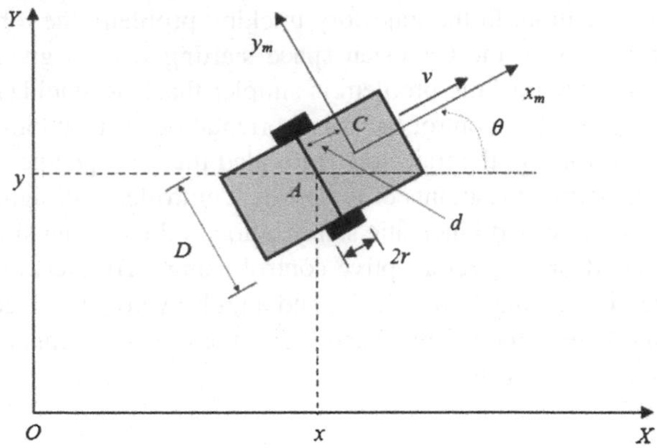

To overcome this problem, we can define two new variables, $\Delta\lambda$ and ϕ. $\Delta\lambda$, is the distance to R, the nearest point from the desired reference that lies on the robot orientation line; ϕ is the angle of the vector that binds the robot position to the desired reference. We can also define $\Delta\phi$ as the difference between the ϕ angle and the robot orientation: $\Delta\phi = \phi - \theta$. We can now easily conclude that:

$$\Delta l = \frac{\Delta\lambda}{\cos(\Delta\phi)} \tag{3}$$

So, if $\Delta\lambda \to 0$ and $\Delta\phi \to 0$ then $\Delta l \to 0$. That is, if we design a control system that assures the $\Delta\lambda$ and $\Delta\phi$ converges to zero, then the desired reference, x_{ref} and y_{ref} is achieved. Thus, the robot positioning control problem can be solved by applying any control strategy that assures such convergence. Ensure the converge of $\Delta\lambda$ and $\Delta\phi$ to zero, as required by Eq. 3. In other words, we want $e_s = \Delta\lambda$ and $e_\theta = \Delta\phi$. Thus, if the controller assures the errors convergence to zero.

$$\theta_{ref} = \tan^{-1}\left(\frac{y_{ref} - y}{x_{ref} - x}\right) = \tan^{-1}\left(\frac{\Delta y_{ref}}{\Delta x_{ref}}\right) \tag{4}$$

To calculate e_s is generally not very simple, because s output signal cannot be measured and we cannot easily calculate a suitable value for s_{ref}. If we define the point R as the reference point for the s controller as shown in Fig. 2, only in this case it is true that $e_s = s_{ref} - s = \Delta\lambda$. So,

$$e_s = \Delta\lambda = \Delta l \cdot \cos(\Delta\phi) = \sqrt{(\Delta x_{ref})^2 + (\Delta y_{ref})^2} \cdot \cos\left[\tan^{-1}\left(\frac{\Delta y_{ref}}{\Delta x_{ref}} - \theta\right)\right] \tag{5}$$

Equations 4 and 5, is presented on Fig. 3. It can be used as a stand-alone robot control system if the problem is just to drive to robot to a given position $(x_{ref}; y_{ref})$ regardless of the final robot orientation. Figure 4 shows the linear and angular errors

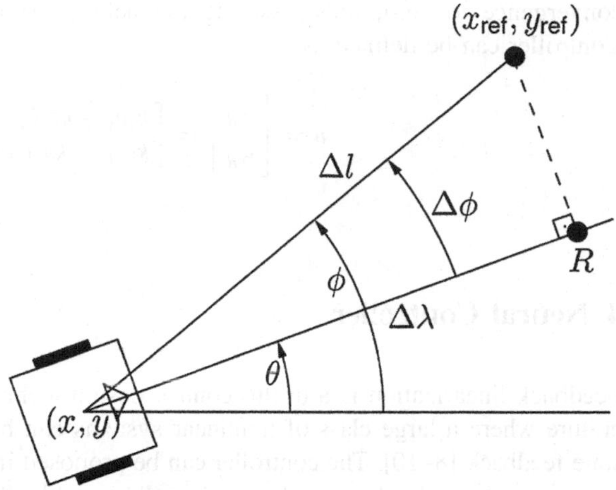

Fig. 2 Robot positioning problem

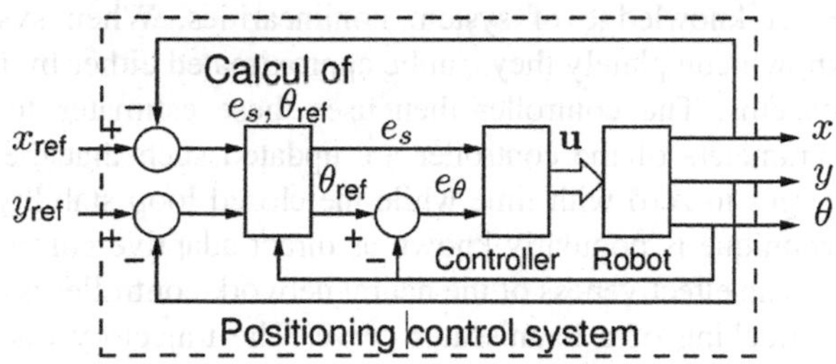

Fig. 3 Robot positioning controller

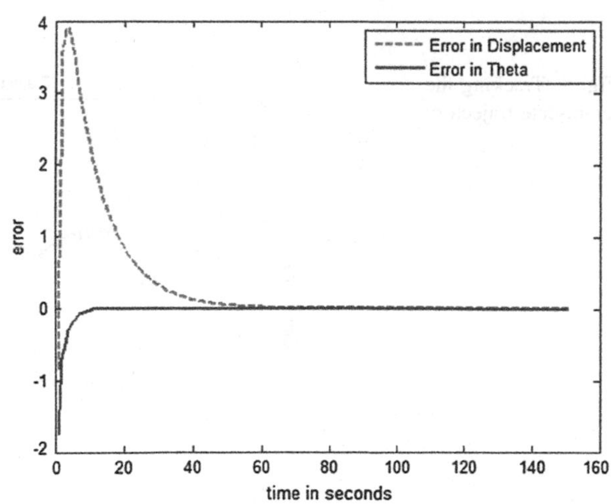

Fig. 4 Linear and angular errors

convergence to zero, thus, assuring the achievement of the control objective. Controller can be defined as

$$u = \begin{bmatrix} v_d \\ w_d \end{bmatrix} = \begin{bmatrix} k_s e_s + k_{sd} \dot{e}_s \\ k_\theta e_\theta + k_{\theta d} \dot{e}_\theta \end{bmatrix} \tag{6}$$

4 Neural Controller

Feedback linearization is a useful control design technique in control systems literature where a large class of nonlinear systems can be made linear by nonlinear state feedback [8–10]. The controller can be proposed in such a way that the closed loop error dynamics become linear as well as stable. The main problem with this control scheme is that cancellation of the nonlinear dynamics depends upon the exact knowledge of system nonlinearities. When system nonlinearities are not known completely they can be approximated either by neural networks or by fuzzy systems. The controller then uses these estimates to linearize the system. The parameters of the controller are updated such that the output tracking error converges to zero with time while the closed loop stability is maintained. The design technique is popularly known as direct adaptive control technique.

The effectiveness of the neural network controller is demonstrated [4] in the case of tracking of a lemniscates curve. The trajectory tracking problem for a mobile robot is based on a virtual reference robot that has to be tracked and trajectory tracking shown in Fig. 5. The overall system is designed and implemented within

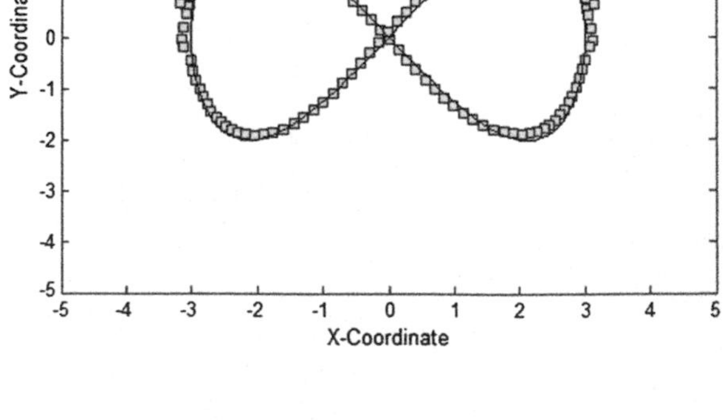

Fig. 5 Tracking the lemniscate trajectory

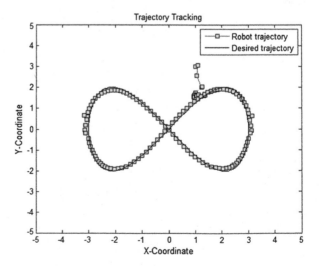

MATLAB environment. The geometric parameters of mobile robot are assumed as $r = 0.08$ m, $D = 0.4$ m, $d = 0.1$ m. $M = 5$ kg, $Ia = 0.05$, $m_0 = 0.1$ kg and $I_0 = 0.0005$. The initial position of robot is $[x_0 \quad y_0 \quad \theta_0] = [1 \quad 3 \quad 30^\circ]$ and the initial robot velocities are $[v, w] = [0.1, 1]$. PD controller gains for kinematic control are $k_s = 0.21$, $k_\theta = 0.6$ and $k_{sd} = k_{\theta d} = 0.01$. We used 6 hidden neurons and set the gain matrix as $K = [1.5\ 0;\ 0\ 1.5]$. The initial values of learning rate, weights, centers and sigma are tuned such a way that it provides good tracking performance.

5 Results and Discussion

The velocities generated from torque control are exactly matched with the values obtained from the kinematic control such that it tracks the trajectory as shown in Fig. 6. The proposed neural controller also ensures small values of the control input torques for obtaining the reference position trajectories. The simulations proved that motor torque of 1 Nm/s is sufficient to drive the robot motion. This mean that smaller power of DC motors is requested. Figure 7 shows that the neural controller is able to stabilize the robot quickly and makes the robot move in the desired path smoothly compared to PD controller. It can be said that the neural controller generated torques is smooth and low. This test is performed to analyze control performance when any disturbance occurred on the robot. In the present paper a sinusoidal trajectory is chosen for this purpose as to prove neural controller performance improves when time increases and applied sudden forces on robot at two

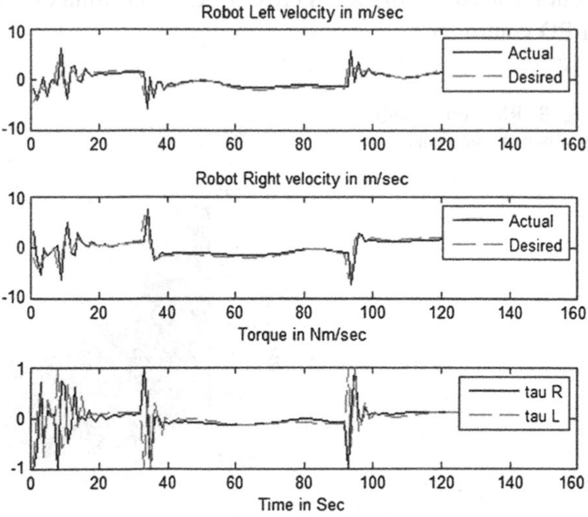

Fig. 6 Inputs to the robot

Fig. 7 Tracking performance when sudden forces applied

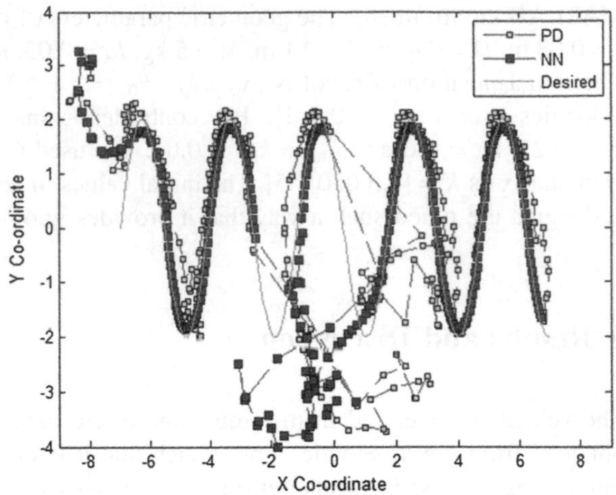

different time instants and observed robot come back to the desired trajectory. As the neural control structure is adaptive, the weights are automatically adjusted using update law such that it tracks the trajectory though any changes happen to dynamics. So the velocity error keeps on reducing with the time and hence the tracking performance improves.

If the control structure uses previous saturated weights as initial weights for the next time reboot of robot makes the error further decreases to lower values. Whereas this is not possible in case of PD controller as the gains are fixed for a particular dynamics and external environments. Figure 8 shows that in case of neural controller, the RMS error in X, Y coordinates decreases faster with time than a PD controller.

Fig. 8 RMS error with number of iterations

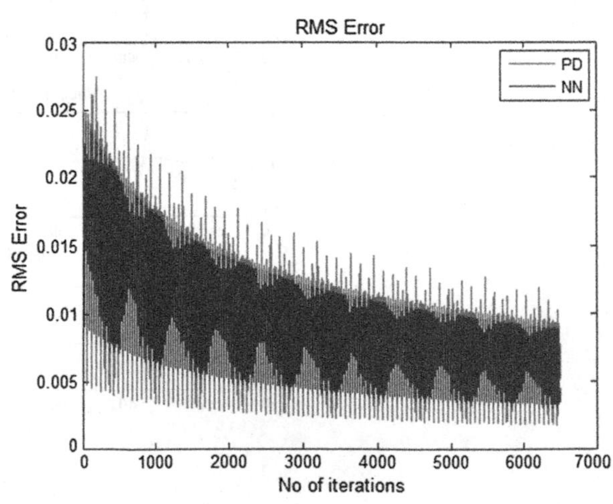

6 Conclusion

In this paper presented a simple method of controlling velocities to achieve desired trajectory by converting x, y, θ into linear displacement (S) and θ which takes care of nonholonomic constraints. Direct adaptive control method using RBF networks to generate motor torques are proposed such that the velocities generated from kinematic control are achieved. From simulation results it is evident that neural controller performance is better than better than PD controller when disturbances occurred. It also converges faster than PD.

References

1. Kolmanovsky, I., McClamroch, N.H.: Development in nonholonomic control problems. IEEE Control Syst. **15**, 20–36 (1995)
2. Fukao, T., Nakagawa, H., Adachi, N.: Adaptive tracking control of a nonholonomic mobile robot. IEEE Trans. Robot. Autom. **16**, 609–615 (2000)
3. Fierro, R., Lewis, F.L.: Control of a nonholonomic mobile robot using neural networks. IEEE Trans. Neural Netw. **9**, 589–600 (1998)
4. Vieira, F.C., Medeiros, A.A.D., Alsinia, P.J., Araujo, A.P. Jr.: Position and orientation control of a two wheeled differentially driven nonholonomic mobile robot
5. Kar, I., Behera, L.: Direct adaptive neural control for affine nonlinear systems. Appl. Soft Comput. **9**, 756–764 (2008)
6. Gholipour, A., Yazdanpanah, M.J.: Dynamic tracking control of nonholonomic mobile robot with model reference adaption for uncertain parameters. Control Intelligent Processing Center for Excellence, University of Tehran
7. Kar, I.: Intelligent control schemes for nonlinear systems. Ph.D. thesis, Indian Institute of Technology, Kanpur, India (2008)
8. Velagic, J., Lacevic, B., Osmic, N.: Nonlinear Motion Control of Mobile Robot Dynamic Model. University of Sarajevo Bosnia and Herzegovina
9. Velagic, J., Osmic, N., Lacevic, B.: Neural Network Controller for Mobile Robot Motion Control, vol. 47. World Academy of Science, Engineering and Technology (2008)
10. Zhong, X., Zhou, Y.: Establishing and maintaining wireless communication coverage among multiple mobile robots using artificial neural network. IEEE 2083–2089 (2011)

Fuzzy C-Means (FCM) Clustering Algorithm: A Decade Review from 2000 to 2014

Janmenjoy Nayak, Bighnaraj Naik and H.S. Behera

Abstract The Fuzzy c-means is one of the most popular ongoing area of research among all types of researchers including Computer science, Mathematics and other areas of engineering, as well as all areas of optimization practices. Several problems from various areas have been effectively solved by using FCM and its different variants. But, for efficient use of the algorithm in various diversified applications, some modifications or hybridization with other algorithms are needed. A comprehensive survey on FCM and its applications in more than one decade has been carried out in this paper to show the efficiency and applicability in a mixture of domains. Also, another intention of this survey is to encourage new researchers to make use of this simple algorithm (which is popularly called soft classification model) in problem solving.

Keywords FCM · Clustering

1 Introduction

Clustering approaches are constructive tools to investigate data structures and have emerged as well-liked techniques for unsupervised pattern recognition and are applied in many application areas such as pattern recognition [1], data mining [2], machine learning [3], etc. Organization of a set of patterns into a cluster may form a low or high degree of both similarity and dissimilarity depending upon the

J. Nayak (✉) · B. Naik · H.S. Behera
Department of Computer Science Engineering and Information Technology, Veer Surendra Sai University of Technology, Burla, Sambalpur 768018, Odisha, India
e-mail: jnayak@gmail.com

B. Naik
e-mail: bnaik@gmail.com

H.S. Behera
e-mail: hsbehera@gmail.com

L.C. Jain et al. (eds.), *Computational Intelligence in Data Mining - Volume 2*,
Smart Innovation, Systems and Technologies 32, DOI 10.1007/978-81-322-2208-8_14

belongingness of patterns into a known or unknown cluster. Generally, clustering can be alienated into two groups like hierarchical and partitioning clustering. Hierarchical clustering results an input into its corresponding output in a nested hierarchy fashion, where as partitioning clustering gives the output by using the objective function which make the partition of the input into a set of fixed number of clusters. Based on the objective function concept, FCM is the best complete technique and is the reason for being widely used.

Clustering can be either hard or fuzzy type. In the first category, the patterns are distinguished in a well defined cluster boundary region. But due to the overlapping nature of the cluster boundaries, some class of patterns may be specified in a single cluster group or dissimilar group. This property limits the use of hard clustering in real life applications. To, reduce such limitations fuzzy type clustering came into the picture [4, 5] and helps to provide more information about the memberships of the patterns. After the fuzzy theory introduced by Zadeh [6], the researchers put the fuzzy theory into clustering. The Fuzzy clustering problems have been expansively studied and the foundation of fuzzy clustering was being proposed by Bellman et al. [7] and Ruspini [8]. Fuzzy clustering problems can be grouped into three branches: (a) Based on fuzzy relation [9]; (b) Based on fuzzy rule learning [10]; and (c) Based on optimization of an objective function. The fuzzy clustering based on the objective function is quite popularly known to be Fuzzy c-means clustering (FCM) [11]. In FCM method, the pattern may belongs to all the cluster classes with a certain fuzzy membership degree [12]. Hoppner et al. [13] have made a good effort towards the survey of FCM.

The main intend of this paper is to present where FCM has been successfully applied from last one decades and make wider the range of its prospective users. The rest of the paper is organized as follows: Sect. 2 describes the main structure of FCM and its different variants. Section 3 provides brief elucidation about the applications areas of FCM. Section 4 discusses some limitations, result discussions and characteristics with its future scope. Finally Sect. 5 concludes our work with some further developments.

2 Fuzzy C-Means Algorithm (FCM)

2.1 Structure of FCM Algorithm

A clustering method that involves in minimizing some objective function [11], belongs to the groups objective function algorithms. When the algorithm is able to minimize an error function [14], it is often called C-Means being c the number of classes or clusters, and if the used classes are using the fuzzy technique or simply fuzzy, then it is known to be FCM. The FCM approach uses a fuzzy membership which assigns a degree of membership for every class. The importance of degree of membership [15] in fuzzy clustering is similar to the pixel probability in a mixture modeling assumption. The benefit of FCM is the formation of new clusters from the data points that have close membership values to existing classes [16]. Basically, there are three basic operators in FCM method: the fuzzy membership function, partition matrix and the objective function.

Let us consider a set of n vectors ($X = (x_1, x_2, \ldots, x_n) 2 \leq c \leq n$) for clustering into c groups. Each vector $x_i \in R^s$ is described by s real valued measurements which represent the features of the object x_i. A membership matrix known as Fuzzy partition matrix is used to describe the fuzzy membership matrix. The set of fuzzy partition matrices ($c \times n$) is denoted by M_{fc} and is defined in Eq. 1.

$$M_{fc} = \{W \in R^{cn} | w_{ik} \in [0,1], \forall i, k; \sum_{i=1}^{c} w_{ik} = 1, \forall k; 0 < \sum_{k=1}^{n} w_{ik} < n, \forall i\} \quad (1)$$

where $1 \leq i \leq c$, $1 \leq k \leq n$.

From the above defined definitions, it can be found that the elements can fit into more than one cluster with different degrees of membership. The total "membership" of an element is normalized to 1 and a single cluster cannot contain all data points. The objective function (Eq. 2) of the fuzzy c-means algorithm is computed by using membership value and Euclidian distance (Eq. 3).

$$J_m(W, P) = \sum_{\substack{1 \leq k \leq n \\ 0 \leq i \leq c}} (w_{ik})^m (d_{ik})^2 \quad (2)$$

where

$$(d_{ik}) = ||x_k - p_i|| \quad (3)$$

where $m \in (1, +\infty)$ is the parameter which defines the fuzziness of the resulting clusters and d_{ik} is the Euclidian distance from object x_k to the cluster center p_i.

The minimization [17] of the objective function J_m through FCM algorithm is being performed by the iterative updation of the partition matrix using the Eqs. 4 and 5.

$$p_i = \sum_{k=1}^{n} (w_{ik})^m x_k / \sum_{k=1}^{n} (w_{ik})^m \quad (4)$$

$$w_{ik}^{(b)} = \sum_{j=1}^{c} 1/[(d_{ik}^{(b)} / d_{jk}^{(b)})^{2/m-1}] \quad (5)$$

The FCM membership function [18] is calculated as:

$$\mu_{i,j} = \left[\sum_{t=1}^{c} \left(\frac{\left\| x_j - v_i \right\|_A}{\left\| x_j - v_t \right\|_A} \right)^{\frac{2}{m-1}} \right]^{-1} \quad (6)$$

$\mu_{i,j}$ is the membership value of *j*th sample and *i*th cluster. The number of clusters is represented by c, x_j is the *j*th sample and v_i cluster center of the *i*th cluster. $\| \|_A$ represents the norm function.

The steps of FCM algorithm are as follows:

1. Initialize the number of clusters c.
2. Select an inner product metric Euclidean norm and the weighting metric (fuzziness).
3. Initialize the cluster prototype $P^{(0)}$, iterative counter b = 0.
4. Then calculate the partition matrix $W^{(b)}$ using (5).
5. Update the fuzzy cluster centers $P^{(b+1)}$ using (4).
6. If $\left\|P^{(b)} - P^{(b+1)}\right\| < \varepsilon$ then stop, otherwise repeat step 2 through 4.

2.2 Studies and Advancements on FCM Algorithm

After the first base concepts of fuzzy set theory by Zadeh [6], Ruspini [8] coined the root concepts of fuzzy partition, more specifically crisp or hard fuzzy clustering algorithm. But in 1974, Dunn [4] extended the hard means clustering to preliminary concepts fuzzy means. In 1981, Bezdek [11] added the fuzzy factor and he proposed FCM. Later some more researchers claim that, FCM is able to solve the partition factor of the classes but unable to give high convergence as like hard clustering. Later, some more variants of FCM Cannon [19], Xie and Liu [20], Kamel and Selim [21], Park and Dagher [22], Pei et al. [23] came into the picture. After a better analysis on the slower convergence speed, Wei and Xie [24] introduced a new competitive learning based rival checked fuzzy c-means clustering algorithm (RCFCM). Ahmed et al. [25] first tailored the FCM algorithm as a bias-corrected FCM (BCFCM) with the regularization of the FCM objective function with a spatial neighborhood regularization term. To establish some more relationships between hard fuzzy c-means (HCM) and FCM, Fan et al. [26] proposed suppressed fuzzy c-means clustering algorithm (S-FCM). Chen and Zhang [27] improved the BCFCM objective function to lower the computational complexity and replaced the Euclidean distance by a kernel-induced distance and proposed kernel versions of FCM with spatial constraints, called KFCM_S1 and KFCM_S2. Y. Dong et al. [28] investigated an algorithm for hierarchical clustering based on fuzzy graph connectedness algorithm (FHC) which performs in high dimensional datasets and finds the clusters of arbitrary shapes such as the spherical, linear, elongated or concave ones. A modified suppressed fuzzy c-means (MS-FCM) algorithm used for both the clustering and parameter selection was proposed by Hung et al. [29]. A Spatial Information based Fuzzy c-means algorithm (SIFCM) used for image segmentation was introduced by Chuang et al. [30]. A Gaussian kernel-based fuzzy c-means algorithm (GKFCM) with a spatial bias correction was introduced by Yang and Tsai [31] to overcome the limitations of KFCM_S1 and KFCM_S2. A new method of partitive clustering called Shadowed c-means clustering, by integrating fuzzy and rough clustering under the framework of shadowed sets is developed by Mitra et al. [32]. An intuitionistic fuzzy hierarchical algorithm was developed Xu [33] is based on hierarchical clustering. Kühne et al. [34] developed a novel fuzzy clustering algorithm by using the observation weighting and context information for the separation of reverberant blind speech. A fuzzy rough semi-supervised outlier detection (FRSSOD) approach proposed by Xue et al. [35] is able to minimize the sum squared errors of the clustering. Geweniger et al. [36] developed a Median Fuzzy c-means (MFCM) method for clustering both similar and dissimilar data. A new method called recursive fuzzy c-means clustering to on-line Takagi–Sugeno fuzzy model identification was introduced by Dovzan and Skrjanc [37]. Ji et al. [38] discussed a modified possibilistic fuzzy c-means algorithm for segmenting the brain MR images. A new fuzzy variant of evolutionary technique was proposed by Horta et al. [39]. A novel fuzzy minmax model along with its applications was broadly discussed by Li et al. [40]. Maraziotis [41] proposed a novel semi supervised fuzzy clustering algorithm (SSFCA) which is used for gene expression profile clustering. A novel strong FCM was introduced by Kannan et al. [42] for segmentation of medical images. Ji et al. [43] proposed a new generalized rough fuzzy c-means (GRFCM) algorithm for brain MR image segmentation. Based on VSs theory and FCM, Xu et al. [44] introduced a new Vague C-means clustering algorithm which is more efficient than the variants of FCM-HDGA, GK-FCM and KL-FCM. A robust clustering approach called F-TCLUST based on trimming a fixed proportion of observations and uses the Eigen value ratio constraint for more feasibility was proposed by Fritz et al. [45]. Integration of pair wise relationships into FCM clustering was successfully introduced by Mei and Chen [46]. A generalized rough fuzzy k-means (GRFKM) algorithm proposed by Lai et al. [47] gives more feasibility solutions in less computational time. Qiu et al. [48] developed a modified interval type-2 fuzzy C-means algorithm for MR image segmentation. An effective comparison between a newly proposed Fuzzy Granular Gravitational Clustering Algorithm (FGGCA) and other FCM techniques was made by Sanchez et al. [49]. Lin et al. [50] developed a size-insensitive integrity-based fuzzy c-means method to deal with the cluster size sensitivity problem. A hybrid clustering method namely a density-based fuzzy imperialist competitive clustering algorithm (D-FICCA) was proposed by Shamshirband et al. [51] for detecting the malicious behaviour in wireless sensor network. The detailed chart of all the developments and variants of FCM is given in Table 1.

Table 1 Various advancements of FCM algorithm

Years	Author(s)	Algorithm	References	Years	Author(s)	Algorithm	References
1965	Zadeh	Fuzzy set	[6]	2010	Mitra et al.	Shadow c-means	[32]
1969	Ruspini	Crisp set	[8]	2010	Xue et al.	FRSSOD	[35]
1974	Dunn	Fuzzy means	[4]	2010	Geweniger et al.	MFCM	[36]
1981	Bezdek	FCM	[11]	2011	Dovžan and Škrjanc	Recursive FCM	[37]
2000	Wei and Xie	RCFCM	[24]	2011	Ji et al.	Possibilistic FCM	[38]
2002	Ahmed	BCFCM	[25]	2012	Maraziotis	SSFCA	[41]
2003	Fan et al.	S-FCM	[26]	2012	Ji et al.	GRFCM	[43]
2004	Chen and Zhang	KFCM-S1 & S2	[27]	2013	Xu et al.	Vague c-means	[44]
2006	Dong et al.	FHC	[28]	2013	Fritz et al.	F-TCLUST	[45]
2006	Hung et al.	MS-FCM	[29]	2013	Lai et al.	GRFKM	[47]
2006	Chuang et al.	SIFCM	[30]	2013	Qiu et al.	Type-2 FCM	[48]
2008	Yang et al.	GKFCM	[31]	2014	Sanchez et al.	FGGCA	[49]
				2014	Shamshirband et al.	D-FICCA	[51]

3 Applications of FCM Algorithm

In order to determine the applications of FCM approach, during the past decade, this paper reviews the literature survey of various articles from 2000 to 2014. We have chosen this period especially for the reason of the wide spread of Information Technology through internet during 2000. A keyword search was performed on some of the major databases like Elsevier, IEEE X-plore, Springer Link, Taylor Francis, Inderscience online databases. For the period of 2000 to 2014 17,751 articles were found. After the filtration of specific title "Fuzzy c-means" in keyword based search, we found 1,463 articles. But majority of the articles are based on some similar applications. For this reason, a no. of distinguishing features of application areas of FCM have been considered in this study. The classification of application areas in this survey has been made in the following manner: Neural Network, Clustering and Classification, Image Analysis, Structural analysis of algorithms together in various application domains.

3.1 Neural Network

A comparative study of the classification accuracy of ECG signals using back propagation learning based MLP have been made by Özbay et al. [52]. They also introduced a new FCM based neural network (FCNN) which is helpful for early medical diagnosis. By taking 10 different training features, they claim that the proposed FCNN architecture can generalize better than ordinary MLP architecture with faster learning.

Mingoti and Lima [53] made a analysis on the performance comparison of the nonhierarchical and hierarchical clustering algorithms including SOM (Self-Organization Map) neural network and FCM. By simulating 2,530 datasets, they found the performance of FCM in all the cases are very encouraging with the stability in presence of outliers.

Staiano et al. [54] proposed a FCM method using the Radial Basis Function Network (RBFN) for organization of data in clustering on the basis of input data and a regression function for improvement of the performance.

A hybridization of SVM and FCM for gene expression data has been done by Mukhopadhyay and Maulik [55]. For the statistical significance ANOVA tool with Tukey–Kramer multiple comparison test has been used. Aydilek and Arslan [56] have proposed a hybrid approach of FCM, GA and SVM for estimation of missing values and optimization of various parameters like cluster size. A PSO trained RBFN has been introduced by Tsekouras and Tsimikas [57] for solving the problem of selection of basis function. The method has three steps of designing which helps in information flow from output to input space.

For determination of classification of partial discharge (PD) events in GasIn-sulated Load Break Switches (GILBS) Sua et al. [58] proposed a hybridization technique of probabilistic neural network and FCM. They used the Discrete wavelet transform (DWT) to suppress noises in the measured signals of high-frequency current transformers (HFCT). Their approach significantly improves the classification correctness ratios of the defect models using conventional observations of phase resolved partial discharge (PRPD). A comparative study between FCM and SVM has been performed by Hassen et al. [59] for Classification of Chest Lesions. It also elaborates the possibility to increase the interpretability of SVM classifier by hybridizing with FCM.

Kai and Peng [60] developed a generalized entropy fuzzy *c*-means (FCM) algorithm, called GEFCM to solve Lagrange multipliers' assignment problem. By the continuous change of index values they obtain good results with the algorithm.

3.2 Clustering and Classification

Fan and Li [61] used a selected suppressed rate for suppressed FCM algorithm for clustering applications. Neha and Aruna [62] introduced a Random Sampling Iterative Optimization Fuzzy c-Means (RSIO-FCM) clustering algorithm which partitions the big data into various subsets and results in formation of effective clusters for elimination of the problem of overlapping cluster centers. An Intuitionistic Fuzzy Representation (IFR) scheme for numerical data set with the hybridization of modified fuzzy clustering algorithm was developed by Visalakshi et al. [63] and compared the results with the crisp as well as fuzzy data. Looney [64] introduces an interactive clustering algorithm with the addition of fuzzy clustering and addressed some issues like optimality, validity, of clusters, prevention of the selection of seed vectors as initial prototypes etc. Miyamoto [65] used fuzzy multiset model for information clustering with application to WWW. Based on three methods like hard c-means, fuzzy c-means, and an agglomerative method they observed that fuzzy multisets give an appropriate model of information retrieval on the WWW.

Fan et al. [26] describe a new suppressed FCM clustering algorithm which able to avoid the limitations of rival checked fuzzy c-means clustering algorithm and helps to establish the relation between hard c-means and fuzzy c-means algorithm. Kim et al. [66] developed a novel initialization scheme for the fuzzy c-means (FCM) algorithm to solve the color clustering problems. They show the effectiveness and reliability of the proposed method by taking various color clustering examples. An improved firefly based fuzzy c-means algorithm (FAFCM) for data clustering have been developed by Nayak et al. [67]. The performance of the proposed algorithm has been tested with various real world data sets and compared with other two algorithms FCM and PSO-FCM, to prove the convergence property in less no. of iterations.

Table 2 Image analysis applications

Image type	Task	References
Brain MR	Segmentation	[43, 48, 73–81]
Breast and brain MRI	Segmentation	[42, 82, 83]
Carotid artery	Segmentation	[84, 85]
Noisy color	Segmentation	[86]
Remote sensing	Classification	[87–89]
Training image	Compression	[68]

3.3 *Image Analysis*

Kong et al. [68] used FCM helps to reduce the resolution and applied for various image compression applications. To overcome the sensitivity of fuzzy c-means clustering Noordam et al. [69] developed a technique called conditional FCM for unequal cluster sizes in multivariate images. Chuang et al. [30] presented a fuzzy c-means (FCM) algorithm that incorporates spatial information into the membership function for clustering which helps to reduce the spurious blobs, increases the regions homogeneity and less sensitive to noise.

Cai et al. [70] developed a fast and robust FCM method by adding local spatial and gray information for image segmentation. They claim for the performance enhancement of the proposed method than original FCM algorithm. Halberstadt and Douglas [71] developed a FCM technique to detect tuberculous meningitis-associated hyper density in CT images to enhance the radiological feature. Kannan et al. [72] proposed an effective robust fuzzy c-means by introducing a specialized center initialization method for execution of the proposed algorithm for a segmentation of breast and brain magnetic resonance images. Some more image analysis applications are listed in Table 2.

3.4 *Structural Analysis of Algorithms*

A novel clustering algorithm called fuzzy J-means algorithm was introduced by Belacel et al. [90] and the performance of the proposed method is compared with the traditional FCM to show the effectiveness as well as efficiency of the proposed algorithm. Wu and Yang [91] developed a new metric to replace the Euclidean norm in c-means clustering procedures which is alternative to FCM and proposed two new algorithms such as alternative hard c-means (AHCM) and alternative fuzzy c-means (AFCM) clustering algorithms. Pedrycza and Vukovich [92] proposed a novel standard FCM technique for vector quantization by extending the original objective function by the supervision component. Pedrycz [93] discusses an issue of exploiting and effectively incorporating auxiliary problem dependent hints in FCM and adds some knowledge based supervision techniques. Wang et al. [94] improved the performance of FCM by the appropriate selection of feature weight vectors and

Table 3 Other applications

Application area	References	Application area	References
Video segmentation	[38, 96]	Forecasting	[104]
Sonar returns	[97]	Speech recognition	[34]
Color clustering	[98]	Forest fire detection	[105]
Real-time applications	[14]	Load frequency	[106]
Signal analysis	[99]	Supply chain management	[107]
Internet portal	[100]	Document analysis	[108]
ECG arrhythmias	[52]	Disease analysis	[109]
Spike detection	[101]	Fluid detection	[110]
Biology	[102]	Mathematics	[111–114]
Damage detection	[103]	Software Engg.	[115]
Gene expression	[41, 55]	Communication	[116]

applying the gradient descent method. Pianykh [95] introduced a simplified clustering algorithm intending for closed-form analytical solution to a cluster fit function minimization problem. Inspite of the above mentioned applications, FCM has also been applied in some other areas like intrusion detection, health sector, video segmentation etc. are listed in Table 3.

4 Analytical Discussions, Limitations and Suggestions

This literature review surveys the applications of FCM in diversified fields in connection with the author's background, the application interest and expertise knowledge in the particular field. Some authors have been repeated for different applications. As we have earlier mentioned some part of the applications may be relevant to other areas of interest like image analysis which include image compression, image segmentation etc. The paper discusses the FCM method applied in a mixture of application areas including engineering, mathematical problems, medical and health issues, earth science, chemistry, agricultural science, computer science, physics, pharmaceutics, Psychology, Finance, Neuro science et., which were retrieved from the databases like Elsevier, IEEE X-plore, Springer Link, Taylor Francis, Inderscience.

The use of FCM methods can be realized in some other areas of both science and engineering. On the other hand, studies and some more applications of FCM in different research fields must be published, to facilitate the wide broaden scope of FCM, in the academic and practical fields. However, many researchers have pointed some drawbacks those are actively associated with FCM like: (1) Traditional fuzzy clustering algorithms is able to utilize only point based membership [44], which may lead to inaccurate description of data vagueness, (2) Loss of information for calculation of fuzzy similarity degree, (3) Slower Convergence speed, (4) Highly sensitive to initialization, (5) Testing for fuzziness, (6) If the points in the datasets are equal [31], then the number of points in the clustering are also equal. FCM is a diversified research area, so in future it can be integrated with various other methodologies. This assimilation of methodologies and cross-disciplinary research may put forward new insights for problem solving with FCM. This paper reviews a no. of major applications using FCM, but still inclusion of some other areas like social and behavioral science etc. are needed. Also, the qualitative and quantitative aspects of FCM technique are to be included in our future work.

5 Conclusion

Since its inception, FCM method is very popular and been widely used in various application domains. This paper presents a detailed survey report concerned with FCM techniques and applications from 2000 to 2014. A use of both keyword and title search has been used for the review of literature in various online databases. Now-a-days, there are no any practical domains where FCM method has not been used. However, as development of new technologies and working domains are changing day by day. So, FCM can be a dynamic method for the new challenging application areas. From Fig. 1 it can be concluded that FCM has been widely used in different Image analysis applications than the other application areas. It is suggested that, diverse applications like cognitive science, physiology etc. might use and find FCM as an alternative method.

This survey has some boundaries. Firstly, an extensive literature survey of FCM and its applications presents a complicated task, because of the widespread background knowledge that is essential, when collecting, studying and classifying these articles. Although describing restricted background knowledge, this paper makes a succinct survey concerned with FCM, from 2000 to 2014 in order to conclude how FCM and its applications have developed, during this period. In reality, the classification of all the methodologies and their applications are based on the FCM keyword index search and abstracts of the articles, gathered for this research. In some other articles, the authors may have used FCM method, but may not have used FCM as an index term or keyword. Hence, this article is oblivious of such

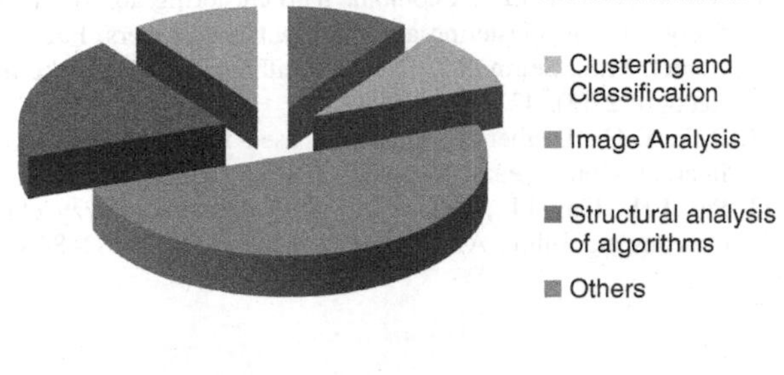

Fig. 1 Applications areas of FCM

reference sources. Another drawback of this article is the inclusion of non English language publications related to FCM. Lastly, it is assumed that some other research articles published in various non-English languages may have used FCM techniques, which are beyond the scope of this article.

References

1. Webb, A.: Statistical Pattern Recognition. Wiley, New Jersey (2002)
2. Tan, P.N., Steinbach, M., Kumar, V.: Introduction to Data Mining. Addison-Wesley, Boston (2005)
3. Alpaydin, E.: Introduction to Machine Learning. MIT Press, Cambridge (2004)
4. Dunn, J.C.: A fuzzy relative ISODATA process and its use in detecting compact well-separated clusters. J. Cybern. **3**(3), 32–57 (1974)
5. Bezdek, J.C.: Fuzzy mathematics in pattern classification, Ph.D. Dissertation. Applied Mathematics, Cornell University. Ithaca. New York (1973)
6. Zadeh, L.A.: Fuzzy sets. Inf. Control **8**(3), 338–353 (1965). doi:10.1016/S0019-9958(65) 90241-X. ISSN 0019-9958
7. Bellman, R.E., Kalaba, R.A., Zadeh, L.A.: Abstraction and pattern classification. J. Math. Anal. Appl. **13**, 1–7 (1966)
8. Ruspini, E.H.: A new approach to clustering. Inf. Control **15**(1), 22–32 (1969)
9. Guoyao, F.: Optimization methods for fuzzy clustering. Fuzzy Sets Syst. **93**, 301–309 (1998)
10. Ravi, V., Zimmermann, H.J.: Fuzzy rule based classification with feature selector and modified threshold accepting. Eur. J. Oper. Res. **123**, 16–28 (2000)
11. Bezdek, J.C.: Pattern Recognition with Fuzzy Objective Function Algorithms. Plenum Press, New York (1981)
12. Ferreiraa, M.R.P., Carvalho, F.A.T.: Kernel fuzzy c-means with automatic variable weighting. Fuzzy Sets Syst. **237**, 1–46 (2014)
13. Höppner, F., Klawonn, F., Kruse, R., Runkler, T.: Fuzzy Cluster Analysis. Wiley (1999)
14. Lazaro, J., Arias, J., Martın, J.L., Cuadrado, C., Astarloa, A.: Implementation of a modified Fuzzy C-Means clustering algorithm for real-time applications. Microprocess. Microsyst. **29**, 375–380 (2005)
15. Icer, S.: Automatic segmentation of corpus collasum using Gaussian mixture modeling and Fuzzy C means methods. Comput. Methods Programs Biomed. **112**, 38–46 (2013)
16. Asyali, M.H., Colak, D., Demirkaya, O., Inan, M.S.: Gene expression profile classification: a review. Curr. Bioinform. **1**, 55–73 (2006)
17. Runkler, T.A., Katz, C.: Fuzzy clustering by particle swarm optimization. In: Proceedings of 2006 IEEE International Conference on Fuzzy Systems, pp. 601–608. Canada (2006)
18. Huang, M., Xia, Z., Wang, H., Zeng, Q., Wang, Q.: The range of the value for the fuzzifier of the fuzzy c-means algorithm. Pattern Recogn. Lett. **33**, 2280–2284 (2012)
19. Cannon, R.L., et al.: Efficient implementation of the fuzzy c-means clustering algorithms. IEEE Trans. Pattern Anal. Mach. Intell. **8**(2), 248–255 (1986)
20. Xie, W.X., Liu, J.Z.: A combine hard clustering algorithm and fuzzy clustering algorithm—fuzzy c-means clustering algorithm with two layers. Fuzzy Syst. Math. **2**(6), 77–85 (1991)
21. Kamel, M.S., Selim, S.Z.: New algorithms for solving the fuzzy clustering problem. Pattern Recogn. **27**(3), 421–428 (1994)
22. Park, D.C., Dagher, I.: Gradient based fuzzy c-means (GBFCM) algorithm. Proc. IEEE Internat. Conf. Neural Networks **3**, 1626–1631 (1994)
23. Pei, J.H., Fan, J.L., Xie, W.X.: A new efficient fuzzy clustering method: cutset of fuzzy c-means algorithm. Acta Electronica Sinica **26**(2), 83–86 (1998). (in Chinese)

24. Wei, L.M., Xie, W.X.: Rival checked fuzzy c-means algorithm. Acta Electronica Sinica **28** (7), 63–66 (2000). (in Chinese)
25. Ahmed, M.N., Yamany, S.M., Mohamed, N., Farag, A.A., Moriarty, T.: A modified fuzzy c-means algorithm for bias field estimation and segmentation of MRI data. IEEE Trans. Med. Imaging **21**, 193–199 (2002)
26. Fan, J.L., Zhen, W.Z., Xie, W.X.: Suppressed fuzzy c-means clustering algorithm. Pattern Recogn. Lett. **24**, 1607–1612 (2003)
27. Chen, S.C., Zhang, D.Q.: Robust image segmentation using FCM with spatial constrains based on new kernel-induced distance measure. IEEE Trans. Syst. Man Cybern. Part B **34**, 1907–1916 (2004)
28. Dong, Y., Zhuang, Y., Chen, K., Tai, X.: A hierarchical clustering algorithm based on fuzzy graph connectedness. Fuzzy Sets Syst. **157**, 1760–1774 (2006)
29. Hung, W.L., Yang, M.S., Chen, D.H.: Parameter selection for suppressed fuzzy c-means with an application to MRI segmentation. Pattern Recogn. Lett. **27**, 424–438 (2006)
30. Chuang, K.S., Tzeng, H.L., Chen, S., Wu, J., Chen, T.J.: Fuzzy c-means clustering with spatial information for image segmentation. Comput. Med. Imaging Graph. **30**, 9–15 (2006)
31. Yang, M.S., Tsai, H.S.: A Gaussian kernel-based fuzzy c-means algorithm with a spatial biascorrection. Pattern Recogn. Lett. **29**, 1713–1725 (2008)
32. Mitra, S., Pedrycz, W., Barman, B.: Shadowed c-means: integrating fuzzy and rough clustering. Pattern Recogn. **43**, 1282–1291 (2010)
33. Xu, Z.S.: Intuitionistic fuzzy hierarchical clustering algorithms. J. Syst. Eng. Electron. **20**(1), 90–97 (2009)
34. Kuhne, M., Togneri, R., Nordholm, S.: A novel fuzzy clustering algorithm using observation weighting and context information for reverberant blind speech separation. Signal Process. **90**, 653–669 (2010)
35. Xue, Z., Shang, Y., Feng, A.: Semi-supervised outlier detection based on fuzzy rough C-means clustering. Math. Comput. Simul. **80**, 1911–1921 (2010)
36. Geweniger, T., Zulke, D., Hammer, B., Villmann, T.: Median fuzzy c-means for clustering dissimilarity data. Neurocomputing **73**, 1109–1116 (2010)
37. Dovžan, D., Škrjanc, I.: Recursive fuzzy c-means clustering for recursive fuzzy identification of time-varying processes. ISA Trans. **50**, 159–169 (2011)
38. Ji, Z.X., Sun, Q.S., Xia, D.S.: A modified possibilistic fuzzy c-means clustering algorithm for bias field estimation and segmentation of brain MR image. Comput. Med. Imaging Graph. **35**, 383–397 (2011)
39. Horta, D., Andrade, I.C., Campello, R.J.G.B.: Evolutionary fuzzy clustering of relational data. Theoret. Comput. Sci. **412**, 5854–5870 (2011)
40. Li, X., Wong, H.S., Wu, S.: A fuzzy minimax clustering model and its applications. Inf. Sci. **186**, 114–125 (2012)
41. Maraziotis, I.A.: A semi-supervised fuzzy clustering algorithm applied to gene expression data. Pattern Recogn. **45**, 637–648 (2012)
42. Kannan, S.R., Ramathilagam, S., Devi, R., Hines, E.: Strong fuzzy c-means in medical image data analysis. J. Syst. Softw. **85**, 2425–2438 (2012)
43. Ji, Z., Sun, Q., Xia, Y., Chen, Q., Xia, D., Feng, D.: Generalized rough fuzzy c-means algorithm for brain MR image segmentation. Comput. Methods Programs Biomed. **108**, 644–655 (2012)
44. Xu, C., Zhang, P., Li, B., Wu, D., Fan, H.: Vague C-means clustering algorithm. Pattern Recogn. Lett. **34**, 505–510 (2013)
45. Fritz, H., Escudero, L.A.G., Iscar, M.: Robust constrained fuzzy clustering. Inf. Sci. **245**, 38–52 (2013)
46. Mei, J.P., Chen, L.: Link FCM: relation integrated fuzzy c-means. Pattern Recogn. **46**, 272–283 (2013)
47. Lai, J.Z.C., Juan, E.Y.T., Lai, F.Z.C.: Rough clustering using generalized fuzzy clustering algorithm. Pattern Recogn. **46**, 2538–2547 (2013)

48. Qiu, C., Xiao, J., Yu, L., Han, L., Iqbal, M.N.: A modified interval type-2 fuzzy C-means algorithm with application in MR image segmentation. Pattern Recogn. Lett. **34**, 1329–1338 (2013)
49. Sancheza, M.A., Castillo, O., Castro, J.R., Melin, P.: Fuzzy granular gravitational clustering algorithm for multivariate data. Inf. Sci. **279**, 498–511 (2014)
50. Lin, P.L., Huang, P.W., Kuo, C.H., Lai, Y.H.: A size-insensitive integrity-based fuzzy c-means method for data clustering. Pattern Recogn. **47**, 2042–2056 (2014)
51. Shamshirband, S., Amini, A., Anuar, N.B., Kiah, L.M., The, Y.W., Furnell, S.: D-FICCA: a density-based fuzzy imperialist competitive clustering algorithm for intrusion detection in wireless sensor networks. Measurement **55**, 212–226 (2014)
52. Özbay, Y., Ceylan, R., Karlik, B.: A fuzzy clustering neural network architecture for classification of ECG arrhythmias. Comput. Biol. Med. **36**, 376–388 (2006)
53. Mingoti, S.A., Lima, J.O.: Comparing SOM neural network with Fuzzy c-means, K-means and traditional hierarchical clustering algorithms. Eur. J. Oper. Res. **174**, 1742–1759 (2006)
54. Staiano, A., Tagliaferri, R., Pedrycz, W.: Improving RBF networks performance in regression tasks by means of a supervised fuzzy clustering. Neurocomputing **69**, 1570–1581 (2006)
55. Mukhopadhyay, A., Maulik, U.: Towards improving fuzzy clustering using support vector machine: application to gene expression data. Pattern Recogn. **42**, 2744–2763 (2009)
56. Aydilek, I.B., Arslan, A.: A hybrid method for imputation of missing values using optimized fuzzy c-means with support vector regression and a genetic algorithm. Inf. Sci. **233**, 25–35 (2013)
57. Tsekouras, G.E., Tsimikas, J.: On training RBF neural networks using input–output fuzzy clustering and particle swarm optimization. Fuzzy Sets Syst. **221**, 65–89 (2013)
58. Sua, M.S., Chia, C.C., Chen, C., Chen, J.F.: Classification of partial discharge events in GILBS using probabilistic neural networks and the fuzzy c-means clustering approach. Electr. Power Energy Syst. **61**, 173–179 (2014)
59. Hassen, D.B., Taleb, H., Yaacoub, I.B., Mnif, N.: Classification of chest lesions with using fuzzy c-means algorithm and support vector machines. Adv. Intell. Syst. Comput. **239**, 319–328 (2014). doi:10.1007/978-3-319-01854-6_33
60. Li, K., Li, P.: Fuzzy clustering with generalized entropy based on neural network. Lect. Notes Electr. Eng. **238**, 2085–2091 (2014). doi:10.1007/978-1-4614-4981-2_228
61. Fan, J., Li, J.: A fixed suppressed rate selection method for suppressed fuzzy c-means clustering algorithm. Appl. Math. **5**, 1275–1283 (2014). doi:10.4236/am.2014.58119
62. Bharill, N., Tiwari, A.: Handling big data with fuzzy based classification approach. Adv. Trends Soft Comput. Stud. Fuzziness Soft Comput. **312**, 219–227 (2014). doi:10.1007/978-3-319-03674-8_21
63. Karthikeyani, N., Visalakshi, S., Parvathavarthini, S., Thangavel, K.: An intuitionistic fuzzy approach to fuzzy clustering of numerical dataset. Adv. Intell. Syst. Comput. **246:** 79–87 (2014). doi:10.1007/978-81-322-1680-3_9
64. Looney, C.G.: Interactive clustering and merging with a new fuzzy expected value. Pattern Recogn. **35**, 2413–2423 (2002)
65. Miyamoto, S.: Information clustering based on fuzzy multisets. Inf. Process. Manage. **39**, 195–213 (2003)
66. Kim, W.D., Lee, K.H., Lee, D.: A novel initialization scheme for the fuzzy c-means algorithm for color clustering. Pattern Recogn. Lett. **25**, 227–237 (2004)
67. Nayak, J., Nanda, M., Nayak, K., Naik, B., Behera, H.S.: An improved firefly fuzzy c-means (FAFCM) algorithm for clustering real world data sets. Smart Innov. Syst. Technol. **27**, 339–348 (2014). doi:10.1007/978-3-319-07353-8_40
68. Kong, X., Wang, R., Li, G.: Fuzzy clustering algorithms based on resolution and their application in image compression. Pattern Recogn. **35**, 2439–2444 (2002)
69. Noordam, J.C., van den Broek, W.H.A.M., Buydens, L.M.C.: Multivariate image segmentation with cluster size insensitive fuzzy C-means. Chemometr. Intell. Lab. Syst. **64**, 65–78 (2002)

70. Cai, W., Chen, S., Zhang, D.: Fast and robust fuzzy c-means clustering algorithms incorporating local information for image segmentation. Pattern Recogn. **40**, 825–838 (2007)
71. Halberstadt, W., Douglas, T.S.: Fuzzy clustering to detect tuberculosis meningitis-associated hyper density in CT images. Comput. Biol. Med. **38**, 165–170 (2008)
72. Kannan, S.R., Ramathilagam, S., Sathya, A., Pandiyarajan, R.: Effective fuzzy c-means based kernel function in segmenting medical images. Comput. Biol. Med. **40**, 572–579 (2010)
73. Zhao, F., Jiao, L., Liu, H., Gao, X.: A novel fuzzy clustering algorithm with non local adaptive spatial constraint for image segmentation. Signal Process. **91**, 988–999 (2011)
74. He, Y., Hussaini, M.Y., Ma, J., Shafei, B., Steidl, G.: A new fuzzy c-means method with total variation regularization for segmentation of images with noisy and incomplete data. Pattern Recogn. **45**, 3463–3471 (2012)
75. Ji, Z., Liu, J., Cao, G., Sun, Q., Chen, Q.: Robust spatially constrained fuzzy c-means algorithm for brain MR image segmentation. Pattern Recogn. **47**, 2454–2466 (2014)
76. Jun, X., Yifan, T.: Research of brain MRI image segmentation algorithm based on FCM and SVM. In: The 26th IEEE Chinese Control and Decision Conference (2014 CCDC), pp. 1712–1716. doi:10.1109/CCDC.2014.6852445
77. Balafar, M.A.: Fuzzy C-mean based brain MRI segmentation algorithms. Artif. Intell. Rev. **41**(3), 441–449 (2014). doi:10.1007/s10462-012-9318-2
78. Laishram, R., Kumar, W.K., Gupta, A., Prakash, K.V.: A novel MRI brain edge detection using PSOFCM segmentation and Canny Algorithm. In: IEEE International Conference on Electronic Systems, Signal Processing and Computing Technologies (ICESC) (2014), pp. 398–401. doi:10.1109/ICESC.2014.78
79. Noordam, J. C., van den Broek, W.H.A.M., Buydens, L.M.C.: Geometrically guided fuzzy c-means clustering for multivariate image segmentation. In: Proceedings on 15th International Conference on Pattern Recognition, 2000, vol. 1. IEEE (2000)
80. Liu, Q., Zhou, L., Sun, X.Y.: A fast fuzzy c-means algorithm for colour image segmentation. Int. J. Inf. Commun. Technol. **5**(3/4), 263–271 (2013)
81. Xiao, K., Ho, S. H., Bargiela, A: Automatic brain MRI segmentation scheme based on feature weighting factors selection on fuzzy c-means clustering algorithms with Gaussian smoothing. Int. J. Comput. Intell. Bioinform. Syst. Biol. **1**(3):316–331 (2010)
82. Zeng, Z., Han, C., Wang, L., Zwiggelaar, R.: Unsupervised brain tissue segmentation by using bias correction fuzzy c-means and class-adaptive hidden markov random field modelling. Lect. Notes Electr. Eng. **269**, 579–587 (2014). doi:10.1007/978-94-007-7618-0_56
83. Lai, D.T.C., Garibaldi, J.M.: A preliminary study on automatic breast cancer data classification using semi-supervised fuzzy c-means. Int. J. Biomed. Eng. Technol. **13**(4), 303–322 (2013)
84. Hassan, M., Chaudhry, A., Khan, A., Kim, J.Y.: Carotid artery image segmentation using modified spatial fuzzy c-means and ensemble clustering. Comput. Methods Programs Biomed. **108**(20–12), 1261–1276
85. Hassana, M., Chaudhry, A., Khan, A., Iftikhar, M.A.: Robust information gain based fuzzy c-means clustering and classification of carotid artery ultrasound images. Comput. Methods Programs Biomed. **113**, 593–609 (2014)
86. Vargas, D.M., Funes, F.J.G., Silva, A.J.R.: A fuzzy clustering algorithm with spatial robust estimation constraint for noisy color image segmentation. Pattern Recogn. Lett. **34**, 400–413 (2013)
87. Zhu, C.J., Yang, S., Zhao, Q., Cui, S., Wen, N.: Robust semi-supervised Kernel-FCM algorithm incorporating local spatial information for remote sensing image classification. J. Indian Soc. Remote Sens. **42**(1), 35–49 (2014)
88. Yu, X.C., He, H., Hu, D., Zhou, W.: Land cover classification of remote sensing imagery based on interval-valued data fuzzy c-means algorithm. Sci. China Earth Sci. **57**(6), 1306–1313 (2014). doi:10.1007/s11430-013-4689

89. He, P., Shi, W., Zhang, H., Hao, M.: A novel dynamic threshold method for unsupervised change detection from remotely sensed images. Remote Sens. Lett. **5**(4), 396–403 (2014). (Taylor & Francis)
90. Belacel, N., Hansen, P., Mladenovic, N.: Fuzzy J-Means: a new heuristic for fuzzy clustering. Pattern Recogn. **35**, 2193–2200 (2002)
91. Wu, K.L., Yang, M.S.: Alternative c-means clustering algorithms. Pattern Recogn. **35**, 2267–2278 (2002)
92. Pedrycz, W., Vukovich, G.: Fuzzy clustering with supervision. Pattern Recogn. **37**, 1339–1349 (2004)
93. Pedrycz, W.: Fuzzy clustering with a knowledge-based guidance. Pattern Recogn. Lett. **25**, 469–480 (2004)
94. Wang, X., Wang, Y., Wang, L.: Improving fuzzy c-means clustering based on feature-weight learning. Pattern Recogn. Lett. **25**, 1123–1132 (2004)
95. Pianykh, O.S.: Analytically tractable case of fuzzy c-means clustering. Pattern Recogn. **39**, 35–46 (2006)
96. Park, D.-C.: Intuitive fuzzy C-means algorithm for MRI segmentation. In: 2010 International Conference on Information Science and Applications (ICISA), pp. 1–7. doi:10.1109/ICISA. 2010.5480541
97. Yang, M.S., Tsai, H.S.: A Gaussian kernel-based fuzzy c-means algorithm with a spatial bias correction. Pattern Recogn. Lett. **29**, 1713–1725 (2008)
98. Kim, W.D., Lee, K.H., Lee, D.: A novel initialization scheme for the fuzzy c-means algorithm for color clustering. Pattern Recogn. Lett. **25**, 227–237 (2004)
99. ŁeRski, J.M., Owczarek, A.J.: A time-domain-constrained fuzzy clustering method and its application to signal analysis. Fuzzy Sets Syst. **155**, 165–190 (2005)
100. Ozer, M.: Fuzzy c-means clustering and Internet portals: a case study. Eur. J. Oper. Res. **164**, 696–714 (2005)
101. Inan, Z.H., Kuntalp, M.: A study on fuzzy C-means clustering-based systems in automatic spike detection. Comput. Biol. Med. **37**, 1160–1166 (2007)
102. Ceccarelli, M., Maratea, A.: Improving fuzzy clustering of biological data by metric learning with side information. Int. J. Approximate Reasoning **47**, 45–57 (2008)
103. Silva, S., Junior, M.D., Junior, V.L., Brennan, M.J.: Structural damage detection by fuzzy clustering. Mech. Syst. Signal Process. **22**, 1636–1649 (2008)
104. Chen, S.M., Chang, Y.C.: Multi-variable fuzzy forecasting based on fuzzy clustering and fuzzy rule interpolation techniques. Inf. Sci. **180**, 4772–4783 (2010)
105. Iliadis, L.S., Vangeloudh, M., Spartalis, S.: An intelligent system employing an enhanced fuzzy c-means clustering model: application in the case of forest fires. Comput. Electron. Agric. **70**, 276–284 (2010)
106. Sudha, K.R., Raju, Y.B., Sekhar, A.C.: Fuzzy C-Means clustering for robust decentralized load frequency control of interconnected power system with generation rate constraint. Electr. Power Energy Syst. **37**, 58–66 (2012)
107. Yin, X.F., Khoo, L.P., Chong, Y.T.: A fuzzy c-means based hybrid evolutionary approach to the clustering of supply chain. Comput. Ind. Eng. **66**, 768–780 (2013)
108. Yan, Y., Chen, L., Tjhi, W.C.: Fuzzy semi-supervised co-clustering for text documents. Fuzzy Sets Syst. **215**, 74–89 (2013)
109. Azara, A.T., El-Said, S.A., Hassaniend, A.E.: Fuzzy and hard clustering analysis for thyroid disease. Comput. Methods Programs Biomed. **11**(1), 1–16 (2013)
110. Liu, L.F., Sun, Z.D., Zhou, X.Y., Han, J.F., Jing, B.,Pan, Y.Y., Zhao, H.T., Neng, Y.: A new algorithm of modified fuzzy c means clustering (FCM) and the prediction of carbonate fluid. In: 76th EAGE Conference and Exhibition (2014). doi:10.3997/2214-4609.20140801
111. Hathaway, R.J., Bezdek, J.C., Hu, Y.K.: Generalized fuzzy c-means clustering strategies using L_p norm distances. IEEE Trans. Fuzzy Syst. **8**(5), 576–582
112. Wang, H.K., Hwang, J.C., Chang, P.L., Hsieh, F.H.: Function approximation using robust fuzzy-Grey CMAC method. Int. J. Model. Ident. Control **14**(4), 227–234

113. Yang, M.S.: On asymptotic normality of a class of fuzzy c-means clustering procedures. Int. J. Gen. Syst. **22**(4) (2007)
114. Yang, M.S., Yu, K.F.: On stochastic convergence theorems for the fuzzy c-means clustering procedure. Int. J. Gen. Syst. **16**(4), 397–411 (2007)
115. Kaushik, A., Soni, A.K., Soni, R.: Radial basis function network using intuitionist fuzzy C means for software cost estimation. Int. J. Comput. Appl. Technol. **47**(1), 86–95 (2013)
116. Juang, C.F., Hsieh, C.D.: Fuzzy C-means based support vector machine for channel equalization. Int. J. Gen. Syst. **38**(3) (2009)

Character Recognition Using Firefly Based Back Propagation Neural Network

M.K. Sahoo, Janmenjoy Nayak, S. Mohapatra, B.K. Nayak and H.S. Behera

Abstract The use of artificial neural network technique has significantly improved the quality of pattern recognition day-by-day with better performance. After the evolution of optimization algorithms, it has given new directions for achieving efficient result. In this paper a firefly based back-propagation network has proposed for character recognition. The firefly algorithm is a nature inspired optimization algorithm and it is simulated into back-propagation algorithm to achieve faster and better convergence rate within few iteration. The characters are collected from system through mouse that are used for training and characters are collected from MS-Paint that are used for testing purpose. The performance is analyzed and it is observed that proposed method performs better and leads to converge in less number of iteration than back-propagation algorithm.

Keywords ANN · Back propagation · Firefly · Back propagation algorithm

1 Introduction

Artificial Neural Networks have emerged as an attractive field of study within engineering via the collaborative efforts of engineers, physicists, mathematicians, computer scientists. A neural network is first and foremost a graph, with patterns

M.K. Sahoo · J. Nayak (✉) · S. Mohapatra · B.K. Nayak · H.S. Behera
Department of Computer Science and Engineering, Veer Surendra Sai University of Technology, Burla, Sambalpur 768018, Odisha, India
e-mail: mailforjnayak@gmail.com

M.K. Sahoo
e-mail: mk186.sahoo@gmail.com

S. Mohapatra
e-mail: sabya1992@gmail.com

B.K. Nayak
e-mail: nayak.bikram1@gmail.com

L.C. Jain et al. (eds.), *Computational Intelligence in Data Mining - Volume 2*,
Smart Innovation, Systems and Technologies 32, DOI 10.1007/978-81-322-2208-8_15

represented in terms of numerical values attached to the nodes of the graph and transformations between patterns achieved via simple message passing algorithms [1]. Some of the nodes in the graph are generally distinguished as being input or output nodes and the graph can be viewed as a depiction of linking from inputs to outputs in terms of a multivariate function. They are error tolerant and can learn and generalize from training data, so there is no need for enormous feats of programming.

From a practical point of view, an ANN is just a parallel computational system consisting of many simple processing elements connected together in a specified way in order to perform a particular task. After the creation of ANN, it must go through the process of updating the weights in the connection links between network layers with the objective of achieving the desired output which is the training of a neural network. There are two approaches for training—supervised and unsupervised. In supervised training, based on the given inputs, the network processes and compares its resulting outputs against the desired outputs and error is calculated. In unsupervised training, the network is provided with inputs but not with desired outputs and the system itself must then decide what features it will use to group the input data.

During the last decades, Character recognition is one of the most attractive area of pattern recognition applications [2]. Many researchers have contributed a good number of research papers for character recognition of different languages, particularly for English language as it is treated as common universal language for all, around the globe. Devireddy and Apparao [3] has proposed a back propagation network capable of recognizing handwritten characters or symbols, inputted by the mouse and for creating new patterns by the user. Espana-Boquera et al. [4] designed a model for recognizing unconstrained offline handwritten texts and presented various techniques for removing slope and slant in handwritten text and the size of text images were normalized with supervised learning methods. Yang et al. [5] proposed a BP network to recognize the English characters which shows superior recognition capability with good convergence speed. Tuli (2012) developed a neural network for character recognition [6] by using back propagation algorithm to recognize the twenty-six characters from A to Z and a to z and studied the effect of variations with error percentage with number of hidden layers. Barve [7] developed an optical character recognition system using artificial neural network, that uses feature extraction step of optical character recognition. Yi et al. [8] proposed a back propagation network based on various optical parameters which are strongly influenced by the performance of the network for character recognition. Sharma and Chaudhary [9] proposed a neural network for recognition of isolated handwritten English character by using binarization, thresholding and segmentation method. Choudhary et al. [10] has proposed a model to extract features obtained by binarization technique for recognition of handwritten characters of English language. Richarz et al. [11] has developed a semi supervised learning method tested on MNIST database of handwritten digits for reducing the human effort to build a character recognizer. De Stefano et al. [12] presented a GA based feature selection approach for handwritten character recognition with efficient discrimination capability of samples belonging to different nonlinearly separable classes.

But one of the major challenging task and open problem discussion in this area is the high recognition accuracy and minimum training time for recognizing English characters using neural network. The recognition of character in a document [2] becomes difficult due to noise, distortion, various character fonts and size, writing styles as handwriting of different persons is different. The performance of character recognition system is primarily depends on effective feature extraction [13, 14] and suitable classifier selection. Some of the important recognition methods including simple template matching, outer contour matching, projection sequence feature matching and external contour projection matching [15] have been developed for accurate recognition which are performing well for standard fonts. But it is quite difficult to achieve the accuracy of the desired recognition results when there are interference and noise. In this paper a firefly based back propagation neural network has been proposed to recognize the alpha numeric characters like characters, numbers and special symbols and a brief analysis report on the effect of variations with error percentage with number of iteration in the neural network is also being studied. The proposed Neural network is able to recognize twenty-six uppercase and lowercase characters, special case letters like '@', '#', '$', '%', '&', '*', '^' etc. and numerals correctly with higher degree of recognition accuracy.

2 Back Propagation Network

The aim of neural network is to mimic the human ability to adapt to changing circumstances and the current environment. This depends heavily on being able to learn from events that have happened in the past and to be able to apply this to the future situation [16]. Back propagation neural network (Fig. 1) is used in many systems and the technique is acting as a crucial factor despite of its limitations [17]. Back propagation is a learning rule [18] for the training of multi-layer feed-forward neural network. The back propagation learning involves propagation of the error

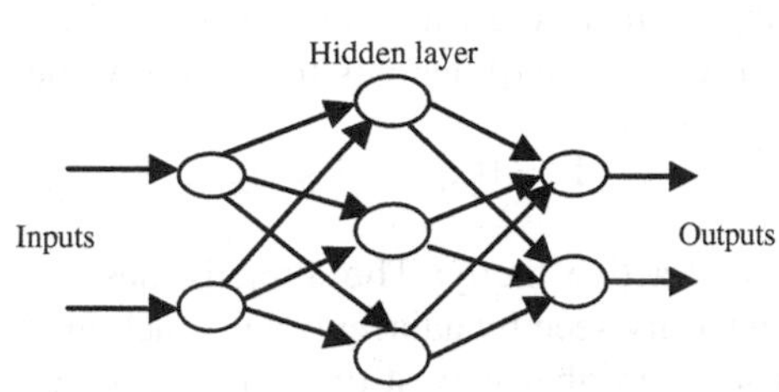

Fig. 1 Back propagation network

backwards [19] from the output layer to the hidden layers in order to determine the update for the weights leading to the units in a hidden layer. In most of the articles from various researches we surveyed that the back propagation technique was the commonly used model for classification problems. The reason behind this is the back propagation is the best technique for the ability to cope with noisy data and for larger inputs as comparison to other networks. In back propagation network, the output values are compared with the correct answer to compute the value of some predefined error function. The goal of back propagation, is to iteratively adjust the synaptic weights in the network to produce the target output by minimizing the output and to train the net to achieve a balance between the ability to respond correctly to the input pattern that are used for training and the ability to provide good response to the input those are similar.

3 Firefly Algorithm

The Firefly Algorithm (FA) is a recent nature inspired metaheuristic algorithm, developed by Yang [20] that has been used for solving nonlinear optimization problems. It is a population-based algorithm [21] to find the global optima of objective functions based on swarm intelligence, investigating the foraging behavior of fireflies. This algorithm is based on the behavior of social insects (fireflies) [22]. For simplicity three idealized rules are to be followed for implementation of the algorithm: (i) Fireflies are unisex so they are attracted towards other fireflies regardless of their sex. (ii) Attractiveness is directly proportional to brightness, so the less brighter firefly will be attracted towards the brighter firefly. (iii) The brightness or light intensity of a firefly is affected or determined by the landscape of the objective function to be optimized [23]. The firefly algorithm has two important issues: the variation of light intensity and formulation of the attractiveness. The attractiveness of the firefly can be determined by its brightness or light intensity which in turn is associated with the encoded objective function. Based on the objective function, initially, all the fireflies are randomly dispersed across the search space. The two phases [24, 25], of firefly algorithm are as follows:

1. Variation of light intensity: The objective function values are used to find the light intensity. Suppose there exist a swarm of n fireflies and x_i represents a solution for firefly i, whereas $f(x_i)$ denotes the fitness value.

$$I_i = f(x_i), \quad 1 \leq i \leq n. \tag{1}$$

2. Movement towards attractive firefly: The attractiveness of a firefly is proportional to the light intensity seen by adjacent flies. Each firefly has its distinctive attractiveness β which describes how strong a firefly attracts other members of

the swarm. But the attractiveness β is relative, it will vary with the distance r_{ij} between two fireflies i and j located at x_i and x_j, respectively, is given as

$$r_{ij} = \left\|x_i - x_j\right\| \quad (2)$$

The attractiveness function $\beta(r)$ of the firefly is determined by

$$\beta(r) = \beta_0 e^{-\gamma r^2} \quad (3)$$

where β_0 is the attractiveness at r = 0 and γ is the light absorption coefficient. The movement of a firefly i at location x_i attracted to a more attractive firefly j at location x_j is given by

$$x_i(t+1) = x_i(t) + \beta_0 e^{-\gamma r^2}\left(x_j - x_i\right) \quad (4)$$

The Pseudocode of Firefly algorithm has been illustrated as follows.

```
Objective function f(x), x = (x1, ..., xd)^T
Generate an initial population of fireflies  xik, i = 1, 2, . . . , n and k = 1, 2, . . . , d
where d=number of dimensions
Maxgen: Maximum no of generations
Evaluate the light intensity of the population Iik which is directly proportional to f (xik)
Initialize algorithm's parameters
While(i<=n)
  While(j<=n)
      If (Ij<Ii)
    Move firefly i toward j in d-dimension using Eq. (4)
     End if
   Attractiveness varies with distance r via
exp[−r^2]
   Evaluate new solutions and update light intensity using Eq. (1)
            j=j+1
    End while
        Rank the fireflies and find the current best
  if stopping criteria is satisfied then stop
          else i=i+1
End while
Post process results and visualization
```

4 Proposed Approach

The proposed network recognizes characters i.e. uppercase and lowercase English letters, numerals and special case letters. The network is trained using firefly based back propagation algorithm and it is observed that mean square error is converged in less number of iteration than using only back propagation algorithm. First error is calculated using back propagation algorithm and then this error is feed as input to firefly algorithm. The brighter firefly is considered having minimum error. Distance between firefly is calculated using Eq. (2), attractiveness among fireflies is calculated by Eq. (3) and movement of firefly is determined by Eq. (4). Various parameters used in the training algorithms are listed as follows: The neural network is trained by using the following steps:

$I = (i_1, i_2, i_3, i_4, \ldots, i_n)$: input training vector	$\$_k$ = error at output unit M_k
$T = (t_1, t_2, t_3, t_4, \ldots, t_m)$: output target vector	$\$_j$ = error at output unit M_j
$\beta_0 \rightarrow$ attractiveness at r = 0	B_j = bias at hidden unit
$\gamma \rightarrow$ light absorption coefficient	B_k = bias at output unit
$E_j \rightarrow$ sum of square error for each input	M_k = output unit k
$\beta \rightarrow$ any constant parameter whose value <l	M_j = output at hidden unit
V_{ij}, $W_{jk} \rightarrow$ weights at output layer and hidden layer	p $\rightarrow$ population of firefly
	α = learning rate

Step 1: Assign small random values to the weights.
Step 2: For j = 1 to n (no. of inputs)
Feed Forward from Input signals to hidden layer can be described as

$$M_{in_j} = B_j + \sum_{i=1}^{n} i_i v_{ij} \tag{5}$$

Actual output is calculated by applying activation function

$$M_j = f(M_{in_j}) \tag{6}$$

which is further propagated to above layer i.e. output units.
Step 3: Each output unit sums its weighted input signals.
For k = 1 to m (no. of inputs)

$$M_{in_k} = B_k + \sum_{i=1}^{m} m_j w_{jk} \tag{7}$$

and by applying the activation function to calculate the output signals

$$M_k = f(M_{in_k}) \tag{8}$$

Step 4: In output layer error can be calculated by using (9)

$$\delta_k = (T_k - M_k)[f(M_{in_k}) * (1 - f(M_{in_k}))] \quad (9)$$

Error information term at hidden layer can be calculated by (10)

$$\delta_j = \left[f\left(M_{in_j}\right) * \left(1 - f\left(M_{in_j}\right)\right)\right] \sum \delta_j w_{jk} \quad (10)$$

This error is the fitness value and is used for calculation of intensity of firefly.

Step 5: Weight and bias correction terms are given by (11) and (12).

$$\Delta W_{jk} = \alpha \delta_k M_j \quad \Delta B_k = \alpha \delta_k \quad (11)$$

Hence,

$$W_{jk(new)} = W_{jk} + \Delta W_{jk} \quad \text{and} \quad B_{k(new)} = B_k + \Delta B_k \quad (12)$$

Each hidden unit also updates its bias and weights.
The weight and bias correction terms are given by (13)

$$\Delta V_{ij} = \alpha \delta_j I_j \quad \text{and} \quad \Delta B_j = \alpha \delta_j \quad (13)$$

Hence,

$$V_{ij(new)} = V_{ij} + \Delta V_{ij} \quad \text{and} \quad B_{j(new)} = B_j + \Delta B_j V_{ij(new)} \quad (14)$$

Step 6: Sum of square error can be calculated by using (15)

$$E_j = \sum_{k=1}^{m} (T_k - M_k)^2 \quad (15)$$

Repeat steps 2 to 6 for p times

Step 7: Mean square error (MSE) can be determined by using (16)

$$E_{av} = 1/p \sum_{j=1}^{n} E_j \quad (16)$$

Step 8: As the sum of square of error (SSE) is inversely proportional to intensity of illumines of firefly, so each SSE value is considered as one firefly in the proposed work.

Step 9: Determine minimum error from SSE list and assign it to E_j (brighter firefly)

For i = 1 to p

For j = 1 to I

If ($E_i > E_j$)

Move firefly i towards j by using Eq. (2)
Attractive function $\beta(r)$ is determined by using (3)
Movement of firefly is determined by using (4)
Update weight and bias at output layer by using the Eqs. (17) and (18)

$$W_{jk}(new) = W_{jk}(new) - \Delta E_i(t) \tag{17}$$

$$B_{k(new)} = B_{k(new)} - \Delta E_i(t) \tag{18}$$

At hidden layer updated weight and biases are given by (19) and (20)

$$V_{ij}(new) = V_{ij}(new) - \Delta E_i(t) \tag{19}$$

$$B_j(new) = B_j(new) - \Delta E_i(t) \tag{20}$$

end for j

end for i

Step 10: Repeat steps 2 to 9 until the MSE is converged.

4.1 Pseudocode of the Proposed Approach

```
Input : Assign all network inputs and output. Initialize all weights with small random
numbers[-1,1],learning parameter and light absorption coefficient.
Output: Updated weight and bias ,sum of squared error(SSE),mean square error(MSE)
Begin
Repeat
For each layer in the network
    For each node in the layer
        Calculate the sum of the inputs to the node .
        Apply activation function to find output value of the node
        Calculate error signal of the node.
        Update weight and bias of each node.
    End
End
Calculate sum of squared error for each generated output using eq. (15)
Each SSE value is considered as one firefly.
    Calculate MSE from SSE list using eq. (16)
Determine minimum error from SSE list and assign it to fj(brighter firefly)
While k < (length of SSE list)
    for i = 1 : n (all n fireflies)
        for j = 1 : i
          if(E_j < E_i)
Calculate the distance between E_i and E_j using(9)
 Move the firefly E_i towards E_j using eq. (4)
Modify corresponding weight and bias value
          End if
        Else
              pass
      End j
    End i
End while
while ((Iterations(max) < Iteration(specified)) OR (mean square error is converged))
```

5 Implementation

The proposed firefly based back propagation algorithm for Character recognition has been implemented in MATLAB 9.0. The implementation consists of two phases like training phase and testing phase.

(i) **Training Phase**: Various English alphabets, numerals, and special characters are taken from the system for training purpose. Imread('*.bmp') is used for reading image from graphics file. A = imread(filename,fmt) reads a gray scale or color image from the file specified by the string filename. We have

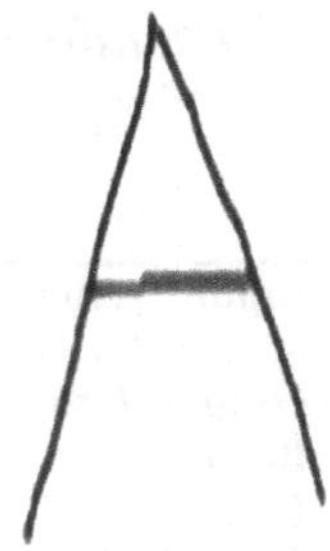

Fig. 2 Input by MS-paint

arbitrarily given the size of image pixels as a matrix using a '1' to represent a pixel that is on and a '0' to represent a pixel that is off. The pixels are then stored in a file for the experimental purpose.

(ii) **Testing Phase**: In this phase, the system learned data patterns are compared with the patterns to be recognized. The character is given as input for testing using MS paint tool. The pixels are extracted and are stored in form of a matrix. The matrix pattern, which is equal to that of training pattern, is given as the output. For example 'A' (Fig. 2) is given as input using MS paint for testing purpose.

6 Experimental Analysis and Results

The proposed algorithm was implemented in MATLAB version 9.0 and results are recorded during each simulation. In the proposed work 1,224 characters (uppercase, lowercase and special characters) and numerals has been tested. We have tested each letter by taking 17 test cases. So, total number of test cases is 26 + 26 + 10 + 10 = 72 * 17 = 1,224. All inputs for test cases are given using MS paint version 6.1. After the experimental analysis, it is found that the letter '^', the number '3', the uppercase alphabet 'H' and the lower case alphabet 'd' have higher recognition accuracy of 94.11, 82.35, 88.23 and 82.35 % respectively (Tables 1, 2, 3 and 4).

Table 1 Accuracy recognition of special case symbols

Special case letter	Accepted	Rejected	Recognition accuracy (%)
!	14	3	82.35
@	12	5	70.58
#	13	4	76.47
$	13	4	76.47
%	13	4	76.47
^	16	1	94.11
&	13	4	76.47
*	15	2	88.23
(	11	6	64.70
)	15	2	88.23

Table 2 Accuracy recognition of digits

No.	Accepted	Rejected	Recognition accuracy (%)
1	12	5	70.58
2	10	7	58.82
3	14	3	82.35
4	12	5	70.58
5	11	6	64.70
6	13	4	76.47
7	13	4	76.47
8	12	5	70.58
9	11	6	64.70
0	9	8	52.94

Table 3 Accuracy recognition of upper case alphabets

Upper case alphabets	Accepted	Rejected	Recognition accuracy (%)
A	13	4	76.47
B	10	7	58.82
C	13	4	76.47
D	11	6	64.70
E	12	5	70.58
F	13	4	76.47
G	13	4	76.47
H	15	2	88.23
I	11	6	64.70
J	14	3	82.35
K	14	3	82.35
L	17	6	64.70
M	17	3	82.35
N	17	4	76.47
O	11	6	64.70
P	12	5	70.58
Q	12	5	70.58
R	14	3	82.35
S	11	6	64.70
T	13	4	76.47
U	12	5	70.58
V	10	7	58.82
W	14	3	82.35
X	13	4	76.47
Y	13	4	76.47
Z	12	5	70.58

Table 4 Accuracy recognition of lower case alphabets

Lower case alphabets	Accepted	Rejected	Recognition accuracy (%)
a	11	6	64.70
b	13	4	76.47
c	13	4	76.47
d	14	3	82.35
e	12	5	70.58
f	13	4	76.47
g	12	5	70.58
h	11	6	64.70
i	9	8	52.94
j	12	5	70.58
k	13	4	76.47
l	12	5	70.58
m	13	4	76.47
n	13	4	76.47
o	12	5	70.58
p	13	4	76.47
q	12	5	70.58
r	11	6	64.70
s	10	6	64.70
t	12	5	70.58
u	11	6	64.70
v	12	5	70.58
w	12	5	70.58
x	13	4	76.47
y	13	4	76.47
z	12	5	70.58

7 Conclusion and Future Work

In this paper, a firefly based back propagation network has been proposed for recognizing the alpha numeric characters. It is a hybridized method, where the nature inspired firefly algorithm is merged with back propagation for optimization in neural network training. The experimental result (Fig. 3) reveals that the proposed method gives better efficiency for various cases of different characters, as it requires less number of iteration for convergence of error in the neural network. The character recognition is fully dependent on the type of inputs received. Since handwritten characters are of various size and style, the recognition method should be much more efficient. The proposed method is designed to recognize the English

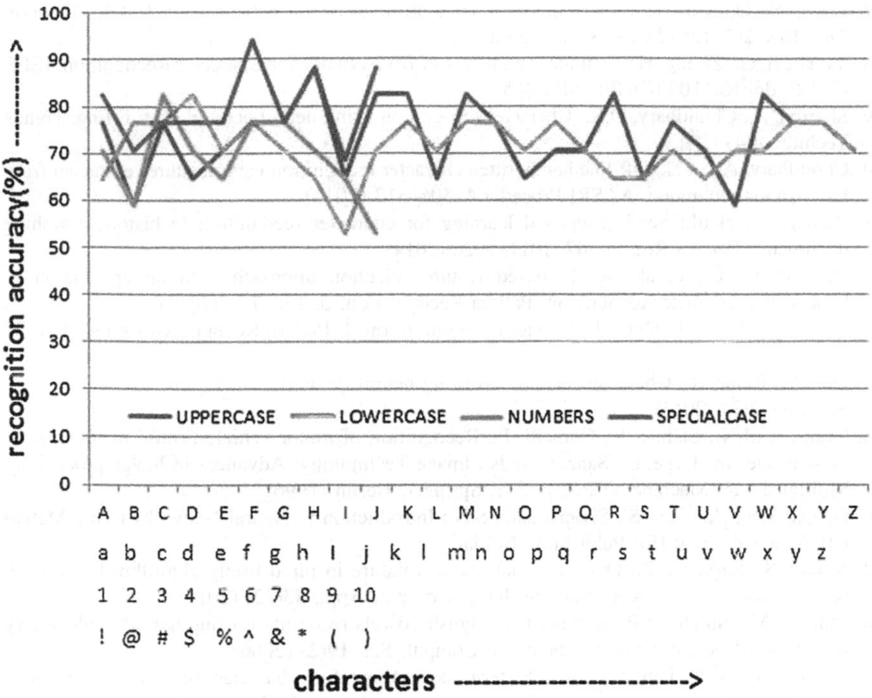

Fig. 3 Recognition accuracy of all alpha numeric characters

alphabets. It can be extended for other languages as well as documents, words etc. Also future research may be done for text documents and various images with better accuracy by tuning the feature extraction parameters.

References

1. Hassoun, M.H.: Fundamentals of Artificial Neural Networks. The MIT Press, Cambridge (1995)
2. Kumar, G., Bhatia, P.K.: Neural network based approach for recognition of text images. Int. J. Comput. Appl. **62** (14), 0975–8887 (2013)
3. Devireddy, S.K., Apparao, S.: Hand written character recognition using back propagation network. J. Theoret. Appl. Inf. Technol. (2009)
4. Espana-Boquera, S., Castro-Bleda, M.J., Gorbe-Moya, J., Zamora-Martinez, F.: Improving offline handwritten text recognition with hybrid HMM/ANN models. IEEE Trans. Pattern Anal. Mach. Intell. **33**(4) (2011)
5. Yang, Y., et al.: English character recognition based on feature combination. Procedia Eng. **24**, 159–164 (2011)
6. Tuli, R.: Character recognition in neural network using back propagation. In: 3rd IEEE International Advance Computing Conference (2013)

7. Barve, S.: Optical character recognition using artificial neural network. Int. J. Adv. Technol. Eng. Res. **2**(2) (2012). ISSN 2250-3536
8. Yi, H., Ji, G., Zheng, H.: Optimal Parameters of Bp Network for Character Recognition. IEEE (2012). doi:10.1109/ICICEE.2012.465
9. Sharma, A., Chaudhary, D.R.: Character recognition using neural network. Int. J. Eng. Trends Technol. **4**(4) (2013)
10. Choudhary, A., et al.: Off-line handwritten character recognition using features extracted from binarization technique. AASRI Procedia **4**, 306–312 (2013)
11. Richarz, J., et al.: Semi-supervised learning for character recognition in historical archive documents. Pattern Recogn. **47**, 1011–1020 (2014)
12. De Stefano, C., et al.: A GA-based feature selection approach with an application to handwritten character recognition. Pattern Recogn. Lett. **3**, 130–141 (2014)
13. Impedovo, S., et al.: Optical character recognition. Int. J. Pattern Recogn. Artif Intell. **5**, 1–24 (1991)
14. Das, V., Rajni, R.: Character recognition using neural network. Int. J. Adv. Trends Comput. Sci. Eng. **2**(3) (2013)
15. Granieri, M.N., Stabile, F., Comelli, P.: Recognition of motor vehicle license plates in rear-view image. In: Jorge, L., Sanz, C. (eds.) Image Technology: Advances in Image processing. Multimedia an Machine Vision, p. 231. Springer, Berlin (1996)
16. Deepa, S.N., Sumathi, S., Sivanandam, S.N.: Introduction to Neural Networks Using Matlab 6.0. Tata Mc-Graw Hill Publishing, Noida
17. Nandy, S., Sarkar, P.P., Das, A.: Analysis of a nature inspired firefly algorithm based back propagation neural network training. Int. J. Comput. Appl. **43**(22) (2012)
18. Mangal, M., Singh, M.P.: Handwritten english vowels recognition using hybrid evolutionary feed-forward neural network. Malays. J. Comput. Sci. **19**(2) (2006)
19. Kosbatwar, S.P., Pathan, S.K.: Pattern association for character recognition by back propagation algorithm using neural network approach. Int. J. Comput. Sci. Eng. Surv. **3**(1) (2012)
20. Yang, X.S.: Nature-Inspired Metaheuristic Algorithms. Luniver Press, Frome (2008)
21. Yang, X.S.: Firefly algorithms for multimodal optimization in: Stochastic algorithms: foundations and applications. In: Watanabe, O., Zeugmann, T. (eds.) SAGA 2009. Lecture Notes in Computer Science 5792, pp. 169–178. Springer, Berlin (2009)
22. Abshouri, A.A., Meybodi, M.R.: New Firefly Algorithm based on Multi Swarm & Learning Automata in Dynamic Environments. IEEE X-plore (2011)
23. Yang, X.S.: Firefly algorithm, stochastic test functions and design optimisation. Int. J. Bio-Inspired Comput. **2**(2), 78–84 (2010)
24. Senthilnath, J., Omkar, S.N., Mani, V.: Clustering using firefly algorithm: performance study. Swarm Evol. Comput. **1**, 164–171 (2011)
25. Nayak, J., Nanda, M., Nayak, K., Naik, B., Behera, H.S. :An improved firefly fuzzy c-means (FAFCM) Algorithm for clustering real world data sets. In: Smart Innovation, Systems and Technologies, vol. 27, pp. 339–348. Springer, Berlin (2014)

Analyzing Data Through Data Fusion Using Classification Techniques

Elizabeth Shanthi and D. Sangeetha

Abstract Knowledge is the ultimate output of decisions on a dataset. Applying classification rules is one of the vital methods to extract knowledge from dataset. Knowledge in a very distributed approach is derived by combining or fusing these rules. In a very standard approach this may generally be done either by combining the classifiers outputs or by combining the sets of classification rules. In this paper, we tend to do a new approach of fusing classifiers at the extent of parameters using classification rules. This approach relies on the fused probabilistic generative classifiers using multinomial distributions for categorical input dimensions and multivariable normal distributions for the continual ones. These distributions are used to produce results like valid/invalid data, error rate etc. Fusing two (or more) classifiers may be done by multiplying the hyper-distributions of the parameters. The main advantage of this fusion approach is that it requires less time to classify the data and is easily extensible for large dataset.

Keywords Data fusion · Classification · Multinomial distribution · Hyper-distribution

1 Introduction

In most of the data mining applications, the task of extracting knowledge (e.g., classification rules) from sample data is divided into a number of subtasks. Typical examples are smart sensor networks, robot teams or software agents that learn locally in their environment. At some point, there is the necessity to fuse or to

E. Shanthi (✉) · D. Sangeetha
Department of Computer Science, Avinashilingam Deemed University,
Coimbatore 641043, India
e-mail: elizabeth_cs@avinuty.ac.in

D. Sangeetha
e-mail: sangeethaprabha@ymail.com

L.C. Jain et al. (eds.), *Computational Intelligence in Data Mining - Volume 2*,
Smart Innovation, Systems and Technologies 32, DOI 10.1007/978-81-322-2208-8_16

combine the knowledge that is now *contained* in a number of *classifiers* in order to apply it to new data. *Probabilistic classifiers* provide outputs that can be interpreted as conditional probabilities as they model the conditional distribution of classes given an input sample. *Generative classifiers* aim at modeling the processes from which the sample data are assumed to originate. *Probabilistic generative classifiers* are usually based on Bayes' theorem [1] given by

$$P(c|\mathrm{x}) = \frac{p(\mathrm{x}|c) \cdot p(c)}{P(\mathrm{x})}. \tag{1}$$

where x is a (multivariate) random variable which models the input space of the classifier (e.g., x € IR^D where D € IN is the input space dimension) and c is a random variable representing a class (e.g., c € $\{1...C\}$). In contrast to generative classifiers, *discriminative classifiers* such as support vector machines, for instance, are *only* expected to have an optimal classification performance on new data (generalization). Compared to the probabilistic generative classifiers the discriminative classifiers take several advantages as well as drawbacks [2, 3]. There are also many application areas where both approaches can successfully be used in combination [4, 5]. *Advantages* are, the class posterior probabilities p $(c|\mathrm{x})$ are very useful to weigh single decisions when several classifiers are combined, e.g., in form of ensembles. A rejection criterion could easily be defined which allows to refuse a decision if none of the class posteriors reaches a pre-specified threshold. Possible *drawbacks* are that, these classifiers are more likely to over-fit to sample data as the (effective) number of parameters is typically quite high. The classification performance is sometimes worse if the data do not (at least nearly) meet the distribution assumptions. Altogether, it depends on the type of application whether such classifiers can successfully be applied [6].

2 Related Work

Knowledge fusion means the knowledge represented by components of classifiers fused at a parameter level. Fusion can take place at various levels or categories viz, data (e.g., sensor measurements or observations) or information extracted from databases can be fused to come to more certain conclusions. Models or parts of models trained from sample data or information can be fused if the models were constructed in a distributed fashion. The outputs of models can be fused to get more certain decisions or as in the case of temporal and spatial data mining to derive conclusions for certain points in space and time.

Here, two main fields can be identified: On one hand, knowledge is often equated with constraints and there is some work focusing on fusion of constraints discussed in [7–9]. On the other hand, knowledge is often represented by graphical models that are subject to fusion, e.g., Bayesian networks, (intelligent) topic maps, or the like as indicated in [10–13]. Paper [14] describes a Bayesian fusion approach based on hyper-parameters and it also exploits the concept of conjugate priors. Parallelization approaches can be found in [15, 16] for instance. It could even be shown that exact approaches are feasible in the sense that they give the same results as if the data were not processed in distributed chunks [17]. These techniques typically assume some shared resources and allow for an exchange of intermediate results with the corresponding communication overhead.

3 Methodology

This paper describes the way of fusing data using one or more classifier and/or combination steps. A classification is a task that begins with a given dataset to do probabilistic generative classifier (CMM), generating classification rule, knowledge fusion and classification, probabilistic classification, fusion techniques for classification, a similarity measure for hyper-distributions and fusion training and analysis.

3.1 Probabilistic Generative Classifier and Generating Classification Rule

A CMM classifier consists of several components each of which represents the knowledge of the classifier about one process 'generating' data in the input space. Here a new fusion classifier at the level of parameters of classification rule generates rules namely RULE 1 and RULE 2 for grouping of real values in that dataset to find a positioner value. In general, a classifier is a function mapping an input value x to an output class c € {1, C} of C possible classes. A probabilistic classifier takes the form p(c|x) which denotes the probability for class c given an input sample x.

According to [17],

$$P(c|X) = \frac{P(X|c)\,p(c)}{p(X)} = \frac{p(c)_{i=1}^{Jc}\,p(j|c)\,p(x|c,j)}{p(X)}. \qquad (2)$$

The classifier is split into C parts one for each class. Here p(c) is a multinomial distribution specifying the prior probability of class c, the conditional densities p(x|| c, j) are called components of the classifier and p(j||c) is another multinomial distribution whose parameters π c, j are called mixture coefficients and weight the components in their respective part of the mixture model. The overall classifier, which is called a classifier based on a mixture model (CMM), consists of $J = \sum_{c=1}^{C} J_c$ components each of which is described by a (usually multivariate) distribution P(x|c, j). As the input data can have both, categorical and continuous dimensions, the distributions p(x|c, j) must be chosen in a way such that both cases can be handled by the classifier.

3.2 Knowledge Fusion and Classification

The fusion mechanism uses the hyper-distributions obtained in the variance inference training process. Doing so, these hyper-distributions are retained throughout the fusion process which has several advantages over a simple linear combination of CMM parameters. The classifier resulting from all fusion and combination steps is also called overall classifier. The following algorithm from [1] gives an overview of the classification and describes explicitly how two CMM classifiers can be merged (fused/combined) but, as all proposed operators are associative, multiple CMM classifiers can easily be merged by iteratively merging pairs of classifiers until only one overall classifier remains.

Algorithm1: Fusion and Combination of CMM

```
Input: Two sets of hyper- distributions C1 and C2
Output: Fused/combined overall classifier
C'<- Ø
foreach c1 in C1 do
    found<- false
    foreach c2 in C2 do
        // similarity evalution
        dist<-ΔH (c1, c2)
        // check for consistent classes
        If dist< H and class (c1) == class (c2) then
            // Fusion
          C'.add(fuse (c1,c2))
            C2.remove (c2)
            found <- true
            break
      if not found then
            // combination
            C'.add (c1)
foreach c2 in C2 do
            // combination
           C'.add (c2)
classifier < - Ø
return classifier
```

3.3 Probabilistic Classification

The main work here is to divide the classifier into four divisions based on the probability of each classification. Probabilistic classifier provides output that are interrupt to an conditional probabilistic that is we are going to classify data based on the input and output of the data. This is done with different folds of probability namely (1, 1), (1, 2), (1, 3), (1, 4). Based on the likelihood and the positioner value the classifier gets fusion based on the formula.

3.4 Fusion Techniques for Classification

If two classifiers model similar processes they are likely to contain many similar components. We now want to detect such a situation in order to fuse all pairs of similar components. When two CMM are trained separately, each with a distinct part of the training data, we have two likelihood functions derived from the two sets of training data and two prior distributions. We now assume that the two priors are equal because in both cases we make use of the same prior knowledge or want to express the same amount of uncertainty about the parameters we want to estimate. Nevertheless, this leaves us with two posterior distributions.

$$\text{Posterior1} \propto \text{likelihood1.prior} \quad \text{and} \quad \text{Posterior2} \propto \text{likelihood2.prior.} \tag{3}$$

Each likelihood is itself a product over all data points in the respective training set. To fuse the posteriors they simply could be multiplied to obtain one overall posterior. This would lead to

$$\frac{\text{likelihood}_1.\text{Prior.}}{\text{posterior}_1}\,\frac{\text{likelihood}_2.\text{Prior}}{\text{posterior}_2} = (\text{likelihood}_1.\text{Likelihoodd}_2).\text{prior}^2. \tag{4}$$

If we had used Eq. (4) for the overall dataset with the same prior, the result would have been

$$\text{Likelihood.Prior} = (\text{likelihood}_1.\text{Likelihoodd}_2).\text{Prior.} \tag{5}$$

Comparing (4) and (5) we see that there is an additional 'prior' factor. As the prior is known, we can compensate for this fact by dividing by this prior which finally leads to our new fusion approach.

$$\begin{gathered}\frac{\text{likelihood}_1.\text{Prior.likelihood}_2.\text{Prior}}{\text{prior}} \\ \text{Posterior} \propto = \frac{\text{Posterior}_1.\text{Posterior}_2}{\text{Prior}}\end{gathered} \tag{6}$$

We cannot simply cancel out the prior here because it is only implicitly contained in the posterior distributions which are the result of the training algorithm. Instead, we multiply the two posteriors and then divide the result by the prior. Finally, we can make the assumption that the resulting overall posterior has the same functional form as the two posteriors that are fused. This is important for two reasons: First, this allows us to easily determine the parameters of the overall posterior and second, we can derive the parameters of the CMM classifier from the fused posterior. This fusion technique is implemented for both classification1 and classification2.

3.5 A Similarity Measure for Hyper-Distributions

The similarity measure should be symmetric such that the order in which two components are compared does not matter. A simple measure ΔH that fulfills this restriction can be derived from the Hellinger distance [18]. The similarity measure ΔH directly operates on the normal and multinomial distributions of the classifier. Theoretically, it would also be possible to compute the Hellinger distance of the hyper-distributions to evaluate the similarity of components but in that case the integral in Equation could not be computed in a closed form.

3.6 Fusion Training and Analysis

In this module we fuse two classifiers based on the likelihood and positioner value. The entire data is viewed and from the fusion value generated an error rate for each classifiers with the generated value gets formulated. The conjugate prior distribution that must be used to estimate the parameters of a multinomial distribution is a Dirichlet distribution. In order to fuse two Dirichlet distributions their density functions are multiplied and then divide the result by the prior. The knowledge that we have a certain distribution type implicitly gives us a suitable normalizing factor for the fused distribution.

4 Experimental Results

4.1 Dataset Used

This data was extracted from the census bureau database found at http://www.census.gov/ftp/pub/DES/www/welcome.html. The above mentioned data was pre-processed in WEKA. Convert the dataset into .arff or .csv format and extract into WEKA. To find a classification technique such as Navie Bayes to do the followings (Correctly classified instances, Incorrectly classified instances, Kappa statistic, Mean absolute error, Root mean squared error, Relative absolute error, Coverage of cases, Mean rel. region size, Total number of instances). To get detailed accuracy by class (TP Rate, FP Rate, Precision, Recall, F-Measure, ROC Area, class) and it formed a confusion matrix. This is depicted in Tables 1, 2 and 3.

Table 1 Evaluation on training set

Correctly classified instances	1,300	81.607 %
Incorrectly classified instances	293	18.393 %
Kappa statistic	0.5299	
Mean absolute error	0.2102	
Root mean squared error	0.3665	
Relative absolute error	56.8115 %	
Root relative squared error	85.2391 %	
Coverage of cases (0.95 level)	96.108 %	
Mean rel. region size (0.95 level)	71.0923 %	
Total number of instances	1,593	

Table 2 Detailed accuracy by class

	TP rate	FP rate	Precision	Recall	F-measure	ROC area	Class
Weighted	0.85	0.29	0.901	0.85	0.875	0.881	≤50K
avg.	0.71	0.15	0.606	0.71	0.654	0.881	>50K
	0.816	0.255	0.828	0.816	0.821	0.881	

Table 3 Confusion matrix

a	b	<– Classified as
1,023	180	a = ≤50K
113	277	b = >50K

4.2 Results Observed

The basic idea of this work is how two or more classifier and thus, the represented knowledge can be combined by means of several fusion and/or combination steps.

Step 1 *Dataset extraction is done before we start our process.*

Step 2 *Generating Classification Rule*

Here Probabilistic Generative Classifier is used to generate classification rules.

Step 3 *Probabilistic Generative Classifier (CMM)*

For rule1 and rule2 the continuous dimensions are modeled with multivariate Gaussian distribution and results noted.

Step 4 *Knowledge fusion and classification*

Classification1 and classification2 are carried out for different dimensions and data fusion occurs.

Step 5 *Probabilistic generative classifier*

The Mahalanobis distance [1] and mixture coefficient value [1] is calculated for classification.

Step 6 *Probabilistic classification1 and classification2*
The classifier here is divided into various folds based on the probability of classification. Data gets fused based on the likelihood and positioner value.

Step 7 *Fusion techniques for classification1 and classification2*
All pairs of similar components are fused for classification1. Similar Fusion technique for classification2 may also be obtained.

Step 8 *A Similarity Measure for Hyper-Distributions*

Here two distributions namely Dirichlet distribution is applied to find parameters namely Hellinger distance, continuous dimension and hyper Distribution, whereas Normal-Wishart distribution is used for detecting error rate. The experiments have shown that this new way of fusing and combining CMM classifiers can successfully be applied in given datasets. The number of components and the classification performance of the overall classifiers obtained with the fusion/combination algorithm depend on the similarity threshold that has to be adjusted by the user depending on the application. It is influenced by parameters such as the type of the dimensions (categorical/continuous), the number of dimensions, or the number of categories in the case of categorical dimensions.

5 Conclusion and Outlook

This work discusses a new technique to fuse two probabilistic generative classifiers (CMM) into one. To identify components of two classifiers that shall be fused, a similarity measure that operates on the distributions of the classifier is suggested. The actual fusion of two components works one level higher on the hyper-distributions which are the result of the Bayesian training of a CMM. Formulas to fuse both Dirichlet and normal Wishart distributions which are the conjugate prior distributions of the multinomial and normal distributions of a CMM are used to obtain a more certain decision of a dataset. Applying data fusion approach to more than two CMM classifiers is straight forward as it is possible to apply the technique iteratively. It will certainly be possible to use the same parameter values (fusion threshold) for all single fusions. While being trivial from a technical point of view, the actual advantages for real applications have still to be pointed out in our future work. If the number of classifiers is known in advance it would also be possible to modify the fusion formulas accordingly. We can also generalize the approach to other distributions, in particular members of the exponential family of distributions and investigate how different prior distributions can be handled. We can find a more intuitive way to parameterize the fusion threshold and we will investigate the weighting of categorical and continuous dimensions in more detail. The proposed techniques could be used in the field of distributed data mining, where datasets have to be split to cope with huge amounts of data and where the communication costs have to be low. It is also possible to use fusion in distributed environments where data are locally processed as they arise (e.g., in smart sensor networks). The work can be applied for a specific application like collaborative learning and intrusion detection.

References

1. Fisch, D., Kalkowski, E., Sick, D.: Knowledge fusion for probabilistic generative classifier with data mining application. IEEE Trans. Knowl. Data Eng. **26**, 652–666 (2014)
2. Bishop, C.M.: Pattern Recognition and Machine Learning. Springer, New York (2006)
3. Fisch, D., Kühbeck, B., Sick, B., Ovaska, S.J.: So near and yet so far: new insight into properties of some well-known classifier paradigms. Inf. Sci. **180**(18), 3381–3401 (2010)
4. Bouguila, N.: Hybrid generative/discriminative approaches for proportional data modeling and classification. IEEE Trans. Knowl. Data Eng. (2011). Accepted for publication doi:10.1109/TKDE.2011.162
5. Hospedales, T.M., Gong, S., Xiang, T.: Finding rare classes: active learning with generative and discriminative models. IEEE Trans. Knowl. Data Eng. (2011). Accepted for publication. doi:10.1109/TKDE.2011.231
6. Fisch, D., Gruber, T., Sick, B.: Swiftrule: mining comprehensible classification rules for time series analysis. IEEE Trans. Knowl. Data Eng. **23**(5), 774–787 (2011)
7. Gray, P., Preece, A., Fiddian, N., Gray, W., Capon, T.B., Have, M., Azarmi, N., Wiegand, I., Ashwell, M., Beer, M. et al.: KRAFT: knowledge fusion from distributed databases and knowledge bases. In: Proceedings of the 8th International Workshop on Database and Expert Systems Applications, pp. 682–691 (1997)
8. Hui, K.Y., Gray, P.: Constraint and data fusion in a distributed information system. In: Embury S., Fiddian N., Gray W., Jones A. (eds.) Advances in Databases, Ser. Lecture Notes in Computer Science, vol. 1405, pp. 181–182. Springer, Berlin
9. Hui, K.Y.: Knowledge fusion and constraint solving in a distributed environment. Ph.D. Dissertation, Department of Computing Science, University of Aberdeen (2000)
10. Pavlin, G., De Oude, P., Maris, M., Nunnink, J., Hood, T.: A multi agent systems approach to distributed Bayesian information fusion. Inf. Fusion **11**(3), 267–282 (2010)
11. Santos Jr., E., Wilkinson, J., Santos, E.: Bayesian knowledge fusion. In: Proceedings of the 22nd International FLAIRS Conference, pp. 559–564 (2009)
12. Wang, Y., Wu, B., Hu, J.: A semantic knowledge fusion method based on topic maps. In: Workshop on Intelligent Information Technology Application, pp 74–76 (2007)
13. Smirnov, A., Pashkin, M., Chilov, N., Levashova, T.: KSNET—approach to knowledge fusion from distributed sources. Comput. Inform. **22**(2), 105–142 (2003)
14. Foina, A.G., Planas, J., Badia, R.M., Ramirez-Fernandez, F.J.: P-means, a parallel clustering algorithm for a heterogeneous multi-processor environment. In: Proceedings of the international conference on high performance computing and simulation (HPCS), pp. 239–248 (2011)
15. Li, Y., Zhao, K., Chu, X., Liu, J.: Speeding up k-means algorithm by GPUs. In: Proceedings of the 10th IEEE International Conference on Computer and Information Technology, pp. 115–122 (2010)
16. Chu, C.T., Kim, S.K., Lin, Y.A., Yu, Y., Bradski, G., Ng, A.Y., Olukotun, K.: Map-reduce for machine learning on multicore. In: Proceedings of NIPS (2006)
17. Fisch, D., Ovaska, S.J., Kalkowski, E., Sick, B.: In your interest objective interestingness measures for a generative classifier. In: Proceedings of the 3rd International Conference on Agents and Artificial Intelligence, pp. 414–423 (2011)
18. Le Cam, L., Yang, G.: Asymptotics in statistics: some basic concepts, 2nd edn. Springer, Berlin (2000)

Multi-objective Particle Swarm Optimization in Intrusion Detection

Nimmy Cleetus and K.A. Dhanya

Abstract In this paper, we proposed particle swarm optimization using multi-objective functions. Intrusion detection system has a significant role in research methodology. Intrusion detection system identifies the normal as well as abnormal behavior of a system. Swarm intelligence plays an essential role in intrusion detection. Random forest classifier is used for detecting attacks. Intrusion detection mechanism based on particle swarm optimization which has a strong global search capability is used for dimensionality optimization. Weighted aggregation method is employed as multi-objective functions. The proposed system has the high intrusion detection accuracy of 97.54 % with a detection time is 0.20 s.

Keywords Particle swarm optimization · Fitness function · Multi-objective functions · Swarm intelligence

1 Introduction

Researchers focus on Intrusion Detection Systems (IDS) to improve the accuracy. They are an unavoidable part of the security management systems for computers and networks that attempt to detect abnormal activities. Thus we proposed multi-objective particle swarm optimization which upgrades the performance of IDS.

Swarm intelligence is the collective behavior of animals such as birds, ants etc. Evolutionary algorithm is having population of solutions instead of a single solution to obtain optimal result. This population of solution can easily identify intrusions in a network.

N. Cleetus (✉) · K.A. Dhanya
Department of Computer Science and Engineering, SCMS School of Engineering and Technology, Ernakulam, Kerala, India
e-mail: cnimmy2008@gmail.com

K.A. Dhanya
e-mail: dhannyashibu@gmail.com

L.C. Jain et al. (eds.), *Computational Intelligence in Data Mining - Volume 2*, Smart Innovation, Systems and Technologies 32, DOI 10.1007/978-81-322-2208-8_17

Optimization problem consistently have more than one objective functions in every field. The target is to optimize the functions with respect to other. This denotes that there is no single solution for these problems. We aim to discover best possible solutions from the objectives.

Particle Swarm Optimization (PSO) [1] is a heuristic search technique. It is based on the behavior of a flock of birds migrating to reach an unknown destination. In PSO, each solution is a bird known as a particle, which is a member of population. Accordingly, each bird speeds towards the best bird using a velocity that depends on its current position. The location of a particle constitutes one solution of a problem in multi-dimensional problem space. When the particles move from one place to another, separate problem solution is generated. This solution is evaluated using fitness function which furnishes a significant value for the solution. The detection rate is varied with respect to the strength of fitness function. Single-objective PSO is found to be not an efficient detection criterion for IDS [2].

The remainder of the paper is organized as follows. Section 2 of the paper depicts the related works done in this area. Section 3 provides a brief description about the feature selection techniques chosen for this study. Section 4 of the paper gives a detailed explanation of the study and the followed method. Section 5 discusses the experiment and results obtained from our study. Sections 6 and 7 explains the inferences and conclusion of the study.

2 Related Work

Zhou et al. [3] implemented a cloud model for intrusion detection system using multi-objective particle swarm optimization. This cloud model generates optimal feature length to increase the efficiency. The result depict that proposed system has low false alarm and high true positive rate.

Malik and Khan [4] proposed a vector evaluated PSO approach for intrusion detection. They proposed a binary version of algorithm to increase the true positive rate. Experiment is conducted on KDD cup 99 dataset. Random forest algorithm is used for classification. Proposed system has high detection and low false alarm.

Bratton and Kennedy [1] defined a standard particle swarm optimization. Particle swarm optimization is a heuristic approach for optimization. To improve the performance of PSO different benchmarks are applied. The study is initiated with local ring topology. The experiment is conducted with 50 particles.

Authors in [5] introduced particle swarm optimization for constrained optimization problems. Authors evaluated the PSO using different benchmark functions. PSO is a good alternative for tackling constrained problems.

In [2] authors proposed particle swarm optimization in multi-objective problems. Weighted aggregation method is applied for multi-objective functions. There are five different fitness functions used in this study. A modified version of PSO is also developed. Result depict that the multi-objective functions is better than single objective functions.

Kennedy and Eberhart [6] examined particle swarm optimization based on heuristic approach for global optimization. The result depicted that the method is simple and it needs only fewer parameters.

Authors in [7] proposed intrusion detection based on particle swarm optimization and neural networks. They developed a network structure and performed the algorithm. The detection rate is higher in PSO using neural network. Experiment is performed in KDD cup 99 dataset. Accuracy of 85 % is attained for the proposed method.

Aljarah and Ludwig [8] proposed map reduce intrusion detection system using particle swarm optimization. Parallel clustering method is applied in intrusion detection. Clustering method is applied to improve the detection rate in particle swarm optimization.

3 Feature Selection Techniques

Evolutionary algorithms are highly sensitive to the dimensionality of feature space. Efficient dimensionality reduction methods eliminate irrelevant features used to mine feature subset that has the potential to represent the quality of entire dataset. Reducing the dimension of feature vector space assists sophisticated algorithms to perform well. The following paragraph describes the feature selection methods used in our study.

3.1 Information Gain (IG)

IG is to measure relevant features of classification using entropy value. The strength of a feature is directly proportional to its entropy value. Variables with higher entropy would be the informative attribute.

$$IG = -\sum_{k=1}^{|C|} \frac{N_{C_k}}{N} \ln \frac{N_{C_k}}{N} + \frac{N_F}{N} \sum_{k=1}^{|C|} \frac{N_{F,C_k}}{N_F} \ln \frac{N_{F,C_k}}{N_F} + \frac{N_{\overline{F}}}{N} \sum_{k=1}^{|C|} \frac{N_{\overline{F},C_k}}{N_{\overline{F}}} \ln \frac{N_{\overline{F},C_k}}{N_{\overline{F}}} \quad (1)$$

where N_F and $N_{\overline{F}}$ are the number of samples containing the feature and number of samples where feature is absent. N_{F,C_k} and $N_{\overline{F},C_k}$ are the number of samples in class C_k consisting of feature F and samples in which feature F is absent. N_{C_k} is the number of samples in class C_k and N is the total number of samples.

3.2 Maximizing Information Gain Feature Selection

Information gain is maximized using Eq. 2,

$$x_i = \begin{cases} 1 + IG(x_i), & IG(\mathrm{x_i}) > \mathrm{m} \\ -(1 - IG(x_i)), & \text{otherwise} \end{cases} \tag{2}$$

where x_i is the ith feature of the feature subset. m corresponds to the trimmed mean (In our case 10 % was considered) obtained from information gain of all features in the dataset. Trimmed mean is used to avoid the impact of extreme values.

4 Proposed Methodology

Investigation involves the following steps (1) dataset preparation (2) prepare train and test set (3) implementation of multi-objective PSO fitness functions and (4) prediction.

4.1 Dataset Preparation

NSL-KDD [9] dataset is used in our study. Dataset is less biased and more refined than KDD cup 99 dataset. It consists of lesser number of records. Training and test samples are not redundant.

4.2 Prepare Train and Test Set

Entire dataset is divided into train and test set (ratio 60:40). Robust features are acquired from the training set using Information gain feature selection. These synthesized features are depicted to generate legitimate models. Models are evaluated using the test data. Dataset preparation is described in Table 1.

Table 1 Dataset preparation

Attacks	Train samples	Test samples
Normal	9,998 (46.93 %)	2,152 (10.10 %)
Neptune	5,001 (23.47 %)	1,579 (7.41 %)
Satan	691 (2.90 %)	727 (3.41 %)
Smurf	529 (2.48 %)	627 (2.94 %)
Total	16,219	5,085

4.3 Particle Swarm Optimization

Particle Swarm optimization is always varies from other evolutionary algorithms. The dataset contain set of values known as particles. They are randomly picked from the dataset. The selection of global best particles (*gbest*) from initial population is the concern of the initial steps. The quality of global best particles are determined using multi-objective fitness functions. Another interest of our study is to find the quality of personal best particles (*pbest*) using this multi-objective fitness functions. After the each iteration *pbest* and *gbest* value is updated according to the fitness functions. The process is repeated for certain iterations. Figure 1 shows the architecture of multi-objective particle swarm optimization.

There are different types of fitness functions for specified problems. The fitness function is based on the problem being optimized (dimension). It can be represented as $f(x_i)$. Fitness function describes how well the particle x_i in the multi-dimension space. Initially the positions and velocity is updated. The particle positions are represented as $x_i(x_{i,0} \ldots x_{i,d})$ and the velocity denoted as $v_i(v_{i,0} \ldots v_{i,d})$. The position and velocity is updated using Eqs. 3 and 4.

$$x_{i,d}(it+1) = x_{i,d}(it) + v_{i,d}(it+1) \tag{3}$$

Fig. 1 Proposed architecture for multi-objective particle swarm optimization

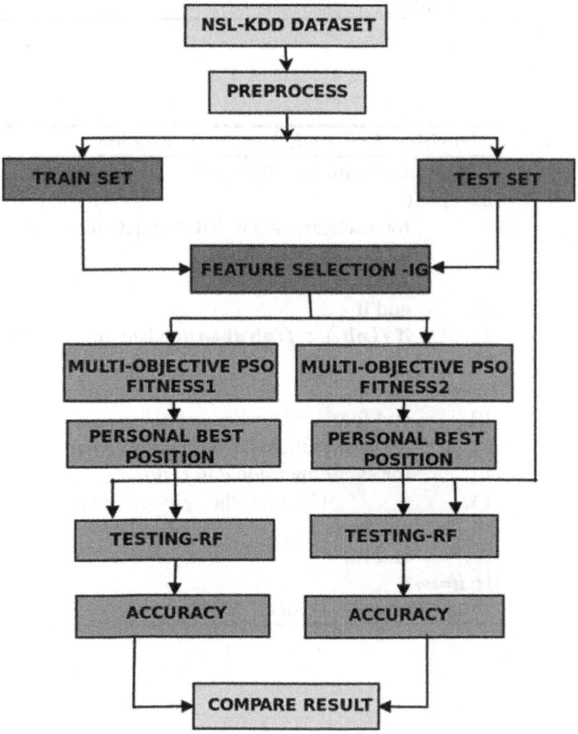

$$v_{i,d}(it+1) = \omega v_{i,d}(it) + C_1 Rnd(0,1)[pb_{i,d}(it) - x_{i,d}(it)] + C_2 Rnd(0,1)[gb_{i,d}(it) - x_{i,d}(it)] \quad (4)$$

where i is the index of particle, d is the considered dimension and it is the iteration number. $x_{i,d}$ and $v_{i,d}$ are the position and velocity of particle i in dimension d. ω is the inertia weight and set as 0.4. C_1 and C_2 are the acceleration constants for cognitive and social component and both are set to 0.5. Rnd is a random value between 0 and 1. $pb_{i,d}$ is the location in dimension d with fitness of all visited location in that dimension of particle.

Algorithm1: Initialize

```
1.for eachparticle i in Sdo
2.     for each dimension d in Ddo//initialize all particle's position and velocity
3.       Initially set the dataset values as position.
4.       v_{i,d} = Rnd[0,1]
5.end for
6.//initialize Particle best position
7.pb_i = x_i
8.//update the global best position
9.if f(pb_i) < f(gb)then
10.      gb = pb_i
11.end if
12. end for
```

Algorithm 2 : Particle Swarm optimization

```
1. Initialize//initialize all particles
2. repeat
3.        for eachparticle i in S do//update the particle's best position
4.        iff(x_i) < f(pb_i)then
5.        pb_i = x_i
6.        end if
7.        if f(pb_i) < f(gb)then//update the global best position
8.        gb = pb_i
9.        end if
10.       end for
11.       for each particle i in S do//update particles velocity and position
12.       for each dimension d in D do
13.                 Calculate the velocity and position.
14        end for
15.       end for
16.it=it+1
17.untilit>MAX-ITERAT IONS
```

Equation 4 is the sum of multiple components. First term of equation, $(\omega v_{i,d}(it))$ is the momentum component, which is the previous velocity. Second term $(C_1 Rnd(0,1)[pb_{i,d}(it) - x_{i,d}(it)]$ is the cognitive component which modifies the particle current distance to the best position that ever visited. At last, the third component $(C_2 Rnd(0,1)[gb_{i,d}(it) - x_{i,d}(it)]$ is social component, it depends the distance of the particle to the best position. Pseudo code for particle swarm optimization is given above.

4.4 Fitness Function [2]

The strength of algorithm depends on its fitness value. Where n is the extracted feature using feature selection method. We considered two different fitness functions,

Fitness 1 [2]:

$$f_1 = \frac{1}{n}\sum_{i=1}^{n} x_i^2 \tag{5}$$

$$f_2 = \frac{1}{n}\sum_{i=1}^{n} (x_i - 2)^2 \tag{6}$$

Fitness 2 [2]:

$$f_1 = x_1 \tag{7}$$

$$g = 1 - \frac{9}{n-1}\sum_{i=2}^{n} x_i \tag{8}$$

$$f_2 = g(1 - (\mathrm{f}_1/g)^2) \tag{9}$$

4.5 Weighted Aggregation Method

Weighted aggregation method [2] is used as a multi-objective function. It is the most frequently used technique. It is integrated with the weighted functions.

$$F = \sum_{i=1}^{k} w_i f_i(\mathrm{x}) \tag{10}$$

where, f is the fitness function, w_i is the weight for i = 1, 2, … k which is non-negative weights. It is defined as,

$$\sum_{i=1}^{k} w_i = 1 \tag{11}$$

Weighted aggregation method is demonstrated using Bang-Bang Weighted Approach (BWA) [2]. The weights are measured using Eqs. 12 and 13.

$$w_1(t) = sign(\sin(2\pi t/F)) \tag{12}$$

$$w_2(t) = 1 - w_1(t) \tag{13}$$

where, t is the iteration number and F is the change in frequency which is always set as 100.

4.6 Classification

The process of predicting the output of an unknown sample is known as classification. Classification is obtained from the developed classification model. Classification algorithm used for this experiment is Random Forest [10] implemented in WEKA [11]. Random forest [10] is an ensemble classifier that contains many decision trees and outputs the class that is the mode of the class's output by individual trees. The method is generated on collection of decision trees with controlled variations by using a combination of bagging idea and the random selection of features.

5 Experiments and Result

The proposed system is designed and validated using NSL KDD dataset [9]. Experiment is conducted with variable particle size of (40, 50, 75, 100, 125 … 300) and iterations (50, 100, 150 … 300). Features are ranked with respect to their gain value and feature with higher entropy are selected. Thus 25 robust features are extracted from the dataset. The dataset is normalized using min-max normalization and multi-objective PSO is applied to it. We have used two set of fitness functions. Finally the model is tested using WEKA [11] with Random Forest (RF) classifier [10].

5.1 Performance Measure

The performance of classification is measured in terms of accuracy, True Positive Rate and False Positive Rate. They are computed as follows,

$$Accuracy\ (Acc) = TP + TN/(TP + FN + TN + FP) \tag{14}$$

$$True\ Positive\ Rate\ (TPR) = TP/(TP + FN) \tag{15}$$

$$False\ Positive\ Rate\ (FPR) = FP/(TN + FP) \tag{16}$$

TP is the number of normal samples classified as normal, *FN* is the misclassified instances of the normal class, *TN* is correctly classified instances of abnormal class, and *FP* is the number misclassified instance of abnormal class.

5.2 Results

Experiment using multi-objective PSO using fitness 1: Table 2 shows the detection rate of fitness function 1. From the table we depict that the efficiency of model is based on the ability of a fitness function. It is tabulated from the table that accuracy is higher in 175 particles at 150 iterations. The highest accuracy obtained is 90.96 %. The detection time for this experiment is 0.19 s.

We cannot rely on accuracy to determine performance of an algorithm as it gives overall accuracy. Thus TPR, FPR is considered for measuring the performance. Good IDS have always high TPR and low FPR. It gives the rate of how many

Table 2 Percentage of accuracy obtained in multi-objective functions using fitness 1

Particles	Iteration						
	50	100	150	200	250	300	350
40	84.38	80.22	78.73	80.88	77.23	79.96	81.54
50	75.11	79.75	78.83	83.04	82.10	87.17	84.71
75	85.53	82.78	81.52	78.33	84.22	84.58	89.78
100	83.69	81.07	83.28	88.20	82.62	80.80	83.31
125	88.80	84.23	90.39	88.31	85.63	89.05	88.10
150	84.84	87.89	86.05	87.19	86.61	85.28	86.06
175	87.67	88.62	**90.96**	87.46	88.02	90.03	87.19
200	86.40	89.46	86.74	88.10	85.67	88.34	87.48
225	90.41	85.49	85.63	88.03	90.27	87.47	89.89
250	89.82	88.80	89.25	87.83	90.93	87.90	88.91
275	90.74	90.79	90.18	87.73	89.47	87.81	88.76
300	89.18	87.04	89.40	89.63	89.46	87.40	90.76

Table 3 Percentage of accuracy obtained in multi-objective functions using fitness 2

Particles	Iteration						
	50	100	150	200	250	300	350
40	96.08	96.11	92.58	96.15	95.67	95.44	94.77
50	96.35	95.38	92.54	96.10	96.07	96.87	93.96
75	96.80	95.74	95.40	96.35	92.88	96.54	95.87
100	96.43	96.60	95.89	95.79	96.41	96.42	96.59
125	96.53	95.98	96.07	96.70	96.45	96.81	96.63
150	96.17	96.21	96.35	96.92	96.73	96.05	96.54
175	96.59	97.04	95.57	97.11	96.70	96.60	96.48
200	96.59	96.72	96.76	96.47	96.58	96.78	96.19
225	96.79	96.79	96.60	96.98	96.02	96.44	96.01
250	96.78	96.75	96.98	97.22	97.28	96.32	96.57
275	97.26	96.03	96.61	96.97	96.97	**97.54**	96.98
300	96.61	96.73	96.81	96.67	96.55	96.59	96.32

normal samples are classified as normal and how many abnormal classes are misclassified as normal. From Table 2, good particles reached in 175 and calculated TPR, FPR rate. From the experiments a high TPR rate of 0.78 and low FPR of 0.22 is established.

Experiment using multi-objective PSO using fitness 2: From the Table 3 we observed that the detection rate is high compared to the fitness 1. In this experiment we obtained that the detection rate is high in 275 particles at 300 iterations. The accuracy depicted is 97.54 %. The detection time measured as 0.20 s. This gives that fitness 2 is efficient for detection.

Detection rate is high in 275 particles form Table 3. High TPR of 0.90 and low FPR of 0.10 is achieved by this particle. From the study, we analyzed that performance of fitness 2 is higher than fitness 1.

6 Inference

The following are the inferences obtained from our study,

- The efficiency of a particle swarm optimization is based on the fitness function.
- The fitness function plays a vital role in intrusion detection system.
- Particle swarm optimization using multi-objective functions can detect the attack in better accuracy of 97.54 % in a particle size of 275 with 300 iterations.
- Perfect IDS have high TPR and low FPR rate.

7 Conclusion

In this paper, we propose an intrusion detection system using particle swarm optimization. We have used multi-objective functions for improving the detection rate. The study compares two different multi-objective functions. Weighted aggregation method is considered as multi-objective function. Relevant features are extracted using information gain method. From the study we observed that the detection rate is higher in fitness 2. Overall accuracy of 97.54 % is obtained for our proposed method in a particle size of 275 at 300 iterations. The detection time is 0.20 s.

References

1. Bratton, D., Kennedy, J.: Defining a standard for particle swarm optimization. In: IEEE Swarm Intelligence Symposium (2007). No. 1-4244-0708
2. Parsopoulos, K.E., Vrahatis, M.N.: Particle Swarm Optimization Method in Multiobjective Problems. ACM (2003). No. 1-58113-445
3. Zhou, L.-H., Liu, Y.-H., Chen, G.-L.: A feature selection algorithm to intrusion detection based on cloud model and multi-objective particle swarm optimization. In: IEEE, Fourth International Computational Intelligence and Design, pp. 182–185 (2011). 978-1-4577-1085-8
4. Malik, A.J., Khan, F.A.: A hybrid technique using multi-objective particle swarm optimization and random forest for probe attacks detection in a network, pp. 2473–2478. In: IEEE International Conference on Systems, Man and Cybernetics (2013)
5. Parsopoulos, K.E., Vrahatis, M.N.: Particle Swarm Optimization Method for Constrained Optimization Problems, LNAI5212, pp. 188–203. Springer, Berlin (2009)
6. Kennedy, J., Eberhart, J.: Particle Swarm Optimization. IEEE (1995). 0-7803-2768
7. Tian, W.J., Liu, J.C.: Network Intrusion Detection Analysis with Neural Network and Particle Swarm Optimization Algorithm. IEEE (2010). ISBN 978-1-4244-5182-1
8. Aljarah, I., Ludwig, S.A.: Map reduce intrusion detection system based on a particle swarm optimization clustering algorithm. In: IEEE Congress on Evolutionary Computation (2013). ISBN 978-1-4799-0454
9. NSL-KDD intrusion detection dataset. Available on http://iscx.ca/NSLKDD/ (2009). Last accessed on 10 Feb 2014
10. Liaw, A., Wiener, M.: Classification and regression by random forest. R News **2/3**, 18–22 (2002)
11. Hall, M., Frank, E., Holmes, G., Pfahringer, B., Reutemann, P., Witten, I.H.: The WEKA data mining software: an update. SIGKDD Explor. **11**(1) (2009)

Vision Based Traffic Personnel Hand Gesture Recognition Using Tree Based Classifiers

R. Sathya and M. Kalaiselvi Geetha

Abstract Human hand gestures can be used as an important communication tool for human computer interaction. As a scientific discipline, computer vision is concerned with the theory behind artificial systems that extract information from images. This paper presents a novel and efficient framework for traffic personnel gesture recognition based on Cumulative Block Intensity Vector (CBIV) of *n*-frame cumulative difference. The experiment carried out on the real time traffic personnel action dataset using Random Forests (RF) and Decision Tree (J48). Experimental results denote the higher performance 97.83 % of the Random Forests classification, compared to the Decision Tree using 5-frame cumulative difference. The main contribution of this paper is the application of incremental tree based classifier techniques to the problem of identification of traffic personnel hand signals in video surveillance, based only on person hand movement.

Keywords Gesture recognition · Video surveillance · Traffic hand signals · Random forests · Decision tree (J48)

1 Introduction

Human action recognition in video is an important topic in computer vision applications such as human computer interaction, automated surveillance, etc., First investigations about this topic began in the seventies with pioneering studies accomplished by Johansson [1].

R. Sathya (✉) · M.K. Geetha
Department of Computer Science and Engineering, Annamalai University, Chidambaram, Tamil Nadu, India
e-mail: rsathyamephd@gmail.com

M.K. Geetha
e-mail: geesiv@gmail.com

L.C. Jain et al. (eds.), *Computational Intelligence in Data Mining - Volume 2*,
Smart Innovation, Systems and Technologies 32, DOI 10.1007/978-81-322-2208-8_18

Naturalistic and intuitiveness of the hand gesture has been a great motivating factor for the researchers in the area of Human Computer Interaction (HCI) to put their efforts to research and develop the more promising means of interaction between human and computers. This paper concentrates on Indian traffic personnel hand gesture recognition. Earlier studies [2–6] on gesture-based interfaces have focused on improving gesture recognition technology. Traffic management on roadway is a challenging task which is increasingly being augmented with automated system.

1.1 Related Work

Human gesture recognition is an active topic in computer vision technique. As established in [7], background subtraction is the more accurate method for traffic monitoring and it is widely exploited in many other applications. In [8], the Regions of Interest (ROI) obtained from color-based segmentation are classified using the Histogram of Oriented Gradients (HOG) feature. Obtaining hand gestures with the Nearest Neighbor (NN) classification has proven to be a promising approach when dealing with depth data [9]. However, recent work uses features that are not specifically designed for depth data. Ye et al. [10] proposed a stratified sampling method to select the feature subspaces for random forests with high dimensional data. In [11] block based human model for real time monitoring. Activity recognition approach proposed in [12, 13] extracted motion information from the difference image.

The rest of this paper is organized as follows. Section 2 discusses the Indian traffic signals. Section 3 describes a proposed approach. Section 4 describes data classification methods. Experimental results are discussed in Sect. 5. Finally, Sect. 6 concludes the paper.

2 Indian Traffic Personnel Signals

In a human traffic control environment, drivers must follow the directions given by the traffic personnel officer in the form of human body gestures. The twelve traffic personnel hand signals are listed as follows, to start one side vehicles, to stop vehicles coming from front, to stop vehicles approaching from back, to stop vehicles approaching simultaneously from front and back, to stop vehicles approaching simultaneously from right and left, to start vehicle approaching from left, to start vehicles coming from right, to change sign, to start one side vehicles, to start vehicles on T-point, to give VIP salute, to manage vehicles on T-point.

3 Proposed Approach

The overall block diagram of the proposed approach is shown in Fig. 1. The input videos are processed at 25 frames per second. The video sequence is converted into frames in .jpg format. To begin with the first frame is compared with the consecutive frames to compute frame differencing followed by the extraction of n-frame cumulative frame difference. ROI is extracted from the video sequence and Cumulative Block Intensity Vector (CBIV) features are extracted. The extracted feature is fed to the tree based classifier for hand gesture recognition. Real time traffic personnel actions are used for experimental purpose as discussed in Sect. 5.

3.1 *Frame Differencing*

Motion information in a video sequence is extracted by pixel-wise differencing of consecutive frames. Figure 2 shows the two consecutive frames and their motion information. Motion information T_k or difference image is calculated using

$$T_k(i,j) = \begin{cases} 1, & \text{if } D_k(i,j) > t; \\ 0, & \text{otherwise}; \end{cases} \tag{1}$$

where D_k is the difference image calculated as,

$$D_k(i,j) = |I_k(i,j) - I_{k+1}(i,j)| \quad 1 \leq i \leq w,\ 1 \leq j \leq h \tag{2}$$

where $I_k(i, j)$ is the intensity value of the pixel (i, j) in the kth frame, t is the threshold, w and h are the width and height of the image respectively. The value of $t = 30$ is used in the experiments.

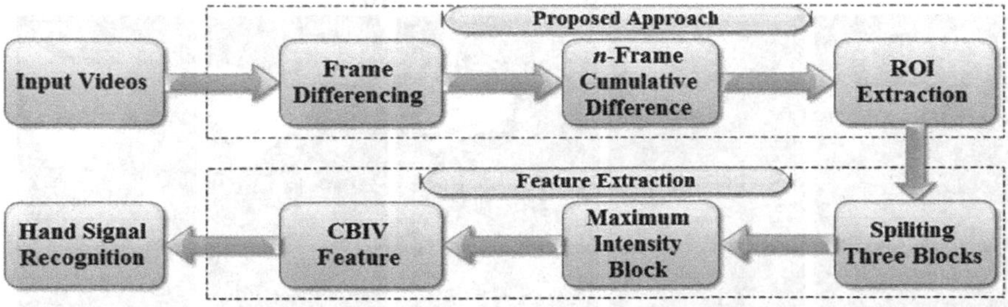

Fig. 1 Block diagram of the proposed approach

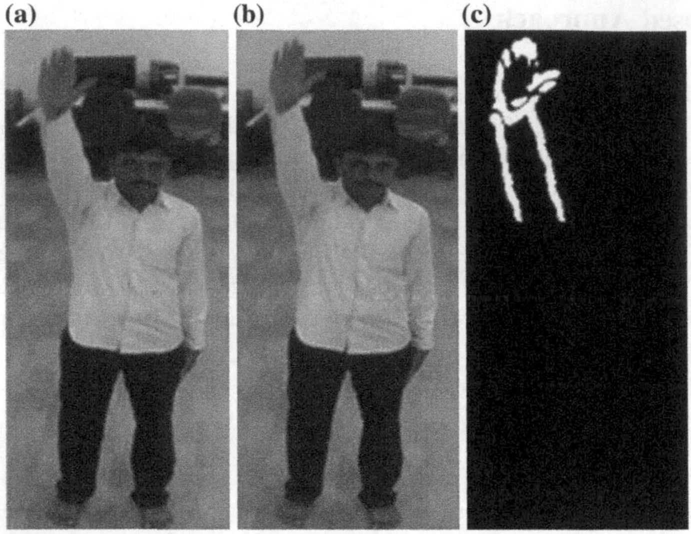

Fig. 2 **a**, **b** Two consecutive frames. **c** Motion information of **a** and **b**

3.2 n-Frame Cumulative Differencing

For identifying the region showing maximum intensity, *n*-frame cumulative differencing is applied. Figure 3a shows 3-frame cumulative difference image. Figure 3b shows 4-frame cumulative difference image. Figure 3c shows 5-frame cumulative difference image. Figure 3d shows 7-frame cumulative difference image. Figure 3e shows 10-frame cumulative difference image.

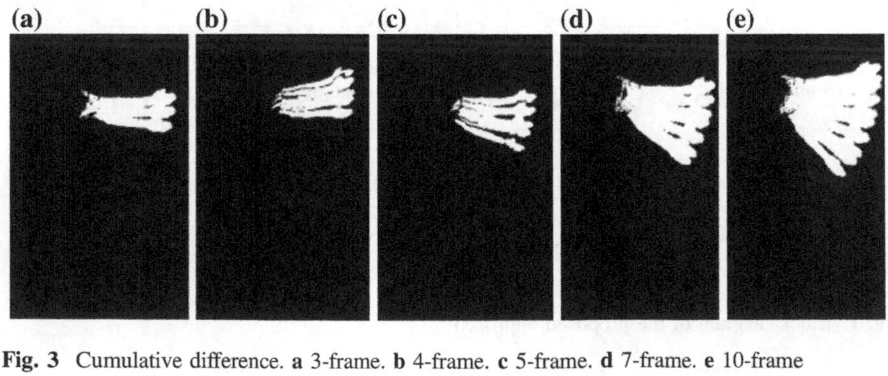

Fig. 3 Cumulative difference. **a** 3-frame. **b** 4-frame. **c** 5-frame. **d** 7-frame. **e** 10-frame

Consecutive difference images are calculated as follows:

$$
\begin{aligned}
D_n(x,y) &= I_p(x,y) - I_{p+1}(x,y) \\
D_{n+1}(x,y) &= I_{p+1}(x,y) - I_{p+2}(x,y) \\
D_{n+2}(x,y) &= I_{p+2}(x,y) - I_{p+3}(x,y) \\
&\cdots\cdots\cdots\cdots\cdots\cdots \\
D_{n+k} &= I_{p+k}(x,y) - I_{p+k+1}(x,y)
\end{aligned} \tag{3}
$$

3.3 ROI Extraction

Once the foreground image is extracted, the next step is to identify the ROI for further analysis. For ROI extraction, the approach used in [14] is utilized. The height of the bounding box $H(t)$ for ROI extraction is calculated using $Height_{(t)} = H_{(t)}/H_{(max)}$, where $H_{(t)}$ is the height of the bounding box in the video frame at time '*t*', H(max) is the maximum value that $H_{(t)}$ has for the entire video sequence. Width of the bounding box is fixed similarity using $Width_{(t)} = W_{(t)}/W_{(max)}$. Finally, ROI is extracted as $ROI = Height_{(t)}/Weight_{(t)}$. For the purpose of the uniformity, the ROI region is considered to be of size 60 × 40 for all actions without any loss in information.

3.4 Cumulative Block Intensity Vector (CBIV)

A novel and efficient feature called Cumulative Block Intensity Vector (CBIV) is proposed in this work. The extracted ROI as discussed in Sect. 3.3 is identified as motion regions. So, in order to minimize computation, the work divided the ROI into three regions. The average intensity of pixels in these regions R1, R2 and R3 are calculated. For all the actions, it is seen that maximum motion is identified either in the R1 or in R2 blocks. Hence in order to minimize computation, only the blocks showing maximum movement are considered for further analysis. The approach uses ROI of size 60 × 40 with each region of size 20 × 40. So, for further analysis the region having maximum intensity of size 20 × 40 only is utilized. This region is further divided into 5 × 5 block for further analysis.

4 Data Classification Methods

Classification is the last stage in the recognition system. Tree based classifiers are one of the much known classifier. The random forests (RF) and Decision Trees (DTs) are utilized to discriminate the traffic personnel hand gestures in this paper.

4.1 Random Forests (RF)

A Random Forests (RF) is a classifier consisting of a collection of tree-structured classifiers first proposed by Ho [15] and further developed by Breiman [16]. It is an effective tool in prediction. It is a combination of tree predictors such that each tree depends on the values of a random vector sampled independently and with the same distribution for all trees in the forest. All trees in the forests are unpruned. RF takes advantages of two powerful machine learning techniques: bagging [17] and random feature selection. The theoretical and practical performance of ensemble classifiers is well document [18, 19] used a novel online-adaptation method to allow random forests adapt to non-stationary financial time series, while [20] demonstrated the random forest algorithm's ability to select features for trend prediction in stock prices.

4.2 Decision Tree (J48)

Decision trees are commonly used methods for pattern classification. Decision tree is a common and intuitive approach to classify a pattern through a sequence of questions in which the next question is depends upon the answer to the current question. Decision tree analysis is a formal, structured approach to making decisions. Many algorithms are available for constructing decision tree models, such as classification and regression tree (CART) stands in [21], Iterative Dichotomiser 3 (ID3) developed by Quinlan [22], Quick, unbiased, efficient, and statistical tree (QUEST) in [23], Chi-squared automatic integration detection (CHAID) introduced in [24] and classifier 4.5 (C4.5, J48) in [25]. In this study, J48 algorithm decision tree were applied to Traffic personnel hand features. J48 classifier is a standard model in C4.5 decision tree for supervised classification.

5 Experimental Results

In this section, proposed method is evaluated using real time traffic hand action datasets. The experiments carried out in C++ with OpenCV 2.2 in Ubuntu 12.04 operating system on a computer with Intel CORE^{TM} I5 processor 2.30 GHz with 4 GB RAMS. The obtained proposed features are fed to tree based classifiers such as, Random Forests and Decision Tree (J48) using open source machine learning tool. WEKA [26] tool to develop the model for each action and these models are used to test the performance, classification trees are used for experimental purpose.

5.1 Dataset

Action recognition has become a very important topic in computer vision. In real time traffic personnel datasets twelve actions. The surveillance system, a surveillance camera mounted on the wall is employed to collect the traffic action videos. They are performed by eight actors. In total, the data consists of 192 video samples. The datasets are shown in Fig. 4. Video clips are at 25 fps. The backgrounds are relatively static. In this work, the samples are taken into a training set of 6 persons at 360 min video and testing set of two persons at 30 min video.

5.2 Evaluation Criteria

To compute accuracy, F-measure is defined as,

$$\text{F-measure} = \frac{2\text{PR}}{(\text{P}+\text{R})} \tag{4}$$

where, P and R are precision and recall. The F-measure computes some average of the information retrieval precision and recall metrics. P is the proportion of the predicted positive cases that were correct, as calculated using the equation:

$$\text{Precision} = \frac{\text{TP}}{(\text{TP}+\text{TR})} \tag{5}$$

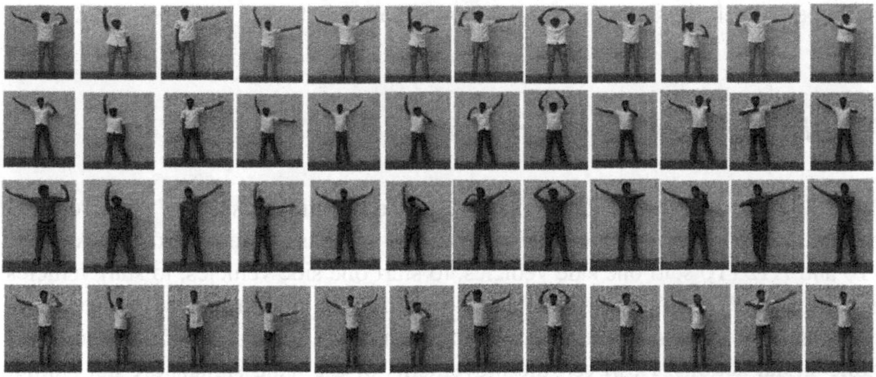

Fig. 4 Indian traffic personnel dataset. *A1* To start one side vehicles, *A2* To stop vehicles coming from *front*, *A3* To stop vehicles approaching from *back*, *A4* To stop vehicles approaching simultaneously from *front* and *back*, *A5* To stop vehicles approaching simultaneously from *right* and *left*, *A6* To start vehicles approaching from *left*, *A7* To start vehicles coming from *right*, *A8* To change sign, *A9* To start one side vehicles, *A10* To start vehicles on T-point, *A11* To give VIP salute *A12* To manage vehicles on T-point

where, TP and FP are True Positive and False Positive. Recall (R) or Sensitivity or True Positive Rate (TPR) is the proportion of positive cases that were correctly identified, as calculated using the equation:

$$\text{Recall} = \frac{\text{TP}}{(\text{TP} + \text{FN})} \tag{6}$$

where, TP and FN are True Positive and False Negative.

5.3 Performance of the Proposed Method

In this section presents hand action recognition problem using open source machine learning tool WEKA [25]. For this purpose, traffic personnel hand action datasets are used shown in Fig. 4. The performances of the classifiers were measured using 10-fold cross validation model. The performance evaluation F-measure for *n*-frame cumulative difference with 25 dimensional CBIV feature using Random Forests classifier is given in Table 1 and Decision Tree (J48) is given in Table 2. Table 3 gives the confusion matrix of the 5-frame cumulative difference using Random Forests algorithm.

Figure 5 shows the average value of the precision and recall of *n*-frame cumulative difference using Random Forests algorithm. Figure 6 shows the average value of the precision and recall of *n*-frame cumulative difference using Decision Tree (J48) algorithm. The proposed approach gives higher precision and recall average value of 5-frame cumulative difference with 25 dimension CBIV feature using Random Forests algorithm. There is no existing work done, related to Indian traffic personnel dataset with in this proposed approach.

5.4 Discussion on the Classification

Table 3 shows the confusion matrix of the recognition results of the proposed method. While observing the table most of the misclassifications exist in A1, A9, A10 and A12 (To start one side vehicles, to start one side vehicles, to start vehicles on T-point and to manage vehicles on T-point). Some misclassifications are due to the similarities of these gestures. For example in Fig. 7 shows the user is doing their traffic signal actions. However, the movements of the hand signals are not similar and the rotation of the hand position is moreover similar. An efficient traffic personnel hand gesture recognition method, by analyzing the misclassification samples find out which gestures is easy confused with others. The performances of these actions are below average.

Table 1 Gesture recognition results for *n*-frame cumulative difference using random forests algorithm (%)

	A1	A2	A3	A4	A5	A6	A7	A8	A9	A10	A11	A12
3-frame	72.2	99.4	96.2	99.2	99.2	98.6	97.9	99.0	71.0	77.8	72.7	91.5
4-frame	76.7	95.7	100	95.9	94.5	97.5	93.5	99.5	70.7	92.2	95.2	88.1
5-frame	**95.5**	**99.4**	**99.4**	**97.6**	**99.2**	**100**	**99.2**	**100**	**96.6**	**98.3**	**98.4**	**90.5**
7-frame	84.4	98.2	99.2	98.2	98.7	99.2	98.4	100	78.1	99.4	98.6	97.6
10-frame	78.8	100	95.6	100	94.8	96.1	97.5	97.0	74.3	87.6	97.7	60.1

Table 2 Gesture recognition results for *n*-frame cumulative difference using decision tree (J48) algorithm (%)

	A1	A2	A3	A4	A5	A6	A7	A8	A9	A10	A11	A12
3-frame	65.2	100	93.3	91.6	91.5	94.6	94.0	97.0	61.2	78.8	80.8	86.1
4-frame	76.6	99.1	98.7	98.4	97.9	98.0	98.0	100	64.6	65.0	98.9	50.6
5-frame	**91.2**	**91.8**	**99.4**	**94.2**	**98.4**	**97.7**	**97.0**	**98.0**	**92.7**	**96.6**	**97.9**	**90.1**
7-frame	79.7	97.8	99.4	98.6	96.7	98.1	96.5	99.3	71.9	93.3	97.7	86.9
10-frame	17.0	95.6	99.4	92.7	79.3	93.1	76.5	99.0	50.0	63.7	85.7	49.1

Table 3 Gesture recognition results at 5-frame cumulative difference for traffic personnel dataset using random forests algorithm—confusion matrix (97.83 %)

	A1	A2	A3	A4	A5	A6	A7	A8	A9	A10	A11	A12
A1	**98.76**	0	0	0.62	0	0	0	0	0	0	0	0.62
A2	0	**98.82**	1.2	0	0	0	0	0	0	0	0	0
A3	0	0	**100**	0	0	0	0	0	0	0	0	0
A4	1.18	0	0	**97.64**	0	0	0	0	0	1.18	0	0
A5	0	0	0	0	**100**	0	0	0	0	0	0	0
A6	0	0	0	0	0	**100**	0	0	0	0	0	0
A7	1.69	0	0	0	0	0	**98.31**	0	0	0	0	0
A8	0	0	0	0	0	0	0	**100**	0	0	0	0
A9	4.5	0	0	0	0	0	0	0	**95.5**	0	0	0
A10	0	0	0	0	0	0	0	0	0	**100**	0	0
A11	1.05	0	0	1.05	0	0	0	0	0	0	**97.9**	0
A12	8.35	0	0	0	1.4	0	0	0	2.78	1.39	0	**86.11**

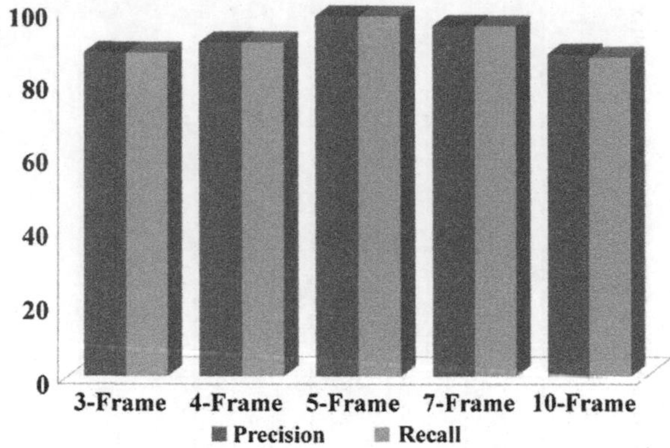

Fig. 5 Average value of *precision* and *recall* value of all actions using random forests

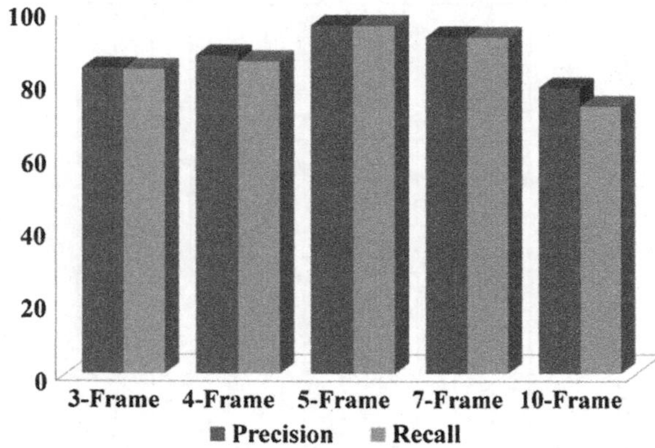

Fig. 6 Average value of *precision* and *recall* value of all actions using (J48)

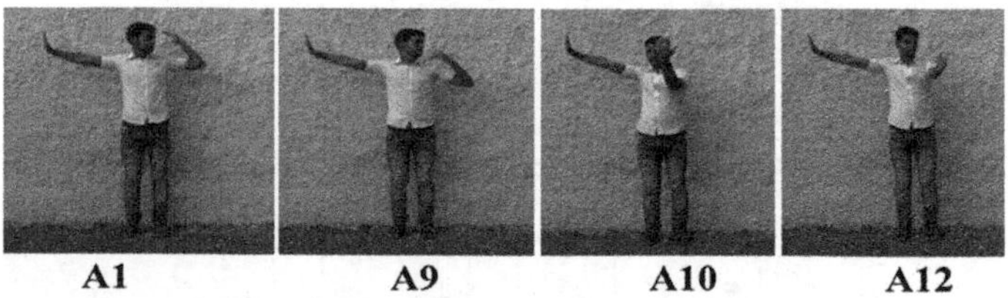

Fig. 7 An example in which proposed method misclassification to recognize the gestures

6 Conclusion and Future Work

This work presented for traffic personnel action recognition for traffic surveillance using Cumulative Block Intensity Vector (CBIV) as feature. Indian traffic personnel performs 12 actions are taken performing the experiment. The ROI extracted from the various cumulative frame difference images are used for classification. These features evaluate the performance using tree based Random forests and decision tree (J48). This approach gives good classification accuracy of 97.83 % for 5-frame cumulative difference with 25 dimension CBIV feature using Random Forests algorithm. And 95.22 % of accuracy for 5-frame cumulative difference with 25 dimension CBIV feature using Decision Tree (J48) algorithm. Future work, intend to enhance the flexibility of this approach by under complex environment.

References

1. Johansson, G.: Visual perception of biological motion and a model for its analysis. Atten. Percept. Psychophys. **14**, 201–211 (1973). doi:10.3758/BF03212378
2. Kang, B., Choi, E., Kwon, S., Chung, M.K.: UFO-zoom: a new coupled map navigation technique using hand trajectories in the air. Int. J. Ind. Ergon. **43**(1), 62–69 (2013)
3. Chaquet, J.M., Carmona, E.J., Fernandez-Caballero, A.: A survey of video datasets for human action and activity recognition. J. Comput. Vis. Underst. **117**(6), 633–659 (2013)
4. Raheja, J.L., Chaudhary, A., Maheshwari, S.: Hand gesture pointing location detection. Optik **125**, 993–996 (2014)
5. Suau, X., Alcoverro, M., López-Méndez, A., Ruiz-Hidalgo, J., Casas, J.R.: Real-time fingertip localization conditioned on hand gesture classification. Image Vis. Comput. **32**, 522–532 (2014)
6. Chaudhary, A., Raheja, J.L., Das, K., Raheja, S.: Intelligent approaches to interact with machines using hand gesture recognition in natural way: a survey. Int. J. Comput. Sci. Eng. Surv. **2**, 111–122 (2011)
7. Mandellos, N.A., Keramitsoglou, I., Kiranoudis, C.T.: A background subtraction algorithm for detecting and tracking vehicle. Expert Syst. 1619–1631 (2011)
8. Qingsong, X., Juan, S., Tiantian, L.: A detection and recognition method for prohibition traffic signs. In: International Conference on Image Analysis and Signal Processing (IASP), 2010, pp. 583–586. IEEE (2010)
9. Ren, Z., Yuan, J., Zhang, Z.: Robust hand gesture recognition based on finger-earth movers distance with a commodity depth camera. In: ACM-MM, pp. 1093–1096 (2011)
10. Ye, Y., Wu, Q., Zhexue Huang, J., Ng, M.K., Li, X.: Stratified sampling for feature subspace selection in random forests for high dimensional data. Pattern Recognit. **46**, 769–787 (2013)
11. Chaaraoui, A.A., Climent-acrez, P., Flarez-Revuelta, F.: A review on vision techniques applied to Human behaviour analysis for ambient-assisted living. Expert Syst. Appl. 10873–10888 (2012)
12. Sathya, R., Geetha, M.K.: Human action recognition to understand hand signals for traffic surveillance. Elixir Int. J. Comput. Sci. Eng. **63**, 18149–18156 (2013)
13. Arunnehru, J., Geetha, M.K.: Motion intensity code for action recognition in video using PCA and SVM. In: Mining Intelligence and Knowledge Exploration, pp. 70–81. Springer International Publishing, Berlin (2013)
14. Liu, Q., Zhuang, J., Ma, J.: Robust and fast pedestrian detection method for far infrared automotive driving assistance systems. Infrared Phys. Technol. **60**, 288–299 (2013)

15. Ho, T.K.: Random decision forests. In Proceedings of the 3rd International Conference on Document Analysis and Recognition, pp. 278–282 (1995)
16. Breiman, L.: Random forests. Mach. Learn. **45**, 5–32 (2001)
17. Breiman, L.: Bagging predictors, machine learning research: four current directions. AIM Mag. **18**, 97–136 (1997)
18. Zbikowski, K., Grzegorzewski, P.: Stock trading with random forests, trend detection tests and force index volume indicators. Artif. Intell. Soft Comput. 441–452 (2013)
19. Xu, Y., Li, Z., Luo, L.: A study on feature selection for trend prediction of stock trading price. In: 2013 International Conference on Computational and Information Sciences, pp. 579–582 (2013)
20. Han, J., Kamber, M.: Data mining; concepts and techniques. Morgan Kaufmann Publishers, Burlington (2000)
21. Svoray, T., Michailov, E., Cohen, A., Rokah, L., Sturm, A.: Predicting gully initiation: comparing data mining techniques, analytical hierarchy processes and the topographic threshold. Earth Surf. Process **37**, 607–619 (2011)
22. Quinlan, J.R.: Introduction of decision trees. Mach. Learn. **1**, 81–106 (1986)
23. Loh, W.Y., Shih, Y.S.: Split selection methods for classification trees. Stat. Sinica **7**, 815–840 (1997)
24. Kass, G.: An exploratory technique for investigating large quantities of categorical data. Appl. Stat. **29**(2), 119–127 (1980)
25. Quinlan, J.R.: C4.5: Programs for Machine Learning. Morgan Kaufmann, San Mateo (1993)
26. Witten, I.H., Frank, E.: Data Mining: Practical Machine Learning Tools and Techniques with Java Implementations. Morgan Kaufmann, Burlington (1999)

Optimization of the Investment Casting Process Using Genetic Algorithm

Sarojrani Pattnaik and Sutar Mihir Kumar

Abstract This paper presents a study in which an attempt has been made to improve the quality characteristic (surface finish) of the wax patterns used in the investment casting process. The wax blend consists of paraffin wax (20 %), carnauba wax (10 %), microcrystalline wax (20 %), polyethylene wax (10 %) and teraphenolic resin (40 %), which provided an improved pattern wax composition. The process parameters considered are injection temperature, holding time and die temperature. The injection process parameters are optimized by genetic algorithm. Further, verification test have been conducted at the obtained optimal setting of process parameters to prove the effectiveness of the method. Finally, a good agreement between the actual and the predicted results of surface roughness of the wax patterns has been found.

Keywords Investment casting · Wax pattern · Surface roughness · Genetic algorithm · Optimization

1 Introduction

The investment casting (IC) or lost wax process is a method of producing high quality precision castings [1]. Dimensional accuracy and excellent surface finish are the major advantages of this process. The major applications of the IC process are in the aircraft and aerospace industries, especially turbine blades and vanes cast in cobalt and nickel-base superalloys [2]. The major steps involved in the IC process are as follows: injection molding of a pattern, ceramic coating, dewaxing, drying and metal casting, followed by minor finishing operations. Wax, plastic, polystyrene or frozen

S. Pattnaik (✉) · S. Mihir Kumar
Department of Mechanical Engineering, VSSUT,
Burla, Sambalpur 768018, Odisha, India
e-mail: rani_saroj7@yahoo.co.in

S. Mihir Kumar
e-mail: mihirsutar05@gmail.com

L.C. Jain et al. (eds.), *Computational Intelligence in Data Mining - Volume 2*,
Smart Innovation, Systems and Technologies 32, DOI 10.1007/978-81-322-2208-8_19

mercury are the pattern materials. However, the wax is the most widely used pattern material for the IC process [3]. These waxes undergo shrinkage on solidification. Most of the researchers have worked on reducing the shrinkage properties of the wax patterns [4–6]. On the other hand, surface finish is also an important quality characteristic of the wax pattern, which cannot be neglected.

The modeling of manufacturing processes using artificial intelligence (AI) technique has been immensely increasing over the last few years. The main purpose of AI is to simulate the human behavior to predict the desired output. Artificial neural network (ANN), fuzzy logic (FL), genetic algorithm (GA), etc., are some basic areas of artificial intelligence. Now a day's, GA is widely used to solve a variety of optimization problems [7–9]. In the present study, an attempt has been made to optimize the wax injection process parameters using GA to obtain the optimum surface finish of a stepped rectangular wax pattern. The obtained optimal condition is verified by confirmatory experiments. The rest of the section is organized as follows: Sect. 2 provides an overview of GA. The experimental work is furnished in Sect. 3. Optimization using GA is described in Sect. 4. Results are presented in Sect. 5, followed by conclusion in Sect. 6.

2 Overview of Genetic Algorithm

As per Zhou and Sun [10], GA is a global optimization technique which is based on Darwinian biological evolution principle. The flowchart for performing simulations by GA is shown in Fig. 1. The solution of a problem that GA attempts to solve is coded into a string of binary numbers known as chromosomes. Each chromosome contains the information in a set of possible process parameters. Initially, a population of chromosomes is formed randomly. The fitness of each chromosome is then evaluated using an objective or fitness function after the chromosome has been decoded. Selected individuals are then reproduced, usually in pairs, through the application of genetic operators. The three operators of GA are reproduction, crossover, and mutation. These operators are applied to the pairs of individuals with a given probability and it results in new offspring. The offspring from reproduction are then further perturbed by mutation. These new individuals then make up the next generation. These processes of selection, reproduction and evaluation are repeated until some termination criteria are satisfied [11].

3 Experimental Work

3.1 Materials

The wax blend was made from base waxes such as petroleum waxes, natural waxes and synthetic waxes. The petroleum waxes were paraffin and micro-crystalline. Carnauba and polyethylene were the natural and synthetic waxes. Teraphenolioc

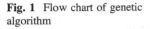

Fig. 1 Flow chart of genetic algorithm

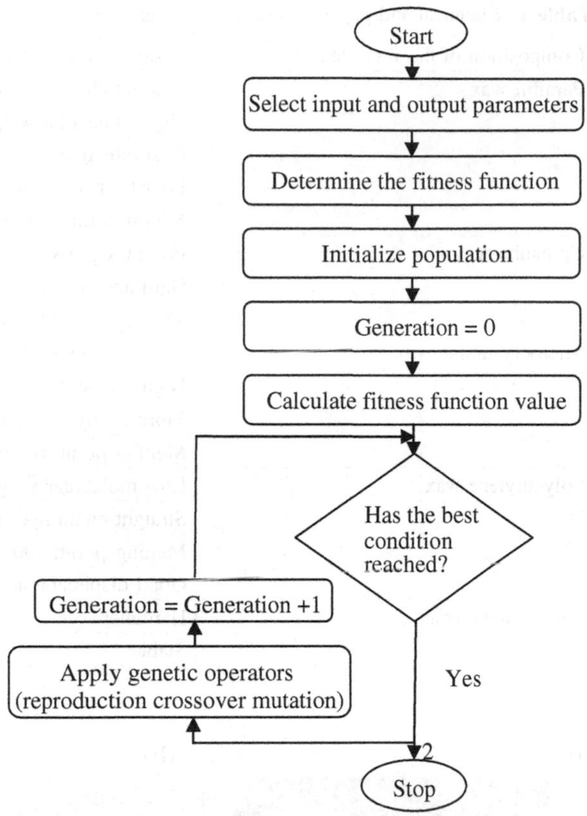

resin was added to the blend to increase the strength of the pattern wax composition. Chemical and physical properties of ingredients are presented in Table 1.

Different blends were made by varying the proportion of all the selected chemicals and wax patterns were made from each of the blends. It was found that the blend containing 20 % of paraffin wax, 10 % of carnauba wax, 20 % of microcrystalline wax, 10 % of polyethylene wax and 40 % of teraphenolic resin showed least shrinkage and surface roughness as compared to other blends and thus, the above mentioned blend was used for carrying out further experiments. The melting point of the blend was found to be 64 °C. Two-dimensional plot of surface roughness and the micro structure of the selected wax blend measured by Atomic force microscope (AFM) and scanning electron microscope (SEM) are depicted in Fig. 2 and an even surface finish and almost a uniform grain structure for the chosen wax blend could be clearly visualized from it.

An aluminum die (Fig. 3) was used for making the stepped wax patterns. The stepped design of the die was used as the wax patterns produced from it acts as a versatile tool for the accurate examination of their quality in terms of dimensional

Table 1 Chemical and physical properties of ingredients

Composition of the wax blend	Chemical and physical properties
Paraffin wax	Straight-chain hydrocarbon
	High molecular weight
	Cost-effective
	Poor heat resistance
	Melting point: 49–71 °C
Carnauba wax	Provides glossy and slippy surface
	Hard and brittle
	Melting point: 82–86 °C
Microcrystalline wax	Branch-chain hydrocarbon
	High molecular weight than paraffin wax
	More costly than paraffin wax
	Melting point: 60–89 °C
Polyethylene wax	Low molecular weight
	Straight-chain hydrocarbon
	Melting point: 100–110 °C
	Good dispersion and fluidity
Teraphenolic resin	Hardener
	Stable

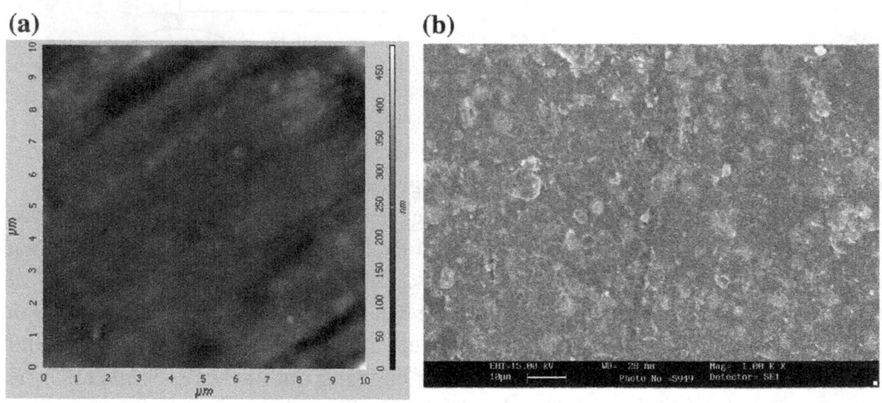

Fig. 2 Surface morphology of the selected wax blend **a** AFM image, **b** SEM image

accuracy and surface finish. Besides an assessment of wax pattern quality, the results of the investigations of the IC process made with a stepped pattern could serve in testing of new binders, ceramic materials, etc.

Fig. 3 An aluminum die used for making stepped wax patterns

3.2 Conduct of Experiments

The process parameters considered in this study were wax injection temperature, holding time and die temperature. Each selected process parameter was analyzed at three levels as shown in Table 2. The selection of parameters and their levels was based on literature review and some preliminary experiments conducted in the laboratory. The experiments were conducted as per Taguchi's L_9 orthogonal array (OA). For each experimental run, the experiments were replicated thrice and the results are furnished in Table 3.

Table 2 Process parameters and their values at different levels

Symbol	Process parameters	Unit	Range	Level 1	Level 2	Level 3
A	Injection temperature	°C	70–80	70	75	80
B	Holding time	min	40–50	40	45	50
C	Die temperature	°C	30–40	30	35	40

Table 3 L_9 OA

Run No.	Input process parameter						Surface roughness (nm)			
	A (°C)		B (min)		C (°C)		R1	R2	R3	Average
1	1	70	1	40	1	30	34.5098	38.0024	35.8084	36.10687
2	1	70	2	45	2	35	32.816	35.4724	38.4272	35.57187
3	1	70	3	50	3	40	46.8169	44.4139	46.7294	45.98673
4	2	75	1	40	2	35	39.1926	38.8481	38.6445	38.89507
5	2	75	2	45	3	40	58.5907	59.6365	59.3761	59.2011
6	2	75	3	50	1	30	62.6494	65.1735	63.9784	63.93377
7	3	80	1	40	3	40	73.4143	73.6743	72.9757	73.35477
8	3	80	2	45	1	30	81.5523	82.1853	82.5779	82.10517
9	3	80	3	50	2	35	86.5739	89.2759	87.3497	87.7332

4 Optimization Using GA

The optimization of the process parameters by GA begins with the construction of a fitness function and the fitness function used in this study is based on regression analysis. The regression model based on experimental data to predict the surface roughness of the wax patterns was developed using Minitab 14 software and it is given by Eq. 1.

$$\begin{aligned} y = 5625.62 - 43.6847x_1 + 20.1914x_2 - 189.49x_3 + 0.245325x_1^2 \\ - 0.0907954x_2^2 + 1.93620x_3^2 - 0.135743x_1x_2 + 0.195781x_1x_3 \end{aligned} \quad (1)$$

where, y is the surface roughness (nm) and x_1, x_2 and x_3 are injection temperature, holding time and die temperature, respectively.

The objective of the present study is to minimize the surface roughness of the wax patterns created by the IC process. Hence, the fitness function is as follows:

Minimize y, subject to constraints:

$$70 \leq x_1 \leq 80 \quad (2)$$

$$40 \leq x_2 \leq 50 \quad (3)$$

$$30 \leq x_3 \leq 40 \quad (4)$$

The various simulation parameters used for optimizing surface roughness using constrained GA for various combinations of process parameters are furnished in Table 4. The optimal condition for minimizing surface roughness as determined by GA is injection temperature at 70 °C i.e. A_1, holding time at 40 min i.e. B_1 and die temperature at 34.89 °C i.e. C_2 and the predicted optimal value of surface roughness is 27.4724 nm. Figure 4a–c shows the best fitness plot, current best individual plot and score diversity plot.

Best fitness plots the best function value in each generation verses iteration number and the optimal settings of process parameters is obtained for the best fitted value of surface roughness. The mean value of surface roughness is 27.4728 nm and the best fitted value is 27.4724 nm. Best individual plots the vector entries of the individual with the best fitness function value in each generation and it shows that injection temperature is the current best process parameter. Score diversity plots a histogram of the scores at each generation. The rationale behind the use of GA lies in the fact that GA has the capability to find the global optimal parameters.

Table 4 Simulation parameters used for optimization by GA

Population size	50
Probability of cross over	0.8
Probability of mutation	0.01
Maximum generations	100

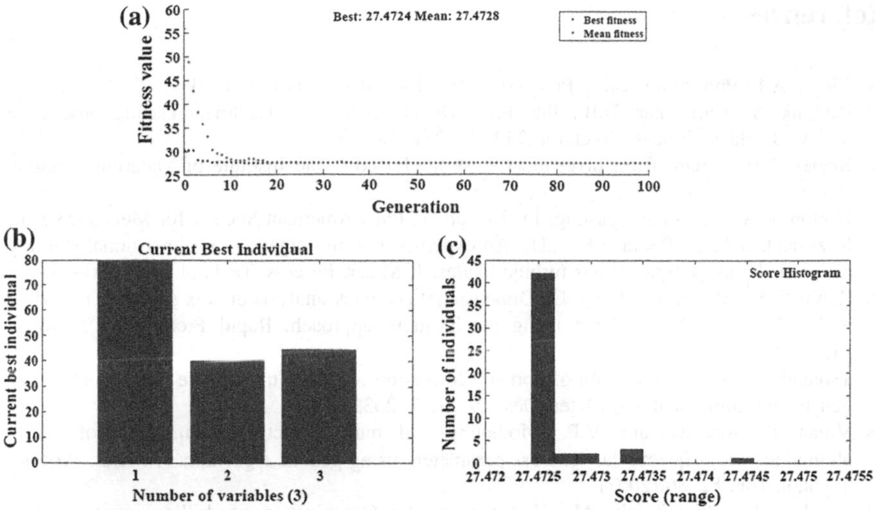

Fig. 4 **a** Best fitness plot; **b** current best individual plot; **c** score diversity plot

5 Confirmatory Experiments

As the predicted optimal condition by GA was not included in the experimental run, three additional experiments were performed at the same and the mean of the surface roughness values was found to be 28.8635 nm. Then, this actual experimental result was compared with the result predicted by GA. The comparison was based on mean absolute percentage error (MAPE) between them as given by Eq. (5) and it was found to be 4.8 %.

$$\text{MAPE} = \left| \left(\frac{MPI_{Exp} - MPI_{Taguchi}}{MPI_{Exp}} \right) \right| \times 100\,\% \tag{5}$$

6 Conclusions

In the present study, the optimization of the surface roughness of the wax patterns used in the IC process was done using GA. The error between the actual and the predicted result by GA was found to be less than 10 %, which shows that GA can be used successfully to solve real problems in the IC process. It may be adopted by the investment casting industries to increase their overall productivity.

References

1. Clegg, A.J.: Precision Casting Processes. Pergamon Press, Oxford (1991)
2. Pattnaik, S., Karunakar, D.B., Jha, P.K.: Developments in investment casting process: a review. J. Mater. Process. Technol. **212**, 2332–2348 (2012)
3. Beeley, P.R., Smart, R.F.: Investment Casting, 1st edn. The Institute of Materials, London (1995)
4. Horton, R.A.: Investment casting. In: Lyman, T. (ed.) American Society for Metals (1987)
5. Rezavand, S.A.M., Behravesh, A.H.: An experimental investigation on dimensional stability of injected wax patterns of gas turbine blades. J. Mater. Process. Technol. **182**, 580–587
6. Rahmati, S., Akbari, F., Barati, E.: Dimensional accuracy analysis of wax patterns created by RTV silicone rubber molding using the Taguchi approach. Rapid Prototyping J. **13**(2), 115–122
7. Tsoukalas, V.D.: Optimization of porosity formation in $AlSi_9Cu_3$ pressure die castings using genetic algorithm analysis. Mater. Des. **29**, 2027–2033 (2008)
8. Vijian, P., Arunachalam, V.P.: Modelling and multi objective optimization of LM24 aluminium alloy squeeze cast process parameters using genetic algorithm. J. Mater. Process. Technol. **186**, 82–86 (2007)
9. Kilickap, E., Huseyinoglu, M., Yardimeden, A.: Optimization of drilling parameters on surface roughness in drilling of AISI 1045 using response surface methodology and genetic algorithm. Int. J. Adv. Manuf. Technol. **52**, 79–88 (2011)
10. Zhou, M., Sun, S.D.: Genetic algorithms: theory and applications. National Defense Industry Press (2002)
11. Reddy, N.S.K., Rao, P.V.: A genetic algorithmic approach for optimization of surface roughness prediction model in dry milling. Mach. Sci. Technol. **9**, 63–84 (2005)

Cyclostationary Feature Detection Based Spectrum Sensing Technique of Cognitive Radio in Nakagami-*m* Fading Environment

Deborshi Ghosh and Srijibendu Bagchi

Abstract The main function of Cognitive Radio network is to sense the spectrum band to check whether the primary user is present or not in a given spectrum band at a place. One of the most efficient ways of spectrum sensing technique is cyclostationary feature detection. Though the computational complexity is very high in case of cyclostationary feature detection, still it is very effective in case of unknown level of noise. Here, spectral correlation function (SCF) of the received signal is determined. In this paper, SCF is calculated using Hanning, Hamming and Kaiser windows and also we have determined the cyclic periodogram of the input signal. In each case, we have considered Nakagami-*m* channel fading. Likelihood ratio test has been performed by varying the parameters of the channel fading distribution. Finally, the effects of the windows have been studied in the numerical section.

Keywords Cognitive radio (CR) · Cyclostationary feature detection · Likelihood ratio test · Nakagami-*m* fading · Spectral correlation function (SCF)

1 Introduction

A Cognitive Radio (CR) is a transceiver designed to use spectrum efficiently. A CR can be programmed and configured dynamically. It has the capability to automatically detect available channels in wireless spectrum and to change accordingly its transmission and reception parameters to allow more number of concurrent wireless communications in a given spectrum band at a place. This process is known as dynamic spectrum management [1]. A CR is basically a software defined radio with

D. Ghosh (✉) · S. Bagchi
RCC Institute of Information Technology, Canal South Road,
Beliaghata, Kolkata 700015, West Bengal, India
e-mail: ghoshdeborshi3@gmail.com

S. Bagchi
e-mail: srijibendu@gmail.com

L.C. Jain et al. (eds.), *Computational Intelligence in Data Mining - Volume 2*,
Smart Innovation, Systems and Technologies 32, DOI 10.1007/978-81-322-2208-8_20

a cognitive engine brain. According to the operator's command, cognitive engine can configure radio-system parameters such as waveform, protocol, operating frequency etc.

Using the limited natural resources efficiently is one of the greatest challenges of our society. The natural frequency spectrum is limited [2, 3] just like coal and petroleum and we have to use it more efficiently in order not to use up all. For this reason, CR is proposed to be a new technology for providing maximum satisfaction of user requirements [4].

Now, in case of CR, the changing of parameters is based on the active monitoring of external and internal radio environment such as network state, user behaviour state etc. The idea behind the CR paradigm is to utilize the idle frequency bands allocated to primary/licensed users by the secondary/unlicensed users without any interference to licensed users' communication. If we scan a portion of radio spectrum, we will find that some frequency bands are largely unoccupied most of the time, some bands are partially occupied and the only remaining bands are heavily used. CR aims to detect idle frequency bands in the spatial and frequency domain and allocate these bands to secondary users.

To detect unused spectrum and sharing it, without harmful interference with the other users; it is very important for a CR network to detect empty spectrum [5]. The most efficient way to detect empty spectrum is detecting the primary user. Now, for spectrum sensing technique of CR, we will apply cyclostationary feature detection method instead of energy Detection and matched Filter Detection.

In case of cyclostationary feature detection, we have determined the Spectral Correlation Function (SCF) using three types of window—Hanning, Hamming and Kaiser windows. Next, we have calculated the cyclic periodogram of the input signal. Now, with this cyclic periodogram and SCF, a test statistic has been formulated. Finally likelihood ratio test has been performed. For each case, we have taken all three types of window and Nakagami-m fading with four combinations of the distribution parameters m and σ and we have found twelve set of observations.

The remainder of this paper is organized as follows. In the next section, an overview of the cyclostationary feature detection has been presented. Section 3, discusses about the bi frequency plot using the three windows. In Sect. 4 Numerical results are given. Likelihood radio test has been performed here to infer about the presence of a primary user in a frequency band. Section 5 concludes the paper.

2 Cyclostationary Feature Detection

Cyclostationary feature detection [6–8] method deals with the inherent cyclostationary properties of a modulated signal based on the fact that the signals are generally coupled with sine wave carriers, repeating spreading, pulse trains or cyclic prefixes which result in periodicity and their statistics like mean and autocorrelation also exhibit periodicity in wide sense. This periodicity trend is used to identify the presence of primary users and that's why this method performs satisfyingly well under low SNR regimes. The noise rejection capability is very high in case of cyclostationary feature detection because noise is random in nature and does not have periodicity property. In practical life, when we don't have any prior knowledge [9–11] of primary user's waveform, then cyclostationary feature detection is the best method to detect the primary user. It only has the drawback of high computationally complexity and that's why the observation time becomes longer than other detection methods.

Here we consider a signal is transmitted over a Nakagami-*m* fading channel and the p.d.f. is given by

$$f(x; m, \sigma) = \frac{2m^m}{\Gamma(m)\sigma^m} x^{2m-1} \exp\left(-\frac{m}{\sigma}x^2\right) \tag{1}$$

The received signal is denoted by $c(t)$.

Now, $c(t)$ is considered to be periodic only if after an interval T_0 it shows the some statistics like mean and autocorrelation.

So, if $c(t)$ is cyclostationary signal, then in case of mean

$$m_c(t + T_0) = m_c(t) \tag{2}$$

And in case of autocorrelation

$$R_c(t + T_0, u + T_0) = R_c(t, u) \tag{3}$$

Now, from Eq. (3) i.e. the autocorrelation equation, Cyclic Autocorrelation (CA) can be derived as

$$R_c\left(t + \frac{\tau}{2}, t - \frac{\tau}{2}\right) = \sum_{\alpha} R_c^{\alpha}(\tau) e^{j2\pi\alpha t} \tag{4}$$

Here α is the cyclic frequency. Now, the Fourier Transform of CA function is the Cyclic Spectral Density (CSD), which is given by

$$S_c^{\alpha}(f) = \int_{-\infty}^{\infty} R_c^{\alpha}(\tau) e^{-j2\pi f\tau} d\tau \tag{5}$$

This CSD is also known as Spectral Correlation Function (SCF). This SCF shows the features of cyclostationary signal c(t). SCF can be plotted on a bi-frequency plane.

3 Implementation of Cyclostationary Feature Detection

In this section, we will do SCF bi-frequency plot using three type of windows—Hanning, Hamming and Kaiser windows.

For implementation, first of all, we have to determine some points like cyclic frequency, carrier frequency, overlap number, window size and fft size. Then we need to compute the SCF for each frame. For this purpose, we shift the signal $c(t)$ by $\alpha/2$ and $-\alpha/2$ in the time domain and thus obtain $c_1(t)$ and $c_2(t)$.

At first, we consider the process of windowing [6]. The window function for Hanning window is given by

$$window(n) = \begin{cases} 0.5\left(1 - \cos\left(\frac{2\pi n}{nwind-1}\right)\right), & 0 \le n \le nwind \\ 0, & otherwise \end{cases} \tag{6}$$

where *nwind* is the window size. Now, the windowed signals are as follows

$$c_{1i}(t) = c_1(t) \cdot window \tag{7}$$

$$c_{2i}(t) = c_2(t) \cdot window \tag{8}$$

Next, we consider $c_{1i}(t)$ and $c_{2i}(t)$ to frequency domain and compute the SCF for each frame as follows

$$S_{xi}^{\alpha}(f) = c_{1i}(f) \cdot conj(c_{2i}(f)) \tag{9}$$

For implementation of SCF, we consider QPSK modulation. Here we consider the signal to noise ratio is 6 dB. The SCF is shown in Fig. 1.

Next for Hamming window, the window function is given by

$$window(n) = \begin{cases} 0.54 - 0.46\left(\frac{2\pi n}{nwind-1}\right), & 0 \le n \le nwind \\ 0, & otherwise \end{cases} \tag{10}$$

The SCF using Hamming is shown in Fig. 2.

Now, we will consider Kaiser window. The window function of Kaiser window is given by

$$window(n) = \begin{cases} \frac{I_0\left(\pi\alpha\sqrt{1-\left(\frac{2n}{nwind-1}-1\right)^2}\right)}{I_0(\pi\alpha)}, & 0 \le n \le nwind \\ 0, & otherwise \end{cases} \tag{11}$$

where I_0 is the zeroth order modified Bessel function of 1st kind and usually $\alpha = 3$.

In Figs. 1, 2 and 3, we have done the bi-frequency plot SCF of the input signal. In Figs. 1 and 2 i.e. in case of Hanning and Hamming windows, we can see a number of small and high peaks in the bi-frequency plot. But in case of Kaiser window, we can see some high peaks present in the SCF, but small peaks are

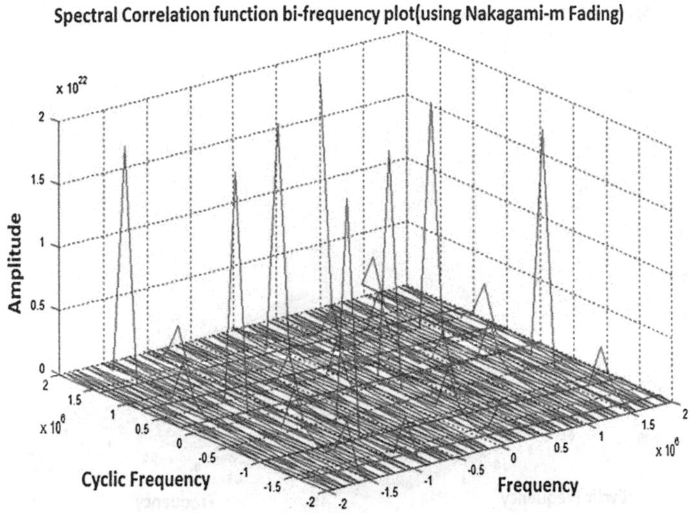

Fig. 1 SCF bi-frequency plot using Hanning window and Nakagami-*m* fading with $m = 1$, $\sigma = 1$

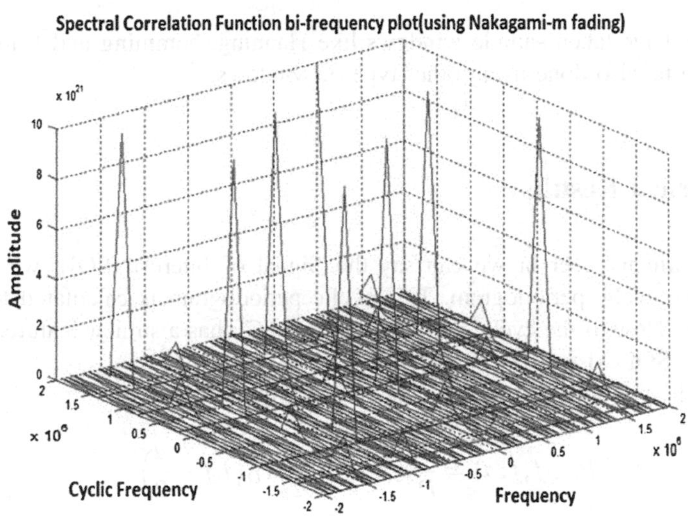

Fig. 2 SCF bi-frequency plot using Hamming window and Nakagami-*m* fading with $m = 1$, $\sigma = 1$

missing as the previous windows. As peaks indicates the presence of primary users, so more number of peaks indicates more number of detected primary users. Now, from these 3 figures, we can conclude that Hanning and Hamming windows are efficient as compares to Kaiser window for detecting primary user. Hanning and Hamming windows are almost in case of primary user detection as we can see that the SCF's are almost same.

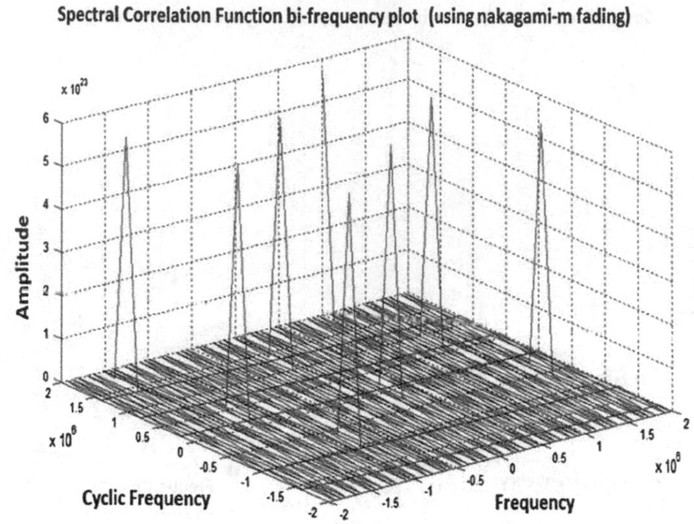

Fig. 3 SCF bi-frequency plot using Kaiser window and Nakagami-*m* fading with $m = 1$, $\sigma = 1$

Here we have taken sample windows like Hanning, hamming and Kaiser window. It can be also done using other type of windows.

4 Numerical Results

To detect primary user or we can say the Signal of Interest (SOI), we have to calculate the cyclic periodogram. This cyclic periodogram is calculated from the input signal. Now, if the cyclic periodogram and SCF has a similar feature, we can say that the SOI exists.

The cyclic periodogram can be calculated by the equation

$$P_{y,D}(f,f;\alpha) = \frac{1}{D} Y_D\left(t, f + \frac{\alpha}{2}\right) Y_D\left(t, f - \frac{\alpha}{2}\right)^* \tag{12}$$

where

$$Y_D(t, v) = \int_{t-\frac{D}{2}}^{t+\frac{D}{2}} y(t) e^{-i2\pi vt} dt \tag{13}$$

Now, from this cyclic periodogram and SCF, we can generate a test statistic, given as

$$W_D(t;\alpha) = \int_{-\infty}^{\infty} S_x(f;\alpha)P_{y,D}(t,f;\alpha)^* df \tag{14}$$

Now, we have to compare the test statistic with a certain threshold to decide whether the SOI exists or not [11]. The SOI in the input signal can be any randomly shifted version of the signal $c(t)$, which is a cyclostationary signal. Now, under H_0 hypothesis the primary user is not communicating and thus the band is considered to be free. Under H_1 hypothesis the primary user is communicating and so the band is considered congested. Now, the hypothesis testing is carried out at the kth decision epoch based on the observations $V_1, V_2, \ldots, V_K$. The observation V_j is the magnitude of W_j i.e. $V_j = |W_j|$

$$L_k = \frac{\pi \prod_{j=1}^{k} f_{v,1}(V_j)}{(1-\pi)\prod_{j=1}^{k} f_{v,0}(V_j)} \tag{15}$$

where $f_{v,m}$ is the probability density function of V_k under the hypothesis H_m and π denotes the prior probability of H_1.

Finally we have different cases of primary user detection for different values of m and σ. In case of each of the three windows, we have taken three values of prior probability i.e. $\pi = 0.3$, 0.4 and 0.5 and also we have considered different combinations of m and σ in case of Nakagami-m fading. In case of 0.5, the probability terms get cancelled from numerator and denominator of likelihood ratio. In all the six cases, we can see that the number of detected primary user increases with the increase of prior probability π.

So, we have observed that in case of Hanning window, for $m = 1$ and $\sigma = 1$, number of user detected gradually increases with prior probability (Fig. 4). For

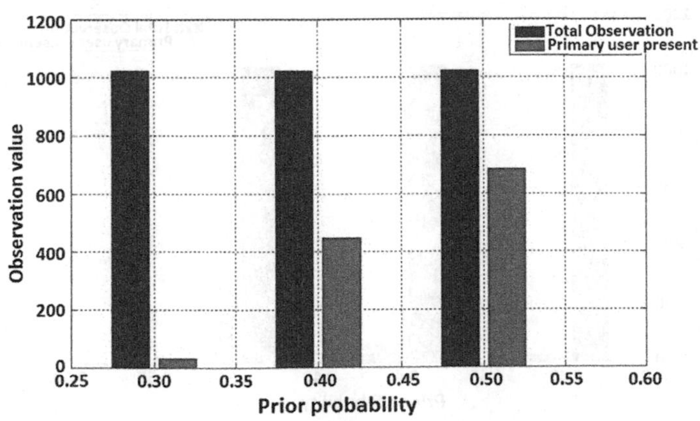

Fig. 4 Number of detected primary users for prior probability $\pi = 0.3$, 0.4, 0.5 in case of Hanning window and Nakagami-m fading with $m = 1$ and $\sigma = 1$

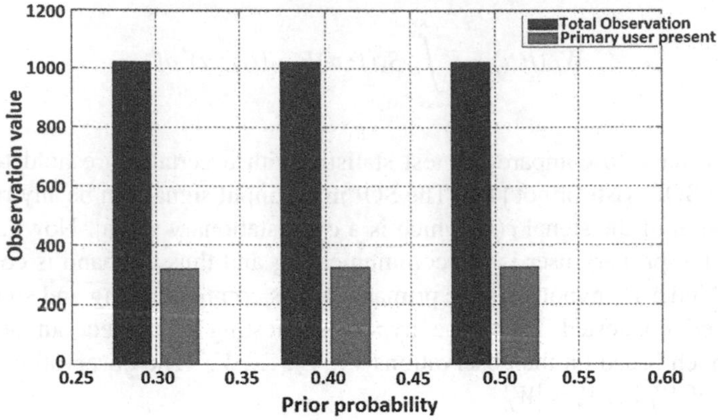

Fig. 5 Number of detected primary users for prior probability $\pi = 0.3, 0.4, 0.5$ in case of Hanning window and Nakagami-m fading with $m = 1$ and $\sigma = 2$

$m = 1$ and $\sigma = 2$, this detection result is more consistent and also not very bad, but not as good as previous case with 0.5 probability (Fig. 5). For $m = 2$ and $\sigma = 1$, number of user detected gradually increases with prior probability (Fig. 6) but the result is not as good as $m = 1, \sigma = 1$. For $m = 2$ and $\sigma = 2$, number of user detected is same for prior probability 0.3 and 0.4 and increases in case of 0.5 prior probability (Fig. 7).

Now in case of Hamming window, for $m = 1$ and $\sigma = 1$, number of user detected gradually increases with prior probability as in case of Hanning window but result is little bit different (Fig. 8). For $m = 1$ and $\sigma = 2$, this detection result is more consistent and also very good, at least better than the previous case with 0.5 probability, just like Hanning window (Fig. 9). For $m = 2$ and $\sigma = 1$, number of user

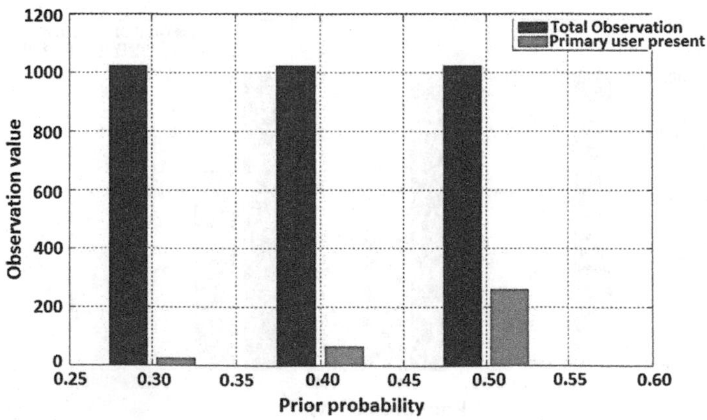

Fig. 6 Number of detected primary users for prior probability $\pi = 0.3, 0.4, 0.5$ in case of Hanning window and Nakagami-m fading with $m = 2$ and $\sigma = 1$

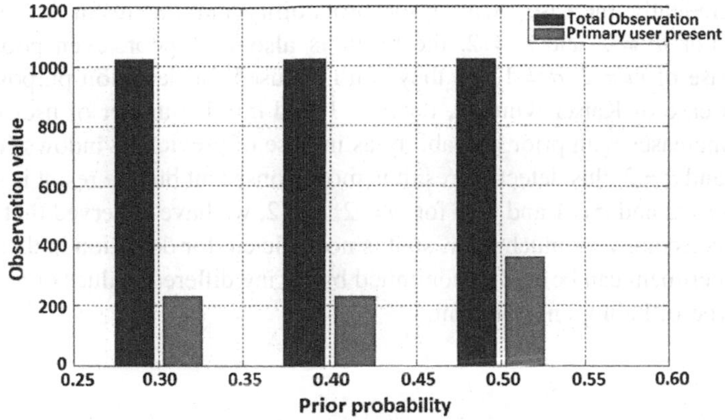

Fig. 7 Number of detected primary users for prior probability $\pi = 0.3, 0.4, 0.5$ in case of Hanning window and Nakagami-*m* fading with $m = 2$ and $\sigma = 2$

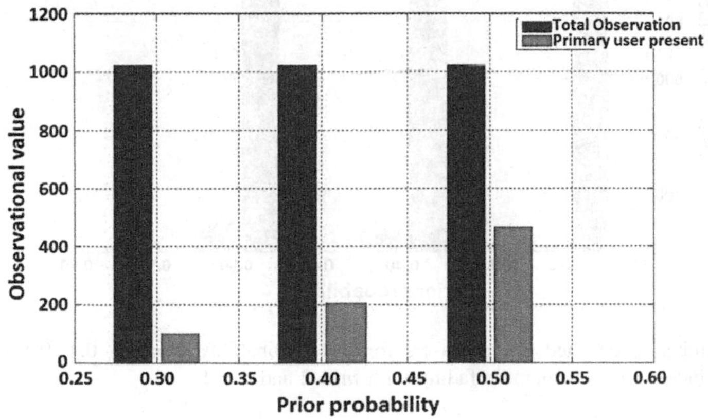

Fig. 8 Number of detected primary users for prior probability $\pi = 0.3, 0.4, 0.5$ in case of Hamming window and Nakagami-*m* fading with $m = 1$ and $\sigma = 1$

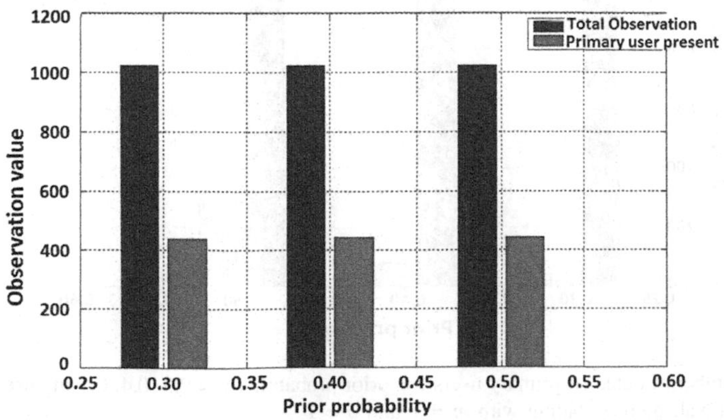

Fig. 9 Number of detected primary users for prior probability $\pi = 0.3, 0.4, 0.5$ in case of Hamming window and Nakagami-*m* fading with $m = 1$ and $\sigma = 2$

detected gradually increases with prior probability but the result is very poor (Fig. 10). For $m = 2$ and $\sigma = 2$, the result is also very poor, even poorer than previous case of $m = 2$, $\sigma = 1$, we they can't be used for detection purpose.

Now in case of Kaiser window, for $m = 1$ and $\sigma = 1$, number of user detected gradually increases with prior probability as in case of previous windows (Fig. 11). For $m = 1$ and $\sigma = 2$, this detection result is more consistent but the result is so much poor. For $m = 2$ and $\sigma = 1$ and also for $m = 2$, $\sigma = 2$, we have observed that number of user detected are very much poor, so it is not efficient for detection in these cases.

This experiment can be further continued by taking different values of m and σ or different type of fading environment.

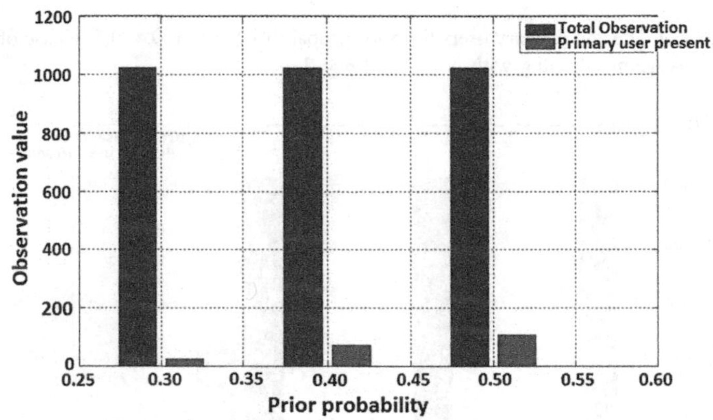

Fig. 10 Number of detected primary users for prior probability $\pi = 0.3, 0.4, 0.5$ in case of Hamming window and Nakagami-m fading with $m = 2$ and $\sigma = 1$

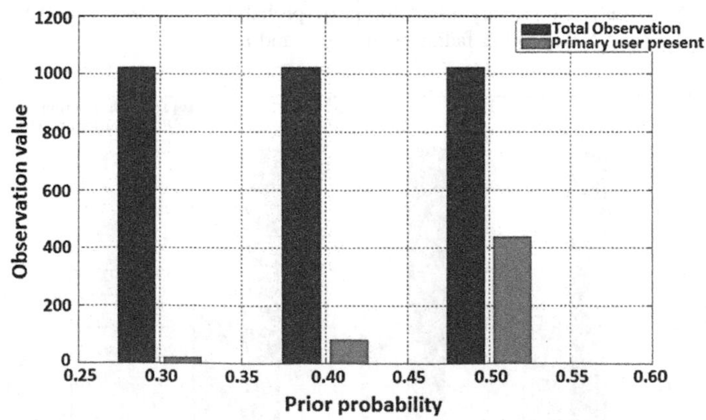

Fig. 11 Number of detected primary users for prior probability $\pi = 0.3, 0.4, 0.5$ in case of Kaiser window and Nakagami-m fading with $m = 1$ and $\sigma = 1$

5 Conclusion

In this paper, SCF of the received signal has been observed with three standard windows under Nakagami-*m* fading environment and efficiency for each of them has been justified by the frequency of detection of primary user. These measurements are taken with different distribution parameters and it has been observed that efficiencies of some windows are better than that of others. The methodology discussed here can be extended further for other window functions as well as different distribution parameters to find the best one for a particular situation to secure primary transmission in the most efficient way.

References

1. Haykin, S.: Cognitive radio: brain-empowered wireless communications. IEEE J. Sel. Areas Commun. **23**(2), 201–220 (2005)
2. Federal Communications Commission (FCC): In the Matter of Facilitating Opportunities for Flexible, Efficient and Reliable Spectrum Use Employing Cognitive Radio Technologies. ET Docket No. 03-108 (2003)
3. Akyildiz, I.F., Lee, W.Y., Mohanty, S.: Next generation dynamic spectrum access/cognitive radio wireless networks. Comput. Netw. **50**, 2127–2159 (2006)
4. Cabric, D., Mishra, S.M., Willkomm, D., Brodersen, R.W., Wolisz, A.: A cognitive radio approach for usage of virtual unlicensed spectrum. In: Proceedings of 14th IST Mobile and Wireless Communications Summit (2005)
5. Ghesami, A., Sousa, E.S.: Spectrum sensing in cognitive radio networks: requirements, challenges and design trade-offs. IEEE Commun. Mag. 32–39 (2008)
6. Gardner, W.A.: Signal interception: a unifying theoretical framework for feature detection. IEEE Trans. Commun. **36**(8) (1988)
7. Cochran, D., Enserink, S.: A cyclostationary feature detector. In: 28th Asilomar Conference (1994)
8. Ustok, R.F.: Spectrum sensing techniques for cognitive radio systems with multiple antennas. MS thesis, Electronics and Communication Engineering, IZMIR Institute of Technology (2010)
9. Tkachenko, A., Cabric, D., Brodersen, R.W.: Cyclostationary feature detector experiments using reconfigurable BEE2. In: Proceeding of New Frontiers in Dynamic Spectrum Access Networks. DySPAN (2007)
10. Sutton, P.D., Nolan, K.E., Doyle, L.E.: Cyclostationary signatures in practical cognitive radio applications. IEEE J. Sel. Areas Commun. **26**(1), 13–24 (2008)
11. Turunen, V., Kosunen, M., Huttunen, A., Kallioinen, S., Ikonen, P., Parssinen, A., Ryynanen, J.: Implementation of cyclostationary feature detector for cognitive radios. In: Proceeding of Cognitive Radio Oriented Wireless Networks and Communications, CROWNCOM (2009)
12. Choi, K.W., Jeon, W.S., Jeong, D.G.: Sequential detection of cyclostationary signal for cognitive radio. IEEE Trans. Wirel. Commun. **8**(9) (2009)

A Modified Real Time A* Algorithm and Its Performance Analysis for Improved Path Planning of Mobile Robot

P.K. Das, H.S. Behera, S.K. Pradhan, H.K. Tripathy and P.K. Jena

Abstract This paper proposed an online path planning of mobile robot in a grid-map environment using modified real time A* algorithm. This algorithm has implemented in simulated and Khepera-II environment and find the optimized path from an initial predefine position to a predefine target position by avoiding the obstacles in its trajectory of path. The path finding strategy is designed in a grid-map and cluttered environment with static and dynamic obstacles with quadrant concept. The optimization the path is found using this algorithm as the goal is present in any of the four quadrant and restricted the movement of the robot to only one quadrant. Robot will plan an optimal path by avoiding obstructions in its way and minimizing time, energy, and distance as the cost, but the original A* algorithm find the shortest path not optimized. Finally, it is compared with other heuristic algorithms.

Keywords Navigation · Improved real time A* algorithm · Path planning · Unknown environment · Khepera-II

P.K. Das (✉) · H.S. Behera
Department of Computer Science and Engineering and Information Technology, VSSUT, Burla, Odisha, India
e-mail: daspradipta78@gmail.com

H.S. Behera
e-mail: hsbehera_india@yahoo.com

S.K. Pradhan
Department of Mechanical Engineering, CET, Bhubaneswar, Odisha, India
e-mail: johndocument@gmail.com

H.K. Tripathy
Department of Computer Science and Engineering, KIIT University, Bhubaneswar, Odisha, India
e-mail: hrudayakumar@gmail.com

P.K. Jena
Department of Mechanical Engineering, VSSUT, Burla, Odisha, India
e-mail: prabirkumarjena07@gmail.com

L.C. Jain et al. (eds.), *Computational Intelligence in Data Mining - Volume 2*,
Smart Innovation, Systems and Technologies 32, DOI 10.1007/978-81-322-2208-8_21

1 Introduction

Path planning for mobile robot is a very complex optimization problem with the constraint [1, 2]. Depending on the model, whether the environment is known or not, the path planning is usually categorized into two groups. The first one is called path planning based on the model of the environment as the mobile robot knows all information about the environment or world map prior to plan. Second one is based on sensors as the mobile robot does not have any information of environment or world map before the execution of the path. The trajectory of the path should be optimal based on the certain criteria such as path length, no of turn the heading direction and time to reach at the target. However, many researchers apply evolutionary computing and intelligent based path planning such as Artificial Neural network, Artificial Immune system, Genetic Algorithm, Ant Colony Optimization etc. into mobile robot, and have made some progress [3].

Motion planning is one of the difficult tasks in intelligent control of a mobile robot which should be performed efficiently. It is often decomposed into online path planning and offline path planning. Offline Path planning is to produce a collision free path in a grid environment by avoiding obstacles and optimize it with respect to some criterion [4, 5]. However, this environment may be imprecise, vast, dynamical and either partially non-structured [6]. In such cluttered environment, path planning problem depends on the sensory information of the environment, which might be associated with imprecision and uncertainty. Thus, for an appropriate motion planning scheme in a cluttered environment, the controller of such kind of robots must have to be adaptive in nature. Online path planning is to schedule the movement of a mobile robot along the planned path. Several methods have been proposed to address the problem of motion planning of a mobile robot. If the path is generates in advance in static environment, then it said to be off-line algorithm. Path is said to be on-line, if it is capable of generating a dynamic path in response to environmental changes. Path planning is the skill of deciding which route to take for navigation under dynamic environment. It involves many computations for continuous movement and sequences, while moving between the predefine start state and goal.

The A* algorithm [7–9] was used in these works for planning purposes and the algorithm was applied only for an environment in which the locations of the obstacles are known in advance. In online path planning, time metric is also an important parameter that cannot be avoided in designing the cost function. Recent researches [10–12] have also considered the Genetic Algorithm (GA) in path planning for a static environment for control of robot motion using GA. The author in [10, 13] also proposed the use of GA in the path planning of mobile robot in static environment with different fitness function. Some of the proposed techniques in [12, 14, 15] suffer from many problems. They include (1) computational cost is more (2) requires large memory spaces, when dealing with dynamic and large sized environments, (3) time consuming process. In [14] authors used Simulated Annealing and Neural Network to handle the path planning problem, respectively.

The author in [16] has proposed the path planning techniques using artificial Immune system.

In this paper the path planning problem of mobile robot is solved using a modified version of the real time A* algorithm in an unknown grid map environment. since the A* algorithm is generally considered as superior to other GA based solution with respect to the constraint stated above, we modified the real time A* algorithm to solve the path planning problem instead of GA based approach to optimize energy in the terms of turn of heading direction, execution time and path length.

The paper has summarized in six sections. The problem formulation with pre-assumptions is presented in the Sect. 2. We provide existing real time A* algorithm in Sect. 3.We have proposed the solution of the path planning with modified real time heuristic A* algorithm by considering Quadrant based world map for simulation and also implemented in Khepera environment, pseudo-code of the path planning is presented in Sect. 4. In Sect. 5, the Experimental result snapshot of the Khepera environment and simulation result is presented and conclusions are narrated in Sect. 6.

2 Formulation of the Problem

In a grid map environment, the mobile robot has to plan the path from its predefine initial state to the goal state by optimizing path length, no of turn of the heading direction and time to reach in the target by avoiding the presence of obstacle in the environment in its trajectory path. The formulation of the problem is to consider the evaluation of the next position of the robot from their current position in a given world map by avoiding the presence of obstacle. The algorithm is considered the following assumptions to compute the optimal path from predefine initial position to the goal.

2.1 Pre-assumptions

1. Initially, robot is placed in the origin of the grid map and Current position of the robot is known with respect to a given reference coordinate system.
2. The goal may be present at any position of any quadrant in the grid map which is known and fixed. Now, we can calculate the position of the goal present in quadrant of the grid map with reference to the initial co-ordinate of the robot position and co-ordinate of the goal.
3. The obstacle may be present at any point of the junction in the grid map. Since the quadrant is known so robot has only two choices for movement. Here we assume that the obstacle may be present at any one of the two possible choices in the path, but not at both.

4. The robot has unique set of actions for motion. A robot can choose anyone action at a given time out of different exist action.
5. The robot is executed its path in steps until it is reached at the goal position.

3 Existing Real Time A* Algorithm

Typical A* algorithm works on a specialized search space represented by a tree or a graph. The objective of this algorithm is to find a specified goal in the process of generating new states(nodes), ultimately terminating at the goal state. The path planning problem of the mobile robot can be solved by modified A* algorithm, well known as real time A* algorithm [13]. To justify the importance of the algorithm, we consider the path planning of a mobile robot on a 2-D grid structure where a grid may contain an obstacle or the robot. Some of the grids in the workspace are empty and the robot has to plan its trajectory path through these empty grid points in each iteration of the algorithm, so as to construct a trajectory of motion towards the prescribed goal. The significance of the real time A* algorithm lies in identifying the vacant until distant grid point which has the shortest distance of the given goal. This is usually done by employing a heuristic function that keeps track of the Euclidean distance of a given neighborhood grid position from the given goal position. The real time A* algorithm is presented below.

//Algorithm Real-Time-A*

1. Set a NODE to be the start state.
2. Create the successor of NODE. If anyone of the successors is a goal state, then exit.
3. Calculate the value of each successor by performing a fixed-depth search starting at that successor. Evaluate all leaf nodes using the A* heuristic function $f = g + h'$, where g is the distance from root node to the leaf node and h′ is the predicted distance to the goal. Pass heuristic estimates up in the search tree such a way that f value of each intermediate node is set to the minimum of the values of its children.
4. Put the NODE to successor with the lowest score, and take the corresponding action in the world map. Store the old NODE in a table along with the heuristic cost of the second-best successor. (With this strategy, we will not enter into a fixed loop, because we never make the same decision at the same node twice.) If this node is generated again in step 2, simply look up the heuristic estimate in the table instead of redoing the fixed-depth search of Step 3.
5. Go to step 2.

Figure 1 presents grid environment with a given obstacles and the trajectory of the planned path of the robot is shown in the form of arrow marks. Here, the robot traversal one of the neighborhood grid points in one iteration, but cannot move diagonally to the nearest grid points.

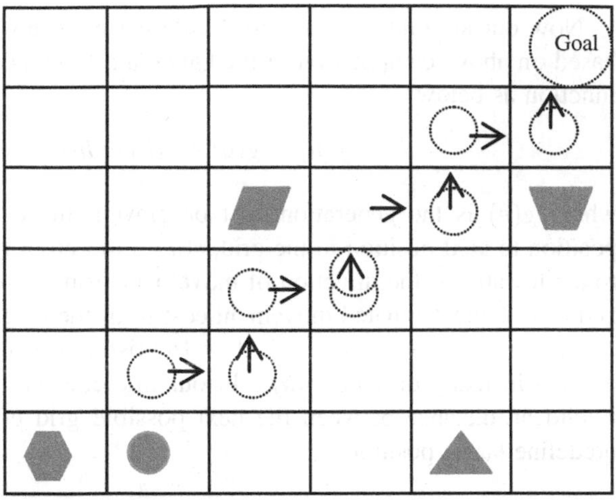

Fig. 1 The theoretical motion planning of a robot in grip map and *circle* represent a robot

4 Proposed Modified Real Time A* Algorithm

Consider the grid environment map, where the map is constructed as set of states is called a grid map and a grid map is built for the navigation of the mobile robot. In a grid map the mobile robot has to find the trajectory path from its predefine initial position to goal position with a cost optimal way. In generally, the robot moves in four neighbor directions vertically or horizontally in grid map to reach at the predefine goal.

But, instead of moving in all the possible four directions to find the trajectory path to the goal, we have modified the A* algorithm in such a way that the robot will first check in which quadrant the goal is present. Hence it will skip the movement of the robot in other three quadrants in each iteration for its movement and it considers only one quadrant every time for finding the trajectory path to the goal. In this way, we have considered only two possible neighbours in the axis point. We have optimized the computational time, no of turn the heading direction and path length for finding the path to the goal as we have restricted the movement of the robot at each step to only one quadrant.

The following equation is used to find the position of the goal in which quadrant it is present. Let us consider the coordinate point of the goal is (x_g, y_g) and the initial coordinate point of the robot is (x_c, y_c). The robot computes the quadrant of the goal by subtracting the goal coordinate point and initial coordinate point of the robot. Let (C_x, C_y) be the difference value of (x_g, y_g) and (x_c, y_c).

Now if,

$$\begin{cases} C_x \geq 0 \; and \; C_y \geq 0 \Rightarrow goal \; is \; in \; First \; Quadrant \\ C_x \leq 0 \; and \; C_y \geq 0 \Rightarrow goal \; is \; in \; Second \; Quadrant \\ C_x \leq 0 \; and \; C_y \leq 0 \Rightarrow goal \; is \; in \; Third \; Quadrant \\ C_x \geq 0 \; and \; C_y \leq 0 \Rightarrow goal \; is \; in \; First \; Quadrant \end{cases} \quad (1)$$

Now our algorithm is used to decide the next movement of the robot position based on above computation of the Quadrant. Each position has an associated cost function as below.

$$f(n) = g(n) + r(n) + h(n) + S(n) \tag{2}$$

where $g(n)$ is the generation cost or movement cost from the predefine initial position to next position in the grid, $r(n)$ is the no of turn required for the robot to rotate towards in the direction of movement from current position to next position and $h(n)$ is the estimated movement cost from the neighbor position of the robot to the target position. This is called as Heuristic cost and the heuristic cost can be defined in many different ways. In our implementation, the Heuristic cost is the Euclidean distance between the next possible grid position of the robot and the predefine target position.

$$h(n) = \sqrt{(x_l - x_g)^2 + (y_l - y_g)^2} \tag{3}$$

Here, (x_l, y_l) is the co-ordinate of the next possible position. $S(n)$ is the direction in which the mobile robot will move around the obstacle is expressed in Tables 1, 2 and 3. The robot selects the next position having minimum $f(n)$ value from the two choices, if there is no obstacle. If there is an obstacle, then robot moves in the other direction. A* is a graph search algorithm that finds a path from given initial node to a given goal node (or one passing a given goal test). The A* algorithm stands by

Table 1 Horizontal obstacle and heading

Conditions	Action
$x_g - x_c \prec 0,\ y_g - y_c \succ 0$	Clockwise
$x_g - x_c \succ 0,\ y_g - y_c \succ 0$	Anti-clockwise
$x_g - x_c \prec 0,\ y_g - y_c \prec 0$	Anti-clockwise
$x_g - x_c \succ 0,\ y_g - y_c \prec 0$	Clockwise

Table 2 Vertical obstacle and heading

Conditions	Action
$x_g - x_c \prec 0,\ y_g - y_c \succ 0$	Anti-clockwise
$x_g - x_c \succ 0,\ y_g - y_c \succ 0$	Anti-clockwise
$x_g - x_c \prec 0,\ y_g - y_c \prec 0$	Clockwise
$x_g - x_c \succ 0,\ y_g - y_c \prec 0$	Anti-clockwise

Table 3 Non-horizontal and non-vertical obstacle relative to mobile robot

Conditions	Action
$x_g \succ x_c,\ y_g \succ y_c\ or\ \lvert x_g - x_c \rvert \succ \lvert y_g - y_c \rvert$	Anti-clockwise
$x_g \prec x_c,\ y_g \prec y_c\ \lvert x_g - x_c \rvert \succ \lvert y_g - y_c \rvert$	Clockwise
$x_g \succ x_c,\ y_g \prec y_c\ or\ \lvert x_g - x_c \rvert \succ \lvert y_g - y_c \rvert$	Clockwise
$x_g \prec x_c,\ y_g \succ y_c\ \lvert x_g - x_c \rvert \succ \lvert y_g - y_c \rvert$	Anti-clockwise

combining the greedy search and uniform-cost search (Dijkstra algorithm). Greedy search minimizes the estimated cost to the goal (that is the heuristic, $h(n)$), and thereby cuts the search cost considerably; but it is neither optimal nor complete. In the other hand, the cost of the path is minimize in uniform-cost search so far (that is elapsed cost, $g(n)$); it is optimal and complete, but can be very inefficient. These two methods can be collectively joined by taking sum of two evaluation functions to get advantages of both methods.

$$f(n) = g(n) + h(n) \tag{4}$$

By using the above heuristic cost of the nodes, the heuristic cost of more than one node may be same, so it cannot give the optimal path to goal. Hence, we have modified Eq. (4) by considering the heading direction as the cost in the heuristic function and direction of move around the obstacle. The modified heuristic function is expressed in Eq. (2).

By using the evaluation function, A* algorithm can enhance the search speed of the best-first algorithm; Overcome the shortcoming on the searching precision of the local optimal search algorithm. A* algorithm uses less memory space than Dijkstra algorithm, the average out degree of one node is marked as 'b', the search depth of the shortest path from start point to end point noted as 'd'. then, it searched all out degree nodes 'b' as next step for moment by considering the shortest distance from the next moment position to the end point. Number of nodes examined is $1 + b + b^2 + b^3 \cdots + b^d$. So, the time complexity of A* algorithm is represented as $O(b^d)$.

Suppose there are a number of obstacles in the environment, where the mobile robot will work and the circumference of each obstacle is limited. The mobile robot does not have any prior knowledge about the environment. The cost of the grid neighbourhood from the current position of the mobile robot can only be found out by its sensor system. The mobile robot should also be able to locate the positions of itself and the goal, and be able to move in any direction. The trajectory path of the mobile robot is expressed in the terms of direct lines towards the goal and set of curved lines around the obstacles. The robot can move around the obstacle either in clockwise or anti-clockwise direction. Table 1 represent the of moment of the heading direction, if obstacle is Horizontal. Similarly, Tables 2 and 3 represent the of moment of the heading direction, if obstacle is Vertical and Non-horizontal and non-vertical respectively. This is identifying by the topological properties of an obstacle relative to the mobile robot heading direction, and the position of the goal relative to the position of the mobile robot is presented in Fig. 2.

In the path planning problem for finding the optimal path, A* algorithm starts from a point, CURRENT (source), to another point, GOAL (destination). The modified A* keeps a list of two possible next steps, called the OPEN list. Then, it chooses a next step that is most likely to lead us to the goal in the minimum time, minimum distance and minimum turn of the heading direction. In order to accomplish this we need to have a heuristic cost that will determine "most likely" position among all possible position. Once that step has decided to move, then, it will add to the CLOSED list.

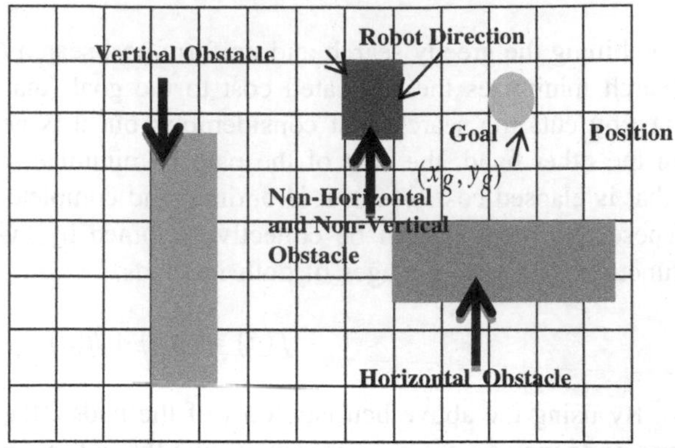

Fig. 2 Properties of obstacle for moving

The modified version of A* algorithm for the path planning of mobile robot is presented below.

Pseudo Code for Modified A* PLANNING ()

Input: Robot initial position as (x_c, y_c), goal position as (x_g, y_g)

Output: trajectory path from (x_c, y_c) to (x_g, y_g)

BEGIN

CURRENT = (x_c, y_c);
OPEN=CURRENT;
CLOSED=EMPTY;
GOAL= (x_g, y_g);

Find the Quadrant of the Goal position using Eq(1);
WhileOPEN $\neq$ EMPTY Or CURRENT $\neq$ GOAL
BEGIN
CURRENT=DELETE_BEST(OPEN)
MOVE_TO (CURRENT)
CLOSED=CURRENT
OPEN=CURRENT
END
END WHILE
END

Procedure DELETE_BEST(OPEN)

BEGIN
S_P=DELETE_QUEUE(OPEN)

Generate Two neighbors S_x and S_y of S_P using Eq(1) ;

Calculate $f(S_x)$ *and* $f(S_y)$ using Eq(2)

If ($f(S_x) \prec f(S_y)$ and ISOBSTACLE (S_x)) THEN return S_y

Else return S_x

END

5 Simulation and Experiment Result

The experiment has been conducted in C environment with Pentium V processor and implemented in the Khepera II environment. Gird map is created with the size of 315 × 315 and origin co-orientate is 15 × 15 pixel value of computer screen. Initially the robot is placed 165 × 165 pixel value of the computer screen and heading direction of the robot is along x-axis, environment is presented in Fig. 6 Extensive experiments were conducted to test the robot performance in the path planning between any two positions i.e. START and GOAL position in the unknown environment with various unexpected obstacles. Figure 7 shows the snapshots taken at different stages of the mobile robot path planning in one such case study. In Fig. 8, we have compared all the possible paths to the goal from the robot position and we found five number of paths. From this we conclude that there exist two optimal paths i.e. Path3 and Path5, but as per our proposed algorithm the next neighbor position will be selected according to the minimum cost which is on the basis of Eq. (2). So the optimal path is the Path 5 in Fig. 8. Similarly in another experiment, the optimal path is Path 6 shown in Fig. 10 out of possible paths shown in Fig. 9. The optimal path selected was shown in Fig. 11. In the other hand, the experiment is also conducted in the Khepera environment, Fig. 3 shows the Khepera network and accessories. Figure 4 shows the setup of the environment for the conduct of the environment and Fig. 5a–c shows the intermediate path during the trajectory of the motion planning towards the predefine goal and Fig. 5d shows the optimal path from predefine initial position to goal position by avoiding the obstacles during the plan of the path. Finally the comparisons of different algorithm with modified A* algorithm in the form of average path length and average time taken is presented in Table 4. Different algorithms trajectory path result is presented in Figs. 12 and 13 of two different simulated environments.

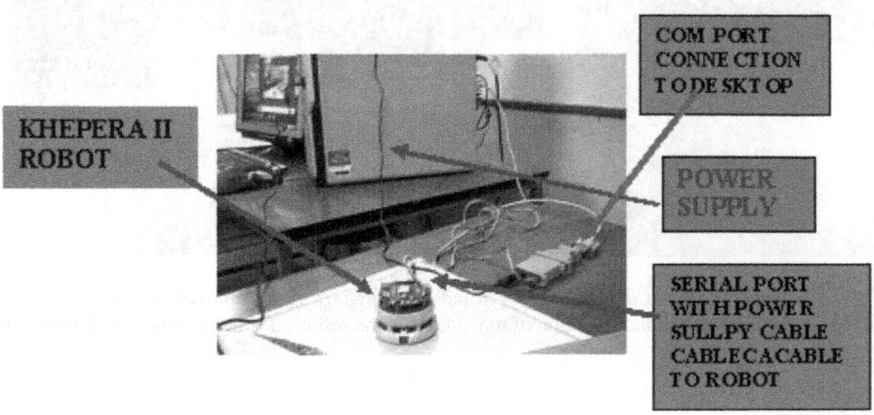

Fig. 3 Khepera environment and its accessories

Fig. 4 Environment setup for robot path planning

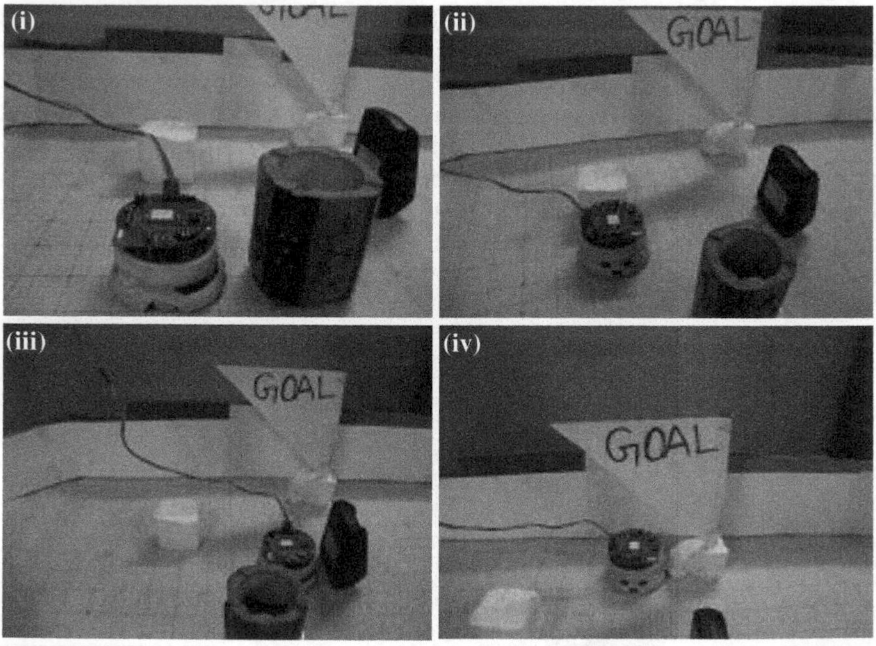

Fig. 5 Snapshots taken at different stages of path planning. (**a**) Initial movement of the robot to next grid. (**b**)–(**c**) Intermediate position of movement of the robot. (**d**) Final position of the robot at the goal

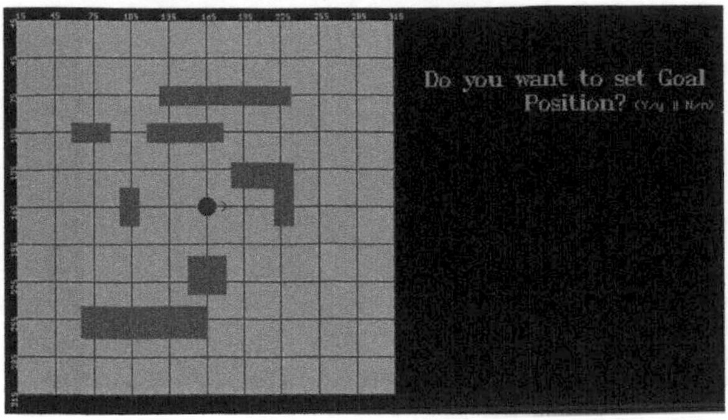

Fig. 6 Environment setup for robot

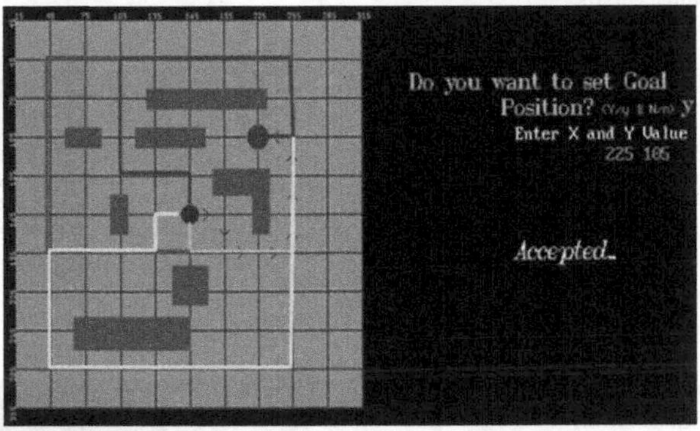

Fig. 7 Possible paths to goal position are shown path planning in different *colored lines* and optimal path is indicated by *arrow marks* (color figure online)

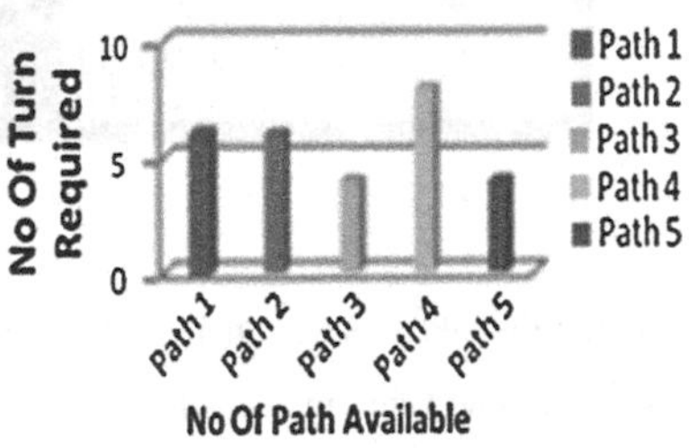

Fig. 8 No. of paths versus No. of turns

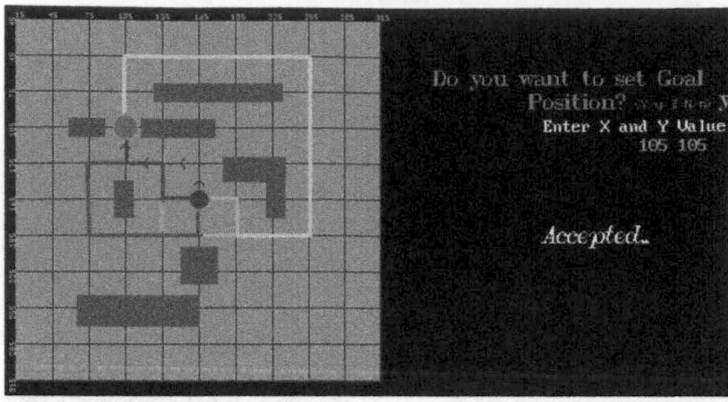

Fig. 9 Possible paths to goal position are shown in different *colored lines* and optimal path is indicated by *arrow symbol* (color figure online)

Fig. 10 No. of paths versus No. of turns

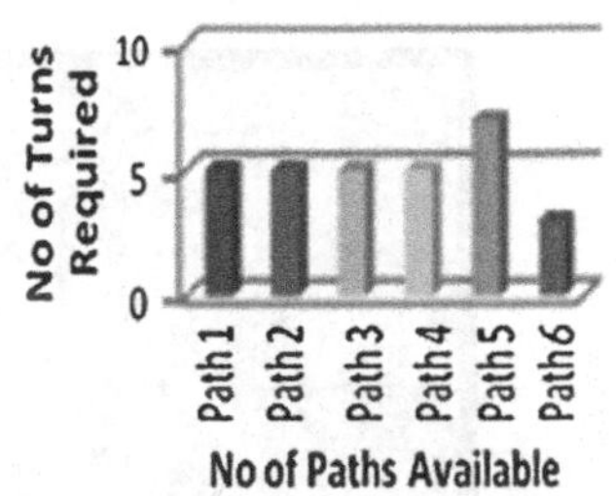

Fig. 11 *Arrow mark* shows optimal path to goal position

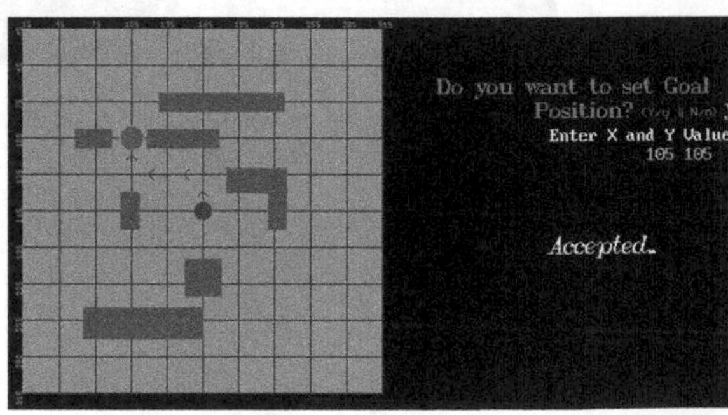

Table 4 Path lengths and average time taken using different algorithm

Different algorithm for path planning	Average time taken in seconds	Average path length in pixel
A* algorithm	5.27	330
Generating algorithm	4.31	270
Dijkstra algorithm	9.10	570
Modified A* algorithm	2.87	180

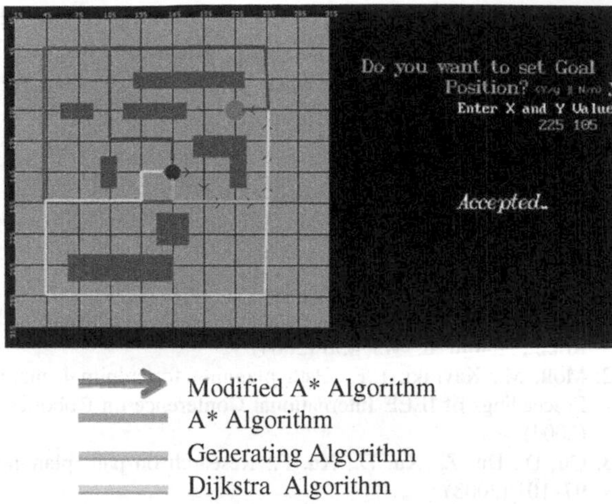

Fig. 12 Different algorithms path represented in different *color code* (color figure online)

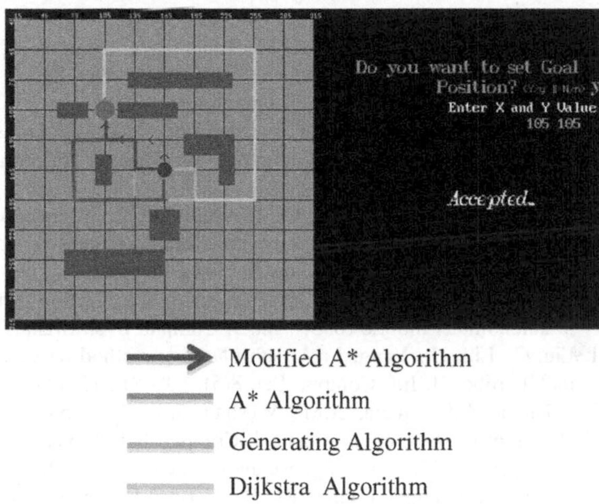

Fig. 13 Different algorithms path represented in different *color code* (color figure online)

6 Conclusions

In this paper, online path planning of mobile robot in unknown environment with static obstacles is presented in simulated environment as well as in Khepera environment. The path planning is performed using modified Heuristic real time A* algorithm. The use of A* algorithm can meet the rapid and real-time requirements of path planning which is otherwise not possible in some path planning using advanced algorithms However only static and dynamic obstacles were considered in the present study.

The result of this work provides a platform for developing robot control and also provides a useful tool for robotics education. For future work, we plan to include visual perception based on static and dynamic obstacles in the study of online path planning of mobile robot.

References

1. Bien, Z., Lee, J.: A minimum-time trajectory planning method for two robots. IEEE Trans. Robot. Autom. **8**, 443–450 (2004)
2. Moll, M., Kavraki, L.E.: Path planning for minimal energy curves of constant length. In: Proceedings of IEEE International Conference on Robotics and Automation, pp. 2826–2831 (2004)
3. Qu, D., Du, Z., Xu, D., Xu, F.: Research on path planning for a mobile robot. Robot **30**, 97–101 (2008)
4. Peng, Q.J., Kang, X.M., Zhao, T.T.: Effective virtual reality based building navigation using dynamic loading and path optimization. Int. J. Autom. Comput. **6**, 335–343 (2009)
5. Florczyk, S.: Robot Vision Video-Based Indoor Exploration with Autonomous and Mobile Robots. WILEY-VCH Verlag GmbH & Co.KGaA, Weinheim (2005)
6. Xiao, J., Michalewicz, Z., Zhang, L., Trojanowski, K.: Adaptive evolutionary planner/navigator for mobile robots. IEEE Trans. Evol. Comput. **1**, 18–28 (1997)
7. LaVall, S.M.: Planning Algorithms. Cambridge University Press, Cambridge (2010). Available: http://msl.cs.uiuc.edu/planning
8. Liu, X., Gong, D.: A comparative study of A-star algorithms for search and rescue in perfect maze. In: International Conference on Electric Information and Control Engineering (ICEICE), pp. 24–27 (2011)
9. Ma, C., Diao, A., Chen, Z., Qi, B.: Study on the hazardous blocked synthetic value and the optimization route of hazardous material transportation network based on A-star algorithm. In: 7th International Conference on Natural Computation, vol. 4, pp. 2292–2294 (2011)
10. Castillo, O., Trujillo, L.: Multiple objective optimization genetic algorithms for path planning in autonomous mobile robots. Int. J. Comput. Syst. Signals **6** (2005)
11. Liu, C., Liu, H., Yang, J.: A path planning method based on adaptive genetic algorithm for mobile robot. J. Inf. Comput. Sci. **8**(5), 808–814 (2011)
12. Taharwa, A.L., Sheta, A.M., Weshah, A.I.: A mobile robot path planning using genetic algorithm in static environment. J. Comput. Sci. **4**, 341–344 (2008)
13. Konar, A.: Artificial Intelligence and Soft Computing: Behavioral and Cognitive Modeling of the Human Brain, 1st edn. CRC Press, Boca Raton (1999)
14. Wei, X.: Robot path planning based on simulated annealing and artificial neural networks. Res. J. Appl. Sci. Eng. Technol. **6**, 149–155 (2013)
15. Das, A., Mohapatra, P., Mishra, P., Das, P.K., Mandhata, S.C.: Improved real time A* algorithm for path planning of mobile robot in quadrant based environment. Int. J. Adv. Comput. Theory Eng. **1**, 25–30 (2012)
16. Das, P.K., Pradhan, S.K., Patro, S.N., Balabantaray, B.K.: Artificial immune system based path planning of mobile robot. In: Soft Computing Techniques in Vision Science, vol. 395, pp. 195–207. Springer, Berlin (2012)

Optimum Design and Performance Analysis of Dipole Planar Array Antenna with Mutual Coupling Using Cuckoo Search Algorithm

Hrudananda Pradhan, Biswa Binayak Mangaraj and Iti Saha Misra

Abstract In this paper, a recently developed meta-heuristic optimization algorithm namely Cuckoo Search (CS) is introduced for analyzing a 4 × 4 dipole planar array (DPA) taking mutual coupling into account. The CS algorithm is employed to optimize the DPA using the multi-objective fitness function considering the Directivity (D), Half Power Beam Width (HPBW), and Front to Side Lobe Level (FSLL) of the antenna. The array is simulated using self developed Method of Moments (MOM) codes using MATLAB (7.8.0.347) and verified using standard MININEC (14.0) software package. Finally, the optimized results are compared with the results of other antenna arrays obtained using various other popular optimization techniques to evaluate the performance of the DPA. The computation time using CS is much less than PSO and 4 × 4 DPA provides higher D with low FSLL.

Keywords CS · DPA · Directivity · FSLL · HPBW · Mutual coupling · MININEC

1 Introduction

Presently, among the varieties of antenna structures planar array is quite popular in offering the benefits of superior performance parameters compared to large linear antenna array or Yagi array having comparable size and cost. They can be used to

H. Pradhan (✉) · B.B. Mangaraj
Department of Electronics and Telecommunication Engineering,
Veer Surendra Sai University of Technology, Sambalpur 768018, Odisha, India
e-mail: mail2hruda@gmail.com

B.B. Mangaraj
e-mail: bbmangaraj@yahoo.co.in

I.S. Misra
Department of Electronics and Telecommunication Engineering,
Jadavpur University, Kolkata 700032, India
e-mail: itisahamisra@yahoo.co.in

L.C. Jain et al. (eds.), *Computational Intelligence in Data Mining - Volume 2*,
Smart Innovation, Systems and Technologies 32, DOI 10.1007/978-81-322-2208-8_22

scan the main beam of the antenna toward any point in space and are used in tracking radars, remote sensing communications. By varying the length of the radiating elements and spacing between the elements the antenna output parameters can be varied. The purpose of optimization is to find out best parameters to satisfy the desired objective i.e. to maximize D, FSLL, and to minimize HPBW. Yagi-Uda arrays are very practical radiator in HF, VHF, and UHF ranges, because they are simple to build, low-cost, and provide desirable characteristics for many applications [1]. But, it is shown in this paper, that a DPA with same or less number of elements gives much better directivity than conventional Yagi-Uda arrays.

Some researches on optimization of linear array have been found in the literature. But same in case of DPA is hardly found. Robinson and Sammi in [2] have introduced PSO for optimization in electromagnetic. Baskar et al. [3] have applied Comprehensive learning PSO (CLPSO) for design of Yagi-Uda array gives superior performance than PSO. Khodier and Al-Aqeel in [4] have applied PSO for linear and circular array optimization and found the considerable enhancements in radiation pattern. Zuniga et al. [5] used PSO for adaptive radiation pattern optimization for linear antenna arrays by Phase Perturbations. Wang et al. in [6] have proposed a complex valued GA which enhances the searching efficiency to synthesize the linear antenna array for confirming current amplitude of elements. Mangaraj et al. [7] have used BFO for the optimization of a multi-objective Yagi Uda array. Joshi et al. in [8] used GA for reduction in SLL and improving D based on modulating parameter M. Kadri et al. in [9] proposed an adaptive PSO and GA where the inertia weights and acceleration coefficient are adjusted dynamically according to the particle's best memories to overcome the limitations of the standard PSO for the synthesis of linear antenna array. Enhancement in D with reduced SLL of linear antenna array applied for WLAN has done by Vishwakarma and Cecil in [10].

Though numerous optimization algorithms are used in antenna arrays, but to the best of our knowledge, Cuckoo Search (CS) algorithm [11, 12] introduced by Yang and Deb has not been used in antenna array problems. This meta-heuristic have been found to be more efficient, because of its ability to find the global solutions in multi-dimensional search space for solving optimization problems with higher success rates. In this paper we use the CS optimization algorithm to optimize the 4 × 4 DPA to achieve better D, low HPBW and high FSLL. It can be seen that the CS is a very promising algorithm which could outperform existing algorithms such as GA, PSO, CLPSO, BFO, CGA, and adaptive PSO and best fit for the design of planar array.

In order to justify our optimum design the result is verified with MININEC software package (14.0). Finally, results of 4 × 4 optimized DPA using CS are compared with PSO technique and also with different types of Yagi-Uda arrays which are optimized using various optimization techniques to evaluate the performance of 4 × 4 DPA.

2 Planar Array and Mutual Coupling

Uniform arrays are usually preferred in design of direct-radiating planar array with a large number of radiating elements. Here, a 4 × 4 DPA shown in Fig. 1 is considered. The total electric field of an array can be formed by multiplying the array factor of the 4 × 4 isotropic sources by the field of a single dipole element [1]. But the array factor method used for calculating far-zone electric field does not hold good for array of dipoles as it doesn't takes mutual coupling into account. Mutual coupling is generally avoided by many researchers as it complicates the analysis and design of antenna. But in practical application it must be taken into account because of its significant contribution. Thus in this paper we consider the mutual coupling into account for calculating the electric field of the planar array.

Each element is fed with a voltage of 1 V. The radius (a) of each wire is assumed to be very small (0.001 λ) and is same for all the elements. The total electric field of the antenna is due to the field contribution from all the radiating elements. To calculate the electric field radiation, Method of Moment (MOM) is used where mutual coupling is taken into account automatically [1]. Each wire is divided into number of segments and each segment's effect on other antenna is calculated using MOM. The process is repeated for all the radiating elements and fields are summed up vectorially and final electric field is calculated.

2.1 Method of Moments Analysis for Planar Array

The total electric field generated by electric current source radiating in an unbounded free space is based on Pocklington's Integral equation [1] is given by

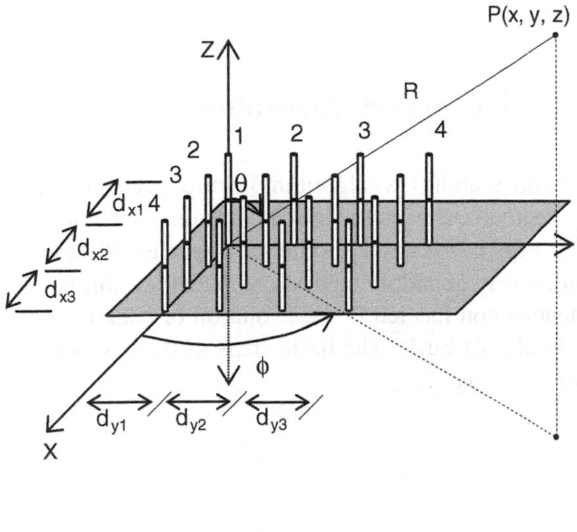

Fig. 1 The geometry of 4 × 4 dipole planar array antenna

$$\int_{-l/2}^{+l/2} I(z')\left(\frac{\partial^2}{\partial z^2}+k^2\right)\frac{e^{-jkR}}{R}dz' = j4\pi\omega\varepsilon_0 E_z^t \quad (1)$$

where l = Length of the one dipole, $I(z')$ = Line-source current, ω = Angular frequency, $k^2=\omega^2\mu\varepsilon$, ε_0 = Permittivity of free space, $R=\sqrt{(x-x')^2+(y-y')^2+(z-z')^2}$, E_z^t = Z-component of total electric field.

For small radius wires (a = 0.001 λ) the current on each element can be approximated by a finite series of odd-ordered even modes. The current in the nth element can be written as Fourier series expansion of the form [1]

$$I(z')=\sum_{m=1}^{M} I_{nm}\cos\left[(2m-1)\frac{\pi z'}{l_n}\right] \quad (2)$$

where I_{nm} represents the complex current coefficient of mode m on element n (the term I_{nm} is the current due to mutual effect, description of which is available in [1]), l_n represents the corresponding length of the nth element and M represents number of current modes.

The total electric field is obtained by summing the field contribution from each radiating element.

$$E_\theta=\frac{j\omega\mu e^{-jkr}}{4\pi r}\sin\theta\sum_{n=1}^{N}\left\{e^{jk(x_n\sin\theta\cos\phi+y_n\sin\theta\sin\phi)}\times\sum_{m=1}^{M} I_{nm}\left[\frac{\sin(Z^+)}{Z^+}+\frac{\sin(Z^-)}{Z^-}\right]\right\}\frac{l_n}{2} \quad (3)$$

where μ = Permeability of the medium, N = No. of dipoles in the array, r = Distance between center of dipole and point of observation, x_n, y_n = X-coordinate and Y-coordinate of the center of dipole respectively, $Z^+=[\frac{(2m-1)\pi}{l_n}+k\cos\theta]\frac{l_n}{2}$ and $Z^-=[\frac{(2m-1)\pi}{l_n}-k\cos\theta]\frac{l_n}{2}$.

3 Cuckoo Search Algorithm

Cuckoo Search (CS) algorithm is introduced by Yang and Deb [11–13]. It has been conceptualized from the brood parasitism behavior of the bird Cuckoo. Cuckoos lay their eggs in the nests of other bird species. If these eggs are discovered by the host bird, it may abandon the nest completely or throw away the alien eggs. This natural phenomenon has led to the evolution of cuckoo eggs to mimic the egg appearance of local host birds. The basic steps of the CS can be summarized in the flow chart shown in Fig. 2.

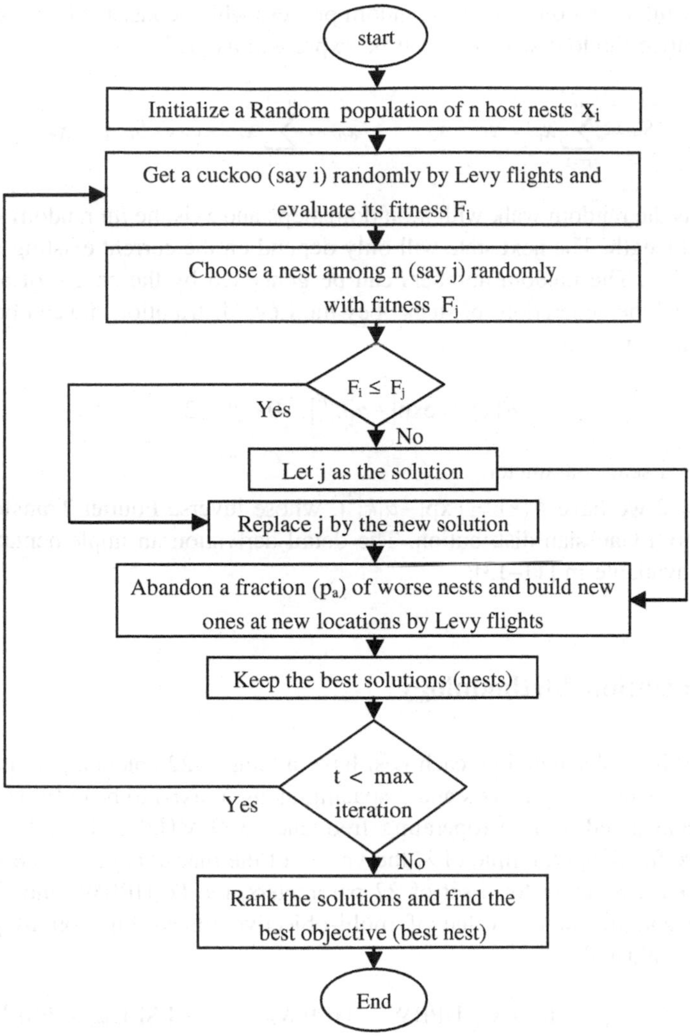

Fig. 2 Flow chart of cuckoo search algorithm

According to the flow chart one of the randomly selected nests (except the best one) is replaced by a new solution $x_i^{(t+1)}$ for the ith cuckoo produced by random walk with Lévy flight performed using the equation

$$x_i^{(t+1)} = x_i^{(t)} + \alpha S \tag{4}$$

where $\alpha > 0$ is the step size which should be related to the scales of the problem of interests and is set to unity in [10]. In this work the step size is taken as 0.01. The parameter S is the length of random walk with Lévy flights.

A Lévy flight random walk is a random process which consists of taking a series of consecutive random steps and can be expressed as [13]

$$S_n = \sum_{i=1}^{n} x_i = x_1 + x_2 + \cdots x_n = \sum_{i=1}^{n-1} x_i + x_n = S_{n-1} + x_n \tag{5}$$

where S_n is the random walk with n random steps and x_i is the ith random step with predefined length. The next state will only depend on the current existing state and the motion x_n. The random numbers can be generated by the choice of a random direction and the generations of steps obey the Lévy distribution and can be defined in terms of Fourier Transform as

$$F(k) = \exp[-\alpha|k|^{\beta}] \quad 0 < \beta \leq 2 \tag{6}$$

where α is a scale parameter.

For $\beta = 2$ we have $F(k) = \exp[-\alpha|k|^2]$, whose inverse Fourier Transform corresponds to a Gaussian distribution. The detail derivation an implementation procedure is available in [11–13].

4 Optimization Methodology

In MATLAB implementation, each nest have a total of 22 antenna parameters (16 lengths, 3 spacing along X-axis and 3 spacing along Y-axis) to be optimized. Since the λ is considered as 1 m (operating frequency 300 MHz) and the lengths and spacing are fractional multiple of λ, the values of the matrix are taken between 0.1 and 1. In each iteration for a set of 22 parameters, the D, HPBW, and FSLL are calculated and the fitness value of multi-objective fitness function as given in Eq. (7) is evaluated.

$$\mathrm{F} = \mathrm{C_1}|\mathrm{D_{des}} - \mathrm{D}| + \mathrm{C_2}|\mathrm{HPBW} - \mathrm{HPBW_{des}}| + \mathrm{C_3}|\mathrm{FSLL_{des}} - \mathrm{FSLL}| \tag{7}$$

where the subscript 'des' stands for the desired. The constants C1, C2, and C3 are taken as 0.5, 0.3, and 0.2 respectively to obtain the best performance. The process is repeated for the number of iterations specified by the CS algorithm and finally the best objective (the best nest) is found by ranking the current best solutions.

5 Result and Discussion

The parameters considered during optimization of DPA at an operating frequency of 300 MHz are D (dBi), HPBW (°), and FSLL (dBi). These optimized parameters are obtained using CS optimization code linked to the structure code of DPA in MATLAB and the radiation pattern is generated. For verification purpose, the parameters of the optimized DPA were considered for the design and simulation of the same DPA using MININEC and radiation patterns are also generated. It is observed that radiation pattern obtained using MATLAB code matches with the pattern obtained using MININEC. In Figs. 3 and 5, the E-plane and H-plane patterns are shown which are obtained using MININEC. In Figs. 4 and 6, E-Plane and H-plane pattern are obtained using MATLAB. Figures 3, 4, 5 and 6 are obtained

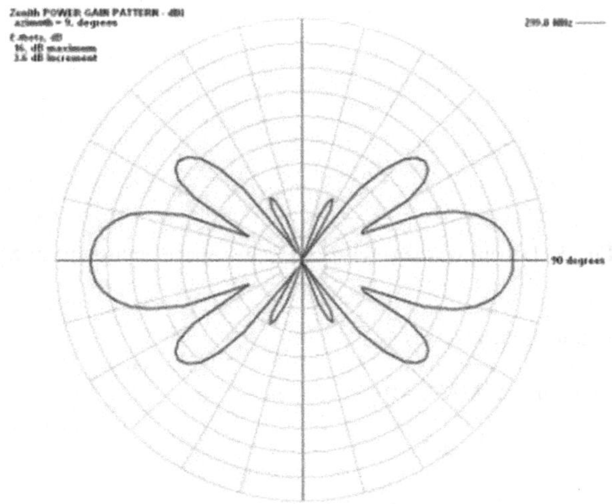

Fig. 3 E-theta versus θ (at $\phi = 0°$: free space environment: MININEC)

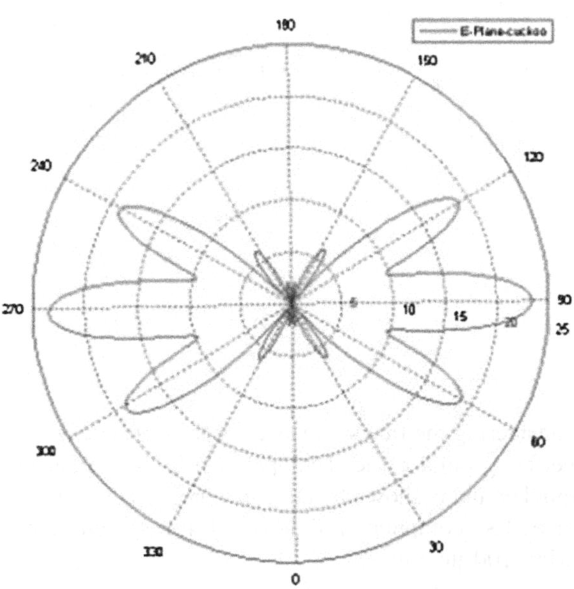

Fig. 4 E-theta versus θ (at $\phi = 0°$: free space environment: MATLAB)

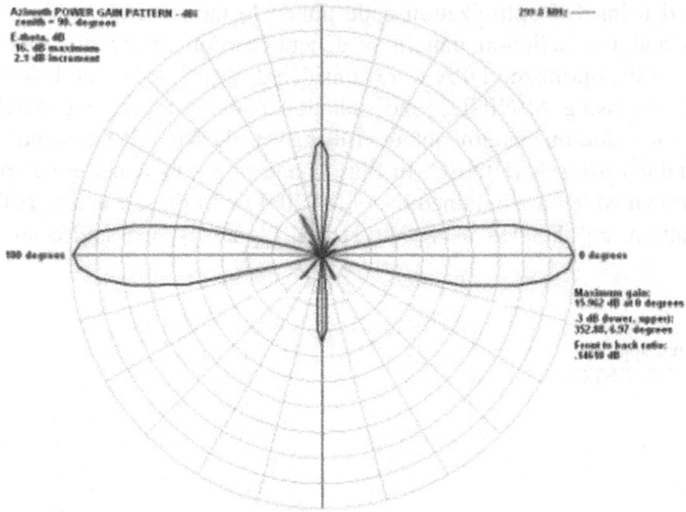

Fig. 5 E-theta versus φ (at $\theta = 90°$: free space environment: MININEC)

Fig. 6 E-theta versus ϕ (at $\theta = 90°$: free space environment: MATLAB)

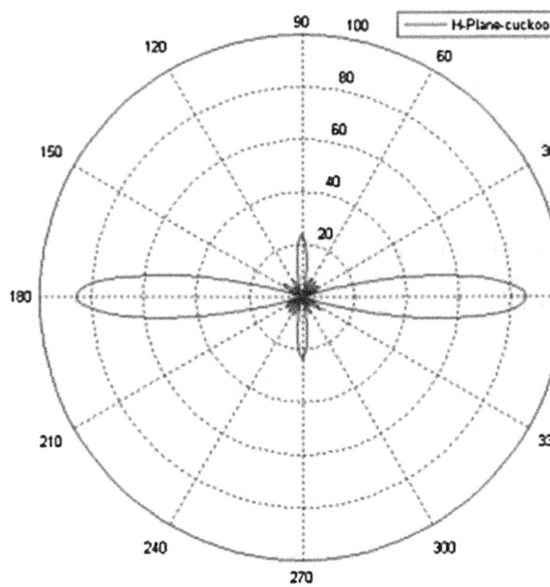

considering the free space as environment, whereas Fig. 7 is obtained considering a perfect ground plane. The approach to the global optimal solution by CS is much quicker and is shown by the convergence curve in Fig. 8. It is also observed that CS takes less computation time (i.e. 1,463 s) than that (i.e. 2,123 s) for PSO considering 100 iterations.

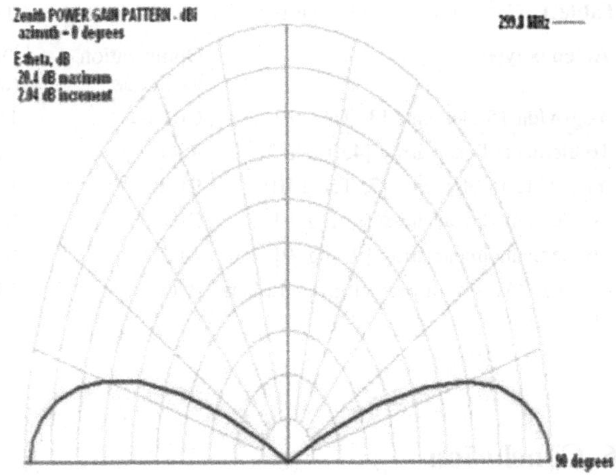

Fig. 7 E-theta versus θ (at $\phi = 90°$: free space environment: MININEC)

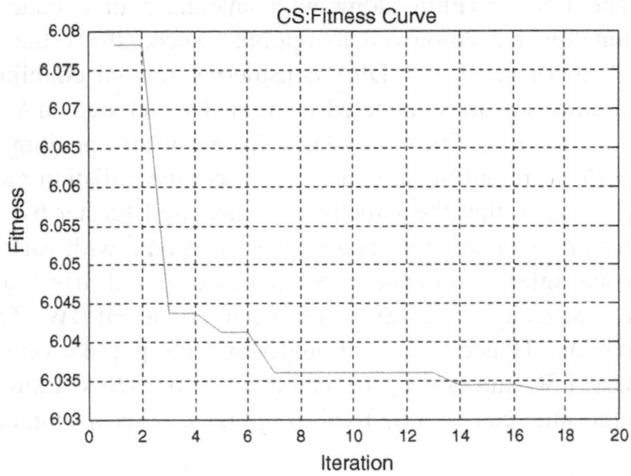

Fig. 8 Convergence characteristics of CS optimization

In Table 1 our designed 4 × 4 DPA is compared with different types of Yagi-Uda, linear, and DPA antennas optimized using different optimization techniques like GA, PSO, CLPSO, BFO, CGA, and adaptive PSO. The optimized result available from various referred journal as in Table 1 indicates very few optimized parameters of the antenna. But in case of ours we have optimized three parameters. The data shows the parameters D and HPBW obtained for our case are better than others. However, the parameter FSLL is slightly less than the others.

Table 1 Comparison of different types of antenna with 4 × 4 DPA optimized with CS

Antenna type	Optimization technique	Directivity (dBi)	HPBW (°)	FSLL (dBi)
Yagi-Uda 15 elements [3, Table 4]	CLPSO	16.40	NA	NA
16 elements linear array [4, Table 2]	PSO	NA	NA	30.7
Yagi-Uda 15 elements [7, Table III]	BFO	17.134	24.259	NA
16 elements linear array [9, Fig. 9]	GA	NA	NA	22
16 elements linear array [9, Fig. 9]	APSO	NA	NA	30
4 × 4 DPA using a perfect ground plane [14]	PSO	19.6	13	31.28
4 × 4 DPA using a perfect ground plane	CS	20.4	10.5	28.36

6 Conclusion

The CS algorithm along with antenna source code and multi-objective fitness function are employed to optimize successfully the parameters D, HPBW, and FSLL of the 4 × 4 DPA considering mutual coupling among all elements. The simulations are conducted using self developed MATLAB code and are verified with aid of MININEC 14.0 software at an operating frequency of 300 MHz. A perfect ground plane is used to reflect the radiation pattern to the other half of the pattern, so that the directivity is increased further by significant amount. The performance of our designed DPA is compared with various antenna arrays optimized using other optimization techniques and is clearly found that the same DPA optimized using CS gives much better D and HPBW. The parameter FSLL is comparable. Hence CS could outperform the popular optimization algorithms such as GA, PSO and best fit for the design of our DPA. Using the same procedure the CS algorithm can also be used to optimize other antenna arrays.

References

1. Balanis, C.A.: Antenna Theory: Analysis and Design, 3rd edn. Wiley, New York (2005)
2. Robinson, J., Sammi, Y.R.: Particle swarm optimization in electromagnetics. IEEE Trans. Antenna Propag. **52**(2), 397–407 (2004)
3. Baskar, S., Alphones, A., Suganthan, P.N., Liang, J.J.: Design of Yagi-Uda antennas using comprehensive learning particle swarm optimization. IEE Proc. Microwaves Antennas Propag. **152**(5), 340–346 (2005)
4. Khodier, M., Al-Aqeel, M.: Linear and circular array optimization a study using particle swarm intelligence. Prog. Electromagnet. Res. B **15**, 347–373 (2009)
5. Zuniga, V., Erdogan, A.T., Arslan, T.: Adaptive radiation pattern optimization for antenna arrays by phase perturbations using particle swarm optimization. In: IEEE Proceedings of NASA/ESA Conference on Adaptive Hardware and Systems, pp. 209–214 (2010)
6. Wang, Y., Gao, S., Yu, H., Tang, Z.: Synthesis of antenna array by complex-valued genetic algorithm. Int. J. Comput. Sci. Netw. Secur. **11**(1) (2011)

7. Mangaraj, B.B., Misra, I.S., Sanyal, S.K.: Application of bacteria foraging algorithm for the design optimization of multi-objective Yagi-Uda array. Int. J. RF Microwave Comput. Aided Eng. **21**(1), 25–35 (2011)
8. Joshi, P., Jain, N., Dubey, R.: Optimization of linear antenna array using genetic algorithm for reduction in side lobs levels and improving directivity based on modulating parameter M. Int. J. Innovative Res. Comput. Commun. Eng. **1**(7), 1475–1482 (2013)
9. Kadri, B., Brahimi, M., Bousserhane, I.K., Bousahla, M., Bendimerad, F.T.: Patterns antennas arrays synthesis based on adaptive particle swarm optimization and genetic algorithms. Int. J. Comput. Sci. Issues **10**(1)(2), 21–26 (2013)
10. Vishwakarma, M., Cecil, K.: Synthesis of linear antenna array using PSO to reduce side lobe level for WLAN. Int. J. Eng. Innovative Technol. (IJEIT) **3**(7), 118–120 (2014)
11. Yang, X.-S., Deb, S.: Engineering optimisation by cuckoo search. Int. J. Math. Model. Numer. Optim. **1**(4), 330–343 (2010)
12. Yang, X.S., Deb, S.: Cuckoo Search via Lévy Flights Proceeings of World Congress on Nature & Biologically Inspired Computing (NaBIC 2009, India), pp. 210–214. IEEE Publications, USA (2009)
13. Yang, X.S., Deb, S.: Nature-Inspired Metaheuristic Algorithms. Luniver press, Bristol (2008)
14. Pradhan, H., Mangaraj, B.B., Misra, I.S.: Designing and optimizing a mutually coupled planar array antenna using PSO for UHF application. In: IEEE Conference on Applied Electromagnetic Conference (AEMC, 2013), pp. 18–20, India (2013)

Model Based Test Case Generation from UML Sequence and Interaction Overview Diagrams

Ajay Kumar Jena, Santosh Kumar Swain and Durga Prasad Mohapatra

Abstract Test case generation is the most crucial job of testing paradigm. Unified Modelling Language (UML) model, give a lot of information for testing which is accepted widely by both the academia and industry. By using UML artifacts the early detection of faults can be achieved during designing while the architectural overview of the software is considered. Using a specific diagram of UML certainly helpful in detecting the flaws. But combining different UML components, more test cases can be generated and different types of faults can be captured by reducing the redundants. In this paper, we propose a method using Sequence Diagram (SD) and Interaction Overview Diagram (IOD) to generate the test cases. An intermediate graph is generated known as Sequence Interaction Graph (SIG) by combining Message Sequence Dependency Graph (MSDG) generated from the sequence diagram and Interaction Graph (IG) generated from Interaction Overview Diagram of the models. By combining the IG and MSDG, different scenarios are generated and it follows the test cases. We use the dominance concept for generating the test cases from these scenarios.

Keywords Sequence diagram · Interaction overview diagram · Testing · Dominance

1 Introduction

Testing is an important phase of software development which aims at producing highly efficient software systems while maintaining the quality and increasing the reliability of the software system [1]. In the procedure of testing, test case

A.K. Jena (✉) · S.K. Swain
KIIT University, Bhubaneswar, Odisha, India
e-mail: ajay.bbs.in@gmail.com

S.K. Swain
e-mail: sswainfcs@kiit.ac.in

D.P. Mohapatra
National Institute of Technology, Rourkela, Rourkela, Odisha, India
e-mail: durga@nitrkl.ac.in

L.C. Jain et al. (eds.), *Computational Intelligence in Data Mining - Volume 2*,
Smart Innovation, Systems and Technologies 32, DOI 10.1007/978-81-322-2208-8_23

generation is the most difficult task [2]. Test cases are mainly designed from the source code of the program [2]. Generally, the code can be generated after the analysis and design of the software. So, it is very difficult to test the software at the early phase of development. To avoid wasting of time consumption and cost utilized in the testing process in case of the code based systems, it is desirable to generate the test cases at the design level so that reliability of the software will be increased. Model based testing is more efficient and effective than code based approach as it is the mixed approach of source code and specification requirements for testing the software [3].

The Unified Modeling Language (UML) is a visual modeling language used to specify, visualize, construct, and document the artifacts of software. By using UML, rather waiting till the end of coding, the user can detect the faults in the designing level of the software which helps in saving the time and effective planning of the software. We propose a method to generate the test cases using UML sequence diagram and interaction overview diagram. Sequence diagrams are used for describing the behaviour by modeling the flow of messages in a system. Sequence diagram describes how an object, or group of objects, interacts within a system [4]. Sequence diagram shows how the messages are exchanged among the objects [4]. Sequence diagram may not be able to decide the components of precondition, post condition, input and the expected output of the test case in many cases. So, IOD will be helpful in picturing the control flow which visualise the sequence of activities. UML interaction overview diagrams combine elements of activity diagrams with sequence diagrams to show the flow of program execution. By using IOD one can deconstruct a complex scenario that would otherwise require multiple if-then-else paths to be illustrated as a single sequence diagram. It shows dependency between the important sequences of a system, which can be presented by an activity diagram. We use dominance concept of a tree to generate the test cases. A node x in a flow graph G dominates node y iff every path in G from the initial node s to y contains x.

The rest of the paper is organised as follows. In Sect. 2, necessary background is presented. The proposed model for test case generation is described in Sect. 3 of our work. A case study of Library book issue use case of Library Information System (LIS) is discussed in Sect. 4 and implementation is given in Sect. 5. Comparison with related work is presented in Sect. 6. Finally, we conclude our work with Sect. 7 with conclusion and future work.

2 Background

In this section, we are briefly describing some necessary concepts graph modeling [5], UML diagrams required for the understanding of the remaining sections.

2.1 Sequence Diagram

Sequence diagrams are well suited for objected oriented software. It shows how the objects interacting each other and sequence of messages exchanged between the objects. An interaction in a sequence diagram is a sequence of messages between the objects to perform a specific task. There are several Interaction operators defined in UML 2.x which are: alternatives (alt), break (break), parallel (par), weak sequence (seq), strict sequence (strict), negative (neg), critical region (region), ignore/consider (ignore/consider), assertion (assert), and loop (loop). Figure 1 shows the sequence diagram of library book issue use case of LIS.

Alternatives (alt): It provides a choice of alternatives, out of which only one alternative will be executed. The interaction operands are evaluated on the basis of specified guard expression. An else guard is provided that evaluates to TRUE if and only if all guards of the other Interaction Operands evaluate to FALSE.

Break (break): It is an Alternative operator where one operand is given and the other assumed to be the rest of the enclosing Interaction Fragment. In the course of processing an interaction, if the guard of the break is satisfied, then the containing interaction abandons its normal execution and instead performs the clause specified by the break fragment.

Parallel (par): It supports the parallel execution of a set of Interaction Operands.

Loop (loop): It indicates that the interaction operand will be repeatedly executed for some number of times which may be indicated with a guard condition and also includes a mechanism to stop the iteration when the guard condition is false.

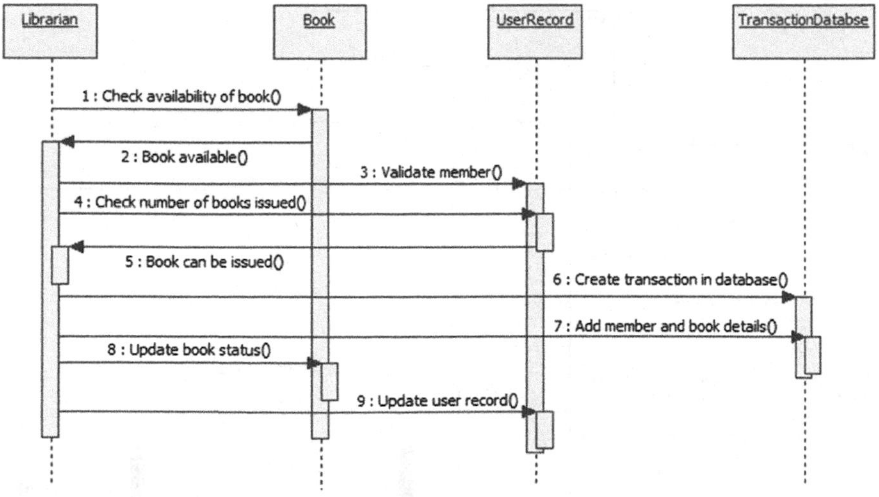

Fig. 1 Sequence diagram of library book issue use case of LIS

2.2 Interaction Overview Diagram (IOD)

Interaction overview diagram appeared in version 2.0 due to the need to complete the faults disadvantages of activity and sequence diagrams. An interaction overview diagram is a UML behavioral diagram that defines interactions and is a variant of the activity diagram, which emphasizes the high-level control flow.

Interaction overview diagrams illustrate an overview of a flow of control in which each node can be an interaction diagram. The nature of the IOD is used to describe the higher level of abstraction of the components within the system. The IOD can show dependence between the important sequences of a system, which can be presented by an activity diagram. The used notation incorporates constructs from sequence diagrams with fork, join, decision and merge nodes from activity diagrams. IODs are special kinds of activity diagrams where the activity nodes are interactions and the activity edges denote the control flow of the interaction. According to the specification of UML 2.x the object flow cannot be represented by an IOD. Figure 2 shows the IOD of the Book issue use case of LIS.

More formally, an IOD can be defined by the tuple IOD = $\langle n_0, N_f, B, D, I, E, Ed \rangle$ where :

n_0 initial node.
N_f $\{nf_1, \ldots, nf_n\}$ is a set of final nodes.
I $\{I_1, \ldots, I_n\}$ is a set of interaction nodes.
B $\{b_1, \ldots, b_n\}$ is a set of join/fork nodes.

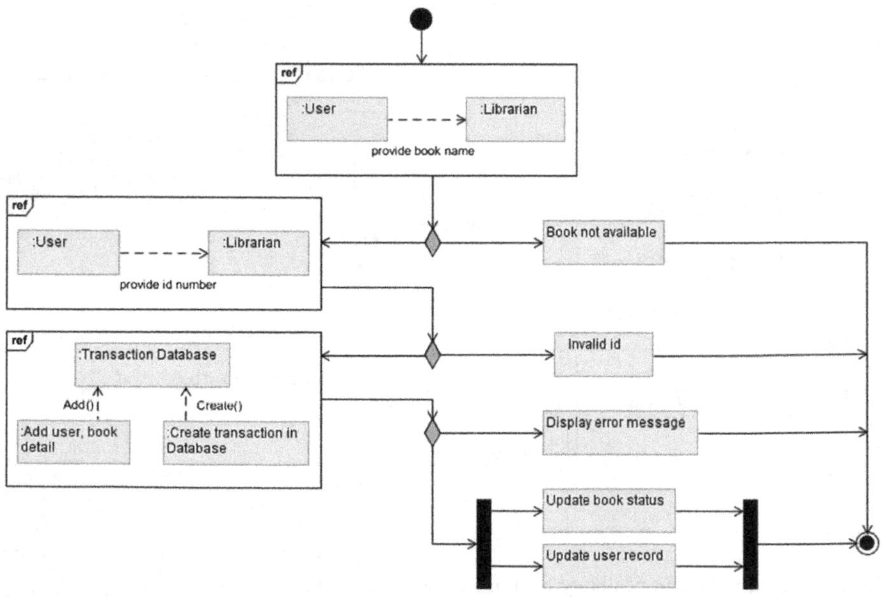

Fig. 2 Interaction overview diagram of book issue use case of LIS

D $\{d_1, \ldots, d_n\}$ is a set of decision/merge nodes.
E $\{e_1, \ldots, e_n\}$ is a set of edges connecting the IOD nodes.
Ed $\{n_0\} \cup Nf \cup I \cup B \cup D \times \{n_0\} \cup Nf \cup I \cup B \cup D \rightarrow E$ is a function which connects two IOD nodes by an edge.

2.3 Dominance

Suppose G = (V, E) be a digraph with two distinguished nodes n_0 and n_k. A node n dominates a node m if every path P from the entry node n_0 to m contains n. Several algorithms are given in the literature to find the dominator nodes in a digraph. By applying the dominance relations between the nodes of a digraph G, we can obtain a tree (whose nodes represent the digraph nodes) rooted at n_0. This tree is called the dominator tree. Figure 7b shows the dominance tree (DT) of G for Book Issue use case of Library Information System based on the intermediate graph. A (rooted) tree DT(G) = (V, E) is a digraph in which one distinguished node n_0, called the root, is the head of no arcs; in which each node n excepts the root n_0 is a head of just one arc and there exists a (unique) path from the root node n_0 to each node n; we denote this path by dom(n) or dominance path. Tree nodes having zero out degree are called the leaves. Here input variables are the functionalities of an application.

3 Proposed Model for Test Case Generation

We have proposed the following model to generate the test cases. Figure 3 represents the proposed model for our approach. The system we are proposing will work as follows.

i. Construct the Sequence and Interaction Overview Diagrams for the system.
ii. Convert the Sequence diagram into Message Sequence Dependency Graph (MSDG) and Interaction Overview Diagram to Interaction Graph (IG) as described in Sect. 4.
iii. An intermediate graph Sequence Interaction Graph (SIG) generated by integrating MSDG and IG.

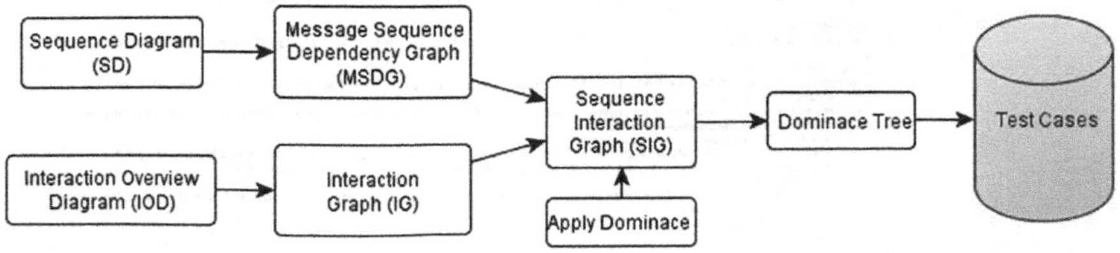

Fig. 3 Proposed model for test case generation

iv. By applying dominance on SIG, a dominance tree is created.
v. The test scenarios are generated from the tree by all possible test paths using Depth First Search (DFS) method.
vi. Finally, the test cases are generated from the test scenarios.

4 Case Study

The Sequence Diagram and Interaction Overview Diagram of Library Book Issue System are created using StarUML and presented in Figs. 1 and 2 respectively. The users coming to the library for issuing books are considered in our use case. The user may or may not be a valid user of the library. The books in the library may or not be available. The books can be issued to the user if he/she is a valid member, books are available and he/she has not issued the no. of books he/she entitled for. Accordingly the error messages will be displayed. If the books will be issued to the user the book status and member records will be updated. The transaction will be also recorded. From Figs. 1 and 2 we generate the XMI code out of which one snapshot is given in Fig. 4.

By using Table 1 and the XMI code in Fig. 4 we generate MSDG of the sequence diagram as presented in Fig. 5. Then, we try to get the ID of the objects from the XMI code and the sequence of the objects. Similarly, by using Table 2 and the XMI code of IOD, the control flow graph is prepared as given in Fig. 6.

4.1 Intermediate Representation

For generating the test scenarios, it is necessary to transform the diagrams into suitable intermediate representations. The intermediate Sequence Interaction Graph (SIG) generated by using the message dependency graph given in Fig. 5 and the


```
<?xml version="1.0" encoding="UTF-8"?>
- <XMI timestamp="Sat Jul 12 12:13:23 2014" xmlns:UML="href://org.omg/UML/1.3" xmi.version="1.1">
  - <XMI.header>
    - <XMI.documentation>
        <XMI.owner/>
        <XMI.contact/>
        <XMI.exporter>StarUML.XMI-Addin</XMI.exporter>
        <XMI.exporterVersion>1.0</XMI.exporterVersion>
        <XMI.notice/>
      </XMI.documentation>
      <XMI.metamodel xmi.version="1.3" xmi.name="UML"/>
    </XMI.header>
  - <XMI.content>
    - <UML:Model xmi.id="UMLProject.1">
      - <UML:Namespace.ownedElement>
        - <UML:Model xmi.id="UMLModel.2" isAbstract="false" isLeaf="false" isRoot="false" namespace="UMLProject.1" isSpecification="false" visibility="public" name="Use
          Case View">
          - <UML:Namespace.ownedElement>
            - <UML:Actor xmi.id="UMLActor.3" isAbstract="false" isLeaf="false" isRoot="false" namespace="UMLModel.2" isSpecification="false" visibility="public"
              name="Librarian">
              - <UML:Namespace.ownedElement>
                - <UML:Collaboration xmi.id="UMLCollaborationInstanceSet.4" isSpecification="false" visibility="public" name="CollaborationInstanceSet1">
                  - <UML:Collaboration.interaction>
                    - <UML:Interaction xmi.id="UMLInteractionInstanceSet.5" isSpecification="false" visibility="public" name="InteractionInstanceSet1"
                      context="UMLCollaborationInstanceSet.4">
                      - <UML:Interaction.message>
                        - <UML:Message xmi.id="UMLStimulus.6" isSpecification="false" visibility="public" name="Check availability of book"
                          interaction="UMLInteractionInstanceSet.5" receiver="UMLObject.25" sender="UMLObject.24">
                          - <UML:Message.action>
                              <UML:CallAction xmi.id="UMLCallAction.7" isSpecification="false" visibility="public" name="" stimulus="UMLStimulus.6"
                                isAsynchronous="false"/>
                            </UML:Message.action>
                          </UML:Message>
```


Fig. 4 Snapshot of XMI code of SD

Table 1 Message dependency from sequence diagram

Symbols	Message passed between objects
S1	Check book availability()
S2	Book available()
S3	Validate member()
S4	Check no of books issued()
S5	Book issued()
S6	Create Transactions()
S7	Add member and Book details()
S8	Update Book Status()
S9	Update User Record()
S10	Disp_Err_Mess1("Book not available")
S11	Disp_Err_Mess2("Not a valid member")
S12	Disp_Err_Mess3("Books entitled completed")
S13	End

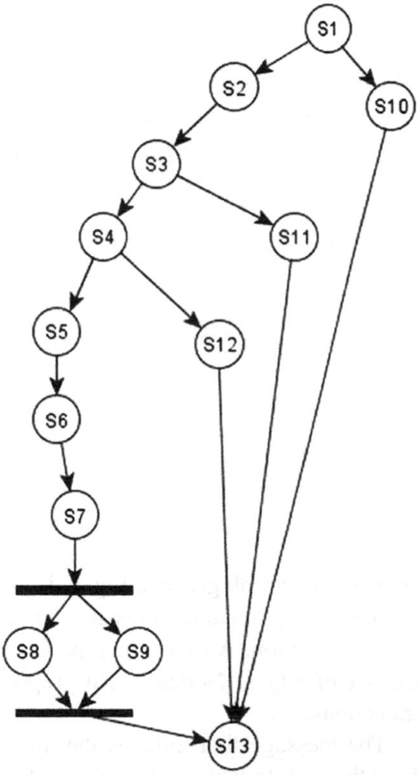

Fig. 5 Message sequence dependency graph of sequence diagram

Table 2 Nodes of the IOD with Id_Number

Identification number	Interaction nodes
I1	Check book
I2	Book not available
I3	Insert user no.
I4	Invalid user
I5	Check no. of books
I6	Display error message
I7	Update book status
I8	Update user record
I9	Issue book
I10	Stop

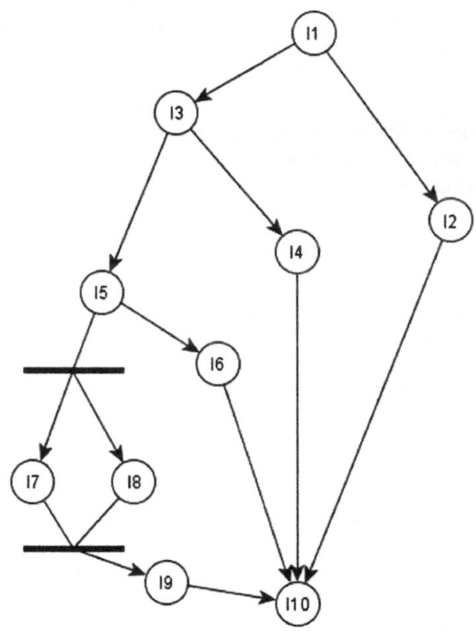

Fig. 6 CFG of the IOD given in Fig. 2

control flow graph given in Fig. 6, is presented in Fig. 7a. SIG is generated from the sequence diagram and represents the possible message/methods between sequences in an interaction. A SIG is a graph G = (V, E), where V is the set of nodes, and E is the set of edges. Nodes of SIG represent the messages and edges represent the transitions.

The message that initiates the interaction is made the root of the graph. In an IOD the each transition is labeled with a guard condition. The conditional predicate corresponding to the guard condition might trivially be an empty predicate which is always true. The initial nodes in the sequence and IOD diagrams are same. For the

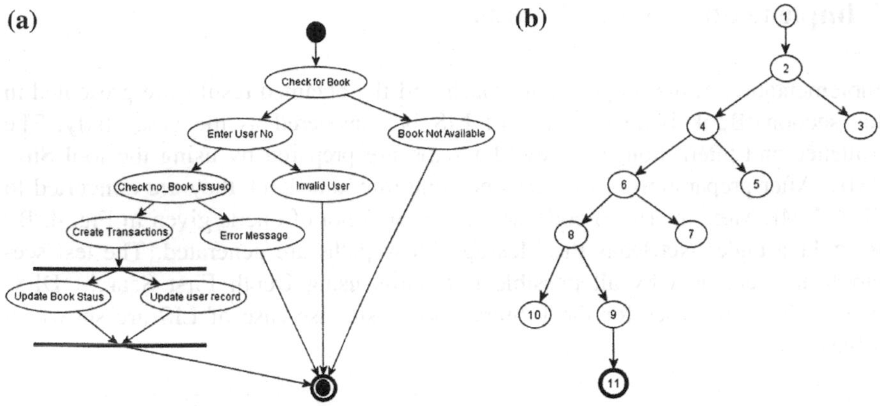

Fig. 7 **a** SIG of library book issue use case. **b** Dominance tree of (**a**)

Table 3 Generated test cases from the dominance tree

Test case no.	User no.	Books issued	Book name	Books after transaction	Expected result	Actual result
1	15104	–	CG	–	Invalid User	Invalid User
2	13104	5	OS	5	Entitled completed	Entitled completed
3	12016	4	PC	4	Error Message	Book Not Available
4	12307	3	AOS	4	Book Issued successfully	Book Issued successfully
5	11509	4	SE	4	Not_in_stock	Not_in_stock

intermediate representation, all the interactions from the IOD are considered first from Fig. 6 and all sequences of interaction are considered from MSDG of Fig. 5. The dominance tree of the SIG is given in Fig. 7a, b. By using the dominance tree the dominance leaf nodes are calculated.

Here, the dominance tree leaf nodes are:

Dom(3) = 1, 2, 3
Dom(5) = 1, 2, 4, 5
Dom(7) = 1, 2, 4, 6, 7
Dom(10) = 1, 2, 4, 6, 8, 10
Dom(11) = 1, 2, 4, 6, 8, 9, 11

By using the above dominance tree we generate the test cases which are presented in Table 3.

5 Implementation and Results

Implementation of the proposed approach and the obtained results are presented in this section. Book issue use case of LIS is considered as the case study. The sequence and interaction overview diagrams are prepared by using the tool Stra-UML. After preparation of the sequence diagram, and IOD, they are converted to XMI (XML Meta Interface) code where the snapshot of one is given in Fig. 4. By using Java under NetBeans the Message Flow paths are generated. The test scenarios are generated by all possible test paths using Depth First Search (DFS) method. The test cases for the Library book issue use case of LIS are shown in Table 3.

6 Comparison with Related Works

A comparison of our work with other related work is presented in this section. To the best of our knowledge no work has been reported till date using combination of diagrams of sequence and Interaction Overview Diagrams.

Model based test cases has been discussed with a large scope in the literature, but by using UML models it is scare [2, 3, 6, 7]. Nayak and Samanta [8] proposed an approach of synthesizing test data from the information embedded in model elements such as class diagrams and sequence diagrams. They used OCL constraints. In their approach, they annotated a sequence diagram with attribute and constraint information derived from class diagram and OCL constraints and mapped it onto a structured composite graph called SCG. The test specifications are then generated from SCG. Sarma and Mall [9] transformed use case diagram to a graph called use case diagram graph (UDG) and sequence diagram into sequence diagram graph (SDG) and then integrated UDG and SDG to form a system testing graph (STG). The STG is then traversed to generate test scenarios by using state-based transition path coverage criteria. Pechtanun and Kansomkeat [10] presented a method to generate test cases from UML activity diagram. They transformed the activity diagram into a grammar, called Activity Convert (AC) grammar. Then the AC grammar is used to generate the test cases. Panthi and Mohapatra [11] proposed a technique for test sequence generation using UML sequence diagram. First they constructed the sequence diagram of the case study. The sequence diagram is then traversed, conditional predicates are selected and these are transformed to source code. Then they generated the test cases by using function minimization method. Our approach is also comparable to the approach of Sarma and Mall [9] in the contrast that sequential messages among the objects and interactions between the objects are explored using sequence and interaction overview diagram of UML 2.x. But, the authors considered the states of the objects within a scenario of the use case.

7 Conclusion and Future Work

In this work, we have proposed an approach to generate the test cases using sequence and interaction overview diagrams. By using multiple diagrams of UML, our approach is significant to diagnose the location of a fault in the earlier stages, thus reducing the testing effort and making better planning. This can also detect more numbers and different types of faults. An intermediate graph called Sequence Interaction Graph is generated by using Message Sequence Dependency Graph of the sequence diagram and Control Flow Graph of the Interaction Overview diagrams. We have used dominance concept in the intermediate graph for generating the test cases. By using our approach, minimum presence of errors are guaranteed in the test case generation. This approach may be extended other diagrams of UML. In our future work, we will consider to prioritize the test cases and to optimize them.

References

1. Mall, R.: Fundamentals of Software Engineering, 3rd edn. Prentice Hall, New Delhi (2009)
2. Abdurazik, A., Offutt, J.: Using UML collaboration diagrams for static checking and test generation. In: Proceedings of the 3rd International Conference on the UML. Lecture Notes in Computer Science, vol. 1939, pp. 383–395. Springer, New York (2000)
3. Ali, S., Briand, L.C., Jaffar-ur-Rehman, M., Asghar, H., Zafar, Z., Nadeem, A.: A state based approach to integration testing based on UML models. J. Inf. Softw. Technol. **49**(11–12), 1087–1106 (2007)
4. Rumbaugh, J., Jacobson, I., Booch, G.: The UML Reference Manual. Addison-Wesley, Reading (2001)
5. Das, H., Jena, A.K., Rath, P.K., Muduli, B., Das, S.R.: Grid computing-based performance analysis of power system: a graph theoretic approach. In: Intelligent Computing, Communication and Devices, pp. 259–266 (2014)
6. Shirole, M., Suthar, A., Rajeev Kumar, R.: Generation of improved test case from UML state diagram using genetic algorithm. In: ACM ISEC'11, pp. 125–134 (2011)
7. Patel, P.E., Patil, N.N.: Test cases formation using UML activity diagram. In: International Conference on Communication Systems and Network Technologies. IEEE (2013)
8. Nayak, A., Samanta, D.: Automatic test data synthesis using UML sequence diagrams. J. Object Technol. **09**(2), 75–104 (2010)
9. Sarma, M., Mall, R.: Automatic test case generation from UML models. In: International Conference on Information Technology, pp. 196–201. IEEE Computer Society (2007)
10. Pechtanun, K., Kansomkeat, S.: Generation of test case from UML activity diagram based on AC grammar. In: International Conference on Computer and Information Science (2012)
11. Panthi, V., Mohapatra, D.P.: Automatic test case generation using sequence diagram. Int. J. Appl. Inf. Syst. (IJAIS) **2**(4), 22–29 (2012)

Enhancing HMM Based Malayalam Continuous Speech Recognizer Using Artificial Neural Networks

Anuj Mohamed and K.N. Ramachandran Nair

Abstract Improving discrimination in recognition systems is a subject of research in recent years. Neural network classifiers are naturally discriminative and can be easily applied to real-world problems. This paper examines the use of multilayer perceptrons as the emission probability estimator in a hidden Markov model based continuous speech recognizer for Malayalam language. The performance of the system has been compared with a recognizer using Gaussian mixture model as the emission probability estimator. Experimental results show that the proposed neural network based acoustic scoring yields significant gains in recognition accuracy and system compactness.

Keywords Continuous speech recognition · Malayalam speech recognition · Hidden Markov model · Gaussian mixture model · Artificial neural network

1 Introduction

Speech is the most direct form of human communication and efficient speech interfaces can make human-computer interaction more effective. A continuous speech recognition (CSR) system converts a sequence of words into a text message independent of the device, speaker or the environment. In continuous speech the words in the utterances are generally spoken without pauses between the words. Therefore, the words in the utterance are strongly co-articulated, which makes the recognition considerably difficult compared to isolated word recognition. Highly

A. Mohamed (✉)
School of Computer Sciences, Mahatma Gandhi University, Kottayam, Kerala, India
e-mail: anujmohamed@mgu.ac.in

K.N. Ramachandran Nair
Department of Computer Science and Engineering, Viswajyothi College of Engineering and Technology, Vazhakulam, Muvattupuzha, Kerala, India
e-mail: knrn@hotmail.com

L.C. Jain et al. (eds.), *Computational Intelligence in Data Mining - Volume 2*,
Smart Innovation, Systems and Technologies 32, DOI 10.1007/978-81-322-2208-8_24

reliable CSR systems are required in many rapidly growing application areas such as transcription of recorded speech, searching audio documents, interactive spoken dialogues and web-enabling via voice.

Many research groups have been working in the field of automatic speech recognition (ASR) and in the past few years there have been proposed a number of ASR systems especially for European languages. Recently, there have also been various efforts towards ASR in Indian languages like Tamil, Telugu, Marathi and Hindi. Malayalam is a Dravidian language spoken mainly in the South West of India. Research in Malayalam speech recognition has started recently and the area of CSR is relatively less investigated. Recent research works on continuous Malayalam speech recognition have been reported in [1–3].

Hidden Markov models (HMM), a statistical framework that supports both acoustic and temporal modeling, are extensively and successfully applied in the state-of-the-art ASR systems. In an ASR system, an acoustic model is intended to capture the characteristic features of each class (acoustic unit) from the given training data and returns the probability of observing/emitting an acoustic unit from a particular state. Multivariate Gaussian mixture model (GMM) is typically used to model the likelihood of an acoustic unit being generated by a given HMM state. Most practical data encountered in speech has complex distribution in the feature space, and hence cannot adequately be described by a GMM, which uses only the first and second order statistics and mixture weights [4]. Also, estimation of the parameters of the GMM is based on maximum likelihood (ML) criterion which is not discriminative. The maximum likelihood approach models each class individually without considering the overall classification performance. The outputs of discriminatively trained artificial neural networks (ANN) in classification mode can be interpreted as estimates of a posteriori probabilities of output classes conditioned on the input. Using Bayes' rule, these state posteriors can be converted to emission probabilities required by the HMM framework. Therefore, an ANN is useful in acoustic scoring of speech patterns, but they are not powerful to deal with the temporal and sequential nature of speech. This fact lead to the idea of using ANNs as emission probability estimators within an HMM, broadly known as hybrid HMM/ANN [5, 6]. Such systems can take advantages of the complementary capabilities of connectionist networks (in particular their discriminative power) and of HMM models (in particular their capacity of handling time). Use of ANN as emission probability estimator has shown good performance for ASR systems for various western languages [7, 8]. The work reported here explores the use of ANN as the emission probability estimator in an HMM based Malayalam CSR system.

The rest of the paper is organized as follows. Section 2, briefs the Bayesian classification theory and its role in ASR. Section 3 describes the HMM modeling approach in continuous speech recognition. Section 4 looks into GMMs and MLPs as emission probability estimators. The experiments conducted and the results are given in Sect. 5. Concluding remarks with a discussion of promising avenues of work to improve the performance of the system are slated in Sect. 6.

2 Mathematical Formulation of Speech Recognition Problem

Typical probabilistic representation of the ASR problem is as follows:

If $O = o_1, o_2, \ldots, o_T$ is the acoustic observation sequence given to the ASR system and $W = w_1, w_2, \ldots, w_N$ is a sequence of words, then the goal of the system is to find the most likely word sequence $\hat{W}$.

$$\hat{W} = \arg\max_{w} P(W|O). \quad (1)$$

P(W|O) is known as the a posterior probability. It is difficult to directly compute the above maximization and this problem can be simplified by applying the Bayes' rule:

$$\hat{W} = \arg\max_{w} \frac{P(O|W)P(W)}{P(O)}. \quad (2)$$

The denominator P(O) in the above equation can be neglected, because the acoustic observation *O* is the same for all competing hypotheses *W*, to give the equation:

$$\hat{W} = \arg\max_{w} P(O|W)P(W). \quad (3)$$

Thus a central issue in the design of a speech recognizer is the accurate estimation of the probabilities given in Eq. 3. The probability, P(O|W), is usually referred to as the likelihood (class-conditional probability) that the speaker produces the acoustic data *O* if he utters the word sequence *W*. This value is typically provided by an acoustic model. The prior probability, P(W), that the speaker utters the word sequence *W* is determined by a language model. A decoder uses these two probability scores and information from the lexicon (pronunciation dictionary) to sort through all possible hypotheses and outputs the most likely word sequence. The input feature vector required by the recognizer is provided by the acoustic front-end. This is the framework under which most statistical ASR systems operate.

3 Continuous Speech Recognition Using HMMs

HMM is the state-of the art technology in speech recognition. The major advantages that support their application in the speech recognition area include their computationally efficient learning and decoding methods for temporal sequences, good sequence handling capabilities and a rich mathematical framework with flexible topology. An HMM is typically defined and represented as a stochastic finite state automation usually with a left-to-right topology as proposed in [9] to reflect the temporal flow, when used for speech. An HMM models an utterance $O = \{o_1o_2\ldots o_t\ldots o_T\}$ as a succession of discrete stationary states $Q = \{q_1q_2\ldots q_k\ldots q_K\}$ with instantaneous transitions between the states and generates an observation. The sequence of states, which represents the spoken utterance, is "hidden" and the parameters of the probability density functions of each state are needed in order to associate a sequence of states $Q = \{q_1q_2\ldots q_k\ldots q_K\}$ to a sequence of observations $O = \{o_1o_2\ldots o_t\ldots o_T\}$. The emission probability distribution estimates the probability with which the given observation O has been generated.

In CSR systems, a sentence is modeled as a sequence of sub word units, usually phonemes, and each phoneme is represented by a continuous density HMM with transition probability $\{a_{ij}\}$ (from state i to state j) and emission (observation) density $\{b_j(o_t)\}$ for observation o_t at state j. The transition probabilities introduce a temporal structure in the modeling of dynamic speech sequence. When a state is entered according to the transition probabilities defined, the HMM generates observations according to the state output distribution.

Once the HMM has been trained with the speech samples it can be used for decoding or recognition. Decoding finds out the most alike path of states within the model and assigns each individual observation to a given state within the model.

4 Methodology

4.1 Speech Corpus and Feature Extraction

The speech databases developed consist of naturally and continuously read Malayalam sentences from both male and female speakers who speak various dialects of Malayalam. 25 phonemes were randomly selected from the frequently used phoneme set to provide a balanced coverage of the phoneme base. After selecting the phonemes to be modeled, 3 phonetically balanced sentences with 15 words were identified to provide optimal coverage of these units. 9 male and 11 female native speakers of Malayalam in the age group of 20–25 with different dialects were identified for reading the sentences. Each speaker read the sentences multiple times. The Windows sound recorder along with a headset microphone was used for recording the speech input. Two databases (Dataset1 and Dataset2) for training and one (Dataset3) for testing were developed.

The input to a speech recognizer is the raw audio signal. The feature extraction stage parameterizes this signal into a discrete sequence of acoustic feature vectors. Most ASR systems use Mel frequency cepstral coefficients (MFCC) as its input feature vector. These coefficients are normally appended with its first (delta) and second (acceleration) derivatives in time to reflect the dynamic nature of the speech signal. To compute the input feature vectors, the above databases were digitized at 8 kHz sampling rate with a precision of 16 bits. After pre-emphasis and application of a Hamming window, 12 MFCC features were extracted from each speech sample along with their delta and acceleration coefficients. These parameters were computed every 8 ms on an analysis window of 16 ms and were used as the input feature vector.

4.2 Selection and Modeling of Basic Speech Units

The quality of the basic units used for speech recognition and their modeling affects the recognition performance of the system using them. Indian languages at large and Malayalam in particular are very phonetic in nature. Its orthography is largely phonemic having a written form (grapheme) that has direct correspondence to the spoken form (phoneme). For reasons of efficiency of representation and also because of the orthography of Malayalam, a phoneme was selected as the basic unit of speech in this work. Context independent modeling was used to model the phonemes because of two reasons. First, context independent nature of phoneme reduces the number of parameters and consequently, the required amount of training material. Second, the ANN based approach requires context independent phoneme modeling. A set of 25 unique context independent phoneme labels were extracted from the lexicon and was used as the basic speech unit to provide full phoneme coverage for all the vocabulary words identified.

The lexical knowledge was modeled using the pronunciation dictionary (lexicon). The lexicon provides a mapping between the words in the vocabulary and the corresponding phoneme representation and is an important aspect for phoneme based systems where the mapping between a word and its lexicon is governed by the pronunciation. A lexical entry was created for each word in the spoken sentences. Malayalam being a phonetic language with one-to-one mapping between grapheme and phoneme, the phonetic decomposition was done first by separation of words into graphemes and then into phonemes. Each lexical entry was characterized by a linear sequence of phoneme units.

For constructing the HMM based sentence model from phonemes, first each sentence was expressed as a series of specified words according to the word transcription of the utterance. Then each word was replaced by its sequence of phonemes according to word pronunciations stored in the lexicon.

4.3 Acoustic Scoring: GMM

The state-of-the-art HMM based ASR systems use GMM to estimate the emission probabilities of HMM states. The baseline system reported in this paper uses continuous density HMMs (CDHMM) where the probability density functions usually used to characterize each state is a mixture of Gaussians (which is a combination of a finite number of Gaussian distributions),

$$b_j(o_t) = \sum_{m=1}^{M} c_{jm} N(o_t; \mu_{jm}, \sum_{jm}), \quad 1 \leq j \leq S \tag{4}$$

where o is the observation vector being modeled, c_{jm} is the weight of the mth mixture component in state j, M is the total number of mixtures and $N(o; \mu, \Sigma)$ is a multivariate Gaussian with mean vector μ_{jm} and covariance matrix Σ_{jm} for the mth mixture component in state j. S represents the number of states in the HMM.

The parameters of the GMM (means and covariance of the Gaussians and GMM weights) were initialized from the training data. The density functions were trained individually for each phoneme. First, the training data was uniformly divided into equal-length segments equivalent to the number of phonemes in the transcription. Then all the acoustic segments corresponding to a phoneme unit were collected. K-means clustering algorithm was applied on all frames in the same state of the unit to partition the set of frames into a number of clusters, where each cluster represents one of the M mixtures of the $b_j(o_t)$. Mean and covariance were computed for each cluster. To reduce the computational complexity a diagonal covariance matrix was used. The covariance matrices were calculated as the sample covariance of the points associated with (i.e. closest to) the corresponding centre. The initial values obtained were then re-estimated using the Baum-Welch algorithm [10] which is based on the maximum likelihood criterion. The transition probabilities were initialized to be equally probable over the possible state transitions. The prior probability Π was set to 1 in the first state, and 0 for the rest of the states.

As a result of the above procedure, a set of HMM phoneme models was created. These phoneme units were then joined together, using the transition probabilities, to form the HMM sentence models.

4.4 Acoustic Scoring: MLP

The neural network used in the proposed hybrid CSR system is a multilayer perception (MLP) which is the most widely used neural network classifier for speech recognition. Each output unit of the MLP is trained to compute the a posteriori probability of a CDHMM state given the acoustic observation.

The network used was a two layer MLP with a single hidden layer. The number of nodes in the input and output layers correspond to the number of acoustic features and the number of phoneme classes to be represented. The number of hidden units was determined by optimizing the performance of the net on a cross validation set, which consisted of utterances that were not used for training. The network training was performed using the standard back-propagation algorithm [11] with cross entropy error criterion. The weights and biases of the network were initialized randomly. The weight and bias values were updated according to gradient descent momentum and an adaptive learning rate.

MFCC feature vectors extracted from the frame to be classified along with features from three surrounding frames were given as input to the network. Such a measure can be interpreted as context-dependency on the frame level and can be a certain substitute for the missing context-dependency on the phoneme level. This also incorporates the dynamic nature of speech. The output layer had 25 neurons (with softmax non-linearity) representing the output phoneme classes. The activation function used in the hidden layer was the sigmoid function. The batch mode of training in which weights were adjusted after each epoch was used to train the system. The network was tested after each epoch on a separate cross-validation set. The posterior probabilities obtained were divided by the prior probabilities to produce the emission likelihoods for the HMM. Evaluation was performed by using the probabilities obtained from the MLP within the HMM framework.

4.5 Decoding

Decoding refers to the process of hypothesizing the underlying hidden state sequence, given the sequence of observations. The Viterbi [12] algorithm was used in the decoding phase to return the hypothesized transcription. Good performance requires that we incorporate basic speech knowledge within the model. The method and complexity of modeling language varies with the speech application. Finite State Network (FSN) is well adapted for Indian languages and was used in this work to represent the valid phoneme sequence to restrict the search space. The sentence model being constructed by concatenating the basic phoneme units, the language model probabilities define which phoneme to follow at each point in the model and were incorporated in the transition probabilities. This value was used by the Viterbi algorithm to update the estimate of a path probability. The hypothesized transcription and the reference transcription were compared and the performance of the system was evaluated using the popular word error rate (WER) metric and sentence recognition rate (SRR).

5 Results and Discussions

Table 1 shows the recognition performance of the continuous Malayalam speech recognizer with GMM and ANN as emission probability estimators. The WERs with both the acoustic scoring approaches and the percentage reduction in the WER obtained with the HMM/ANN hybrid approach are given. A strong improvement has been observed when the HMM emission probabilities were computed using the discriminatively trained MLPs i.e., a relative improvement of 33.25 and 41.65 % on DB1 and on DB2 respectively. This can be explained by the fact that, being a

Table 1 Performance comparisons between HMM and HMM/ANN hybrid approaches

Dataset	WER		SRR		GMM → ANN	
	GMM	ANN	GMM	ANN	Percentage of reduction in WER	Percentage of increase in SRR
Dataset1	4	2.67	88.89	93.33	33.25	4.75
Dataset2	5.33	3.11	86.67	91.11	41.65	4.87

In the HMM approach emission probabilities were estimated using GMMs and in the HMM/ANN hybrid emission probabilities were estimated using MLPs. The hybrid approach has provided better performance

Table 2 Comparison of trainable parameters: GMM versus ANN as emission probability estimators

No. of trainable parameters		Percentage of reduction in trainable parameters
GMM	ANN	GMM → ANN
70,200	10,475	85.08

discriminative approach, more information is stored by the MLP during the training phase. In the GMM each phoneme class is trained independently using the maximum likelihood approach with in-class data only. But, the MLP is trained with both in-class and out-of-class data which helps it to discriminate among classes more efficiently.

Numbers of free parameters to be trained in the case of both the approaches are given in Table 2. The value given is for the best performance obtained. A relative reduction of 85.08 % in the number of trainable parameters has been observed in the case of ANN acoustic models. As the likelihoods model the surface of distribution, GMMs require more parameters compared to discriminant functions which model boundaries.

6 Conclusions

This paper has reported a comparison of the GMM and ANN based acoustic models in the context of a small vocabulary, speaker independent Malayalam continuous speech recognizer. Experimental results have shown that the use of ANN increases the recognition performance of the system. There was also a relative reduction of 85.08 % in the number of free parameters compared to the GMM based approach. The use of contextual information and discriminative training has been found to enhance the performance of the system. The incorporation of ANN in an HMM based CSR system also provides advantages like context sensitivity, a natural structure for discriminative training, model accuracy etc., over the use of traditional mixtures of Gaussian as acoustic model.

In this work, the MLP was trained in the multi-class classification approach. In classification problems like ASR, the high number of classes to be separated makes the boundaries between classes complex. Computationally expensive learning algorithms learn many small problems much faster than a few large problems. Therefore, the use of posteriors derived from MLP trained in pairwise classification mode may be helpful to improve the performance of the system.

References

1. Anuj, M., Nair, K.N.R.: Continuous Malayalam speech recognition using hidden Markov models. In: Proceedings of 1st Amrita-ACM-W Celebration on Women in Computing in India (2010)
2. Anuj, M., Nair, K.N.R.: HMM/ANN hybrid model for continuous Malayalam speech recognition. Procedia Eng. **30**, 616–622 (2012)
3. Kurian, C.: Analysis of unique phonemes and development of automatic speech recognizer for malayalam language. Ph.D. Thesis, Cochin University of Science and Technology, Kerala (2013)
4. Yegnanarayana, B., Kishore, S.P.: AANN: an alternative to GMM for pattern recognition. Neural Networks **15**, 459–469 (2002)
5. Richard, M.D., Lippmann, R.P.: Neural network classifiers estimate Bayesian a posteriori probabilities. Neural Comput. **3**, 461–483 (1991)
6. Bourlard, H., Morgan, N.: Continuous speech recognition by connectionist statistical methods. IEEE Trans. Neural Networks **4**, 893–909 (1993)
7. Seid, H., Gamback, B.: A speaker independent continuous speech recognizer for Amharic. In: INTERSPEECH 2005, 9th European Conference on Speech Communication and Technology, Lisbon (2005)
8. Meinnedo, H., Neto, J.P.: Combination of acoustic models in continuous speech recognition hybrid systems. In: Proceedings of ICSLP, Beijing (2000)
9. Bakis, R.: Continuous speech recognition via centiseconds acoustic states. J. Acoust. Soc. Am. **59**(S1) (1976)
10. Baum, L.E.: An inequality and associated maximization technique in statistical estimation for probabilistic functions of Markov processes. Inequalities **3**, 1–8 (1972)
11. Rumelhart, D.E., Hinton, G.E., Williams, R.J.: Learning representations by back-propagating errors. Nature **323**, 533–536 (1986)
12. Viterbi, A.J.: Error bounds for convolutional codes and asymptotically optimum decoding algorithms. IEEE Trans. Inf. Theory **13**, 260–269 (1982)

Classification of Heart Disease Using Naïve Bayes and Genetic Algorithm

Santosh Kumar and G. Sahoo

Abstract Data mining techniques have been widely used to mine knowledgeable information from medical data bases. In data mining Classification is a supervised learning that can be used to design models describing important data classes, where class attribute is involved in the construction of the classifier. Naïve Bayes is very simple, most popular, highly efficient and effective algorithm for pattern recognition. Medical data bases are high volume in nature. If the data set contains redundant and irrelevant attributes, classification may produce less accurate result. Heart disease is the leading cause of death in India as well as different parts of world. Hence there is a need to define a decision support system that helps clinicians to take precautionary measures. In this paper we propose a new algorithm which combines Naïve Bayes with genetic algorithm for effective classification. Experimental results shows that our algorithm enhance the accuracy in diagnosis of heart disease.

Keywords Naïve Bayes · Genetic algorithm · Heart disease · Data mining

1 Introduction

Data mining is the process of automatically extracting knowledgeable information from huge amounts of data. It has become increasingly important as real life data enormously increasing [1]. Data mining is an integral part of KDD, which consists of series of transformation steps from preprocessing of data to post processing of data mining results. The basic functionality of data mining involves classification,

S. Kumar (✉) · G. Sahoo
Department of Computer Science and Engineering, Birla Institute of Technology, Mesra, Ranchi, Jharkhand, India
e-mail: san77j@gmail.com

G. Sahoo
e-mail: gsahoo@bitmesra.ac.in

L.C. Jain et al. (eds.), *Computational Intelligence in Data Mining - Volume 2*,
Smart Innovation, Systems and Technologies 32, DOI 10.1007/978-81-322-2208-8_25

association and clustering. Classification is a pervasive problem that encompasses many diverse applications. To improve medical decision making data mining techniques have been applied to variety of medical domains [2–4] etc. Many health care organizations are facing a major challenge is the provision of quality services like diagnosing patients correctly and administering treatment at reasonable costs. Data mining techniques answer several important and critical questions related to health care.

Naïve Bayes is one of the most popular classification technique introduced by Reverend Thomas Bayes [5, 6]. Without any additional data, classification rules are generated by the training samples themselves.

Evolutionary algorithms are used for the problem that can't be solved efficiently with traditional algorithms. Genetic algorithms (GA) are computing methodologies constructed with the process of evolution [7]. GA have played a vital role in many engineering applications. Heart disease is the leading cause of death in developed countries and is one of the main contributors to disease burden in developing countries like India. Several studies reveal that heart disease was the leading cause of mortality accounting 32 % deaths, as high as Canada and USA. Hence there is a need to design and develop a clinical decision support for classification of heart disease. In this paper we propose a classification algorithm which combines Naïve Bayes and genetic algorithm, to predict heart disease of a patient. This paper is organized as follows: Sect. 2 we review the concepts of datasets, feature selection and classification. Sections 3 and 4 explains research methodology and our proposed classifier respectively. Results are discussed in Sect. 5 followed by concluding remarks in Sect. 6.

2 Datasets, Feature Selection and Classification

In this section we review the concepts like datasets, feature selection, classification, Naïve Bayes, Genetic algorithm and heart disease.

In the process of prediction, the accuracy of the predicted results in data mining depends mainly on how well the classifier is being trained [8]. A data set is a collection of data. Feature selection is selecting the relevant features from the data set. Classification is finding the unknown target value based on the known target values in the data set.

The training of the classifier is done mainly based on the selection of classification algorithm and the data sets, which are given as input to the classifier. Some of the data in the data sets may not be useful for the prediction, which if eliminated would reduce the burden on the classification algorithms. This can be achieved with the help of feature selection, which is also called as filtering. This process of filtering the data helps to obtain better features to be selected among many features. For example, before the feature selection process, if there are features like id, age, gender, blood group, blood pressure for a patient in the heart disease data set, after the filtering process the irrelevant features like 'id' are removed and the necessary features are selected.

2.1 Datasets

A dataset is a collection of data that is related to one category. All the datasets for heart disease, hypothyroid, breast cancer, primary tumor, heart stalog and lymph were collected from the UCI machine learning repository [9, 10]. Since WEKA takes an input data set in the Attribute Relationship File Format (ARFF) format, all the data sets were converted to the ARFF file format. This format has attributes and instances (data). Two portions of the data for each disease were used, one is for training the classifier and the other is to test the trained classifier. Also the test dataset output is generated in the ARFF format.

ARFF is the text format file used by WEKA to store data. The ARFF file contains two sections, one is the header and the other is the data section. The first line of the header defines the relation name, which is usually the dataset name. Then, there is the list of the attributes. Each attribute is associated with a unique name and a type. The type describes the kind of data contained in the variable and what values it can have. The variables types are: numeric, nominal, etc. The class attribute can be changed and depends on the researcher/user and datasets.

2.1.1 Heart Disease

Heart disease is not only related to heart attack but may also include functional problems such as heart-valve abnormalities. These kinds of problems can lead to heart failure. Heart disease is also known as cardiovascular disease (CVD) in medical terms.

This dataset is collected from UCI machine learning repository [10]. In this data set (HDD) 13 attributes were used as shown in Table 1.

Table 1 Attributes of heart disease

No.	Field name	Class label
1	Age	Real
2	Sex	Nominal
3	CP	Nominal
4	TRESTBPS	Real
5	Cholesterol	Real
6	FBS	Number
7	RESTECG	Number
8	THALACH	Number
9	EXANG	Number
10	Old peak	Number
11	Slope	Number
12	CA	Number
13	THAL	Number

2.2 Feature Selection

Feature selection is the process of selecting a subset of relevant input variables for use in model construction from large datasets. Most of the times the data in the datasets contain many redundant or irrelevant attributes or features, which are not useful for the model construction. Redundant attributes are those, which provide no more information than the already selected features, and irrelevant features provide no useful information and sometimes make the training data less feasible [11]. By using feature selection, we can reduce the irrelevant and redundant features and hence it takes less time to train the model, which can help to improve the performance of the resulting classifiers. The reduced attributes is shown in Table 2. It is known that the machine learning methods themselves will automatically select the most appropriate attributes and delete the irrelevant ones. But in practical cases, the performances of those algorithms are still affected and can be improved by pre-processing. So by using some of the WEKA provided filtering methods to pre-process the data set, and possibly improve the final prediction results. Feature selection techniques are often used in domains where there are many features and comparatively few samples.

Feature selection algorithms are divided into three categories; filters, wrappers and embedded. Filters evaluate each feature independent from the classifier, rank the features after evaluation and take the superior one. The filter model is usually chosen when the no. of features becomes very large due to its computational efficiency. Some heuristic algorithms can be used for subset selection such as genetic algorithm, greedy stepwise, best first or random search. So the filters are more efficient, but it doesn't take into account the fact that selecting the better features may be relevant to classification algorithms. In the following subsections we will discuss feature subset selection through genetic algorithm (search).

2.2.1 Genetic Algorithm

Genetic algorithms have played a major role in many applications of the engineering science, since they constitute powerful tool for optimization. A simple genetic

Table 2 List of reduced attributes

Predictable attribute
Diagnosis
Value 0: No heart disease; Value 1: Has heart disease
Reduced input attributes
Type: Chest pain type
Rbp: Reduced blood pressure
Eia: Exercise induced angina
Oldpk: Old peak
Vsal: No. of vessels colored
Thal: Maximum heart rate achieved

algorithm is a stochastic method that performs searching in global search spaces, depending on some probability values. For these reasons it has the ability to converge to the global minimum or maximum depending on the specific application and to skip possible local minima or maxima. Basically there are three fundamental operators in GA, selection, crossover and mutation within chromosomes [12]. As in nature, each operator occurs with a certain probability. There must be a fitness function to evaluate individual's fitness. The fitness function is a very important component of the selection process since offspring for the next generation are determined by the fitness value of the present population. Selection is an operator applied to the current population in a manner similar to the one of the natural selection found in biological systems. The fittest individuals are promoted to the next population and poorer individuals are discarded. Crossover allows solutions to exchange information in the same way the living organism use in order to reproduce themselves. Mutation operator applied to an individual single bit of an individual binary string can be flipped with respect to a predefined probability.

2.3 Classification

Classification is a data mining technique, which is used to predict the unknown values by training one of the classifiers using known values. The concept of using a "training set" is to produce the model. The classifier takes a data set with known output values and uses this data set to build the classification model. Then, whenever there is a new data point with test data, with an unknown output value, the already trained classification model produces the output.

The following subsections introduce a classification techniques such as naïve Bayes which are used to build the model.

2.3.1 Naïve Bayes

The Naïve Bayes classifier is a probabilistic classifier based on the Bayes rule of conditional probability. It means, the Naive Bayes classifier uses probability to classify the new instance. It makes use of all the attributes contained in the dataset, and analyzes them individually as though they are equally important and independent of each other [13]. The Naïve Bayes classifier considers each of these attributes separately when classifying a new instance. It works under the assumption that the presence or absence of a particular feature of a class is unrelated to the presence or absence of another feature [14]. An advantage of the Naïve Bayes classifier is that it does not require large amounts of data to train the model, because independent variables are assumed; only the variances of the attributes for each class need to be determined, not the entire attributes.

3 Methodology

For building efficient and effective Decision support system we used the standard feature selection algorithms involving genetic search to identify important reduced input features. The reduced data sets are further classified by Naïve Bayes classifier on discretized values. Since results using discretized features are usually more compact, shorter and accurate than using continuous values.

The attribute to be predicted “Diagnosis” uses a value “1” for patients with heart disease and value “0” for patients with no heart disease. The key used is Patient ID. These attributes along with input attributes are shown in Table 3. Bayes’ Rule states that a conditional probability is the likelihood of some conclusion, C, given some evidence/observation, E, where a dependency relationship exists between C and E. The probability is denoted by P(C/E) given by

$$P(C/E) = \frac{P(E/C)P(C)}{P(E)} \tag{1}$$

Naïve Bayes classification algorithm works as follows:

1. Let D be a training set of tuples and their associated class labels. Each tuple is represented by an n-dimensional attribute vector, $T = (t_1, t_2\ldots, t_n)$, depicting n-measurements made on the tuple from n-attributes, respectively $A_1, A_2, A_3 \ldots A_n$.
2. Suppose that there are m-classes, $C_1, C_2\ldots C_m$. Given a tuple, T, the classifier will predict that T belongs to the class having the highest posterior probability, Conditioned on T. That is the Naïve Bayes’ classifier predicts that tuple t belongs to the class C_i if and only if $P(C_i/T) > P(C_j/T)$ for $1 \le j \le m, j \ne i$ Thus, we have to maximize $P(C_i/T)$. The class for which $P(C_i/T)$ is maximized is called the maximum posteriori hypothesis. By Bayes’ theorem $P(C_i/T) = P(T/C_i)$ $P(C_i)$ $P(T)$ As $P(T)$ is constant for all classes, only $P(T/C_i)P(C_i)$ needs to be maximized. If the class’ prior probabilities are not known, then it is commonly assumed that the classes are equally likely, that is, $P(C_1) = P(C_2) = \cdots = P(C_n)$ and we therefore maximize $P(T/C_i)$. Otherwise, we maximize $P(T/C_i)$ $P(C_i)$.
3. In order to calculate $P(T/C_i)$, the Naïve assumption of class conditional independence is made. There are no dependence relationships amongst the attributes. Thus,

Table 3 Description of various datasets

Dataset	Instances	Attributes
Hypothyroid	3,770	30
Breast cancer	286	10
Primary tumor	339	18
Heart stalog	270	14
Lymph	148	19

$$P(T/C_i) = \prod_{k=1}^{n} P\left(\frac{tk}{ci}\right) = P(t_1/C_i) \times P(t_2/C_i) \times \cdots \times P(t_m/C_i) \quad (2)$$

4. In order to predict the class level of T, $P(T/C_i)\ P(C_i)$ is evaluated for each class C_i. The classifier predicts that the class label of tuple T is the class C_i if $P(T/C_i)\ P(C_i) > P(T/C_j)\ P(C_j)$ for $1 \leq j \leq m$, $j \neq i$. The predicted class label of tuple T is the class C_i for which $P(T/C_i)\ P(C_i)$ is the maximum.

Example:
Let us consider a tuple T.

T = (age > 60, sex = female, cp = 2, trestbps > 130, Chol > 200, Fbs = 0, Restecg = 2, Thalch > 150, Exang = 0, Oldpeak < 3, Slope = 2, Ca = 2, Thal = 6).

The 13 attributes used are as mentioned in Table 1. The risk levels have been rated as 0, 1, 2, 3 and 4. This is the final outcome that we have to find out. C_1, C_2 … C_5 are the corresponding classes. Thus,

C_1 correspond to Risk-level 0.
C_2 correspond to Risk-level 1.
C_3 correspond to Risk-level 2.
C_4 correspond to Risk-level 3.
C_5 correspond to Risk-level 4.

We need to maximize $P(T/C_i)\ P(C_i)$ for i = 1, 2, 3, 4, 5.
The prior probability of each class can be calculated using training tuples.

$P(C_1) = 164/303 = 0.54$
$P(C_2) = 55/303 = 0.18$
$P(C_3) = 36/303 = 0.118$
$P(C_4) = 35/303 = 0.115$
$P(C_5) = 13/303 = 0.043$

To compute $P(T/C_i)$, the conditional probabilities have been listed in Table 4.
Similarly, other probabilities could be calculated. Consider all the values required for $P(T/C_1)$ listed in Table 5.

Now,

$$\begin{aligned} P(T/C_1) &= P(A_1/C_1) \times P(A_2/C_1) \times P(A_3/C_1) \times P(A_4/C_1) \times P(A_5/C_1) \times P(A_6/C_1) \\ &\quad \times P(A_7/C_1) \times P(A_8/C_1) \times P(A_9/C_1) \times P(A_{10}/C_1) \times P(A_{11}/C_1) \\ &\quad \times P(A_{12}/C_1) \times P(A_{13}/C_1) \\ &= 0.21 \times 0.44 \times 0.25 \times 0.39 \times 0.82 \times 0.86 \times 0.41 \times 0.725 \times 0.86 \times 0.98 \\ &\quad \times 0.29 \times 0.042 \times 0.036 \\ &= 6.9 \times 10^{-7} \end{aligned}$$

Table 4 Description of conditional probabilities

P(age > 60/C_1)	35/164 = 0.21
P(age > 60/C_2)	14/55 = 0.25
P(age > 60/C_3)	14/36 = 0.38
P(age > 60/C_4)	10/35 = 0.286
P(age > 60/C_5)	6/13 = 0.46
P(sex = female/C_1)	72/164 = 0.44
P(sex = female/C_2)	9/55 = 0.16
P(sex = female/C_3)	7/36 = 1.99
P(sex = female/C_4)	7/35 = 0.2
P(sex = female/C_5)	2/13 = 0.15
P(cp = 2/C_1)	0.25
P(cp = 2/C_2)	0.11
P(cp = 2/C_3)	0.027
P(cp = 2/C_4)	0.057
P(cp = 2/C_5)	0

Table 5 Description of required value for conditional probabilities

A_1	P(age > 60/C_1)	0.21
A_2	P(sex = 0/C_1)	0.44
A_3	P(cp = 2/C_1)	0.25
A_4	P(Trestbps > 130/C_1)	0.39
A_5	P(Chol > 200/C_1)	0.82
A_6	P(fbs = 0/C_1)	0.86
A_7	P(Resteccg = 2/C_1)	0.41
A_8	P(Thalch > 150/C_1)	0.725
A_9	P(Exang = 0/C_1)	0.86
A_{10}	P(OldPeak < 3/C_1)	0.98
A_{11}	P(Slope = 2/C_1)	0.29
A_{12}	P(Ca = 2/C_1)	0.042
A_{13}	P(thal = 6/C_1)	0.036

In the same way calculation is done for respective $P(T/C_i)$, which is shown in Table 6.

To find a classic C_i, multiply the above mentioned values with respective $P(C_i)$.

$$P(T/C_2)\ P(C_2) = 8.4 \times 10^{-7} \times 0.18 = 1.15 \times 10^{-7}$$
$$P(T/C_1)\ P(C_1) = 6.9 \times 10^{-7} \times 0.54 = 3.7 \times 10^{-7}$$
$$\text{Max}\ (P(T/C_i)\ P(C_i)) = P(T/C_1)\ (C_1)$$

Thus Naïve Bayes classifier predicts Class C_1 i.e. Risk-level = 0 for tuple T.

Table 6 Description of calculated values for respective $P(T/C_i)$

$P(X/C_i)$	Value
$P(X/C_1)$	6.9×10^{-7}
$P(X/C_2)$	8.4×10^{-7}
$P(X/C_3)$	6.6×10^{-8}
$P(X/C_4)$	1.6×10^{-7}
$P(X/C_5)$	0

3.1 Parameters of a Genetic Algorithm

A genetic algorithm operates iteratively and tries to adapt itself progressively over the iterations. At every iteration the genetic algorithm evaluates the population of chromosomes on the basis of its fitness function and ranks them according to the fitness value. Genetic algorithms also apply the crossover and the mutation operator to explore new traits in chromosomes. There are various parameters defining a genetic algorithm. These parameters can be varied to obtain better performance. In this section we discuss some of the parameters that are provided as input to the genetic algorithm. The extent to which each of the factors affects the performance of the genetic algorithm and the optimum values of the parameters for the dataset are showed in Fig. 2.

- **Crossover rate**: The crossover rate specifies how often a crossover operator would be applied to the current population to produce new offspring. The crossover rate, mutation rate and selection rate determine the composition of the population in the next generation.
- **Mutation rate**: The mutation rate specifies how often mutation would be applied after the crossover operator has been applied.
- **Selection rate**: This parameter comes into play if elitism is applied to the genetic algorithms. When creating a new population there is a chance that the best chromosome might get lost. Elitism is a method which prevents the best chromosome from getting lost by copying a fixed percentage of the best chromosomes directly into the next generation. The selection rate specifies how often the chromosomes from the current generation would be carried over to the next generation directly by means of elitism.
- **Population size and generation**: This parameter sets the number of chromosomes in the population at a given instance of time (i.e., in one generation). It also determines whether the initial population is generated randomly or by heuristics.
- **Number of iterations**: This parameter dictates the stopping criteria for the genetic algorithm. Generally the stopping criteria used in genetic algorithms is the number of iterations. In some cases genetic algorithms are halted if the average fitness value crosses a certain threshold value.
- **Selection type**: This parameter dictates the type of selection mechanism to be used.
- **Crossover type**: This parameter dictates the type of crossover to be used.

4 Proposed Method

Our proposed approach combines Naïve Bayes and genetic algorithm to improve the classification accuracy of heart disease data set. We used genetic search as a goodness measure to prune redundant and irrelevant attributes, and to rank the attributes which contribute more towards classification. Least ranked attributes are removed, and classification algorithm is built based on evaluated attributes. This classifier is trained to classify heart disease data set as either healthy or sick. Our proposed algorithm consists of two parts.

1. First part deals with evaluating attributes using genetic search.
2. Part two deals with building classifier and measuring accuracy of the classifier.

4.1 Proposed Algorithm

Proposed algorithm is shown below:

Step 1. Load the dataset.
Step 2. Apply genetic search on dataset.
Step 3. Attributes are ranked based on the value.
Step 4. Select the subset of higher ranked dataset.
Step 5. Apply (Naïve Bayes + GA) on the subset of attributes that maximize classification accuracy.
Step 6. Calculate accuracy of classifier, which measures the ability of classifier to correctly classify unknown sample.

Step 1 to 4 comes under part 1 which deals with attributes and their ranking. **Step 5** is used to build the classifier and **step 6** records the accuracy of the classifier.

Accuracy of the classifier is computed as

$$\text{Accuracy} = \frac{\text{Number of samples correctly classified}}{\text{Number of total samples}} \tag{3}$$

Or in a two-classes case

$$\text{Accuracy} = \text{TP} + \text{TN}/\text{TP} + \text{TN} + \text{FP} + \text{FN} \tag{4}$$

Given two classes, we usually use a special terminology describing members of the confusion matrix. Terms *Positive* and *Negative* refer to the classes. *True Positives* are positive instances that were correctly classified, *True Negatives* are also correctly classified instances but of the negative class. On the contrary, *False Positives* are incorrectly classified positive instances and *False Negatives* are incorrectly classified negative instances.

5 Results and Discussion

The performance of our proposed approach has been tested with 6 medical data sets. All 6 data sets were chosen from UCI Repository [15] and attributes are selected based on opinion from expert doctor's advice. Information about these attributes is listed in Table 3. A way to validate the proposed method, we have tested with emphasis on heart disease besides other machine learning data sets also taken from UCI repository. In Table 3, column 1 represents name of the attribute, column 2 no. of instances, and column 3 represents no. of attributes in each dataset. The comparison of our proposed algorithm with 2 algorithms is listed in Table 7. Accuracy of 5 data sets is improved by our approach. Attributes of heart disease and their corresponding data type is shown in Table 1. Our method (Naïve Bayes + GA) was not successful for breast cancer and primary tumor. This may due to our proposed approach could not account for irrelevant and redundant attributes present in above mentioned data sets. Cross over rate for GA should be high and 60 % is preferable, so we set the value at 60 %. Mutation rate should be low and we set mutation value at 0.033. Population size should be good to improve the performance of GA.

The results acquired by Naïve Bayes and GA reveals that by integrating GA with Naïve Bayes will improve the classification accuracy for many data sets and especially heart disease.

Parameters of Naïve Bayes and genetic search are described Figs. 1 and 2. Figure 3 shows accuracy comparison with and without GA. Table 8 shows accuracy of various

Table 7 Accuracy comparison with various algorithms

Dataset name	NB + GA (our approach)	NN + PCA	GA + ANN
Primary tumor	74.2	80	82.1
Hypothyroid	100	97.06	97.37
Lympography	100	99.3	99.3
Heart stalog	100	98.14	99.6
Breast cancer	91	97.9	95.45
Heart disease	100	100	100

Fig. 1 Naïve Bayes parameters

<u>Naïve Bayes Parameters</u>
1) P=1, 2, 3……N
2) Cross Validate=True
3) Variance =class probability
4) Maximum likelihood =True
5) No normalization=False

Genetic Search Parameters

1) Cross over probability=60%
2) Mutation probability=0.033
3) Maximum generation=20
4) Report frequency=20
5) Seed=1

Fig. 2 Genetic search parameters

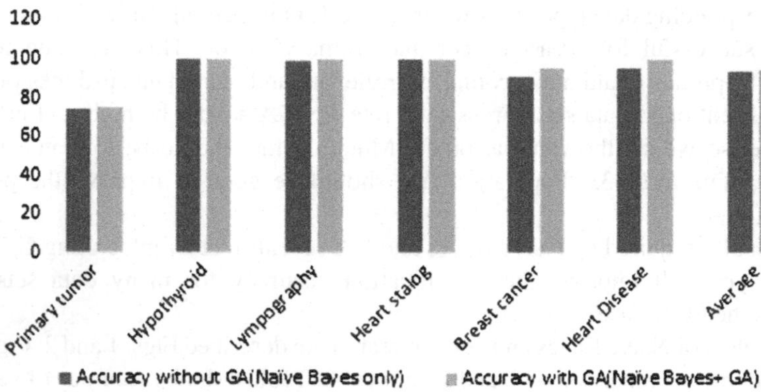

Fig. 3 Accuracy comparison with and with and without GA

Table 8 Accuracy comparison with and without GA

Dataset name	Accuracy without GA (Naïve Bayes only)	Accuracy with GA (Naïve Bayes + GA)
Primary tumor	74	75.2
Hypothyroid	100	100
Lympography	99	100
Heart stalog	100	100
Breast cancer	91	95.3
Heart disease	96	100
Average	93.5	95.08

data sets with and without GA. Average accuracy of our approach is higher than Naïve Bayes approach without GA. Accuracy of heart disease is improved 5 % over classification algorithm without GA. From the results it is also observed that integrating GA with Naïve Bayes outperforms the other methods with greater accuracy.

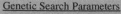

6 Conclusion

In this paper we have presented a novel approach for classifying heart disease. As a way to validate the proposed method, we have tested with emphasis on heart disease besides other machine learning data sets also taken from UCI repository. Experimental results carried out on 6 data sets show that our approach is a competitive method for classification. This prediction model helps the doctors in efficient heart disease diagnosis process with fewer attributes. Heart disease is the most common contributor of mortality in India, USA and other countries as well. Identification of major risk factors and developing decision support system, and effective control measures and health education programs will decline in the heart disease mortality.

References

1. Berry, M.W., et al.: Lecture Notes in Data Mining. World Scientific, Singapore (2006)
2. Sahoo, A.J., Kumar, Y.: Seminal quality prediction using data mining methods. Technol. Health Care (2014)
3. Kumar, Y., Sahoo, G.: Prediction of different types of liver diseases using rule based classification model. Technol. Health Care **21**(5), 417–432 (2013)
4. Yadav, G., Kumar, Y., Sahoo, G.: Predication of Parkinson's disease using data mining methods: a comparative analysis of tree, statistical, and support vector machine classifiers. Indian J. Med. Sci. **65**(6), 231 (2011)
5. Lewis, D.D.: Naive (Bayes) at forty: the independence assumption in information retrieval. In: Machine Learning ECML-98, pp. 4–15. Springer, Berlin (1998)
6. Han, J, Kamber, M.: Data Mining Concepts and Techniques. Morgan Kaufman Publishers, San Francisco
7. Goldberg, D.E.: Genetic Algorithm in Search Optimization and Machine Learning. Addison Wesley, Boston (1989)
8. Hall, M., et al.: The WEKA data mining software: an update. ACM SIGKDD Explor. Newslett. **11**(1), 10–18 (2009)
9. Fabrice, G., Hamilton, H.J.: Quality Measures in Data Mining, vol. 43. Springer, Heidelberg (2007)
10. www.ics.uci.edu/~mlearn
11. Powers David, M.W.: Evaluation: from precision, recall and F-measure to ROC, informedness, markedness and correlation. J. Mach. Learn. Technol. **2**(1), 37–63 (2011)
12. Sivanandam, S.N., Deepa, S.N.: Introduction to Genetic Algorithms. Springer, Berlin (2008)
13. Hand, D.J., Yu, K.: Idiot's Bayes—not so stupid after all? Int. Stat. Rev. **69**(3), 385–399 (2001)
14. Fakhraei, S., et al.: Confidence in medical decision making application in temporal lobe epilepsy data mining. In: Proceedings of the Workshop on Data Mining for Medicine and Healthcare. ACM (2011)
15. Newman, D.J., Hettich, S., Blake, C.L., Merz, C.J.: UCI Repository of Machine Learning Databases. Department of Information and Computer Science, University of California, Irvine (1998)
16. Jabbar, M.A., Deekshatulu, B.L., Chandra, P.: Heart Disease Prediction System Using Associative Classification and Genetic Algorithm, pp. 183–192. Elsevier, New York (2012)

17. Gansterer, W.N., Ecker, G.F.: On the relationship between feature selection and classification accuracy. Work **4**, 90–105 (2008)
18. Yan, H., Zheng, J., Jiang, Y., Peng, C., Xiao, S.: Selecting critical clinical features for heart diseases diagnosis with a real-coded genetic algorithm. Appl. Soft Comput. **8**, 1105–1111 (2008)
19. Dash, M., Liu, H., Motoda, H.: Consistency based feature selection. In: Proceedings of the 4th International Conference on Knowledge Discovery and Data Mining 'ICKDDM00', pp. 98–109 (2000)
20. Palaniappan, S., Awang, R.: Intelligent Heart Disease Prediction System Using Data Mining Techniques. IEEE (2008)

Solution for Traversal Vulnerability and an Encryption-Based Security Solution for an Inter-cloud Environment

S. Kirthica and Rajeswari Sridhar

Abstract Techniques of borrowing resources from external clouds to satisfy a user's need paves way to several security threats. This is tackled by securing the data available in the borrowed resource. Data Security is provided using the traditional Advanced Encryption Standard (AES) algorithm. The traversal vulnerability issue poses a threat to security. This is handled in this work thereby, protecting every virtual machine (instance) created in the cloud from access by other malicious instances available in the same cloud. Security is evaluated based on cost. The results show that a secure cloud is designed.

Keywords Cloud security · Inter-cloud security · Intra-cloud security · Traversal vulnerability

1 Introduction

Cloud computing is a technology that provides hardware, networks, storage, services and interfaces as a service [1] through a network. This service includes providing software, infrastructure, and storage, either as separate components or a complete platform to meet the user's demand.

Security plays a highly critical role in cloud environments as such environments act as huge store-houses for sensitive information and thus tend to attract hackers who might find ways to get at them [2].

In this paper, *internal private cloud* (*IPC*) or *private cloud* refers to the private cloud to which security is to be provided. Also, the term *external cloud* (*EC*) refers

---

S. Kirthica (✉) · R. Sridhar
College of Engineering, Guindy, Anna University, Chennai 600025, Tamil Nadu, India
e-mail: s.kirthica@gmail.com

R. Sridhar
e-mail: rajisridhar@gmail.com

L.C. Jain et al. (eds.), *Computational Intelligence in Data Mining - Volume 2*,
Smart Innovation, Systems and Technologies 32, DOI 10.1007/978-81-322-2208-8_26

to the cloud (private or community or public or hybrid) whose resources are used by the IPC to satisfy its user's demand.

A cloud environment being a shared environment, it is possible for a user of the cloud to access data of another user in the same cloud. Such an intra-cloud security threat is termed as traversal vulnerability [3]. In this work, a solution for this threat has been provided. Thus, intra-cloud security is provided by securing a user's instance from access by another user i.e. the virtual machine allocated by the cloud for every user is secured.

Security between clouds is handled in this work by securing the portion of an EC which has been allocated to the user of the IPC. This inter-cloud security is provided by using a traditional encryption algorithm to encrypt the data to be stored in the portion of the EC. *Instance* (*EBIn*) is the resource allocated to an IPC's user for satisfying the user's demand. *Volume* is the resource borrowed by the IPC from EC to satisfy the requirement of the IPC's user.

This paper is organized as follows: Sect. 2 discusses literature survey of the identified security threats. Sections 3 and 4 deals with the flow of activities involved in providing inter- and intra-cloud security respectively. In Sect. 5, the results of proposed solution are given. Section 6 concludes the paper by giving the overall contribution of the work and future works.

2 Related Work

While providing elasticity there arises certain security challenges [3] which are as follows:

- providing a distinction between virtual environments created for cloud users for easier administration
- loss of important data stored in the cloud
- access of a user's data by another without necessary permission
- Traversal Vulnerability which is the ability to access data of one user (including the encryption carried out) by another under the same hypervisor in which case encryption will be of no use.

A few major security issues addressed [4] and which are handled in this work are:

- users of the cloud share a common infrastructure but run different instances of applications
- traditional security algorithms require a local copy of the data to verify integrity, hence they cannot be used to secure data correctness

Cloud users are offered security as a service by means of an On-Demand Security Perimeter whenever connecting to the Internet [5]. In our work, this concept is used for providing a secure boundary for the resource allocated to a user.

In the work by Sudhir and Akassh [6], security in every cloud environment is provided by a trusted third party called "Cloud Certifier". The cloud certifier provides a unique key for the cloud for which it is in charge. This key is used to secure inter-cloud communication by exchanging it with the other cloud involved in the communication using the traditional "Diffie-Hellman Key Exchange" technique. This ensures that the communication between the two cloud environments is secure. However, this work depends on third party for security services which is not reliable.

In a Trusted Federation model [4], a server accepts a connection from a peer only if the peer supports TLS and presents a digital certificate issued by a root certification authority (CA) that is trusted by the server. In this model, a cloud provider establishes secure communication with another cloud provider by requesting the trust provider service (Intercloud Root) for a trust token. The trust provider service sends two copies of secret keys, the encrypted proof token of the trust service along with the encrypted requested token. An additional system, Intercloud root, is used to effect secure inter-cloud communication by establishing trust. This causes an overhead and is unreliable.

All the solutions given above depend on a third party. This, in turn, poses an additional threat. Hence, in this paper we propose a security solution which is not reliant on a third party to cater to intra and inter-cloud security. The security threat targeted, traversal vulnerability, is solved by providing a dynamic secure boundary for each instance of the IPC and inter-cloud security is provided by securing the data stored in the EC. The solutions being provided here have not been handled in any of the earlier works.

3 Intra-cloud Security

Security is provided to a private cloud by securing the virtual machine or the instance allocated to a user. Any instance in any cloud is subject to threats such as traversal vulnerability from other malicious instances in the cloud. The aim is to prevent traversal vulnerability thus preventing access of data of one instance from another instance. An instance should be accessible only by the user who created that instance and not by any other user of the cloud.

Figure 1 shows how security is provided within the private cloud. When a request of a user is satisfied by the private cloud, a dynamic secure boundary is established for the resource allocated to the user. This secure boundary restricts the access of the resource by other users of the cloud by providing credentials to only the user of that resource.

When the demand of the user increases and if it is satisfied by the private cloud itself, the limit of the secure boundary is extended. If the demand of the user decreases, the extra resource is released and the footprint of the boundary is destroyed. Thus the traversal vulnerability problem discussed by Owens [3] is solved.

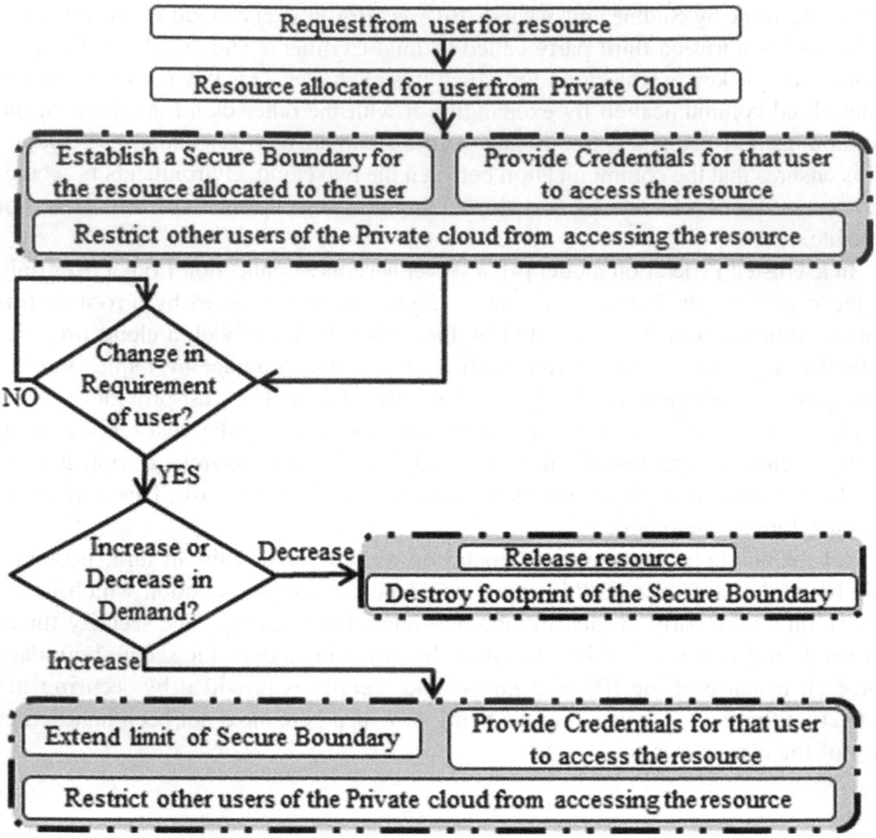

Fig. 1 Intra-cloud security flow

In order to protect an instance (EBIn) in the private cloud (IPC) from the threats, algorithm 1 is used. The public IP addresses of all instances in the cloud except that of the instance to be protected (EBIn) are listed in the denied hosts list which the user of EBIn cannot access. Algorithm 1 lists the process that would achieve intra-cloud security.

Algorithm to provide Intra-cloud Security

```
provide_intra_cloud_security
input: Identifier of Instance (EBIn) to be protected
output: EBIn protected from attack by other instances
begin

    PK_EBIn = private key of EBIn
    IP_EBIn = IP address of EBIn
    IPAddressSet = set of all public IP addresses provided
    by IPC to create instances
```

```
    IPAddressSet = IPAddressSet—IP_EBIn
    add IPAddressSet to list of denied hosts in EBIn using
    PK_EBIn and IP_EBIn
end
```

4 Inter-cloud Security

When there is a need to access an External Cloud (EC) from the Internal Private Cloud (IPC), the additional resource provided to a user from it should not be accessible by:

- any other user of IPC
- any user of EC

This necessitates the need for inter-cloud security in addition to the already provided intra-cloud security. It is assumed that, in the EC, one user can access another user's data i.e. Traversal Vulnerability is possible. Hence, it is the duty of the controller of IPC to defend itself from the threat in EC.

Figure 2 shows how the portion of resource allocated to the user of IPC from EC is secured. The activities of the IPC controller are not made known to any host in the network in which the IPC has been setup including its users. This makes the IPC controller fit for carrying out encryption since the encryption key and the technique used will not be revealed to any host in the network.

For securing the data in the external cloud, the data is encrypted by the IPC controller and then stored in EC. Since encryption is carried out in the secured IPC which is not accessible by any host in the network, there is no possibility of any user of EC to know about the data or the encryption technique used. This overcomes the issue of traditional cryptographic algorithms not being able to work in cloud environments as discussed by Ren et al. [4].

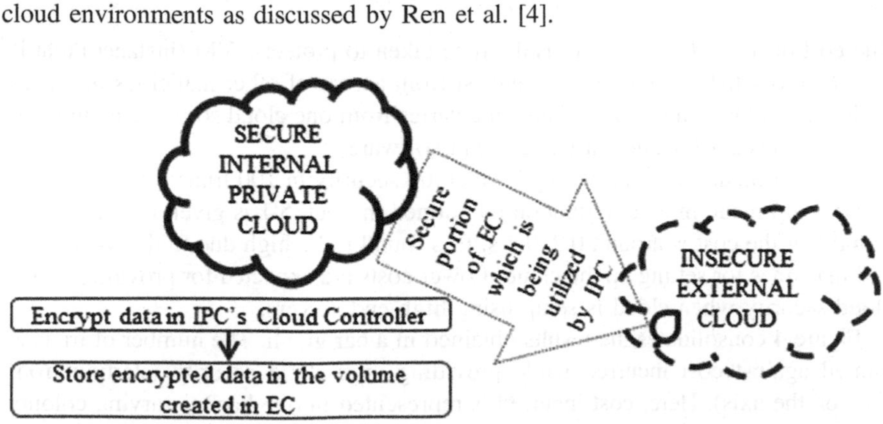

Fig. 2 Inter-cloud security. The internal private cloud (IPC) secures the resource allocated to its user from the external cloud (*EC*) by following the process given under the arrow

5 Implementation and Evaluation

For evaluating security, the parameter "Cost of Cloud Security" is used. It is the time overhead caused by implementing the security algorithm.

5.1 Machine Configuration

The open source computer software, Eucalyptus [7], for building cloud computing environments is used for demonstrating this work. For setting up the internal private cloud, Eucalyptus Faststart v3.4.1 is installed in Frontend and Node Controllers configuration i.e. a Eucalyptus cloud with all Frontend components on a single system, and two Node Controllers (NCs) on separate machines. For setting up external cloud (EC), Eucalyptus Faststart v3.4.1 is installed in Cloud-in-a-box configuration i.e. a Eucalyptus cloud with frontend and node controllers in a single machine. Two such ECs are setup. The NCs control virtual machine instances.

To set up the IPC, the Frontend and two NCs are installed in separate partitions in three machines, each having Intel (R) Core™ i7-3770 processor. The partition in each machine has disk capacity of 300 GB, RAM capacity of 3.48 GB, RAM frequency of 3.39 GHz and CPU frequency of 3.40 GHz.

One of the two ECs is set up on a single machine with the same configuration as above and the other EC is set up on a single node of a three-node six core Intel Xeon E5-2630 processor with disk capacity of 500 GB + 3 TB extended storage and RAM capacity of 96 GB.

5.2 Intra-cloud Security

The cost of intra-cloud security is the time taken to protect a VM (instance) that is provided to satisfy a cloud user's request from access of other malicious instances in the same cloud and is O(1). This time varies from one cloud software to another. But it is constant for one particular cloud software.

The cost incurred for providing intra-cloud security in 100 trials in a eucalyptus cloud setup made in a configuration mentioned in Sect. 5.1 is given in Fig. 3. It is found that the cost is around 0.2–0.4 s. It is found to be high due to the use of low end machines for setting up the cloud. Lower costs are expected for providing intra-cloud security when cloud is setup using high end servers.

Figure 4 consolidates the results obtained in a bar graph. The number of trials is plotted against cost incurred while providing intra-cloud security (starting from 0.15 on the axis). Here, cost incurred is represented in one bar by varying colours meaning that its value occurred for "n" different trials. For instance, the values of cost 0.20, 0.23, 0.25, 0.36 and 0.37 have occurred for 4 trials each. This is

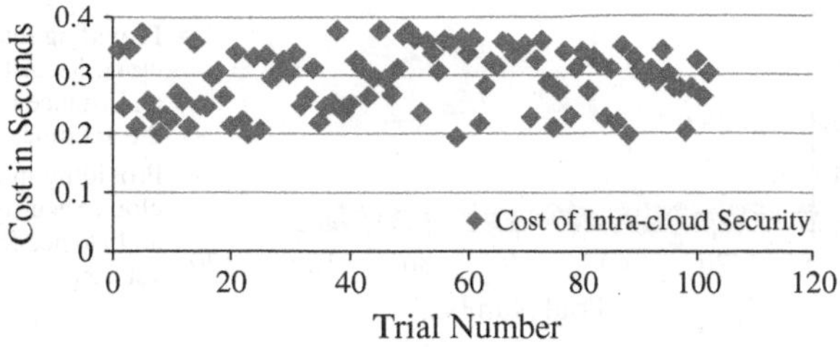

Fig. 3 Cost of intra-cloud security

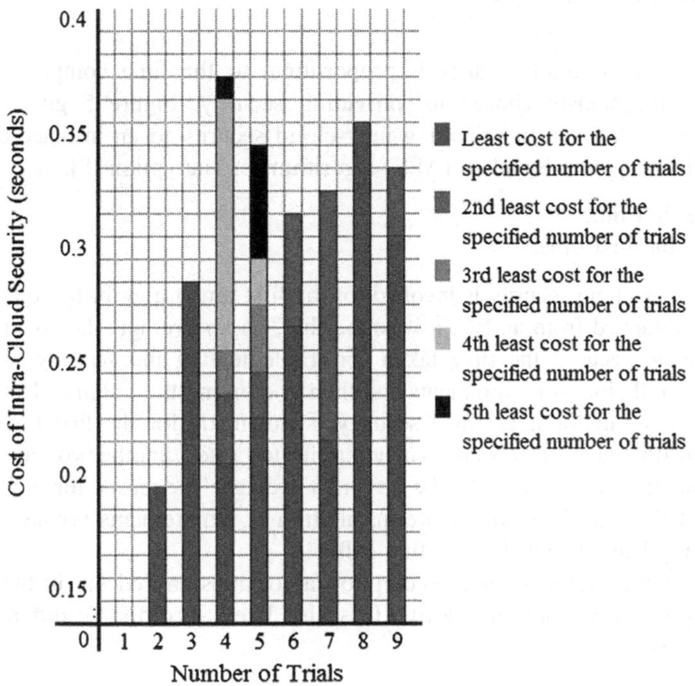

Fig. 4 Cost of intra-cloud security—consolidated version (Color figure online)

represented by blue, red, green, yellow and black colours which indicate the 1st, 2nd, 3rd, 4th and 5th least cost of security for the specified number of trials respectively. For example, in the graph shown in Fig. 4, from 0 to the least cost secured for 4 trials (0.2 s) blue colour is used, from 0.2 to the next least cost for the same number of trials (0.23 s) red colour is used and so on.

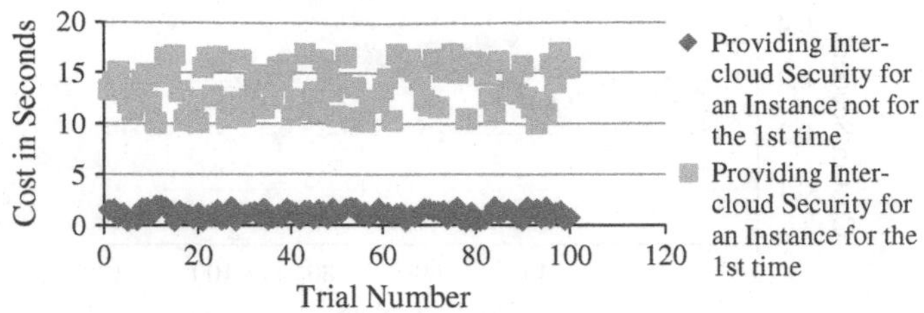

Fig. 5 Cost of inter-cloud security

5.3 Inter-cloud Security

The cost of inter-cloud security is proportional to the time complexity of the encryption mechanism chosen for providing security. Figure 5 gives the cost incurred for 100 trials for providing inter-cloud security to an instance using the Advanced Encryption Standard (AES) algorithm for encryption [8]:

- for the first time
- not for the first time

When inter-cloud security is invoked for the first time for an instance to which a volume is attached from an EC, it is found that, on an average, the cost is around 15 s. This is because the time taken for implementing the security algorithms includes installation of requirements and then encrypting the volume. The samples taken when a call for inter-cloud security is not made for the first time for the instance in IPC, i.e., it has been invoked earlier to protect another volume attached to the same instance, are found to be 1 s on an average. The reason for such reduced cost is that this call does not require installation of requirements because it would have already been installed in the first call.

This cost differs for different encryption algorithms carried out in the IPC for encrypting the data stored in the EC. It is also likely to differ for different cloud environments.

6 Conclusion and Future Work

While working in a cloud environment, a user must feel secure about the data being stored. High security is thus essential to handle data—confidential or not. A security solution for traversal vulnerability (ability to access one instance or virtual machine from another in the same cloud) has been proposed. The security threats posed by interoperating with external clouds are also overcome by providing cloud data security. The solutions provided handle the shortcomings identified in earlier works. The evaluation parameters used show the effectiveness of the techniques employed.

In this work, only eucalyptus cloud has been used for testing the solutions proposed. Other real-time cloud software like OpenStack, OpenNebula, etc., and public cloud environments like Amazon Web Services, Windows Azure, etc., have not been explored. They can be used as private cloud providers and external clouds respectively. Other possible areas for expanding this work are preventing traversal vulnerability in public cloud environments and providing Security-as-a-Service (SeaaS).

References

1. Miller, M.: Cloud Computing Web-Based Applications that Change the Way You Work and Collaborate Online, 1st edn. Que publishing, Upper Saddle River (2008)
2. Srinivasamurthy, S., Liu, D.: Survey on cloud computing security. In: Proceedings of Conference on Cloud Computing, CloudCom, vol. 10 (2010)
3. Owens, D.: Securing elasticity in the cloud. Commun. ACM **53**(6), 46–51 (2010)
4. Ren, K., Wang, C., Wang, Q.: Security challenges for the public cloud. IEEE Internet Comput. **16**(1), 69–73 (2012)
5. Chaudhry, J.: The Essential Guide to Cloud Security
6. Sudhir, D. Akassh, M.A.: Security in inter-cloud communication. In: EIE's 2nd International Conference on Computing, Energy, Networks, Robotics and Telecommunication, pp. 69–71 (2012)
7. Johnson, D., Murari, K., Raju, M., Suseendran, R.B., Girikumar, Y.: Eucalyptus Beginner's Guide-UEC Edition. CSS Corp (2010)
8. Wenbo, M.: Modern Cryptography: Theory and Practice. Prentice Hall, Englewood Cliffs (2004)

Efficient Spread of Influence in Online Social Networks

Gypsy Nandi and Anjan Das

Abstract Influence maximization in Online Social Networks (OSNs) is the task of finding a small subset of nodes, often called as seed nodes that could maximize the spread of influence in the network. With the success of OSNs such as Twitter, Facebook, Flickr and Flixster, the phenomenon of influence exerted by such online social network users on several other online users, and how it eventually propagates in the network, has recently caught the attention of computer researchers to be mainly applied in the marketing field. However, the enormous amount of nodes or users available in OSNs poses a great challenge for researchers to study such networks for influence maximization. In this paper, we study efficient influence maximization by comparing the general Greedy algorithm with two other centrality algorithms often used for this purpose.

Keywords Online social networks · Influence maximization · Greedy algorithm · High-degree heuristic algorithm · Eigenvector centrality algorithm

1 Introduction

Nowadays, as Online Social Networks (OSNs) are attracting millions of people, the latter rely on making decisions based on the influence of such sites [1]. For example, influence maximization can help a user make a decision on which movie to watch, which online community to join, which product to purchase, and so on. Thus, influence maximization has become essential for effective viral marketing, where marketing agencies or companies can convince online customers to buy

---

G. Nandi (✉)
Assam Don Bosco University, Guwahati 781017, Assam, India
e-mail: gypsy.nandi@gmail.com

A. Das
St. Anthony's College, Shillong 793001, Meghalaya, India
e-mail: anjan_sh@rediffmail.com

L.C. Jain et al. (eds.), *Computational Intelligence in Data Mining - Volume 2*,
Smart Innovation, Systems and Technologies 32, DOI 10.1007/978-81-322-2208-8_27

products through the help of those influential Online Social Network (OSN) users who can play a major role in influencing several others. This generates the demand for researchers to make an in-depth study of efficient influence maximization in OSNs which is currently a hot topic related to data mining OSNs.

The basic computational problem in influence maximization is that of selecting a set of initial users that are more likely to influence the other users of an OSN [2]. Domingos and Richardson [3] was the first to provide an algorithmic treatment to the problem of influence maximization by providing a heuristic solution to this problem. Kempe et al. [4] have discussed the greedy algorithm for influence maximization and also further proposed two elementary influence maximization models namely, the Linear Threshold (LT) model and the Independent Cascade (IC) model. Leskovec et al. [5] developed the Cost-Effective Lazy Forward selection (CELF) algorithm for influence maximization which provides a much faster result than a simple greedy algorithm, yet takes long time to find even 50 most influential users in an OSN. Chen et al. [6] proposed the Maximum Influence Arborescence (MIA) model for spread of influence under the Independent Cascade model which also provides a comparatively efficient solution than the general greedy algorithm. Chen et al. [7] also proposed a scalable heuristic called Local Directed Acyclic Graph (LDAG) for the LT model. However, their model is not suitable for various social media sites for influence maximization. Goyal and Lakshmanan [8] proposed a Credit Distribution model for studying the problem of learning influence probabilities. Recently, few papers [9, 10], have also studied influence maximization in OSNs from the topic modeling perspective, i.e., analyzing the influence strength of a user on other users based on a specific topic.

Several algorithms and techniques are thus being developed in the recent years to improve the running time and the output of the influence maximization problem. To summarize, our contributions in this paper is to make a comparative analysis of some of the basic standard influence maximization algorithms that paves a way to understand for the researchers in this field to work for improvement and for new techniques and algorithms.

The rest of the paper is organized as follows. Section 2 discusses the basic standard influence maximization algorithms, namely, the general greedy algorithm, the high-degree heuristic algorithm, and the eigenvector centrality algorithm. Section 3 shows our experimental results. We conclude the paper in Sect. 4.

2 Basic Standard Influence Maximization Algorithms

In this section, we discuss the three standard influence maximization algorithms that are often used to find the most influential users in an OSN, namely the general greedy algorithm, the high-degree heuristic algorithm, and the eigenvector centrality algorithm, for the IC model.

Table 1 Notations used in various algorithms

Variables	Descriptions
k	Number of seed sets or influential users to be selected from G
n	Number of nodes or vertices in G
m	Number of edges in G
S	Set containing the influential nodes
S_{im}	Number of rounds of simulations
d_v	Degree of a node in G
v_{di}	Vertex having top ith highest degree centrality value
V_{csi}	Vertex having top ith highest centrality score value
$C_S(v_i)$	Centrality score of node v_i
W_Sum()	weighted sum of centralities of all nodes in a node's neighborhood
V_{large}	Largest value of v in V

An OSN is modeled as an undirected or directed graph G = (V, E), where V denotes a set of nodes or vertices in the network and E denotes a set of edges or links, indicating the relationship between vertices. For convenience, Table 1 lists few important variables and their meanings that are mentioned for the algorithms below in this paper.

2.1 The General Greedy Algorithm

The general greedy algorithm relies on the computation of influence spread and is designed using specific properties of any influence propagation models, such as the Independent Cascade (IC) model or the Linear Threshold (LT) model [4]. Both these models are probabilistic in nature. In the IC model, coin flips decide whether an active node will succeed in activating its subsequent neighbours. In the LT model the activation of a node is determined by the node threshold chosen uniformly at random, together with the influence weights of active neighbors. A considerable amount of research effort has been devoted to developing algorithms for influence maximization under the above two propagation models.

In this section, Algorithm 1 describes the general greedy algorithm which takes as input the OSN graph G and the required number of seed sets or influential users k. Let Random (S) denote the random process of influence cascade from the seed set S using the concept of IC model, and the output is a random set of vertices influenced by S. The final output contains the set S with k most influential seeds.

In each round i, the above algorithm adds one vertex v into the set S in such a way that this vertex v together with the current set S maximizes the spread of influence in

Algorithm 1: GeneralGreedy (G, k) [12]

```
1:  Initialize S = ϕ, S_im = 20000
2:  for i = 1 to k do
3:    for each vertex v ∈ V \ S do
4:          s_v = 0
5:          for i = 1 to S_im do
6:                s_v += |Random(S U { v })|
7:          end for
8:          s_v = s_v / S_im
9:    end for
10:   S = S U { arg max_{v∈V\S} {s_v}}
11: end for
12: output S
```

G (Line 10). Thus, for each vertex v ∈ S, the simulation process is carried out S_{im} times repeatedly to estimate the influence spread of S U {v} (Lines 3–9).

2.2 The High-Degree Heuristic Algorithm

Another approach for finding the set of influential users using the IC model can be considered by finding the degree of each node. Much of the experimental results have shown that selecting vertices with maximum degrees as seeds results in larger influence spread than other heuristics. However, the output generated by implementing the high-degree heuristic algorithm results in a lesser influence spread as compared to the output produced by the general greedy algorithm. But, the time complexity of the high-degree heuristic algorithm is much better than the general greedy algorithm.

In this section, Algorithm 2 describes the basic process of using the concept of high degree heuristic algorithm to select k seeds for influence propagation in an OSN. The high-degree centrality measure is proposed by Nieminen in [11]. For a node v, its degree centrality can be defined as

$$\mathrm{DC}(v) = \sum_{i=1}^{n} \sigma(u_i, v) \tag{1}$$

where, $\sigma(u_i, v)$ is defined as follows:

$$\sigma(u_i, v) = 1, \quad \text{iff } u_i \text{ and } v \text{ are connected}$$
$$= 0, \quad \text{otherwise}$$

Algorithm 2: HighDegree (G, k)

1: Initialize $S = \phi$
2: **for** each vertex v **do**
3: compute its degree d_V
4: **end for**
5: **for** i = 1 to k **do**
6: $S = S \cup \{v_{di}\}$
7: **end for**
8: output S

Once the k seeds are selected (as explained in Algorithm 2), a centrality measure for IC model is carried out, which is based on propagation probability and degree centrality to estimate the spread of influence.

2.3 The Eigenvector Centrality Algorithm

The eigenvector centrality algorithm can also be used as a measure for the spread of influence of a node in an OSN. This technique assigns relative scores to all nodes in an OSN based on the concept that connections to high-scoring nodes add more to the score of the node in question than equal connections to low-scoring nodes. Thus, in case of eigenvector centrality, we need to compute the centrality of a node as a function of the centralities of its neighbors. Algorithm 3 describes the Eigenvector centrality algorithm to select the top k influential nodes in G.

Algorithm 3: EigenVector (G, k)

1: Initialize $C_S(v_i)= 1$ for each $v_i \in V$
2: **do**
2: **for** each vertex v_i **do**
3: $W\_Sum((v_i) = 0$
4: **for** each neighborhood x_j of v_i **do**
5: $W\_Sum(v_i)= W\_Sum(v_i) + C_S(v_i)^* C_S(x_j)$
6: **end for**
7: $C_S(v_i)= W\_Sum(v_i)$
8: **end for**
9: **for** each vertex v_i **do**
10: $v_i = v_i / v_{large}$
11: **end for**
10: **until** no change in all v_i
11: **for** m = 1 to k **do**
12: $S = S \cup \{ v_{csi} \}$
13: **end for**
14: output S

In the above algorithm, steps 2–8 discuss about the scores of each node being recomputed as the weighted sum of centralities of all nodes in a node's neighborhood. In steps 9–11, the nodes are normalized by dividing each value of node by the largest value.

3 Experimental Evaluation

In our experiment, we use a real-world and publicly available dataset containing a social graph G. This dataset comes from the *Flixster* website (www.flixster.com) which is one of the main players in the mobile and social movie rating business. The *Flixster* dataset contains 1,45,135 distinct users and 70,588,19 edges which indicates the friendship ties of all users. In our experiment, some standard data cleaning was performed by removing all users that do not appear at least 15 times in the data set.

We conduct experiments on the three different influence maximization algorithms namely, the general greedy algorithm, the high-degree heuristic algorithm and the eigenvector centrality algorithm. Our experiments aim at illustrating the performance of each of these algorithms based on the amount of spread of influence using the Independent Cascade model.

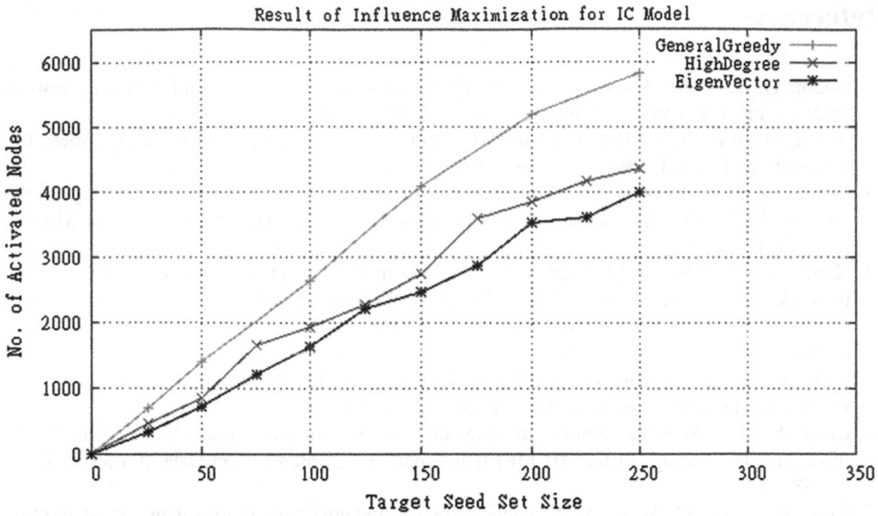

Fig. 1 Comparing influence spread using (a) general greedy, (b) high-degree, (c) eigenvector centrality algorithms

As can be seen from Fig. 1, the general greedy algorithm outperforms the high-degree heuristic algorithm as well as the eigenvector centrality algorithm which indicates that relying exclusively on the structural properties of a graph is not a better marketing strategy but we need to also explicitly consider the dynamics of information in an OSN.

4 Conclusions

This paper gives a comparative analysis of the three basic standard techniques used for effective influence maximization in OSNs and also provides algorithms and experimental results to analyze the same. We believe that these algorithms will remain the simplest and basic algorithms for studying and analyzing the concept of influence maximization in OSNs and will hence help a researcher in this field to get a preliminary idea about the same. As a future work, some comparative study of the latest algorithms with these basic algorithms will be studied for maximizing the spread of influence in OSNs.

References

1. Nandi, G., Das, A.: A survey on using data mining techniques for online social network analysis. Int. J. Comput. Sci. Issues **10**(6), 162–167 (2013)
2. Nandi, G., Das, A.: Online social network mining: current trends and research issues. Int. J. Res. Eng. Technol. **03**(4), 346–350 (2014)
3. Domingos, P., Richardson, M.: Mining the network value of customers. In: Proceedings of the 7th ACM SIGKDD International Conference on Knowledge Discovery and Data Mining, pp. 57–66 (2001)
4. Kempe, D., Kleinberg, J.M., Tardos, E.: Maximizing the spread of influence through a social network. In: Proceedings of the 9th ACM SIGKDD International Conference on Knowledge Discovery and Data Mining (KDD'03), pp. 137–146 (2003)
5. Leskovec, J., Krause, A., Guestrin, C., Faloutsos, C., Briesen, J., Glance, N.S.: Cost-effective outbreak detection in networks. In: Proceedings of the 13th ACM SIGKDD Conference on Knowledge Discovery and Data Mining, pp. 420–429 (2007)
6. Chen, W., et al.: Scalable influence maximization in social networks under the linear threshold model. In: Proceedings of the 2010 IEEE International Conference on Data mining (ICDM), pp. 88–97 (2010)
7. Chen, W., Wang, C., Wang, Y.: Scalable influence maximization for prevalent viral marketing in large-scale social networks. In: Proceedings of the 16th ACM SIGKDD International Conference on Knowledge Discovery and Data mining (KDD), pp. 1029–1038 (2010)
8. Goyal, F.B., Lakshmanan, L.V.S.: A data-based approach to social influence maximization. Proc. VLDB Endowment **5**(1), 73–84 (2011)
9. Liu, L., Tang, J., Han, J., Jiang M., Yang, S.: Mining topic-level influence in heterogeneous networks. In: Proceedings of the 19th ACM International Conference on Information and Knowledge Management (CIKM), pp. 199–208 (2010)
10. Barbieri, N., Bonchi, F., Manco, G.: Topic-aware social influence propagation models. In: Proceedings of the 2012 IEEE 12th International Conference on Data Mining (ICDM), pp. 81–90 (2012)
11. Nieminen, J.: On the centrality in a graph. Scand. J. Psychol. **15**, 332–336 (1974)
12. Chen, W., Wang, Y., Yang, S.: Efficient influence maximization in social networks. In: The Proceedings of the 15th ACM SIGKDD International Conference on Knowledge Discovery and Data Mining, pp. 199–208 (2009)

Adaptive FIR Filter to Compensate for Speaker Non-linearity

Varsha Varadarajan, Kinnera Pallavi, Gautam Balgovind and J. Selvakumar

Abstract In this paper we have implemented an adaptive filter to compensate for the non-linearity in a speaker. An attempt has been made to minimize the Mean Square Error (MSE) and convergence time using the LMS adaptive algorithm. Two adaptations of the LMS have been considered, the general adaptive LMS algorithm and the Leaky LMS algorithm. The Leaky LMS adaptation is observed to be more efficient with almost a 40 % decrease in convergence time. The filter coefficients for the above objective function are obtained using MATLAB. The target processor for implementing the two algorithms is Tensilica/Xtensa SDK toolkit using 'C' language which enables the codes to be directly dumped on to hardware.

Keywords Adaptive echo cancellation · LMS adaptive algorithm · Leaky LMS algorithm · Variable leaky LMS · Convergence rate

1 Introduction

In any telephone network, echoes and noise at points internal to and near the end of connection are generated. One of the means used to combat this echo for both speech and data transmission is echo cancellation [1]. This usually happens in some low-cost consumer products, here the loudspeaker signal may contain a certain level of nonlinear distortions, this is because the nonlinear signal components are

V. Varadarajan (✉) · K. Pallavi · G. Balgovind · J. Selvakumar
Department of ECE, SRM University, Kattankulathur, Chennai, India
e-mail: varshavarad@hotmail.com

K. Pallavi
e-mail: kinnerapallavi@gmail.com

G. Balgovind
e-mail: gautambalgovind@gmail.com

J. Selvakumar
e-mail: selvakumar.j@ktr.srmuniv.ac.in

L.C. Jain et al. (eds.), *Computational Intelligence in Data Mining - Volume 2*,
Smart Innovation, Systems and Technologies 32, DOI 10.1007/978-81-322-2208-8_28

shrouded by the linear ones, and insignificant distortions in speech signals are usually accepted by telephone customers. However this system cannot cancel the nonlinear echo components, and thus the far end subscriber will surely hear a highly scrambled echo. Also the required echo reduction level cannot be reached in most hands-free telephone configurations [2]. Therefore, additional adaptive techniques are required to cancel these echo components. These measures depend on the kind of nonlinearities present in the transmission chain. In this paper we have considered the case of a nonlinear speaker which is compensated for by an adaptive FIR filter. An adaptive filter is a form of self-learning filter, capable of adapting to a channel, or a particular set of signals, rather than being designed with the characteristics of a single filter in mind. The LMS algorithm is one such approach to adaptive filter design using an iterative weight update algorithm to update to coefficient of an FIR filter to best construct a desired signal from one corrupted with noise. The block representation of generalized FIR adaptive filter is in Fig. 1.

In Fig. 2, input x(n) is fed to both the speaker and the adaptive filter. Due to some non linearity in the speaker, a non linear output z(n) is obtained from the speaker. To this a noise which is approximately 1 % of the feedback is added to the speaker output. The speaker output along with the noise is represented as y(n). This y(n) is sent to the filter. Coefficients h(n) are mixed with y(n). A residual output e(n) which can also be called as error signal is obtained by subtracting y(n) from d(n) wherein d(n) is desired output. This residual output is sent back to the adaptive filter in order to adjust the coefficients such that the desired output is obtained. This forms the basic outline of this paper.

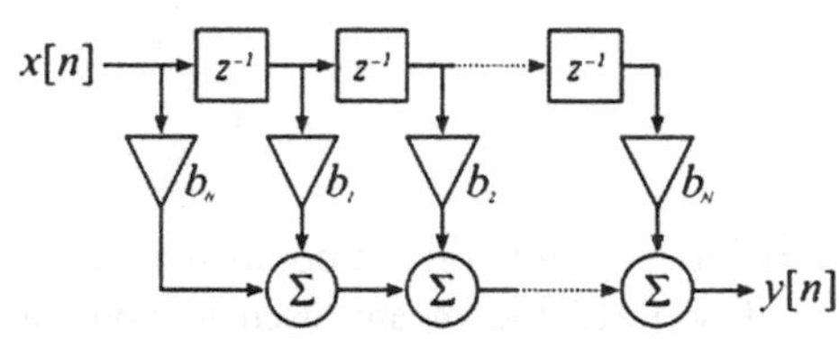

Fig. 1 Diagram of a generalized FIR adaptive filter

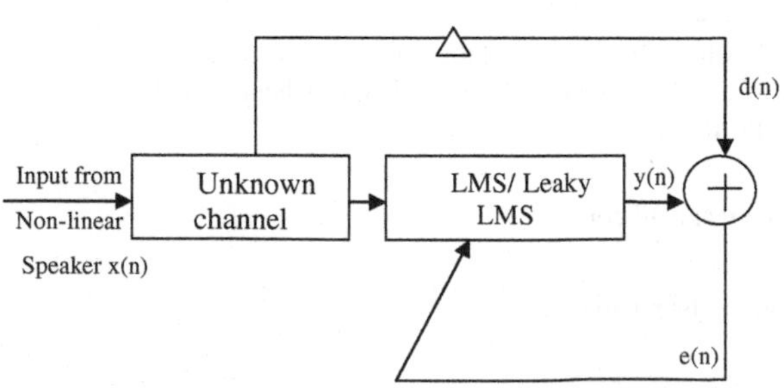

Fig. 2 Block diagram of a non linear speaker compensated for by an adaptive FIR filter

2 Adaptive Filtering

A filter is basically a device that maps the input signal supplied, to its output signal such that it extracts only the user desired information from the initial input real-time signal. Now an adaptive filter is required when either the fixed specifications are unknown or when the given specifications cannot be satisfied by time-invariant filters. Technically an adaptive filter automatically becomes a non-linear filter because its characteristics are independent of the input signal thus ensuring that the homogeneity and additive properties required in a linear system are absent in adaptive filters. The adaptive filters are however time varying since their parameters are continuously changing in order to fulfill system requirement [3]. Filter adaptively in FIR and IIR filters are used mainly for echo and acoustic noise cancellation. In these applications, an adaptive filter tracks changing conditions in the environment and changes filter specifications accordingly. Although in theory, both FIR and IIR structures can be used as adaptive filters, stability problems and the local optimum features in IIR filters makes them less suitable for such uses. Thus FIR filters are used for all practical applications as in this paper. Different algorithms can be used to implement adaptively, such as, Least Mean Square (LMS) algorithm and Recursive Least Square (RLS) algorithm. LMS is usually preferred due to its computational efficiency and design simplicity.

3 Adaptive Noise Cancellation

Adaptive noise cancellation is done purely for the purpose of improving the Signal to Noise Ratio (SNR) and for this, noise has to be removed from the path of transmission of signal. Consider the following example, communication between a pilot who is in air, and a ground control tower. A Jet Engine produces noise of over 140 dB, but normal human speech is much below 50 dB. If one is in the ground control tower, there will definitely be issues in hearing what the pilot has to say. In such a situation adaptive noise cancellation will come in handy and the noise caused by the engine can be removed while retaining the pilot's speech. In Fig. 3, we wish to obtain the signal $s(n)$ wherein $s(n)$ is the pilot's speech signal. However one cannot acquire $s(n)$ directly. Only $s(n) + v_1(n)$ can be obtained, where $v_1(n)$ is the jet engine noise. You cannot separately obtain $v_1(n)$ either. Thus to remove $v_1(n)$ from $s(n) + v_1(n)$, an adaptive filter is used. Firstly one must equip a sensor so as to acquire only the jet engine noise $v_2(n)$ and send this signal into the adaptive filter [4, 5]. On comparing Fig. 3 to Fig. 2 we can presume the signal $s(n) + v_1(n)$ in Fig. 3 corresponds to desired signal $d(n)$ in the diagram of the adaptive filter. $v_2(n)$ corresponds to $x(n)$. And finally e(n) in Fig. 2 denotes the resulting signal that is close to $s(n)$ in Fig. 3.

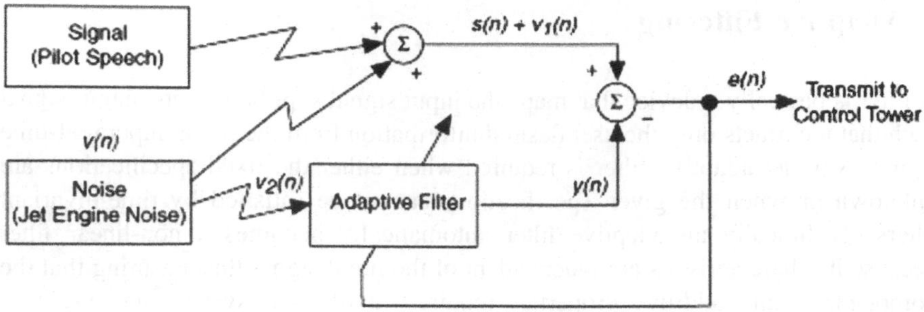

Fig. 3 Block representation of an adaptive noise cancellation system

4 Least Mean Square Algorithm

The most commonly used algorithm for adapting coefficients is the Least Mean Square (LMS) algorithm. This is generally used for adaptive echo and noise cancellation. The output response of an adaptive LMS filter is of the form as in (1)

$$\text{FIR filter: } y_n = \sum_{i=0}^{N-1} b_i x_{n-i} \tag{1}$$

Figure 4 depicts the functioning of a LMS adapted system wherein the coefficients are continuously getting adapted based on the value of residue.

When the output response is corrupted by noise or echo the LMS algorithm allows for updating of the coefficients b_i given by (2)

$$\forall i \in [0, N-1] b_i(n+1) = b_i(n) + \delta E(e_n x_{n-i}). \tag{2}$$

where 'e_n' is denoted as the error signal or residue value calculated as ($d_n - y_n$), where 'd_n' is the desired response. Also 'δ' is the convergence factor greater than or

Fig. 4 Depiction of coefficient updating using LMS

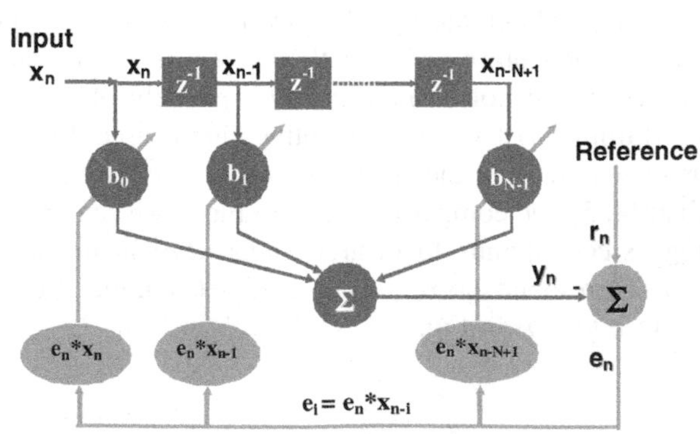

equal to zero. Larger the convergence factor the convergence speed but lower the precision [6].

The LMS algorithm can also be adapted into two other forms, Leaky LMS and Normalized LMS. In a previous paper that we have referred to [7], the authors have proposed a variable leaky LMS by using eigen values. According to that paper if $\{\lambda_1, \lambda_2,\ldots,\lambda_n\}$ are the eigen values seen by the general LMS algorithm then those that are seen by the leaky LMS algorithms are $\{\lambda_{1+\gamma},\ldots,\lambda_{N+\gamma}\}$. The γ value is required to be as large as possible in order for them to achieve that greatest possible reduction in eigen value spread. The variable leaky LMS algorithm used by them is given as in (3). Here γ_k is a time varying parameter.

$$\omega_{k+1} = (1 - 2\mu\gamma_k)\omega_k + 2\mu\varepsilon_k x_k \quad (3)$$

The leak factor here $1 - 2\mu\gamma_k$ is adjusted based on the formula of the a posteriori LMS error E_k^{LMS}:

$$\begin{aligned} E_k^{LMS} &= d_k - \omega_{k+1}^T x_k \\ &= d_k - \left(\omega_k^T + 2\mu \in_k \chi_k^T\right) x_k \\ &= \varepsilon_k\left(1 - 2\mu\chi_k^T x_k\right) \end{aligned} \quad (4)$$

This is the error when the new weight ω_{k+1} together with the old desired response d_k and input vector x_k. In this paper we aren't utilizing Eigen vectors as done in the reference paper [7] instead we directly apply a leaky factor to the coefficient update algorithm. Hence our convergence time does not depend on the Eigen value spread as above. In this paper we have simulated our programs using the general LMS algorithm and the leaky LMS algorithm. The evidence of this can be seen in the next section.

5 Simulations and Testing

The LMS algorithm was simulated and tested using Tensilica/Xtensa SDK toolkit. Apart from this, MATLAB has been used for the generation of audio signals and initial supply of filter coefficients.

5.1 Input Audio Signal and Filter Coefficients Generation

Using MATLAB, we initially generated a filter with sampling rate of 16 kHz using the Hamming window technique. In order to obtain the coefficients we used the Digital Signal Processing (DSP) application [8, 9]. These coefficients were then used in ourfirst program in Xtensa/Tensilica SDK toolkit in order to generate our

Table 1 Details of chosen coefficients for desired output

Specification categories	Specifications chosen
Response type	Lowpass
Design method	FIR (window)
Filter order	20
Window	Hamming
Frequency	f_a = 16,000 Hz, f_c = 8,000 Hz

desired output for echo cancellation. This desired output was obtained by generating the digital format of the pure signal without noise and echo and putting these values in our first program along with the initial filter coefficients. We first read the original signal using the same third party tool using command (wavread) and noted its digital format. Next step is to use a similar process to obtain the digitalized version of the corrupted and attenuated signal. This was the real signal for which noise cancellation was performed using adaptive filtering in our second program in Xtensa/Tensilica SDK toolkit. Table 1 represents the specifications of our filter coefficient generation and the impulse response of the proposed filter respectively.

5.2 Introduction to XTENSA/TENSILICA SDK Simulation

Using the XTENSA SDK Kit we have set up the simulation for comparing the algorithms in two steps:-

Step 1: In this proposed work we have utilized file handling to supply the program with the digitalized format of the original signal without noise or echo from a text file talked about in Sect. 5.1. We also supplied the filter coefficients obtained in the Sect. 5.1 in order to get an output response based on the general direct FIR implementation structure of a 13-tap FIR filter. We utilized multiply and accumulate strategy to compute the output response each time an input sample entered the register. This generated output response was then written into a text file which was to be the desired response of the adaptive filter.

Step 2: In the proposed work we have supplied the corrupted signal in the form of white noise, echo and a 15 dB attenuation using file handling as in the case of the first program. We have also supplied the desired response generated in the first program. We started of the program with filter coefficients initialized to zero and obtained an output response for the first input sample after which it was checked with the desired value to obtain the residue or error value e(n) which was then used to update the coefficients based on the LMS algorithm to minimize this residue to the most optimum value at the maximum possible speed tradeoff.

5.3 *Testing Methodology for Proposed Technique Under the Two Different Scenarios*

Technique 1: Initially we used the original general LMS algorithm and implemented it in our program. However, while this did give a good residue value the convergence time was on the higher side. For the first few sample the convergence time was relatively fast. However as the number of the samples progressed the convergence time decreased.

Technique 2: Next we applied the same program with the leaky LMS algorithm. Here thc filter coefficient is multiplied by a leaky factor 'β' in the range 0–1. This allows for more control over the filter response. To this algorithm we applied various scenarios by changing the convergence factor and the leaky factor and found that the initial convergence time for the first few samples was always higher but after that the convergence time was barely noticeable.

The leaky LMS algorithm used by us is given by (5)

$$b(n) = \beta * b(n) + \mu * e(n) * x(i - n) \tag{5}$$

Tables 2 and 3 are representations for real-time signals of 30 consecutive samples as.

Table 2 Comparison of LMS Adaptive algorithm for various convergence factors

S. No.	Convergence factor (μ)	Run time (min)	Residual error [d(n) − e(n)] range
1	100	3	30×10^{-6} to -30×10^{-6}
2	200	5	30×10^{-6} to -30×10^{-6}
3	300	19	30×10^{-6} to -30×10^{-6}
4	400	26	30×10^{-6} to -30×10^{-6}

Table 3 Comparison of leaky LMS adaptive algorithm for various convergence factors

S. No.	Convergence factor (μ)	Leaky factor (β)	Run time (s)	Residual error [d(n) − e(n)] range
1	100	0.5	65	1×10^{-6} to -1×10^{-6}
2	200	0.5	40	1×10^{-6} to -1×10^{-6}
3	300	0.5	28	1×10^{-6} to -1×10^{-6}
4	400	0.5	23	1×10^{-6} to -1×10^{-6}

6 Conclusion

Based on the comparison discussed in Tables 2 and 3 and the comparison charts based on the two tables it can be seen that using the leaky LMS algorithm is more efficient than the general LMS algorithm. It reduces the convergence time by a large amount of 40 %, while at the same time minimizing the residue error by almost 96 %. In case of the general LMS algorithm, Run time tends to very high for lower convergence factors like 10. The run time slowly begins to decrease till a convergence factor of 10 after which the run time tends to increase again. The convergence factor readings 100–400 have been represented in Table 2 and Fig. 5. In Table 3 representing the Leaky LMS algorithm the leaky factor helps in a tradeoff between the convergence time and the residue error, thus giving more stability to the system. Figure 6 denotes the relation between runtime and convergent factor for Leaky LMS. This results in a run time of seconds along with a lower residual error whereas the general LMS results in a run time of minutes and a much larger residual error range. Leaky factor being the main point of difference between Leaky LMS and general LMS algorithm is shown in comparison to the convergence time in Fig. 7. Also from Table 3 it can be seen that the most optimum residue error and convergence rate is found with a convergence factor of 400 and a leaky factor of 0.5 Apart from this, since our adaptation of leaky LMS doesn't use Eigen values there is no necessity to continuously maintain a low eigen value. This is required because larger eigenvalue increases the convergence time. Since our adaptation doesn't use

Fig. 5 Runtime (min) versus Convergence factor for general LMS

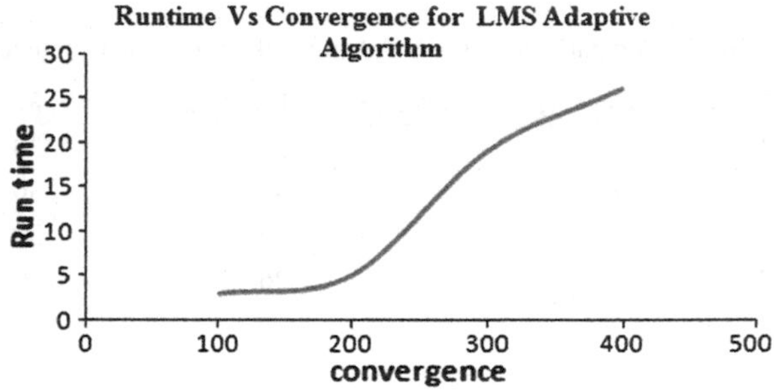

Fig. 6 Runtime (s) versus convergence factor for leaky LMS

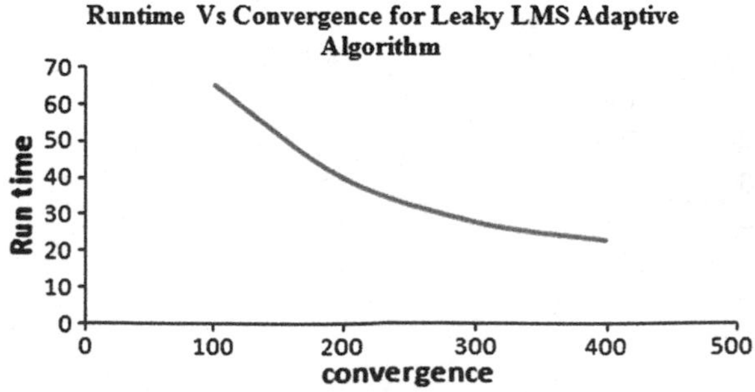

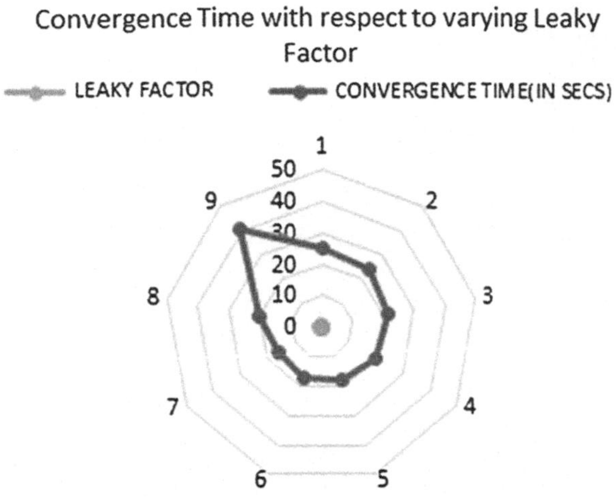

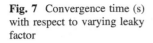

Fig. 7 Convergence time (s) with respect to varying leaky factor

eigen values it saves any extra maintenance effort required to do that. Also unlike previous cases wherein MATLAB is alone used, wc have used XTENSA C to implement our algorithm. The XTENSA C simulator creates a highly accurate representation of an XTENSA hardware processor hence the codes can be directly dumped on to hardware. Thus on comparing these two algorithms our Leaky LMS algorithm definitely seems to be the more efficient one for same range of residual error.

References

1. Messerschmitt, D.G.: Echo cancellation in speech and data transmission. IEEE J. Sel. Areas Commun. **2**(2), 283–297 (1984)
2. Stenger, A., Rabenstein, R.: Adaptive Filters for Nonlinear Acoustic Echo Cancellation. http://citeseerx.ist.psu.edu/viewdoc/summary?doi=10.1.1.108.4589
3. Paulo, S.R.D.: Introduction to Adaptive Filtering, pp. 1–12, 3rd edn. Springer, Berlin (2008)
4. Widrow, B., et al.: Adaptive noise cancelling: principles and applications. Proc. IEEE **63**(12), 1692–1716 (1975)
5. Selvakumar, J., Bhaskar, V., Narandran. S.: FPGA based efficient fast FIR algorithm for higher order digital FIR filter. In: IEEE Computer Society, IEEE Xplore and Proceedings of 3rd International, Symposium on Electronics System Design (ISED, 2012), vol. 2, pp. 43–47. Shibpur, Kolkata (2012)
6. www.ti.com/ww/cn/uprogram/share/ppt/c5000/16adaptive_v104.ppt
7. Kamenetsky, M.: A variable leaky LMS adaptive algorithm. Proc. IEEE **1**, 125–128 (2004)
8. Parhi, K.K.: VLSI Digital Signal Processing Systems Design and Implementation
9. LeelaKrishna, T., Selvakumar, J.: FPGA implementation of high speed fir low pass filter for EMG removal from ECG. Int. J. Res. Eng. Technol. **2**(3), 12–24 (2013)

A Pi-Sigma Higher Order Neural Network for Stock Index Forecasting

S.C. Nayak, B.B. Misra and H.S. Behera

Abstract Multilayer perceptron (MLP) has been found to be most frequently used model for stock market forecasting. MLP is characterized with black-box in nature and lack of providing a formal method of deriving ultimate structure of the model. Higher order neural network (HONN) has the ability to expand the input representation space, perform high learning capabilities that require less memory in terms of weights and nodes and have been utilized in many complex data mining problems. To capture the extreme volatility, nonlinearity and uncertainty associated with stock data, this paper considered a HONN, called Pi-Sigma Neural Network (PSNN), for prediction of closing prices of five real stock markets. The tunable weights are optimized by Gradient Descent (GD) and a global search technique, Genetic Algorithm (GA). The model proves its superiority when trained with GA in terms of Average Percentage of Errors (APE).

Keywords Stock index forecasting · Multilayer perceptron · Higher order neural network · Pi-sigma neural network · Genetic algorithm

1 Introduction

Mining stock market trend is challenging due to the extreme nonlinearity, uncertainty and non-stationary characteristics of the stock market. The stock market is very complex and dynamic by nature, and has been a subject of study for modeling

S.C. Nayak (✉) · H.S. Behera
Veer Surendra Sai University of Technology, Burla, India
e-mail: sarat_silicon@yahoo.co.in

H.S. Behera
e-mail: hsbehera_india@yahoo.com

B.B. Misra
Silicon Institute of Technology, Bhubaneswar, India
e-mail: misrabijan@gmail.com

L.C. Jain et al. (eds.), *Computational Intelligence in Data Mining - Volume 2*,
Smart Innovation, Systems and Technologies 32, DOI 10.1007/978-81-322-2208-8_29

its characteristics by researchers. Factors such as gold rate, petrol rate, foreign exchange rate as well as the economic and political situation of the country, trader's expectation and investor's psychology are influencing the behavior of stock market. Hence, an accurate forecasting is both necessary and beneficial for all investors in the market including investment institutions as well as small individual investors. Hence there is a need to developing an automated forecasting model which can accurately estimate the risk level and the profit gained in return.

Traditionally statistical models can be applied on stationary data sets and can't be automated easily. At every stage it requires expert interpretation and development. They cannot be employed to mapping the nonlinearity and chaotic behavior of stock market. The most used statistical method is autoregressive moving average (ARMA) and autoregressive integrated moving average (ARIMA).

During last two decades there are tremendous development in the area of soft computing which includes artificial neural network (ANN), evolutionary algorithms, and fuzzy systems. This improvement in computational intelligence capabilities has enhances the modeling of complex, dynamic and multivariate nonlinear systems. These soft computing methodologies has been applied successfully to the area data classification, financial forecasting, credit scoring, portfolio management, risk level evaluation etc. and found to be producing better performance. The advantage of ANN applied to the area of stock market forecasting is that it incorporates prior knowledge in ANN to improve the prediction accuracy. It also allows the adaptive adjustment to the model and nonlinear description of the problems. ANNs are found to be good universal approximator which can approximate any continuous function to desired accuracy.

It has been found in most of the research work in financial forecasting area used ANN, particularly multilayer perceptron (MLP). Suffering from slow convergence, sticking to local minima are the two well known lacuna of a MLP. In order to overcome the local minima, more number of nodes added to the hidden layers. Multiple hidden layers and more number of neurons in each layer also add more computational complexity to the network.

In the other hand, HONN are type of feed forward network which provide nonlinear decision boundaries, hence offering better classification capability as compared to linear neuron [1]. They are different from ordinary feed forward networks by the introduction of higher order terms into the network. HONN have fast learning properties, stronger approximation, greater storage capacity, higher fault tolerance capability and powerful mapping of single layer trainable weights [2]. In most of neural network models, neural inputs are combined using summing operation, where in HONN, not only summing units, but also units that find the product of weighted inputs called as higher order terms. Due to single layer of trainable weights needed to achieve nonlinear separability, they are simple in architecture and require less number of weights to capture the associated nonlinearity [3, 4]. As compared to networks utilizing summation units only, higher order terms in HONN can increase the information capacity of the network. This representational power of higher order terms can help solving complex nonlinear problems with small networks as well as maintaining fast convergence capabilities [5].

Evolutionary training algorithms are capable of searching better than gradient descent based search techniques. The Neuro-Genetic hybrid networks gain wide application in nonlinear forecasting due to its broad adaptive and learning ability [6]. The new hybrid iterative evolutionary learning algorithm is more effective than the conventional algorithm in terms of learning accuracy and prediction accuracy [7]. With the increase in order of the network, there may exponential growth in tunable weights in HONN and hence more computation time. However, there is a special type of HONN called as Pi-Sigma neural network (PSNN) using less number of weights has been introduced by Shin and Ghosh [8] in 1991. The PSNN has been successfully employed solving several difficult problems including polynomial factorization [9], zeroing polynomials [10], classification [11, 12], time series forecasting [13, 14]. In this paper, a PSNN is considered for the task of short-term prediction of closing prices of some real stock market. The network has been trained by both GD and GA.

The rest of the paper is organized as follows. The brief description and the related studies of PSNN have been described in Sect. 2. The model architecture and the training algorithms are described by Sect. 3. Section 4 gives the experimental results and discussion followed by a concluding remark.

2 Pi-Sigma Neural Network Architecture

The Pi-Sigma Neural Network has architecture of fully connected two-layered feed forward network. It is a class of HONN first introduced by Shin and Ghosh [8]. The two layers are termed as summing and product layer respectively. The input layers are connected to the summing layer and the output of this layer is feed to the product unit. The weight set between input and summing layer is trainable and the weight set between summing and product unit is non-trainable and set to unity. Hence, this network having only one tunable weight set reduces the training time of the network drastically [15]. The summing units use a linear activation where as the product unit uses a nonlinear activation function to produce the output of the network. Incorporation of extra summing unit increases the order by one. The product units give the networks higher order capabilities by expanding input space into higher dimension space offering greater nonlinear separability without suffering the exponential increase in weights. The architecture of Pi-Sigma neural network forecasting model trained by GA is shown by Fig. 1.

As shown in Fig. 1, the output at *jth* summing unit in the hidden layer is computed by summation of product of each input x_i with the corresponding weight w_{ij} between *ith* input and *jth* hidden unit and represented by Eq. 1.

$$y_j = \sum_{i=1}^{n} w_{ij} * x_i \tag{1}$$

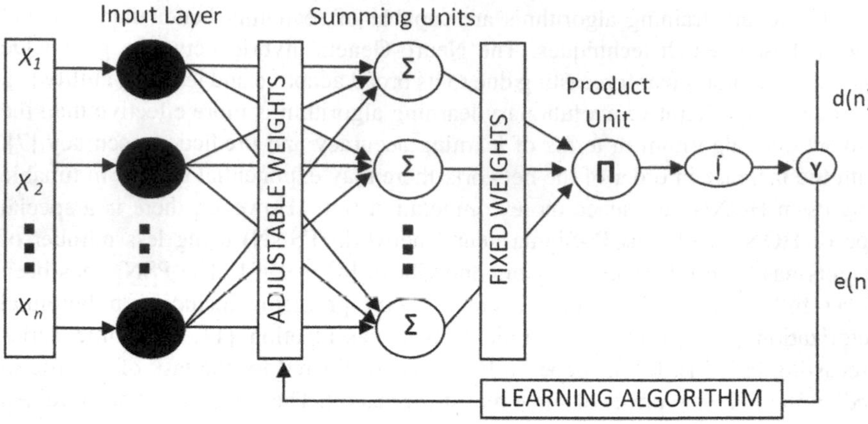

Fig. 1 Pi-sigma neural network forecasting model

where n is the number of input signals, i.e. number of closing prices here. The output of the network is now computed by making the product of the output of each summing units in the hidden layer and then passing it to a nonlinear activation function i.e. sigmoid function here. The output Y is represented by Eq. 2.

$$Y = \sigma\left(\prod_{j=1}^{k} y_j\right) \tag{2}$$

where k represents the order of the network same as the number of summing unit in the hidden layer.

3 Methodology and Experimental Results

This section presents the forecasting results obtained by employing the above models. The closing prices are collected for each transaction day of the stock exchange for the year 2012. In order to scale these data sigmoid normalization method has found to be superior. The normalized values are now considered as the input vector to the model. The sliding window method has been used to decide the input pattern for the model. Let $\boldsymbol{X(n) = (x_1, x_2, x_3, \ldots, x_n)}$ be the normalized closing prices. Let $\boldsymbol{W(n) = [w_{11}, w_{12}, w_{13}, \ldots, w_{n1}, w_{n2}, \ldots, w_{nm}]}$ represent the elements of a weight vector associated with the input vector $\boldsymbol{X(n)}$ at each summing unit neuron. Each input pattern $\boldsymbol{X(n)}$ is applied to the model sequentially and the desired closing prices value is supplied at the output neuron. Given the input, the model produces an output Y, which acts as an estimate to the desired value.

The error signal ***e(n)*** is calculated as the difference between the desired response and the estimated output of the model.

$$e(n) = d(n) - y(n) \tag{3}$$

The error signal ***e(n)*** and the input vectors are employed to the weight update algorithm to compute the optimal weight vector. The well known back propagation algorithm is employed for training the model. The GD based training algorithm for PSNN is as follows:

Algorithm 1. GD based Training Algorithm

---

1. Calculate the output Y of the model using Eq. 2.
2. Supply the desired output to compute the error signal by minimizing the error function computed by Eq. 4.

$$E(n) = \frac{1}{N}\sum_{i=1}^{N}(d_n - Y(n))^2 \tag{4}$$

3. Compute the weight changes by gradient descent method using Eq. 5 where learning rate is η.

$$\Delta W = \eta\left(\prod_{j=1}^{k} y_j\right)X(n) \tag{5}$$

4. Update the weight using change in weight and momentum factor α described as in Eq. 6.

$$W = W + \propto \Delta W \tag{6}$$

5. If stopping condition satisfies
   Stop training
   Else go to step 1.

To overcome the demerits of GD based back propagation, we employed the GA which is a popular global search optimization.

We adopted the binary encoding for GA. Each weight and bias value constitute of 17 binary bits. For calculation of weighted sum at output neuron, the decimal equivalent of the binary chromosome is considered. A randomly initialized population with 40 genotypes is considered. GA was run for maximum 150 generations with the same population size. Parents are selected from the population by elitism method in which first 20 % of the mating pools are selected from the best parents and the rest are selected by binary tournament selection method. A new offspring is generated from these parents using uniform crossover followed by mutation operator. In this experiment the crossover probability is taken as 0.6 and mutation probability is taken as 0.001. In this way the new population generated replaces the current population and the process continues until convergence occurs.

The fitness of the best and average individuals in each generation increases towards a global optimum. The uniformity of the individuals increases gradually leading to convergence.

The major steps of the GA based PSNN models can be summarized as follows.

Algorithm 2. GA Based Training Algorithm

---

1. Setting training data, i.e. choosing number of closing prices as input vector for the network.
2. Random initialization of search spaces, i.e. populations.

   Initialize each search space, i.e. chromosome with values from the domain [0, 1].

3. While (termination criteria not met)

   For each chromosome in the search space

   i. Calculate the weighted sum at each summing unit node of hidden layer.

   ii. Compute the product of output of summing unit at the output unit and passing it through a sigmoid transformation..

   iii. Present the desired output, calculate the error signal and accumulate it.

   iv. Fitness of the chromosome is equal to the accumulated error signal.

   End

   Apply crossover operator.

   Apply mutation operator.

   Select better fit solutions.

   End

4. Present the testing input vector, immediate to the training vectors.

   Calculated the estimated signal and calculate the error value.

5. Repeat the steps 1-5 for all training and testing patterns, calculate the total error signals.
6. Calculate the average percentage of errors (APE) for the whole financial time series.

Extensive experiments have been conducted and average results of 10 simulations have been collected. APE has been considered for performance metric in order to have a comparable measure across experiment with different stocks. The formulas are represented by Eq. 7.

$$APE = \frac{1}{N}\sum_{i=1}^{N}\frac{|x_i - \hat{x}_i|}{xi} \times 100\% \qquad (7)$$

The prediction performance of the GD based and GA based PSNN model experimented on BSE, NASDAQ, TAIEX, FTSE and DJIA closing prices of 2012 are shown in Fig. 2 for one-day-ahead forecasting. It can be observed that the error signal reduced drastically when the tunable weights are optimized by GA. The GD based model performs best for DJIA generating APE 1.0675. Similarly, GA based model has the best performance for TAIEX data generating APE value of 0.5866. In case of NASDAQ data, performance improvement is nominal, where as for TAIEX and FTSE it is significant. However, overall performance of the GA based PSNN is quite better for all data sets. Prediction performance of these models for one-week-ahead is presented by Fig. 3. In this case also, the GA based model performs better. The best result is repeated for TAIEX which is 0.5898. Similarly, for one-month-ahead prediction the results are shown by Fig. 4.

Similarly, for one-month-ahead prediction the results are shown by Fig. 4.

From Fig. 4, it can be observed that the APE value has been growing as with number of days ahead forecasting. The GA based PSNN model shows best performance for DJIA generating APE value 0.8533 and worst for FTSE with APE of 0.9906. But its performance over GD based PSNN is superior for all stock market data sets.

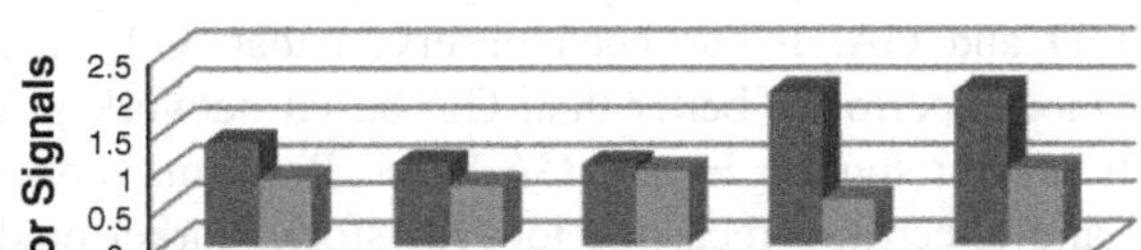

	BSE	DJIA	NASDAQ	TAIEX	FTSE
GD Based PSNN	1.3462	1.0675	1.0508	2.0075	2.0144
GA Based PSNN	0.8591	0.7753	0.9672	0.5866	0.9785

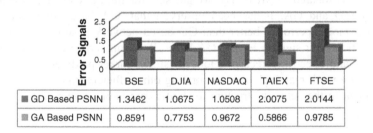

Fig. 2 Comparison of one-day-ahead forecasting error signals of five different stock indices for GD based PSNN and GA based PSNN models

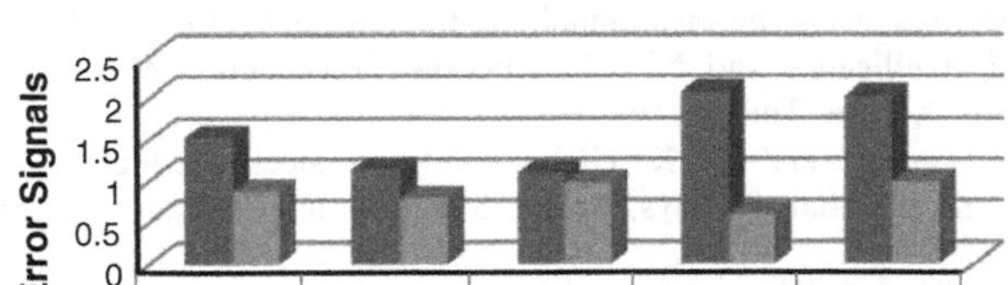

	BSE	DJIA	NASDAQ	TAIEX	FTSE
GD Based PSNN	1.5212	1.1325	1.0886	2.0683	2.0177
GA Based PSNN	0.8735	0.7882	0.9695	0.5898	0.9798

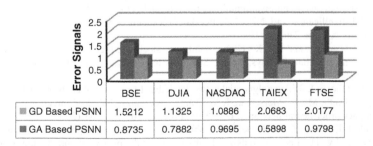

Fig. 3 Comparison of one-week-ahead forecasting error signals of five different stock indices for GD based PSNN and GA based PSNN models

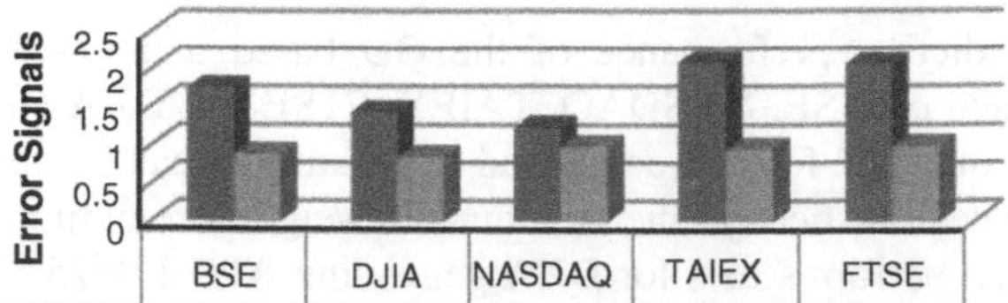

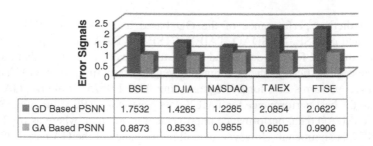

	BSE	DJIA	NASDAQ	TAIEX	FTSE
GD Based PSNN	1.7532	1.4265	1.2285	2.0854	2.0622
GA Based PSNN	0.8873	0.8533	0.9855	0.9505	0.9906

Fig. 4 Comparison of one-month-ahead forecasting error signals of five different stock indices for GD based PSNN and GA based PSNN models

4 Conclusion

A Pi-Sigma Neural Network has been developed and employed for short as well as long term closing prices prediction of five fast growing global stock markets. The stock data considered for experimentation includes BSE, DJIA, NASDAQ, TAIEX and FTSE. The model has simple and computationally less complex network structure as compared to MLP. The weight sets for the network have been optimized by GD and GA. It has been observed that, GA based Pi-Sigma neural forecasting model performs better than GD based network in terms of generating less prediction error signals. Hence, GA based Pi-Sigma model can be adopted as an efficient and effective forecasting tool for stock market prediction.

The future work may include incorporating other efficient training methods and enhancement in the network structure for better forecasting accuracies.

References

1. Guler, M., Sahin, E.: A new higher-order binary-input neural unit: learning and generalizing effectively via using minimal number of monomials. In: Third Turkish Symposium on Artificial Intelligence and Neural Networks Proceedings, 51–60. Middle East Technical University, Ankara, Turkey (1994)
2. Wang, Z., Fang, J., and Liu, X.: Global stability of stochastic high-order neural networks with discrete and distributed delays, Chaos, Solutions and Fractals, (2006). doi:10.1016/j.chaos. 2006, 06.063
3. Shin, Y., Ghosh, J.: Ridge polynomial networks. IEEE Trans. Neural Netw. **6**, 610–622 (1995)
4. Park, S., Smith, M.J.T., Mersereau, R.M.: Target recognition based on directional filter banks and higher-order neural networks. Dig. Sign. Process. **10**, 297–308 (2000)
5. Leerink, L.R., Giles, C.L., Horne, B.G., Jabri, M.A.: Learning with product units. In: Tesaro, G., Touretzky, D., Leen, T. (eds.) Advances in neural information processing systems 7, pp. 537–544. MIT Press, Cambridge (1995)
6. Kwon, Y.K., Moon, B.R.: A hybrid neuro-genetic approach for stock forecasting. IEEE Trans. Neural Netw **18**(3), 851–864 (2007)

7. Yu, L., Zhang, Y. Q.: Evolutionary fuzzy neural networks for hybrid financial prediction. IEEE Trans. Syst Man Cyber Part C Appl Rev **35**(2), 244–249 (2005)
8. Shin, Y., Ghosh, J.: The pi–sigma network: an efficient higher-order neural network for pattern classification and function approximation. In: International Joint Conference on Neural Networks (1991)
9. Perantonis, S., Ampazis, N., Varoufakis, S., Antoniou, G.: Constrained learning in neural networks: application to stable factorization of 2-d polynomials. Neural Process. Lett. **7**(1), 5–14 (1998)
10. Huang, D.S., Ip, H.H.S., Law, K.C.K., Chi, Z.: Zeroing polynomials using modified constrained neural network approach. IEEE Trans. Neural Netw. **16**(3), 721–732 (2005)
11. Shin, Y., Ghosh, J.: Efficient higher-order neural networks for classification and function approximation. Int. J. Neural Syst. **3**, 323–350 (1992)
12. Epitropakis, M.G., Plagianakos, V.P., Vrahatis, M.N.: Hardware-friendly higher-order neural network training using distributed evolutionary algorithms. Appl. Soft Comput. **10**, 398–408 (2010)
13. Ghazali, R., Husaini, N.H., Ismail, L.H., Samsuddin, N.A.: An Application of jordan pi-sigma neural network for the prediction of temperature time series signal. Recurrent Neural Networks and Soft Computing, (2012) Dr. Mahmoud ElHefnawi (Ed.), ISBN: 978-953-51-0409-4
14. Ghazali, R., Hussain, A.J., Liatsis, P.: Dynamic ridge polynomial neural network: forecasting the univariate non-stationary and stationary trading signals. Expert Syst. Appl. **38**, 3765–3776 (2011). (Elsevier)
15. Shin Y., Ghosh, J.: Realization of boolean functions using binary pi-sigma networks. In: Dagli, C.H., Kumara, S.R.T., Shin, Y.C. (eds.) Intelligent engineering systems through artificial neural networks, pp. 205–210. ASME Press, New York (1991)

Comparison of Statistical Approaches for Tamil to English Translation

R. Rajkiran, S. Prashanth, K. Amarnath Keshav and Sridhar Rajeswari

Abstract This work proposes a Machine Translation system from Tamil to English using a Statistical Approach. Statistical machine translation (SMT) is a machine translation paradigm where translations are generated on the basis of statistical models whose parameters are derived from the analysis of bilingual text corpora. It is the most widely used machine translation paradigm for the tradeoff between efficiency and implementation feasibility and due to its partial language independency. In syntax based approach, a phrase table is created which identifies the most probabilistically likely English translation of each Tamil phrase in the input sentence. In hierarchical phrase based approach, a rule table is used to reduce the input Tamil sentence into the output English sentence. We evaluated the two approaches based on different parameters like corpus size, gram size of language model and achieved a BLEU score of 0.26.

Keywords Statistical machine translation · Syntax based · Hierarchical phrase based · BLEU score

---

R. Rajkiran (✉) · S. Prashanth · K.A. Keshav · S. Rajeswari
Department of Computer Science and Engineering, Anna University,
College of Engineering, Guindy, Chennai 600025, Tamilnadu, India
e-mail: rajkiran2507@gmail.com

S. Prashanth
e-mail: prashanths1992@gmail.com

K.A. Keshav
e-mail: keshav.amarnath@gmail.com

S. Rajeswari
e-mail: rajisridhar@gmail.com

L.C. Jain et al. (eds.), *Computational Intelligence in Data Mining - Volume 2*,
Smart Innovation, Systems and Technologies 32, DOI 10.1007/978-81-322-2208-8_30

1 Introduction

Machine Translation (MT), on an overview, can be done with a source-target language dictionary. The process of translation is a tedious and time consuming one necessitating the importance to be automated. Translation is typically done on a sentence level basis to preserve the meaning of sentences. Important information required for translation is the grammar of the target language, known as the language model. Some of the main problems faced in MT are word sense disambiguation, anaphora resolution, idiomatic expressions etc. MT is an important and upcoming field for the past 2 decades, since one of its applications includes Cross Lingual Information Retrieval (CLIR) which is the need of the hour.

There are different approaches to machine translation like rule based machine translation [1] which includes dictionary based, transfer based and Interlingua machine translation, example based machine translation [2] and statistical machine translation [3].

In this paper, we have discussed about a Machine Translation system for Tamil to English based on statistical machine translation approaches [3].

This paper is organized as follows: Sect. 2 deals with the different approaches to Machine Translation and the advantages and disadvantages in each approach. Section 3 explains the system architecture with detailed module design. Section 4 discusses the implementation details and the different algorithms used in each module. Section 5 deals with the results of the implementation and the various inferences that can be made from the test data. Section 6 is the conclusion of the paper, in which we summarize our work with criticism and suggest improvements to the existing implementation that can improve translation quality.

2 Related Work

Machine Translation is an important field in computational linguistics and hence different approaches exist to MT. Some of them are discussed in the following sections.

2.1 Rule Based Machine Translation

Rule Based MT systems, also known as Knowledge Based Machine Translation systems, are described as the 'classical approach' to MT. Dictionary Based Machine Translation is one of the most primitive approaches to machine translation. It involves word by word translation using a source-target language dictionary [4]. The sentence structure is not preserved in this approach. In the transfer Based and Inter-lingual Machine Translation approach the source language is converted to an intermediate representation which is then translated into the target language. The difference between the two is that in interlingual machine translation, the intermediate representation is completely language independent. Transfer based MT can achieve a very high percentage of efficiency (up to 90 %), but is dependent on the source-target language pair. Superficial transfer (syntactic transfer) and Deep transfer (semantic transfer) are the two subclasses in transfer based MT.

Universal Networking Language (UNL) is the most predominant of the Interlingua based approaches. The main aim of Interlingua based approaches is the development of 2n pairs of language to interlingua (en-conversion) and interlingua to language (de-conversion) pairs rather than n * (n − 1) pairs of translation systems for given 'n' languages. Inter-lingua based approaches are highly scalable and independent of source-target language pairs, unlike transfer based MT systems.

2.2 Example Based Machine Translation

Example-based machine translation (EBMT) is a method of machine translation often characterized by its use of a bilingual corpus with parallel texts as its main knowledge base, at run-time. Example based machine translation system is very efficient in handling phrasal verbs but is not useful in deep linguistic analysis where one source language sentence can have multiple correct target language sentences [2].

2.3 Statistical Machine Translation

Statistical Machine Translation uses parameters that are derived from the analysis of bilingual text corpora to generate translations. The probability distribution P(e|f) denotes the probability that a string 'e' in the target language can be the translation for a string 'f' in the source language. Various heuristic approaches based on Hidden Markov Models have been developed for efficient search of translation strings.

2.4 Hybrid Approaches

Hybrid approaches to machine translation involve combining multiple translation approaches to obtain better results. Some of the approaches to hybrid MT are—Multi Engine MT, Statistical rule generation and Multi pass MT [5]. Researchers at Carnegie Mellon University have combined example based, transfer based and statistical translation sub systems into one Multi Engine MT system [6]. The main advantage of systems built using Multi pass paradigm is that it removes the need for the rule-based system, significantly reducing the amount of human effort and labor necessary to build the system.

2.5 Important Works

Statistical Machine Translation is described in detail by Koehn [3]. The training data set is the result of independent mining work described in [7] and ZdenekŽabokrtský [8]. The data set consists of sentence pairs such that for each sentence in the Tamil part of the corpus, the English part of the corpus contains the translated sentence in the same position (line number). The Och and Ney [9] text alignment algorithm is used to weed out any misalignments present in the training data set. Misaligned sentences are those which are present in the respective line numbers but may not be translations of each other. The statistical machine translation systems are implemented using an open source toolkit, Moses [10]. The post processing of output sentences for grammar correction is described using a morphological tagger and sentence structure rules [11].

2.6 Observations from the Survey

Given all the different approaches to MT, we have chosen the statistical machine translation paradigm because of the versatility of the system. Tamil is a free word-order language and English is fixed word-order language, hence the reordering of a given input sentence to the proper sentence structure in the target language output is considerably efficient when done using a statistical MT system. Hence, over time, the errors in translation gradually decrease and we move towards more perfect translation in the SMT approach to Machine Translation.

3 System Design

The entire block diagram of the system is shown in Fig. 1.

The parallel bilingual text corpus consists of English and Tamil sentences from Wikipedia articles from different domains, Biblical verses and day-to-day news from the newspaper. The bilingual corpus is first tokenized to separate clubbed punctuation and even out the effect of type casing on translation. The output of this phase is sent for alignment, where Och and Ney [9] text alignment algorithm is used to align the corpus. The target language part (English) of the corpus is used to generate the language model. The aligned corpus is sent to the training module where in syntax based SMT, a phrase table is created and in hierarchical phrase based SMT, a rule table is created. The phrase table contains the following fields,

- Tamil phrase.
- Corresponding English translation phrase.
- inverse phrase translation probability φ (f |e).
- inverse lexical weighting lex(f|e).

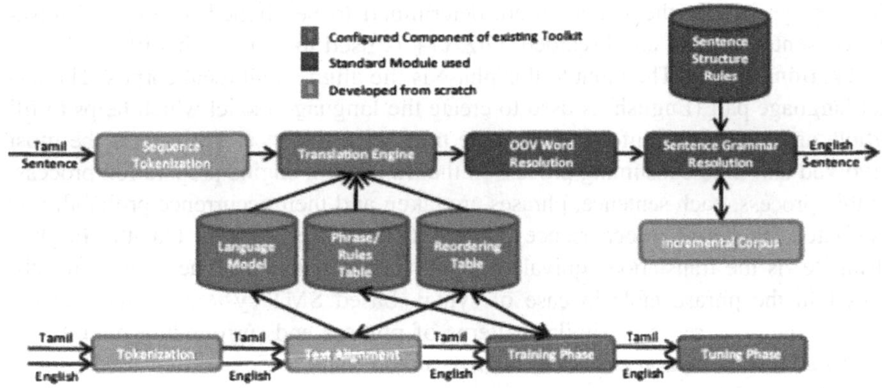

Fig. 1 System architecture

- direct phrase translation probability φ (e| f).
- direct lexical weighting lex(e|f).

The rule table contains the following fields,

- source string and source left-hand-side
- target string and target left-hand-side
- alignment between non-terminals (using word positions)
- score(s): here only one, but typically multiple scores are used
- frequency counts of source and target phrase (for debugging purposes, not used during decoding)

The probability values in the phrase table and rule table are weighted in the tuning phase to improve the quality of translation. This part of the software is a one-time process and creates the necessary lookup information for the translation engine. The actual working of the translation algorithm is discussed in detail in the following sections.

3.1 Module Design

Tokenization. The input bilingual text corpora are tokenized based on punctuation and spacing. Each word is treated as an individual translatable unit, a token. Then, the text is case adjusted. This is necessary because unless required in the case of proper nouns, the type case of a given word should not affect the translation.

Text Alignment. Text alignment is performed using the Och and Ney Text alignment algorithm. It is based on the concept that sentences of approximately the same length are alignable. Hence, in the first pass, the sentences of both languages are parsed and the average length is found and based on the standard deviation from the average length, the sentences are determined to be aligned or not, and misaligned sentence pairs are discarded. GIZA++ is used to achieve this [9].

Training Phase. The input to this phase is the aligned bilingual corpus. The target language part (English) is used to create the language model which helps to fill article and supporting information to the translated sentence. Training is the most involved and time consuming process in the translation engine preparation process. In this process, each sentence, phrases are taken and their occurrence probability is estimated (P(e|f)). The occurrence probability is the probability that the English string 'e' is the translation equivalent of the Tamil string 'f'. The values are tabulated in the phrase table in case of syntax based SMT system. In hierarchical system, the occurrence of similar patterns of phrases and surrounding phrases are identified and reduction rules are created. These values/rules will be used to perform the translation.

Tuning Phase. In addition to training phase, the allocated values must be checked and weighted. Hence a small portion of the dataset is used to tune the phrase/rule table and reordering tables. Actual translation is performed and the results are used to adjust the assigned weights to the phrase table values and rule table reduction rules. The output of this module is a tuned phrase table/rule table and reordering table which can be used to perform translation from Tamil to English.

Sentence Tokenization. Since the translation of each sentence is independent of the previous or next (it is assumed that the input sentence contains information for anaphora resolution), if more than 1 sentence is provided, then it is split and tokenized before passing it to the translation engine.

Translation Engine. This is the main module of the software which translates the input Tamil sentence to English sentence without loss of meaning by referring the phrase/rule table, language model and reordering tables. The input tokenized sentence is translated, annotated with article information and reordered based on the sentence structure [10].

OOV Word Resolution. The translation engine translates input sentences based on the words that are present in the training corpus. Hence, if a word doesn't occur in the training corpus and occurs in the input sentence, it is called an Out Of Vocabulary (OOV) word. To resolve this, we use an online dictionary lookup using the Agarathi dictionary for Tamil to English [12]. Since it is a word level dictionary look up, words that do not contain inflections will be resolved properly and words with inflections may or may not get resolved.

Sentence Grammar Resolution. After OOV word resolution, the output sentence is checked for grammatical correctness and a second reordering is performed to en-sure the grammatical correctness of the output sentence. This output sentence, along with the input sentence is stored in the incremental corpus which serves as the feedback loop for improvement of phrase/rule table leading to improvement in the translation quality [11].

Table 1 Corpus statistics

Domain	Sentences	#English tokens	#Tamil tokens
Bible	26,792 (15.77 %)	703,838	373,082
Cinema	30,242 (17.80 %)	445,230	298,419
News	112,837 (66.43 %)	2,834,970	2,104,896
Total	169,871	3,984,038	2,776,397

4 Data Set

The training data set for creation of the translation engine is derived from 2 sources. It contains translation of Wikipedia articles using MTurkers on 8 different domains such as Technology, Religion, Sex, Things, Language and Culture, People, Places and Events [7]. The BLEU score of the parallel corpus is 9.81 % (case insensitive BLEU-4). The second is described in [8]. The data for the corpus comes from 3 major sources namely, the bible, news and cinema. The statistics of the corpus is given in Table 1.

The combination of these 2 datasets was used as the training and tuning dataset for the syntax based and hierarchical phrase based SMT systems.

5 Results and Discussion

5.1 Dataset for Testing

The dataset for testing consists of a random sampling of sentences from the training dataset. About 1,000 sentences are chosen randomly and excluded from the training dataset. This is done so that the training and testing data consist of the same information domain and hence minimizes OOV words. The results of testing are summarized below.

5.2 Performance Evaluation

The statistical machine translation approaches have been evaluated for the various parameters as discussed in the following sub-sections.

BLEU Score. Bilingual Evaluation Understudy (BLEU) score is a measure to determine the translation quality of the given sentence. It is used to determine how close the machine translation output is similar to a human translation output. The output of the software is evaluated against a reference translation. If a word occurs in the candidate translation many times and does not occur in the reference translation, then it does not contribute any weight. If a word is present in the reference translation many times and not in the candidate translation, it brings the weight down. Hence, BLEU gives an accurate evaluation of the translation quality of the software.

Based on similarity, a score between 0 and 1 is assigned for each sentence. Overall average of all sentences gives the BLEU score of the translation. For better projection, BLEU scores are represented in percentage i.e., a BLEU score of 0.261 is represented as 26.1 %.
Algorithm for BLEU

- A candidate translation and reference translation are taken. A portion of the training set is given as input to the system. The Tamil portion of the training set is candidate translation and the English portion is reference translation.
- The maximum frequency of occurrence of a word in the reference translation is found. Let that be countmax.
- For each word in the candidate translation, the number of occurrences of the word in the candidate translation is counted. Let that be countword[i]. If it is greater than countmax, it is clipped to countmax.
- The sum over all the countword[i] is taken and divided by the total number of words in the reference translation. The resultant output is the BLEU score of the given translation system.

The variation in BLEU score based on different parameters like corpus size, gram size of the language model, and increase in quality after post processing. The readings are tabulated in Table 2.

Variation of BLEU score against the gram size of the language model for a corpus size of 1,60,000 sentences is shown in Fig. 2 and in Table 3.

Variation of BLEU score by introduction of post processing to resolve OOV words is shown in Fig. 3 and in Table 4.

Word Error Rate. It is defined as the Levenshtein distance on a sentence level between the candidate and reference translation. Word Error Rate,

$$\mathrm{WER} = \frac{\mathrm{S} + \mathrm{D} + 1}{\mathrm{N}} \tag{1}$$

Table 2 BLEU scores of translation systems

Corpus size	Syntax based SMT	Hierarchical phrase based SMT
70,000	4.71	3.54
1,60,000	14.11	13.63
2,30,000	26.10	25.99

Fig. 2 Variation of BLEU score-corpus size

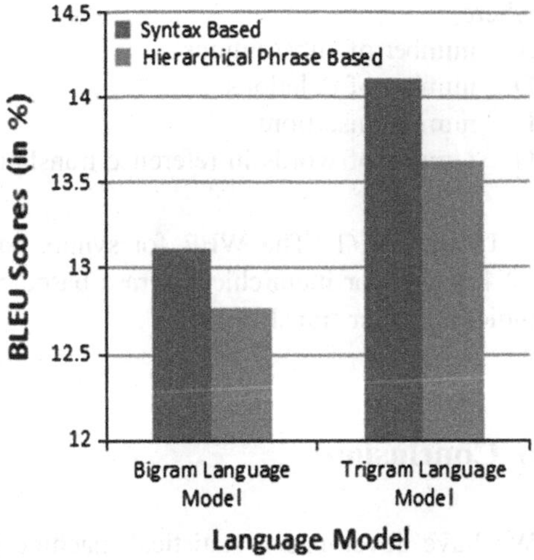

Table 3 BLEU scores of Bigram and Trigram LM

MT system	Bigram LM	Trigram LM
Syntax based	13.12	14.11
Hierarchical phrase based	12.77	13.63

Fig. 3 Variation of BLEU score—post processing

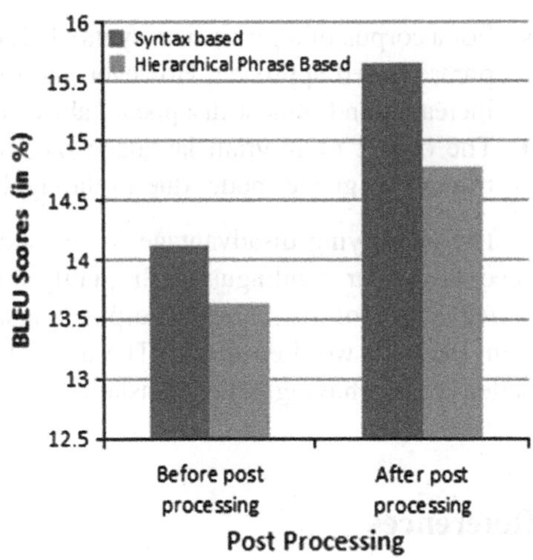

Table 4 BLEU scores before and after post processing

MT system	Before post-processing	After post-processing
Syntax based	14.11	15.65
Hierarchical phrase based	13.63	14.77

where,

S number of substitutions
D number of deletions
I number insertions
N number of words in reference translation

Using Eq. (1) The WER for syntax based SMT system was evaluated to be 39.4 % and for hierarchical phrase based SMT system was 54.3 % (Lower value indicates better translation).

6 Conclusion

We have developed a statistical machine translation for Tamil to English using syntax based and hierarchical phrase based approaches. We used different corpus sizes to train the system. In a corpus of 2,30,000 sentences (1 million words approximately), we achieved a BLEU score of 26.10 % for syntax based approach and 25.99 % for hierarchical phrase based approach. The output produced more natural translations than other MT approaches due to the usage of statistical machine translation.

Based on the implementation of multiple SMT systems on different approaches, we have come to the following inferences:

- For a corpus of a given size, syntax based approach fares better than hierarchical phrase based approach. However, the gap be- comes smaller as the corpus size increases and almost disappears above a corpus of more than 1 million words.
- The output of bi-gram language model is lesser in translation quality than a trigram language model due to the agglutinative nature of Tamil language.

The underlying disadvantage of statistical machine translation paradigm is the necessity of large bilingual high quality text corpora. Hence, due to the less than perfect nature of the reference corpus, translation quality is affected. This is evident from the high word error rate. The translation engine can be improved by using a better corpus, having better translations and increasing the corpus size.

References

1. Forcada, M.L., Ginest´ı-Rosell, M., Nordfalk, J., ORegan, J., Ortiz- Rojas, S., Pe´rez-Ortiz, A., Sa´nchez-Mart´ınez, F., Ramırez- Sa´nchez, G., Tyers, F.M.: Apertium: a free/open-source platform for rule-based machine translation. Mach. Transl. **25**(2), 127–144 (2011)
2. Somers, H.: Review article: example-based machine translation. Mach. Transl. **14**(2), 113–157 (1999)
3. Koehn, P.: Statistical Machine Translation. Cambridge University Press, Cambridge (2010)

4. Muegge, U.: An excellent application for crummy machine translation: automatic translation of a large database. In: Proceedings of the Annual Conference of the German Society of Technical Communicators, pp. 18–21 (2006)
5. Aleksic, V., Thurmair, G., Will, T.: Hybrid machine translation system. US Patent App. 11/885,688, 7 Mar 2005
6. Hogan, C., Frederking, R.E.: An evaluation of the multi-engine MT architecture. In: Machine Translation and the Information Soup, pp. 113–123. Springer, Heidelberg (1998)
7. Post, M., Callison-Burch, C., Osborne M.: Constructing parallel corpora for six Indian languages via crowdsourcing. In: Proceedings of the Seventh Workshop on Statistical Machine Translation, pp. 401–409. Association for Computational Linguistics (2012)
8. ZdenekŽabokrtský, L.O.: Morphological processing for English–Tamil statistical machine translation. In: 24th International Conference on Computational Linguistics, pp. 113–122 (2012)
9. Och, F.J., Ney, H.: A systematic comparison of various statistical alignment models. Comput. Linguist. **29**(1), 19–51 (2003)
10. Koehn, P., Hoang, H., Birch, A., Callison-Burch, C., Federico, M., Bertoldi, N., Cowan, B., Shen, W., Moran, C., Zens, R., et al.: Moses: open source toolkit for statistical machine translation. In: Proceedings of the 45th Annual Meeting of the ACL on Interactive Poster and Demonstration Sessions, pp. 177–180. Association for Computational Linguistics (2007)
11. Stymne, S., Ahrenberg, L.: Using a grammar checker for evaluation and postprocessing of statistical machine translation. In: LREC, pp. 2175–2181 (2010)
12. Parthasarathi, R., Karky, M.: Agaraadhi: a novel online dictionary framework. In: 10th International Tamil Internet Conference of International Forum for Information Technology in Tamil, pp. 197–200

An Integrated Clustering Framework Using Optimized K-means with Firefly and Canopies

S. Nayak, C. Panda, Z. Xalxo and H.S. Behera

Abstract Data Clustering platform is used to identify hidden homogeneous clusters of objects to analyze heterogeneous data sets based upon the attribute values in the domain of Information Retrieval, Text Mining, Web Analysis, Computational Biology and Others. In this work, a hybrid clustering algorithm for K-Means called Optimized K-Means with firefly and canopies, has been proposed by integration of two meta-heuristic algorithms: Firefly algorithm and Canopy pre-clustering algorithm. The result model has been applied for classification of breast cancer data. Haberman's survival dataset from UCI machine learning repository is used as the benchmark dataset for evaluating the performance of the proposed integrated clustering framework. The experimental result shows that the proposed optimized K-Means with firefly and canopies model outperforms traditional K-Means algorithm in terms of classification accuracy and therefore can be used for better breast cancer diagnosis.

Keywords K-means · Firefly algorithm · Canopy pre-clustering algorithm · Breast cancer · Classification · Medical data

S. Nayak (✉)
Department of Computer Application, Veer Surendra Sai University of Technology, Burla 768018, Odisha, India
e-mail: vssut.sanjib@gmail.com

C. Panda · Z. Xalxo · H.S. Behera
Department of Computer Science Engineering and Information Technology, Veer Surendra Sai University of Technology, Burla 768018, Odisha, India
e-mail: chinmoyxyz@gmail.com

Z. Xalxo
e-mail: zarinaxalxo16@gmail.com

H.S. Behera
e-mail: mailtohsbehera@gmail.com

L.C. Jain et al. (eds.), *Computational Intelligence in Data Mining - Volume 2*, Smart Innovation, Systems and Technologies 32, DOI 10.1007/978-81-322-2208-8_31

1 Introduction

Clustering techniques have a lot of important applications in several diverse fields and problems which are computationally intensive. So to address these multi-objective, multi-modal complex applications and solving many inter-disciplinary real world problems a variety of clustering mechanisms are available. As the data set can be large in many cases as with large number of elements in the data set or with each element with many features or with a large number clusters to be formed [1–5]. Clustering is used to partition a set of objects which have associated multi-dimensional attribute vectors into homogeneous groups such that the patterns within each group are similar [6]. Several unsupervised learning partitions the set of objects into a given number of groups according to an optimization criterion [7, 8]. One of the most popular and widely studied clustering methods is K-Means algorithm. In this work, an emphasis on understanding and analysis of Firefly based K-Means with canopy clustering view is presented. Although the K-Means algorithm is relatively straight-forward and converges in practice quickly but it is computationally expensive. Using canopies for the first stage of clustering and then feeding the output to the Firefly based K-Means remarkably reduces the computing time of K-Means algorithm [9]. Using Firefly based K-Means algorithm enhancement in accuracy of K-Means algorithm is achieved. They are implemented by mapping and reducing the steps in the algorithm.

Clustering based on canopies can be applied to many different underlying clustering algorithms like Greedy Agglomerative Clustering and Expectation-Maximization [1, 2]. Another algorithm efficiently performs K-means clustering by finding good initial starting points, but is not efficient when the number of clusters is large [3]. There has been almost no work on algorithms that work efficiently when the data set is large in all three senses at once when there are a large number of elements, thousands of features, and hundred thousands of clusters [10, 11].

This paper is organized as follows. Section 2 describes the basic preliminaries. Section 3 gives an overview of proposed hybridized K-Means algorithm. Section 4 discusses experimental results followed by concluding remarks in Sect. 5.

2 Preliminaries

2.1 Firefly Algorithm

Introduced by Yang in 2008, the firefly algorithm is based upon three idealized rules: (1) All fireflies will be attracted to each other. (2) The sense of attraction is directly proportional to the degree of brightness and this proportionality decreases with increase in distance. Thus, for any two flashing fireflies, the less bright one will move towards the brighter one. If there is no brighter one than a particular firefly, it will move randomly. (3) The measure of brightness for a firefly is determined by the landscape of the objective function defined for the optimization problem.

For a maximization problem, the brightness can simply be proportional to the value of the objective function [12]. The pseudo code for the algorithm is as follows.

Firefly Algorithm

1. Initialize algorithm parameters
   I. Max Gen: maximum number of generation
   II. γ : Light absorption coefficient
   III. r : Particular distance from light source
   IV. d : Domain Space
2. Objective function $f(x)$ where $x = (x_1 \quad x_2 \quad -- \quad x_d)^T$
3. Generate initial population of fireflies $x_i = (x_1 \quad x_2 \quad -- \quad x_n)$
4. Define light intensity of I_i at x_i determined by $f(x_i)$
5. While $(t < MaxGen)$
6. For $i = 1:n$
7. For $j = 1:n$
8. IF $(I_j > I_i)$
9. Move firefly I towards j in d-dimension via L'evy flight
10. End if
11. Attractiveness varies with distance r via
12. Evaluate new solutions and update light intensity
13. End for j
14. End for i
15. Rank fireflies and find the current best
16. End while
17. Post process result and visualization

End.

In the firefly algorithm there are two issues which include variation of light intensity and the attractiveness. For simplicity, it is assumed that the attractiveness of a firefly is determined by its brightness which associated with the objective function of the optimization problem. As light intensity decreases with the distance from its source and light is also absorbed in the media, so we should allow the attractiveness to vary with the degree of absorption.

2.2 K-means Algorithm

The k-means algorithm can work very well for compact and hyper spherical clusters. The time complexity of K-Means algorithm is O(N*K_i) [4]. K-means can be used to cluster high dimensional data sets. It uses a two-phase iterative algorithm to minimize the sum of point-to-centroid distances, summed over clusters.

The first phase consists of iterations for placing the points nearest to their respective cluster centroids with repositioning of cluster centroids. This phase is generally less useful for small datasets as the convergence rate is poor. Although this batch phase is fast but it approximates a solution as a starting point for the second phase.

The second phase uses online updates, where points are individually reassigned if doing so will reduce the sum of distances, and cluster centroids are recomputed after each reassignment. The second phase consists of one pass though all the points. The second phase will converge to a local minimum, although there may be other local minima with lower total sum of distances. The problem of finding the global minimum can only be solved in general by an exhaustive choice of starting points, but using several replicates with random starting points typically results in a solution that is a global minimum. Formally, the k-means clustering algorithm is described as follows.

K-Means Algorithm:

1. Set k: Choose a number of desired clusters, k.
2. Initialization: Choose k starting points to be used as initial estimates of the cluster centroids. These are the initial starting values.
3. Classification: Examine each point in the data set and assign it to the cluster whose centroid is nearest to it.
4. Centroid Calculation: When each point is assigned to a cluster, recalculate the new k centroids.
5. Convergence condition: Repeat steps 3 and 4 until no point changes its cluster assignment, or until a maximum number of passes through the data set is performed changes clusters or until a maximum number of passes through the data set is performed.

3 Proposed Clustering Algorithms

3.1 Efficient Clustering with Canopies

The key idea of the canopy algorithm is that one can greatly reduce the number of distance computations required for clustering by first cheaply partitioning the data into overlapping that one can greatly reduce the number of distance computations required for clustering by first cheaply partitioning the data into overlapping subset, and then only measuring distances among pairs of data points that belong to a common subset.

The basic objective of the canopy algorithm begins with setting up of points and removing one at random. It creates a canopy which contains this point and it iterates through the remainder of the point set. At each point if the distance from the first point is less than T1, then adds the point to the cluster. If after addition the point is less than T2 then removes the point from the set. This way the point that are very close to the original points will avoid all further processing the program loops until the set is empty.

We divide the clustering process into two stages. In the first stage we use the cheap distance measure in order to create some new number of overlapping subsets, called canopies. A canopy is simply a subset of the elements that, according to the approximate similarity measure, are within some distance threshold from a central point, an element may appear under more than one canopy, but at least should be present in a single canopy (Fig. 1).

3.2 Firefly Based K-means with Canopies

The proposed Firefly based K-Means with Canopies algorithm improves and increases performance accuracy of traditional K-Means algorithm. The hybridized algorithm is formulated as follows:

Pseudo code for firefly based K-Means with Canopies

```
Initialize algorithm parameters:
MaxGen: the maximum number of generations
Objective function of f(x) where x=(x1 x2 -- xd)^T
Generate initial population of clusters k or k (i=1, 2,..., n)
  While (t<MaxGen)
   For i = 1 to k;
   end for i;
Inputs:
I={i1,...,ik}(Instances to be clustered)
N(Number of clusters)
Outputs:
C={c1,..,cn}(cluster centroids)
M:  I→ (cluster membership)
Procedure Kmeans
   Set C to initial value(e.g. random selection of I)
   End
  boolean pointStronglyBoundToCanopyCenter = false
     for (Canopy canopy : canopies) {
     double centerPoint= canopyCenter.getPoint();
     if(distanceMeasure.similarity(centerPoint, x) > T1)
     pointStronglyBoundToCanopyCenter = true
}
     if(!pointStronglyBoundToCanopyCenter){
                    canopies.add(new Canopy(0.0d));
   While m has changed
        For each j ∈ {1 ...
            Recompute   as the centroid of
            {i|m(i)=j}
        End
```

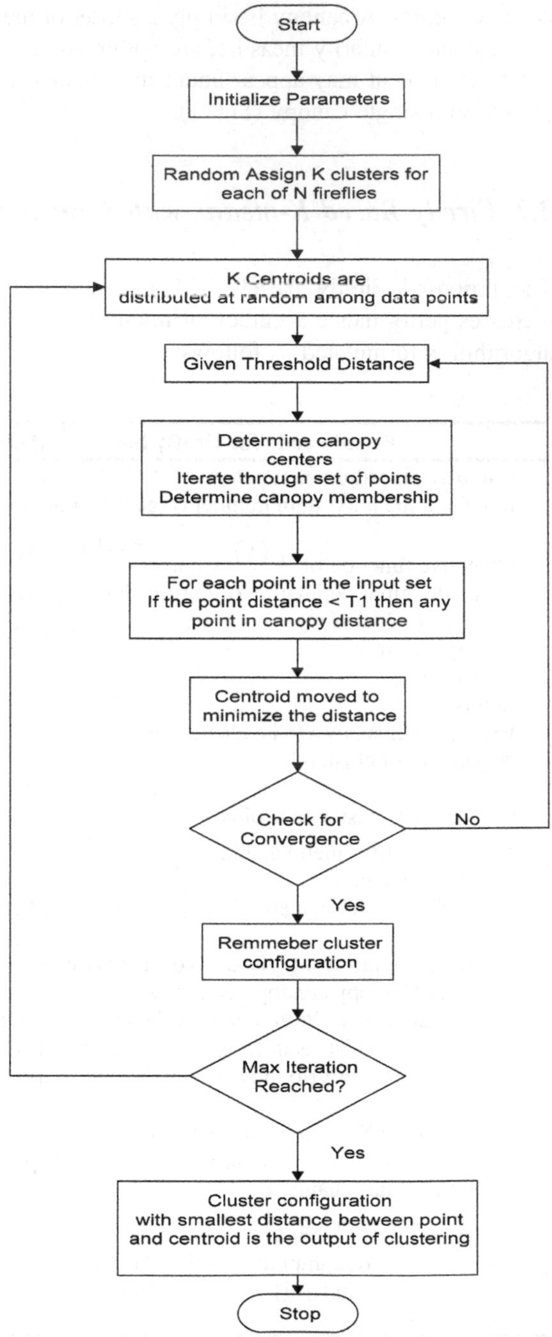

Fig. 1 Flow chart for clustering with canopies

3.3 Map-Reduce Implementation of K-means Algorithm with Canopy Algorithm

This sub-section discusses about the Map-Reduce implementation of K-Means algorithm with Canopy algorithm. The flow chart and algorithm steps are explained as follows (Fig. 2).

The major steps are as follows:

1. Data Preparation: The input data needs to be converted into a format suitable for distance and similarity measures. The data may need some pre-processing if it is of text format.
2. Picking Canopy Centers: The input points are split into data for the individual mappers. The input data received is then converted to canopies sequentially. The canopy generation algorithm is repeated by using the canopy centers using the reducer.
3. Assign Points to Canopy Centers: The canopy assignment step would simply assign points to generated canopy centers.
4. Pick K-Mean Cluster Centers and Iterate Until Convergence: The k-means center for each point is found out in the mapper stage of the map reduces procedure. The distance between the k-mean center and the point is computed only if they share a canopy, hence the computation is reduced. The average of

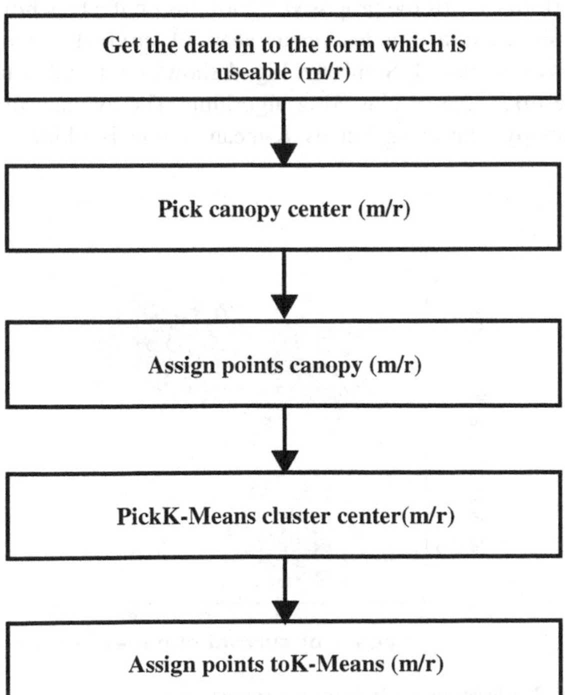

Fig. 2 Flow chart of map-reduce implementation of K-means algorithms with canopy algorithm

both the coordinates is computed by the reducer. The resultant is the new k-center. The iteration is repeated until convergence is achieved.

5. Assign Points to the K-Mean Centers: The points are assigned to the k-means center and are called cluster groups.

3.4 Data Sets for Experimentation

For testing of K-means with canopy clustering, the data sets used for experimentation here are the Haberman's Survival Data. The dataset contains cases from a study that was conducted at the University of Chicago's Billings Hospital on surgery for breast cancer. It contains 4 attributes of numerical type (Age of the patient, Year of the survival of patients who had undergone operation, No. of positive axillary nodes detected and Survival Status) out of which survival status was taken performing operations on the auxiliary nodes detected.

4 Experimental Results

Extensive simulations have been conducted in order to show the efficiency and performance of the proposed algorithms on the benchmark data sets. The number of positive axillary nodes versus years of survival of patient undergone operation is shown by Fig. 3. Similarly, Fig. 4 shows the result obtained by applying canopy at the first stage of clustering algorithm. The average time for firefly K-means with Canopy Clustering versus K-means alone is plotted and the execution time for

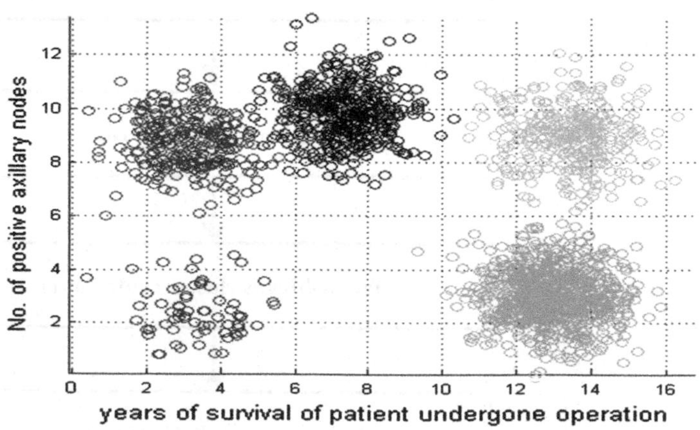

Fig. 3 Clustering on Haberman's survival data set

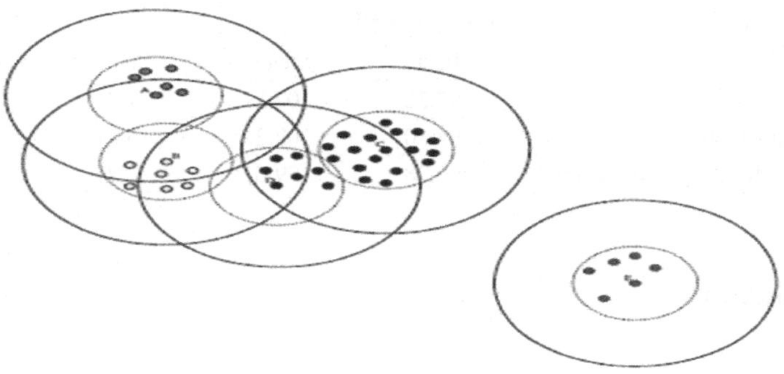

Fig. 4 Results applying canopy at the first stage of a clustering algorithm

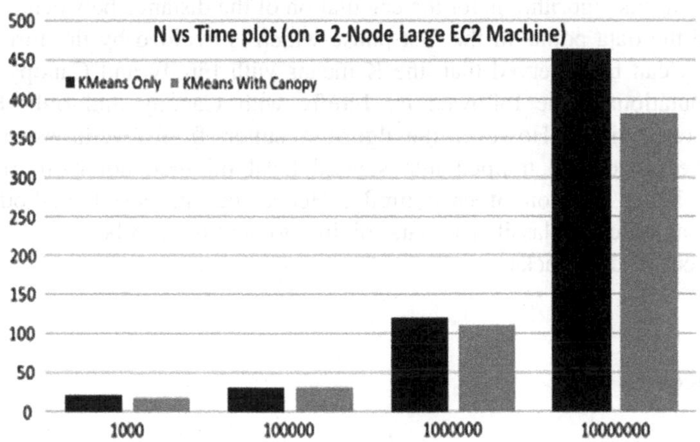

Fig. 5 Average time for K-means and K-means with canopy

canopy based K-means shows better trade-off for larger data sets as using canopies decrease the distance measure time of the K-means algorithm. This is shown by Fig. 5.

From the experimental studies it can be observed that, the total sum of distances decreases on each iteration, as K-means reassigns points between clusters and re-computes cluster centroid. Here the second phase does not works on reassigning and hence tells the first phase to have reached the optimal value. In some problems, the first phase might not reach a minimum.

The above experiment shows a good trade-off between using the Firefly and K-means algorithm with the use of overlapping canopies and the K-means algorithm itself in terms of execution time of both the algorithms as a majority of time used by K-means algorithm is for the calculation of the distance between the cluster center and the data points in the first phase which is avoided by the formation of canopies. Table 1 represents the result obtained when the above algorithm applied on Haberman data set.

Table 1 Computational time for Haberman data set

Algorithm	Best	Mean	Standard deviation
K-means	97.32	102.57	11.34
Firefly with canopy	97.10	102.26	5.81
K-means with firefly and canopy	96.78	99.61	7.21

5 Conclusion

The above experiment shows a good tradeoff between using the Firefly and K-means algorithm with the use of overlapping canopies and the K-means algorithm itself in terms of execution time of both the algorithms as a majority of time used by K-means algorithm is for the calculation of the distance between the cluster center and the data points in the first phase which is avoided by the formation of canopies. It can be observed that, the K-means with Firefly and Canopy requires less computational time followed by Firefly with Canopy and only K-means algorithm respectively. However, the algorithm suffers from several issues such as slow convergence rate, trapped into several local minima, not memorizing the history of better position of each firefly. Hence, our proposed method can be utilized as an efficient classifier in data mining domain but can be studied more for improvement of drawbacks.

References

1. McCallum, A., Nigam, K., Ungar, L.H.: Efficient clustering of 'high-dimensional data sets with application to reference matching. In: Proceedings of the Sixth ACM SIGKDD International Conference on Knowledge Discovery and Data Mining, pp. 169–178 (2000). doi:10.1145/347090.347123
2. Moore, A.W.: Very fast EM-based mixture model clustering using multi resolution kd-trees. Adv. Neural Inf. Process. Syst. 543–549 (1999)
3. Bradley, P.S., Fayyad, U., Reina, C.: Scaling clustering algorithms to large databases. In: Proceedings of 4th International Conference on Knowledge Discovery and Data Mining (KDD-98). AAAI Press, Menlo Park, Aug 1998
4. Xu, R., Wunsch, D.: Survey of clustering algorithms. IEEE Trans. Neural Netw. **16**(3), 645–678 (2005)
5. Anderberg, M.R.: Cluster Analysis for Application. Academic Press, Waltham (1973)
6. Kanungo, T., Mount, D.M.: An efficient K-means clustering algorithm: analysis and implementation. IEEE Trans. Pattern Anal. Mach. Intell. **24**(7), 881–892 (2002)
7. Garcia, J., Fdez-Valdivia, J., Cortijo, F., Molina, R.: Dynamic approach for clustering data. Signal Process. **44**(2), 181–196 (1994)
8. Judd, D., McKinley, P., Jain, A.: Large-scale parallel data clustering. In: Proceedings of International Conference on Pattern Recognition, Aug 1996
9. Hassanzadeh, T., Meybodi, M.R.: A new hybrid approach for data clustering using firefly algorithm and K-means, May 2012

10. Ester, M., Kriegel, H., Sander, J., Xu, X.: A density-based algorithm for discovering clusters in large spatial databases with noise. In: Proceedings of the 2nd International Conference on Knowledge Discovery and Data Mining, vol. 96, pp. 226–231, Aug 1996
11. Ester, M., Kriegel, H., Xu, X.: Knowledge discovery in large spatial databases: focusing techniques for efficient class identification. In: Advances in Spatial Databases. Springer, Berlin, pp. 67–82 (1995)
12. Dubes, R.C., Jain, A.K.: Algorithms for Clustering Data. Prentice Hall, Upper Saddle River (1988)

A Cooperative Intrusion Detection System for Sleep Deprivation Attack Using Neuro-Fuzzy Classifier in Mobile Ad Hoc Networks

Alka Chaudhary, V.N. Tiwari and Anil Kumar

Abstract This paper proposed a soft computing based solution for a very popular attack, i.e. sleep deprivation attack in mobile ad hoc networks (MANETs). As a soft computing solution, neuro-fuzzy classifier is used in binary form to detect the normal and abnormal activities in MANETs. The proposed detection scheme is based on distributed and cooperative architecture of intrusion detection system and simulations have been carried out through Qualnet simulator and MATLAB toolbox that shows the results of proposed solution in respect of performance metrics very effectively.

Keywords Mobile ad hoc networks (MANETs) · Security issues · Intrusion detection system (IDS) · Soft computing · Adaptive neuro-fuzzy inference system (ANFIS) · Neuro-fuzzy · Sleep deprivation attack

1 Introduction

Mobile Ad hoc Networks (MANETs) are really attractive and flexible during the communication because MANETs do not take in any predefined infrastructure or any centralized management points. In MANETs, each the mobile node acts as a router as well as the mobile client to broadcast the data packages across the network. MANETs uses in many applications i.e. disaster relief management,

A. Chaudhary (✉) · A. Kumar
Department of Computer Science and Engineering, Manipal University Jaipur, Jaipur, India
e-mail: alka.chaudhary0207@gmail.com

A. Kumar
e-mail: dahiyaanil@yahoo.com

V.N. Tiwari
Department of Electronics and Communication Engineering, Manipal University Jaipur, Jaipur, India
e-mail: vivekanand.tiwari@jaipur.manipal.edu

L.C. Jain et al. (eds.), *Computational Intelligence in Data Mining - Volume 2*,
Smart Innovation, Systems and Technologies 32, DOI 10.1007/978-81-322-2208-8_32

military areas, communications among groups of people [1, 2]. Similarly, conventional networks, MANETs also prone to some types of attacks due to communication via wireless connections, resource constraints, cooperativeness between the mobile clients and dynamic topology. In terms of security of MANETs, intrusion detection system (IDS) is an essential component so that it is known as the second line of defence [3].

When any attempt to compromise with the security attributes such as confidentiality, repudiation, integrity, and availability of resources, then these attempts said to be the intrusions and detection of such intrusions are known as intrusion detection system (IDS) [3]. Usually there are three basic types of IDS architectures in literature: stand-alone or local intrusion detection systems, distributed and cooperative intrusion detection systems, hierarchical Intrusion Detection Systems although for better intrusion detection, IDS architecture should be distributed and cooperative in MANETs [3]. This paper emphasized on distributed and cooperative intrusion detection system based on neuro-fuzzy classifier for detection of sleep deprivation attack in MANETs that are discussed in Sect. 3. AODV routing protocol is used very commonly in MANETs so that I have considered AODV routing protocol in this research [4]. The targeted attack in this research is sleep deprivation attack, but for a detailed description of attacks in MANETs readers can adopt the reference [5]. Here, in this research the targeted attack is sleep deprivation attack so during simulation believed as, in route discovery phase of AODV an attacker node broadcasts the route request (RREQ) packets to their neighbor nodes with a destination IP address that is existed in these network address ranges but in actually, it does not present in the network.

Soft computing is a modern approach towards intelligent systems that can easily treat the uncertainty [6]. Here, a new scheme designed based on neuro-fuzzy classifier, which is in binary form for distinguishing the normal and abnormal activities so that Adaptive neuro-fuzzy inference system (ANFIS) is used as a neuro-fuzzy classifier in binary form and, subtractive clustering technique is use for deciding the fuzzy rules.

The subsequent sections are as follows: Sect. 2, the data extraction on specific features has been done with the help of Qualnet simulator 6.1 for sleep deprivation attack. Section 3, presents the proposed scheme for mobile ad hoc networks. Section 4, evaluate the performance of system in distributed and cooperative environment and Sect. 5, presents the conclusion.

2 Selection of Features and Dataset

Features are eminent attributes that are applied as an input for our suggested system. For developing the system fundamentals, selection of attributes are most significant. Our system has considered particularly on sleep deprivation attack based features. Table 1 show the list of selected features that are observed from each node in

Table 1 List of features

Features abbreviations	Explanations
num_hops	Aggregate sum of the hop counts of all active routes
num_req_initd	No. of RREQ packets initiates by this node
num_req_receivd	No. of RREQ packets received to this node
num_req_recvd_asDest	No. of RREQ packets received as a destination for this node
num_rep_initd_asDest	No. of RREP packets initiated from the destination by this node
num_rep_initd_asIntermde	No. of RREP packets initiated from the an intermediate node
num_rep_fwrd	No. of RREP packets forwarded by intermediate nodes
Num_rep_recvd	No. of RREP packets received by this node
Num_rep_recvd_asSrce	No. of RREP packets received as source by this node
num_dataPks_Initd	No. of data packets sent as source of the data by this node
num_dataPks_fwrd	No. of data packets forwarded by this node
num_dataPks_recvd	No. of data packets sent as destination of the data by this node
Consumed_battery	Calculates the consumed battery to perform any operation by this node
num_nbrs	No. of neighbors of node during simulation time
num_addNbrs	No. of added neighbors of node during simulation time
num_rmveNbrs	No. of remove neighbors of node during simulation time

MANETs in respect of AODV routing protocol and data are collected periodically by each node based on the selected features [7]. The unusual decrements in the battery power that is used up by each node may be signals of sleep deprivation attack.

Here in this research, Qualnet simulator 6.1 [8] is used to draw out the dataset based on selected features for measuring the results of Neuro-fuzzy classifier so that in Table 2, presents the segmentation between the normal and attacks type of data for training, checking and testing at the simulation time 1,000 s. Table 3 depicted the simulation parameters that are used during simulation.

Table 2 Distribution of data samples in training and checking phase

Distributions of data samples	Classes	
	Normal	Attack (sleep deprivation attack)
Training	18,000	18,000
Checking	4,700	4,700
Testing	4,800	4,800
Total	27,500	27,500

Table 3 List of simulation parameters

Simulation parameters	
Simulator	Qualnet 6.1
Routing protocol	AODV
Packet size	512 bytes
Radio type	802.11b
Energy model	Generic
Pause time	30 s
Battery model	Linear model
Simulation time	1,000 s (training and testing)
Batter charge monitoring	Interval 60 s
Traffic type	CBR
Simulation area	1,500 m × 1,500 m
Number of nodes	15 and 35 nodes
Mobility	Random way point
Mobility speeds	0–25 mps
Malicious nodes	4

3 Proposed Scheme

This section describes the suggested method of binary neuro-fuzzy classifier based IDS for MANETs. Here, ANFIS is applied as a binary neuro-fuzzy classifier and fuzzy rules are selected based on subtractive clustering technique. ANFIS suggests a process in terms of learning the information from a presented dataset for fuzzy modeling, where the parameters of membership functions are computed that take into account the related FIS to track or manage in the skillful way of given input and output information. The related parameters of membership functions will be changed according to the procedure of learning [9]. Subtractive clustering [10] is immediate, single pass algorithm for calculating the clusters and cluster centers from a presented dataset. It is an extension of mountain clustering technique that is suggested by Yager.

The proposed scheme based intrusion detection system can detect the attacks in distributed and cooperative manner so that in the network, each node has an IDS agent that is communicated with their one hop away nodes for detecting the attacks. During the communication, neighbour nodes exchange the information and finally reach on decisions regarding malicious activities that are present in MANETs or not and then give responses [5]. Here, list of features that are given in Table 1, are used as the input for proposed binary neuro-fuzzy classifier based IDS and data patterns are labeled with 0 and 1 where 0 presents the normal patterns and 1 presents the attack patterns. Figure 1 shows the process steps of our proposed system. Here in this research, a subtractive clustering technique employed with neighbourhood radius $r_a = 0.5$ for segmenting the training data and builds the automatic fuzzy rules to make the structure of fuzzy inference system (FIS) for ANFIS training. Figure 2

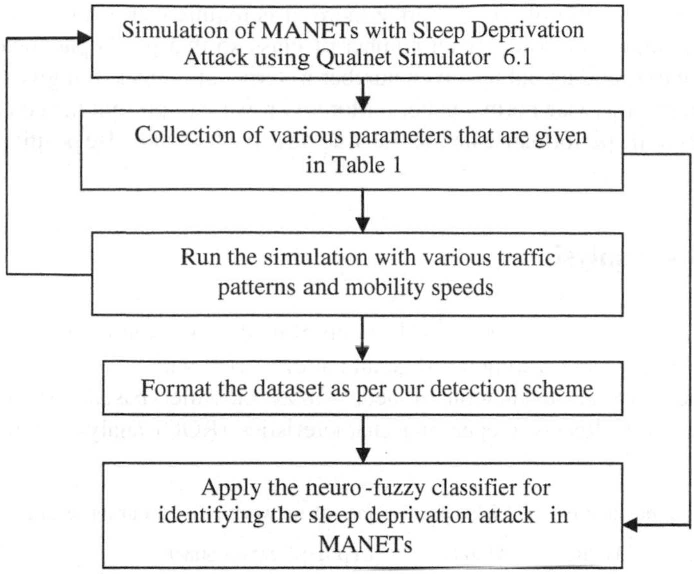

Fig. 1 Flow the process of our proposed system

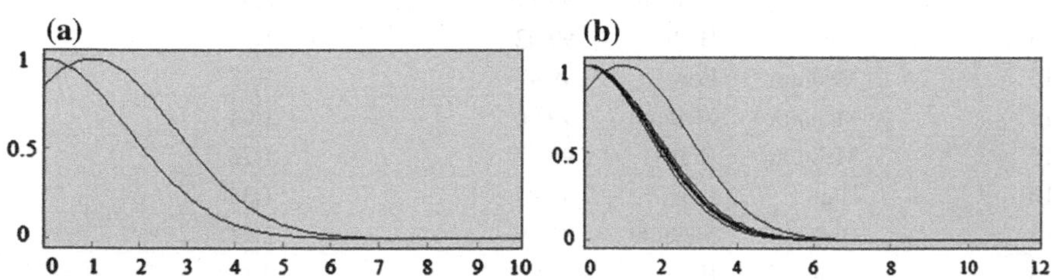

Fig. 2 **a** Before training MFs on input feature 9. **b** After training MFs on input feature 9. **a**, **b** Shows the initial and final membership functions for some input features

shows the initial and final membership functions for some input features during ANFIS training phase. Checking data are utilized for model validation because during the training phase, later on a certain time model starts over fitting so that FIS may act biased with other data set.

The applied ANFIS contains 302 nodes and total fitting parameters are 411, whereas 270 are premise parameters and 141 are consequent parameters. After 50 epochs of learning, root mean square error (RMSE) of training is 0.3261 and checking is 0.3523.

The architecture of ANFIS produce exclusively single output so that here in this paper ANFIS output is assigned by the class number of Table 1 input features, i.e. 0 presents the normal class and 1 presents the attacks. It is not necessary that ANFIS gives output in respect of 0 or 1. Furthermore, it is require estimating class number through rounding out of the given number of class so that μ is a parameter that is responsible to rounding out the given number in terms of integer. If it gives 1 means attack pattern otherwise normal pattern. Moreover, we will present the consequence of μ based on the performance metrics i.e. true positive rate and false positive rate [7].

4 Results Analysis

From the results aspects, testing has been employed to analyze the performance of our proposed IDS. For testing our IDS, all patterns used with actual labeled dataset. For the performance evaluation of neuro-fuzzy classifier based distributed and cooperative IDS, Receiver operating characteristics (ROC) analysis is utilized in

Table 4 True positive rate and false positive rate of network size 15 during testing at $\mu = 0.5$

No. of nodes	Traffic	Mobility	Sleep deprivation attack	
			Cooperative detection	
			True positive rate (%)	False positive rate (%)
15	Low	Low	99.34	0.68
15	Low	Medium	99.79	0.87
15	Low	High	99.02	1.1
15	Medium	Low	99.47	1.18
15	Medium	Medium	99.56	1.64
15	Medium	High	98.97	1.78
15	High	Low	97.94	1.34
15	High	Medium	98.39	2.21
15	High	High	98.91	2.81

Table 5 True positive rate and false positive rate of network size 35 during testing at $\mu = 0.5$

No. of nodes	Traffic	Mobility	Sleep deprivation attack	
			Cooperative detection	
			True positive rate (%)	False positive rate (%)
35	Low	Low	99.53	0.71
35	Low	Medium	99.83	0.90
35	Low	High	99.11	1.08
35	Medium	Low	99.58	1.21
35	Medium	Medium	99.69	1.68
35	Medium	High	99.14	1.74
35	High	Low	97.86	1.53
35	High	Medium	98.41	2.37
35	High	High	98.97	2.94

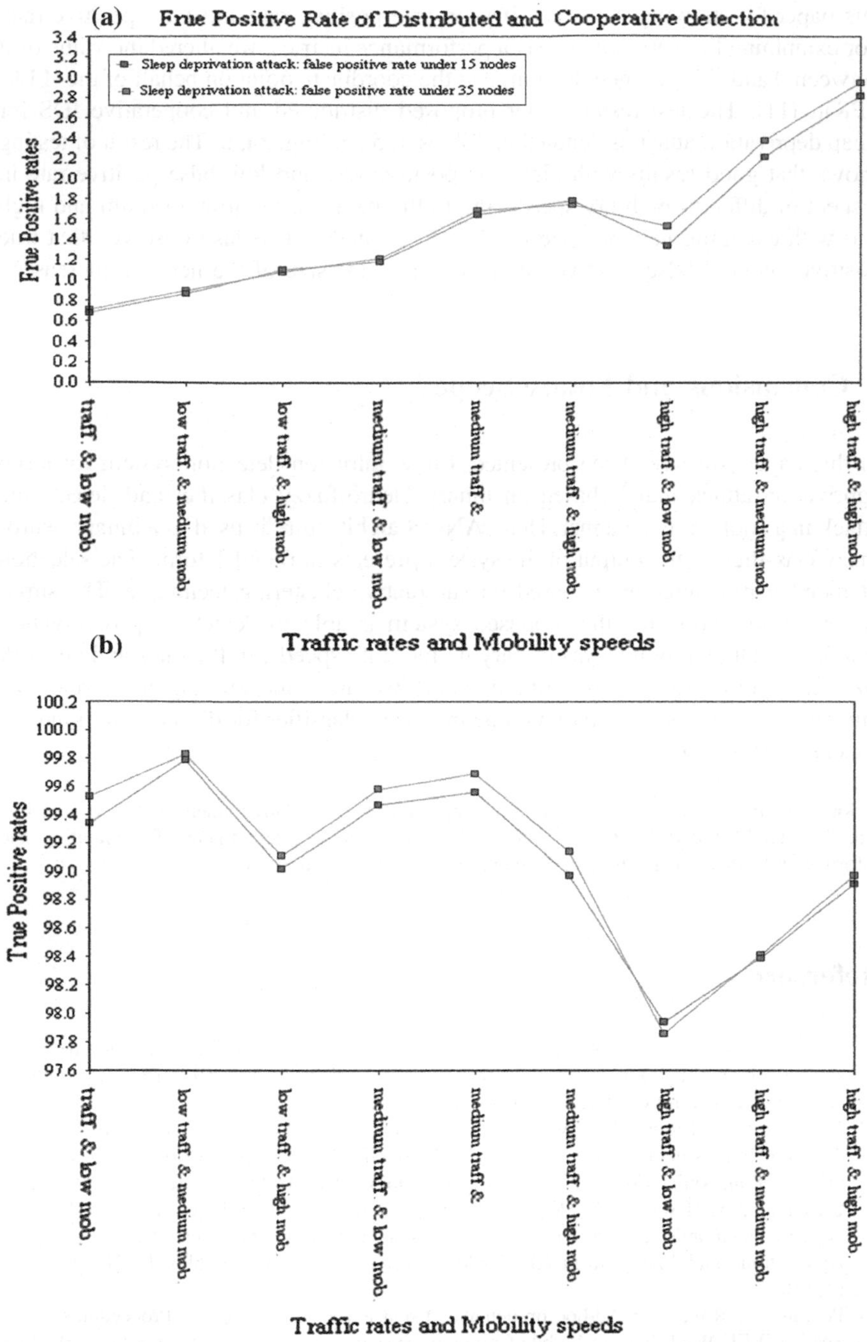

Fig. 3 **a** True positive rates for cooperative IDS in respect sleep deprivation attack under 15 and 35 nodes. **b** False positive rates for cooperative IDS in respect sleep deprivation attack under 15 and 35 nodes

this paper for showing the effect of μ on true positive rate and false positive rate. For examining the variation between performance metrics, we altered the value of μ between 0 and 0.5 and then diagrammed the coordinate point on behalf of the (FPR, TPR)μ [11]. The test result of our proposed distributed and cooperative IDS for sleep deprivation attack is denoted in Tables 4, 5 and Fig. 3a, b. The result of testing shows that good results with high true positive rate and low false positive rate in respect of different mobility speeds and traffic patterns, i.e. low, medium and high and with changing network size i.e. 15 and 35 nodes. It is also observed that true positive rate and false positive rate increases if the size of the network increases.

5 Conclusions and Future Scope

In this paper, we have been presented a new intrusion detection system for sleep deprivation attack that is based on binary Neuro-fuzzy classifier and detects the attack in a cooperative manner. Here, ANFIS architecture is used as a binary neuro-fuzzy classifier so the output of the system presents in 0 and 1 form. The selection of membership functions is based on subtractive clustering technique. The simulation results show that the proposed system is able to detect sleep deprivation attack very efficiently in terms of varying mobility speeds, traffic rates and network size. In future work, In our future work, we are concentrating to develop an intrusion detection system based on neuro-fuzzy classifier for detecting all kinds of attacks in MANETs.

Acknowledgments This work is partially supported by DST (Government of India) vide File No. DST/TSG/NTS/2012/106, we acknowledge A.N.TOOSI (Department of Computing and Information System), University of Melbourne for his useful suggestions.

References

1. Chaudhary, A., Tiwari, V.N., Kumar, A.: Design an anomaly based fuzzy intrusion detection system for packet dropping attack in mobile ad hoc network. In: 2014 IEEE International Advance Computing Conference (IACC), pp. 256–261 Feb 2014
2. Chaudhary, A., Kumar, A., Tiwari, V.N.: A reliable solution against packet dropping attack due to malicious nodes using fuzzy logic in MANETs. In: 2014 International Conference on Optimization, Reliability, and Information Technology (ICROIT), pp. 178–181 (2014)
3. Chaudhary, A., Tiwari, V.N., Kumar, A.: Analysis of fuzzy logic based intrusion detection systems in mobile ad hoc networks. In: Bharati Vidyapeeth's Institute of Computer Applications and Management (BVICAM), vol. 6, no. 1, pp. 690–696. ISSN 0973-5658 (2014)
4. Perkins, C., Royer, E.: Ad-hoc on-demand distance vector routing. In: Proceedings of the Second IEEE Workshop on Mobile Computer Systems and Applications, pp. 90–100. IEEE (1999)

5. Sen, S., Clark, J.A., Tapiador, J.E.: Security Threats in Mobile Ad Hoc Networks. Security of Self-Organizing Networks: MANET, WSN, WMN, VANET, pp. 127–147. Auerbach Publications, Boca Raton (2010)
6. Zadeh, L.A.: Roles of soft computing and fuzzy logic in the conception, design and deployment of information/intelligent systems. In: Computational Intelligence: Soft Computing and Fuzzy-Neuro Integration with Applications, pp. 1–9. Springer, Berlin (1998)
7. Şen, S., Clark, J.A.: A grammatical evolution approach to intrusion detection on mobile ad hoc networks. In: Proceedings of the Second ACM Conference on Wireless Network Security, pp. 95–102. ACM (2009)
8. QualNet Network Simulator. Available http://www.Scalable-networks.com
9. Jang, J.S.R., Sun, C.T., Mizutani, E.: Neuro-Fuzzy and Soft Computing: A Computational Approach to Learning and Machine Intelligence, 1st edn. Prentice Hall of India, New Delhi (1997)
10. Chiu, S.: Fuzzy model identification based on cluster estimation. J. Intell. Fuzzy Syst. **3**, 267–278 (1994)
11. Gomez, J., Dasgupta, D.: Evolving fuzzy classifiers for intrusion detection. In: Proceedings of the 2002 IEEE Workshop on Information Assurance, vol. 6, no. 3, pp. 321–323. IEEE Computer Press, New York (2002)

Improving the Performance of a Proxy Cache Using Expectation Maximization with Naive Bayes Classifier

P. Julian Benadit, F. Sagayaraj Francis and U. Muruganantham

Abstract The Expectation Maximization Naive Bayes classifier has been a centre of attention in the area of Web data classification. In this work, we seek to improve the operation of the traditional Web proxy cache replacement policies such as LRU and GDSF by assimilating semi supervised machine learning technique for raising the operation of the Web proxy cache. Web proxy caching is utilized to improve performance of the Proxy server. Web proxy cache reduces both network traffic and response period. In the beginning section of this paper, semi supervised learning method as an Expectation Maximization Naive Bayes classifier (EM-NB) to train from proxy log files and predict the class of web objects to be revisited or not. In the second part, an Expectation Multinomial Naïve Bayes classifier (EM-NB) is incorporated with traditional Web proxy caching policies to form novel caching approaches known as EMNB-LRU and EMNB-GDSF. These proposed EMNB-LRU and EMNB-GDSF significantly improve the performances of LRU and GDSF respectively.

Keywords Web caching · Proxy server · Cache replacement · Classification · Expectation Maximization Naive Bayes classifier

---

P. Julian Benadit (✉) · F. Sagayaraj Francis
Department of Computer Science and Engineering,
Pondicherry Engineering College, Pondicherry 605014, India
e-mail: benaditjulian@gmail.com

F. Sagayaraj Francis
e-mail: fsfrancis@pec.edu

U. Muruganantham
Department of Computer Science and Engineering,
Pondicherry University, Pondicherry 605014, India
e-mail: mahianandh18@gmail.com

L.C. Jain et al. (eds.), *Computational Intelligence in Data Mining - Volume 2*,
Smart Innovation, Systems and Technologies 32, DOI 10.1007/978-81-322-2208-8_33

1 Introduction

As the World Wide Web and users are maturing at a very rapid rate, the performance of web systems becomes rapidly high. Web caching is the one of the best methods for improving the performance of the proxy host. The idea behind in web caching is to maintain the most popular web log data that likely to be re-visited in the future in a cache, such that the performance of web system can be improved since most of the user requests can directly access from the cache. The main idea of web caching algorithm is the cache replacement policy. To ameliorate the execution of web caching, researchers have proposed a number of cache replacement policies Table 1. Many of these conventional replacement algorithms take into account several factors and assign a key value or priority for each web document stored in the cache. However, it is difficult to have to have an omnipotent replacement policy that performs well in all places or for all time because each replacement policy has a different structure to optimize the different resources. Moreover, several factors can influence the cache replacement policy to have a better replacement decision and it is not an easy task because one parameter is more important than the other one.

Due to this restriction, there is a need for an efficient method to intelligently handle the web cache by satisfying the objectives of web caching requirement. Thus the motivation, for incorporating the intelligent methods in the web caching algorithms. Another motivation to intelligent web caching algorithms is to train the availability of web access log files. In our previous surveys, the intelligent techniques have been applied in web caching algorithm. These studies typically build a prediction model by training the web log files. By progressing to use of the prediction model, the caching algorithms become more effective and adaptive to the web cache environment compared to the conventional web caching algorithms. However, these studies didn't take into account the user design and feature request when making the prediction model. Since the users are the origin of web access, it is necessary to establish a prediction model, whether the object can be revisited in future or not. In this paper, we use the web logs file to train by the Expectation

Table 1 Cache replacement policies

Policy	Key parameters	Eviction
LFU	No. of references	The least frequently accessed
LRU	Time since last access	The least recently accessed first
GDS	Document size Sp	Least valuable first according to value $p = Cp/Sp + L$
	Document cost Cp	
	An inflation value L	
GDSF	Document size Sp	Least valuable first according to value $p = (Cp \cdot fp/Sp)\, \beta + L$
	Document cost Cp	
	Number of non-aged references fp time since last access	
	An inflation value L	

Maximization Naïve Bayes classifier (EM-NB) [1] and to classify whether the web objects can be revisited in future or not. Based on this method, we can obtain user interest web objects that can be revisited in future or not. We then proposed the semi supervised learning mechanism EM-NB to classify the web log files and then it is incorporated with traditional caching algorithm called EMNB-LRU and EMNB-GDSF to improve its web caching performance.

The organization of this paper is as follows. In the following part, we survey the related study in web caching. In Sect. 3 we give a brief introduce of Expectation Maximization Naive Bayes classifier model. Section 4 we introduce the proposed novel web proxy caching approach and show how it integrates with the caching Algorithms. Experiment Results are described in Sect. 5. Performance Evaluation is presented in Sect. 6. Finally, we conclude our paper in Sect. 7.

2 Related Works

Web caching plays a vital role in improving web proxy server performance. The essence of web caching is so called "replacement policy", which measure the most popular of previously visited documents, retaining it in the cache those popular documents and replaces rarely used ones. The basic idea of the most common caching algorithms is to assign each document a key value computed by factors such as size, frequency and cost. Using this key value, we could rank these web documents according to corresponding key value. When a replacement is to be carried, the lower ranked web documents will be evicted from the cache. Among these key value based caching algorithms; GDSF [2] is the most successful one. It assigns a key value to each document in the cache as $K(h) = L + F(h) * C(h)/s(h)$, where L is an inflation factor to avoid cache Pollution, C(h) is the cost to fetch, F(h) is the past occurrence frequency of h and S(h) is the size of h. Accessibility of web log files that can be used as training data promotes the growth of intelligent web caching algorithms [3–5].

In preceding papers exploiting supervised learning methods to cope with the matter [3, 6–8]. Most of these recent studies use an Adaptive Neuro Fuzzy Inference System (ANFIS), Naïve Bayes (NB), Decision tree (C4.5), in worldwide caching Table 2. Though ANFIS training might consume a wide amount of time and need further process overheads. Also Naïve Bayes, result in less accuracy in classifying the large web data sets and similarly Decision Tree also result in less prediction accuracy in training large data sets. So In this paper, we attempted to increase the performance of the web cache replacement strategies by integrating semi supervised learning method of Expectation Maximization Naive Bayes classifier (EM-NB). In conclusion, we achieved a large-scale evaluation compared with other intelligent classifier like ANFIS, Naïve Bayes, and Decision Tree (C4.5) in terms of precision and recall on different log files and the proposed methodology has enhanced the performance of the web proxy cache in terms hit and byte hit ratio.

Table 2 Summary of intelligent web caching approaches

Name	Principle	Limitation
NB [3]	Constructed from the training data to estimate the probability of each class given the document feature values of a new instance	• Violation of independence assumption
		• Zero conditional probability problem
C4.5 [6]	A model based on decision trees consists of a series of simple decision rules, often presented in the form of a graph	• Not good for predicting the values of a continuous class attribute
		• Low prediction accuracy, high variance
ANFIS [7]	Neuro-fuzzy system (ANFIS) has been employed with the LRU algorithm in cache replacement	• The training process requires a long time and extra computations
		• The byte hit ratio is not good enough

3 Expectation Maximization Naive Bayes Classifier

One of the LU learning techniques uses the Expectation–Maximization (EM) algorithm [9]. EM is a popular iterative algorithm for maximum likelihood estimation problems with missing data. The EM algorithm consists of two steps, the Expectation step (or E-step), and the Maximization step (or M-step). The E-step basically fills in the missing information based on the current approximation of the parameters. The M-step, which maximizes the likelihood, re-estimates the parameters. This leads to the next iteration of the algorithm, and so on.

The ability of EM to work with missing information is exactly what is needed for learning from labelled and unlabelled examples. The web document in the labelled set (denoted by L) all have class labels (or values). The web document in the unlabeled.

Set (denoted by U) can be considered as having missing class labels. We can use EM to estimate them based on the current model, i.e., to assign probabilistic class labels to each document *di* in *U*, (i.e., Pr(*cj*|*di*)). Subsequently a number of iterations, all probabilities will converge. Notice that the EM algorithm is not really a specific "algorithm", but is a framework or strategy. It only carries a base algorithm iteratively. We will use the naïve Bayesian (NB) algorithm as the base algorithm, and run it iteratively. The parameters that EM estimates the class prior probabilities (see Eq. 1). In this paper, we use a NB classifier in each iteration of EM, (Eqs. 1 and 2) for the E-step, and (Eq. 3) for the M-step. Specifically, we first build a NB classifier *f* using the labeled examples in *L*. We then use of to classify the unlabeled examples in U, more accurately, to ascribe a probability to each class for every unlabelled example (i.e., Pr(*cj*|*di*)), which takes the value in [0, 1] instead of {0, 1}.

Let the set of classes be $C = \{c1, c2 \ldots c|C|\}$. That is, it assigns *di* the class probabilities of $\Pr(c1|di)$, $\Pr(c2|di)$, …, $\Pr(c|C||di)$. This is different from the example in the labelled set L, where each web document belongs to only a single class c_k (i.e. $\Pr(c_k|d_i) = 1(\text{Revisited again})$ and $\Pr(c_j|d_i) = 0(\text{Not Revisited again})\, j \neq k$.

3.1 Implementation of Expectation Maximization Naive Bayes Classification Algorithm

3.1.1 Training Phase

Input: Collection of training documents; Set of target values (categories, topics)

Output: Files containing the probabilities of $\Pr(c_j)$ and $\Pr(w_r|c_j)$

Algorithm EM (L, U)

1. Take an initial naïve Bayesian classifier f from only the labelled set L using Equation 1and 2

$$\mathrm{pr}(w_t \mid c_j;\hat{\Theta}) = \frac{\lambda + \sum_{I=1}^{|D|} N_{ti}\,\mathrm{pr}(c_j \mid d_i)}{\lambda \mid V \mid + \sum_{s=1}^{|V|} \sum_{I=1}^{|D|} N_{si}\,\mathrm{pr}(c_j \mid d_i)} \tag{1}$$

$$\mathrm{pr}(c_j;\hat{\Theta}) = \frac{\sum_{I=1}^{|D|} \mathrm{pr}(c_j \mid d_i)}{\mid D \mid} \tag{2}$$

2. **Repeat // E-step**
3. For Each Example d_i in U do
4. Using the current classifier f to compute $\Pr(c_j \mid d_i)$ using equation 3

$$\mathrm{pr}(c_j \mid d_i;\hat{\Theta}) = \frac{\mathrm{pr}(c_j \mid \hat{\Theta})\pi_{k=1}^{|di|}\mathrm{pr}(wd_{i,k} \mid c_j;\hat{\Theta})}{\sum_{r=1}^{|c|} \mathrm{pr}(c_j|\hat{\Theta})\pi_{k=1}^{|di|}\mathrm{pr}(wd_{i,k}|c_j;\hat{\Theta})} \tag{3}$$

5. **End //M-step**
6. Learn a new initial naïve Bayesian classifier f from L $\cup$ U .
7. Until the classifier parameters stabilize return the classifier f from the last Iteration.

4 Proposed Novel Web Proxy Caching Approach

The proposed system will present a framework (Fig. 1) for novel Web proxy caching approaches based on machine learning techniques [3–5]. In Our Proposed work we use the semi supervised mining algorithm for classifying the datasets that can be revisited again or not in the future. The mining steps consist of two different phases for classifying the data sets; in the first phase (Fig. 1) we preprocess to remove the irrelevant information in the proxy log data sets, Different techniques are given at the preprocessing stage such as data cleansing, data filtering and

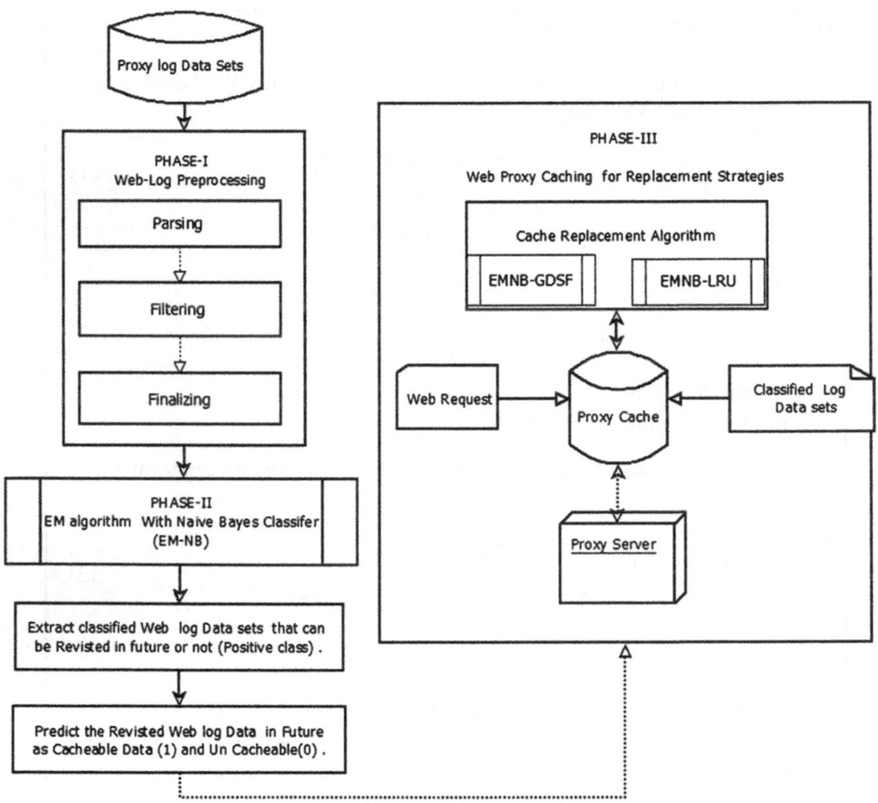

Fig. 1 Working flow of web proxy caching approach based on Expectation Maximization Naïve Bayes classifier

information consolidation. Once this task has been accomplished, the proxy data sets have been trained in the second phase (Fig. 1). By the classifier EM-NB, which predicts whether the web objects that can be revisited once more in the future or not. In the third phase (Fig. 1), the Predicted web object has been integrated to the traditional web proxy caching algorithm like LRU and GDSF for replacement strategies in order to improve the hit and byte hit ratio of the conventional replacement algorithm.

4.1 Expectation Naïve Bayes Classifier—Greedy Dual Size Frequency

The main advantage of the GDSF principle is that it executes well in terms of the hit ratio. But, the byte hit ratio of GDSF principle is too reduced. Thus, the EM-NB classifier is integrated with GDSF for advancing the performance in terms of the hit and byte hit ratio of GDSF [2]. The suggested novel proxy caching approach is called (EM-NB)—GDSF. In (EM-NB)—GDSF, a trained EM-NB classifier is used to predict the classes of web objects either objects may be re-visited later or not. After this, the classification, assessment is integrated into cache replacement policy (GDSF) to give a key value for each object in the cache buffer; the lowest values are removed first. The proposed (EM-NB)—GDSF.

4.2 Expectation Naïve Bayes Classifier—Least Recently Used

LRU policy is the most common web proxy caching scheme among all the Web proxy caching algorithms [10]. But, LRU policy suffers from cache pollution, which means that unpopular data's will remain in the cache for a long period. For reducing cache pollution in LRU, an EM-NB classifier is joint with LRU to form a novel approach Called (EM-NB)—LRU.

4.2.1 Algorithm for Greedy Dual Size Frequency—Least Recently Used

Input: Requested objects from proxy trace **Output:** Total Caching cost using LRU mechanism **Steps:** 1. Receive request for object P 2. IF P is already present in cache THEN 3. Server request internally from cache 4. Goto Step 16 5. ELSE IF P is not in present in cache 6. Serve P request from origin server 7. Increment total cost by p 8. IF P can be accommodated in cache THEN 9. Bring P into cache 10. ELSE 11. P cannot be accommodated 12.Evict least recently requested object from cache and replace it with P. 13. END IF 14. Incremental total cost by c 15. END IF 16. Repeat steps 1 to 14 for next object request of trace data. **5 Experimental Results**	**Input:** Requested objects from proxy trace **Output:** Total Caching cost using GDSF mechanism **Steps:** 1. Receive request for object P 2. IF P is already present in cache THEN 3. Serve request P internally from cache 4. ELSE IF P is not in present in cache 5. Serve request P from origin server 6. Increment total cost P; $H(p) \leftarrow L + f(p) \times c(p) / s(p)$ 7. IF P can be accommodated in cache THEN 8. Bring P into cache 9. ELSE 10. While there is not enough free cache for P 11. do $L \leftarrow \min\{H(q) \mid q$ is in cache Evict q which satisfies $H(q) = L$ and replace it with P. 12. END IF 13. $H(p) \leftarrow L + f(p) \times c(p) / s(p)$ 14. END IF 15. Repeat steps 1 to 14 for next object request of trace data.

5 Experimental Results

5.1 Proxy Log File Collection

We received data from the proxy log files of the Web object requested in some proxy Servers found in UC, BO2, SV, SD, and NY nearby the United States of the IR cache network for 15 days [11]. An access proxy log entry generally consists of the consequent fields: timestamp, elapsed time, log tag, message protocol code, size, user identification, request approach, URL, hierarchy documents and host-name, and content type.

5.2 Web Pre-processing

Web pre-processing is a usage mining technique that involves transforming web log data into a structured format. WWW log information is frequently incomplete, inconsistent and likely to contain many errors. Web preprocessing prepares log data for further classification using machine learning classifier. It takes three different steps such as Parsing, Filtering and Finalizing. After the pre-processing, the final format of our Web log files consists of a URL-ID, Timestamp, Delay Time, and Size as presented in Table 3.

5.3 Training Phase

The training datasets are prepared the desired features of Web objects are taken out from pre-processed proxy log files. These features comprise of URL-ID, Time-stamp, Delay Time, size. The sliding window of a request is that the period, a far and later once the demand were created. In additional, the sliding window ought to be about the signify time that the information usually stays during a cache (SWL is 15 min).

Once the dataset is prepared Table 4, the machine learning techniques (MNB) is applied depending on the concluded dataset to categorize the World Wide Web objects that will be re-visited or not. Each proxy dataset is then classified into training data (75 %) and testing data (25 %). Therefore, the dataset is normalized according into the series [0, 1]. When the dataset is arranged and normalized, the machine learning methods are applied using WEKA 3.7.10.

Table 3 Sample of pre-processed dataset

URL-ID	Timestamp	Frequency	Delay time (ms)	Size (bytes)	No. of future requests
1	1082348905.73	1	1,088	1,934	2
2	1082348907.41	1	448	1,803	3
4	1082349578.75	2	1,488	399	0
1	1082349661.61	2	772	1,934	0
6	1082349688.90	1	742	1,803	1
4	1082349753.72	1	708	1,233	3

Table 4 Sample of training dataset

Recency	Frequency	Size	SWL frequency	No. of future request	Target output
900	1	1,934	1	2	1 (cacheable)
900	1	1,803	1	3	1 (cacheable)
900	2	399	2	0	0 (uncacheable)
900	2	1,934	2	0	0 (uncacheable)
1,226.15	1	1,803	1	1	1 (cacheable)
1,145.08	1	1,233	1	3	1 (cacheable)
900	1	2,575	2	1	1 (cacheable)

5.4 Web Proxy Cache Simulation

The simulator WebTraff [12] can be modified to rendezvous our suggested proxy caching approaches. WebTraff simulator is used to evaluate distinct replacement Policies such as LRU, LFU, GDS, GDSF, FIFO and RAND policies. The trained, classified datasets are integrated with WebTraff to simulate the suggested novel World Wide Web proxy caching approaches. The WebTraff simulator receives the arranged log proxy document as input and develops file encompassing performance measures as outputs.

6 Performance Evaluation

6.1 Classifier Evaluation

Precision (Eq. 4) and recall (Eq. 5) are more suitable in such applications because they measure how accurate and how complete the classification is on the positive class (re-visited object). It is convenient to introduce these measures using a confusion matrix Table 5. A confusion matrix contains information about actual and predicted results given by a classifier.

Based on the confusion matrix, the precision (p) and recall (r) of the positive class are defined as follows:

$$\text{Precision (p)} = \frac{\text{TP}}{\text{TP} + \text{FP}} \tag{4}$$

Table 5 Confusion matrix for a two-class problem

	Predicted positive	Predicted negative
Actual positive	True positive (TP)	False negative (FN)
Actual negative	False positive (FP)	True negative (TN)

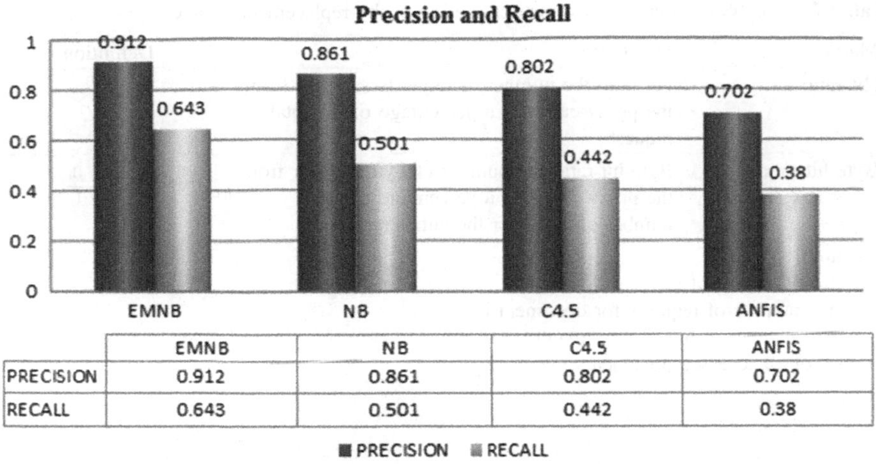

Fig. 2 Comparison of recall and precision

$$\text{Recall (r)} = \frac{\text{TP}}{\text{TP} + \text{FN}} \quad (5)$$

In words, precision p is the number of correctly classified positive examples divided by the total number of examples that are classified as positive. Recall r is the number of correctly classified positive examples divided by the total number of actual positive examples in the test set. From the (Fig. 2) apparently displays that the EMNB accomplishes the best Precision and recall for all datasets.

In summation, the computational time for training EMNB is faster than NB and C4.5, ANFIS for all datasets Table 6. Thus, we can conclude that the applications of EMNB in web proxy caching are more valuable and effective when associated with other machine learning algorithm.

Table 6 The training time (in seconds) for different datasets

Datasets	Training time (s)			
	EMNB	NB	C4.5	ANFIS
BO2	0.11	0.12	0.36	20.39
NY	0.23	0.35	0.85	22.66
UC	0.36	0.59	1.03	18.54
SV	0.11	0.33	0.69	16.18
SD	0.55	1.32	2.90	16.92
AVG	0.272	0.542	1.232	18.93

Table 7 Examples of performance metrics used in cache replacement policies

Metric	Description	Definition
Hit ratio	Hit ratio the number of requests satisfied from the proxy cache as a percentage of the total request	$\frac{\sum_{i \in R} h_i}{\sum_{i \in R} f_i}$
Byte hit ratio	Byte hit ratio the number of byte transfer from the proxy cache as a percentage of total number of bytes for the entire request	$\frac{\sum_{i \in R} s_i \cdot h_i}{\sum_{i \in R} s_i \cdot f_i}$

Notation

s_i = size of document i

f_i = total number of requests for document i

h_i = total number of hits for document i

R = set of all accessed documents

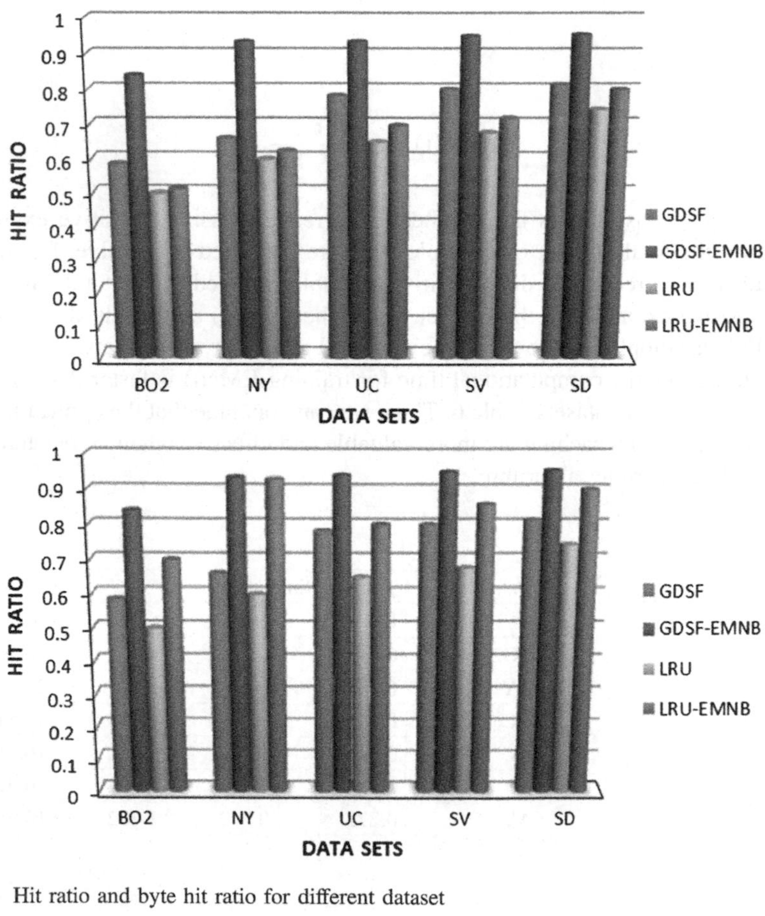

Fig. 3 Hit ratio and byte hit ratio for different dataset

6.2 Evaluation of Integrated Web Proxy Caching

6.2.1 Performance Measure

In web caching, hit ratio (HR) and byte hit ratio (BHR) Table 7 are two commonly utilized metrics for assessing the performance of web proxy caching strategies [2, 4, 9].

HR is well-defined as the ratio of the number of demands served from the proxy cache and the complete number of demands. BHR denotes to the number of bytes assisted from the cache, riven up by the complete number of byte assisted. The results in Fig. 3 Specify that EMNB-GDSF increases GDSF performance in terms of HR and EMNB-LRU over LRU is in terms of HR and in terms of BHR.

7 Conclusion

This work proposes two new web proxy caching approaches, namely EMNB-LRU, and EMNB-GDSF for improving the operation of the conventional World Wide Web proxy caching algorithms. Primarily, EMNB discovers from World Wide Web proxy log file to forecast the categories of objects to be revisited or not. Experimental results have revealed that EMNB achieves much better Precision and performance much faster than the other classifiers. In addition, in future we can consider incorporating the clustering approach to process web logs, so that a more accurate user interest model could be obtained by the EMNB and other intelligent classifiers.

References

1. Han, J., Kamber, M.: Data Mining: Concepts and Techniques, pp. 101–103. Morgan Kaufmann, Burlington (2001)
2. Cherkasova, L.: Improving WWW Proxies Performance with Greedy-Dual-Size-Frequency Caching Policy. Technical Report HPL-98-69R1. Hewlett-Packard Laboratories, Nov 1998
3. Ali, W., Shamsuddin, S.M., Ismail, A.S.: Intelligent Naïve Bayes-based approaches for web proxy caching. Knowl. Based Syst. **31**, 162–175 (2012)
4. Romano, S., ElAarag, H.: A neural network proxy cache replacement strategy and its implementation in the squid proxy server. Neural Comput. Appl. **20**, 59–78 (2011)
5. Kumar, C., Norris, J.B.: A new approach for a proxy-level web caching mechanism. Decis. Support Syst. **46**, 52–60 (2008)
6. Quinlan, J.R.: C4.5: Programs for Machine Learning. Morgan Kaufmann, Burlington (1993)
7. Ali Ahmed, W., Shamsuddin, S.M.: Neuro-fuzzy system in partitioned client side web cache. Expert Syst. Appl. **38**, 14715–14725 (2011)
8. Chen, H.T.: Pre-fetching and re-fetching in web caching system. Algorithms and Simulation, Master thesis, Trent University, Peterborough, Ontario (2008)
9. Liu, B.: Web Data Mining: Exploiting Hyperlinks, Contents, and Usage Data, pp. 173–176. Springer, Berlin (2007)

10. Podlipnig, S., Boszormenyi, L.: A survey of web cache replacement strategies. ACM Comput. Surv. **35**, 374–398 (2003)
11. NLANR.: National Lab of Applied Network Research (NLANR), and Sanitized Access Logs. Available at http://www.ircache.net/2010
12. Markatchev, N., Williamson, C.: WebTraff: a GUI for web proxy cache workload modeling and analysis. In: Proceedings of the 10th IEEE International Symposium on Modeling, Analysis, and Simulation of Computer and Telecommunications Systems, pp. 356–363. IEEE Computer Society (2002)
13. Kin-Yeung, W.: Web cache replacement policies a pragmatic approach. IEEE Netw. **20**, 28–34 (2006)

Comparative Study of On-Demand and Table-Driven Routing Protocols in MANET

G. Kumar Pallai, S. Meenakshi, A. Kumar Rath and B. Majhi

Abstract In this paper, we have studied a comparative analysis on three MANET routing protocols namely AODV, DSR and DSDV. The performances of protocols are evaluated with respect to packet delivery ratio, throughput, average end-to-end delay, routing overhead, normalized routing load, and packet drop and loss metrics. Experiments has been conducted with 80 numbers of nodes using random waypoint mobility model for constant bit rate traffic using NS-2.35 simulator as a function of varying speed. Simulation results conclude that the throughput performance of both the on-demand routing protocols is fairly comparable. However, DSR protocol outperforms in terms of routing overhead, normalized routing load and packet drop metrics than AODV. Overall, AODV protocol exhibits superior performance in terms of packet delivery ratio and packet loss.

Keywords AODV · CBR · DSR · DSDV · Mobile ad hoc networks (MANETs) · Routing protocols · Routing table (RT)

G.K. Pallai (✉)
Department of Computer Science and Engineering,
NM Institute of Engineering and Technology, Bhubaneswar, India
e-mail: gyan.pallai@gmail.com

S. Meenakshi
Department of Humanities and Social Sciences,
National Institute of Technology, Rourkela, Rourkela, India
e-mail: mee_kshi@rediffmail.com

A.K. Rath
Department of Computer Science and Engineering,
Veer Surendra Sai University of Technology, Burla, India
e-mail: amiyaamiya@rediffmail.com

B. Majhi
Department of Computer Science and Engineering,
National Institute of Technology, Rourkela, Rourkela, India
e-mail: bmajhi@nitrkl.ac.in

L.C. Jain et al. (eds.), *Computational Intelligence in Data Mining - Volume 2*,
Smart Innovation, Systems and Technologies 32, DOI 10.1007/978-81-322-2208-8_34

1 Introduction

Mobile ad hoc network (MANET) [1] is based on self configuring, that enables mobile devices to communicate with each other without any centralized administration. Each node in a MANET has the freedom to move randomly in any direction. MANET is adaptive; it establishes connections among mobile devices by easily adding and removing them to and from the network. The applications of MANET include military operations, emergency operations, mining, rescue operations, industrial monitoring law enforcement, and wireless and sensor networks. The random movement of mobile nodes generates a frequent network topology changes. Due to network decentralization, each mobile node must act as both host and router. Therefore, the most vital and demanding issue in MANET is routing. A lot of routing protocols has been designed to route the packets between any pair of nodes. Establishing an efficient routing path and reliable data transmission are the main objectives of a routing protocol. MANET protocols are broadly categorized into three types namely reactive (on-demand), proactive (table-driven) and hybrid routing protocols. A proactive routing protocol [2] maintains consistent and up-to-date routing information for the complete network. Each node maintains one or more routing tables to store network topology information. Some of the examples of proactive routing protocols are: Destination sequence distance vector (DSDV), Wireless routing protocol (WRP), Cluster head gateway switch routing (CGSR), Global state routing (GSR), Fisheye state routing (FSR), Source tree adaptive routing (STAR) and Optimized link state routing (OLSR). Reactive routing protocols [2] have an advantage over proactive routing protocol. In this approach, a route discovery process is initiated only when it is needed by the source node. Reactive protocols maintain routing information for the active routes only. This mechanism reduces the unnecessary routing overheads unlike the table-driven protocols. Some of the examples of reactive routing protocols include Ad hoc on-demand distance vector routing (AODV), Dynamic source routing (DSR), Temporarily ordered routing algorithm (TORA). A hybrid protocol combines the advantages of reactive and proactive routing protocols.

The paper is structured as follows: Sect. 2 presents the related work. Overview of AODV, DSR and DSDV protocols are described in Sect. 3. The simulation environment, simulation parameters, performance metrics and performance analysis are discussed in Sect. 4. Section 5 concludes the paper.

2 Related Work

Several quantitative simulation studies were done to appraise the performance of routing protocols in MANET. Taneja et al. [3], have presented the experimental analysis of AODV and DSR routing protocols by varying speed using NS-2 simulator. The result summarizes that AODV found to be reliable as it achieves better

PDR in denser mediums than DSR. The performance comparison of DSDV, AODV and DSR routing protocols by varying pause time and maximum speed has been done by Tonk et al. [4]. The simulation results show that AODV exhibits highest PDR and NRL, whereas DSR attains the highest average end-to-end delay. Huang et al. [5] have studied the performance of DSDV, DSR and AODV protocols for network of size 50 nodes by varying speed. Under higher-mobility, AODV performs better than DSR in terms of PDR. Baraković et al. [6], performed a simulation based study on AODV, DSR and DSDV by varying load and speed. The study concludes that DSR outperforms all the protocols with increase in load and mobility. The performance comparison of DSR, DSDV and AODV protocols under CBR traffic was evaluated by Malhotra and Sachdeva [7]. The results conclude that DSR performs better than DSDV whereas AODV outperforms the other two protocols.

3 An Overview of AODV, DSR and DSDV Protocols

3.1 Ad Hoc On-Demand Distance Vector Routing (AODV)

AODV [8, 9] is a pure on-demand routing protocol. In this approach, nodes belong to active routes are only required to maintain the routing information. When a source node desires to send packets to the destination it initiates a route discovery process [10]. A unique node sequence number is generated by the destination node in order to find latest optimal path information to the destination. AODV utilizes broadcast-id and node sequence number to ensure loop freedom [9]. In order to find a path to the destination, a route request (RREQ) packet is broadcasted by the source node. The request packet carries the broadcast-id and node sequence numbers for both source and destination which also includes time to live field. The destination node or an intermediate node that has the latest path information responds the source node by generating a route reply (RREP) packet. Otherwise, the receiving node rebroadcast the RREQ packet to its surrounding neighbor nodes. Periodical hello messages are communicated among neighboring nodes in order to indicate their continual presence.

3.2 Dynamic Source Routing (DSR)

DSR [11] protocol utilizes source routing mechanism. In this approach, the source determines the complete path to the destination node. DSR does not communicate hello-messages among the mobile nodes unlike AODV. In this algorithm, every node maintains the optimal path information to all the possible destinations. For any network topology changes, the whole network gets informed by means of flooding [10].

DSR RREQ packet employs the address of both destination and source node, and a unique id-number. A node after receiving the RREQ packet first verifies for the route availability, if not found then it append its own address and then forward the packet along its outgoing links. The destination node or the node that has the route information for the destination, responds by sending a RREP message. The node that generates the RREP packet appends the complete path information contained in the RREQ packet.

3.3 Destination Sequence Distance Vector (DSDV)

DSDV [12] is a proactive routing protocol. In DSDV, every node maintains a routing table that holds the hop count and next hop information to their destinations. A unique sequence number is generated by the destination node to avoid loops [8]. Periodical updates are required to exchange the latest topological information changes among the nodes. This periodical updates generate a large overhead in the network. Routing table updates follow two mechanisms. The first method is called "full dump" that includes all available routing information. This kind of updates must be used only when there is complete change in network topology. Incremental updates are carried out only when there is a smaller change in network topology. All the updates should be compacted in a single network protocol data unit (NPDU) [12].

4 Simulation Results and Analysis

4.1 Simulation Environment

NS-2.35 [13] simulator was used for simulation to evaluate the performance of three routing protocols by varying nodes' speed. A network size of 80 nodes was used and randomly distributed in a 500 m × 500 m square simulation area. The random way point mobility [14] model was used to generate the node mobility. The running time of simulation is 250 s. Nodes were made to travel at different speeds such as 5, 10, 15, 20, 25 m/s respectively. To exchange data traffic between the nodes, CBR sources have been used. The sending rate of CBR is set to 5 packets/s. The size of the data packet is 512 bytes. Maximum 10 numbers of nodes were chosen as source-to-destination pair to represent the traffic load. The parameters taken for simulation are shown in Table 1.

Table 1 Simulation parameters

Parameter type	Value
Simulation duration	250 s
MAC layer protocol	802.11
Topology size	500 m × 500 m
Number of nodes	80
Traffic sources	Constant-bit-rate (CBR)
Packet size	512 bytes
Max number of connections	10
Sending rate	5 packets/s
Pause time	5 s
Speed	5, 10, 15, 20, 25 m/s

4.2 *Performance Metrics*

The performances of routing protocols are evaluated using the following metrics: packet delivery ratio, throughput, average end-to-end delay, normalized routing load, and packet loss.

4.2.1 Packet Delivery Ratio

This performance metric refers to the ratio of number of delivered data packets to the destinations. The performance of the protocol is better if the packet delivery ratio is higher, which implies that how successful the packets have been deliver.

$$\text{Packet Delivery Ratio}(\text{PDR}[\%]) = \frac{\sum_{1}^{n} \text{Received Packets}}{\sum_{1}^{m} \text{Sent Packets}} * 100$$

4.2.2 Throughput

Throughput refers to the rate of successful data packets transmitted over a communication channel in a given amount of time. This metric measures the effectiveness of a routing protocol in the network.

$$\text{Throughput (bps)} = \frac{\sum_{1}^{n} \text{Received packets} * \text{Packet size} * 8}{\text{Total simulation time}}$$

4.2.3 Average End-to-End Delay

This metric represents the average time taken by a data packet to arrive at the destination. This delay can be caused by many reasons like buffering during route discovery latency, queuing at the interface queue, retransmission delays at the MAC, and propagation and transfer times.

$$Avg\,EtE\,Delay\,(\mathrm{s}) = \frac{\sum_1^n (Packet\,sent\,time - Packet\,receive\,time)}{\sum_1^n Received\,packets}$$

4.2.4 Normalized Routing Load

Normalized routing load is the number of routing packets transmitted per data packet delivered at the destination. Each hop-wise transmission of a routing packet is counted as one transmission. It evaluates the efficiency of the routing protocol.

$$\text{Normalized Routing Load (NRL)} = \frac{\sum_1^k \text{Routing Packets}}{\sum_1^n \text{Received Packets}}$$

4.2.5 Packet Loss

Packet loss occurs when one or more data packets unable to reach their destination. Once a packet reaches the network layer it is forwarded to the destination when a valid route is available; otherwise it is buffered until it reaches the destination. A packet is lost when the buffer is full or the time that the packet has been buffered, exceeds the limit.

$$Packets\,loss = \sum_1^m Sent\,packets - \sum_1^n Received\,packets$$

4.3 Results and Analysis

The graphical representation of three routing protocols with respect to packet delivery ratio is illustrated in Fig. 1. The result shows that the reactive protocols achieve high packet delivery ratio and proven to be more reliable than proactive protocol. DSR shows better performance than DSDV, whereas AODV outperforms the other two protocols. AODV attains the highest packet delivery ratio of 99 %, and found to be more stable among other two protocols. The packet delivery rate of AODV is 10 % higher than DSR and 28 % higher than DSDV. The reason being

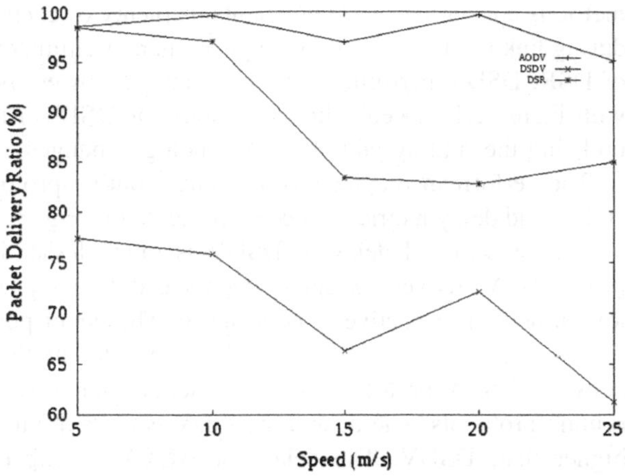

Fig. 1 Comparative analysis of packet delivery ratio w.r.t. speed

that the AODV protocol utilizes hop by hop routing technique and has employs enhanced path repair mechanism. At node speed 10 m/s, the packet delivery rate of DSR is 97 % and its performance degrades by 15 % with speed of 15 m/s. However, later it delivers consistent uniform delivery rate with speed variation. However, the DSDV performance suffers poorly and maintains the delivery rate between 77 and 70 % with speed variation.

Figure 2 illustrates the throughput performance results for three routing protocols. From the simulations results, it is very obvious that reactive routing protocols AODV and DSR are superior to a proactive one DSDV. Both the reactive protocols exhibits fairly high throughput with varying speed. They start with 3.7 kbps and maintain consistently uniform delivery rate till speed of 20 m/s. However, AODV performance declines a bit under higher speed as it holds only one route per destination. On the other hand, DSR proved to be reliable choice when throughput

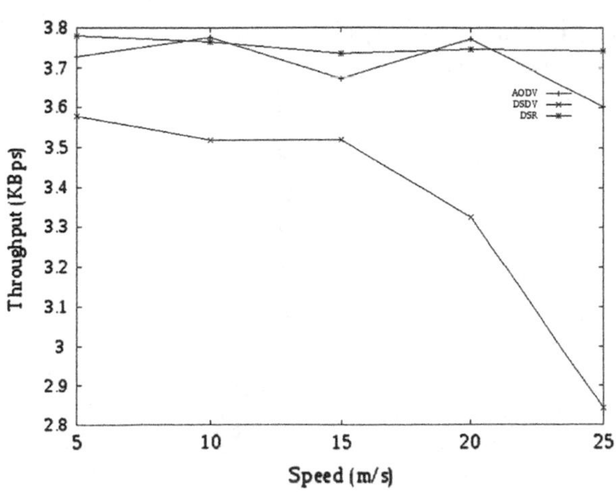

Fig. 2 Comparative analysis of throughput w.r.t. speed

metric is concerned. As DSR maintains plenty of cache routes the reaction of DSR during link failures is mild. Such phenomenon ultimately improves the performance of DSR. DSDV performs better especially at lower speed and decreases gradually with increase in speed. This is obvious for DSDV, since it takes longer time for updating the routing table for any topology changes.

The performance comparison of three routing protocols with respect to average end-to-end delay metric has been represented in Fig. 3. It has been observed that the average end-to-end delay of DSDV is much lesser in comparison to other two protocols. Moreover, except at node speed 15 m/s, DSDV consistently gives the lowest delay irrespective of node speed. The better performance of DSDV is quite obvious, as it proactively holds the routes to all the destinations for the entire network. DSDV protocol does not generate route discovery process unlike reactive routing protocols. The delay for AODV is 89 % higher than DSR and about 96 % higher than DSDV. The delay for AODV is expected since for any network topology changes, a fresh route discovery process is initiated by sending route request packets. However, DSR performs better than AODV since it uses source routing and route cache mechanism and also do not involve in periodical updates.

The experimental results of normalized routing load for three routing protocols have been illustrated in Fig. 4. Experimental results demonstrate that DSR consumes significantly less routing overhead among three routing protocols. The performance for DSR is quite obvious as it need not have to depend on periodical activities. The utilization of route cache mechanism and multiple routes per destination reduces the routing overhead DSR. It can be well noticed that the NRL of AODV is always at high peak rate than the other two protocols because it keeps only one route per destination. AODV periodically sends route request and route reply packets in order to update the routing tables with latest routing information. DSDV has the second highest NRL since it proactively keeps routes to all destinations in routing table, regardless of topology changes.

Fig. 3 Comparative analysis of end-to-end delay w.r.t. speed

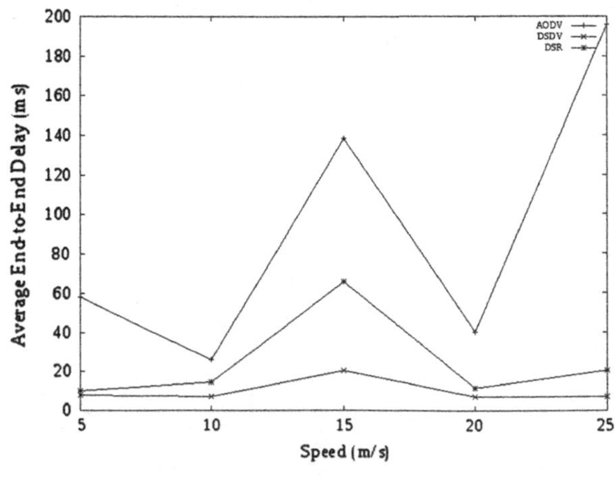

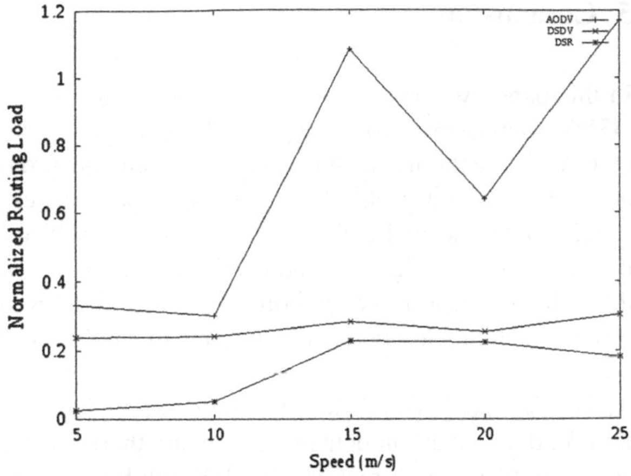

Fig. 4 Comparative analysis of normalized routing load w.r.t. speed

As represented in Fig. 5, AODV outperforms all protocols in terms of packet loss followed by DSR while DSDV exhibits the worst loss. It is clearly proved that both reactive routing protocols perform better than proactive protocol with respect to packet drop. The reason is obvious for DSDV, since it requires longer time to update the routing tables for the whole network. With increase in number of broken links, time to update increases which ultimately results in dropping more number of packets. However, in AODV, if a node finds link break, then it immediately reports to the source by generating an error message. In terms of packet drop, AODV performs much better than DSDV but DSR outperforms all.

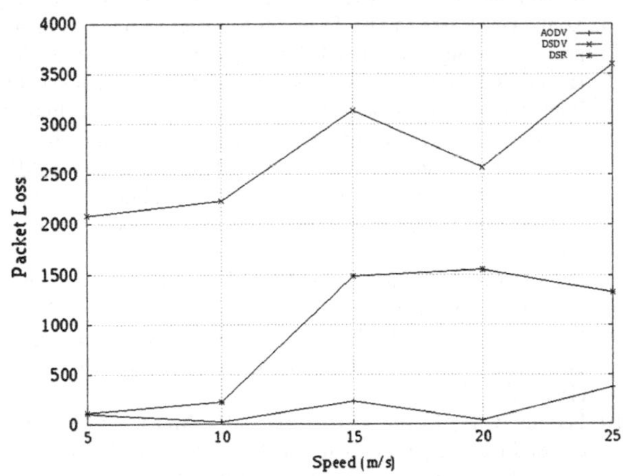

Fig. 5 Comparative analysis of packet loss w.r.t. speed

5 Conclusion

In this paper, we have assessed the performance differentials of AODV, DSR and DSDV routing protocols using NS-2.35 simulator. Simulation analysis concludes that AODV exhibits its superiority over all the protocols with respect to packet delivery ratio and packet loss metrics. However, severely suffers with respect to routing overhead and end-to-end delay since AODV broadcasts periodic HELLO messages to its neighbors more frequently to find and repair routes. However, throughput performance of both AODV and DSR are fairly comparable. DSR protocol exhibits its supremacy in terms of routing overhead, NRL and packet drop but achieves second highest performance in terms of PDR, end-to-end delay and packet loss. AODV and DSR maintained a high packet delivery ratio compared to DSDV, due to their on-demand nature and their fast recovery when the nodes move at moderate and high mobility. DSDV exhibits considerably less end-to-end delay due to its table-driven characteristic. DSDV performs averagely in terms of routing overhead and NRL, whereas badly suffers with respect to all other performance metrics. Overall, on-demand routing protocols have better performance than the table-driven protocol.

References

1. Chlamtac, I., Conti, M., Liu, J.: Mobile ad hoc networking: imperatives and challenges. Ad-Hoc Netw. J. **1**, 13–64 (2003)
2. Royer, E.M., Toh, C.K.: A review of current routing protocols for ad-hoc mobile wireless networks. IEEE Pers. Commun. Mag. **6**, 46–55 (1999)
3. Taneja, S., Kush, A., Makkar, A.: Experimental analysis of DSR, AODV using speed and pause time. Int. J. Innov. Manage. Technol. **1**, 453–458 (2010)
4. Tonk, G., Kashyap, I., Tyagi, S.S.: Performance comparison of ad-hoc network routing protocols using NS-2. Int. J. Innov. Technol. Exploring Eng. (IJITEE) **1** (2012)
5. Huang, R., Zhuang, Y., Cao, Q.: Simulation and analysis of protocols in ad hoc network. In: International Conference on Electronic Computer Technology, pp. 169–173 (2009)
6. Baraković, S., Kasapović, S., Baraković, J.: Comparison of MANET routing protocols in different traffic and mobility models. Telfor J. **2**, 8–12 (2010)
7. Malhotra, R., Sachdeva, B.: Multilingual evaluation of the DSR, DSDV and AODV routing protocols in mobile ad hoc networks. SIJ Trans. Comput. Netw. Commun. Eng. (CNCE) **1** (2013)
8. Raju, J., Garcia-Luna-Aceves, J.J.: A comparison of on demand and table driven routing for ad hoc wireless networks. In: Proceedings of IEEE ICC (2000)
9. Perkins, E., Royer, E.M.: Ad-hoc on-demand distance vector routing. In: Proceedings of the 2nd IEEE Workshop on Mobile Computing Systems and Applications, vol. 3, pp. 90–100 (1999)
10. Perkins, C.E., Royer, E.M., Das, S.R., Marina, M.K.: Performance comparison of two on demand routing protocols for ad hoc networks. IEEE Pers. Commun. Mag. **8**, 16–29 (2001). (Special Issue on Mobile Ad Hoc Networks)

11. Johnson, D.B., Maltz, D.A.: Dynamic source routing in ad hoc wireless networks. In: Imielinski, T., Korth, H.F. (eds.) Mobile Computing, pp. 153–181. Kluwer Academic Publishers, Berlin (1996)
12. Perkins, C.E., Begat, P.: Highly dynamic destination-sequenced distance-vector routing (DSDV) for mobile computers. In: Proceedings of the ACM SIGCOMM'94 Conference on Communications Architectures, Protocols, and Applications, pp. 234–244, London (1994)
13. Greis, M.: Tutorial for the UCB/LBNL/VINT Network Simulator "ns". http://www.isi.edu/nsnam/ns/tutorial/
14. Camp, T., Boleng, J., Davies, V.: A survey of mobility models for ad hoc network research. Wireless Commun. Mobile Comput. (WCMC) **2**, 483–502 (2002). (Special Issue on Mobile Ad Hoc Networking. Research, Trends and Applications)

A Novel Fast FCM Clustering for Segmentation of Salt and Pepper Noise Corrupted Images

B. Srinivasa Rao and E. Srinivasa Reddy

Abstract The conventional fuzzy C-means (FCM) is most frequently used unsupervised clustering algorithm for image segmentation. However, it is sensitive to noise and cluster center initialization. In order to overcome this problem, a novel fast fuzzy C-means (FFCM) clustering algorithm is proposed with the ability to minimize the effects of impulse noise by incorporating noise detection stage to the clustering algorithm during the segmentation process without degrading the fine details of the image. This method also improves the performance of the FCM algorithm by finding the initial cluster centroids based on histogram analysis, reducing the number of iterations. The advantages of the proposed method are: (1) Minimizes the effect of impulse noise during segmentation, (2) Minimum number of iterations to segment the image. The performance of the proposed approach is tested on different real time noisy images. The experiment results show that the proposed algorithm effectively segment the noisy image.

Keywords Clustering · Image segmentation · Histogram · Salt-and-pepper noise · Fuzzy C-means · Image processing

1 Introduction

Image segmentation classically is defined as the process by which an original image is partitioned into some homogeneous regions with respect to some characteristics such as gray value or texture. The technique is commonly used by many consumer

B. Srinivasa Rao (✉)
Department of Information Technology, GITAM Institute of Technology, GITAM University, Visakhapatnam 530045, Andhra Pradesh, India
e-mail: sreenivas.battula@gmail.com

E. Srinivasa Reddy
Department of Computer Science and Engineering, ANUCET, Acharya Nagarjuna University, Guntur 522510, Andhra Pradesh, India
e-mail: esreddy67@gmail.com

L.C. Jain et al. (eds.), *Computational Intelligence in Data Mining - Volume 2*,
Smart Innovation, Systems and Technologies 32, DOI 10.1007/978-81-322-2208-8_35

electronic products or for some specific application. The algorithms generally based on similarity and particularity, these can be divided into different categories: thresholding [1], template matching [2, 3], region growing [4, 5] and edge detection [6, 7]. Each technique has its own advantages and limitations in terms of suitability, performance and computational cost. In edge-based methods, the local discontinuities are detected first and then connected to form longer, hopefully complete boundaries. In region-based methods, areas of an image with homogeneous properties are found, which in turn give the boundaries, but it suffers from time-consuming and over-segmentation problems. On the other hand the threshold technique is simplest in segmenting methods. To set two thresholds on the histogram of the image, we can classify between the two thresholds in the histogram as the same region and classify the others as the second region. This technique produces a good quality and fast segmentation, but it is sensitive to noise.

Clustering is an unsupervised classification of patterns into groups of similar objects; widely used in medical diagnostic studies, image analysis, image processing, decision making, machine learning situation etc. [8–13]. The goal of clustering is descriptive and is to discover a new set of categories, the new groups are of interest in themselves, and their assessment is intrinsic. In image segmentation, clustering algorithms iteratively computes the characteristics of each cluster and segment the image by classifying each pixel in the closest cluster according to a distance metric, segmentation results that can be obtained are better but over segmentation is one of the major problems that must be faced.

Segmenting images with clustering algorithms have been applied in numerous applications including medical applications, specifically in the biomedical image analysis. Several previous studies have proven that clustering algorithms is capable in segmenting and determining certain regions of interest on medical images [14, 15]. It is because in the biomedical image segmentation task, clustering algorithm is often suitable since the number of clusters for the structure of interest is usually known from its anatomical information [16]. Digital images acquired through many consumer electronics products are often corrupted (faulty memory locations or impaired pixel sensors) by salt-and-pepper noise [17]. FCM is one of the most frequent clustering-based segmentation methods used for image segmentation. FCM algorithms with spatial constraints have been proven effective for image segmentation. However, it is very sensitive to noise, outliers and other imaging artifacts [18].

The rest of this paper is organized as follows: Sect. 2 presents the FCM clustering algorithm, Sect. 3 presents Fast FCM clustering algorithm, Sect. 4 presents the proposed algorithm, Sect. 5 presents Experimental results and finally Sect. 6 report conclusions.

2 Fuzzy C-means (FCM)

The Fuzzy C-means [19] is an unsupervised clustering algorithm. The main idea of introducing fuzzy concept in the Fuzzy C-Means algorithm is that an object can belong simultaneously to more than one class and does so by varying degrees called memberships. It distributes the membership values in a normalized fashion. It does not require prior knowledge about the data to be segmented. It can be used with any number of features and number of classes. The fuzzy C-means is an iterative method which tries to separate the set of data into a number of compact clusters. The Fuzzy C-means algorithm is summarized as follows (Fig. 1).

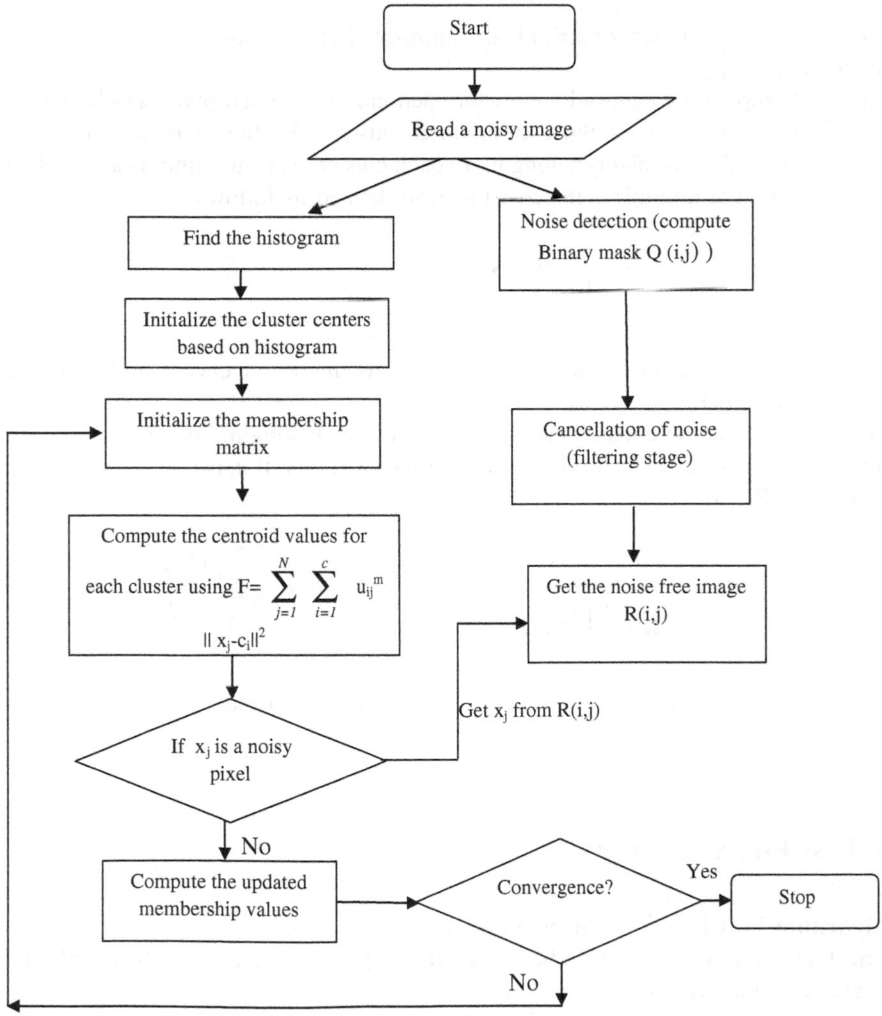

Fig. 1 Block diagram of the proposed method

Algorithm Fuzzy C-Means (x, n, c, m)
Input: n = number of pixels to be clustered; x = $\{x_1, x_2, \ldots, x_n\}$: pixels of real time image; c = number of clusters; m = 2: the fuzziness parameter;
Output: u: membership values of pixels and segmented Image
Begin
Step_1: Initialize the membership matrix u_{ij} is a value in (0,1) and the fuzziness parameter m(m = 2). The sum of all membership values of a pixel belonging to clusters should satisfy the constraint expressed in the following.

$$\sum_{j=1}^{c} u_{ij} = 1 \tag{1}$$

for all i = 1, 2, … n, where c(=2) is the number of clusters and n is the number of pixels in the image.
Step_2: Compute the centroid values for each cluster c_j. Each pixel should have a degree of membership to those designated clusters. So the goal is to find the membership values of pixels belonging to each cluster. The algorithm is an iterative optimization that minimizes the cost function defined as follows:

$$F = \sum_{j=1}^{N} \sum_{i=1}^{c} u_{ij}^{m} || x_j - c_i ||^2 \tag{2}$$

where u_{ij} represents the membership of pixel x_j in the ith cluster and m is the fuzziness parameter.
Step_3: Compute the updated membership values u_{ij} belonging to clusters for each pixel and cluster centroids according to the given formula. If x_j is noisy pixel get the pixel from R(i, j)

$$\mathbf{u_{ij}} = \frac{1}{\sum_{k=1}^{c} \left(\left(\frac{x_j - v_i}{x_j - v_k} \right)^{2/(m-1)} \right)} \quad \text{and} \quad \mathbf{v_i} = \frac{\sum_{j=1}^{N} u_{ij}^{m} x_j}{\sum_{j=1}^{N} u_{ij}^{m}} \tag{3}$$

Step_4: Repeat steps 2–3 until the cost function is minimized.
End.

3 Fast Fuzzy C-means

Algorithm Fast Fuzzy C-Means (x,n,c,m)
Input: N = number of pixels to be clustered; x = $\{x_1, x_2, \ldots, x_n\}$: pixels of real time image; c = number of clusters;
Output: u: membership values of pixels and segmented Image

Begin
Step_1: Find the histogram of the image.
Step_2: Based on the number of clusters divide the histogram bins into l/c parts where l is the maximum gray value and c is the number of clusters. (e.g.: for 3 clusters 1–85, 86–170, 171–256).
Step_3: Consider the highest peak intensity value from each part (excluding noise pixels), get the pixels with these intensity values initialize these values as initial centroids.
Step_4: Start FCM algorithm with these initialized centroids.
End.

4 The Proposed Novel Fast FCM Clustering Technique

We propose a new method of clustering based segmentation technique, specifically for images corrupted with impulse noise. The novel Fast FCM clustering is introduced to overcome the problem of noise sensitivity in the segmentation process which increases the robustness of the segmentation process with respect to noise. This proposed method is a two stage process, in the first stage the detection of salt-and pepper noise and its locations. The second stage will perform thc actual clustering process. The 'noise-free' pixels will be totally considered as the input data and they will give full contribution on the clustering process. Otherwise, for the 'noise' pixels, the fuzzy concept is applied to determine the degree of contributions of these 'noise' pixels on the clustering process. The combination of noise detection, cancellation and the clustering allows more versatile and powerful methods to achieve a better segmentation especially on noisy images.

4.1 Salt-and-Pepper Noise Detection and Noise Cancellation

For a gray scale digital image the intensity is stored in an 8-bit integer, giving a possible 256 gray levels in the interval [0,255]. The salt-and-pepper noise takes on the minimum and maximum intensities. It can be either minimum intensity value near 0 i.e. L_{lower} (appears black i.e. pepper) or maximum intensity value near i.e. 255 L_{upper} (appears white i.e. salt). The histogram of the image is used to identify these two types of noise intensities. If an image corrupted with salt-and-pepper noise would peak at the ends of the noisy image histogram [20]. These two salt-and-pepper noise intensities will be used to identify possible 'noise-pixels' in the image. According to [21], a binary noise mask Q(i, j) will be created to mark the location of 'noise-pixels' by using;

$$Q(i,j) = \begin{cases} 0, & X(i,j) = L_{Upper} \text{ or } L_{Lower} \\ 1, & \text{Otherwise} \end{cases} \tag{4}$$

where X(i, j) is the pixel at the location (i, j) with intensity X, Q(i, j) = 1 represents the ‘noise-free’ pixel to be retained in the next clustering stage while Q(i, j) = 0 represents ‘noise’ pixels.

4.2 Noise Cancellation and Clustering

After the binary mask M(i, j) is created, in order to allow more versatile methods of clustering-based segmentation in noisy images all “noise pixels” marked with will be replace by an estimated correction term

$$X^1(i,j) = (1 - F(i,j)\, X(i,j) + F(i,j)\, M(i,j) \tag{5}$$

where M(i, j) is the median of in the 3 × 3 window given by:

$$M(i,j) = \text{median}\{X(i+k, j+l) \quad \text{with } k, l \in (-1, 0, 1) \tag{6}$$

After the median pixel is found, the absolute luminance difference, d(i, j), is computed by using;

$$d(i+k, j+l) = \{X(i+k, j+l) - X(i,j) \quad \text{with } (i+k, j+l) \neq (i,j) \tag{7}$$

Then the local information of the ‘noise’ pixels in 3 × 3 window is calculated by taking the maximum value of the absolute luminance difference given by;

$$D(i,j) = \max\{(i+k, j+l)\} \tag{8}$$

According to [21], “noise pixels” will be set to the maximum intensity 255 while “noise-free pixels” will assume other values in the dynamic range. Based on this the choice of using maximum operator rather than the minimum operator is justified in [21]. Next the fuzzy reasoning is applied to the extracted local information D(i, j). The fuzzy set defined by the fuzzy membership function F(i, j) is defined by:

$$F(i,j) = \begin{cases} 0; & D(i,j) < T1 \\ D(i,j) - T1/T2 - T1; & T1 < D(i,j) < T2 \\ 1; & D(i,j) \geq T2 \end{cases} \tag{9}$$

whereby for optimal performance, the threshold value *T*1 and *T*2 are set to 10 and 30 respectively as described in [21]. Then the corrected value of noise pixel is calculated using (5).

To improve the efficaciousness of the FFCM clustering towards noise, these corrected values (i.e., for the noise pixels) are used to replace original pixels values during the process of assigning the data to their nearest centre. Then the new position for each cluster is calculated using (2). The term x_j in (2) is substituted by:

$$x_j = \begin{cases} X(i,j) & \text{if } Q(i,j) = 1 \\ X^1(i,j) & \text{if } Q(i,j) = 0 \end{cases} \tag{10}$$

By the end of this stage we get a image R(i, j) which is noise free image, if x_j is noise free then the value of x_j will be Original image pixel value (i.e. X(i, j)) otherwise consider the pixel from R(i, j). Integrating this method in the fast FCM clustering algorithm, a novel Fast fuzzy C-Means clustering for segmentation of noisy images is proposed.

5 Experimental Results

In this section, the performance of the proposed novel Fast FCM is compared with conventional FCM. We present the experimental results on several standard images. In this experiment, we have used images corrupted with the salt-and-pepper noise to test the effectiveness and the efficiency of the proposed algorithm. The experiments were performed in a 2.99 GHz Intel Core 2 Duo processor, Windows XP with 3.21 GB RAM, using Matlab R2010a.

We execute the proposed Fast FCM clustering algorithm and convectional FCM clustering with varying number of clusters on total 60 real time images contaminated by different levels of salt-and-pepper noise to investigate the robustness of the algorithm. Two images are chosen to enable the visualization of the proposed algorithm. The images are named *building*, *butterfly* are shown in Figs. 2a and 3a respectively. Figures 2b and 3b, are the same aforementioned images corrupted with 20 and 30 % density of salt-and-pepper noise respectively.

5.1 Initialization of Centroids

Figures 2i and 3i are the histogram of the noisy images, from these histograms based on the number of clusters divide the histogram bins into l/c parts where l is the maximum gray value and c is the number of clusters. For example using our proposed method Fig. 2 with three clusters and the cluster centers are c1 = 59, c2 = 138 and c3 = 231.

On the other hand with random initialization, three cluster centers are c1 = 63, c2 = 55 and c3 = 132 and the final cluster centers with the proposed method are c1 = 53.8544, c2 = 133.4829 and c3 = 209.5001. These final cluster centers are almost nearer to the proposed cluster centers, hence we get convergence quickly.

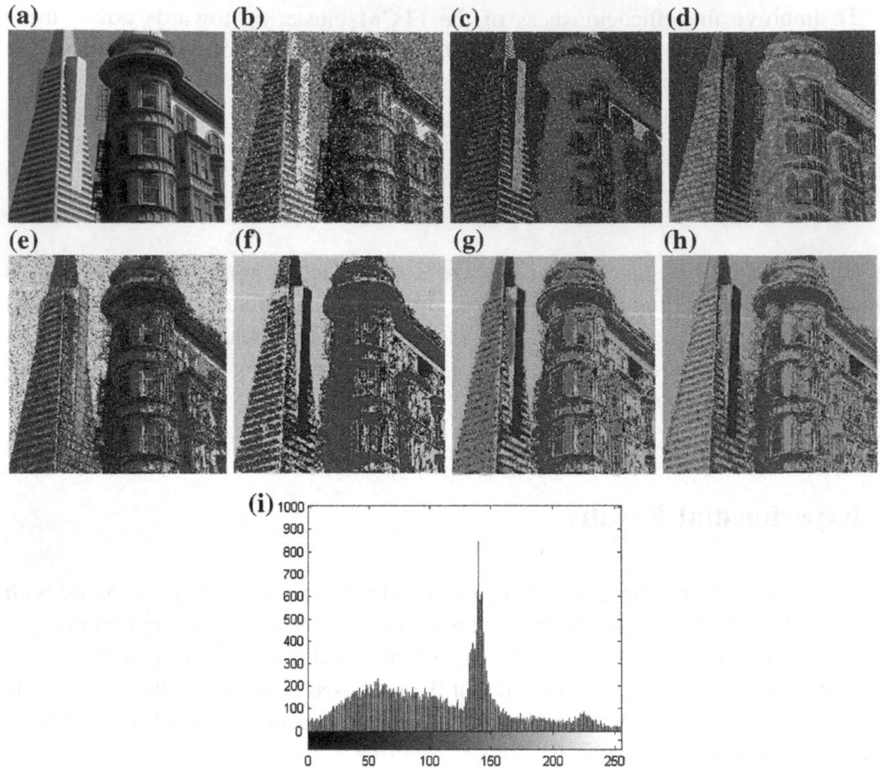

Fig. 2 Segmentation results on *Building* with 20 % density of salt-and pepper noise using; **a** original image, **b** noisy image, **c** conventional FCM with c = 3, **d** conventional FCM with c = 4, **e** conventional FCM with c = 5, **f** proposed FFCM with c = 3, **g** proposed FFCM with c = 4, **h** proposed FFCM with c = 5 and **i** histogram of the noisy image

The proposed method gives much better segmentation results with c = 5 (i.e. number of clusters).

The another advantage of proposed method over conventional method is that, the novel Fast FCM clustering algorithm give better and clearer segmentation results compared to conventional FCM algorithm which is influenced by noise. From Fig. 2f–h are produced better segmentation results where c–e are effected by noise. The black and white particles are significantly able to be reduced. These findings prove that the proposed algorithm is robust with respect to the noise effect.

As for the *Butterfly* image which is contaminated with 30 % of salt-and-pepper noise as illustrated in Fig. 3, the proposed method outperforms the conventional method. The proposed novel Fast FCM clustering algorithm have successfully reduced the black and white particles, created less corrupted image and maintained the shape of the butterfly.

This is due to the ability of the proposed algorithm to ignore the noise pixels during the segmentation process. Furthermore, comparison results in terms of the

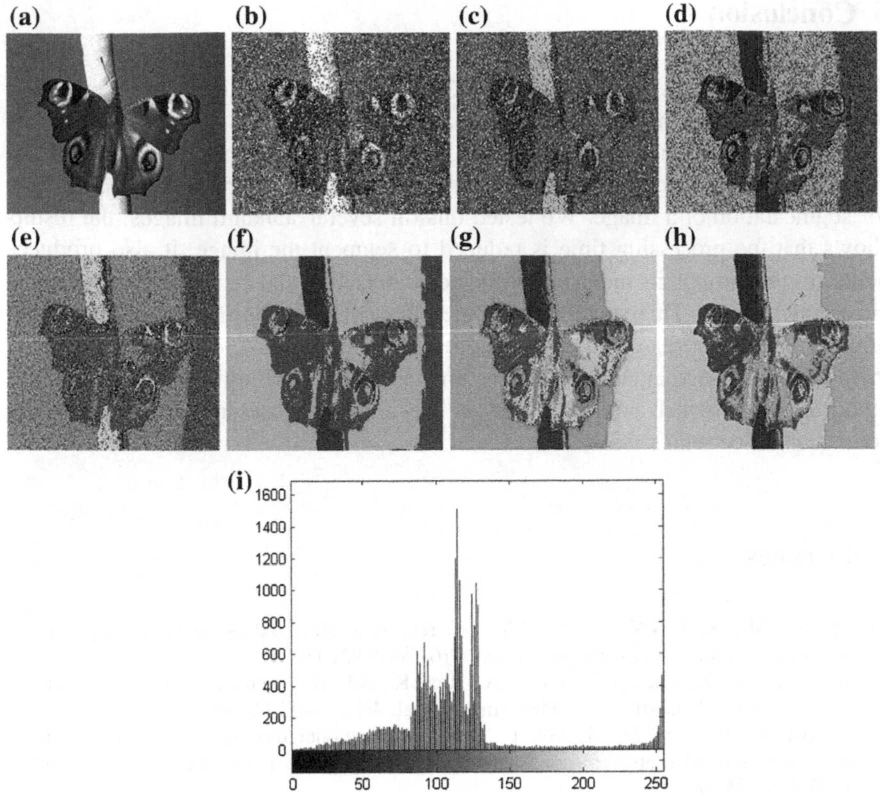

Fig. 3 Segmentation results on *Butterfly* with 30 % density of salt-and pepper noise using; **a** original image, **b** noisy image, **c** conventional FCM with c = 3, **d** conventional FCM with c = 4, **e** conventional FCM with c = 5, **f** proposed FFCM with c = 3, **g** proposed FFCM with c = 4, **h** proposed FFCM with c = 5, **i** histogram of the noisy image

running time (in seconds) until the final segmentation is given in Table 1. Likewise we tested for 60 real time images which gives less processing time for segmentation of a image.

Table 1 Comparison of computation efficiency salt-and-pepper noise density Fig. 2 with 20 % and Fig. 3 with 30 % noise

No. of clusters	CPU time in seconds			
	FCM		Proposed method	
	Figure 2	Figure 3	Figure 2	Figure 3
C = 3	9.24	6.35	3.59	4.58
C = 4	14.88	11.87	7.22	10.22
C = 5	23.27	11.79	17.93	9.67

6 Conclusion

This paper presents a novel Fast Fuzzy-C-Means clustering algorithm is proposed for image segmentation, especially for images corrupted with salt-and-pepper noise. The proposed algorithm produces results faster than the conventional FCM with the novel initialization method based on histogram analysis to start the FCM clustering for segmentation of a image. We tested this on several standard images, the results shows that the processing time is reduced to segment the image. It also produces better results through its inclusion of the noise detection and cancellation stage in its clustering process. This stage reduces the effect of noise during the segmentation process. Furthermore, this finding suggests the proposed clustering as a novel method for the segmentation of noisy images and is very efficient with its computational time, which could be used as pre- or post-processing technique in the consumers' electronics fields.

References

1. Cheriet, M., Said, J.N., Suen, C.Y.: A recursive thresholding technique for image segmentation. IEEE Trans. Image Process. **7**(6), 918–921 (1998)
2. Warfield, S.K., Michael, K., Jolesz, F.A., Ron, K.: Adaptive, template moderated, spatially varying statistical classification. Med. Image Anal. **4**(1), 43–55 (2000)
3. Lalonde, M., Beaulieu, M., Gagnon, L.: Fast and robust optic disc detection using pyramidal decomposition and Hausdorff-based template matching. IEEE Trans. Med. Imaging **21**(11), 1193–1200 (2001)
4. Seunghwan, Y., Rae-Hong, P.: Red-eye detection and correction using in painting in digital photographs. IEEE Trans. Consum. Electron. **55**(3), 1006–1014 (2009)
5. Mat-Isa, N.A., Mashor, M.Y., Othman, N.H.: Automatic seed based region growing for pap smear image segmentation. In: Kuala Lumpur International Conference on Biomedical Engineering, Kuala Lumpur, Malaysia (2002)
6. Paik, J.K., Park, Y.C., Park, S.W.: An edge detection approach to digital image stabilization based on tri-state adaptive linear neurons. IEEE Trans. Consum. Electron. **37**(3), 521–530 (1991)
7. Siyoung, Y., Donghyung, K., Jechang, J.: Fine edge-preserving deinterlacing algorithm for progressive display. IEEE Trans. Consum. Electron. **53**(3), 1654–1662 (2009)
8. Algorri, M.E., Flores-Mangas, F.: Classification of anatomical structures in MR Brain images using fuzzy parameters. IEEE Trans. Biomed. Eng. **51**, 1599–1608 (2004)
9. Doulamis, A.D., Doulamis, N., Kollas, S.: Non-sequential video content representation using temporal variation of feature vectors. IEEE Trans. Consum. Electron. **46**, 758–768 (2000)
10. Nickel, K., Stiefelhagen, R.: Visual recognition of pointing gestures for human-robot interaction. Image Vis. Comput. **25**, 1875–1884 (2007)
11. Setnes, M.: Supervised fuzzy clustering for rule extraction. IEEE Trans. Fuzzy Syst. **8**, 416–424 (2000)
12. Sulaiman, S.N., Mat, N.A.: Isa: adaptive fuzzy-K-means clustering algorithm for image segmentation. IEEE Trans. Consum. Electron. **56**, 2661–2668 (2010)
13. Sulaiman, S.N., Mat, N.A.: Isa : Denoising-based clustering algorithms for segmentation of low level salt-and-pepper noisecorrupted images. IEEE Trans. Consum. Electron. **56**, 2702–2710 (2010)

14. Mat-Isa, N.A., Mashor, M.Y., Othman, N.: Pap smear image segmentation using modified moving k-mean clustering. In: International Conference on Biomedical Engineering, Kuala Lumpur Malaysia (2002)
15. Mat-Isa, N.A., Mashor, M.Y., Othman, N.H., Sulaiman, S.N.: Application of moving k-means clustering for pap smear image processing. In: Proceeding of International Conference on Robotics, Vision, Information and Signal Processing, Penang, Malaysia (2002)
16. Mat-Isa, N.A., Samy, A.S., Ngah, U.K.: Adaptive fuzzy moving K-means algorithm for image segmentation. IEEE Trans. Consum. Electron. **55**(4), 2145–2153 (2009)
17. Toh, K.K.V., Ibrahim, H., Mahyuddin, M.N.: Salt-and-pepper noise detection and reduction using fuzzy switching median filter. IEEE Trans. Consum. Electron. **54**(4), 1956–1961 (2008)
18. Cai, W., Chen, S., Zhang, D.: Fast and robust fuzzy C-means clustering algorithms incorporating local information for image segmentation. Pattern Recogn. **40**(3), 825–838 (2007)
19. Zhang, J.-H., Ha, M.H., Wu, J.: Implementation of rough fuzzy K-means clustering algorithm in Matlab. In: Proceedings of Ninth International Conference on Machine Learning and Cybernetics, July 2010
20. Luo, W.: Efficient removal of impulse noise from digital images. IEEE Trans. Consum. Electron. **52**(2), 523–527 (2006)
21. Toh, K.K.V., Isa, N.A.M.: Noise adaptive fuzzy switching median filter for salt-and-pepper noise reduction. IEEE Signal Process. Lett. **17**(3), 281–284 (2010)

Non Linear Autoregressive Model for Detecting Chronic Alcoholism

Surendra Kumar, Subhojit Ghosh, Suhash Tetarway, Shashank Sawai, Pillutla Soma Sunder and Rakesh Kumar Sinha

Abstract In this study, the Non Linear Autoregressive model of the resting electroencephalogram (EEG) was examined to address the problem of detecting alcoholism in the cerebral motor cortex. The EEG signals were recorded from chronic alcoholic (n = 20) conditions and the control group (n = 20). Data were taken from motor cortex region. The dimension of the extracted features are reduced by linear discrimination analysis (LDA) and classified by Support Vector Machine (SVM). The 600 sample from each group gave the best result using Support Vector Machine classifier. The maximum classification accuracy (90 %) with SVM clustering was achieved with the EEG Fz channel. More alterations are identified in the left hemisphere. Considering the good classification accuracy with SVM on Fz electrode, it can be suggested that the non-invasive automated online diagnostic system for the chronic alcoholic condition can be developed with the help of EEG signals.

S. Kumar (✉) · P.S. Sunder · R.K. Sinha
Department of Bio Engineering, Birla Institute of Technology, Mesra, India
e-mail: sikuranchi@gmail.com

P.S. Sunder
e-mail: somasunderpillutla@gmail.com

R.K. Sinha
e-mail: rksinhares@gmail.com

S. Ghosh
Department of Electrical Engineering, National Institute of Technology, Raipur, Raipur, India
e-mail: aceghosh@gmail.com

S. Tetarway
Department of Physiology, Rajendra Institute of Medical Sciences, Ranchi, India
e-mail: suhashtetarway@gmail.com

S. Sawai
Department of Electrical Engineering, Birla Institute of Technology, Mesra, India
e-mail: shashanksawai@yahoo.in

L.C. Jain et al. (eds.), *Computational Intelligence in Data Mining - Volume 2*,
Smart Innovation, Systems and Technologies 32, DOI 10.1007/978-81-322-2208-8_36

Keywords Alcohol · Cerebral motor cortex · Electroencephalogram · LDA · Non linear autoregressive model · Support vector machine

1 Introduction

Alcohol consumption is a major risk factor leading to disability, illness and mortality issues. This accounts for approximately 9 % of global disease in developing countries. Also in our modern society, alcohol abuse and dependence are the cause of major health issues [1]. The brain signal or electroencephalogram (EEG) is one of the most complicated tools available amongst the various bio signals, which are being used in medical acumen. EEG provides a direct determination of cortical behavior with millisecond temporal resolution. The constantly changing EEG patterns depends on large number of factors, associated with both internal and external environment and it can also provide long-term insight of the psycho-physiological dynamics of many chronic disorders including alcoholism.

It is often difficult for clinicians and researchers to identify the subject for an alcohol problem. The subjects may complain about their digestion, pain or weakness, but due to social issues, hardly reveal their abuse of alcohol. A doctor who suspects the alcohol problem with the subject may ask a series of questions but in general denial is a hallmark in alcoholism [2]. On the other hand, information extracted from the cerebral cortical activities from the alcoholic subjects may provide concrete platform for the establishment of alcoholism [3]. Advance digital signal processing and soft computing tools can be considered as very important in setting definite EEG spectral variations as a marker of alcoholism as it has been demonstrated in various psychopathological conditions.

Thus, the aim of this study is to examine the permanent alteration in a sample of standard database of the EEG and to determine if the alteration in a particular area of brain is a consequence of alcohol use. It is established that alcohol affect the motor system most prominently and thus on the cerebral motor cortex region of the brain is supposed to the most vulnerable to alcoholism [4]. Therefore, the present work examines the EEG spectral changes, if any, on different motor cortex region by using the data extracted from C3, C4, CZ, P3, P4, PZ, F3, F4 and FZ electrodes. Furthermore, with the help of these alcoholic EEG data, a procedure based on the combined framework of Nonlinear Autoregressive Coefficient, Linear Discrimination Analysis (LDA), and Support Vector Machine (SVM) has been proposed for the identification EEG changes due to alcoholism.

2 Materials and Methods

2.1 Subjects

This study include total of 40 male subjects running in age from 32 to 38 years (mean 35 years) were divided in two group, 20 each in alcoholic and control. The alcoholic subjects were started consuming same kind of alcohol known as Mahuwa, fermented from flower of mahuwa (Madura Longifolia) almost every day from the age of 20 year. The maximum alcoholic content is 13.45 % (w/v) [5]. But on the day of recording they had not taken any drugs or alcohol so that the permanent marker of the alcohol can be established. On the other hand the control subjects reported that they had never taken any kind of alcohol or tobacco in their life time.

2.2 Data Recording

EEG data was recorded using RMS System (Recorder and Medicare Systems Pvt. Ltd, India) of 19 Channel with Ag/AgCl electrodes those were placed with standard10-20 system. EEG data were digitized at sampling frequency of 256 Hz. The unipolar reference region was linked at the right and left earlobes and the Nasion electrode is used as ground. Recording is done in radio frequency (RF) shielded soundproof room with controlled temperature (24 ± 1 °C), relative humidity (45 %) and the electrode impedance were kept below 5 kΩ for the recording. The data were exquisite with amplification gain of 10^4. In the present work, Nonlinear Autoregressive model is applied to consecutive waveforms of 2 s epochs. Prior to the transformation, the EEGs are filtered with low-pass Butterworth filter of fourth order having pass band frequency of 0–40 Hz. The complete experimental protocol is illustrated in Fig. 1.

3 Feature Extraction by Non Linear Autoregressive Coefficient

An approach based on non linear modeling has been adopted for the extraction of features from EEG signals. Considering the non stationary and stochastic nature of the signal, the entire time series data has been divided in windows of 2 s each over which the signal characteristics are assumed to be constant [6].

Over a given below, data is fitted to a non linear autoregressive α represented as:

$$y(k) = f\left[y(k-1), \ldots y\left(k-n_y\right)\right] + \epsilon(k) \tag{1}$$

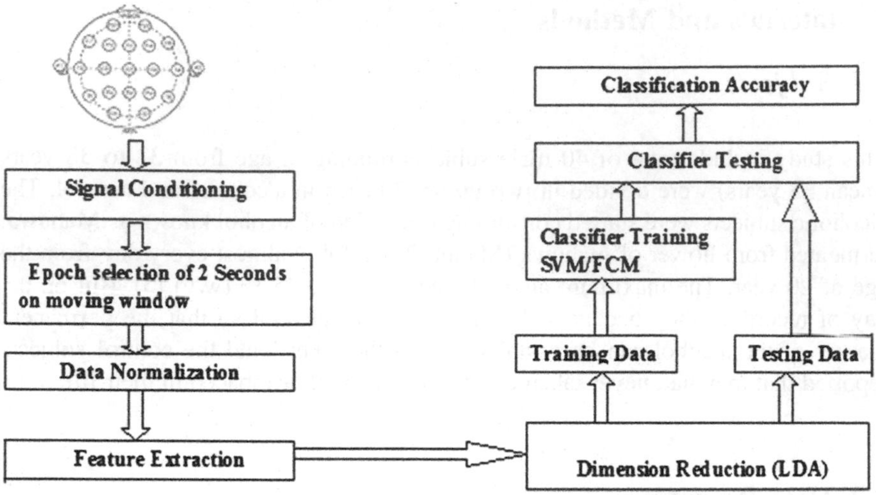

Fig. 1 The illustrative experimental protocol and work plan of the study

where y(k) represents the magnitude of the output at the present sampling instant, f (.) is a non linear function of past outputs to a degree k, n_y is the order of the model and $\epsilon(k)$ is a noise component having zero mean and finite variance. For the present problem, both the order and degree has been taken as 'three'. This has been selected based on the tradeoff between minimizing model complexity and reducing the deviation between the EEG signal and output over the window length, with the assured order and degree Eq. (1) can be written as:

$$\begin{aligned} y(k) = \alpha_0 + \sum_{i=1}^{3} \alpha_i y(k-i) + \sum_{i=1}^{3}\sum_{j=i}^{3} \alpha_{ij} y(k-i)y(k-j) \\ + \sum_{i=1}^{3}\sum_{j=i}^{3}\sum_{l=j}^{3} \alpha_{ijk} y(k-i)y(k-j)y(k-l) \end{aligned} \quad (2)$$

The model involves the obtaining the coefficient of $y(k)(\alpha_0, \alpha_i, \alpha_{ij}, \alpha_{ijk})$ from past and present values of y(k). The unknown parameters are obtained using least squares technique, by reformulating Eq. (2) in linear regression form as discussed in (1).

Since the parameters (model coefficients) represent EEG characteristics over the window length, they can be used as discriminative features for detecting alcoholism. Considering the high dimension (20) of the parameter vector, LDA is applied as a dimension reduction technique. The sample of extracted features from alcoholic and control subjects are presented in Fig. 2.

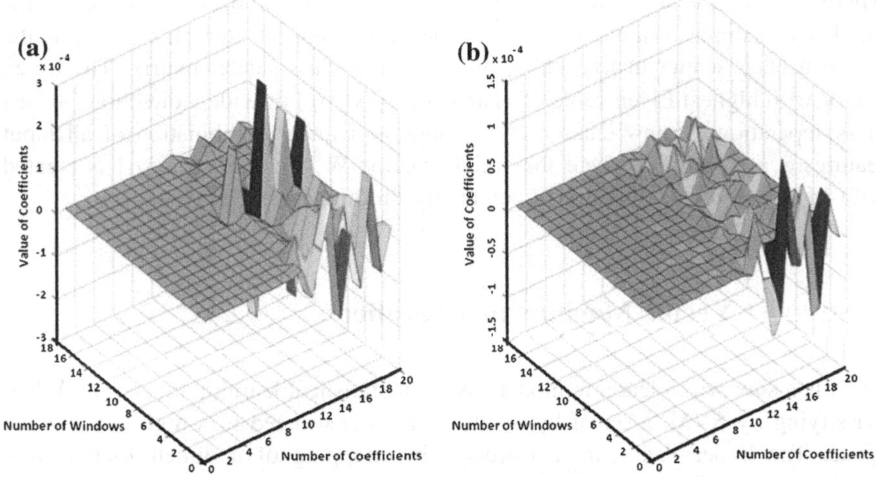

Fig. 2 The sample of feature vector of 20 non linear autoregressive coefficients obtained from 18 Epoch (2 s) EEG data from **a** alcoholic and **b** control subjects

4 Linear Discriminate Analysis as a Dimension Reducer

LDA is a popular dimension reduction algorithm, which aims maximizing the inter class covariance and minimizing the intra class covariance simultaneously, while preserving the class separability. LDA explores for a projection in which the data point of different classes are distant while the data point of the same class are close. The optimized projection of LDA is achieved by the Eigen decomposition of the non-singular scatter matrix derived from the extracted features. To overcome to this singularity problem, which might arise because of linearly related features in high dimension, dimension reduction is done in two stages. In the first stage the dimension is reduced by principal component analysis (PCA) and singular value decomposition (SVD). Following SVD, a diagonal matrix S is obtained of the same dimension as X (data matrix), with non-negative elements in decreasing order as X = U * S * V, where U and V are unitary matrices.

In the second stage the LDA seeks to minimize the following objective function:

$$\mathrm{A} = arg_A \max \frac{A^T S_b A}{A^T S_w A} \quad (3)$$

$$s_b = \sum_{k=1}^{c} m_k (\mu^k - \mu)(\mu^k - \mu)^T \quad (4)$$

$$s_w = \sum_{k=1}^{c} (\sum_{i=1}^{m_k} m_k (x^{i_k} - \mu)(x^{i_k} - \mu)^T) \quad (5)$$

where A is the transformation vector, μ is the total sample mean vector, m_k is the number of sample in the kth class, x^{i_k} is the ith sample in the kth class. s_b is the between class scatter matrix and s_w is the within class scatter matrix. The Eigen vector with highest Eigen value of matrix $\mathrm{S}(\mathrm{S} = s_w^{-1}s_b)$ provide a direction for best class separation. A new dataset y is created as a linear combination of all input features x as $y = x^t\mathrm{W}$; where the weight vector $\mathrm{W} = [w_1, w_2, \ldots w_M,]$ is created with M Eigen vector of matrix S containing the highest Eigen value.

5 Support Vector Machine as Classifier

SVM is a supervised learning model based on statistical learning theory [7]. While classifying by SVM, a set of hyper-planes are constructed in a high dimensional space. The hyper-planes are constructed by mapping of n dimensional feature vector into a k dimensional space via a nonlinear function Φ(x) with the aim of minimizing the margin between two classes of data. For assigning data to two different classes, the hyper-plane equation is given as

$$\mathrm{Y(x)} = W^T\Phi(\mathrm{x}) = \sum_{K=1}^{K} W_k(\mathrm{x}) + W_0 \quad (6)$$

where $\mathrm{W} = [\mathrm{w}_1, \mathrm{w}_2, \mathrm{w}_3 \ldots \mathrm{w}_\mathrm{K}]$, is the weight vector and W_0 represent the bias. Proper separability is achieved by the hyper plane for the large distance between the neighboring data point of both the class by use of kernel function. Kernel function is generally used for mapping the feature space to a high dimensional space in which the classes are linearly separable among the different type of kernel i.e.—linear, quadratic, classical and Gaussian. For the present work linear kernel has been adopted.

6 Results and Discussion

The effectiveness of proposed SVM based approach in classification of EEG data using the feature space obtained in coefficient from non linear autoregressive model has been investigated in this section. Considering the redundancy in the number of samples to be used for the classification. The feature vector can be used as a characteristic feature for clustering EEG into control and alcoholic groups. For the present case, 600 EEG epochs of two second are available for the alcoholic and control groups. Entire dataset is equally divided into training and testing sets. For generating the results, the training dataset should contain data points spanning the entire feature space. For the same size of the training dataset, classification accuracy on testing dataset is found to be dependent on selected data points for training.

Table 1 The class accuracy obtained using SVM clustering based with the number of sample parameters of different channels

Electrode position	Number of samples used for classification				
	270	230	200	180	160
F3	72	74	80	86.11	81.5
Fz	55	67.39	90	88.89	87.5
F4	57	67.39	70	58.33	71.88
C3	66	71.74	72.5	77.78	75
Cz	59	76.09	72.5	72.22	75
C4	51	69.57	65	63.89	71.88
P3	68	82.61	75	75	87.5
Pz	66	71.74	67.50	72.22	75
P4	62	69.57	70	72.22	62

In this study, the classifier is tested with 270, 230, 200, 180 and 160 samples from each group. The class accuracy obtained using SVM clustering based with the number of sample parameters of different channels is summarized in Table 1. The average time required for clustering algorithm to classify a dataset of 600 EEG epochs on a 2 GHz, core 2 duo processor with 2 GB RAM under Windows XP is about 1.5 s.

Best classification performance using SVM clustering (90 % accuracy) is obtained using the feature sample of 200 in the EEG acquired from the Fz electrode position. The result shows that the left frontal electrodes (F3, Fz) are relatively higher accuracy than other electrode positions. It is also observed that even number of electrodes (F3, C3 and P3) have more accuracy then odd no of electrodes (F4, C4 and P4).

It has been observed that the class accuracy is dependent on the number of samples used for training and testing. The accuracy gradually increased with the no of samples from 160 to 200 and then started reducing from 220 to 270. The higher value of classification accuracy is achieved, when the 200 samples are used as a classifying feature. In other words, 200 samples are coming out to be a more distinguishable parameter for detecting alcoholism as compared to the other number of samples. For the extraction of maximum information from acquired data, the selection of optimal channel locations is very important. Few works have been reported using visual evoked response (VEP) for the selection of optimal number of channels [8]. Conversely, in this work, considering the diminished limbic control in alcoholic subjects, we have selected nine channels from cerebral motor cortex region for the analysis. Fz area of brain is having highest classification accuracy, that can be explain in a way that the person with chronic alcoholism is having hyper active Fz zone in comparison to control. Conclude to this result, it is also observed that F3 is following the path of Fz, therefore, it is assumed that the right part of motor cortex area is active or fail to present synchronizing activity in chronic alcoholism (Fig. 3).

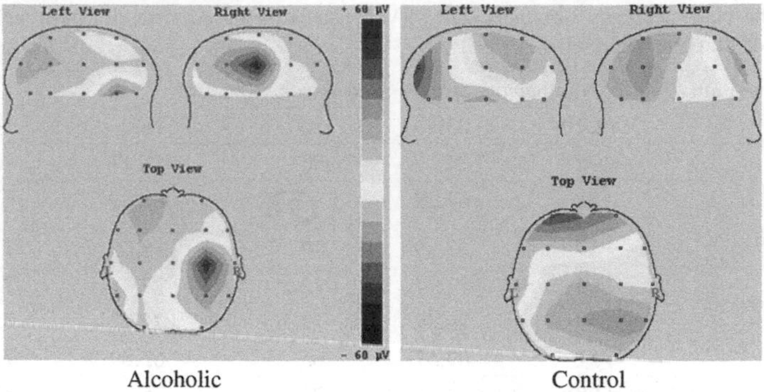

Fig. 3 Distribution of EEG amplitude between alcoholic and control subjects

Dependence on alcohol and craving for alcohol reflects an irresistible urge to drink and is characterized by anticipation and preoccupation with the needed product. In many studies, it has been reported that the alcohol abuse results in the activation of a region in the brain called nucleus accumbens, which is the 'reward centre' of the brain and is situated in the frontal cortex [9]. The frontal cortex is highly responsible for integrating the incoming sensory information such as sight, smell and sound and is also connected to the brain regions, which controls emotions. On examining our results, it can be analyzed that the maximum alteration are in the frontal cortex region (F3, Fz and F4) that can be correlated with report that the chronic alcoholic person are higher craving rating for alcohol [10].

In terms of left and right hemisphere electrode position, F3, C3 and P3 electrodes are having higher classification accuracy then F4, C4 and P4 electrodes. Many pharmacological studies have dealt with the neural mechanism in alcoholism, which express alcohol's ability to modulate wide variety of ion channels, neurotransmitter receptor and transporters [11]. Among neurotransmitter systems, gamma amino butyric acid (GABA) and glutamate are highly sensitive to the effect of alcohol resulting in decreased overall brain excitability, [12]. Animal studies on the effects of alcohol also suggest altered level of histamine and dopamine, which leads to changes in the cortical excitability [13]. The anatomical studies with magnetic resonance imaging (MRI) have also supported the effects of alcohol on nervous system and analyzed reduction in cortical gray matter volume in alcoholic subjects in comparison to the control ones. Furthermore, the functional imaging studies using positron emission tomography (PET) and signal photon emission computed tomography (SPECT) have shown more distinct activation in right cerebral hemisphere [10]. This is also in accordance to our results, which shows more alterations in left sided electrodes (F3, C3 and P3) with respect to right side electrodes (Table 1 and Fig. 3).

7 Conclusion

With the ultimate aim of developing a real time monitoring for detecting alcoholism and study the effect of alcohol on EEG, a technique based on the combination of Non Linear Autoregressive analysis and SVM has been proposed to classify EEG patterns into control and chronic alcoholic groups. The Non Linear Autoregressive analysis decomposes the EEG into time-coefficient features. The feature space obtained is used as a population of data points for classification. The superiority of certain area of brain in detecting alcoholism over others is established by the study. The proposed algorithm is able to achieve a maximum classification accuracy of 90 %. Furthermore, the results obtained from this work can be considered as a platform for considering EEG alterations as a marker for chronic alcoholism, in which the chronic alcoholic conditions can be efficiently and non-invasively detected without any clinical and/or psychological questionnaire. Conversely, the research studies reveal that there are variety of parameters involved that can be used as efficient prognostic factor for chronic alcoholic conditions. Thus, it can be suggested that to obtain a better result and classification accuracy, some other electrophysiological and biochemical parameters may also be added. The results obtained from this study will definitely help in the development of a real time diagnostic system for the chronic alcoholism. However, the observations and diagnosis of the expert clinician cannot be ignored in any case.

References

1. Rehm, J., Mathers, C., Popova, S., Thavorncharoensap, M., Teerawattananon, Y., Patra, J.: Global burden of disease and injury and economic cost attributable to alcohol use and alcohol-use disorders. Lancet **373**(9682), 2223–2233 (2009)
2. Patkar, A.A., Sterling, R.C., Gottheil, E., Weinstein, S.P.: A comparison of medical symptoms reported by cocaine-, opiate-, and alcohol-dependent patients. Subst. Abuse. **20**, 227–235 (1999)
3. Daskalakis, Z.J., Christensen, B.K., Fitzgerald, P.B., Roshan, L., Chen, R.: The mechanisms of interhemispheric inhibition in the human motor cortex. J. Physiol. **543**, 317–326 (2002)
4. Cohen, H.L., Porjesz, B., Begleiter, H.: EEG characteristics in males at risk for alcoholism. Alcohol. Clin. Exp. Res. **15**, 858–861 (1991)
5. Benerji, D.S.N., Rajini, K., Rao, B.S., Banerjee, D.R.N., Rani, K.S., Rajkumar, G., Ayyanna, C.: Studies on physico-chemical and nutritional parameters for the production of ethanol from Mahua flower (*Madhuca indica*) using *Saccharomyces Cerevisiae*—3090 through submerged fermentation (SMF). J. Microb. Biochem. Tech. **2**, 46–50 (2010)
6. Ghosh, S., Maka, S.: Modeling based approach for evaluation of insulin sensitivity. Biomed. Signal Process. Control **4**, 49–56 (2009)
7. Cortes, C., Vapnik, V.: Support-vector networks. Mach. Learn. **20**, 273–297 (1995)
8. Porjesz, B., Rangaswamy, M., Kamarajan, C., Jones, K.A., Padmanabhapillai, A., Begleiter, H.: The utility of neurophysiological markers in the study of alcoholism. Clin. Neurophysiol. **116**, 993–1018 (2005)
9. Anton, R.F.: What is craving? Models and implications for treatment. Alcohol Res. Health. **23**, 165–173 (1999)

10. George, M.S., Anton, R.F., Bloomer, C., Tenback, C., Drobes, D.J., Lorberbaum, J.P., Nahas, Z., Vincent, D.J.: Activation of prefrontal cortex and anterior thalamus in alcoholic subjects on exposure to alcohol-specific case. Arch. Gen. Psychiatry **58**, 345–352 (2001)
11. Lovinger, D.M.: Communication networks in the brain: neurons, receptors, neurotransmitters and alcohol. Alcohol. Res. Health. **31**, 196–214 (2008)
12. Nardone, R., Bergmann, J., Kronbichler, M., Caleri, F., Lochner, P., Tezzon, F., Ladurner, G., Golaszewski, S.: Altered motor cortex excitability to magnetic stimulation in alcohol withdrawal syndrome. Alcohol. Clin. Exp. Res. **34**, 628–632 (2010)
13. Lintunen, M., Hyytiä, P., Sallmen, T., Karlstedt, K., Tuomisto, L., Leurs, R., Kiianmaa, K., Korpi, E.R., Panula, P.: Increased brain histamine in an alcohol-preferring rat line and modulation of ethanol consumption by H_3 receptor mechanisms. FASEB J. **15**, 1074–1076 (2001)

Theoretical Analysis of Expected Population Variance Evolution for a Differential Evolution Variant

S. Thangavelu, G. Jeyakumar, Roshni M. Balakrishnan and C. Shunmuga Velayutham

Abstract In this paper we derive an analytical expression to describe the evolution of expected population variance for Differential Evolution (*DE*) variant—*DE/current-to-best/1/bin* (as a measure of its **explorative** power). The derived theoretical evolution of population variance has been validated by comparing it against the empirical evolution of population variance by *DE/current-to-best/1/bin* on four benchmark functions.

Keywords Differential evolution · Explorative-exploitative balance · Population variance · Explorative power · Empirical evolution of population variance

1 Introduction

Given an optimization problem every practitioner would seek an algorithm that provides acceptably good solutions quicker and has lesser number of control parameters to tune it. Differential Evolution (*DE*), which is one of the most recent additions to the repertoire of Evolutionary Algorithm (*EA*) family, is one such algorithm. *DE*, conceived by Storn and Price in 1995 [1], is a simple yet powerful real-parameter evolutionary algorithm. The robustness of *DE* and its superior

S. Thangavelu (✉) · G. Jeyakumar · R.M. Balakrishnan · C.S. Velayutham
Amrita School of Engineering, Amrita Vishwa Vidyapeetham,
Coimbatore, Tamil Nadu, India
e-mail: s_thangavel@cb.amrita.edu

G. Jeyakumar
e-mail: g_jeyakumar@cb.amrita.edu

R.M. Balakrishnan
e-mail: mroshnib@gmail.com

C.S. Velayutham
e-mail: cs_velayutham@cb.amrita.edu

L.C. Jain et al. (eds.), *Computational Intelligence in Data Mining - Volume 2*,
Smart Innovation, Systems and Technologies 32, DOI 10.1007/978-81-322-2208-8_37

performance in benchmark optimization problems and consequently in real-world applications have been amply demonstrated in the literature [2, 3].

Despite the fact that *DE*, too, is a population based optimizer employing iteratively the variation operators and selection operation like a typical *EA*, its defining characteristic is its *differential mutation* operation that generates mutant vectors by perturbing parent solutions. The multitude of ways by which parent solutions can be perturbed has resulted in many *DE* variants. These variants primarily differ in the way the differential mutation is implemented as well as the type of recombination operator used and thus consequently differing in their efficacy to solve a given optimization problem. Understanding the efficacy of each *DE* variant is crucial as it provides necessary insight to choose the right variant for a given problem thus shortening the lengthy trial-and-error approach.

The extensive empirical analyses of the performance efficacy depicted by different *DE* variants have still not provided with a clear understanding/explanation as to why a particular *DE* variant behave the way it is for a given optimization scenario. This calls for sufficient research focus towards theoretical investigation of *DE* variants. There has been few but significant theoretical research works in *DE* literature, to understand the behavior of *DE* variants [4–11]. Towards this, this paper attempts to directly extend Zaharie's theoretical measure of population diversity, derived for *DE/rand/1/bin* [4] to yet another *DE* variant *DE/current-to-best/1/bin*. Subsequently, the thus derived analytical expression measuring population diversity of above said *DE* variant has been empirically validated.

This paper is organized as follows. Section 2 provides a general description of a typical *DE* algorithm. Section 3 presents a brief review of related works. In Sect. 4, an analytical expression to measure the population diversity of the identified *DE* variant has been derived. Section 5 empirically validates the accuracy of the derived expression and finally, Sect. 6 concludes the paper.

2 Differential Evolution Algorithm

A typical *DE* algorithm begins with uniformly randomized *NP* *D*-dimensional parameter vectors, represented as $X_{i,G} = \left\{x_{i,G}^1, x_{i,G}^2, \ldots, x_{i,G}^D\right\}$ with $(X_{0,G})$ denoting the population of initial points. Subsequently, each generation of *DE* is marked by the formation of a new population through differential mutation, crossover and selection operations. A typical *DE* run comprises of repeated generation of new population by the iterative application of variation and selection operators until a stopping criterion is satisfied.

The differential mutation operation creates one mutant vector $\left(V_{i,G} = \left\{v_{i,G}^1,\right.\right.$ $\left.\left. v_{i,G}^2, \ldots, v_{i,G}^D\right\}\right)$ for each individual $(X_{i,G})$ (the so called target vector in the parlance of *DE*) in the current population by perturbing a random/best population vector with scaled differences of random but distinct population members. Depending on the way a mutant vector is created, a number of *DE* variants have been proposed in the literature. Using the standard notation employed in *DE* literature, the *DE/current-to-best* variant is represented as

$$V_{i,G} = X_{i,G} + K \cdot \left(X_{best,G} - X_{i,G}\right) + F \cdot \left(X_{r_1^i,G} - X_{r_2^i,G}\right) \quad (1)$$

where, $X_{best,G}$, in the above equations, denotes the best solution in the current population i.e. G. The selection of random but distinct population members is ensured by the mutually exclusive random indices $r_1^i, r_2^i, i \in 1, \ldots, NP$ generated anew for each mutant vector.

Subsequent to the differential mutation operation, a crossover operator generates a trial vector (offspring) $U_{i,G}$ for each target vector by mixing the respective target ($X_{i,G}$) and mutant ($V_{i,G}$) vectors. The two crossover schemes typically used in the *DE* literature are binomial (uniform) crossover and exponential crossover. As this paper focuses only on the *DE* variant employing binomial crossover, this section describes only the former scheme. The binomial crossover is defined as follows

$$u_{i,G}^j = \begin{cases} v_{i,G}^j & \text{if}\,(rand_j[0,1) \le C_r) \vee (j = j_{rand}) \\ x_{i,G}^j & \text{otherwise} \end{cases} \quad (2)$$

where $C_r \in (0, 1)$ (the crossover rate) is a user specified positive real number and $rand_j$ [0,1) is the *j*th random number generation. $j_{rand} \in 1, \ldots, D$ is a random parameter index chosen once for each i to ensure the presence of at least one parameter from the mutant vector $V_{i,G}$.

Finally, a selection scheme between the target vector $X_{i,G}$ and trial vector $U_{i,G}$ decides the survivor among the two, based on the objective function values, for next generation as follows (assuming a minimization problem)

$$X_{i,G+1} = \begin{cases} U_{i,G} & \text{if}\, f\left(U_{i,G}\right) \le f\left(X_{i,G}\right) \\ X_{i,G} & \text{otherwise.} \end{cases} \quad (3)$$

3 Related Works

There has been several empirical studies concerning the behavior and efficacy of *DE* variants in the *DE* literature. In retrospect, there are only few but significant theoretical resulting concerning the behavior of *DE* under restricted assumptions can be found in the literature. Zaharie in [4], derived an analytical expression that describes the evolution of population variance of *DE/rand/1/bin* (also applicable to *DE/rand/2/bin*) as a measure of its explorative power. Prior to this work, Zaharie in [5], analyzed the influence of five recombination operators on the convergence properties and exploitative power of a class of *EAs*. The relationship between the control parameters [population size (*NP*), scale factor (*F*) and crossover rate (C_r)] of *DE/rand/1/bin* and the evolution of population variance has been analyzed both from a theoretical and an empirical perspective in [6]. Further, the theoretical analyses in [4, 6] have been extended to *DE/best/1/bin* and *DE/best/2/bin* in [7]. Based on the earlier results on population variance evolution, the idea of population diversity control and parameter adaptation on theoretical grounds was proposed in [8]. Allowing the crossover rate to decide on mutating a given number of components, a comparative analysis of binomial and exponential crossover has been carried out and a relationship between the probability of mutating a given number of components vs. the crossover rate was derived in [9].

The earlier analysis of the influence of variation operators (as well as their parameters) carried out on *DE/rand/1/bin* has been extended to *DE/rand-to-best/1/** (* represent both binomial and exponential crossover) and *DE/current-to-rand/1* variants in [10]. Zaharie in [11], has reviewed the significant theoretical results concerning *DE's* convergence. In addition, the population variance of *DE/either-or* variant has been computed and presented in [8]. It is worth noting that the above mentioned works in *DE* literature have computed the population variance as a function of variation operators for all the classical mutation strategies enumerated in the previous section except *DE/current-to-best/1* strategy. Consequently, this paper directly extends Zaharie's theoretical measure of population diversity to *DE/current-to-best/1/bin* variant. A similar work has also been done in [12] for the variants *DE/best/1bin*, *DE/rand/2/bin* and *DE/best/2/bin.*

4 Population Variance for DE/current-to-best/1/bin

This attempt to derive an expression for population variance is based on the Beyer's statement [13], "the ability of an *EA* to find a global optimal solution depends on its ability to find the right relation between exploitation of the elements found so far and exploration of the search space" and by following [4]. It has been accepted in the Evolution Strategies (*ES*) community [14] that the selection operation may be viewed as realizing exploitation of the information about search space by favoring good solutions (based on their objective function values) and variation operators (mutation and crossover) may be viewed as realizing exploration of search space by introducing unexplored new information about search space into the population. Consequently, the explorative power of an *ES* is considered to be implicated by its population variability in *ES* theory. Hence, the population variance may be considered as a useful measure of its variability. In this paper we analyze the explorative power of *DE/current-to-best/1/bin* through analytical representation as well as empirical observation of the evolution of population variance as against the number of generations.

Zaharie in [4], derived a theoretical relationship between the expected population variance after mutation-crossover and the initial population variance for *DE/rand/1/bin*. The main result of [4] is the following theorem.

Theorem *Let x = {x_1,…,x_m} be the current population, Y = {Y_1,…,Y_m} the intermediate population obtained after applying the mutation and Z = {Z_1,…,Z_m} the population obtained by crossing over the population x and Y. If F is the parameter of the mutation step (i.e. scale factor) and p_c is the parameter of the crossover step (i.e. crossover rate) then*

$$E(Var(Z)) = \left(2F^2 p_c + 1 - \frac{2p_c}{m} + p_c^2\right) Var(x)$$

where Var(x) is the initial population variance before crossover and mutation and E(Var(Z)) is the expected variance for the population obtained after crossover and mutation. The standard DE algorithm has a couple of constraints as in the original DE proposed in [1] *viz. the randomly generated indices $r_1^i \neq r_2^i \neq i$, for all $i \in 1, \ldots, NP$ are mutually exclusive and at least one parameter from mutant vector $V_{i,G}$ must find its place in trial vector $U_{i,G}$ from mutant vector $V_{i,G}$ during the binomial crossover. The above theorem has been derived by ignoring the above said constraints and by assuming a population (of size m) of scalar elements as against n-dimensional vectors (since the transformation by mutation and crossover operators are made similarly and independently to all components in a vector).*

Following Zaharie's work, the derivation of expected population variance for *DE/current-to-best/1/bin* has been carried out, in this paper, in two steps viz. computation of expected population variance

(i) after applying *DE/current-to-best/1* mutation
(ii) after applying binomial crossover.

In the following derivations, the random populations (i.e. *Y* obtained after applying mutation and *Z* obtained after applying crossover) and their elements are denoted with uppercase and the deterministic population (i.e. initial population *x*) is denoted with lowercase. In the subsequent discussions *Var*(*X*), representing variance of a population *X*, is considered a random variable if its elements are influenced by some random elements. The variance of the (deterministic) population $x = \{x_1, \ldots, x_m\}$ of *m* scalar elements ($x_l \in R$) is

$$Var(x) = \frac{1}{m}\sum_{l=1}^{m}(x_l - \bar{x})^2 = \overline{x^2} - \bar{x}^2 \quad (4)$$

where $\overline{x^2}$ is the quadratic population mean and $\bar{x} = \frac{1}{m}\sum_{l=1}^{m} x_l$ is the population mean. If a population is a random variable, then its expected variance *E*(*Var*(*X*)) can be taken as a measure of explorative power, where mean *E* is computed with respect to all random elements which influence the population *X*.

As a first step, the influence of differential mutation *DE/current-to-best/1* on the expected variance has to be determined. Each element Y_l of the intermediate population *Y* (after mutation) is obtained as (Refer Eq. 1)

$$V_l = X_l + K \cdot (X_{best} - X_l) + F \cdot \left(x_{\alpha_1^l} - x_{\alpha_2^l}\right) \tag{5}$$

where $x_{\alpha_1^l}$ and $x_{\alpha_2^l}$ with $l \in \{1, \ldots, m\}$ are random distinct elements selected for each l. The random distinct elements can also be viewed as random variables with property $P(\alpha_i^l = k) = \frac{1}{m}$ with $i \in \{1, \ldots, m\}$. Hence

$$E\left(x_{\alpha_i^l}\right) = \sum_{k=1}^{m} P(\alpha_i^l = k) x_k = \frac{1}{m} \sum_{k=1}^{m} x_k = \bar{x} \tag{6}$$

and

$$E\left(x_{\alpha_i^l}^2\right) = \sum_{k=1}^{m} P(\alpha_i^l = k) x_k^2 = \frac{1}{m} \sum_{k=1}^{m} x_k^2 = \overline{x^2}. \tag{7}$$

Also, the random distinct elements are mutually dependent since their values must be distinct thus one must depend on others. Hence, for $i \neq j$, $E\left(x_{\alpha_i^l} x_{\alpha_j^l}\right) \neq E\left(x_{\alpha_i^l}\right) E\left(x_{\alpha_j^l}\right)$. Extending the property $P(\alpha_i^l = k) = \frac{1}{m}$, it follows that

$$P\left(\alpha_i^k = k,\ \alpha_j^l = l\right) = \frac{1}{m} \cdot \frac{1}{m-1} = \frac{1}{m^2 - m}.$$

So

$$E\left(x_{\alpha_i^l} x_{\alpha_j^l}\right) = \sum_{\substack{k=1 \\ k \neq l}}^{m} P\left(\alpha_i^k = k, \alpha_j^l = l\right) x_k x_l = \frac{1}{m^2 - m} \sum_{k \neq l} x_k x_l$$

$$= \frac{1}{m^2 - m} \left[\left(\sum_{k=1}^{m} x_k \right)^2 - \sum_{k=1}^{m} x_k^2 \right] = \frac{1}{m^2 - m} \left[m^2 \bar{x}^2 - m \overline{x^2} \right]$$

$$= \frac{m}{m-1} \bar{x}^2 - \frac{1}{m-1} \overline{x^2}$$

$$E\left(x_{\alpha_i^l} x_{\alpha_j^l}\right) = \frac{m}{m-1} \bar{x}^2 - \frac{1}{m-1} \overline{x^2}. \tag{8}$$

Now, the expected variance of the intermediate population Y can be computed as

$$E(Var(Y)) = E\left(\overline{Y^2}\right) - E\left(\bar{Y}^2\right) \tag{9}$$

where

$$E\left(\overline{Y^2}\right) = \frac{1}{m}E\left(\sum_{l=1}^{m} y_l^2\right) = \frac{1}{m}E\left(\sum_{l=1}^{m}\left(x_l + K\cdot(x_{best} - x_l) + F\cdot(x_{\alpha_1^l} - x_{\alpha_2^l})\right)^2\right)$$

$$E\left(\overline{Y^2}\right) = \overline{x^2}(1-K)^2 + K^2x_{best}^2 + 2Kx_{best}\bar{x}(1-K) + 2F^2\frac{m}{m-1}Var(X). \tag{10}$$

Since x_{best} could be any element from the population, accepting the approximation $x_{best} \approx \bar{x}$, to simplify the analysis, yields

$$E\left(\overline{Y^2}\right) = \overline{x^2} + \left(K^2 - 2K + \frac{m}{m-1}2F^2\right)Var(x). \tag{11}$$

The next term to be computed is $E(\bar{Y}^2)$. It is known that

$$\begin{aligned}
E(\bar{Y}^2) &= E\left[\left(\frac{1}{m}\sum_{l=1}^{m} y_l\right)^2\right] \\
&= \frac{1}{m^2}E\left[\left(\sum_{l=1}^{m}\left(x_l + K\cdot(x_{best} - x_l) + F\cdot(x_{\alpha_1^l} - x_{\alpha_2^l})\right)\right)^2\right] \\
&= \frac{1}{m^2}E\left[\sum_{l=1}^{m}\left(x_l + K\cdot(x_{best} - x_l) + F\cdot(x_{\alpha_1^l} - x_{\alpha_2^l})\right)^2\right. \\
&\quad \left. + \sum_{k\neq l}\left(x_lK\cdot(x_{best} - x_l) + F\cdot\left(x_{\alpha_1^l} - x_{\alpha_2^l}\right)\right)\left(x_k + K\cdot(x_{best} - x_k) + F\cdot(x_{\alpha_1^k} - x_{\alpha_2^k})\right)\right] \\
&= \frac{1}{m}E\left(\overline{Y^2}\right) + \frac{m-1}{m}\bar{x}^2(1-K)^2 + \frac{m-1}{m}2\bar{x}Kx_{best}(1-K) + \frac{m-1}{m}K^2x_{best}^2
\end{aligned}$$

$$E(\bar{y}^2) = \frac{1}{m}E\left(\overline{Y^2}\right) + \left(\frac{m-1}{m}\right)\bar{x}^2. \tag{12}$$

Thus from Eqs. (11) and (12)

$$\begin{aligned}
E(Var(Y)) &= E\left(\overline{Y^2}\right) - \frac{1}{m}E(\bar{Y}^2) = \left(\frac{m-1'}{m}\right)E\left(\overline{Y^2}\right) - \frac{m-1}{m}\bar{x}^2 \\
&= \left(\frac{m-1}{m}\right)\overline{x^2} + \left(\frac{m-1}{m}\right)\left(K^2 - 2K + \frac{m}{m-1}2F^2\right)Var(x) - \left(\frac{m-1}{m}\right)\bar{x}^2 \\
&= \left(\frac{m-1}{m}\right)Var(x) + \left(K^2 - 2K + \frac{m}{m-1}2F^2\right)\left(\frac{m-1}{m}\right)Var(x)
\end{aligned}$$

$$E(Var(Y)) = \left(\left(\frac{m-1}{m}\right)(1-K)^2 + 2F^2\right)Var(x). \tag{13}$$

Having computed the influence of *DE/current-to-best/1* mutation on the expected population variance, the next step involves the computation of the expected population variance after applying the binomial crossover. Each element Z_l of the population Z (after applying binomial crossover) is obtained as

$$Z_l = \begin{cases} Y_l & \textit{with probability } p_c \\ x_l & \textit{with probability } 1 - p_c \end{cases} \quad \text{where} \quad l = 1, \ldots, m. \tag{14}$$

The expected population variance after applying crossover is computed as follows

$$E(Var(Z)) = E\left(\overline{Z^2}\right) - E(\bar{Z}^2). \tag{15}$$

Using Lemma 4.1 from [1], the terms $E\left(\overline{Z^2}\right)$ and $E(\bar{Z}^2)$ are obtained as follows

$$E\left(\overline{Z^2}\right) = \frac{1}{m}\sum_{l=1}^{m} E(Z_l^2) = \frac{1}{m}\sum_{l=1}^{m}\left[(1 - p_c)x_l^2 + p_c E(y_l^2)\right]$$
$$E\left(\overline{Z^2}\right) = (1 - p_c)\overline{x^2} + p_c E\left(\overline{y^2}\right)$$

and

$$E(\bar{Z}^2) = \frac{1}{m^2} E\left[\left(\sum_{l=1}^{m} Z_l\right)^2\right] = \frac{1}{m^2}\left(E\left(\sum_{l=1}^{m} Z_l^2\right) + E\left(\sum_{k\neq l} Z_k Z_l\right)\right)$$
$$E(\bar{Z}^2) = \frac{1}{m} E\left(\overline{Z^2}\right) + \frac{1}{m^2}\sum_{k\neq l} E(Z_k Z_l)$$

where

$$\begin{aligned} \sum_{k\neq l} E(Z_k Z_l) &= \sum_{k\neq l} E(Z_k)E(Z_l) \\ &= \sum_{k\neq l}((1 - p_c)x_k + p_c E(Y_k))((1 - p_c)x_l + p_c E(Y_l)) \end{aligned} \tag{16}$$

However

$$E(Y_l) = E\left[x_l + K \cdot (x_{best} - x_l) + F \cdot \left(x_{\alpha_1^l} - x_{\alpha_2^l}\right)\right] = Kx_{best} + (1 - K)\bar{x}.$$

So

$$\sum_{k\neq l} E(Z_k Z_l) = \left(\sum_{l=1}^{m} Z_l\right)^2 - \sum_{l=1}^{m} Z_l^2$$
$$= \left[\sum_{l=1}^{m}((1-p_c)x_l + p_c(Kx_{best} + (1-K)\bar{x}))\right]^2$$
$$- \sum_{l=1}^{m}((1-p_c)x_l + p_c(Kx_{best} + (1-K)\bar{x}))^2$$

Now

$$E(\bar{Z}^2) = \frac{1}{m}E\left(\overline{Z^2}\right) + \frac{1}{m^2}(m^2(1-P_c)^2\bar{x}^2 + m^2P_c^2(Kx_{best} + (1-K)\bar{x})^2$$
$$+ 2m^2P_c(1-P_c)x_l(Kx_{best} + (1-K)\bar{x})$$
$$- m(1-P_c)^2\overline{x^2} - mP_c^2(Kx_{best} + (1-K)\bar{x})^2$$
$$- 2mP_c(1-P_c)\bar{x}(Kx_{best} + (1-K)\bar{x})$$

and

$$E(Var(Z)) = E(\overline{Z^2}) - E(\bar{Z}^2)$$
$$= \frac{m-1}{m}E\left(\overline{Z^2}\right)\frac{1}{m^2}(m^2(1-P_c)^2\bar{x}^2 + m^2P_c^2(Kx_{best} + (1-K)\bar{x})^2$$
$$+ 2m^2P_c(1-P_c)x_l(Kx_{best} + (1-K)\bar{x}) - m(1-P_c)^2\overline{x^2}$$
$$- mP_c^2(Kx_{best} + (1-K)\bar{x})^2 - 2mP_c(1-P_c)\bar{x}(Kx_{best} + (1-K)\bar{x}).$$

Accepting the above said approximation

$$E(Var(Z)) = \left[\left(\frac{m-1}{m}\right)\left(1 + P_c\left(K^2 - 2K + \frac{m}{m-1}2F^2\right)\right) + \frac{(1-P_c)^2}{m}\right]Var(x). \tag{17}$$

Thus the expected population variance after applying *current-to-best/1* mutation and *binomial* crossover on the initial population x is

$$\left[\left(\frac{m-1}{m}\right)\left(1 + P_c\left(K^2 - 2K + \frac{m}{m-1}2F^2\right)\right) + \frac{(1-P_c)^2}{m}\right]Var(x) \tag{18}$$

where F and K are scaling factor, P_c is crossover probability and m is the number of elements in the population.

5 Empirical Evolution of Population Variance

Having derived the analytical expression as mentioned earlier, this section analyze the validity of the expression with simulation experiments.

The experimental study involved four benchmarking functions of dimension 30, with a population size of $NP = 50$, a fixed value of $P_c = 0.5$. The benchmarking functions are chosen with different modality and decomposability (f_1—Schwefel's Problem 2.21 (Unimodal Separble), f_2—Schwefel's Problem 1.2 (Unimodal Nonseparable), f_3—Generalized Restrigin's Function (Multimodal Separble) and f_4—Ackley's Function (Multimodal Nonseparable)) [15]. The details of the benchmarking functions are presented in Table 1.

Interestingly, Zaharie [4] observed that there is a critical value (F_c) for the mutation scaling factor F under which the variance decreases and over which it increases. For the assumed population size (i.e. 50) and P_c value (0.5) the F_c value for *DE/current-to-best/1/bin* is calculated to be 0.67. Accordingly, the empirical and expected population variances for the *DE/current-to-best/1/bin* have been measured for two values of parameter F (one less than F_c and the other greater than F_c). Since in *DE* the transformation to the candidate vectors take place component wise, the empirical population variance for the variants has been computed by averaging the component-wise variance for 100 independent runs.

Table 2 displays the (theoretical) expected and empirical population variances of *DE/current-to-best/1/bin* for the four benchmarking functions. The theoretical and empirical population variances are denoted as *tVar* and *eVar* in Table 2. As can be seen from the Table 2, the *DE* variant's runs for each function display gradual loss of variance or gradual increase in population variance. This can be easily observed by comparing the variances at the beginning and at the end of runs. It is worth mentioning that while the variance decreases invariably for the variant for the first F value, it increases invariably for the second F value as both F values are on either side of the critical value (F_c). As can be seen from Table 2, the decreasing and increasing pattern of theoretical expected variance is matched by that of empirical population variance.

Figure 1 reiterates this similar evolution pattern of theoretical and empirical population variances of *DE/current-to-best/1/bin*, for both F values. However, there is a difference between the theoretical and empirical variances, which may be

Table 1 Description of the benchmarking functions used for the experiment

Fn.	Function description
f_1	$f_{sch3}(x) = \max_i\{\lvert x_i\rvert, 1 \le i \le n\}, -100 \le x_i \le 100$
f_2	$f_{sch2}(x) = \sum_{i=1}^{n} (\sum_{j=1}^{n} x_j)^2, -100 \le x_i \le 100$
f_3	$f_{grf}(x) = \sum_{i=1}^{n} [x_i^2 - 10\cos(2\pi x_i) + 10], -5.12 \le x_i \le 5.12$
f_4	$f_{ack}(x) = 20 + e - 20\exp\left(-0.2\sqrt{\frac{1}{n}\sum_{i=1}^{n} x_i^2}\right) - \exp\left(\frac{1}{n}\sum_{i=1}^{n}\cos(2\pi x_i)\right), -30 \le x_i \le 30$

Table 2 Empirical and theoretical variance measured for *DE/current-to-best/1/bin*

DE/current-to-best/1/bin ($F_{=0.67)c}$)

DG	f_1				G	f_2			
	F = 0.6		F = 0.7			F = 0.6		F = 0.7	
	eVar	tVar	eVar	tVar		eVar	tVar	eVar	tVar
1	3,326.27	3,326.27	3,725.02	3,725.027	1	3,244.25	3,244.25	3,890.69	3,487.67
3	3,215.14	3,190.29	4,827.68	4,916.78	3	2,536.34	2,476.13	5,598.12	5,645.22
6	3,070.6	3,001.38	6,509.62	6,499.66	6	2,034.22	2,013.62	7,916.82	7,934.34
9	2,997.37	2,908.11	8,696.33	8,506.32	9	1,700.45	1,637.49	11,302.97	11,412.23
12	2,827.42	2,799.73	11,570.61	11,490.31	12	1,367.45	1,331.63	15,310.81	15,423.23
15	2,801.76	2,755.71	12,666.67	12,560.24	15	1,122.19	1,082.90	21,364.62	21,123.45
18	2,727.33	2,699.81	16,816.11	16,666.12	18	912.62	880.62	29,351.04	29,402.03
21	2,612.69	2,620.9	21,866.97	21,763.06	21	723.23	716.13	40,863.98	40,178.23
24	2,503.82	2,499.59	29,552.12	29,552.34	24	556.43	582.37	56,909.67	57,009.69
27	2,430.66	2,422.03	39,452.99	38,999.92	27	469.23	473.59	77,875.89	77,198.23
30	2,225.77	2,210.13	56,133.63	56,002.12	30	401.23	412.61	94,095.99	95,002.24
G	f_3				G	f_4			
	F = 0.6		F = 0.7			F = 0.6		F = 0.7	
	eVar	tVar	eVar	tVar		eVar	tVar	eVar	tVar
1	8.75	8.75	9.84	9.84	1	337.27	337.27	381.33	381.33
3	8.32	8.45	13.23	12.34	3	334.73	313.72	522.54	519.23
6	8.06	8.26	17.40	16.54	6	324.74	302.23	710.04	709.34
9	7.8	7.93	22.71	21.01	9	312.12	299.16	990.13	982.33
12	7.54	7.76	29.84	29.01	12	298.3	290.23	1,441.95	1,439.12
15	7.29	7.20	38.88	37.97	15	277.44	282.23	2,135.06	2,132.23
18	7.19	7.16	49.71	47.99	18	256.38	262.28	3,213.61	3,199.21

(continued)

Table 2 (continued)

G	f_3				*G*	f_4			
	F = 0.6		*F* = 0.7			*F* = 0.6		*F* = 0.7	
	eVar	*tVar*	*eVar*	*tVar*		*eVar*	*tVar*	*eVar*	*tVar*
21	7.06	6.99	64.95	65.02	21	245.52	259.22	4,943.11	5,003.21
24	6.95	6.89	84.08	83.02	24	236.89	242.19	7,407.12	7,419.22
27	6.81	6.72	107.65	106.32	27	221.51	229.28	11,057.49	11,103.34
30	6.71	6.66	125.61	124.92	30	211.35	212.28	14,369.16	14,432.11

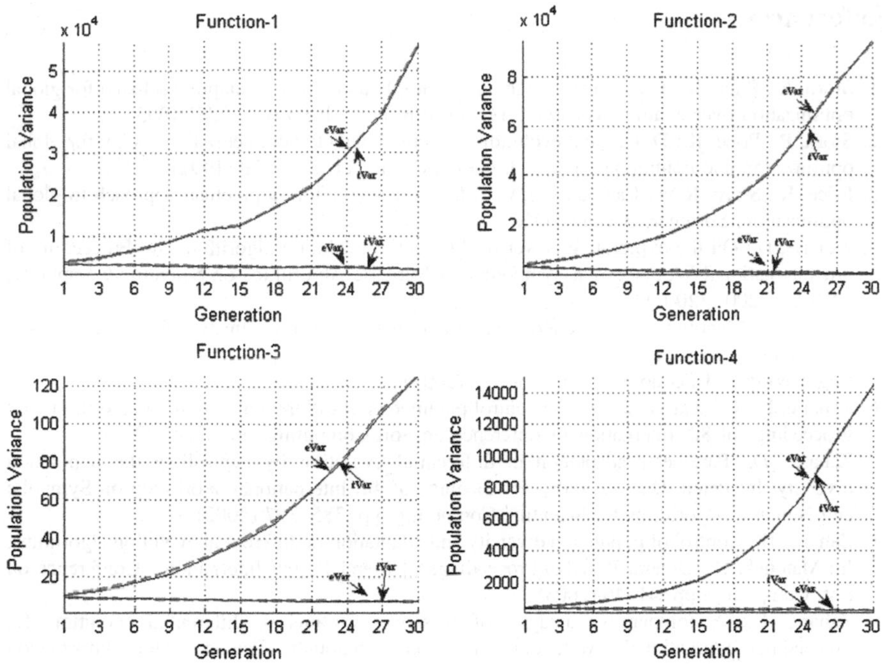

Fig. 1 Evolution of theoretical and empirical population variances for *DE/current-to-best/1/bin*, for both F values (one above and one below the F_c)

attributed to the fact that the theoretical derivation ignores the restriction that the indices of chosen solution vectors should not be equal to the parent vector index. Even then the difference is not significant as can be seen from the figure.

Thus the derivation of analytical expression to describe the evolution of expected population variance for *DE/current-to-best/1/bin* has been validated empirically.

6 Conclusion

This paper intended to analyze the explorative power of *DE/current-to-best/1/bin* from a theoretical perspective. Consequently, a theoretical relationship between the initial population variance and expected population variance after applying variation operators (mutation followed by binomial crossover) has been derived as a measure of explorative power by extending Zaharie's work on *DE/rand/1/bin*. Subsequently, the validity of the derived relationship has been demonstrated with empirical studies.

References

1. Storn, R., Price, K.: Differential evolution—a simple and efficient adaptive scheme for global optimization over continuous spaces. Technical Report TR-95-012, ICSI 1995
2. Storn, R., Price, K.: Differential evolution—a simple and efficient heuristic strategy for global optimization and continuous spaces. J. Global Optim. **11**, 341–359 (1997)
3. Price, K., Storn, R.M., Lampinen, J.A.: Differential evolution: a practical approach to global optimization. Springer, Berlin (2005)
4. Zaharie, D.: On the explorative power of differential evolution algorithms. In: Proceeding of 3rd International Workshop on Symbolic and Numeric Algorithms on Scientific Computing, SYNASC-2001 (2001)
5. Zaharie, D.: Recombination operators for evolutionary algorithms. In: Ivanchev, D., Todorov, M.D. (eds.) Proceedings of the 27th Summer School of Applications of Mathematics in Engineering and Economics, pp. 625–627 (2001)
6. Zaharie, D.: Critical values for the control parameters of differential evolution algorithms. In: Proceedings of 8th International Conference on Soft Computing, pp. 62–67 (2002)
7. Zaharie, D.: Parameter adaptation in differential evolution by controlling the population diversity. In: Petcu, D., et al (eds.) Proceedings of 4th International Workshop on Symbolic and Numeric Algorithms for Scientific Computing, pp. 385–397 (2002)
8. Zaharie, D.: Control of population diversity and adaptation in differential evolution algorithms. In: Matouek, R., Omera, P. (eds.) Proceedings of Mendel's 9th International Conference on Soft Computing, pp. 41–46 (2003)
9. Zaharie, D.: A comparative analysis of crossover variants in differential evolution. In: Proceedings-International Multiconference on Computer Science and Information Technology, vol. 2, pp. 172–181, Wisła, Poland (2007)
10. Zaharie, D.: Statistical properties of differential evolution and related random search algorithms. In: Brito, P. (eds.) Proceedings of the International Conference on Computational Statistics Porto, pp. 473–485 (2008)
11. Zaharie, D.: Differential evolution from theoretical analysis to practical insights. In: Proceeding of 19th International Conference on Soft Computing, Brno, Czech Republic, 26–28 June 2013
12. Jeyakumar, G., Shunmuga Velayutham, C.: A comparative study on theoretical and empirical evolution of population variance of differential evolution variants, SEAL 2010. LNCS 6457, pp. 75–79 (2010)
13. Beyer, H.G.: On the 'Explorative Power' of ES/EP like algorithms. In: Porto, V.W., Saravanan, N., Waagen, D.E., Eiben, A.E. (eds.) Proceedings of the 7th Annual Conference on Evolutionary Programming. Lecture Notes in Computer Science, vol. 1447–1998, pp. 323–334. Springer, Berlin (2008)
14. Beyer, H.G., Deb, K.: On the analysis of self adaptive evolutionary algorithms. Technical Report, CI-69/99, University of Dortmund (1999)
15. Mezura-Montes, E., Velazquez-Reyes, J., Coello Coello, C.A.: A comparative study on differential evolution variants for global optimization. In: Genetic and Evolutionary Computation Conference (GECCO'06), pp. 485–492 (2006)

Quantification and 3D Visualization of Articular Cartilage of Knee Joint Using Image Processing Techniques

M.S. Mallikarjunaswamy, Mallikarjun S. Holi and Rajesh Raman

Abstract The articular cartilage of knee joint plays an important role in smooth movement and lubrication of the joint. Osteoarthritis (OA) is a degenerative disease of the knee joint, commonly affecting the elderly around the world. Visualization and morphological analysis of cartilage plays an important role in the assessment of OA and rehabilitate the affected people. In the present work, thickness and volume of the cartilage was quantified using an edge detection based interactive segmentation method from knee joint magnetic resonance images (MRI) and the joint is visualized in 3D. Volume of interest (VOI) processing approach was used to reduce the number of voxels processed in 3D rendering of articular cartilage. The method reduces the processing time in comparison with manual and other semiautomatic methods. The agreement of thickness and volume measurements was assessed using Bland-Altman plots in comparison with manual method.

Keywords Cartilage volume · Interactive segmentation · 3D visualization · MRI · Osteoarthritis

---

M.S. Mallikarjunaswamy (✉)
Department of Biomedical Engineering, BIET, Davangere 577004, India
e-mail: ms_muttad@yahoo.co.in

M.S. Mallikarjunaswamy
Department of Instrumentation Technology, S.J. College of Engineering, Mysore 570006, India

M.S. Holi
Department of Electronics and Instrumentation Engineering, University B.D.T. College of Engineering, Visvesvaraya Technological University, Davangere 577004, India
e-mail: msholi@yahoo.com

R. Raman
Department of Radio-Diagnosis, J.S.S. Medical College, JSS University, Mysore 570015, Karnataka, India
e-mail: rajeshiyer81@gmail.com

L.C. Jain et al. (eds.), *Computational Intelligence in Data Mining - Volume 2*,
Smart Innovation, Systems and Technologies 32, DOI 10.1007/978-81-322-2208-8_38

1 Introduction

The knee joint plays an important role in mobility and stability of the human body. It supports the body during routine and difficult activities [1]. It is a frequently injured joint of the body in elderly people. The articular cartilage is a thin layer of soft tissue at the end of femur and tibia, which smoothens the movement of the knee joint. Figure 1 shows a schematic of knee joint anatomy. The cartilage gets abraded by the grinding mechanism at the points of contact between the articular surfaces over a period of time. As a consequence, the underlying articular margins of the femur and tibia are exposed and results in formation of bone spurs known as osteophytes. In OA affected knee, the joint is stiff with decreased motion and is painful [2]. The OA causes disability and reduces the quality of life in the elderly population. The OA is a progressive disease and therefore its early detection and cessation of progression are primary concerns of medical doctors. The knee articular cartilage is visible in MRI of knee joints. The cartilage is surrounded by ligaments and other tissues of knee joint hence it is difficult to observe the degradation of the cartilage and quantify the thickness at an early stage of OA.

Progression of OA is investigated using MRI. The thickness and volume visualization of cartilage is useful in identification of the region of cartilage degradation.

2 Earlier Work

The segmentation methods of knee joint cartilage can be classified into manual, semiautomatic and fully automatic methods. The manual methods are operator dependent and time consuming. Therefore semiautomatic and fully automatic methods are preferred. A good number of segmentation methods are reported in the literature. Cohen et al. [3] developed a semiautomatic method based on the region of interest and image gradient evaluation for the detection of cartilage boundary. Cashman et al. [4] developed a method based on edge detection and thresholding

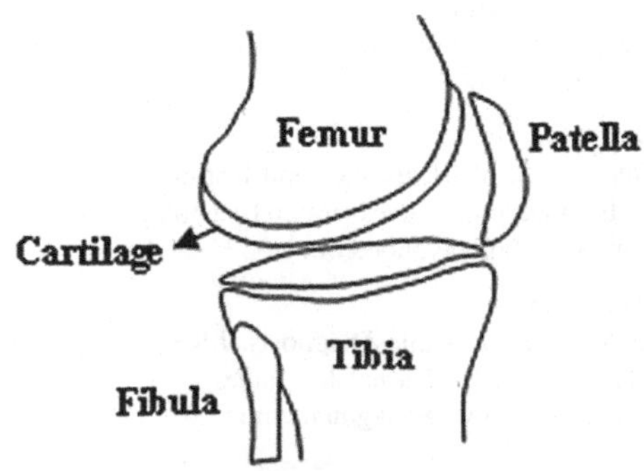

Fig. 1 Knee joint anatomy

and further boundary discontinuities, were bridged using B-spline interpolation. A semiautomatic method based on radial search was developed [5] and the boundary of the femur–cartilage interface was detected starting from the center of the femur. In earlier study [6] a fully automatic method based on voxel classification approach was developed, which reduced the processing time. Kauffmann et al. [7] developed 2D active contour method based on local coordinate system to fix the shape of the cartilage. A method based on gradient vector flow snakes [8] to segment the cartilage from knee joint MRI was also developed. The morphology of the knee joint using direct volume rendering technique is obtained using commercial software [10]. The volume of interest processing approach for medical image segmentation can reduce the computations [10]. The volume segmentation is classified into three categories namely structural, statistical and hybrid techniques. The edge detection based methods of volume segmentation are classified as structural techniques. Ozlem et al. [11] compared the grades of radiologist diagnosis of OA affected knee joints with the cartilage volume measurements obtained using image processing software. The volume segmentation of anatomical regions plays an important role in the diagnosis and treatment of the diseases. The interactive segmentation methods can provide better performance than the fully automatic methods in medical diagnostic applications. In this work, the objective was to quantify and 3D visualize the articular cartilage from MRI and compare the thickness and volume measurements of the articular cartilage using edge detection based method with the manual segmentation method for the assessment of accuracy and agreement between the methods.

3 Methodology

The knee joint MR images are obtained from OA Initiative (OAI), National Institute of Health, which includes normal and OA affected knee joint images. The MR images of this database include water excitation double echo steady-state (DESS) imaging protocol with sagittal slices at 1.5-T. The imaging parameters for the sequence are: TR/TE: 16.3/4.7 ms, matrix: 384 × 384, FOV: 140 mm, slice thickness: 0.7 mm, x/y resolution: 0.365/0.365 mm. The study involves total knee MR images of 55 subjects including 10 normal and 45 OA affected subjects with varying age group from 25 to 85 years. MR images are processed using Matlab 7.1 software for segmentation of cartilage and 3D visualization.

The obtained MR images are preprocessed for noise removal using median filter of (3 × 3). Median filter removes noise without affecting edges and boundary information in an image. An interactive segmentation method based on edge detection [12] was used to segment the cartilage from knee joint MRI. In this segmentation method, Canny edge detection technique [13] was used to obtain the location of femur cartilage boundary and cartilage synovial boundary. The inner boundary, the cartilage femur interface was completely detected in Canny edge detected image and sample points were obtained from the detected boundary. In the

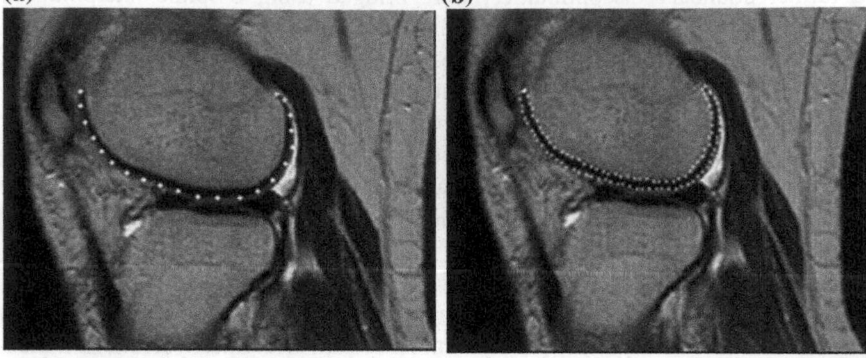

Fig. 2 Segmentation of cartilage **a** marking of outer boundary, **b** interpolated cartilage boundary overlapped on knee joint MRI

cartilage synovial interface the boundary detected is not continuous; the manual sample points were added in addition to sample points of detected boundary of interface. The sample points of inner and outer boundary were further increased using B-spline interpolation. The outer boundary detection and sample points after interpolation are shown in Fig. 2a, b. A mask is developed for cartilage boundary and cartilage is segmented from MR image.

Thickness and local volume of cartilage were computed using segmented cartilage. The segmented cartilage was saved to a 3D array. The processing steps were repeated for all the images of MRI sequence in which cartilage is visible. The edge information of cartilage is obtained from structure of cartilage from MRI using edge detection technique. These edges are the local edges, when they are saved to 3D array it forms contour of cartilage in 3D. The contour boundary is the separation of voxels of cartilage from the voxels of other tissues of knee joint in MRI. The method is a structure based approach of volume segmentation. For 3D volume rendering of cartilage, direct method of volume rendering using stack of 2D images was adopted. The segmented cartilage images of entire sequence of knee MRI were saved to 3D array one after the other. The segmentation algorithm loops till the completion of all images of MRI sequence. The segmented cartilage images are stacked as 3D array as shown in Fig. 3.

The pseudo code of the procedure is as follows

//Segment the cartilage and save the mask

for n=1: N *loop*
Read the knee joint MRI in the sequence as image I
Segment the cartilage and save the mask as image b
*multiply images (I * b)*
Save the resultant image as v_n
End

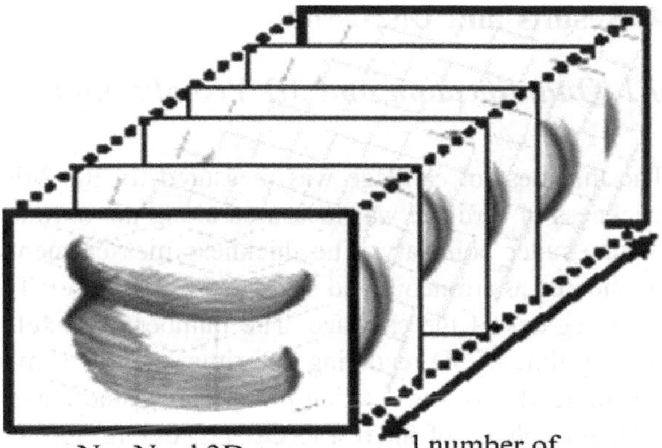

Fig. 3 3D volume rendering of articular cartilage

//Save the segmented image into 3D array

Create 3D array of zeros of size X=*128*×128×1×N
for n = 1 : N *loop*
I = *Read image* v_n *(from the saved sequence of segmented images)*
Save to 3D array [x (:, :, 1, n)]=I (:, :, 1);
End

//Volume render the cartilage

C=squeeze the 3D array (X)
Volume render the 3D array
Visualize the cartilage image in 3D

The saved array is squeezed as 3D image and volume rendered for visualization. The squeeze operation removes all singleton dimensions. In volume rendering, the processing time is dependent on the computational complexity. The total number of voxels processed for volume rendering is the product of pixels in each segmented image and the number of segmented images from MRI. To reduce the computational complexity in volume rendering, volume of interest processing is adopted. In the segmented image, the pixels pertaining to cartilage and its boundary were considered as the volume of interest. From the segmented images the volume of interest is saved to 3D array. The size of the 3D array is based on the number of pixels of the cartilage and the number of 2D images segmented and saved. The volume rendering algorithm adopts texture mapping techniques. The visualized cartilage can be rotated faster for visualization of different regions without any limitations of memory handling capabilities of processor.

4 Results and Discussion

4.1 Quantification and 3D Visualization

The thickness of cartilage was measured for the MRI of the entire data set. The thickness of cartilage was measured along the normal to the inner boundary curve till the outer boundary. The thickness measurements of more than 3.5 mm are considered as erroneous and discarded from the set. The thickness was computed in three regions of the cartilage. The number of voxel processed increases the processing time of 3D rendering algorithm and also imposes limitations of hardware for 3D rendering and rotation of visible 3D cartilage. Therefore, to overcome this problem volume of interest (VOI) processing approach was used for 3D volume rendering. The VOI is the cartilage portion of knee MRI. The reduced size images are saved as a stack of images for volume rendering. The visualized cartilage can be rotated faster for visualization of different regions. Table 1 compares the computational complexity of VOI approach with direct method. Considering the computational complexity of the direct method as 100 % and it is reduced to 11.11 % using VOI.

The cartilage of knee joint MRI was also segmented using manually marking cartilage boundary as region of interest. The segmented cartilage images are loaded to

MIMICS 15.01 (Materialise, Belgium) Medical image processing and editing software. The volume was visualized and thickness and volume are quantified. Figure 4 shows the processed images using MIMICS software. Figure 4a shows an input segmented cartilage image. Figure 4b shows volume rendered cartilage. Figure 4c, d shows 3D visualization of OA affected and eroded cartilages.

4.2 Bland-Altman Analysis of Measurement Methods

The thickness and volume measurements of developed edge detection based interactive segmentation method were compared with measurements of manual method. Table 2 shows the limiting values of measurement (d ± 2 s).

The residual of two measurements and percentage error was calculated. The percentage error found was less than 4.1 %. To asses the agreement between the

Table 1 Comparison of computational complexity

Method of volume rendering	Image size	Number of MR images processed	Total number of voxels processed	Computational complexity (%)
Direct	384 × 384	80	11.79×10^6	100.00
VOI	128 × 128	80	1.31×10^6	11.11

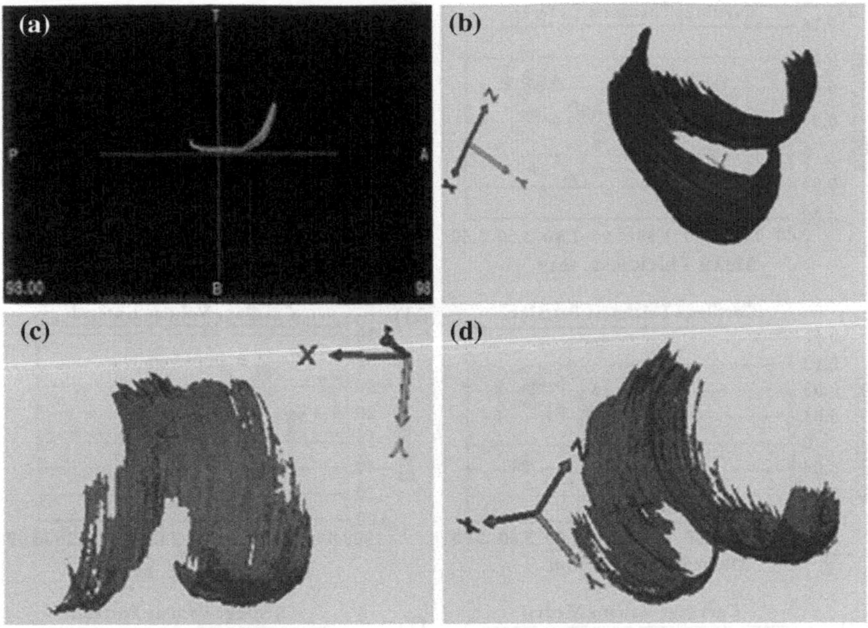

Fig. 4 Visualization of articular cartilage **a** segmented cartilage image, **b** 3D rendered view, **c**, **d** degraded cartilages due to OA

Table 2 Limits of differences for the assessment of agreement between the methods

Cartilage measurements	% error maximum	Mean of difference (d)	Standard deviation of difference (s)	Lower limit d − 2 s	Upper limit d + 2 s
Thickness lateral	<3.5	0.04	0.047	−0.05	0.14
Thickness medial	<4.1	0.05	0.049	−0.05	0.15
Thickness patellar	<4.1	0.03	0.070	−0.11	0.17
Volume lateral	<3.8	30.87	33.25	−35.64	97.37
Volume medial	<4.0	23.63	45.66	−67.69	114.96
Volume patellar	<3.5	18.93	37.86	−56.78	94.64
Total volume	<3.8	19.4	104.2	−227.9	189.1

measurements of developed method with manual method, the Bland-Altman analysis [14] was carried out. The Bland-Altman assessment plots of cartilage thickness and volume measurements using two methods are shown in Fig. 5.

Form the results it was observed that deviations in measurements of developed method are in acceptable limits (d ± 2 s) for both thickness and volume quantification of cartilage. The developed method can be used as an alternative method of thickness and volume measurement of articular cartilage of knee joints. The developed image processing technique renders volume of the cartilage with reduced computational complexity (11.11 %) and hence takes less processing time for

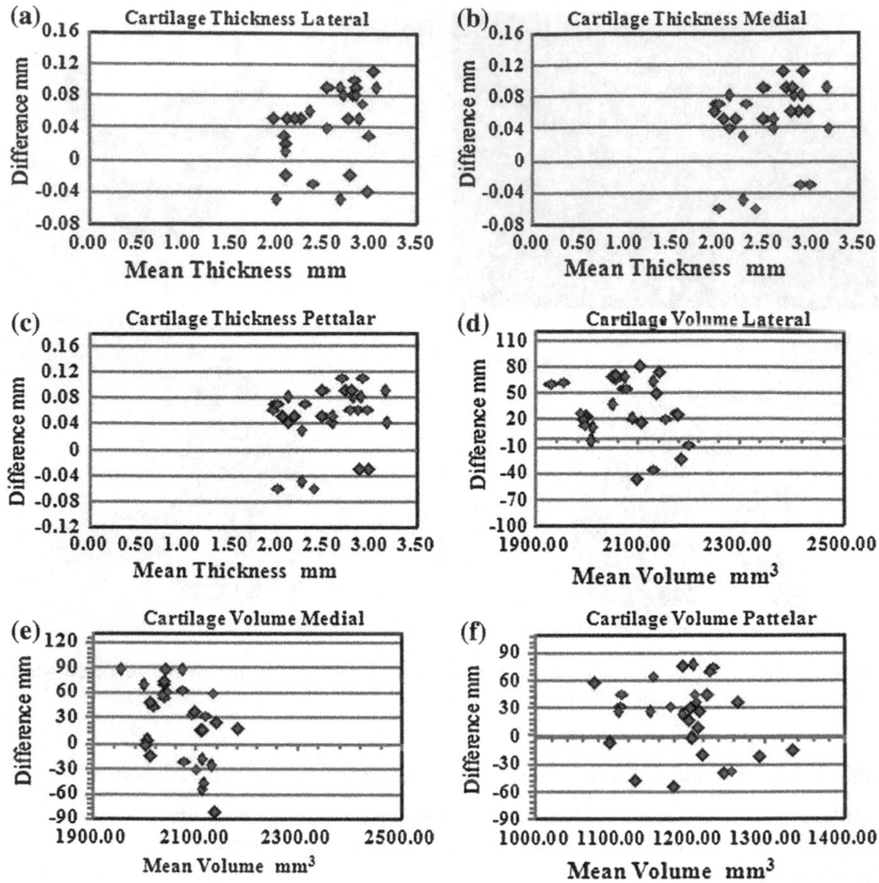

Fig. 5 **a**, **b**, **c** Bland-Altman plots of cartilage thickness measurements, **d**, **e**, **f** Bland-Altman plots of cartilage volume measurement

segmentation and 3D visualization cartilage from knee joint MRI. The developed image processing based method of quantification and 3D visualization of articular cartilage is useful in diagnosis and progressive study of knee joint affected by OA.

Acknowledgments Osteoarthritis Initiative (OAI), National Institute of Health, USA for providing knee MR images.

References

1. Levangie, P.K., Norkin, C.C.: Joint structure and function: a comprehensive analysis, 4th edn. F.A. Davis Company, Philadephia (2006)
2. Joshi, J., Kotwal, P.: Essentials of orthopedics and applied physiotherapy, 1st edn. Elsevier, New Delhi (2008)

3. Cohen, Z.A., McCarty, D.M., Kwak, S.D., Legrand, P., Fogarasi, F., Ciaccio, E.J., Ateshian, G.A.: Knee cartilage topography, thickness, and contact areas from MRI: in-vitro calibration and in-vivo measurements. Osteoarthritis Cartilage **7**, 95–109 (1999)
4. Cashman, P.M.M., Kitney, R.I., Gariba, M.A., Carter, M.E.: Automated techniques for visualization and mapping of articular cartilage in MR images of the osteoarthritic knee: a base technique for the assessment of microdamage and submicro damage. IEEE Trans. Nanobiosci. **1**, 42–51 (2002)
5. Poh, C.L., Kitney, R.I.: Viewing interfaces for segmentation and measurement results. In: Proceedings of 27th Annual Conference on IEEE Engineering in Medicine and Biology, Shanghai, China, pp. 5132–5135 (2005)
6. Folkesson, J., Dam, E.B., Olsen, O.F., Pettersen, P.C., Christiansen, C.: Segmenting articular cartilage automatically using a voxel classification approach. IEEE Trans. Medi. Imaging **26**, 106–115 (2007)
7. Kauffmann, C., Gravel, P., Godbout, B., Gravel, A., Beaudoin, G., Raynauld, J.-P., Pelletier, J. M., de Guise, J.A.: Computer-aided method for quantification of cartilage thickness and volume changes using MRI: validation study using a synthetic model. IEEE Trans. Biomed. Eng. **50**, 978–988 (2003)
8. Tang, J., Millington, S., Acton, S.T., Crandall, J., Hurwitz, S.: Surface extraction and thickness measurement of the articular cartilage from MR images using directional gradient vector flow snakes. IEEE Trans. Biomed. Eng. **53**, 896–907 (2006)
9. Anastasi, G., Bramanti, P., Di Bella, P., Favaloro, A., Trimarchi, F., Magaudda, L., Gaeta, M., Scribano, E., Bruschetta, D., Milardi, D.: Volume rendering based on magnetic resonance imaging: advances in understanding the three-dimensional anatomy of the human knee. J. Anat. **211**, 399–406 (2007)
10. Udupa, J.K., Herman, G.T.: 3D imaging in medicine, 2nd edn. CRC Press, Boca Raton (1999)
11. Baysal, O., Baysal, T., Alkan, A, Altay, Z., Yologlu, S.: Comparison of MRI graded cartilage and MRI based volume measurement in knee osteoarthritis. Swiss Med. Weekly **134**, 283–288 (2004)
12. Swamy, M.S.M., Holi, M.S.: Segmentation, visualization and quantification of knee joint articular cartilage using MR images. In: Swamy, P.P., Guru, D.S. (eds), Multimedia Processing, Communication and Computing Applications. LNEE, vol. 213, pp. 321–332. Springer, Berlin (2013)
13. Canny, J.: A computational approach to edge detection. IEEE Trans. Pattern Anal. Mach. Intell. **8**, 679–698 (1986)
14. Altman, D.G., Bland, J.M.: Measurement in medicine: the analysis of method comparison studies. Statistician **32**, 307–337 (1983)

Application of Particle Swarm Optimization and User Clustering in Web Search

Sumathi Ganesan and Sendhilkumar Selvaraju

Abstract User clustering is the most significant process in web usage mining. This approach tries to generate the clusters of users with the similar travels in the web search. Preprocessing is needed to extract the relevant data which is used for user clustering. Now a day Particle Swarm Optimization (PSO) approach is used in web search applications. This paper applies a Particle Swarm Optimization algorithm to web user grouping in association with the Open Directory Project (ODP) dataset. The experimental result shows that the effectiveness of Particle Swarm Optimization to be a suitable approach for web user clustering as compared to the K-means and DB-Scan clustering methods.

Keywords Web search · User clustering · Swarm intelligence · Particle swarm optimization · ODP taxonomy

1 Introduction

With the huge amount of web pages, there is a difficult for users to get the needed information from the web. So, the information retrieval is a very crucial task to provide the needed information to users with different interests [1]. The user's clustering will group users with similar web travels into clusters which is the most essential task of the web usage mining [2]. Users within the same cluster have the similar interests while users within the diverse clusters have dissimilar interests [3]. So, user clustering is applied to find user groups with similar path directions in the web search.

S. Ganesan (✉) · S. Selvaraju
Department of Information Science and Technology, CEG Campus, Anna University, Chennai 600025, Tamil Nadu, India
e-mail: sumisundhar@auist.net

S. Selvaraju
e-mail: ssk_pdy@yahoo.co.in

L.C. Jain et al. (eds.), *Computational Intelligence in Data Mining - Volume 2*, Smart Innovation, Systems and Technologies 32, DOI 10.1007/978-81-322-2208-8_39

Particle swarm optimization is a Swarm Intelligence (SI) based optimization algorithm which simulates the social activity of bird flocking [4]. Based on the similarity of user searching behavior and bird flocking behavior, this paper uses PSO algorithm for user grouping in association with the ODP dataset. ODP dataset contains the set of URLs which are grouped under various categories [5]. The ODP data set is used for the purpose of finding the categories for clustering the users. This paper attempts to group users based on their interests using PSO approach. It also proves that the PSO technique gives the improved results for web user clustering.

The organization of the paper is as follows. Section 2 provides a review of related works on the application of PSO in web search. Section 3 explains the data preprocessing to remove the unwanted data. Section 4 highlights the PSO based clustering of usage data in association with ODP taxonomy. Section 5 presents the performance evaluation that proves PSO algorithm gives better result for clustering the users. Section 6 gives the discussion about the experimental results. Section 7 concludes the paper and provides directions for future research.

2 Literature Work

Categorization of users is a major task in web search applications. Particle swarm optimization approach has been employed in the web search domain in different ways. They are summarized in this section.

Cagnina et al. [6] adopted a discrete PSO algorithm of CLUDIPSO for clustering short texts on the web. At each iteration this system grouped all particles quickly. They also introduced the CLUDIPSO* approach to reduce the problem with medium size corpora of short texts. These algorithms achieves a maximum of F-measure values. In 2014, Bakshi et al. [7] incorporated the PSO-based alpha estimate method to minimize the error in sparsity and the cold start problem of collaborative filtering approach. Experimental results exposes that this approach will improve the scalability and accuracy of the recommendation systems. The MOPSO-DAR technique [8] is used for searching variable length direct association rules. It improves and explores the indirect association among the users and the items for giving more precise recommendations.

The work in [9] proposes a two-stage term reduction approach that demonstrates a ranker combined with a heuristic search algorithm which is efficient to obtain an optimal subset of terms. It is also minimizing the computational cost. The studies presented in [10] introduced a new hybrid PSOK approach for clustering of web documents. An experiment shows that this approach generates much better results for clustering. In this paper, the PSO technique applies to evolving web user groups of related travels in the web search which is presented in the following section.

3 Data Preprocessing

Crawler data are the raw data. This data contains the largest amount of unwanted or irrelevant data. This data makes the user clustering process inefficient. Preprocessing of the crawler data is the initial step which is needed for the user clustering. The search data were collected by using a crawler. Here a user access data of 50 students were collected from the department of Information Science and Technology, Anna university web usage logs. These access logs contain user requests from 6 August 2012 to 9 November 2012 which is stored as a text file. The size of this data is 214 MB. Each access log entry of the crawler data consists of: URL, IP address, Time Spent (Seconds), Scrolling speed encountered and Average speed (Pixels/second).

Preprocessing is applied to this data for cleaning purpose that has removed the irrelevant data. For extracting the needed information of web user, following steps are done for data cleaning process [11]:

(a) *URLs with images, audio and video.* The information contains file names with GIF, JPEG, CSS, etc. can be found and these fields are eliminated.
(b) *Unsuccessful Paths.* The details with the failure status codes in the crawler data set can be found and these fields are eliminated.

Then from the cleaned data, the users are differentiated with the help of Internet Protocol (IP) addresses. Since the IP addresses of users are same, the users' web travelling activity was used to identify the user. This activity was identified by using the URL or search query. For this purpose, the title tag of each page visited by the 50 users was extracted from the access path and they were stored in a separate file. Figure 1 describes the preprocessed data which contain only needed information for user clustering.

Further, this data is applied for web user clustering.

4 Clustering of Usage Data

Grouping of users is a crucial task in web usage mining process. For user clustering, PSO approach and ODP taxonomy was applied on the preprocessed user data which is derived from the previous section. This is shown in Fig. 2.

4.1 ODP Taxonomy

ODP taxonomy contains the set of URLs which are grouped under various categories. There are a huge number of taxonomies on the web. The ODP dataset includes the top level categories of Adult, Arts, Business, Computers, Games,

URL	IP Address	Time Spent (Seconds)	Average Speed (Pixels/seconds)	Title
http://in.search.yahoo.com/search?p=Sachin+Tendulkar&s=Search&fr2=sb-top&fr=yfp-pmh-704&rd=r1	10.6.156.247	134.453	097.5986055226126	Sachin Tendulkar
http://www.iitm.ac.in/component/events/?eventid=1247&task=viewEvents	10.6.156.247	0.297	097.5986055226126	Admission Tests
http://www.mathworks.com/help/toolbox/fuzzy/fp310.html?nocookie=true#top_of_page	10.6.156.247	1.547	45.92024099048865	Clustering / Fuzzy Clustering
http://www.ebay.com/sch/i.html?_nkw=guitar&_sacat=619&fsb=Search&_odkw=fuzzy+neural+networks&_osacat=267&_trksid=p3286.c0.m270.l1313	10.6.156.247	381.641	350.2293617122634 pixels/second	guitar | eBay

Fig. 1 Preprocessed data

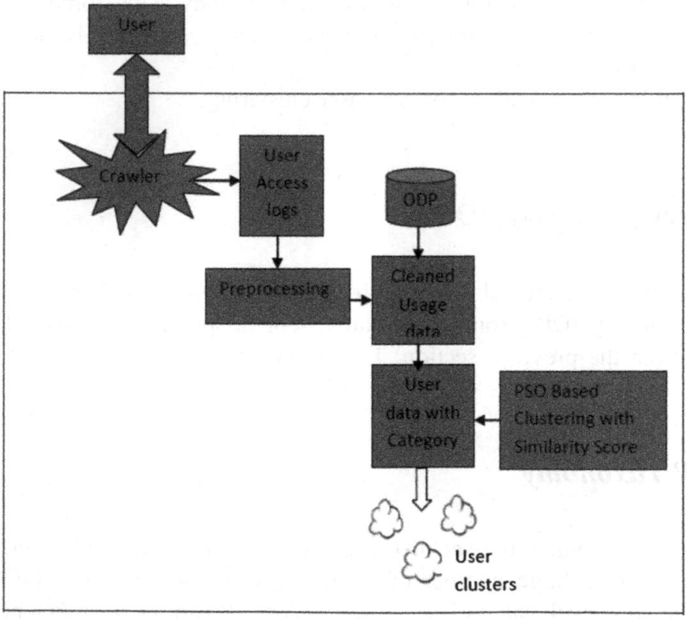

Fig. 2 System architecture

Health, Home etc. In this ODP dataset, the path categories of all the keywords are shown [5]. Dmoz [12] is an Open Directory Project (ODP). The ODP data set is used for the purpose of finding the categories for clustering the users.

The title tag extracted for the user differentiation is compared to the paths of these ODP category datasets and the corresponding relevant path is acquired for all the 50 users. Here ten categories of Education, Sports, Engineering, Computer, Entertainment, News, Travel, Finance, Music and others can be taken as the cluster or category names. Initially, all the paths of the ODP dataset were put in all the 10 categories randomly. Then by applying PSO technique, the best cluster or category of each user was identified by an iterative process.

4.2 PSO Based User Clustering

Particle Swarm Optimization (PSO) is an SI based optimization technique which was developed by Kennedy and Eberhart in the mid 1990s [4]. The optimization technique of Particle Swarm Optimization (PSO) that replicates the movement and flocking of birds. In this technique, each user is referred to as particle and it is initialized with a set of 50 users randomly in the searching space. Here the users' travels with a same rate of interest change (velocity) are considered for grouping. Each user changes its interest from one category to another based on the velocity updates. Here users are denoted by $U = (u_0, u_1, u_2, \ldots, u_{49})$ and the average rate of changes in interest is denoted by the velocity and its rate is $V = (1, 0.74, 0.9, \ldots, 0.82)$. The average rate of changes in interest is evaluated by using the cosine similarity measure. This Similarity score [13] is calculated by using the Eq. (1):

$$\mathrm{Sim\_Score(q, category_i)} = \frac{\sum_i s_{q,j} s_{i,j}}{\sqrt{\sum s_{q\cdot j}^2}\sqrt{\sum s_{i\cdot j}^2}}. \quad (1)$$

This similarity score is calculated between the category of search query of the users and category with the keywords of the travelled pages of the corresponding user sessions. These categories were found by using the ODP dataset. This similarity rate is in the range between 0 and 1. Here $s_{q,j}$ is a score of the query's category. For user u_0 its value is 0.5 and query's category is education. $s_{i,j}$ is a score of the list of categories and its value for user u_0 is 1. From these values the similarity score for the user u_0 is 1. These similarity scores of 50 users are used as the initial value of the velocity for users.

Fitness Function: Each user is evaluated by finding their fitness value. The following equation can be applied for fitness calculation:

$$\mathrm{Fitness} = \alpha * \gamma(\mathrm{S_I(j)}) + \beta * \frac{|\mathrm{N}| - |\mathrm{S}|}{|\mathrm{N}|}. \quad (2)$$

Where $S_l(j)$ is the subset of pages selected by lth user at jth iteration, $\gamma(S_l(j))$ is the clustering quality of the pages selected, |S| is the number of user sessions. |N| is the total number of pages visited. α and β are two parameters that correspond to the importance of clustering quality and subset length, with $\alpha \in [0,1]$ and $\beta = 1 - \alpha$. In our experiment α is set to 0.85 and β is set to 0.15. For user u_0 the calculated fitness value is 5.45 [14]. Based on this fitness function, the best value of the lth user is denoted by I_l. This is known as the individual best. Individual best value for user u_0 is 5.45. From the fitness values of all users the best fitness value is taken as the overall best. It is denoted by O and its initial value is 3.27. In each iteration, the velocity and the new category can be found by using the Eqs. (3) and (4):

$$V_l^{(j)} = \omega * v_l^{(j-1)} + a_1 * d_1\left(I_l - X_l^{(j-1)}\right) + a_2 * d_2\left(O - X_l^{(j-1)}\right). \tag{3}$$

$$X_l^{(j)} = X_l^{(j-1)} + V_l^{(j)}. \tag{4}$$

Where ω is the inertia weight that controls the impact of previous and current velocity of each user. Here the inertia value of 0.9 is taken. This value should be between 0.9 and 1.2 to improve the PSO's performance. d1 and d2 are two independently uniformly distributed random variables in the range [0,1] and d1 = d2 = 1. a1 and a2 are positive constant parameters called acceleration coefficients which control the maximum step size between successive iterations. For the values of a1 = a2 = 2 PSO generates the most optimized clustering result.

According to Eq. (3) the rate of interest change (velocity) of the user is found using three factors: previous rate of interest change of the user, the distance of the user from their best previous category and the distance from the best category of the entire population. Velocity and position of each user are found at each iteration by using the Eqs. (3) and (4) [15]. During the first iteration, the values of velocity and position for user u_0 are 0.94 and 4.94. These values can be changed in successive iterations. The user may change to another group by the value of rate of interest change (velocity) according to Eq. (4). The stop condition takes the maximum number of iterations or minimum error requirements [16]. Here the maximum number of iterations is set to 2,000. It generates the most optimized result for user clustering.

User Clustering is a process of grouping users into a set of clusters with their similarity in categories. For this purpose, PSO algorithm is applied and Fig. 3 shows the clustering of users with their categories.

Education, Sports, Engineering, Computer, Entertainment, News, Travel, Finance, Music and Others can be taken as categories. User-0 and User-31 comes under the category of Education. Similarly, other users are grouped into their respective categories.

PSO CLUSTERING RESULT
Education : User-0 : Reference/Education/Colleges_and_Universities/Asia/India/Tamil_Nadu/An na_University User-31 : Reference/Education/Colleges_and_Universities/North_America/United_Sta tes/Texas/Texas_A&M_University/Admissions/Conferences
Sports : User-1 : Business/Consumer_Goods_and_Services/Sporting_Goods/Cricket User-6 : Business/Consumer_Goods_and_Services/Sporting_Goods/Cricket User-10 : Business/Consumer_Goods_and_Services/Sporting_Goods/Cricket User-11 : Business/Consumer_Goods_and_Services/Sporting_Goods/Cricket User-18 : Business/Consumer_Goods_and_Services/Sporting_Goods/Cricket User-33 : Business/Consumer_Goods_and_Services/Sporting_Goods/Cricket User-36 : Business/Consumer_Goods_and_Services/Sporting_Goods/Cricket
Computer : User-2 : Computers/Software/Databases/Data_Mining User-4 : Computers/Artificial_Intelligence/Fuzzy User-9 : Computers/Software/Databases/Oracle User-12 : Computers/Multimedia/Music_and_Audio/Audio_Formats/MP3/News_and _Media/MP3.com User-13 : Computers/Software/Information_Retrieval/Web_Clustering

Fig. 3 User clustering with PSO algorithm

5 Performance Evaluation

The quality of the user clustering with PSO approach is evaluated by applying the measures of purity and entropy. These measures can be used to show the quality of the user clustering which can be described as follows [17]:

5.1 Purity

Purity of a cluster signifies the percentage of correctly clustered users. On the whole the purity of a clustering is a weighted sum of the purities of all clusters. Far value of purity gives the better clustering result.

Table 1 Performance comparison

Clustering algorithms	Purity	Entropy
PSO	0.975	0.643
DBSCAN	0.745	0.914
K-means	0.840	0.885

5.2 Entropy

Entropy of a cluster is defined as the degree to which each cluster consists of users of a single class. The final entropy of a clustering is a weighted sum of the entropy values of all clusters. Smaller entropy value shows the better clustering result. PSO technique can achieve a high value for the purity and low value for the entropy.

The same data set was given to DBSCAN and K-means algorithms. These algorithms were evaluated for purity and entropy values and the comparative analysis is given in Table 1.

```
====================================================
DBSCAN CLUSTERING RESULT
====================================================
Entertainment :
User-3 : Arts/Animation/Movies/Titles
User-8 : Arts/Movies/Titles/B/Bollywood-Hollywood
User-19 : Adult/Arts/Movies/Databases
User-20 : Adult/Arts/Movies/Databases
User-25 : Adult/Arts/Movies/Databases
User-28 : Adult/Arts/Movies/Databases
Music :
User-12 :
Computers/Multimedia/Music_and_Audio/Audio_Formats/MP3/New
s_and_Media/MP3.com
Others :
User-7 :
Kids_and_Teens/School_Time/Social_Studies/Geography/North_Am
erica/United_States/Government/Elections
User-14 : Recreation/Collecting/Themes/Politics
User-16 :
Health/Conditions_and_Diseases/Wounds_and_Injuries/Bites_and_St
ings/Mammals
User-22 : Health/Organizations/Grant-Making_Foundations
User-27 : Health/Conditions_and_Diseases/Skin_Disorders
User-37 :
Health/Conditions_and_Diseases/Sleep_Disorders/Centers/Australia
```

Fig. 4 User clustering with DBSCAN

From Table 1 it can be observed that PSO technique outperforms than the other two clustering techniques. The PSO optimization technique achieves the high value of purity and low value of entropy compared to other techniques. These results show that in terms of the cluster quality (High Purity and Low Entropy) [17].

6 Discussion

The result of PSO technique when compared with other clustering technique highlights vast differences. For example, User-12 is grouped under the category "Computer" using PSO algorithm, while the same user is categorized under the category "Music" in DBSCAN and Kmeans categorized the same user under the category of "Arts". This is shown in Figs. 3, 4 and 5 respectively.

The reason for this variation in grouping is due to the fact that, PSO technique utilizes the search path used by the user and accordingly computes the relevancy of web pages to user's search queries. Thus the experiment results highlighted in this paper proves that PSO technique performs well for user grouping in web search application than the other traditional clustering algorithm.

```
==============================================================
SIMPLEKMEANS CLUSTERING RESULT
Computer :
User-15 :
Computers/Artificial_Intelligence/Agents/Publications/Software_Engineeri
ng
User-23 : Adult/Computers/Internet/Searching/Directories
User-24 : Computers/Software/Globalization/Operating_Systems/Windows
User-29 : Computers/Data_Formats/Document
User-30 : Computers/Programming/Languages/C
User-34 : Computers/Data_Formats/Document/Text/Word
User-35 : Adult/Computers
Arts :
User-3 : Arts/Animation/Movies/Titles
User-8 : Arts/Movies/Titles/B/Bollywood-Hollywood
User-12 :
Computers/Multimedia/Music_and_Audio/Audio_Formats/MP3/News_an
d_Media/MP3.com
News :
User-32 : News/Breaking_News
Music :
User-5 : Arts/Music/Bands_and_Artists/3/311/Tablature/Guitar
User-21 : Arts/Music
User-26 : Arts/Music/Lyrics/Love_Songs
```

Fig. 5 User clustering with K-means

7 Conclusion and Future Work

This paper proposes an effective way for web user grouping in web search applications. The system consists of two stages. The first stage is the preprocessing stage, where unwanted log data were removed. In the second stage the cleaned data were mapped with ODP dataset to find the category. Then PSO approach was applied for clustering the users based on the similar travels. The performance of the PSO clustering algorithm is compared with the other clustering techniques using the cluster validity measures of purity and entropy. The performance evaluation highlights that the PSO algorithm gives better result for clustering the users based on their similar travels in web search. The results can further be refined by using ontologies. Recommendation and re-ranking will be done in the future based on the initial grouping done using PSO technique.

References

1. Zhu, Z., Chen, X., Zhu, Q., Xie, Q.: A GA based query optimization method for web information retrieval. Appl. Math. Comput. **185**, 919–930 (2007)
2. Vijaya Jumar, T., Guruprasad, H.S.: Clustering and visualization of web usage data using SOM and XML. Int. J. Emerg. Trends Technol. Comput. Sci. **2**(4), 23–29 (2013)
3. Han, J., Kamber, M.: Data Mining Concepts and Techniques, 2nd edn. Morgan Kaufmann Publishers, Burlington (2006)
4. Kennedy, J., Eberhart, R.: Particle swarm optimization. IEEE International Conference Neural Networks, pp. 1942–1948 (1995)
5. Zhang, D., Lee, W.S.: Web taxonomy integration using support vector machines. International Conference World Wide Web, pp. 472–481. ACM (2004)
6. Cagnina, L., Errecalde, M., Ingaramo, D., Rosso, P.: An efficient particle swarm optimization approach to cluster short texts. Inf. Sci. **265**, 36–49 (2014)
7. Bakshi, S., Jagadev, A.K., Dehuri, S., Wang, G.: Enhancing scalability and accuracy of recommendation systems using unsupervised learning and particle swarm optimization. Appl. Soft Comput. **15**, 21–29 (2014)
8. Tyagi, S., Bhardwaj, K.K.: Enhancing collaborative filtering recommendations by utilizing multi-objective particle swarm optimization embedded association rule mining. Swarm Evol. Comput. **13**, 1–12 (2013)
9. Karabulut, M.: Fuzzy unordered rule induction algorithm in text categorization on top of geometric particle swarm optimization term selection. Knowl.-Based Syst. **54**, 288–297 (2013)
10. Jaganathan, P., Jaiganesh, S.: An improved k-means algorithm combined with PSO approach for efficient web document clustering. International Conference Green Computing, Communication and Conservation of Energy, pp. 772–776 (2013)
11. Dixit, D., Kiruthika, M.: Preprocessing of web logs. Int J. Comput. Sci. Eng. **02**(07), 2447–2452 (2010)
12. ODP—Open Directory Project. Available online at: http://dmoz.org/
13. Wibowo, A., Handojo, A., Halim, A.: Application of topic based vector space model with wordnet. International Conferences on Uncertainty Reasoning and Knowledge Engineering, pp. 133–136 (2011)
14. Zahran, B.M., Ghassan, K.: Text feature selection using particle swarm optimization algorithm. World Appl. Sci. 69–74 (2009)

15. Alireza, A. (ed.): Combining PSO and K-means to enhance data clustering. International Symposium on Telecommunications, pp. 688–691 (2008)
16. PSO available online at: www.swarmintelligence.org/tutorials.php
17. Steinbach, M., Karypis, G., Kumar, V.: A comparison of document clustering techniques. KDD Workshop on Text Mining pp. 1–20 (2000)

Analyzing Urban Area Land Coverage Using Image Classification Algorithms

T. Karthikeyan and P. Manikandaprabhu

Abstract In this paper mainly deals with classifying high resolution image of an urban land cover area. It aims to extract the features like texture, shape, size and spectral information in feature extraction process. In this work, various classification algorithms particularly Naïve Bayes, IBk, J48 and Random Tree are implemented. The classification accuracy always depends on the effectiveness of the extracted features. Experimental results show that the accuracy performance obtained by Decision Tree based J48 algorithm is better than other classification algorithms.

Keywords Decision tree · Feature extraction · Image classification · K nearest neighbors · Naïve bayes · Urban land cover

1 Introduction

Satellite images are used in many applications such as astronomy and geosciences information systems. The images received from satellite contain huge amount of data to be deciphered and to be processed. But our human eye is insensitive to realize subtle changes in the image characteristics such as intensity, color, texture or brightness. So the manual human processing is not successful to retrieve the hidden treasures of information in the satellite image. The optimal solution is the processing of satellite images with digital computers.

Landsat imagery is most consistent and vital earth observation instrument. Landsat imagery is with quite high resolution earth observation data system, which is acquired through sensors. The satellite sensors acquire high reliability images of the planet surface in a systematic approach. Land cover is the physical material at the surface of the earth. It includes grassland, asphalt, trees, bare soil, concrete, etc.

T. Karthikeyan (✉) · P. Manikandaprabhu
PSG College of Arts and Science, Coimbatore, Tamil Nadu, India
e-mail: t.karthikeyan.gasc@gmail.com

P. Manikandaprabhu
e-mail: manipsgphd@gmail.com

L.C. Jain et al. (eds.), *Computational Intelligence in Data Mining - Volume 2*,
Smart Innovation, Systems and Technologies 32, DOI 10.1007/978-81-322-2208-8_40

Land cover information plays an important role in sustainable management, development and exploitation of resources, environmental protection, planning, scientific analysis, monitoring and modeling. The data become even more essential when there are rapid changes on the Earth's surface due to dynamic human activities as well as natural factors. Remotely sensed data in particular satellite images, among different advantages such as huge repetitive competencies, several spectral bands or multiple frequency/polarization are more effective tools for land cover mapping and they have been applied extensively for land cover monitoring and classification. Therefore, the challenging tasks are to understand the contribution of each dataset to select the most useful input features and to determine the combined datasets which can maximize the benefits of multi-source remote sensing data and to give the highest classification accuracy [1]. However, limited research has explored ways to determine variables from multi-source data in order to increase the classification accuracy [2].

This paper is organized as follows. Section 2 deals with the feature extraction process. Section 3 elaborates with classification algorithms like Naïve Bayes, K Nearest Neighbors, decision tree and random tree. Section 4 discusses with the experimental results and followed by conclusion. In Fig. 1 discussed with our overall system process.

2 Feature Extraction

It is the process of extracting image features into a distinguishable extent [3]. It is a group of features called image signature. It is carried out by using colors, textures and shapes or spectral. Extraction of knowledge on the built environment from remote sensing imaging may be an advanced task mainly because of them manifold

Fig. 1 Our system process

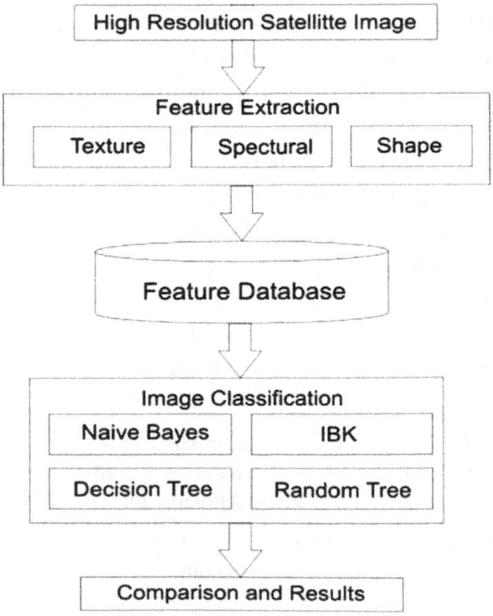

combinations of surface materials and also the diversity of size, shape and placement of the substance creating a distinctive image scene.

It is widely acknowledged that image domains outside spectral information such as temporal, geometrical or image texture fields should be utilized so as to tackle the difficulty of knowledge mining. In recent years, procedures that create use of the prolonged information content of image segments are well fit in the remote sensing community [4].

Image analysis can also be useful once interpreted to build the environment from Medium Resolution (MR) satellite images [5, 6] Wherever urban patterns completely different and primarily defined by the position of buildings, streets and open-spaces cannot sufficiently be represented by the spectral values of a single pixel.

When segments are used as the basic spatial data for analysis, a spacious volume of image structure will be formed to illustrate the objects of interest in an image. Normalized Difference Vegetation Index (NDVI) [7] contains spectral features like mean and standard deviation values of every image band, lowest and highest pixel values, mean and standard deviation of band index.

Texture features, which are taken into consideration of neighborhood relations between pixels, can be capably calculated over the parts. Commonly used in remote sensing analysis, second-order texture descriptors are derived from the Gray Level Co-occurrence Matrix (GLCM), like angular second moment, contrast, correlation or entropy. Haralick et al. [8] proposed GLCM scientific notation and the related second-order textural descriptors.

Shape descriptors will be usually divided into region-based and contour-based. Region-based features are mined from the shape region and contain area, eccentricity or moment features. Contour-based features will be extracted from the contour of an object and convexity, perimeter, compactness or major axis alignment. Reviews of shape representation and description techniques with mathematical explanations of the various mentioned descriptors can be found in [9, 10].

In order to reduce the dimensionality of the feature space and preselect the foremost vital features for a particular classification task, dataset and classifier, few feature selection methods are projected within the literature [11, 12]. Most generally established technique in remote sensing applications could be a manual feature selection with conventional knowledge exploration tools like histograms or scatterplots. This methodology needs better understanding of the classification strategy and also the aspect of the features in study. Conversely, increasing the quantity of features of manual methods becomes unfeasible and additional quantitative feature selection techniques are needed [13]. The feature databases information details in Table 1.

3 Image Classification

The classification algorithms are vital for classification. Appropriate selection of classification algorithms can result in a substantial improvement in the quality of the classification results. The classification algorithms are supported entirely diverse

Table 1 Details of feature database

Attribute name	Image feature description	Feature type
ClassName	Nine classes: asphalt, buildings, cars, concrete, grass, pool, shadows, soil and trees	Class
Area	Area in m^2	Size
Assym	Asymmetry	Shape
BordLngth	Border length	Shape
BrdIndx	Border index	Shape
Bright	Brightness	Spectral
Compact	Compactness	Shape
Con_GLCM	Contrast of NIR band—GLCM	Texture
Cor_GLCM	Correlation of NIR band—GLCM	Texture
Dens	Density	Shape
Ent_GLCM	Entropy of NIR band—GLCM	Texture
LW	Length/Width	Shape
Mean_G	Mean of green	Spectral
Mean_NIR	Mean of near infrared	Spectral
Mean_R	Mean of red	Spectral
NDVI	Normalized difference vegetation index	Spectral
Rect	Rectangularity	Shape
Round	Roundness	Shape
SD_G	Standard deviation of green	Texture
SD_NIR	Standard deviation of near infrared	Texture
SD_R	Standard deviation of red	Texture
ShpIndx	Shape index	Shape

concepts (statistical, nearest neighbor, tree-based). Common to all classifiers is that they generate use of training samples. Earlier, a training dataset is formalized in its vital structure. It is often used to train a classifier on each and every kind of input image type and feature space. The implemented classification algorithms are briefly represented in the following paragraphs.

3.1 Naive Bayes (NB)

A Naive Bayesian classifier based on Bayes theorem is a probabilistic statistical classifier [14]. Here, the term “naive” indicates conditional independence among features or attributes. The “naive” assumption greatly reduces computation complexity to a simple multiplication of probabilities. The major benefit of the Naïve Bayesian classifier is its speediness of use. This speediness occurs through the simplest algorithm among other classification techniques. Because of this simplicity, it can readily handle a data set with many attributes. In addition, the naive

Bayesian classifier needs only small set of training data to develop accurate parameter estimations. Because it requires only the calculation of the frequencies of attributes and attribute outcome pairs in the training data set [15, 16].

3.2 K Nearest Neighbors (IBk)

IBK is an instance-based learning approach. It is the K Nearest Neighbors (KNN) classification model. It segregates each and every unlabeled occurrence of k nearest neighbors. In multi-dimensional feature space, a set of training occurs and assigns a class value and confers to the majority of an exact class within this spanned region [17]. The basic practice of this algorithm is that each unseen instance is always compared with existing ones using a distant metric—most commonly Euclidean distance and the closest existing instance is used to assign the class for the test sample. Previously, a list of nearest neighbors is attained, the forecast based on voting (mass or distance-weighted) [18].

3.3 J48 (Decision Tree)

Quinlan [19] introduced a decision tree algorithm (known as Interactive Dichotomiser (ID3)) in 1979. C4.5 [17], as a successor of ID3, is the most widely-used decision tree algorithm. Decision tree classifiers construct a flowchart-like tree structure in a top to bottom, recursive, divide-and-conquer, manner. The Attribute Selection Method (ASM) is the key process in the construction of a decision tree. The ASM selects a splitting criterion that best splits the given records into each of the class attribute, whose sorting result is closest to the pure partitions by the class in terms of class parameter. Selected attributes become nodes in a decision tree. Decision trees, which consist of IF-THEN rules, are classification models. In other words, constructing a decision tree is the training step of classification. The major advantage to the use of decision trees is the class-focused visualization of data. This visualization is useful in that it allows users to readily understand the overall structure of data in terms of what it attribute mostly affects the class (the root node is always the most significant attribute to the class). Pruning is a useful technology even though there may be some minor errors in the trees generated under this method.

3.4 Random Tree

Random Tree has been proposed by Breiman and Cutler [20]. It is also called as random forests [18]. It is an ensemble of decision tree classifiers. Decision Tree uses a sequence of easy decisions supported outcome of sequential tests for class label assignment. Label assignments implemented at the leaf nodes of the tree are used in allocation strategy [21]. In Random Tree, the feature input vector is classified with apiece tree within the forest anywhere. The final prediction is based on a greater part of voting. The trees are trained with similar parameters. However with entirely different sets of training occurrences, the training sets are chosen by using a bootstrap method. A random subset of the variables is employed at apiece node of the trained trees to discover the most effective split. The size of subsets generated at individual node is predetermined for every nodes and trees by a training constraint. Inaccurate classification of every tree is predictable from the data, which are collected by the vectors to be unseen throughout the training part of the single tree classifiers by sampling with understudy.

4 Experimental Results

4.1 Study Area

The Landsat image of Coimbatore city in India, acquired in the month of May 2014 was used for the analysis. The investigations have been done for the area near Airport and PSGCAS campus located within latitude 11° 01 to 11° 30N and longitude 76° 95 to 77° 30 as it is shown in Fig. 2.

4.2 Parameters for Performance Evaluation

For performance evaluation, normally used accuracy measures (overall accuracy, precision and recall) were estimated using training (66 %) and testing (34 %) for each image type. Overall accuracy is defined like the total number of correctly classified instances divided by the total number of test instances. Accuracy measures were averaged for every classification and also both the training and testing data are used to train the classifiers under various environments. To construct its potentials to be evaluated, the virtual performance and reliability of the procedure

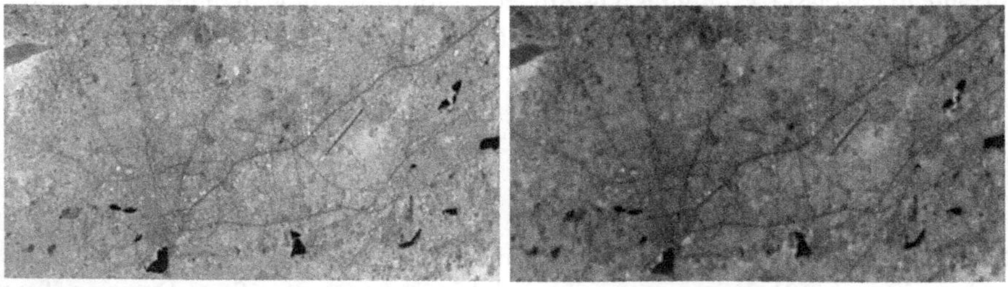

Fig. 2 Landsat image of Coimbatore, Tamil Nadu, India

for a given task should be arrived. Precision is the number of correctly classified positive examples divided by the number of examples labeled by the system as positive. Recall is the number of correctly classified positive examples divided by the number of positive examples in the data. It is also referred to as the sensitivity or true positive rate or producer's accuracy. It means the part of actual positives is classified as positives. Kappa statistics is computed as a degree of non random agreement among observers and/or measurement of a specific categorical variable.

4.3 Graphical Results

The graphical result shown in Figs. 3 and 4 shows the performance analysis related to Accuracy, Kappa Statistic, Precision and Recall of various classification algorithms.

This study has taken various classification methods and compared the results on the basis of accuracy, precision, recall. According to the graphical results shown in Figs. 3 and 4 and Table 2, it can be determined that Decision Tree based J48 algorithm has the highest accuracy percentage of 73.26 %.

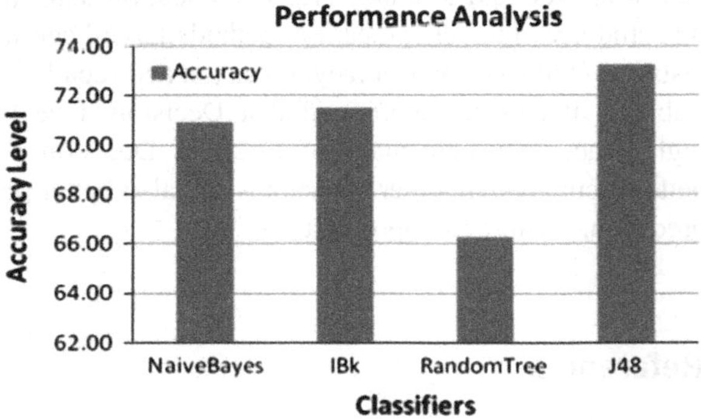

Fig. 3 Performance analysis based on accuracy

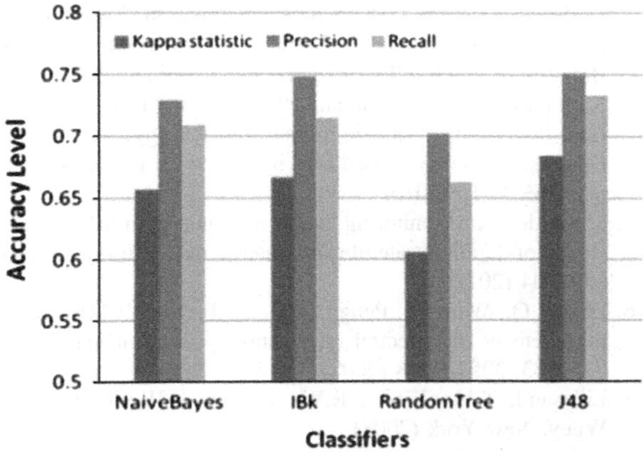

Fig. 4 Performance analysis based on kappa statistic, precision and recall

Table 2 Details of classification statistical results

Classifiers	Accuracy	Kappa statistic	Precision	Recall
NaiveBayes	70.93	0.6569	0.729	0.709
IBk	71.51	0.6666	0.0668	0.715
RandomTree	66.28	0.6056	0.702	0.663
J48	73.26	0.6845	0.751	0.733

5 Conclusion

The main goal of this research is to present an image classification strategy in the problem of urban land-cover data. This paper is just the preliminary attempt to explore probabilistic, nearest neighbor and tree based classification techniques, which are used for land-cover classification in remote sensing imagery. This work can also extend to some other classifier like function-based or Rule based classification and also to reduce the feature database using dimensionality reduction process.

The classification accuracy always depends on the effectiveness of the extracted features and classification algorithms used. Naïve Bayes, K Nearest Neighbors, Decision Tree and Random Tree classification algorithm are used in this study. In this study, various classification methods have been undertaken and compared the results on the basis of accuracy, precision and recall. According to Figs. 3 and 4 and Table 2, it can be determined that Decision Tree based J48 algorithm has the highest accuracy percentage of 73.26 %. Decision Tree based J48 algorithm has better accuracy than other algorithms and also better performance analysis results of precision, recall and kappa statistic.

References

1. Peddle, D.R., Ferguson, D.T.: Optimization of multisource data analysis: an example using evidence reasoning for GIS data classification. Comp. Geosci **28**, 45–52 (2002)
2. Li, G., Lu, D., Moran, E., Hetrick, S.: Land-cover classification in a moist tropical region of Brazil with Landsat TM imagery. Int. J. Remote Sens. **32**(23), 8207–8230 (2011)
3. Karthikeyan, T., Manikandaprabhu, P., Nithya, S.: A survey on text and content based image retrieval system for image mining. Int. J. Engg. Res. Tech. **3**(3), 509–512 (2014)
4. Blaschke, T.: Object based image analysis for remote sensing. ISPRS J. Photogramm. Remote Sens. **65**, 2–16 (2010)
5. Schneider, A.: Monitoring land cover change in urban and peri-urban areas using dense time stacks of Landsat satellite data and a data mining approach. Remote Sens. Environ. **124**, 689–704 (2012)
6. Zhang, Q., Wang, J., Peng, X., Gong, P., Shi, P.: Urban built-up land change detection with road density and spectral information from multi-temporal Landsat TM data. Int. J. Remote Sens. **23**, 3057–3078 (2002)
7. Lillesand, T.M., Kiefer, R.W., Chipman, W.: Remote Sensing and Image Interpretation. Wiley, New York (2008)

8. Haralick, R., Shanmugam, K., Dinstein, I.: Textural features for image classification. IEEE Trans. Sys. Man Cyber. **3**, 610–621 (1973)
9. Zhang, D., Lu, G.: Review of shape representation and description techniques. Pattern Recognit. **37**, 1–19 (2004)
10. Peura, M., Iivarinen, J.: Efficiency of simple shape descriptors. Asp. Vis. Form. 443–451 (1997)
11. Langley, P.: Selection of relevant features in machine learning. In: AAAI Fall Symposium, pp. 127–131 (1994)
12. Koprinska, I.: Feature selection for brain-computer interfaces. In: 13th Pacific-Asia Internatioanl Conference on Knowledge Discovery and Data Mining: New Frontiers in Applied Data Mining, pp. 106–117 (2009)
13. Witten, I.H., Frank, E., Hall, M.A.: Data Mining: Practical Machine Learning Tools And Techniques. Elsevier, Amsterdam (2011)
14. Karthikeyan, T., Thangaraju, P.: PCA-NB algorithm to enhance the predictive accuracy. Int. J. Eng. Tech. **6**(1), 381–387 (2014)
15. Dumitru, D.: Prediction of recurrent events in breast cancer using the Naive Bayesian classification. Ann. Univ. Craiova Math. Comp. Sci. Ser. **36**(2) (2009)
16. Karthikeyan, T., Thangaraju, P.: Analysis of classification algorithms applied to hepatitis patients. Int. J. Comp. Applns. **62**(5), 25–30 (2013)
17. Wu, X., Kumar, V., Quinlan, J.R., Ghosh, J., et al.: Top 10 algorithms in data mining. Knowl. Inf. Syst. **14**, 1–37 (2007)
18. Wieland, M., Pittore, M.: Performance evaluation of machine learning algorithms for urban pattern recognition from multi-spectral satellite images. Remote Sens. **6**, 2912–2939 (2014)
19. Quinlan, J.R.: C4.5: programs for machine learning. Morgan Kaufmann, Burlington (1993)
20. Breiman, L., Cutler, A.: Random forests. Available online: http://www.stat.berkeley.edu/users/breiman/RandomForests. Accessed on 4 June 2014
21. Pal, M., Mather, P.M.: An assessment of the effectiveness of decision tree methods for land cover classification. Remote Sens. Environ. **86**, 554–565 (2003)

CALAM: Linguistic Structure to Annotate Handwritten Text Image Corpus

Prakash Choudhary and Neeta Nain

Abstract In this paper, we report our effort in building a multi linguistic structure Cursive and Language Adaptive Methodology (CALAM) to create, annotate and validate linguistic dataset. CALAM provides a way for fetching and retrieval of information in a scientific and systematic manner through design and development of an annotated corpus of handwritten text image. It is a useful tool to annotate multi-lingual handwritten image dataset (Hindi, English, and Urdu etc.). The annotation is not limited with the grammatical tagging, but structural markup is also done. Annotation of handwritten text image is done in a hierarchical manner starting from handwritten form to segmented lines, words, and components. The component level markup is useful for finding strokes and list of ligatures in Urdu language. Along with a hierarchical access structure, CALAM provides the functionalities of Indexing, Insertion, Searching and Deletion of words and phrases in handwritten form. Apart from dataset fetching and retrieval it also automatically generates XML tagged file for each annotated handwritten text image for all dataset.

Keywords Corpus · Corpus annotation · Handwritten image · Handwritten document labeling

P. Choudhary (✉)
Department of Computer Science and Engineering, National Institute of Technology, Manipur, Imphal, India
e-mail: choudharyprakash87@gmail.com

N. Nain
Department of Computer Engineering, Malaviya National Institute of Technology, Jaipur, Jaipur, India
e-mail: neetanain@yahoo.com

L.C. Jain et al. (eds.), *Computational Intelligence in Data Mining - Volume 2*, Smart Innovation, Systems and Technologies 32, DOI 10.1007/978-81-322-2208-8_41

1 Introduction

Corpus is a structured and large data set collection of real life language, chosen to be as varied as possible to cover a large volume of distinct texts. Annotated Corpus play a significant role in any kind of computer-based linguistics research. Obvious application areas include all features of a language such as grammatical information, style of writing, syntax, lexicography, statistical analysis and testing, checking occurrences or validating linguistic rules within all branches of applied and theoretical linguistics territory. Computer process able corpora facilitates linguistic research, as electronically readable corpora have dramatically reduced the time needed to find a particular information in a Corpus.

In principle, Corpus Linguistics is an approach that aims at investigating nature of language and all its properties by analyzing large collections of text samples. This approach has been used in a number of research areas for ages: from descriptive study of a language, to language education, to lexicography, etc. It broadly refers to exhaustive analysis of any substantial amount of authentic, spoken/written text samples. In general, it covers large amount of machine-readable data of actual language that includes the collections of literary and non-literary text samples to reflect on both the synchronic and diachronic aspects of a language. The uniqueness of corpus linguistics lies in its way of using modern computer technology in the collection of language data, methods used in processing of language databases, techniques used for information retrieval, and strategies used to explore all kinds of language-related research and applications development activities.

This paper describe the structure of the CALAM, as embodied in the filename of the text. It describes the design and development of a multilingual Corpus of large volume of handwritten text and Unicode dataset. The paper explores all the features of a natural language: writing style, grammatical category information, and machine translation, aligned translation for sentence by sentence, phrase by phrase, or word by word. The Corpus is completely labelled for content information as well as content detection and supports the evaluation of systems like linguistic handwriting recognition, writer identification. The database was also experimented for the benchmarking of handwritten text recognition algorithms by generating a XML file of annotated handwritten text image.

The paper first introduces the experimental setup for the collection and distribution of data in a systematic manner, and then report on the process of information fetching and feeding in both handwritten text image and corresponding Unicode text format simultaneously on same screen. The paper is organized as: Sect. 2, details the related work with Corpus and their structure. Section 3, describes the collection and distribution of raw text sample. Section 4 is concerned with the annotation of dataset and the methodology to develop the Corpus. Section 5 describe the validation of dataset and Sect. 6 does the comparatively study of structure. Finally some conclusions are presented in Sect. 7.

2 Related Work

The Corpus methodology though started in 1990, is still a thrust area for linguistic domain. In the era of computation linguistic research Corpora has become a revolutionized area in all branches of linguistics. Standard datasets demand has been increased in recent years for different research area. This provides a platform for researchers to evaluate various linguistic techniques on the same dataset.

The most popular handwritten databases used for linguistic research is IRESTE [1]. It is a handwritten image database of French and English languages containing isolated words and characters without labelling. IAM [2, 3] is the first available annotated dataset of full length English sentences. The database is available in both printing and handwritten image format. Annotation has been done for line as well as word. MNIST [4] is a database of handwritten digits. CENPARMI [5] is the first Urdu handwritten Corpus which includes isolated digits and characters of Urdu language. CMATER [6] is a database of unconstrained handwritten Bangla and English mixed script document images. CENIP-UCCP [7] is the only available Corpus of Urdu handwritten text image with full sentences. Some other widely used databases in the field of handwriting recognition are CEDAR [8], NIST [9], ETL9 (Japan) [10], and PE92 (Korea) [11].

From the survey it was found that a less number of annotated handwritten datasets is available as compared to printed datasets. We do not find any handwritten dataset for Indian languages. CALAM provides a way to develop a large volume of data set for handwritten text images in Indic scripts and their corresponding labelled Unicode texts.

3 CALAM: Design of Experiment

Our Proposed methodology is to design and develop a Corpus consisting of full length text sentences. In order to be representative of all the phenomena across that language the corpus should contain a large verity of text samples. To maintain the balancing among the resources and building a corpus some salient features of the corpus suggested by Dash [12] in a general introduction about corpus linguistic has been considered.

3.1 Category Wise Distribution of Data

To cater to a huge vocabulary and maintain the balancing among the resources throughout the database, domain of the Corpus would be a data collection of six different categories. The categories are further divided into subcategories to capture maximum variance in word collection. List of category and denoted keyword of corresponding category and their subcategory for collection of data is as follows:

1. History—H
   (a) Indian History—IH
   (b) World History—WH
2. Literature—L
   (a) Poetry/Religion—PR
   (b) Gazals/Shyari—GS
   (c) Biography—BI
3. Science—S
   (a) Medical—ME
   (b) Physics—PH
   (c) Chemistry—CH
4. News—N
   (a) International—IN
   (b) National—NA
   (c) Sports—SP
5. Architecture—A
   (a) Rural Architecture—RA
   (b) Urban Architecture—UA
6. Politics—P
   (a) Central Government—CG
   (b) State Government—SG

The corpus development starts with the raw collection of data and ends with appropriate tagging and labelling of the collected text in the database. A form is designed to systematic collection of handwritten text images for corpus. The design and structure of the form is split into four parts as shown in Fig. 1, each part is separated from each other with a horizontal line.

For corpus understanding we are using Urdu language as an example. The same steps are done for Hindi and English scripts. The various parts of the form are organized as:

Part 1: The first part comprises the title for a language in the Dataset and a unique identification number. For example Urdu language and Indian History form 1 will have an id as (URD-H-IH-001). The id of the corresponding form is automatically updated or generated once a language and category/subcategory is selected.
Part 2: The second part of the form consists of 3–5 lines of printed text which is collected from various sources. Where a line can have around 60–70 words.
Part 3: Third part of the form is left blank where the writers replicate the printed text in his/her natural handwriting.

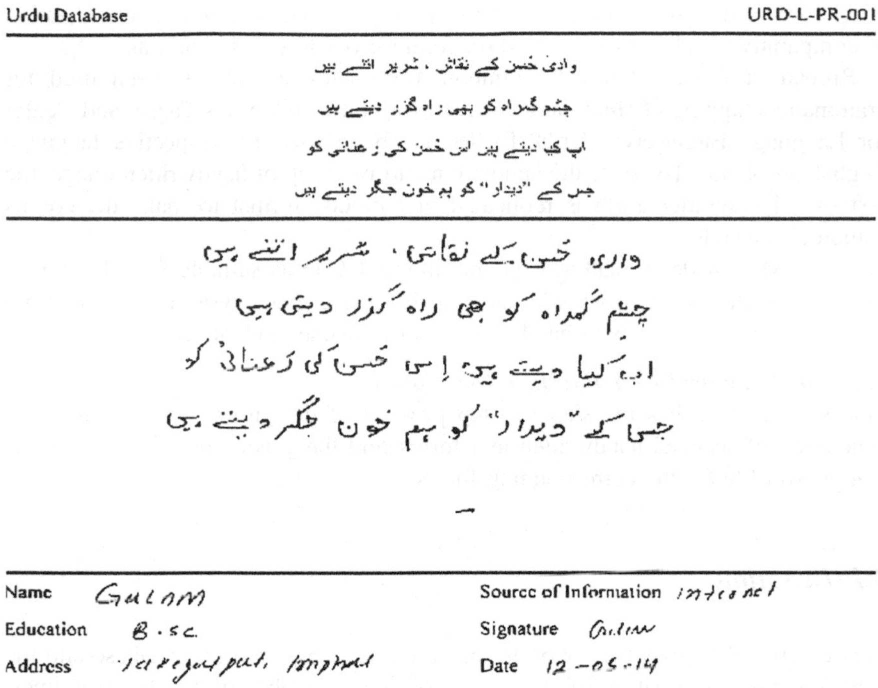

Fig. 1 Layout of a handwritten text form

Part 4: Fourth part of the form have six attributes: Name, Education, Address, Source of Information, Signature and date of form filling, writers can optionally provide this information.

The filled forms are scanned at the resolution of 600 dpi at a grey-level. The images were saved in PNG-format. Each form was completely scanned, including both the printed and handwritten text which can be useful for experiment on machine-printed and handwritten text separation. Transcription coding is stored in Unicode utf-general-ci UTF-8 which provide a fully support for Urdu Unicode.

4 Annotation

Annotation (labeling) of text in a Corpus is prerequisite for any Corpus development process. Annotation is a time consuming and error prone task, so it requires utmost care. Annotation makes a corpus useful to support machine learning and computational linguistic related research. Apart from pure text annotation, CALAM provides some additional linguistics features about the nature of language such as transcription of corpus to other language. Transcription of corpus across languages provide more fruitful resources in term of cross linguistic research, and realization of comparative study and helps in discovering cross linguistic variants.

Bureau of Indian Standards common Tagset framework has been used for grammatical tagging of Hindi language, British National Corpus Tagset and Center for Language Engineering Urdu Tagset has been used for respective language English and Urdu. To create the annotation and mark-up of handwritten image, the text of a handwritten form is replicated in Unicode format to make the corpus computer readable.

CALAM provides a platform for multilingual Corpus suitable for all types of linguistic related research where a large scale of fine grade systematic data across language is provided in both handwritten and machine readable format.

CALAM: Graphical User Interface Description

This section describes the step by step process of designing a corpus after the generation of scanned handwritten text forms and the generation of a meta-information xml file for the corresponding forms.

4.1 Insertion

This functionality gives an option to insert a new image into the database and the corresponding information of the image such as: number of handwritten lines, skew, transcription, date of creating and updating etc. The images are stored in a separate folder while the data is directed towards the respected fields in the database.

4.1.1 Auto-Indexing

The ID of each image inserted is automatically indexed according to the selected language, category and subcategory. The user selects the particulars language of form and the id field is appended accordingly.

For example: As shown in Fig. 1 form id is URD-L-PR-001 where URD as the language URDU, L is Category, PR is subcategory and last 3 digit is form number in respective subcategory.

Auto-indexing is also applicable for the ID of the segmented lines and words of handwritten image.

Each handwritten form will get a unique Id which is as follows:

(a) File name is language (2 bits)—category (3 bits)—subcategory (3 bits)—xxxxxxxx((8 bit form no). The index structure is shown in Fig. 3.
(b) So the Index of form id is 16 bits = Total number of forms (maximum) = 216 = 65,536.

(c) There can be maximum 8 categories so 2,048 forms in each category and there can be 8 subcategories hence 256 forms in each subcategory.
(d) CALAM can have a maximum of 4 languages with 16,384 handwritten forms in each language.

4.1.2 Handwritten Image Storage in Database

To achieve the consistency throughout the database all the handwritten text images stored in the database get the same unique id which was generated during the auto-indexing. The filename consists of a series of codes chained together with hyphen characters. The codes used for languages in the database are drawn from ISO-639. For example Urdu URD, English ENG and Hindi HIN etc.

A unique auto-indexing for word level.
The name of the image file in database is in the following format:
[Language code]-[Subcategory-id]-[FormNo].png
The name of the line_le is generally of the format:
[Language Code]-[Cat]-[Subcat]-[FormNo]-[Line No].png
The name of the word_le is generally of the format:
[Language Code]-[Cat]-[Subcat]-[FormNo]-[Line No]-[WordNo].png

4.1.3 Searching

This functionality facilitates the user to find any particular text/image using keyword, string, image ID, line ID and word ID in any category of the corpus.

Search generates output of all the database entries of images contained in the query string in the search box corresponding to the text fields in the database. For example, when an ID is inputted, the search engine searches the database for the particular ID, and displays the result. There is a link that redirects to the image that is searched. It helps to directly access the needed attributes and annotated information. It also highlights the searched query on the result page.

4.1.4 Deletion or Modification in Existing Form

This functionality gives an option for deletion and updating of an existing image from the database and it automatically deletes lines, words and components corresponding to that particular image.

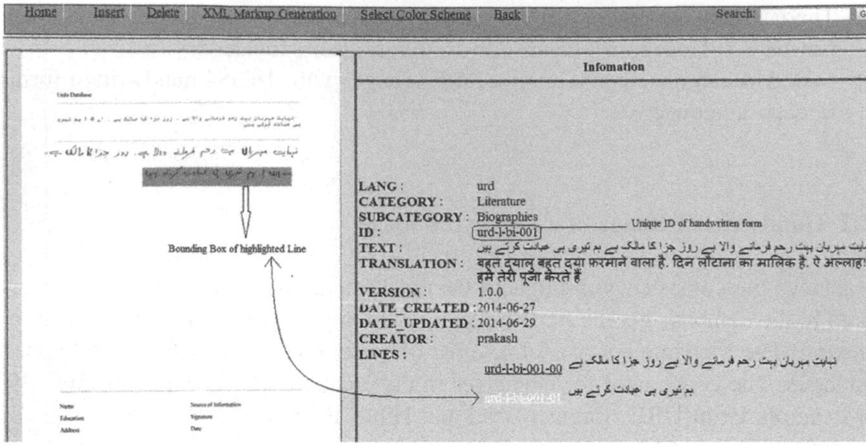

Fig. 2 A graphical interface of handwritten text image and transcription information

4.1.5 Bounding Box

Structural mark-up is done for lines, words and ligatures. This is required for proper benchmarking of segmentation techniques for handwritten text recognition. A bounding box is displayed over the component selected for better visibility, so that one can recognize the path of the image components. A mapping has been done between the window screen and the view port. When cursor points on unique id of lines, words and ligatures a rectangular bounding box appears on the corresponding line, word or ligature of the image in the view port. A sample of graphical user interface is shown in Fig. 2.

4.1.6 XML Mark-Up of Handwritten Image

Each image in the corpus is mark-up with meta-information as shown in Table 1. XML is the mostly used file format to generate ground-truth annotation results of corpus. CALAM provides the functionality of generating an XML file for each image in the database based on data entry description. The user can select an image to generate corresponding XML formatted file and then they can download or directly view the XML file for that image.

All the meta-information of handwritten text image, segmented line, word, component and writers name are formatted in XML format with ground truth data [13]. A sample of XML format meta-information which is automatically extractable from data entry procedure is presented in Fig. 3.

Standard Character Encoding Scheme (CES) under the guidelines of Text Encoding Initiative (TEI) is used for electronic data encoding and XML files meta-information.

Table 1 Meta information of XML formatted

Specification	Meta-information
Handwritten image	Image unique ID Image text Date of creation and updating Writer name Number of line Translation cross language Pixel coordinate of image skew
Segmented line	Line unique ID Pixel coordinates for bounding box Text, transcription Number of words
Segmented word	Word unique ID Pixel coordinates for bounding box Text, transcription Number of component First letter of word
Component	Pixel coordinates for bounding box

5 Validation

Data validation is the process of ensuring that a program operates on clean, correct and useful data. Validation checks are very important to maintain the integrity of any database structure. They are equipped in our corpus by using auto indexing and cross indexing routines, often called validation rules, validation constraints or check routines for correctness, meaningfulness, and security of data that are input to the system. In a nutshell, data needs to be validated at the same stage/level where it is most likely to be erroneous. The different types of data validation techniques applied such as: form level validation, search criteria validation, field level validation and range validation.

6 Comparative Study of CALAM

A comparatively analysis of CALAM with existing structures (Pix Labeler [14], GTLC [15], Truthing Tool [16], APTI [17]) for handwritten text image corpus is shown in Table 2.

As compared to above structures CALAM provides a facility to display handwritten text image file and transcription material of corresponding image on the same screen in a collaboration context. An Automatic XML file of meta-information can be generated on the basic of database entries. It also provides a structural markup information for benchmarking of handwritten text segmentation and OCR techniques.

```xml
<image>
<lang>urd</lang>
<category>Literature</category>
<subcategory>Biographies</subcategory>
<id>urd-1-bi-001</id>
<text>39</text>
<version>1.0.0</version>
<date_created>2014-06-27</date_created>
<date_updated>2014-06-29</date_updated>
<creator>prakash</creator>
<lines>
<id>urd-1-bi-001-00</id>
<text>
نہایت مہربان بہت رحم فرمانے والا ہے روز جزا کا مالک ہے ہم تیری ہی عبادت کرتے ہیں
</text>
<translation>
बहुत दयालु बहुत दया फ़रमाने वाला है. दिन लौटाना का मालिक है. ऐ अल्लाह! हम तेरी पूजा करते हैं
</translation>
<x>170</x>
<y>850</y>
<height>150</height>
<width>2100</width>
<character_width>140</character_width>
<slant>1</slant>
<threshold>.2</threshold>
<nowords>7</nowords>
```


Fig. 3 XML mark-up file of handwritten image

Table 2 Comparative analysis of CALAM

Structure	Language	Annotation	Output	Application
Pix labeler	English	Image	Text, XML	Labelling
GTLC	Chinese	Line, word, character	Image, ground truth text, XML	Annotation
APTI	Arabic	Image	Image, XML	Transcription
Truthing tools	English	Image	Image, text	Retrieval of text from image
CALAM	Multi-lingual	Image, line, word, component	Image, unicode text, auto generated XML file	Annotation, OCR transcription, benchmarking, corpus design

7 Conclusions

In this paper we have presented a structure for developing a standard corpus for various languages, currently it has been experimented with three languages Urdu, Hindi and English. The uniformity of structure provide an appropriate way for annotation of handwritten text images. We describes the data collection methodology and characteristics of the structure to manipulate the data for benchmarking tests.

Structure has a potential to provide researchers all the facilities for linguistic research on same platform. The aim of the structure is to build a resources that would provide ground truth annotation for handwritten text images. Structure would be rich source to design a large volume dataset for natural language processing related research. All the experiments from data collection to validating and XML meta-information has been done with utmost care by following standard procedures and rules. Forth part of handwritten text form having demographic information of writers could be used to train a system for automatic data fetching from handwritten form.

Acknowledgments This work is financially supported by Department of Science and Technology, Government of Rajasthan.

References

1. Christian, V.G., Michel, P., Stefan, K., Philippe, B.: The IRESTE on/off (IRONOFF) dual handwriting database. In: International Conference Document Analysis and Recognition, pp. 455–458 (1999)
2. Marti, U., Bunke, H.: A full English sentence database for off-line handwriting recognition. In: International Conference Document Analysis and Recognition, pp. 705–708 (1999)
3. Marti, U., Bunke, H.: The IAM-database: an English sentence database for off-line handwriting recognition. Int. J. Doc. Anal. Recogn. **5**, 39–46 (2002)
4. Lecun, Y., et al.: The MNIST database of handwritten digits (image) (1999)
5. Waqas, M., Lei, C., Nobile, N., Suen, C.Y.: A new large Urdu database for off-line handwriting recognition. In: International Conference Image Analysis and Processing. Lecture Notes in Computer Science, pp. 538–546, Italy (2009)
6. Sarkar, R., Das, N., Basu, S., Kundu, M., Nasipuri, M., Basu, D.: A database of unconstrained handwritten Bangla and English mixed script document image. Int. J. Doc. Anal. Recogn. (IJDAR) **15**, 71–83 (2012)
7. Raza, A., Siddiqi, I., Abidi, A., Arif, F.: An unconstrained benchmark Urdu handwritten sentence database with automatic line segmentation. In: International Conference Frontiers in Handwritten Recognition (ICFHR), pp. 491–496 (2012)
8. J. Hull: A database for handwritten text recognition research. IEEE Trans. Pattern Anal. Mach. Intell. 550–554 (1994)
9. Wilkinson, R., Geist, J., Janet, S., Grother, P., Burges, C., Creecy, R., Hammond, B., Hull, J., Larsen, N., Vogl, T., Wilson, C.: The first census optical character recognition systems: NISTIR 4912. The U.S. Bureau of Census and the National Institute of Standards and Technology, Gaithersburg (1992)
10. Saito, T., Yamada, H., Yamamoto, K.: On the data base ETL 9 of hand printed characters in JIS Chinese characters and its analysis. IEICE Trans. 757–764 (1985)
11. Dae-Hwan, K.I.M., Hwang, Y., Sang-Tae, P.A.R.K., Eun-Jung, K.I.M., Sang-Hoon, P.A.E.K., Sung-Yang, B.A.N.G.: Handwritten Korean character image database PE92. In: International Conference Document Analysis and Recognition (ICDAR), pp. 470–473 (1993)
12. Dash, N.S.: Corpus Linguistics: A General Introduction. CIIL, Mysore (2010)

13. Agrawal, M., Bali, K., Madhvanath, S.: UPX: a new XML representation for annotated datasets of online handwriting data. In: International Conference Document Analysis and Recognition (ICDAR), vol. 2, pp. 1161–1165, Seoul, Korea (2005)
14. Saund, E., Lin, J., Sarkar, P.: PixLabeler: user interface for pixel-level labeling of elements in document images. In: International Conference Document Analysis and Recognition (ICDAR), pp. 446–450, Spain (2009)
15. Yin, F., Wang, Q.-F., Liu, C.-L.: A tool for ground-truthing text lines and characters in off-line handwritten Chinese documents. In: International Conference Document Analysis and Recognition ICDAR, pp. 951–955 (2009)
16. Elliman, D., Sherkat, N.: A truthing tool for generating a database of cursive words. In: International Conference Document Analysis and Recognition, pp. 1255–1262, USA (2001)
17. Slimane, F., Ingold, R., Kanoun, S., Alimi, M.A., Hennebert, J.: A new arabic printed text image database and evaluation protocols. In: International Conference Document Analysis and Recognition (ICDAR), pp. 946–950, Spain (2009)

A Novel PSO Based Back Propagation Learning-MLP (PSO-BP-MLP) for Classification

Himansu Das, Ajay Kumar Jena, Janmenjoy Nayak, Bighnaraj Naik and H.S. Behera

Abstract Particle swarm optimization (PSO) is a powerful globally accepted evolutionary swarm intelligence method for solving both linear and non-linear problems. In this paper, a PSO based evolutionary multilayer perceptron is proposed which is intended for classification task in data mining. The network is trained by using the back propagation algorithm. An extensive experimental analysis has been performed by comparing the performance of the proposed method with MLP, GA-MLP. Comparison result shows that, PSO-MLP gives promising results in majority of test case problems.

Keywords Data mining · Classification · Particle swarm optimization · Genetic algorithm · Multilayer perceptron

1 Introduction

PSO is a meta-heuristic evolutionary optimization technique which can be directly applied in a continuous global space environment and was proposed by Kennedy and Eberhart [1, 2]. Due to its simplest algorithmic structure, less parameter use and

H. Das (✉) · A.K. Jena
School of Computer Engineering, KIIT University, Bhubaneswar, Odisha, India
e-mail: das.himansu2007@gmail.com

A.K. Jena
e-mail: ajay.bbs.in@gmail.com

J. Nayak · B. Naik · H.S. Behera
Department of Computer Science Engineering and Information Technology, Veer Surendra Sai University of Technology, Burla, Sambalpur 768018, Odisha, India
e-mail: mailforjnayak@gmail.com

B. Naik
e-mail: mailtobnaik@gmail.com

H.S. Behera
e-mail: mailtohsbehera@gmail.com

L.C. Jain et al. (eds.), *Computational Intelligence in Data Mining - Volume 2*,
Smart Innovation, Systems and Technologies 32, DOI 10.1007/978-81-322-2208-8_42

free from gradient use of an objective function, it is quite popular in the swarm intelligence community. The conceptual development of PSO is based upon the behavior of the swarms like bird, fish etc. Strong convergence and global optimization solution makes this algorithm more popular and attracts the attention of the researchers for solving various wide range of diversified problems [3–9].

A hybrid PSO algorithm (DNPSO) with the diversity enhancement and neighborhood search has been proposed by Wang et al. [10]. Neri et al. [11] addressed a novel optimization method called Compact PSO which employs the probabilistic representation and search logic of PSO but does not use either the position or velocity. A fuzzy based hybridized PSO is proposed by Valdez et al. [12] to improve the performance of the modular neural network. A general method for object detection in images based on deformable models and swarm intelligence/ evolutionary optimization algorithms is proposed by Ugolottia et al. [13]. SHIN and KITA [14] used the second global best and second personal best particles to improve the performance of the original PSO. Akay [15] proposed two swarm-intelligence-based global optimization algorithms, namely particle swarm optimization (PSO) and artificial bee colony (ABC), those have been applied to find the optimal multilevel thresholds. A Particle swarm optimization with grey evolutionary analysis for performing a global search over the search space with faster convergence speed has been introduced by Leu et al. [16]. Sun et al. [17] described a novel PSO technique called FEPSO, for reducing the number of fitness evaluations as well as computational cost. Imran et al. [18] gave an brief overview on various variants of PSO and analyzed on the parameters performance of PSO. PAN et al. [19] analyzed the performance of a standard PSO based on Markov chain by defining the state sequence of a single particle or swarm.

In this study, a PSO based back propagation trained multilayer perceptron is proposed for data classification. The proposed method has been tested with various benchmark datasets considered from UCI machine learning repository. The rest of the paper is organized in the following manner. Section 2 introduces some basic concepts like PSO and MLP. Section 3 describes the method of proposed work. Section 4 presents Experimental Setup and Result Analysis. Section 5 concludes our work with future scope.

2 Preliminaries

2.1 Particle Swarm Optimization

Particle swarm optimization (PSO) [1, 2] is a widely used stochastic based algorithm and it is able to search global optimized solution. Like other population based optimization methods, the particle swarm optimization starts with randomly initialized population for individuals and it works on the social behavior of particle to get the global best solution by adjusting each individual's positions with respect to global best position of particle of the whole population (Society). Each individual is adjusting by changing the velocity according to its own experience and by observing the position of the other particles in search space by use of Eqs. 1 and 2. Equation 1 is for social and cognition behavior of particles respectively where c1 and c2 are the constants in between 0 and 2 and rand (1) is random function which produces random number between 0 and 1.

$$V_i(t+1) = V_i(t+1) + c_1 * rand(1) * (lbest_i - X_i) + c_2 * rand(1) * (gbest_i - X_i) \quad (1)$$

$$X_i(t+1) = X_i(t) + V_i(t+1) \quad (2)$$

Basic steps of PSO can be visualized as:

Initialize the position of particles $V_i(t)$ (population of particles) and velocity of each particle $X_i(t)$.
Do

Compute fitness of each particle in the population.
Generate local best particles (LBest) by comparing fitness of particles in previous population with new population.
Choose particle with higher fitness from local best population as global best particle (GBest).
Compute new velocity $V_i(t+1)$ by using Eq. 1.
Generate new position $X_i(t+1)$ of the particles by using Eq. 2.

***While** (iteration <= maximum iteration OR velocity exceeds predefined velocity range);*

2.2 Multilayer Perceptron

MLP (Fig. 1) is the simplest neural network model which is consists of neurons called perceptron (Rosenblatt 1958). From multiple real valued inputs, the perceptron compute a single output according to its weights and non-linear activation functions. Basically MLP network is consists of input layer, one or more hidden layer and output layer of computation perceptron.

MLP is a model for supervised learning which uses back propagation algorithm. This consists of two phases. In the 1st phase, error (Eq. 4) based on the predicted outputs (Eq. 3) corresponding to the given input is computed (forward phase) and in the 2nd phase, the resultant error is propagated back to the network based on that weight of the network are adjusted to minimize the error (Back Propagation phase).

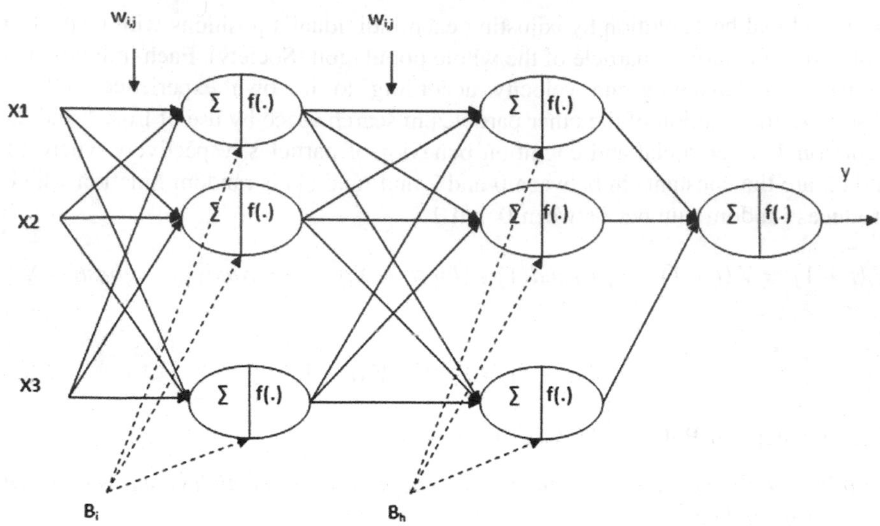

Fig. 1 MLP with input layer, single hidden layer and output layer

$$y = f\left(\sum_{i=1}^{n} w_i x_i + b\right) \quad (3)$$

where w is the weight vector, x is the input vector, b is the bias and $f(.)$ is the non-linear activation function.

$$\delta_k = (t_k - y_k)f(y_{in_k}) \quad (4)$$

where t_k and y_k is the given target value and predicted output value of input kth pattern and δ_k is the error term for kth input pattern.

The popularity of MLP increases among the neural network research community due to its properties like nonlinearity, robustness, adaptability and ease of use. Also it has been applied successfully in many applications [20–28]. It can also be applied to some advanced computing applications [29–32] of grid computing.

3 Proposed Method

In this section, we have proposed a PSO based back propagation learning-MLP (PSO-BP-MLP) for classification. Here basic concepts and problem solving strategy of PSO evolutionary algorithm is used to enhance performance of MLP classifier.

Algorithm PSO based Back Propagation Learning- MLP (PSO-BP-MLP) for classification

---

% Initialization of population

P = round (rand(n,(c-1)*(c-1)));

Where n is the number of weight-sets(chromosomes) in the population 'P' and c is the number of attributes in dataset (excluding class label).

% Initialization of velocity

v = rand (n,(c-1)*(c-1));

% NEURAL SETUP for MLP

Bh =(rand(c-1,1))'; % Bh: Bias of hidden layer.

Bo = rand(1); % Bo: Bias of output layer.

%PSO Iterations

Iter=0; **% Iter: Iteration**

while(1)

1. Selecting local best weight-sets (lbest) by comparing with weight-sets in previous population. If it is first iteration, then initial population (P) is considered to be local best(lbest). Otherwise, new 'lbest' population is formed by selecting best weight-sets from previous population (P) and current local best (lbest).

```
Iter = Iter+1;
If (Iter == 1)
        lbest = P;
else
        [lbest] = lbestselection (lbest, P, tdata, t);
end
```

2. Compute fitness of all weight-sets in local best 'lbest'. Each weight-sets are set individually in MLP and trained with training data 'tdata'. RMSE for each weight-sets are calculated with respect to target 't'. Based on RMSE, fitness of weight-sets are calculated by using 'fitfromtrain' procedure.

```
[F] = fitfromtrain (lbest, tdata, t);
```

3. Select a global best 'gbest' from local best 'lbest' based on their fitness(F) (calculated by using 'fitfromtrain' procedure) by using 'gbestselection' procedure.

```
[gbest] = gbestselection (P, F, rmse);
```

4. Compute new velocity 'vnew' from population (P), velocity (v), local best 'lbest' and global best 'gbest'by using 'calcnewvelocity' procedure.

```
vnew = calcnewvelocity (P, v, lbest, gbest);
```

5. Update next position by using current population (P) and new velocity (vnew).

```
P = P + vnew;
If (Iter == maxIter)
        break;
end
```

end

---

function [lbest]=lbestselection (lbest, P, tdata, t)

1. Compute fitness of all weight-sets in local best 'lbest' and previous population 'P'. Each weight-sets are set individually in MLP and trained with training data 'tdata' by using 'fitfromtrain'. RMSE for each weight-sets are calculated with respect to target 't'. Based on RMSE, fitness of weight-sets is calculated.

```
[F1] = fitfromtrain (lbest, tdata, t);
[F2] = fitfromtrain (P, tdata, t);
```

2. Compare fitness of weight-sets in lbest and P by comparing F1 and F2 , where F1 and F2 are fitness vector of lbest and P respectively. Based on this comparison, generate new lbest for next generation.

```
for i=1:1:number of weight-sets in P or lbset.
    if(F1(i,1)<=F2(i,1))
        lbest(i,:)=P(i,:);
    else
        lbest(i,:)=lbest(i,:);
    end
end
end
```

Function [F] = fitfromtrain(data, tdata, t)

Step-1: Repeat step 1 to 6 for all the weight-set 'w' in 'P'.

Error=0.

Here tdata is the training data without class label and it is considered to be tdata={x_1,x_2,....,x_L}, L is the number of input patterns.

Step-2: Repeat step- to step- for all the patterns $x_k = (x_{k1}, x_{k2}, \ldots, x_{kn})$ in given dataset tdata, where n is the number of attribute value in a single input pattern.

Step-3: Feed forward stage.

$$Z_{in_j} = Bh_j + \sum_{i=1}^{n} x_{ki} w_{ij}$$

Here Z_{in_j} represents j[th] hidden unit and i=1,2..n, where n is the number of input which are connected to a single hidden unit.

$y_{in_k} = B_{ok} + \sum_{j=1}^{m} Z_j Wh_{jk}$, Where j=1,2..m, m is the number of hidden unit connected to one output unit.

$y_k = f(y_{in_k})$, for k=1,2,..p, where p is the number of output unit in output layer and $f(.)$ is the sigmoid activation function and can defined as follows.

$$f(y_{in_k}) = \frac{1}{1 + e^{-y_{in_k}}}$$

Step-4: Calculation of error terms in hidden layer.

$\delta_k = (t_k - y_k) f(y_{in_k})$, for each output unit y_k, k=1,2...m. For each hidden unit Z_j, j=1,2...n, sums of delta input is computed as follows.

$$\delta_{in_j} = \sum_{i=1}^{m} \delta_k\, w_{jk}$$

Error terms are calculated as: $\Delta_j = \delta_{in_j} f_1(z_{in_j})$, where $f_1(.)$ is the activation function which is defined as:

$$f_1(z_{in_j}) = f(z_{in_j})\left(1 - f(z_{in_j})\right) = \frac{1}{1+e^{-z_{in_j}}}\left(1 - \frac{1}{1+e^{-z_{in_j}}}\right)$$

Finally error of the network in calculated as: Error = Error + Δ_j.

Step-5: if all input patterns in dataset are processed then goto step-6.

Else goto step-1.

Step-6: Compute RMSE based on total error of the network and compute fitness of the weight-set as F(i)=1/RMSE. , where F(i) is the fitness of ith weight-set in P.

end

```
function vnew=calcnewvelocity (P,v,lbest,gbest)
  for i = 1:1:number of row in population 'P'.
     for j = 1:1: of column in population 'P'.
        vnew(i,j) = v(i,j) + r1 * c1 * ( lbest(i,j) - P(i,j) ) + r2 * c2 * ( gbest(1,j) - P(i,j) );
     end
  end
end
```

4 Experimental Setup and Result Analysis

In this section, the comparative study on the efficiency of our proposed method has been presented. Benchmark datasets (Table 1) from UCI machine learning repository [33] and KEEL dataset repository [34] have been used for classification and the result of proposed PSO-MLP model is compared with MLP, GA-MLP based on Genetic Algorithm. Datasets information is presented in Table 1. Datasets have been normalized and scaled in the interval −1 to +1 using Min-Max normalization before training and testing is made. Classification accuracy (Eq. 5) of models has been calculated in terms of number of classified patterns are listed in Table 2.

If cm is confusion matrix of order m × n then, accuracy of classification is computed as:

$$Clasification\,Accuracy = \frac{\sum_{i=1}^{n}\sum_{j=1,\ i==j}^{m} cm_{i,j}}{\sum_{i=1}^{n}\sum_{j=1}^{m} cm_{i,j}} \times 100 \tag{5}$$

Table 1 Data set information

Dataset	Number of pattern	Number of features (excluding class label)	Number of classes	Number of pattern in class-1	Number of pattern in class-2	Number of pattern in class-3
Monk 2	256	06	02	121	135	–
Hayesroth	160	04	03	65	64	31
Heart	256	13	02	142	114	–
New thyroid	215	05	03	150	35	30
Iris	150	04	03	50	50	50
Pima	768	08	02	500	268	–
Wine	178	13	03	71	59	48
Bupa	345	06	02	145	200	–

Table 2 Performance comparison in terms of accuracy

Dataset	Accuracy of classification in average					
	MLP		GA-MLP		PSO-MLP	
	Train	Test	Train	Test	Train	Test
Monk 2	86.94648	85.27453	87.23734	87.85732	90.19375	92.44732
Hayesroth	83.48576	82.38657	85.43675	81.04653	88.38271	81.97365
Heart	82.84653	74.77453	85.44937	75.03645	86.23755	75.84651
New thyroid	92.03782	73.26876	92.74834	73.92756	93.02785	75.92784
Iris	90.87365	92. 15368	92. 56873	95. 93158	92. 51736	93. 36158
Pima	73.73645	73.88647	76.82642	77.38275	78.17464	78.28746
Wine	80.69731	80.87294	88.93645	77.37565	90.75389	91.77319
Bupa	68.66251	69.82637	70.97485	71.27459	70.27841	71.21465

4.1 Parameter Setting

During simulation, c1 and c2 constants of PSO has been set to 2 and used throughout the experiment. In MLP, one input layer, one hidden layer and one output layer for the neural network has been set during training and testing.

5 Conclusion

PSO is a popular interesting swarm intelligence technique which is able to find both the global minima and maxima for complex problems. This paper describes the PSO based back propagation neural network for classification of various benchmark datasets. The comparison of performance analysis of the results indicates that the proposed method gas better classification accuracy than the other defined methods. But many researchers have found out some of the major limitations of PSO like: slow convergence, convergence at local minima and large search space etc. In each iteration, the time complexity is more due to the search of the weakest performed particle. In future, our work may extend in this interest by better adjustment of all the parameters along with hybridization of some higher order neural network.

References

1. Kennedy, J., Eberhart, R.: Particle swarm optimization. In: Proceedings of the 1995 IEEE International Conference on Neural Networks, vol. 4, pp. 1942–1948 (1995)
2. Kennedy, J., Eberhart, R.: Swarm intelligence morgan kaufmann, 3rd edn. Academic Press, New Delhi (2001)
3. Cai, J., Pan, W.D.: On fast and accurate block-based motion estimation algorithms using particle swarm optimization. Inf. Sci. **197**(15), 53–64 (2012)
4. Du, W., Li, B.: Multi-strategy ensemble particle swarm optimization for dynamic optimization. Inf. Sci. **178**(15), 3096–3109 (2008)
5. Zhang, Y., et al.: A bare-bones multi-objective particle swarm optimization algorithm for environmental/economic dispatch. Inf. Sci. **192**(1), 213–227 (2012)
6. Chuanwen, J., Bompardb, E.: A hybrid method of chaotic particle swarm optimization and linear interior for reactive power optimization. Math. Comput. Simul. **68**(1), 57–65 (2005)
7. Li, Y., et al.: Dynamic optimal reactive power dispatch based on parallel particle swarm optimization algorithm. Comput. Math Appl. **57**(11–12), 1835–1842 (2009)
8. Li, Y., et al.: Optimal reactive power dispatch using particle swarms optimization algorithm based pareto optimal set. Lect. Notes Comput. Sci. **5553**, 152–161 (2009)
9. Naik, B., Nayak, J., Behera, H.S.: A Novel FLANN with a hybrid PSO and GA based gradient descent learning for classification. In: Proceedings of the 3rd International Conference on Frontiers of Intelligenct Computing (FICTA). Advances in Intelligent Systems and Computing 327, vol. 1, 745–754 (2015). doi:10.1007/978-3-319-11933-5_84

10. Wang, H., et al.: Diversity enhanced particle swarm optimization with neighborhood search. Inf. Sci. **223**, 119–135 (2013)
11. Neri, F., et al.: Compact particle swarm optimization. Inf. Sci. **239**, 96–121 (2013)
12. Valdez, F., et al.: Modular neural networks architecture optimization with a new nature inspired method using a fuzzy combination of particle swarm optimization and genetic algorithms. Inf. Sci. **270**, 143–153 (2014)
13. Ugolottia, R., et al.: Particle swarm optimization and differential evolution for model-based object detection. Appl. Soft Comput. **13**, 3092–3105 (2013)
14. Shin, Y., Kita, E.: Effect of second best particle information for particle swarm optimization. Procedia Comput. Sci. **24**, 76–83 (2013)
15. Akay, B.: A study on particle swarm optimization and artificial bee colony algorithms for multilevel thresholding. Appl. Soft Comput. **13**, 3066–3091 (2013)
16. Leu, M.-S., et al.: Particle swarm optimization with grey evolutionary analysis. Appl. Soft Comput. **13**, 4047–4062 (2013)
17. Sun, C., et al.: A new fitness estimation strategy for particle swarm optimization. Inf. Sci. **221**, 355–370 (2013)
18. Imran, M., et al.: An overview of particle swarm optimization variants. Procedia Eng. **53**, 491–496 (2013)
19. Pan, F., et.al. : Analysis of standard particle swarm optimization algorithm based on markov chain. Acta Automatica Sinica. vol. 39, no. 4. April (2013)
20. Mahyar, H., et al. : Comparison of multilayer perceptron and radial basis function neural networks for EMG-based facial gesture recognition. In: Proceedings of the 8th international conference on robotic, vision, signal processing and power applications. Springer, Singapore (2014)
21. Ndiaye, A., et al.: Development of a multilayer perceptron (MLP) based neural network controller for grid connected photovoltaic system. Int. J. Phys. Sci. **9**(3), 41–47 (2014)
22. Roy, M., et al.: Ensemble of multilayer perceptrons for change detection in remotely sensed images. Geoscience and remote sensing letters. IEEE11.1, pp. 49–53 (2014)
23. Hassanien, A., et al.: MRI breast cancer diagnosis hybrid approach using adaptive ant-based segmentation and multilayer perceptron neural networks classifier. Appl. Soft Comput. **14**, 62–71 (2014)
24. Aydin, K., Kisi, O.: Damage detection in Timoshenko beam structures by multilayer perceptron and radial basis function networks. Neural Comput. Appl. **24**(3–4), 583–597 (2014)
25. Velo, R., et al.: Wind speed estimation using multilayer perceptron. Energ. Convers. Manag. **81**, 1–9 (2014)
26. Lee, S.: Choeh. J. Y. : Predicting the helpfulness of online reviews using multilayer perceptron neural networks. Expert Syst. Appl. **41**(6), 3041–3046 (2014)
27. Azim, S., Aggarwal, S. : Hybrid model for data imputation: Using fuzzy c means and multi layer perceptron. 2014 IEEE International Advance computing conference (IACC). IEEE (2014)
28. Chaudhuri, S., et al.: Medium-range forecast of cyclogenesis over North Indian Ocean with multilayer perceptron model using satellite data. Nat. Hazards **70**(1), 173–193 (2014)
29. Das, H., Mishra, S.K., Roy, D.S.: The topological structure of the Odisha power grid: a complex network analysis. IJMCA **1**(1), 012–016 (2013)
30. Das, H., Roy, D. S. : A grid computing service for power system monitoring. Int. J. Comput. Appl. 62 (2013)
31. Das, H., et.al. : The complex network analysis of power grid: a case study of the West Bengal power network. In: Intelligent Computing, Networking and Informatics, pp. 17–29. Springer India (2014)
32. Das, H., et.al. : Grid computing-based performance analysis of power system: a graph theoretic approach. Intell. Comput. Commun. Dev. 259–266 (2015)

33. Bache, K., Lichman, M.: UCI machine learning repository. University of California, Irvine. CA. School of Information and Computer Science. (2013) [http://archive.ics.uci.edu/ml]
34. Alcalá-Fdez, J., et al.: KEEL data-mining software tool: data set repository. integration of algorithms and experimental analysis framework. J. Multiple-Valued Logic Soft Comput. **17** (2–3), 255–287 (2011)

Quantum Based Learning with Binary Neural Network

Om Prakash Patel and Aruna Tiwari

Abstract In this paper, a quantum based binary neural network learning algorithm is proposed for solving two class problems. The proposed method constructively forms the neural network architecture and weights are decided by quantum computing concept. The use of quantum computing optimizes the network structure and the performance in terms of number of neurons at hidden layer and classification accuracy. This approach is compared with MTiling-real networks algorithm and it is found that there is a significant improvement in terms of number of neurons at the hidden layer, number of iterations, training accuracy and generalization accuracy.

Keywords Quantum computing · Qubit · Binary neural network · Qubit gates

1 Introduction

In recent years, neural network has attracted attention of many researchers not only in field of computer science but also in biomedical, economics, mathematics and many more field. This technique is used to solve wide range of problems like data mining, clustering, classification, series prediction etc. But still finding an optimal neural network for all kinds of problems is an open area of research. To find an optimal result, it depends on several parameters of networks such as network architecture, training time, testing time, connections, number of neurons at hidden layer and number of hidden layers [1]. For optimizing the parameters like connection weight Lu et al. [2] proposed a quantum based algorithm. This algorithm works on the concept of quantum computing for optimizing connections and

O.P. Patel (✉) · A. Tiwari
Department of Computer Science and Engineering, Indian Institute of Technology Indore, Indore, India
e-mail: Phd1301201003@iiti.ac.in

A. Tiwari
e-mail: artiwari@iiti.ac.in

L.C. Jain et al. (eds.), *Computational Intelligence in Data Mining - Volume 2*,
Smart Innovation, Systems and Technologies 32, DOI 10.1007/978-81-322-2208-8_43

weights in MLP based neural network structure. Designing a neural network architecture using MLP requires much iteration for convergence and also causes under-fitting and over-fitting problem.

There are many ways available through which neural network architecture can be formed constructively during learning. One such algorithm based on binary neural network is proposed by Kim and Park named as ETL algorithm.

It classifies data on the basis of hyperplane generation and works for two class problem. Chaudhari et al. proposed mETL which uses geometric concepts of ETL for classification of multiclass data. It start learning by selecting any random sample as core vertex. Thus network formed after learning depends on the core selection; therefore solution is not unique [3]. In this paper, an algorithm is proposed for binary neural network using quantum computing concept. This algorithm constructively forms network structure by updating network weights using quantum processing concept. This paper is organized as follows; Sect. 2 describes literature review. Section 3 describes required preliminaries briefly. Section 4 describes details of proposed methodology. Section 5 is presented with the experimental work which shows the performance of proposed algorithm. Then it is compared with MTiling-real networks algorithm [4]. Section 6 is presented with the concluding remarks.

2 Literature Review

Quantum inspired computing concept is used by many researchers for the learning algorithms [5–10] since last decade. For the classical computers, Narayanan and Moore in the 1996 introduced quantum inspired learning algorithm for classical computers. In 2000 and 2002 Han and Kim [5] proposed quantum inspired algorithm with genetic concept for solving knapsack problem with new termination criteria. In 2006 Mori et al. [6] proposed qubit inspired neural network towards its practical applications for image processing in night vision cameras. In 2008 de Araujo et al. [7] proposed a quantum-inspired intelligent hybrid method for stock market forecasting based on neural network method.

Binary neural network attracted attention of many researchers because implementation of this algorithm is more close to computer system. In 2003 Narendra et al. proposed binary neural network training algorithms based on linear sequential learning [8]. There are many BNN algorithms developed which work faster. One of such is proposed by Wang et al. a fast modified constructive- covering algorithm for binary multi-layer neural networks in 2006. There are many parameters are taken to measure the performance of neural network. One of the important performance issue is the number of hidden layer neurons, this issue is worked out for binary neural network by deriving bounds on the number of hidden layer neurons in [8]. In order to more efficiently realize boolean function by neural network Zhang et al. [9] proposed binary higher order neural networks for realizing boolean functions in 2011.

3 Preliminaries

Here, we are proposing a quantum based binary neural network learning algorithm for finding the network structure. Network formed consist of three layers input layer, hidden layer and output layer. Let $X = (x_1, x_2, x_3..., x_n)$ denotes the input samples which are in binary form, where n is the number of input samples. Let $W = (w_1, w_2, w_3..., w_n)$ be a connection weight which are updated by using quantum concept, quantum gates [5]. In the proposed learning algorithm, we make use of sigmoid activation function for neuron which is given as follows.

$$f(net) = \frac{1}{1 + e^{-net}} \quad (1)$$

where,

$$net = \sum_{i=1}^{n} w_i \times x_i \quad (2)$$

Quantum concept makes use of quantum bits called as qubits. It is smallest unit of data information in quantum computing. It is defined by the pair (σ, β)

$$\psi = \alpha|0\rangle + \beta|1\rangle \quad (3)$$

where α^2 gives the probability that qubit will be found in 0 state and β^2 gives the probability that qubit will be found in 1 state. A qubit may be in any state either in 0 in the 1 or in linear super position of both 0 and 1. A qubits individual as a string of m qubits is defined as:

$$\psi = \left\langle \begin{matrix} \sigma_1 \\ \beta_1 \end{matrix} \cdots \begin{matrix} \sigma_n \\ \beta_n \end{matrix} \right\rangle \quad (4)$$

where

$$\sigma_i^2 + \beta_i^2 = 1, \quad i = 1, 2, 3..., n \quad \text{and} \quad 0 \leq \alpha \leq 1, \ 0 \leq \beta \leq 1$$

Qubit represent any superposition of 0's and 1's state. For example two-qubit system would perform the operation on 4 values, and a three-qubit system on eight thus n qubit will perform operation of 2^n values. A quantum bit individual contains a string of q quantum bits. Let's take an example of two quantum bits, which are represented as follows:

$$\psi = \left\langle \begin{array}{c|c} 1/\sqrt{2} & 1/\sqrt{2} \\ 1/\sqrt{2} & 1/\sqrt{2} \end{array} \right\rangle \quad (5)$$

$$\psi = (1/\sqrt{2} \times 1\sqrt{2})\langle 00\rangle + (1/\sqrt{2} \times 1\sqrt{2})\langle 01\rangle + (1/\sqrt{2} \times 1\sqrt{2})\langle 10\rangle + (1/\sqrt{2} \times 1\sqrt{2})\langle 11\rangle \tag{6}$$

These quantum bits are used later on for representing the quantum weight are as follows.

$$Q^g = (q_1^g, q_2^g, q_3^g \ldots q_n^g); \tag{7}$$

where $q_i^{g,i} = [y_1^{g,i}, y_2^{g,i}]; 0 \leq y_1^{g,i} \leq 1$ and $0 \leq y_2^{g,i} \leq 1$; $i = 1, 2, 3 \ldots n, n$ is number of input sample; $g = 1, 2, 3 \ldots m, m$ is number of iteration required for finding an optimal weight. These quantum bits are updated using quantum gates [5].

Observation Process: Classical computers work on bits therefore it is necessary to convert quantum bits in terms of classical bits 1 and 0. To generate classical bit from qubits a random vector $r = [r_1, r_2, r_3 \ldots r_n]$ is generated where $r_i = (r_1^i, r_2^i)$ and $0 \leq r_{1,2}^i \leq 1$ following operation are performed.

$$if(r_i \leq (\alpha^2)) \text{ then,} \quad s_b = 1 \quad \text{and} \quad s_b = 0, \quad \text{where b} = 1, 2. \tag{8}$$

Here, let us take vector $S^g = [S_1^g, S_2^g, S_3^g \ldots, S_n^g]$, where $S_i^g = [s_1 s_2]$ correspond to qubit vector Q_g. String S_i is of two bit length which corresponding to the quantum weight space. This is then mapped to the final weights in first iteration. Further, the weights for the neuron are selected using Gaussian random generator. To select weights from binary values here, the Gaussian random generator have been used with mean value μ_i^g and variance σ_i^g. Therefore, weight vector is now formalized as follows:

$$W^g = (w_1^g, w_2^g, w_3^g, \ldots, w_n^g), \quad \text{where } w_i^g = N(\mu_i^g, \sigma_i^g) \tag{9}$$

Separability Parameter (λ): To know whether an input sample lie within a particular class, a separability parameter λ is used. This parameter is decided by taking at least 10 % of samples from both the classes and observes values of $f(net)$ by using Eqs. 1 and 2. In our case, 20 instances of class A and 38 instance of class B has been taken for observation. For example if values generated by $f(net)$ for class A lies in range (1.4–2.5) and for class B lies in range (0.3–1.5), then the value of λ can be taken as 1.3.

4 Proposed Approach

A Quantum Based Learning for forming Binary Neural Network algorithm (QL-BNN) is proposed. This approach makes use of quantum concept for constructing neural network. Network structure formed is of three layers consist of input, hidden and output layer, which accepts binary form of inputs and generate binary form of outputs. The approach works in two phases, hidden layer learning and output layer learning.

4.1 Hidden Layer

Hidden layer training starts by taking one neuron first. Weights of a neuron Defined by quantum values $Q^1 = (q_1^1, q_2^1, q_3^1 \ldots q_n^1)$; as shown in Eq. 7. These quantum values are initialized by 0.707. (this is equal probability of any bit in 0 or 1 state.) After initialization of quantum values as weight for first neuron, observation process starts. A random vector $r = [r_1, r_2, r_3, \ldots r_n]$, is generated by using vector $S^1 = [S_1^1, S_2^1, S_3^1 \ldots, S_n^1]$, is generated which is classical bit representation of quantum bits used for first neuron. With the help of vector S^1 and Gaussian random generator, weight matrix $W^1 = (w_1^1, w_2^1, w_3^1, \ldots, w_n^1)$, is generated [5]. These weights are applied as an input to the hidden neuron using Eqs. 1 and 2. For further proceeding with the learning of more samples, a separability parameter λ is used as defined earlier. Now this learning process will continue till number of iterations which is user specified. In this case, number of iterations has been taken 100. Whenever, new samples are coming for learning, it is checked with present neurons. If it is not learnt with the existing ones then a new neurons will be added and processed. New neuron will be added till learning of the sample reaches tolerance limit. Tolerance limit is defined by user. It may be 95 % of total training sample of any value greater than 90 %. Whole process is repeated from initialization of qubits to find out optimal weights for new neuron.

Based on the discussion, QL-BNN in the form of algorithm is presented subsequently. In this algorithm, input layer is just the number of input nodes corresponding to the number of bits representing a particular sample, where one by one input samples are applied and then following algorithm is used for learning of hidden neurons. In this algorithm, some necessary parameters has been used which is described as follows:

$X(j)$ where $j = 1, 2, 3 \ldots, r$; is sample of class A which gives output '1'.

$Y(k)$ where $k = 1, 2, 3 \ldots, s$; is sample of class B which gives output '0'.

Two values *count*1 and *count*2 has been taken, it describes the number of samples of class A and B. Here $S*$ vector is used which store best value from S^g corresponding to best result or value of sum. Let us take a variable specifying user tolerance limit is denoted by tol_limit. The Algorithm 1 is presented next.

Algorithm1 : *QL-BNN Algorithm*

Hidden layer learning:

(a) Select a new neuron at the hidden layer While(iteration ≡ user define number) Do

Step-1: Initialize different parameters

$$Q^g = (q_1^g, q_2^g, q_3^g ......q_n^g);$$

where $g = 1,...,m; m$ is the number of iterations to update loop
$i = 1......n; n$ is number of input sample in binary form.
$sum_0 = 0;$
$S^* = 0;$

Step-2: Initialize $g = 1$

Step-3: Observe Qg and generate binary string S^g Using *Eq.* 8

Step-4: Store initial value of S^g in S^*

Step-5: Select following link for weight selection by formula:
Decimal value of $(S^g) + 1$

Step-6: Generate weight w_i by Gaussian random generator

Step-7: Evaluate for each sample
$net_A(j) = W^g \times X(j)$
$net_B(k) = W^g \times Y(k)$
if$(f(netA(j)) > \lambda)$
increase count1 by 1 endif
if$(f(net_B(k)) \le \lambda)$ inc
rease count2 by 1
endif

Step-8: $sum = count1 + count2$
if$(sum_g) < (sum_g - 1))$ update weight by observing bits in S^g and S^*
endif

Step-9: Update quantum weight for all iterations.

end while

(b) Select the weight corresponding to maximum value of sum_g in all iterations
Finalize this weight matrix for the first neuron

(c) After applying all the samples as counted by variable sum_g perform the following check to add more neurons to converge the learning While (sample learnt ≤ tol_limit)
Add new neuron and finalize it's weight by using step-1 to step-9.
end while.

4.2 Output Layer

Formation of output layer will start by taking one neuron at the output layer for connecting with all the hidden neuron. Quantization weights for the output layer is decided in the same way as for hidden layer neurons. Iterations required to learn output layer neurons is comparatively less with respect to hidden layer neurons. Because number of connection between input layer to hidden generally are more as compared to connection between hidden layer to output layer. Therefore corresponding to connection less iteration will be require to get optimize weights for output layer. It is user defined variable. Since this method handles, only two class problem, one neuron will be required at the output layer.

Working of QL-BNN is explained by taking simple example. Let us consider a database of eight instances with 3 input variable (000, 001, 010, 011, 100, 101, 110, 111).

Data set (000, 010, 111) belongs to class *A* which generate output 1 and data set (100, 101, 110) belongs to class *B* which generate output 0. Now initialize quantum values $Q^1 = (0.707|0.707, 0.707|0.707, 0.707|0.707, 0.707|0.707)$. Initialize random vector r let say $r = (0.8|0.3, 0.69|0.15, 0.37|0.77, 0.75|0.83)$. By using Eq. 7 observation process gives following result *S* = (01, 01, 10, 00). Now according to *step*-5 and *step*-6 in Algorithm 1, Gaussian random generator is used for generating weight, which generate weight values $W^1 = (0.23, -0.078, 0.25, -0.36)$. Now these weight values (Using step 7 $net_A(1)$ and $net_B(1)$) are calculated for given data set. Also, count number of samples which satisfy $f(net_A(1)) > \lambda$ and $f(net_B(1)) \leq \lambda$.

Let say class *A* sample (000, 010) and class *B* samples (100, 101) satisfy the above stated condition for generated weight values. It shows that out of six data set from both the classes, this neuron learnt for four samples in first iteration. Now, update qubits by quantum gate and again check value of sum for all iterations for first neuron only. If value of sum in all iterations found more than 4 then weight corresponding to this value will be considered as final weight, else weight corresponding to *sum* = 4 will be the final weight. The value of angular displacement $\Delta\theta$ will be decided on the basis of sum and bits values of *s* and *s** as described [5]. After deciding the number of neurons required at hidden layer output layer neuron is selected. Same process is applied for deciding weights at output layer. However number of iterations require for deciding weights of output layer is respectively less as compared to iterations require for deciding weights of the hidden layer. If number of neuron at hidden layer is one then there is no requirement of output layer neuron.

5 Experimental Results

In this section QL-BNN has been tested on PIMA Indian diabetes dataset. This dataset is collected from UCI Repository of machine learning [10]. Experimental setup starts with initialization of quantum bits Q_g and random vector *r*. To update

Table 1 Parameters used for experimentation

Parameters	Value
Iteration or generation	100
Quantum bits	Number of attribute in data set
$\Delta\theta$	$0.03 \times \Pi$
ϱ	0.05
λ	0.6

quantum weights, let us take angular value $\Delta\theta = 0.03 * \Pi$ and total 100 iterations. To stop quantum weight to converge from 0 and 1, the value of $\varrho = 0.05$, is taken in account. The value of λ is taken into account 0.6. The Gaussian random value have been taken from both range positive and negative and values corresponding to four weight space are N(−0.15, 0.03), N(−0.65, 0.03), N(0.15, 0.03), N(0.65, 0.03). Experiments have been carried out on Intel core i-5 processor with 4 GB RAM and on windows 7. All necessary parameters have been described in Table 1. The performance of proposed method has been evaluated on PIMA Indian diabetes dataset, which is describe next.

PIMA Indian diabetes data set has total 768 instances consist of 81 attributes after converting it in terms of bits. To test the proposed algorithm the data set have been divided in two subsets 75 % i.e. 576 instances is used for training and 25 % i.e. 192 instances is used for testing. It is observed from Table 2, that best training accuracy is achieved for set-1 which is 100.00 % with training time 1.6064 s and the worst training

Accuracy achieved for set-7 which is 96.3581 % with training time 1.6117 s. On the contrary, the best generalization accuracy is achieved for set-1 which is 98.9572 % with testing time 0.6261 s and the worst generalization accuracy is achieved for set-3 which is 96.2311 % with testing time 0.6280 s. Table 3, shows the classification results.

5.1 Discussion

As result shown in Table 3, the QL-BNN algorithm produces better results than MTiling-real networks algorithm [4] and they compared in terms of iterations for learning, training accuracy, generalization accuracy and timing. The QL-BNN algorithm produces better result with only two neurons in 100 iterations. The selection of angular value is also an important factor because large value results in fast convergence of the algorithm which does not lead to the desired result. Similarly small value may increase the number of iterations. Therefore to produce best result, value of $\Delta\theta$ may vary in the range of $\Delta\theta \in (0.01 \times \Pi, 0.05 \times \Pi)$.

However, for this approach the angular value of $\Delta\theta$ taken as $0.03 \times \Pi$. With all the parameters taken into account as discussed above, Table 3 shows the comparative results with MTiling-real networks, which are reported in terms of four

Table 2 Experimental results for PIMA Indians diabetes data set

	Training time (s)	Training accuracy	Number of iterations at first neuron	Number of iterations at second neuron	Number of neuron at hidden layer	Testing time (s)	Generalization accuracy
Set-1	1.6064	100	85	62	2	0.6261	98.9584
Set-2	1.7415	97.8521	83	65	2	0.6546	97.5287
Set-3	1.6152	99.5246	84	73	2	0.6236	96.2541
Set-4	1.6200	98.0903	88	61	2	0.6258	97.2365
Set-5	1.6079	98.3625	87	67	2	0.6237	98.4583
Set-6	1.6075	99.8562	86	63	2	0.6213	97.3541
Set-7	1.6409	96.7428	84	73	2	0.628	96.2311
Set-8	1.6482	98.7412	83	62	2	0.6245	97.2153
Set-9	1.6117	96.3581	85	69	2	0.6233	96.5742
Set-10	1.6311	97.7852	86	72	2	0.6243	98.8695

Table 3 Comparison of QL-BNN with MTiling-real networks algorithm

S. No.	Parameters	MTiling-real	QL-BNN
1	Number of neuron at hidden layer	–	2
2	Average training time (s)	–	1.636
3	Average testing time (s)	–	0.6276
4	Maximum iteration	500	100
5	Training accuracy	85.3 %	98.1458 %
6	Generalization accuracy	80.6 %	97.3024 %

parameters: iterations, training accuracy, training time (in seconds), generalization accuracy and testing time (in seconds).

6 Conclusion

In this paper, Quantum Based Learning with Binary Neural Network algorithm is presented. It can solve any two class classification problem by forming three layers network structure. The neural network architecture is formed constructively and weights are updated by quantum computing concept. This algorithm is tested on benchmark diabetes dataset and compared with MTiling-real network algorithm. From the experimental results it is observed that QL-BNN algorithm is better in terms of training and generalization accuracy, number of neurons and number of iterations. Since this problem is for two class problem, therefore there is no overhead involved in learning of output layer.

References

1. Simon, H.: Neural Networks and Learning Machines. Prentice Hall, Upper Saddle River (2008)
2. Lu, T.C., Yu, G.R., Juang, J.C.: Quantum-based algorithm for optimizing artificial neural networks. IEEE Trans. Neural Netw. Learn. Syst. **24**, 1266–1278 (2013)
3. Xu, Y., Chaudhari, N.: Application of binary neural networks for classification. In: 2003 International Conference on Machine Learning and Cybernetics, vol. 3, pp. 1343–1348. IEEE (2003)
4. Parekh, R., Yang, J., Honavar, V.: Constructive neural-network learning algorithms for pattern classification. IEEE Trans. Neural Netw. **11**, 436–451 (2000)
5. Han, K.H., Kim, J.H.: Quantum-inspired evolutionary algorithms with a new termination criterion, H_ε gate, and two-phase scheme. IEEE Trans. Evol. Comput. **8**, 156–169 (2004)
6. Mori, K., Isokawa, T., Kouda, N., Matsui, N., Nishimura, H.: Qubit inspired neural network towards its practical applications. In: International Joint Conference on Neural Networks, 2006, IJCNN'06, pp. 224–229. IEEE (2006)
7. de Araujo, R.A., Aranildo, R., Ferreira, T.: A quantum-inspired intelligent hybrid method for stock market forecasting. In: IEEE World Congress on Computational Intelligence Evolutionary Computation, 2008, CEC 2008, pp. 1348–1355. IEEE (2008)
8. Wang, D., Chaudhari, N.S.: Binary neural network training algorithms based on linear sequential learning. Int. J. Neural Syst. **13**, 333–351 (2003)
9. Zhang, C., Yang, J., Wu, W.: Binary higher order neural networks for realizing boolean functions. IEEE Trans. Neural Netw. **22**, 701–713 (2011)
10. Blake, C., Merz, C.J.: {UCI} repository of machine learning databases. Department of Information and Computer Science, University of California Irvine, Irvine, CA (1998) [Online]. Available http://www.ics.uci.edu/mlearn/MLRepository.html

Graphene Nano-Ribbon Based Schottky Barrier Diode as an Electric Field Sensor

Dipan Bandyopadhyay and Subir Kumar Sarkar

Abstract In this paper, an analytical approach has been made to represent graphene nanoribbon based schottky barrier diode as an Electric Field Sensor. Mainly the studies of the relationship between electric field and schottky barrier lowering and the relationship between electric field and current density have been presented. It is observed that electric field increases with the increase of schottky barrier lowering (i.e. the lowering of the schottky barrier potential enhances) which in turn enhances the net current density as more and more carriers can cross the metal-semiconductor barrier owing to the lowered barrier potential. Ultimately this encourages us to predict easily the corresponding electric field either from the schottky barrier lowering value or from the value of the current density. Thus, it can be stated that a GNR based schottky diode can function as an electric field sensor.

Keywords Graphene · Graphene nanoribbon (GNR) · Schottky diode · Schottky barrier lowering · Current density · Electric field sensor

1 Introduction

An Electric field sensor can be defined as a device, which can produce a functionally related output generally in the form of an electrical signal or an optical signal, responding to an Electric Field. In this paper, graphene nanoribbon based schottky barrier diode has been presented as an Electric Field Sensor. Due to the requirement of energy bandgap, graphene nanoribbon (GNR), a monolayer graphene sheet (width of few nanometers) has been introduced in place of graphene

D. Bandyopadhyay (✉) · S.K. Sarkar
Department of Electronics and Telecommunication Engineering, Jadavpur University, Kolkata 700032, India
e-mail: idipan89@gmail.com

S.K. Sarkar
e-mail: su_sircir@yahoo.co.in

L.C. Jain et al. (eds.), *Computational Intelligence in Data Mining - Volume 2*,
Smart Innovation, Systems and Technologies 32, DOI 10.1007/978-81-322-2208-8_44

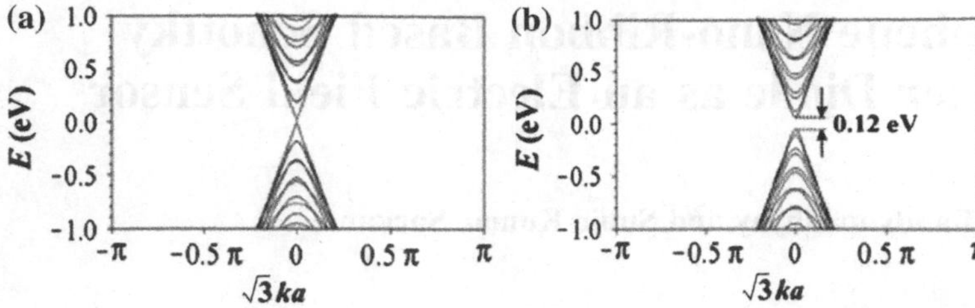

Fig. 1 **a** Band structure of Metallic armchair-GNR ($N = 44$). **b** Band structure of semiconducting ac-GNR ($N = 45$) N. number of hexagonal carbon rings across the width of the GNR, a: lattice constant ($a = 0.246$ nm)

(zero bandgap material) which is a flat monolayer of carbon atoms tightly packed into a 2-D honeycomb lattice. Moreover band-gap engineering is possible in GNR because the band-gap of this material depends on the orientation relative to its crystal structure as well as the width [1, 2]. On the basis of edge types, GNR can be classified into two types (1) zigzag GNR (ZGNR) and (2) armchair GNR (AGNR). Regardless of the width, this zig-zag GNR exhibits metallic behavior whereas on the other hand arm-chair GNR can show both metallic as well as semi-conducting behaviors, depending on widths [3]. When the dimer line of GNR satisfies the equation $n = 3m + 2$, it can exhibit metallic behavior whereas its semiconducting behavior is observed when the dimer line satisfies the equation $n = 3m$ or $n = 3m + 1$, where n is the number of dimer lines and m is an integer. The band structure of metallic armchair GNR is shown in Fig. 1a which clearly shows that the conduction band and the valence band are touching each other, confirming the metallic property. Again Fig. 1b is presenting us the band structure of the semiconducting armchair GNR where it is found that a gap of 0.12 eV is present (for N = 45) between the conduction and the valence band, indicating us the semiconducting property of an armchair GNR [4].

In the schottky barrier diode discussed here, transport of electrons through a semiconductor(Armchair GNR)-metal(Pd) junction is based on the thermionic emission theory which is a semi-classical approach developed by Bethe. There are three basic assumptions of this thermionic emission theory: (i) the energy barrier height at the interface should be much higher than kT (0.0259 eV) (ii) the junction plane should be at thermal equilibrium, (iii) presence of an electrical current should not affect this equilibrium. This paper mostly focuses on the study of the electric field and the schottky barrier lowering along with the current density based on the thermionic theory which will ultimately encourage us to present a GNR based schottky diode as an electric field sensor. Here the responsivity of the device has been defined as change in electric field due to the change in some parameters (viz. temperature). For this proposed sensor it is observed that the responsivity of the device more or less increases with the temperature and the current density which is a sign of a good sensor.

2 Schottky Barrier Lowering, Electric Field and Current Density Calculation

The net-current density of the metal-semiconductor junction in a schottky diode is given by [5],

$$J = J_{sT}[\exp(\frac{qV_a}{kT}) - 1] \quad (1)$$

where V_a is the applied forward bias voltage and J_{sT} is the reverse current saturation density.

The reverse current saturation density (J_{sT}) in a schottky barrier diode is given by

$$J_{sT} = A^* T^2 \exp(\frac{-q\phi_{Bn}}{kT}) \quad (2)$$

where T is the temperature in Kelvin, k is the Boltzmann's constant and A^* is the Richardson's constant which is given by

$$A^* = \frac{4\pi q m_n^* k^2}{h^3} \quad (3)$$

where h is Planck's constant and m_n^* is the effective electron mass in GNR and m_n^* is given by,

$$m_n^* = E_G(A + BE_G) \quad (4)$$

where A = 0.1053 (m_o/eV) B = 0.0339 (m_o/eV^2) are the fitting parameters for 3p + 1 (family) airmchair GNR. Due to image force lowering, the Schottky barrier height (ϕ_{Bn}) changes and now [5]

$$\phi_{Bn} = \phi_{BO} - \Delta\phi \quad (5)$$

Therefore, the new reverse saturation current density will be given by

$$J_{sT} = A^* T^2 \exp(\frac{-q\phi_{BO}}{kT}) \exp(q\frac{\Delta\phi}{kT}) \quad (6)$$

3 Calculation of ϕ_{BO}

The band gap energy (E_G) can be calculated as [6]:

$$E_G = \frac{2v\eta\pi}{W} \quad (7)$$

Here v is the velocity and W is the width of the nanoribbon.

Again, alternatively

$$E_G = \frac{B_I 0.69}{W} \quad (8)$$

where B_I is the band index and $B_I = 0, 1, 2\ldots$

Considering B_I as 1 and W as 10 nm, we get the bandgap energy as $E_G = 0.069$ eV

Putting this value of E_G in Eq. 7, we get,

$$\frac{v\eta\pi}{W} = 0.0345 \quad (9)$$

Now, Electron affinity of GNR is given by [6],

$$q\chi = W_G - \frac{v\eta\pi}{W} \quad (10)$$

where, W_G is the work function of 2D graphene (4.6 eV). Therefore, Electron affinity of GNR is 4.5655 eV. The barrier height of the diode is given ideally by the difference between the work function of the metal (Here, palladium whose work function (W_{Pd}) is 5.12 eV) and the electron affinity of GNR.

Therefore,

$$\phi_{BO} = W_{Pd} - q\chi \quad (11)$$

Therefore, the ideal barrier height (ϕ_{BO}) comes out to be 0.5545 eV which is much larger than kT (0.0259 eV).

4 Derivation of the General Expressions of $\Delta\phi$ and ξ

An electric field will be created when an electron, residing within a dielectric, is at x distance from the metal. The field lines being same and perpendicular to that metal surface, prompts us to consider an image charge, locating inside the metal and at a same distance x from the surface. Now the Coulomb force of attraction on the electron due to this image charge will be given by

$$F(x) = q\xi(x) = \frac{-q^2}{4\pi\varepsilon_G\{x-(-x)\}^2} = \frac{-q^2}{16\pi\varepsilon_G x^2} \tag{12}$$

The corresponding potential in absence of other fields will be given by:

$$\phi(x) = -\int_x^\infty \xi(x)dx = \frac{-q}{16\pi\varepsilon_G x} \tag{13}$$

When the metal surface experiences an electric field, two external forces are being experienced by the carriers that escape from the metal surface. One of these two forces is the image force that arises from Coulomb attraction force as a result of image charges induces inside the metal by the escaping carrier which has been discussed earlier, and the Lorentz force due to the electric field. Due to these two combined external forces, the schottky barrier is lowered and is given by

$$-\Delta\phi = \frac{-q}{16\pi\varepsilon_G x} - \xi x \tag{14}$$

Differentiating both sides of (14) with respect to x, we get

$$-\frac{d\phi}{dx} = \frac{q}{16\pi\varepsilon_G x^2} - \xi \tag{15}$$

Here x which is the position of the maximum barrier, can be found out from the condition $\frac{d\phi}{dx} = 0$.

Thus, using this condition, the above equation can be simplified to:

$$x = \sqrt{\frac{q}{16\pi\varepsilon_G\xi}} \tag{16}$$

Putting the values of x from the above equation into (14), we get

$$\begin{aligned}\Delta\phi &= \sqrt{\frac{q^2 16\pi\varepsilon_G\xi}{q16^2\pi^2\varepsilon_G^2}} + \xi\sqrt{\frac{q}{16\pi\varepsilon_G\xi}} \\ \Delta\phi &= \sqrt{\frac{q\xi}{4\pi\varepsilon_G}} \\ \xi &= \frac{\Delta\phi^2 4\pi\varepsilon_G}{q}\end{aligned} \tag{17}$$

where, permittivity of graphene (ε_G): $2.4 \times 8.85 \times 10^{-14}$ F/cm [7].

Since, the new reverse saturation current density (i.e. after the schottky barrier lowering) is given by

$$J_{sT} = A^*T^2 \exp(\frac{-q\phi_{BO}}{kT}) \exp(q\frac{\Delta\phi}{kT})$$

Therefore,

$$\frac{J_{sT}}{A^*T^2} = \exp\{\frac{-q}{kT}(\phi_{BO} - \Delta\phi)\}$$
$$\frac{-q}{kT}(\phi_{BO} - \Delta\phi) = \ln(\frac{J_{sT}}{A^*T^2})$$
$$\Delta\phi - \phi_{BO} = \frac{kT}{q}\ln(\frac{J_{sT}}{A^*T^2})$$
$$\Delta\phi = \frac{kT}{q}\ln(\frac{J_{sT}}{A^*T^2}) + \phi_{BO}$$

Using the expression of schottky barrier lowering obtained earlier, we have

$$\sqrt{\frac{q\xi}{4\pi\varepsilon_G}} = \frac{kT}{q}\ln(\frac{J_{sT}}{A^*T^2}) + \phi_{BO}$$
$$\frac{q\xi}{4\pi\varepsilon_G} = [\frac{kT}{q}\ln(\frac{J_{sT}}{A^*T^2}) + \phi_{BO}]^2 \qquad (18)$$
$$\xi = \frac{4\pi\varepsilon_G}{q}[\frac{kT}{q}\ln(\frac{J_{sT}}{A^*T^2}) + \phi_{BO}]^2$$

5 Results and Discussions

In this section, the variation of current density with the schottky barrier lowering at different temperatures, the variations of electric field with the current density at different temperatures, the variation of the responsivity of the device with the current density at different temperatures and with the temperature at different current densities and lastly the variation of the electric field with the schottky barrier lowering have been shown.

In Fig. 2a it is observed that with the increase of the schottky barrier lowering (i.e. lowering of the potential barrier) the current density also increases as more and more carriers can cross the potential barrier. It is also seen that the current density increases with the increase in temperature (from 290 to 300 K).

Figure 2b depicts that with the increase of the current density the electric field also increases and thus quite obviously for a particular value of the current density, the value of the corresponding electric field can be easily predicted.

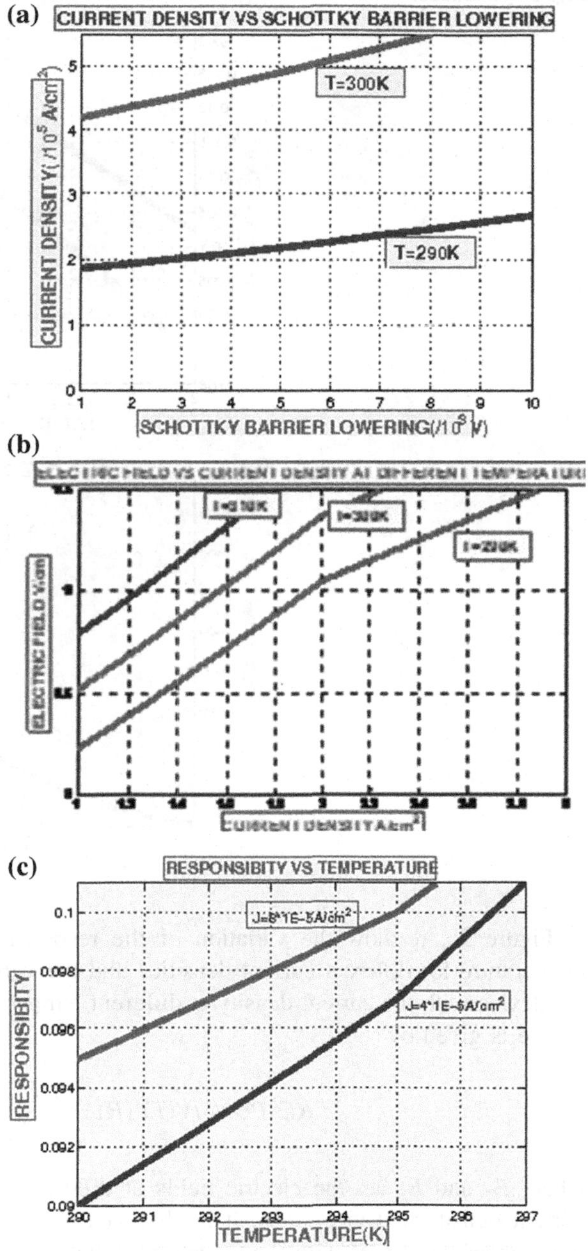

Fig. 2 **a** The variation of the current density with the schottky barrier lowering at different temperatures has been shown. **b** The variation of the electric field with the current density (10^{-5}) at different temperatures has been shown. **c** The variation of the responsivity of the device with the temperature for different current densities has been shown. **d** The variation of the responsivity of the device with the current density (10^{-5}) at different temperatures has been shown. **e** The variation of the electric field with the schottky barrier lowering has been shown. The value of the electric field can be easily predicted noticing the value of schottky barrier lowering

Fig. 2 (continued)

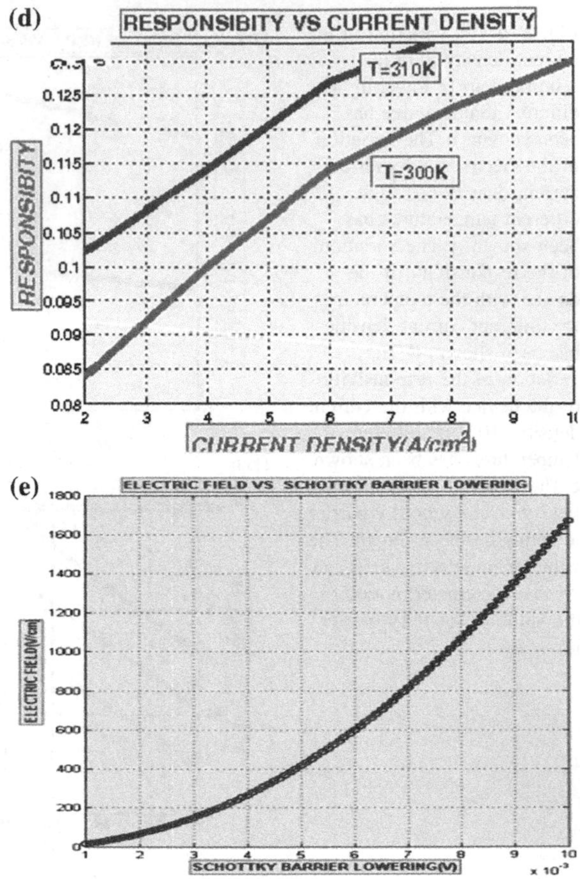

Figure 2c, d show the variation of the responsivity of the device with the temperature for different current densities and the variation of the responsibity of the device with the current density at different temperatures. Responsivity (R) of a device is given by

$$RESPONSIVITY(R) = \frac{E_2 - E_1}{E_1}$$

where E_1 and E_2 are the electric fields at different temperatures for a particular current density and vice versa. It is observed that the responsivity of the device more or less increases with the temperature and the current density.

In Fig. 2e it is observed that as the schottky barrier lowering is increased the electric field will also increase owing to the parabolic relationship between them. The value of the electric field can be easily predicted noticing the value of schottky barrier lowering. Actually, the current density which has been discussed so far is the reverse saturation current density. Since this current density is related to the net-current density of the metal-semiconductor junction [5], $J = J_{sT}[\exp(\frac{qV_a}{kT}) - 1]$, V_a is the applied forward bias voltage, a change in reverse saturation current due to the schottky barrier lowering will obviously change this net current density, thereby changing the current flowing through this device. It is also important to remember that the schottky barrier lowering also changes with the change in temperature. The temperature affects the kinetic energy of the carriers. With the increase of the temperature, the carriers becoming more energetic accumulate near the metal-semiconductor interface. Thus, the schottky barrier lowering is increased as the temperature increases.

6 Conclusion

It can be concluded from our discussion that GNR based schottky barrier diode can be presented as an electric field sensor. The variations of the electric field with the schottky barrier lowering and the current density have been studied along with the variations of the responsivity of the device with current density and temperature.

References

1. Geim, K., Novoselov, K.S.: The rise of graphene. Nat. Mater. **6**, 183–191 (2007). doi:10.1038/nmat1849
2. Kiat, W.K., Ismail, R., Ahmadi, M.T.: Schottky barrier lowering effect on graphene nanoribbon based schottky diode. In: RSM2013 Proceedings of 2013 Langkawi, Malaysia (2013)
3. Zheng, H.X., Wang, Z.F., Luo, T., Shi, Q.W., Chen, J.: Analytical study of electronic structure in armchair graphene nanoribbon. Phys. Rev. B. **75**, 165414. doi:10.1103/PhysRevB.75.165414
4. Xu, C., Li, H., Banerjee, K.: Modeling, analysis, and design of graphene nano-ribbon interconnects. IEEE Trans. Electron Devices **56**(8), 1567–1578 (2009). doi:10.1109/TED.2009.2024254
5. Neamen, D.A.: Semiconductor physics and devices: basic principles. Irwin (1992)
6. Mao, L.-F., Wang, Z.O., Zhang, L.-J., Ji, A.-M., Zhu, C.-Y., Yang, J.: Current-voltage characteristics of graphene nanoribbon schottky diodes. IETE J. Res. **58**(1), 65–71 (2012). doi:10.4103/0377-2063.94084
7. Lemme, M.C., Echtermeyer, T.J., Baus, M., Kurz, H.: A graphene field effect device. IEEE Electron Device Lett. **28**(4), 282–284 (2007). doi:10.1109/LED.2007.891668

Dynamic Slicing of Object-Oriented Programs in Presence of Inheritance

S.R. Mohanty, M. Sahu, P.K. Behera and D.P. Mohapatra

Abstract This paper proposes an *DG traversal dynamic slicing* algorithm for dynamic slicing of object-oriented programs in presence of inheritance, which can facilitate various software engineering activities like program comprehension, testing, debugging, reverse engineering, maintenance etc. This paper creates an intermediate program representation called dynamic graph to represent the execution trace of an object-oriented program. Then the proposed slicing algorithm is applied on the intermediate representation to compute the dynamic slice. The advantage of this approach is that, the intermediate program representation is small, hence needs less memory to store and the proposed algorithm is space efficient as well as time efficient and computes precise dynamic slices.

Keywords Dynamic slice · Dynamic graph · Object-oriented program

1 Introduction

Slicing is a technique of program analysis. This concept was originally developed by Wieser [1]. Generally, program slice can be computed with respect to the slicing criterion ⟨S, V⟩, Where S is the statement number and V is the set of variables used

S.R. Mohanty (✉)
Department of CS, RIMS, Rourkela 769012, Odisha, India
e-mail: swatee_18@rediffmail.com

M. Sahu · D.P. Mohapatra
Department of CSE, NIT Rourkela, Rourkela 769008, Odisha, India
e-mail: madhu_sahu@yahoo.com

D.P. Mohapatra
e-mail: durga@nitrkl.ac.in

P.K. Behera
Department of CSA, Utkal University, Vani Vihar, Bhubneswar 751004
Odisha, India
e-mail: p_behera@hotmail.com

L.C. Jain et al. (eds.), *Computational Intelligence in Data Mining - Volume 2*,
Smart Innovation, Systems and Technologies 32, DOI 10.1007/978-81-322-2208-8_45

or defined at S. Object-oriented technique modularizes the program, but at the same time it is very complex and difficult to debug, test and maintain. Slicing extracts set of statements from a program, which is relevant to a particular computation. Such strategies are usually called filtering techniques. The most important filtering technique is program slicing [2]. This paper computes the dynamic slices of object-oriented programs in presence of inheritance. The rest of the paper is organized as follows. Some basic concepts and definitions are discussed in Sect. 2. Section 3 presents the proposed algorithm for computing dynamic slices of an object oriented program in presence of inheritance. Section 3.1 presents working of the proposed algorithm. This paper compares the proposed work with existing ones in Sect. 4. Section 5 concludes the paper and presents the future work.

2 Basic Concepts and Definitions

This section is organized into two parts. The first part describes the intermediate representation for an object-oriented program in presence of inheritance and the second part describes basic concepts and definitions used in the proposed algorithm.

2.1 Intermediate Program Representation

Program Dependence Graph: Warren et al. [3] proposed Program Dependence Graph (PDG) to represent the intra-procedural programs. Program dependence graph represents the data dependencies and control dependencies within the statements of the programs.
System Dependence Graph: Horwitz et al. [4] proposed System Dependence Graph (SDG) to represent the inter-procedural programs. SDG can nicely represent inter-procedural programs by adding some additional edges like call edge, parameter in and out edge, summary edge etc.
Class Dependence Graph: Larsen and Harrold [5] extended system dependence graph to represent the object-oriented features. They introduced a new representation know as class dependence graph to represents the classes in an object-oriented programs.

To represents the execution trace of the object-oriented programs in presence of inheritance, this paper uses Dynamic Graph (DG) as the intermediate representation of the program. DG is a subset of SDG. DG can be defined as $G_h = (N_d, E_d)$. Where N_d is set of nodes in the graph and E_d is the set of edges in the graph. $N_d = \{n \mid n$ represents statement of a program$\}$, $E_d = \{e \mid e$ represents data and control dependency between n_1 and n_2, $(n_1, n_2) \in N_d\}$.

Consider the example program in Fig. 1.

Figure 2 shows the SDG of the example program given in Fig. 1.

```
CE1 class Person{
S2     String FirstName;
S3     String LastName;
SME4   Person(String fName, String lName){
S5     FirstName = fName;
S6     LastName = lName;}
ME7    void Display(){
S8     System.out.println("First Name : " + FirstName);
S9     System.out.println("Last Name : " + LastName);}}
CE10 class Student extends Person{
S11    int id;
S12    String standard;
S13    String instructor;
SME14  Student(String fName, String lName, int nId, String
stnd, String instr){
SMC15  super(fName,lName);
S16    id = nId;
S17    standard = stnd;
S18    instructor = instr;}
ME19   void Display(){
MC20   super.Display();

S21    System.out.println("ID : " + id);
S22    System.out.println("Standard : " + standard);
S23    System.out.println("Instructor: " + instructor);}}
CE24 class Teacher extends Person{
S25    String mainSubject;
S26    int salary;
S27    String type; //Primary or Secondary School teacher
SME28 Teacher(String fName, String lName, String sub, int
slry, String sType){
S29    super(fName,lName);
S30    mainSubject = sub;
S31    salary = slry;
S32    type = sType;}
ME33   void Display(){
MC34   super.Display();
S35    System.out.println("MainSubject : " + mainSubject);
S36    System.out.println("Salary : " + salary);
S37    System.out.println("Type:"+type);} }
MCE38 class InheritanceDemo{
MME39 public static void main(String args[]){
S40    int i;
S41    i=Integer.parseInt(in.readLine());
S42    if(i==2){
SMC43  Person pObj = new Person("Rayan","Miller");
MC44   pObj.Display();}
         else
SMC45  {Student sObj = new Student("Jacob","Smith",1,"1 -
B","Roma");
MC46   sObj.Display();
SMC47  Teacher tObj = new
Teacher("Daniel","Martin","English","6000","Primary Teacher");
MC48   sObj.Display();}}}
```

Fig. 1 An example program

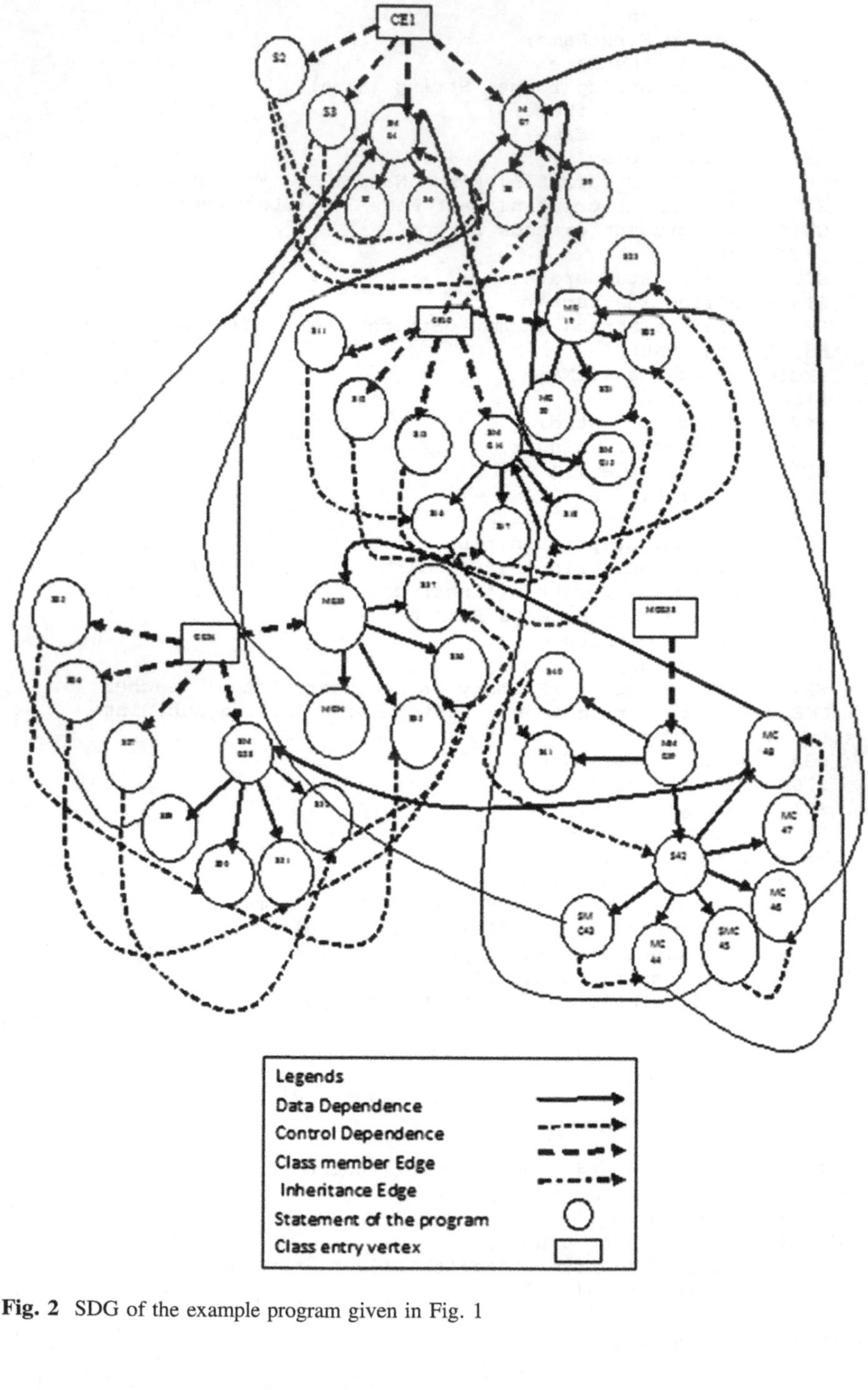

Fig. 2 SDG of the example program given in Fig. 1

2.2 Basic Concepts and Definitions

This section presents some definitions used in the proposed algorithm.

Control Dependence: Let x and y are two different nodes in the SDG. Node y depends on x if there is a directed path from x to y, indicating that execution of y depends on execution of x. Then node y is said to be control dependent on node x. In the example program node SMC43 is control dependent on node S42.

Data Dependence: Let x and y are two nodes in the system dependence graph. Then node y is data dependence on node x, if a variable var defined at x is used at y. There is a directed path exist from x to y. In the example program node S42 is data dependent on node S40.

Def(var): Let var be a variable in a program P. Then a node u is said to be Def(var) node if node u defines variable var. Def(pObj) = SMC40.

Use(var): Let var be a variable in the program P. Then a node u is said to be Use (var)node, if node u uses the variable var. Use(pObj) = S43.

DefVarSet(u): Let u and var be the node and variable respectively. Then DefVarSet (u) = {var | var is a variable of the program P and u is a Def(var) node}. DefVarSet (SMC40) = pObj.

UseVarSet(u): Let var be a variable of a program P. And u be a node. Then UseVarSet(u) = {var | var is a variable of the program P and u is a Use(Var) node}. UseVarSet(S43) = pObj.

Let the example program executed with the input value i = 1. According to the predicate condition, object of class Person is created and it called the Display() method. The dynamic graph of the example program is shown in Fig. 3.

ActiveDataSlice: Let P be a program and var be a variable. Before execution of the program P, ActiveDataSlice(var) = ϕ. Let u be a def(var) node and UseVarSet (u) = var_1, var_2, …, var_K. Let program P be executed with a given set of input value. Then ActiveDataSlice(var) = u U ActiveDataSlice(var_1) U ⋯ U ActiveDataSlice (var_k) U ActiveDataSlice(var_t), where t is the most recently executed node of s in the SDG.

ActiveControlSlice: Let s be the test node in the SDG of the program P and UseVarSet(u) = {var_1,var_2,…,var_K}. Before execution of the program P, Active-ControlSlice = ϕ. After each execution of the node s in an actual run of the program, ActiveControlSlice(s) = {s} U ActiveDataSlice(var_1) U ⋯ U ActiveDataSlice(var_k) U ActiveDataSlice(var_t), where t is the most recently executed predicate node.

DyanSlice(s, var): Let s be a node in the program P, and the variable var be in set DefVarSet(s) U UseVarSet(s). Before execution of the program P DyanSlice (s, var) = ϕ. For each execution of the statement s, DyanSlice(s, var) = ActiveDataSlice (var) U ActiveControlSlice(t), where t is the most recently executed predicate node of s.

ActiveCallSlice: For a call node u ActiveCallSlice(u_{call}) = ActiceDataSlice(var) U ActiveControlSlice(u_{call}), where var is the variable or object used to call the method.

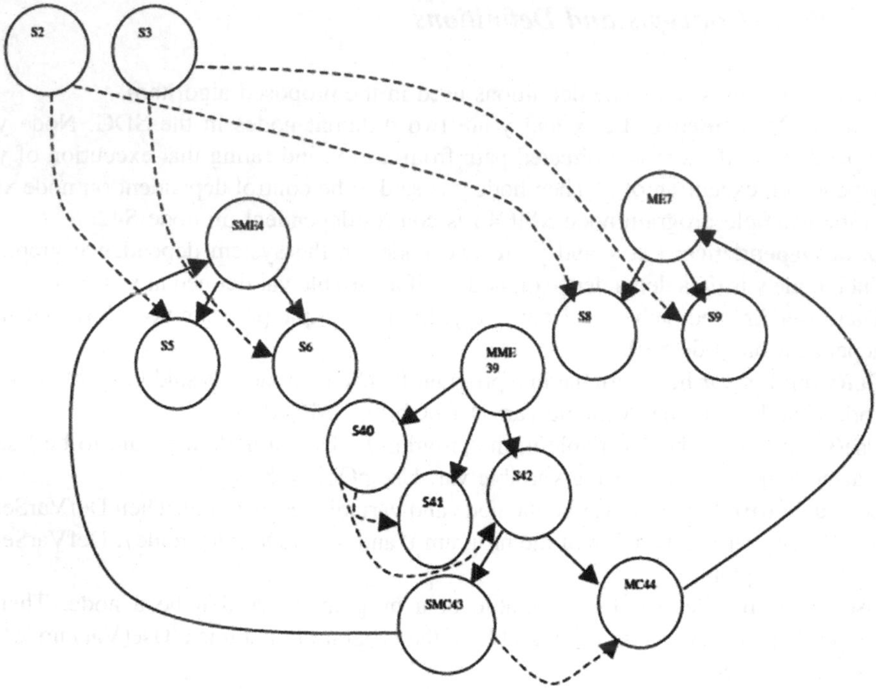

Fig. 3 DG of the example program given in Fig. 1

3 Proposed Algorithm

This section proposes the *DG traversal dynamic slicing* algorithm to compute the dynamic slices of object-oriented programs in presence of inheritance.

The proposed algorithm is given below.

Algorithm—DG traversal dynamic slicing algorithm

1) Consider the program P.
2) *Initialization*: Before execution of the program do the followings:
   a) For each node u do the followings:
      If u is a predicate node ActiveControlSlice(u) = ϕ, For each variable var ϵ DefVarSet(u) U UseVarSet(u), set DyanSlice(u, var) = ϕ
   b) For every variable var of the program P, set ActiveDataSlice(var) = ϕ.
   c) Set ActiveCallSlice = ϕ.
3) Run the program P with given set of input value.
4) Get the execution trace of the program and then construct the dynamic graph (DG).
5) *Dynamic slice computation*: Enter the slicing criterion, node u on which slice has to be computed.

a) If u is a Def(var) node and not a call node, DyanSlice(u, var) = ActiveData Slice(var). Compute ActiveDataSlice(var) by traversing DG through the incoming data dependence edge and list the reached nodes.
b) If u is a call node, DyanSlice(u, var) = ActiveCallSlice(u). Compute ActiveCallSlice(u) by traversing DG through the outgoing control dependences edges and incoming data dependence edges and list the reached node.
c) If u is a test node, DyanSlice(u, var) = ActiveControlSlice(u). Compute ActiveControlSlice(u) by traversing DG through all control dependence edges and list the reached nodes.
d) If u is a Def(var) and Use(var) node, DyanSlice(u, var) = ActiveDataSlice (var) U ActiveControlSlice(t), where t is the most recent executed predicate node.

6) *Slice look up*:
Extract the nodes which are reached during the traversal of the dynamic graph, those constitute the dynamic slice.

3.1 Working of the Proposed Algorithm

The proposed algorithm has been successfully implemented using Java. The example program runs with input value i = 1. The slicing criterion is ⟨SMC43, pObj⟩ The algorithm computes precise dynamic slices more accurately and efficiently. Table 1 shows the framework for the proposed algorithm. Table 1 also shows the slice for node MC44, SME4 and ME7.

4 Comparison with Related Work

This work has been influenced with the work done by Mund [6]. But the proposed representation is a better way to represent the data as well as the control dependencies.

Table 1 Dynamic slices of the example program at different statements

Slice node	Types of node	Dynamic slice
SMC43	Call node	SMC43, SME4, S5, S6, S2, S3
MC44	Call node	MC44, ME7, S8, S9, S2, S3, SMC43
SME4	Call node	SME4, S5, S6, S2, S3
ME7	Call node	ME7, S8, S9, S2, S3

Jain and Garg [7] proposed d-u chain which will be large for the large program. As this paper computes dynamic slice based on the execution trace of the program under consideration, thus the slice generated is more precise and correct.

Du et al. [8] used system dependence graph as the intermediate program representation. This paper proposes an intermediate representation called dynamic graph, which contains the required dependence edges i.e. data and control dependence. Hence, dynamic graph is simple and can be traversed faster. which generates precise dynamic slices for software engineering applications.

More over although number of nodes increases, the time of computing slices does not increase, rather it depends on the number of dependencies (control and/or data) in the representation. This is another advantage of the proposed algorithm.

5 Conclusion and Future Work

This paper presented an algorithm DG traversal dynamic slicing algorithm to compute the dynamic slices of object-oriented programs in presence of inheritance. This paper used dynamic graph to represent the data and control dependencies between the statements of the execution trace.

The proposed algorithm is not influenced by the number of nodes in the dynamic graph but it is influenced by the number of data and control dependencies in the intermediate representation. This paper does not consider polymorphism, dynamic binding and message passing, which are also very important features of object-oriented programs. So in future, our work will focus on the above said features and computing the dynamic slice. Also, we will be focusing on computing slicing of distributed object-oriented and concurrent object-oriented programs.

References

1. Wieser, M.: Program slicing. In: Proceedings of the 5th International Conference on Software Engineering, pp. 439–449. IEEE Press (1981)
2. Zhao, J.: Dynamic slicing of object-oriented programs. J. Nat. Sci. **6**, 391–397 (2001)
3. Warren, J.D., Ferrante, J., Ottenstein, K.J.: The program dependence graph and its use in optimization. ACM Trans. Program. Lang. Syst. **9**, 319–349 (1987)
4. Horwitz, S., Reps, T., Binkley, D.: Interprocedural slicing using dependence graphs. ACM Trans. Program. Lang. Syst. **12**, 26–61 (1990)
5. Larsen, L., Harrold, M.J.: Slicing object-oriented software. In: Proceeding of ICSE, pp. 495–505 (1996)
6. Mund, G.B.: An efficient inter-procedural dynamic slicing method. J. Syst. Softw. **79**, 791–806 (2006)
7. Jain, P., Garg, N.: A novel approach for slicing of object oriented programs. ACM SIGSOFT Softw. Eng. Notes. **38**, 1–4 (2013)
8. Du, L., Xiao, G., Yu, Y.: Research on algorithm for object-oriented program slicing. J. Convergence Inf. Technol. **6** (2011)

9. Ottenstein, K., Ottenstein, L.: The program dependence graph in software development environments. In: Symposium on Practical Software Development Environments, vol. 19, pp. 177–184 (1984)
10. Korel, B., Lask, J.: Dynamic program slicing. Inf. Process. Lett. **29**, 155–163 (1988)
11. Agrawal, H., Horgan, J.: Dynamic program slicing. In: Proceeding of the ACM SIGPLAN Conference on Programming Languages Design and Implementation, vol. 25, pp. 246–256 (1990)
12. Mund, G.B., Mall, R., Sarkar, S.: Computation of intra-procedural dynamic program slices. Inf. Softw. Technol. **45**, 499–512 (2003)
13. Mohapatra, D.P., Mall, R., Kumar, R.: An overview of slicing techniques for object-oriented programs. Informatica **30**, 253–277 (2006)
14. Beszedes, A., Gergely, T., Gyimóthy, T.: Graph-less dynamic dependence-based dynamic slicing algorithm. In: Sixth IEEE International Workshop on Source Code Analysis and Manipulation, vol. 6, pp. 21–30 (2006)
15. Mohanty, S.R., Behera, P.K., Mohapatra, D.P.: Computing dynamic slices of object-oriented program using dependency information. Int. J. Comput. Appl. **80**, 1–7 (2013)

Prediction of Heart Disease Using Classification Based Data Mining Techniques

Sujata Joshi and Mydhili K. Nair

Abstract Data Mining is an interesting field of research whose major objective is to find interesting and useful patterns from huge data sets. These patterns can be further used to make important decisions based on the result of the analysis. Healthcare industry today generates huge amount of data on a day to day basis. This data has to be analysed and hidden and meaningful patterns can be discovered. Data mining plays a promising and significant role in this aspect. Data Mining techniques can be used for disease prediction. In this research, the classification based data mining techniques are applied to healthcare data. This research focuses on the prediction of heart disease using three classification techniques namely Decision Trees, Naïve Bayes and K Nearest Neighbour.

Keywords Data mining · Classification technique · Heart disease · Healthcare · Decision tree · Naïve bayes · K-Nearest neighbor · Dataset

1 Introduction

Heart Disease is a class of diseases that involve the heart, the blood vessels or both. The most common causes of heart disease are atherosclerosis and/or hypertension. Atherosclerosis is a condition that develops when a substance called plaque builds up in the walls of the arteries. This buildup narrows the arteries, making it harder for blood to flow through. If a blood clot forms, it can stop the blood flow. This can

S. Joshi (✉)
Department of Computer Science and Engineering, Nitte Meenakshi Institute of Technology, Bangalore, Karnataka, India
e-mail: sujata_msrp@yahoo.com

M.K. Nair
Department of Information Science and Engineering, M. S. Ramaiah Institute of Technology, Bangalore, Karnataka, India
e-mail: mydhili.nair@gmail.com

L.C. Jain et al. (eds.), *Computational Intelligence in Data Mining - Volume 2*,
Smart Innovation, Systems and Technologies 32, DOI 10.1007/978-81-322-2208-8_46

cause a heart attack or stroke. The major risk factors for heart diseases are age, gender, high blood pressure, diabetes mellitus, tobacco smoking, processed meat consumption, excessive alcohol consumption, sugar consumption, family history, obesity, lack of physical activity, psychosocial factors, and air pollution.

Heart disease is the leading cause of deaths worldwide, however since the 1970s, mortality rate due to heart related diseases have declined in many high-income countries. At the same time, heart related deaths and diseases have increased at a fast rate in low and middle-income countries. Although heart disease usually affects older adults, the symptoms may begin in early life, making primary prevention efforts necessary from childhood. Therefore risk factors may be modified by having healthy eating habits, exercising regularly, and avoiding of smoking tobacco.

In today's world, most of the hospitals maintain their patient data in electronic form through some hospital database management system. These systems generate huge amount of data on a daily basis. This data may be in the form of free text, structured as in databases or in the form of images. This data may be used to extract useful information which may be used for decision making. This requirement has led to the use of Knowledge Discovery in Databases (KDD) which is responsible for transforming data of low-level into high-level knowledge for decision making. Data mining which is one of the KDD process aims at finding useful patterns from large datasets. These patterns can be further analyzed and the result can be used for effective decision making and analysis. The various tasks of data mining are classification, clustering, association analysis and outlier detection. In this paper, various data mining classification techniques are applied to healthcare data related to heart diseases. It has helped to determine the best prediction technique in terms of its accuracy and error rate on the specific dataset.

2 Related Work

There has been an increase in the number of people suffering from heart diseases in the recent years [1]. With the advent of information technology and its applications data mining plays a very important and apt role in early detection of diseases. Data mining is extensively used in all fields and healthcare industry in particular [2–6]. In the healthcare industry, the data mining techniques are used for diagnosis of diseases [7], disease prediction [8], and analysis [9]. Data mining techniques can be applied for predicting the outcome of interest. Hence prediction is a very important task. The issues and guidelines of Predictive data mining in clinical medicine is discussed in [10]. Research work [7, 11, 12] related to heart disease diagnosis using data mining techniques is the motivation for this work. Classification based on Gini index is discussed in [13]. The data mining techniques Decision tree, Naïve Bayes and KNN are discussed in [8, 10, 14, 15]. A model based on Combination of Naïve Bayes Classifier and K-Nearest Neighbor is proposed in [16]. A clinical decision support system using association rule mining is discussed in [17]. A prediction system for lung cancer detection is proposed in [18]. A diagnostic tool is proposed in [19] for skin diseases. In [6, 9], the researchers analyze healthcare data using different data mining techniques. After the extensive literature survey of the dataset, algorithms, methods employed by the authors, results and future work, it is found that there is a lot of scope in discovering efficient methods of medical diagnosis for various diseases and their analysis. This work is an attempt to predict the occurrence of heart diseases using classification data mining techniques namely Decision Tree, Naïve Bayes and K-Nearest Neighbor techniques.

3 Classification

Classification is one of the important data mining tasks. The objective of classification is to assign a class to previously unseen data accurately. Classification consists of two stages:

Stage 1: Model construction
Stage 2: Model usage

Classification creates a model for the attributes of the dataset. A dataset is divided into training set and test set. In the first stage the training set is used to build the classification model using a learning algorithm. In the second stage, the learned model is put into operational use i.e. it is used to validate the test set. If the model performs well, then the model is now ready for prediction.

3.1 Classification Techniques

In this study, the classification techniques, Decision tree, Naïve Bayes and KNN are explored and applied to the dataset.

3.1.1 Decision Trees

The decision tree is a structure that includes root node, branch and leaf node. Each internal node denotes a test on an attribute, each branch denotes the outcome of test and each leaf node holds the class label. The first node in the tree is the root node. First, an attribute is selected and placed at the root node, and a branch is made for each possible value. This splits up the data set into subsets, one for every value of the attribute. Now repeat the process recursively for each branch, using only those instances that actually reach the branch. When all instances at a node have the same classification, the tree development can be stopped. To select the best split the measures used generally are Gini, Entropy or Classification error.

3.1.2 Naïve Bayes Classifier

Classification based on Bayes Theory is known as Bayesian Classification. Naive Bayes classifier is a statistical based classifier which is based on Bayes Theory. It assumes that attributes are statistically independent. This classifier is based on probabilities.

Given two events A and B, P(*A*) is prior probability and *P*(*A*|*B*) is posterior probability, then according to Bayes theorem

$$P(A|B) = P(B/A)P(A)/P(B) \text{ and } P(B|A) \text{ is computed as } P(A \cap B)/P(A)$$

These Bayesian probabilities are used to determine the most likely next event for the given instance given all the training data. Conditional probabilities are determined from the training data.

This classifier yields optimal prediction (given the assumptions). It can also handle discrete or numeric attribute values.

3.1.3 K-Nearest Neighbor

Nearest neighbor method is a instance based classification technique that remembers all the instances. When the new instance is encountered, it uses previous instances as a model and compares it with the new instance. Prediction for the current instance is the one with the most similar previously observed instance. K-NN classifies the instances using the K nearest neighbors. This classifier has faster training rate but is slow when the dataset is large since it has to evaluate all instances.

4 Methodology

4.1 TOOL Used

WEKA [20] Tool (Waikato Environment for Knowledge Analysis), is a set of data mining algorithms and tools which can be used for analysis of data. WEKA is developed in JAVA. WEKA allows analyzing the data sets saved in .arff format using various algorithms. In this study, the Decision tree, Naïve Bayes and K-NN algorithms are applied to heart data set and the results of applying these techniques are shown.

4.2 Data Source

The heart diseases data set from the UCI [21] Learning Repository is used for this study. The heart data set consists of 303 records and 14 attributes. The attributes are listed in Table 1.

Table 1 Attributes of the heart.arff file

No	Attribute	Type
1	age	Real
2	sex	{female, male}
3	cp	{typ_angina, asympt, non_anginal, atyp_angina}
4	trestbps	Real
5	chol	Real
6	restecg	{left_vent_hyper, normal, st_t_wave_abnormality}
7	thalach	real
8	restecg	{left_vent_hyper, normal, st_t_wave_abnormality}
9	exang	{no, yes}
10	oldpeak	real
11	slope	{up, flat, down}
12	ca	real
13	thal	{fixed_defect, normal, reversable_defect}
14	num	{'<50', '>50_1','>50_2', '>50_3', '>50_4'}

4.3 Decision Tree

The decision tree is created by selecting the best split at every node. To select the best attribute for the split, the information gain is computed at each node and the attributes are ranked accordingly. Here the attribute evaluator used is Gain Ratio AttributeEval and the search method used is Ranker method from WEKA Tool. The ranked attributes are listed in Table 2.

Table 2 Attribute ranking based on information gain

Info gain	Rank	Attribute
0.17	12	thal
0.16	13	ca
0.15	9	exang
0.13	8	thalach
0.11	3	cp
0.10	10	oldpeak
0.09	11	slope
0.065	2	sex
0.060	1	age
0.022	7	restecg
0	6	fbs
0	5	chol
0	4	trestbps

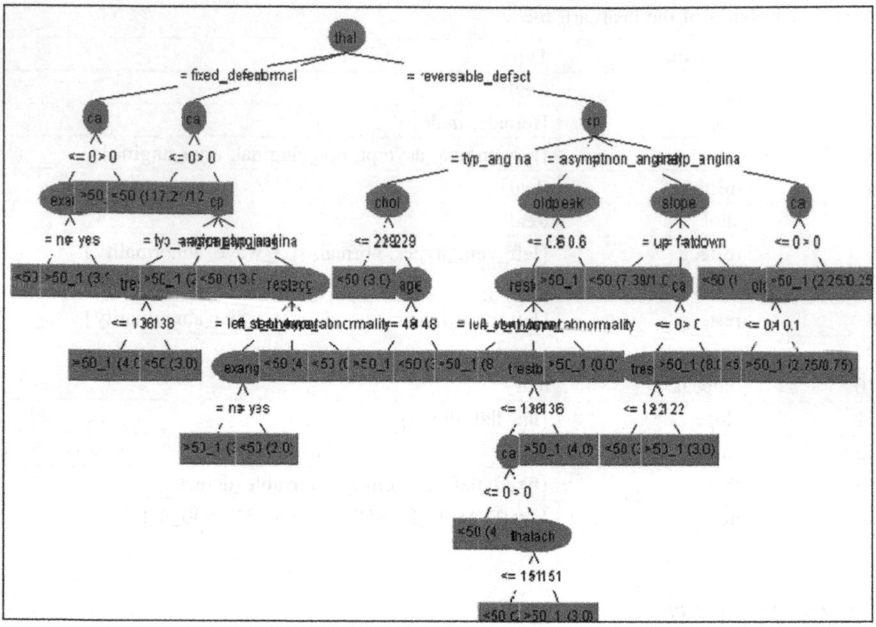

Fig. 1 Decision tree generated using J48 algorithm

Table 3 Results of decision tree algorithm

	No of instances	Percentage (%)
Correctly classified instances	279	92.0792
Incorrectly classified instances	24	7.9208
Total instances	303	

The attributes selected in the order are: **12**, **13**, **9**, **8**, **3**, **10**, **11**, **2**, **1**, **7**, **6**, **5**, **4**.

The Decision Tree algorithm J48 is then applied to the heart data set and the decision tree in Fig. 1 is generated. This decision tree can be used for prediction. The results are shown in Table 3.

4.4 Naïve Bayes

The attribute evaluator used is Gain Ratio AttributeEval and the search method used is Ranker method. The ranked attributes are same as in Decision tree. The Naïve Bayes algorithm is applied to the heart data set and the results of few attributes are shown in Table 4.

The results are shown in Table 5.

Table 4 Results of few attributes using Naïve Bayes technique

Attribute	<50 (0.54)	>50_1 (0.45)	>50_2 (0)	>50_3 (0)	>50_4 (0)
cp					
typ_angina	17.0	8.0	1.0	1.0	1.0
asymp	40.0	105.0	1.0	1.0	1.0
non_anginal	70.0	19.0	1.0	1.0	1.0
atyp_angina	42.0	10.0	1.0	1.0	1.0
[total]	169.0	142.0	4.0	4.0	4.0
restecg					
left_vent_hyper	69.0	80.0	1.0	1.0	1.0
Normal	97.0	57.0	1.0	1.0	1.0
st_t_wave_abnormality	2.0	4.0	1.0	1.0	1.0
[total]	168.0	141.0	3.0	3.0	3.0

Table 5 Results of Naïve Bayes technique

	No of instances	Percentage (%)
Correctly classified instances	255	84.1584
Incorrectly classified instances	48	15.8416
Total instances	303	

Table 6 Results of K-nearest neighbor technique

	No of instances	Percentage (%)
Correctly classified instances	303	100
Incorrectly classified instances	0	0
Total instances	303	

4.5 K-Nearest Neighbor

The KNN algorithm is applied to the heart data set and the results are shown in Table 6.

5 Results and Conclusion

The evaluation measures used are Sensitivity, Specificity and Accuracy

(i) **Sensitivity = TP/P**
(ii) **Specificity = TN/N**
(iii) **Accuracy = (TP + TN)/(P + N)**

Table 7 Summarization of prediction techniques with performance

Prediction technique	Sensitivity	Specificity	Accuracy
Decision tree	0.921	0.085	0.922
Naïve bayes	0.842	0.165	0.842
KNN	1	0	1

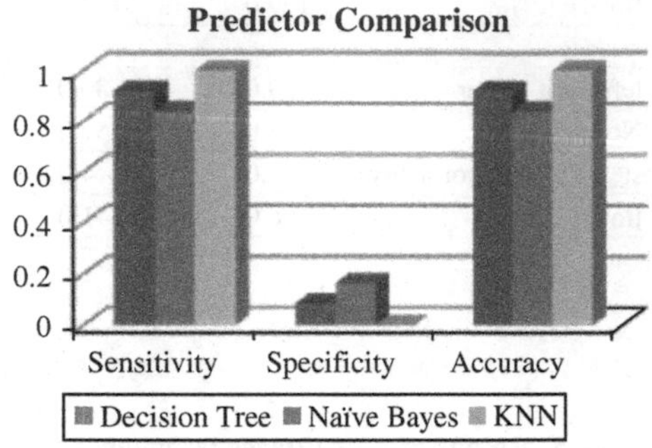

Fig. 2 Comparison of prediction techniques

where TP is true positives, TN is true negatives, P and T are actual positives and actual negatives respectively. A good predictor must have high sensitivity, low specificity and high accuracy. The comparisons of these measures with respect to the three prediction techniques are summarized in Table 7.

The experiments are conducted with WEKA tool and the algorithms applied on the heart dataset. The graph in Fig. 2 reveals that sensitivity and accuracy are high and specificity is low. Hence the predictors perform well on operational use. With respect to model creation the results show that KNN has highest accuracy as expected since KNN remembers all the instances. But when used for prediction the Decision Tree performs well when compared to other two methods for the given heart dataset.

References

1. Heart Disease—General Info and Peer reviewed studies: [Online] Available http://www.aristoloft.com
2. Patka, S., et al.: Recent trends and rapid development of applications in data mining. IOSR J. Comput. Sci. (IOSR-JCE) 73–78. e-ISSN: 2278-0661, p-ISSN: 2278-8727 (2014)
3. Tomar, D., Agarwal, S.: A survey of data mining approaches for healthcare. Int. J. Bio-Science and Bio-Technology **5**(5), 241–256 (2013)
4. El-Sappagh, S.H., et al.: Data mining and knowledge discovery: applications, techniques, challenges and process models in healthcare. Int. J. Eng. Res. Appl. (IJERA) **3**(3), 900–906. ISSN: 2248-9622 www.ijera.com (2013)
5. Koh, H.C., Tan, G.: Data mining applications in healthcare. J. Healthc. Inf. Manag. **19**(2), 65 (2011)

6. Obenshain, M.K.: Application of data mining techniques to healthcare data. Infect. Control Hosp. Epidemiol. **25**(8), 690–695 (2004)
7. Shouman, M., Turner, T., Stocker, R.: Using data mining techniques in heart disease diagnosis and treatment. In: Proceedings in Japan–Egypt Conference on Electronics, Communications and Computers, vol. 2, pp. 174–177. IEEE (2012)
8. Bellazzi, R., Zupan, B.: Predictive data mining in clinical medicine: current issues and guidelines. Int. J. Med. Inf. **77**(2), 81–97 (2006)
9. Gosain, A.: Analysis of healthcare data using different data mining techniques, IEEE, ISBN: 978-1-4244-4711-4 (2009)
10. Milovic, B., Milovic, M.: Prediction and decision making in health care using data mining. Int. J. Public Health Sci. (IJPHS) **1**(2), 69–78 (2012). ISSN: 2252-8806
11. Melillo, P., De Luca, N., Bracale, M., Pecchia, L.: Classification tree for risk assessment in patients suffering from congestive heart failure via long-term heart rate variability. IEEE J. Biomed Health Inf. **17**(3), 727–733 (2013)
12. Rao, R.B., Krishan, S., Niculescu, R.S.: Data mining for improved cardiac care. ACM SIGKDD Explor. Newsl. **8**(1), 3–10 (2006)
13. Suneetha, N., Hari, V.M.K., Kumar, V.S.: Modified gini index classification: a case study of heart disease dataset. Int. J. Comput. Sci. Eng. **2**(6), 1959–1965 (2010)
14. Han, J., Kamber, M.: Data Mining: Concepts And Techniques. Morgan Kaufmann, San Francisco (2001)
15. Tan, P.N., Steinbach, M., Kumar, V.: Introduction to Data Mining, 4th edn. Pearson Publications, Boston
16. Ferdousy, E.Z., Islam, M.M, Matin, M.A.: Combination of Naïve Bayes classifier and K-nearest neighbor in the classification based predictive models. J. Comput. Inf. Sci. **6**(3), 48–56. ISSN: 1913-8989 (2013)
17. Cheng, C., Chanani, N., Vengopalan, J., Maher, K., Wang, D.: icuARM—an ICU clinical decision support system using association rule mining. IEEE J. Transl. Eng. Health Med. **1** (2013)
18. Krishnaiah, V., Narasimha, G., Chandra, N.S.: Diagnosis of lung cancer prediction system using data mining classification techniques. Int. J. Comput. Sci. Inf. Technol. **4**, 39–45 (2013)
19. Cataloluk, H., Kesler, M.: A diagnostic software tool for skin diseases with basic and weighted K-NN, IEEE. ISBN: 978-1-4673-1448-0/12 (2012)
20. WEKA: Data Mining Machine Learning Software. [Online] Available http://www.cs.waikato.ac.nz/ml/weka/
21. UCI Machine Learning Repository. [Online] Available http://archive.ics.uci.edu/ml/datasets.html
22. Cios, K.J., William Moore, G.: Uniqueness of medical data mining. J. Artif. Intell. Med. **26**(1), 1–24 (2002)

An Empirical Analysis of Software Reliability Prediction Through Reliability Growth Model Using Computational Intelligence

Manmath Kumar Bhuyan, Durga Prasad Mohapatra, Srinivas Sethi and Sumit Kar

Abstract The objective of this paper is to predict software reliability using non-parametric neural network of *computational intelligence (CI)*. The study uses data sets containing failure history such as number of failures, failure time interval etc. In this paper, we explore the applicability of feed-forward neural network with back-propagation training as a reliability growth model for software reliability prediction. The prediction result is compared with that of traditional parametric software reliability growth models. The results described in the proposed model exhibits an accurate and consistent behavior in reliability prediction. The experimental results demonstrate that the proposed model provides a significant difference respect to accuracy and consistency.

Keywords Software reliability prediction · Reliability growth models · Neural networks

M.K. Bhuyan (✉) · S. Sethi · S. Kar
Department of Computer Science and Engineering, IGIT, Utkal University, Bhubaneswar, Odisha, India
e-mail: manmathr@gmail.com

S. Sethi
e-mail: ssethi@igit.ac.in

S. Kar
e-mail: sumittalk2u@gmail.com

D.P. Mohapatra
Department of Computer Science and Engineering, NIT, Rourkela, Odisha, India
e-mail: dpk@nitrkl.ac.in

L.C. Jain et al. (eds.), *Computational Intelligence in Data Mining - Volume 2*,
Smart Innovation, Systems and Technologies 32, DOI 10.1007/978-81-322-2208-8_47

1 Introduction

Software reliability growth models (SRGM) have acquired significant role in predicting software reliability. The typical definition of software reliability is the ability of a system or component to perform the failure free intended function without error for a specified exposure period of time in a defined environment (i.e. at different testing phases both within a project and across projects) [1–5]. Day by day, the demand of the software products is growing rapidly. The prediction of software is becoming a major role in the software development process. The impact of the failures produces severe consequences such as environmental impact, inconvenience, economical losses, loss of human life etc. Software reliability is an important facet of software quality characteristic [6].

The time duration between successive failures or the cumulative failure time is a vital factor of software reliability [7, 8]. Though many analytical models are proposed so far, but neural network (NN) produces accurate and significant results.

Reliability prediction gives confidence about software correctness. Reliability prediction balances the additional cost of testing and improves software reliability [5]. At the time of development of any product or system, we need to ensure its reliability and consistency in its performance, because a system's reliability has a major impact on maintenance, repair costs, continuity of service, and customer satisfaction [6]. It is better to continue testing rather than releasing faulty product to the market.

The SRGM assumes that reliability grows after a defect has been detected and fixed [9]. SRGM is applied to guide the test board in taking decision of whether to stop or continue the testing process. Though many methods proposed so far for software reliability engineering (SRE), acceptable software reliability techniques are still strongly needed.

The rest of the paper is organized as follows: In Sect. 2, we discuss software reliability growth model structure. In this section, we describe feed forward network modeling architecture. The application of models to software reliability prediction and architecture construction are presented in Sect. 3. The training on the neural network is also described in this section. The basic terminology and experimental observations are presented in Sect. 4. Section 5 confines a summary, some conclusions, and notes for future direction.

2 Software Reliability Growth Model Structure

Software reliability modeling (SRM) has been used as the most important and successful predictor of software quality and reliability since long time [9]. In order to get an operative powerful system, it is essential to predict software reliability by minimizing software fault. SRGM assumes that the numbers of faults residing in the software system are already observed.

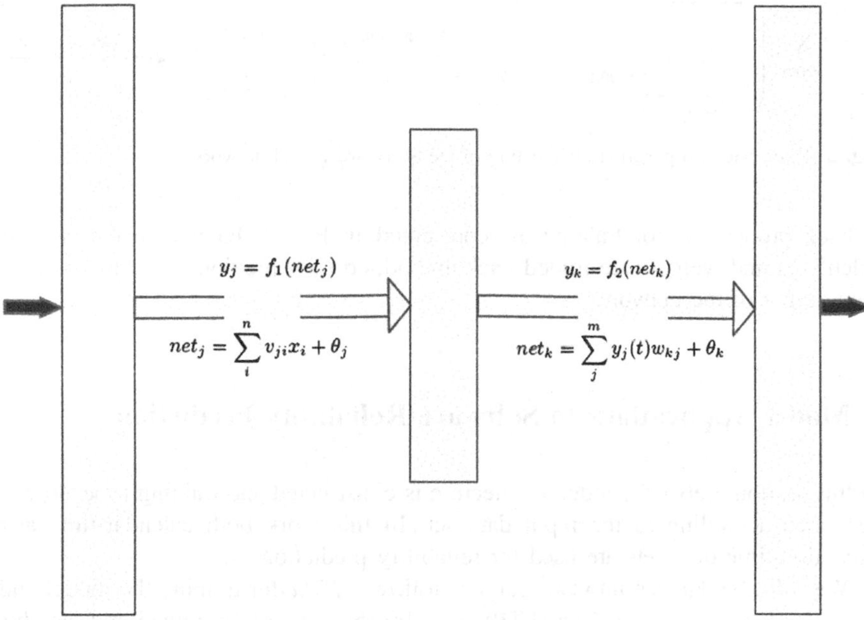

Fig. 1 A simple architecture of feed-forward neural network

In this work, the parameter learning rule is applied. The network model with parameter learning in multilayer architecture is shown in Fig. 1. In this model, the input layer receives x_i (i.e. number of failures) as input. This layer typically performs no function other than buffering the input signal. Any layer between the input and output layer is called a hidden layer, because it is internal to the network and has no direct contact with the external environment. A simple feed forward network assimilates the current activation in memory and generates the output. For each iteration, the errors are back propagated with a set of connection weights even further. It is nothing but the repetition of weights for an arbitrary number of times.

The conventional feed-forward neural network consists of two-layered network. The network comprises of two steps mapping which can represented as

$$y = F_2(F_1(x)) \tag{1}$$

where F_1 is a transfer function between input and hidden node. F_2 is a transfer function between hidden and output node.

The hidden nodes are fully connected with input and output nodes. The basic feed forward back-propagation neural network architecture comprises of two steps: (1) feed forward neural network, (2) back propagation with error from output layer. The network architecture is shown in Fig. 1.

The feed forward network model is a simple supervised learning model. In back-propagation learning, the error is calculated for each pair of input pattern and then it

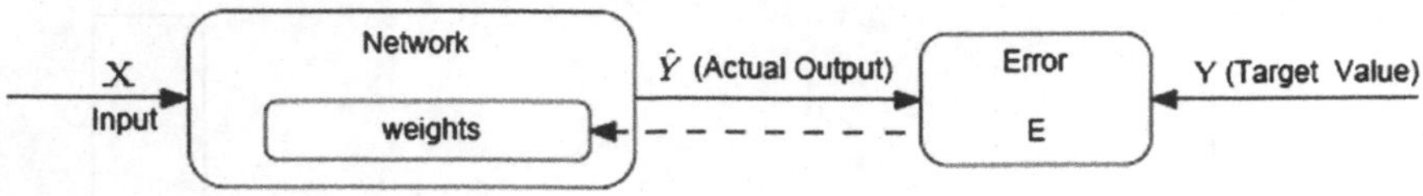

Fig. 2 Basic back-propagation architecture of feed-forward neural network

is back propagated for training as represented in Fig. 2. Each time the error is calculated and weights are folded back and added with previous error to compute new output of the network.

3 Model Applications to Software Reliability Prediction

In this section, network model architecture is constructed and training procedure is described according to the input data set. In this work both calendar-time and execution-time data sets are used for reliability prediction.

We split the data set into two parts as follows: 75 % for training the model and 25 % for *long term prediction* (LTP). Initially the network is trained and weights are fixed.

Based on this fixed weights, the output (i.e. Cumulative failure) is calculated using remaining 25 % data set for LTP.

3.1 Adapting the Network Approach and Training

Back-propagation learning techniques are used in the Eq. 1 to update the weights of the network for training the feed forward network. The input vector 'x' is propagated with a layer associated with weight V as depicted in Eq. 2. Hence, Eq. 3 shows the input to the ith hidden unit at time t.

$$y_j = f_2(net_j) \tag{2}$$

$$net_j = \sum_{i}^{n} v_{ji}x_i + \theta_j \tag{3}$$

where n is the number of input nodes, θ_j is a threshold value, x in input (i.e. number of cumulative failures), v is the weight between input node and hidden node, and f1 is an activation function between input and hidden node that needs to be continuous, differentiable, and non-decreasing. The output of the network is calculated by the state and weight W associated with that output layer, m is the number of 'state' or 'hidden' nodes. The output from hidden node to output node is calculated using Eqs. 4 and 5.

Where f_2 is an activation function between hidden node and output node, w_{kj} is the weight between hidden node and output node, θ_k is the bias.

$$y_k = f_2(net_k) \tag{4}$$

$$net_k = \sum_j^m y_j(t) w_{kj} + \theta_k \tag{5}$$

The feed-forward neural network is a category of static network as it is a network with no output feed-back. We first trained the SRGM on 75 % data set. Then after fixing the weight, the testing phase work on rest 25 % data set. In this paper, the back-propagation training (BPT) algorithm with supervised learning is applied to get the back-propagated error. The error in back-propagation is propagated towards the hidden layer to minimize the error.

The Network Architecture Construction: This section gives a brief discussion about network training using back-propagation learning. In this paper, we consider n-continuous-valued nodes. Output y_i in Eq. 6 is the execution time (i.e. actual failure interval) with activation function f(.) and weight w_{ij} (i.e. from node j to node i) associated with this connection. Here the input nodes xi receives external inputs (i.e. cumulative number of failure after (i − 1)th failure interval).

$$y_i = f(h_i) = f(\sum_i w_{ij} y_i + x_i) \tag{6}$$

where, $h_i = \sum_j w_{ij} + x_i$, x_i is the net input to the node i.

The predicted output y_i is compared with the desired output. Then the network is trained by using input data set when desire output patterns exist as the network model belongs to supervised learning. As per this learning, the actual outputs ($\hat{Y}$) come closer to target output (Y). The summed square error (SSE) is the difference between actual cumulative failure and desired cumulative failure with cost function is calculated using Eq. 7.

$$L = \frac{1}{2}\sum_{k=1}^m (\hat{y}_k - y_k)^2 = \frac{1}{2}\sum_{k=1}^m L_k^2 \tag{7}$$

(m is the total no of output nodes and is an index over training sequence, $\hat{y}$ is the desired cumulative failures, y_k is the actual predicted cumulative failures)

Where the summation ranges over all the output units such as

$$L_k = \begin{cases} \hat{y}_k - y_k, & k\text{th } \textit{output node} \\ 0, & \textit{otherwise} \end{cases}$$

Using Gradient-descent method, the proposed network model gives a weight update rule that requires a matrix inversion at each step.

By applying gradient decent method, the weight changes Δw_{ij} and Δv_{lj} is represent in Eqs. 8 and 9 respectively.

$$\Delta w_{il} = \eta(\hat{y}_i - y_i)a'(f_i)z_l \tag{8}$$

$$\Delta v_{lj} = \eta \sum_{i=1}^{m} [(\hat{y}_i - y_i)a'(f_i)w_{il}]a'(g_l)x_j \tag{9}$$

where η is a scalar parameter that is used to set the rate of adjustment, referred as learning rate.

Feed Forward Network Training: To train this network, we have considered multiple numbers of neurons in input, hidden and output layers. The number of hidden layer is one and number of neuron varies from 9 to 13. Decreasing or increasing beyond the above range does not have any impact on prediction accuracy. The input layer neurons are exempted from computation. The error is computed in output layer and difference is calculated between the actual output and target output value. The whole data set is split into two parts: (1) 75 % data set for training the model and (2) 25 % data set for testing the model.

Normally, there are two numbers of stopping criteria used in the prediction process: (1) the computation process and weight adjustment process (i.e. it is continued until a minimum or a training data close to minimum is found) [10]. (2) the mean square error (MSE) falls below a fixed threshold value.

In our experiment, we have adopted the maximum MSE tolerable error as the stopping criterion. Training the network terminates when the MSE is below a specified tolerance limit. Based on this stopping criterion, the network is back propagated towards hidden layer. This process is continued and weight is up-dated until the MSE is close to maximum tolerable error. In the first epoch, the weights are typically initialized, next onwards, set of weights chosen at random and weights are adjusted in proportion to their contribution to error [11]. There is no restriction for numbers of neuron in each layer. In this work, the logistic Function binary sigmoidal $F(x) = 1/(1 + e^{-\lambda x})$ is used, where λ is the steepness parameter at each step of learning. The range of this transfer function varies from 0.0 to 1.0. The logistic transfer function is used to reduce the computational burden during training [12]. The proposed model is implemented using MATLAB Version 7.10. The network model is trained with initial weights and continues until stopping criteria is satisfied and best weights are recorded for training data calculation and end-point-prediction of the reliability. After the stopping condition is satisfied, the weights are recorded for LTP. The sum square error (SSE) is calculated between the desired outputs and actual output with help of error function using Eq. 7. The target values are not present for hidden units. The error precision is fixed at beginning is taken as $E_{max} = 0.005$.

4 Experimental Results and Observations

The system failure data are collected during testing phase. Normally data set used for reliability growth model having defect severities 2 and 3 as per TANDEM report [13]. In our prediction experiment, we considered software failure data obtained from John D. Musa, Software Reliability Data, Cyber Security and Information Systems Information Analysis Center, January 1980 [14]. The software failure data are collected through the website at Bell Telephone Laboratories. The data set consists of (a) Failure Number, (b) time between failures (TBF) of different medium-sized applications.

We considered y_i as the time between (i − 1)th and ith software failure, we forecast y_{i+1} by using $\{x^{(1)}, y^{(1)}\}$, $\{x^{(2)}, y^{(2)}\}$, …, $\{x^{(i)}, y^{(i)}\}$. This pattern is as follows:

Cumulative failure number	Next failure number
$y_1, y_2 \cdots y_k$	y_{k+1}
$y_2, y_3 \cdots y_{k+1}$	y_{k+2}
	…
$y_{p-k}, y_{p-(k+1)} \cdots y_{p-1},$	y_p

The data sets DS [1] and DS [2] that contain 136 and 191 failures respectively, are considered in our experiment. The number of nodes in the input and output layer is one. The number of hidden layer is one and numbers of neurons in the hidden layer are taken from 9 to 45. We observed that, the result is reasonable good up to 40 neurons, after that the same result is repeated. The training performance shows little bit good result by considering two hidden layers, but not for all data set. Taking more than two hidden layers produces same result as that of single hidden layer.

List of some prediction parameters:

- **The Average Error (*AE*)**: AE is used to compare our model with other models to test the predictive accuracy.

$$\text{Relative Error } (\%)\, RE_i = (|(Forcasted_i - Actual_i)/Actual_i|) * 100$$

$$\text{Average Relative Error } (\%) : 1/n \sum_{1}^{n} RE_i$$

- **The Root Mean Square Error *(RMSE)***: To determine fit to the proposed model with large data set.. The lower is RMSE, the higher is prediction accuracy.

$$\text{RMSE} = \left[\sqrt{\sum_{1}^{n}(Forcasted_i - Actual_i)^2}\right] / n$$

- **Mean Absolute Error *(MAE)***: The MAE and RMSE are used together to analyze the variation In the errors on data set.

$$\left[\sum_{1}^{n}|(Forcasted_i - Actual_i)|\right] / n$$

We perform the goodness of fit (GOF) test to observe the convergence of the curve using the data set taken for this work. R2 is the measuring unit for this curve fitting [15]

$$R^2 = 1 - \frac{\sum_{i=1}^{n}(m_i - m(t_i))^2}{\sum_{i=1}^{n}(m_i - \sum_{j=1}^{n}\frac{m_j}{n})^2} \tag{10}$$

Using data set DS [1] for prediction, we got the following measured data: (1) average relative error (ARE)(%) is 2.8768, Normalized RMSE (NRMSE) is 0.0089 and mean absolute error (MAE) is 0.0484. Figure 3a has shown the desired output and predicted output and Fig. 3b shows the difference between these two. The ARE is 2.8462, RMSE is 0.0092, NRMSE is 0.0607 and MAE is 0.0496 for data set DS [2]. Table 1 shows the different measurement value for DS [1] and DS [2]. The AEs, we found in this experiment are showing good accuracy and consistent result then some well-known method, which is illustrated in Table 2. Our proposed model converging continuously to desire number of failures that indicates the performance of GOF for this data set.

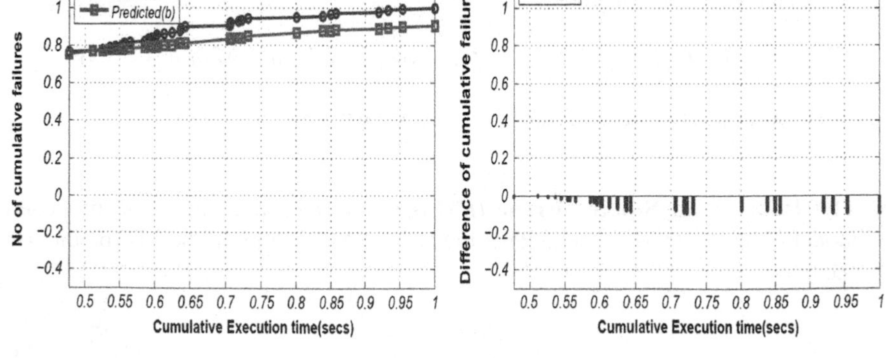

Fig. 3 Graphical representation of end-point prediction and relative for DS [1]. **a** End-point reliability prediction using test data; **b** Relative prediction error

Table 1 Predictive performance on different data set

Measurement parameter	DS [1]	DS [2]
AE	2.8768	2.8462
RMSE	0.008	0.0092
NRMSE	0.058	0.0607
MAE	0.0484	0.0496

Table 2 A comparison on predictive performance on DS [1]

Model	AE	Normalization error
Logarithm	16.84	6.29
Exponential	28.35	8.20
Power	18.35	11.39
FFN-generalization	24.73	2.80
Elman-net-prediction	3.05	2.09
Proposed model	2.8768	0.0585

Table 3 RMSE and NRMSE results of LTP for different analysis

Data set	Predicted value	Measured parameter
DS	0.135522	NRMSE
DS [1, 16]	0.119402	NRMSE
DS [10, 17]	0.076335	NRMSE

In the first trial, we fixed the number of neurons in input and output layers. Consequently, we increased the number of neurons in hidden layer. We found that, the AE is also decreasing accordingly, but at certain number of neurons in hidden layer, the AE as well as other measured value remain on unaltered.

The NRMSE estimated in this model is compared with the result of [16], which is shown in Table 3. We found it is producing a significant result using this model for DS [1] (Fig. 4).

4.1 Observation

Table 1 depicts the predicted value of AE, RMSE, and MAE after satisfying the stopping criteria. The comparison between some analytical models and the proposed feed forward neural network model for software reliability prediction are summarized for DS [1] in Table 2. It is observed that the proposed model is a better predictor than that of other models.

Table 2 is considered for data set of the project "Real Time Command and Control System DS [1]" with various models from [18].

Normally the reliability of software is measure of GOF and end-point-predictability. The accuracy and consistency of software are measured by normalized root

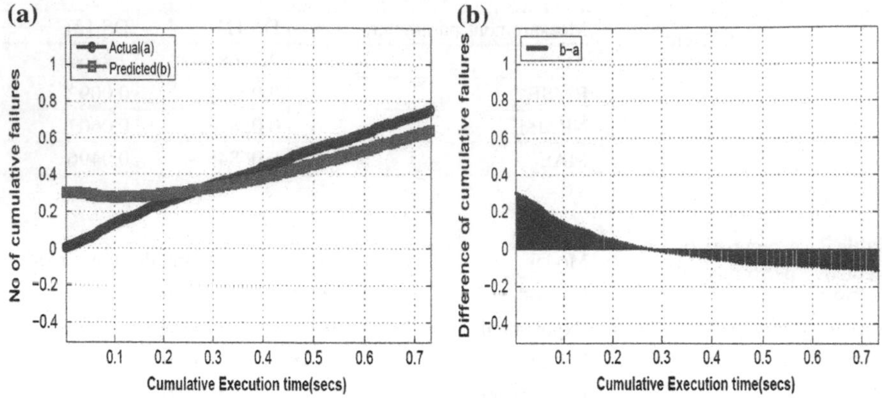

Fig. 4 Graphical representation of end-point prediction and relative for DS [2]. **a** End-point reliability prediction using test data; **b** Relative prediction error

mean square error (NRMSE) on remaining test data set. So NRMSE is also used as software measurement criteria [17]. The end-point prediction in Table 2 shows comparing this AE with many work [18–20], it is observed that it is giving better result. Besides this, another measurement criteria NRMSE is also found minimum then various reliability prediction model [16, 17, 21] and RMSE value with [19, 20, 22].

Model quality is observed if its predictions are close to the ideal line passing through the zero error [23]. Figures 3b and 4b shows the prediction closeness between the actual value and prediction value. It is found that the difference between the two values are significantly close to each other, which indicates that lesser the AE, higher is the accuracy in prediction.

Some observations using feed forward neural network with back-propagation learning are listed below:

- The overall performance in terms of accuracy, consistency and application potential is better than the analytical models.
- If the data set is having large fluctuation, then it will effect the prediction accuracy. So the performance result of the network is effected. Such fluctuations may not expect from software manager's point of view.
- It is easy to design and implement for various data sets. However, it produces less accurate and inconsistent behavior for fluctuate data set.
- Weights are taken randomly, so every time expecting the same on same number of epoch is an issue.
- It may not produce satisfactory result for insufficient training data size.
- The model cannot manage well with major changes that were not reflected in training phase.

The proposed model is a powerful technique because of its nonlinear and logistic properties and being less constraint on the number of observations. The proposed model shows that it yields a lower average relative prediction error compared to other neural network [23, 24] and recurrent neural network [8, 18] approaches. This approach is illuminated as a compliment to the repository of software reliability models, rather than as a substitution to them.

5 Conclusions

In this model, we presented a novel technique for software reliability prediction using feed forward neural network of CI through reliability growth model with BPT technique. The use and applicability of feed forward network model during software testing phase was also discussed. A methodology for developing a model from failure data was proposed and illustrated by means of data set collected from different projects. Our proposed technique accurately predicts the software reliability. We presented experimental evidences showing that feed forward network using back propagation through reliability growth model is giving the accurate result comparable to other methods. Experimental results show that this approach is computationally feasible and can significantly reduce the cost of testing the software and estimating software reliability. This model is easily compatible with different smooth trend data set and projects.

References

1. Boland, P.J.: Challenges in software reliability and testing. Technical report, Department of Statistics National University of Ireland, Dublin Belfield—Dublin 4 Ireland (2002)
2. IEEE: Standard glossary of software engineering terminology. Standards Coordinating Committee of the IEEE Computer Society (1991)
3. Khatatneh, K., Mustafa, T.: Software reliability modeling using soft computing technique. Eur. J. Sci. Res. **26**, 154–160 (2009)
4. Musa, J.D., Okumoto, K.: A logarithmic poisson execution time model for software reliability measurement. In: Straeter, T.A., Howden, W.E., Rault, J.C., (eds) ICSE, Orlando, Florida, NJ, Proceedings of the 7th International Conference on Software Engineering, pp. 230–238, IEEE Computer Society, USA (1984)
5. Goel, A.L.: Software reliability models: assumptions, limitations, and applicability. IEEE Trans. Soft. Eng. **11**, 1411–1423 (1985)
6. Bhuyan, M.K., Mohapatra, D.P., Sethi, S.: A survey of computational intelligence approaches for software reliability prediction. ACM SIGSOFT Softw. Eng. Notes **39**, 1–10 (2014)
7. Tian, L., Noor, A.: Software reliability prediction using recurrent neural network with bayesian regularization. Int. J. Neural Syst. **14**, 165–174 (2004)
8. Tian, L., Noore, A.: On-line prediction of software reliability using an evolutionary connectionist model, Science Direct. J. Syst. Softw. **77**, 173–180 (2005)
9. Ullah, N., Morisio, M.: An empirical study of reliability growth of open versus closed source software through software reliability growth models. In: Leung, K.R.P.H., Muenchaisri, P. (eds.) APSEC. Hong Kong, vol. 1, pp. 356–361. China, IEEE (2012)
10. Hush, D.R., Herne, B.G.: Progress in supervised neural networks. Sig. Process. Mag. IEEE **10**, 8–39 (1993)

11. Karunanithi, N., Malaiya, Y., Whitley, D.: Prediction of software reliability using neural networks. In: Proceedings IEEE International Symposium on Software Reliability Engineering, pp. 124–130, TX, IEEE, Austin (1991)
12. Sivanandam, S.N., Deepa, S.: Principles of soft computing, 1st edn. Wiley, India (2008)
13. Wood, A.: Software reliability growth models. Technical Report 96.1, Tandem Computers (1996)
14. Musa, J.D.: Software reliability data. Data & Analysis Center for Software (1980)
15. Pillai, K., Nair, V.S.S.: A model for software development effort and cost estimation. IEEE Trans. Softw. Eng. **23**, 485–497 (1997)
16. RajKiran, N., Ravi, V.: Software reliability prediction using wavelet neural networks. In: International Conference on Computational Intelligence and Multimedia Applications, IEEE, vol. 1, pp. 195–199. Savakis, Tamil Nadu (2007)
17. Mohanty, R., Ravi, V., Patra, M.R.: Hybrid intelligent systems for predicting software reliability. Appl. Soft Comput. **13**, 189–200 (2013)
18. Karunanithi, N., Whitley, D., Malaiya, Y.K.: Prediction of software reliability using connectionist models. IEEE Trans. Softw. Eng. **18**, 563–574 (1992)
19. Singh, Y., Kumar, P.: Prediction of software reliability using feed forward neural networks. In Conference, I. (ed.): Computational intelligence and software engineering (CiSE), IEEE, pp. 1–5 (2010)
20. Singh, Y., Kaur, A., Malhotra, R.: Empirical validation of object-oriented metrics for predicting fault proneness models. J. Softw. Qual. Control **18**, 3–35 (2009). Springer Science Business Media, LLC
21. RajKiran, N., Ravi, V.: Software reliability prediction by soft computing techniques. J. Syst. Softw. **81**, 576–583 (2008)
22. Costa, E.O., Aurora R.S., V., Souza, P.G., eds.: Modeling software reliability growth with genetic programming, Chicago, Illinois. In: Proceedings of the 16th IEEE International Symposium on Software Reliability Engineering (2005)
23. Karunanithi, N., Whitley, D.: Prediction of software reliability using feed forward and recurrent neural nets. In: Baltimore, M.D. Neural networks, IJCNN IEEE, vol. 1, pp. 800–805 (1992)
24. Karunanithi, N., Whitley, D., Malaiya, Y.K.: Using neural networks in reliability prediction. IEEE Softw. **9**, 53–59 (1992)

A Harmony Search Based Gradient Descent Learning-FLANN (HS-GDL-FLANN) for Classification

Bighnaraj Naik, Janmenjoy Nayak, H.S. Behera and Ajith Abraham

Abstract The Harmony Search (HS) algorithm is meta-heuristic optimization inspired by natural phenomena called musical process and it quite simple due to few mathematical requirements and simple steps as compared to earlier meta-heuristic optimization algorithms. It mimics the local and global search procedure of pitch adjustment during production of pleasant harmony by musicians. Although HS has been used in many application like vehicle routing problems, robotics, power and energy etc., in this paper, an attempt is made to design a hybrid FLANN with Harmony Search based Gradient Descent Learning for classification. The proposed algorithm has been compared with FLANN, GA based FLANN and PSO based FLANN classifier to get remarkable performance. All the four classifier are implemented in MATLAB and tested by couples of benchmark datasets from UCI machine learning repository. Finally, to get generalized performance, 5 fold cross validation is adopted and result are analyzed under one-way ANOVA test.

Keywords Data mining · Machine learning · Classification · Harmony search · Functional link artificial neural network · Gradient descent learning

B. Naik (✉) · J. Nayak · H.S. Behera
Department of Computer Science Engineering and Information Technology, Veer Surendra Sai University of Technology, Burla, Sambalpur 768018, Odisha, India
e-mail: mailtobnaik@gmail.com

J. Nayak
e-mail: mailforjnayak@gmail.com

H.S. Behera
e-mail: mailtohsbehera@gmail.com

A. Abraham
Machine Intelligence Research Labs (MIR Labs), Washington, USA
e-mail: ajith.abraham@ieee.org

A. Abraham
IT4Innovations—Center of Excellence, VSB—Technical University of Ostrava, Ostrava, Czech Republic

L.C. Jain et al. (eds.), *Computational Intelligence in Data Mining - Volume 2*, Smart Innovation, Systems and Technologies 32, DOI 10.1007/978-81-322-2208-8_48

1 Introduction

The Harmony Search (HS) (Geem et al. [1]) is a meta-heuristic algorithm which is inspired by musical process of searching for a perfect shape of harmony. The algorithm is based on natural musical processes in which a musician searches for a better state of harmony by tuning pitch of each musical instrument, such as during jazz improvisation. Music improvisation by pitch adjustment is analogous to local and global search process to find better solution in any optimization techniques.

Many researchers are attracted towards the study of harmony search and its applications due to the fact that, HS algorithm has few mathematical requirements as compared to earlier meta-heuristic optimization algorithms and can be easily used for optimization problems. The founder of HS have used harmony search to solve various standard benchmark engineering optimization problems which includes vehicle routing problems [2, 3], design of water distribution networks [4] etc. In recent years, HS has been used in Robotics [5–7], Control [8, 9], Power and energy [10–13], Medical [14–16] and different engineering applications [17–19].

Manjarres et al. [20] surveyed about 160 published papers during the year 2012 in the scientific databases of Elsevier, IEEE and Springer. 23 % of papers are contributed to variants of the HS algorithms and rest 77 % paper are published in the area of different application of HS which includes robotics (2 %), medical (3 %), engineering (31 %), control (1 %), power and energy (14 %), water system management (5 %), cross-application (2 %) and others (19 %). In this paper, an attempt is made to use harmony search to design Harmony Search based Gradient Descent Learning for FLANN classifier. Prior to this, from 1990 to 2013, a good number of FLANN model with hybrid learning scheme [21–47] are adopted for solution to many engineering problems.

Rest of this paper is organized as follow: Preliminaries are introduced in Sect. 2 followed by the Proposed Method in Sect. 3. Experimental Setup is illustrated in Sect. 4 and Simulation Results is presented in Sect. 5. Statistical Analysis is provided in Sect. 6 followed by Conclusions in Sect. 7.

2 Preliminaries

2.1 Harmony Search

In general, harmony search is composed of 4 major steps as follows:

Step 1. *Initialize a harmony memory (HM) with randomly generated solution vectors.*
Step 2. *Improvise HM to get New Harmony Memory (NHM).*
Step 3. *Update the HM based on comparison between solution vectors of HM and NHM in terms of fitness. If any harmony in HM is less fitter than harmony in NHM, then harmony in HM will be excluded by adding harmony from NHM.*
Step 4. *Repeat Steps 2 and 3 until no further significant growth in fitness of solution vector is noticed or the maximum number of iterations is reached.*

The harmony memory (HM) is a collection of pre-defined amount of solution vectors similar to a population of a genetic algorithm. Initially HM is initialized with random solution vectors and gradually, solution vectors in HM improved by using step-2 of harmony search procedure known as improvisation step. This step is completed controlled by parameters: Harmony Memory Consideration Rate (HMCR), Pitch Adjustment Rate (PAR) and Bandwidth (BW).

The HMCR controls the balance between exploration and exploitation and it is set between 0 and 1.if HMCR set to 0, searching procedure behaves as purely random search and a value 1 for HMCR indicates 100 % of previous solution vectors from HM are taken into consideration for next generation, means, there is no chance to improve the harmony from outside the HM. In this manner, HMCR maintains the balance between exploration and exploitation. Another parameter PAR determines rate adjustment of solution vectors based on bandwidth (BW), which is usually a variable behaves as step size. The HMCR and PAR determines Memory Consideration Probability (MCP), Pitch Adjustment Probability (PAP) and Random Probability (RP) as follows.

$$\begin{aligned} MCP &= HMCR * (1 - PAR) * 100 \\ PAP &= HMCR * PAR * 100 \\ RP &= 100 - MCP - PAP \end{aligned}$$

Basically, Improvisation of HM is governed by these parameters (MCP, PAP, and RP).

For example: if HMCR = 0.99 and PAR = 0.45 then MCP = 0.9 * (1 − 0.45) * 100 = 49.5 and PAP = 0.9 * 0.45 * 100 = 40.5 and RP = 100 − 49.5 − 40.5 = 10. Which means, during harmony improvisation phase (step-2), from previous harmony memory (HM) 49.5 % of solution vectors are migrated to New Harmony Memory (NHM), 40.5 % of solution vectors are gone through pitch adjustment and then included into NHM and 10 % of solution vectors are gone through modification by adding randomly generated values with existing solution vector in HM.

2.2 Functional Link Artificial Neural Network

Functional Link Neural Network (FLANN, Fig. 1) [48] is a class of Higher Order Neural Networks that utilize higher combination of its inputs and is much more modest than MLP since it has a single-layer network compared to the MLP but still is able to handle a non-linear separable classification task.

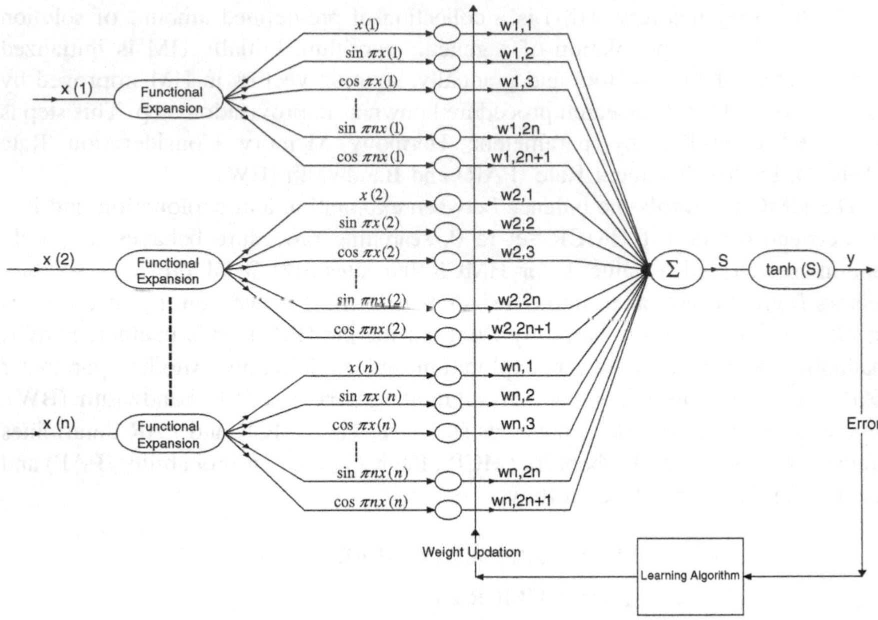

Fig. 1 Functional link artificial neural network

If x is a dataset with data in a matrix of order m × n then functionally expanded values can be generated by using Eq. 1.

$$\begin{aligned}\varphi(x_i(j)) = \{&x_i(j), \cos \Pi x_i(j), \sin \Pi x_i(j),\\ &\cos 2\Pi x_i(j), \sin 2\Pi x_i(j)\ldots\\ &\cos n\Pi x_i(j), \sin n\Pi x_i(j)\}\end{aligned} \tag{1}$$

for i = 1,2…m and j = 1,2…n, Where m and n are number of input pattern and number of input values of each input pattern respectively except class level (Probably last column of dataset x).

$$\begin{aligned}\varphi = \{&\{\varphi(x_1(1), \varphi(x_1(2)\ldots\varphi(x_1(n))\}^T,\\ &\{\varphi(x_2(1), \varphi(x_2(2)\ldots\varphi(x_2(n))\}^T\ldots\\ &\{\varphi(x_m(1), \varphi(x_m(2)\ldots\varphi(x_m(n))\}^T\}\end{aligned}$$

W_i is the weight vector Initialized randomly for a single input value of a input pattern as $W_i = \{w_{i,1}, w_{i,2}, \ldots w_{i,2n+1}\}$, Where i = 1,2…n. For the set of input value of a single pattern weight vector is $W = \{W_1, W_2 \ldots W_m\}^T$ Net output of FLANN network is obtained as follows.

$$Y = \tanh(S) = \{\tanh(s_1), \tanh(s_2) \ldots \tanh(s_m)\} = \{y_1, y_2 \ldots y_m\},$$

where S is calculated as $S = \varphi XW = \{s_1, s_2 \ldots s_m^T\}$.

2.3 *Gradient Descent Learning*

Error of kth input pattern is generated as $e(k) = Y(k) - t(k)$ which is used to compute error term in Eq. 2.

$$\delta(k) = \left(\frac{1 - y_k^2}{2}\right) \times e(k) \tag{2}$$

If $\varphi = (\varphi_1, \varphi_2 \ldots \varphi_L)$, $e = (e_1, e_2 \ldots e_L)$ and $\delta = (\delta_1, \delta_2 \ldots \delta_L)$ are vector which represent set of functional expansion, set of error and set of error term respectively then weight factor of w 'ΔW' can be computed by using Eq. 3.

$$\Delta W_q = \left(\frac{\sum_{i=1}^{L} 2 \times \mu \times \varphi_i \times \delta_i}{L}\right) \tag{3}$$

where $= 1, 2 \ldots L, q = 1, 2 \ldots L \times (2n + 1)$ and L is the number of input pattern. Weight updating is done as $w_{new} = w + \Delta W$ where $w = (w_1, w_2 \ldots w_{L\times(2n+1)})$ and $\Delta W = (\Delta W_1, \Delta W_2 \ldots \Delta W_{L\times(2n+1)})$.

3 Proposed Method

In this section, a FLANN model with a hybrid gradient descent-learning scheme based on Harmony Search have been presented. Basic problem solving strategy of Harmony Search is used to design a fast and accurate learning method as illustrated in Algorithm 1.

Algorithm 1. HS-GDL-FLANN

```
% HMS : Harmony Memory Size , HMCR : Harmony Memory Consideration Rate, PAR : Pitch
Adjustment Rate, BW : Bandwidth.
Step – 1 : Create and randomly initialize a harmony memory (HM) with size HMS.
        % Initialization of HMS, HMCR , PAR and BW.
        HMS=50; HMCR=0.9; PAR=0.45; BW=0.0001;
        % Initialize Population of HM
        HM =-0.1 + (0.1 - -0.1).*rand(HMS, (c-1)*(2*n+1));
Step – 2 : Improvise a new harmony from the HM
        % Computes MCP (memory consideration probability), PAP (pitch adjustment
        probability) and RP (randomization probability).
            MCP=HMCR*(1-PAR)*100;
            PAP=HMCR*PAR*100;
            RP=100-MCP-PAP;
 while(1)
    for i=1:1:HMS
     r=rand(1)*100;
    % Select jth weight-set randomly from harmony memory with memory consideration
    probability (MCP) which serve as new Harmony memory (NHM).
            If (1<=r && r<=MCP)
                    j =f loor ( mod((rand(1)*1000), HMS ) )+1;
                    NHM(i,:)=HM(j,:);
            End if
    % Select jth weight-set randomly from harmony memory for pitch adjustment to improve
    quality of weight-set in HM which serves as new harmony memory (NHM). The PAR and
    appreciate bandwidth (bw) serve this purpose. It is similar to the local search method with
    step size of variable distance bandwidth.
            If (MCP+1<=r && r<=PAP)
                    j=floor(mod((rand(1)*1000),HMS))+1;
                    r1=rand(1);
                    if (r1<=0.5)
                            cbw=(rand(1)*BW);
                            NHM(i,:)=HM(j,:)-cbw;
                    else
                            cbw=(rand(1)* BW);
                            NHM(i,:)=HM(j,:)+cbw;
                    End if
            end if
    % Select jth weight-set randomly from harmony memory with random probability which
    serve as new harmony memory(NHM). In this phase, a jth weight-set is selected randomly
    from HM and added to NHM by suitably adding or subtracting a random value from it.
            If (PAP+1<=r && r<=RP)
                    j=floor(mod((rand(1)*1000),HMS))+1;
                    NHM(i,:)=HM(j,:) + (-0.1 + (0.1 - -0.1).*rand(1));
            End if
    end for
    % Update the HM. If the new harmony (weight-sets) in NHM is better than the harmony
    in the HM, then add the new harmony into the HM by excluding the worst harmony from
    the HM.
    for i =1:1:HMS
            cw=HM(i,:);
            F1(i,1)=fitfromtrain (φ,cw, t, μ);
    End for

    for i=1:1:HMS
            cw=NHM(i,:);
            F2(i,1)=fitfromtrain (φ,cw, t, μ);
    End for
    lf=length(F1);
    for i=1:1:lf
            if (F1(i,1)<F2(i,1))
                    HM(i,:)=NHM(i,1);
            end if
    end for
    % Repeat Step 2 until the maximum number of iterations is reached.
    if (iter <= MAX_ITERATION)
            break;
    end if
  iter=iter+1;
end While
```

Algorithm – 2. Fitfromtrain Procedure

```
function F=fitfromtrain (φ, w, t, μ)
    S = φ X w
    Y = tanh(S);
    If φ = (φ_1, φ_2 ... φ_L), e = (e_1, e_2 ... e_L) and δ = (δ_1, δ_2 ... δ_L) are vector which represent set
    of functional expansion, set of error and set of error tern respectively then weight factor of w
    'ΔW' is Computed as follow ΔW_q = (Σ_{i=1}^{L} 2×μ×φ_i×δ_i / L).
    compute error term δ(k) = ((1-y_k^2)/2) × e(k), for k=1,2...L where L is the number of pattern.
    e = t - y;
    Compute root mean square error (RMSE) from target value and output.
    F=1/RMSE, where F is fitness of the network instance of FLANN model.
end
```

4 Experimental Setup

The proposed method has been implemented using Matlab 9.0 on a system with the following configuration: Intel Core Duo CPU T2300, 1.66 GHz, 2 GB RAM and Microsoft Windows XP Professional 2002 OS. Dataset (Table 1) used for classification are originated from UCI learning repository [49] and 5-folds cross-validated dataset (Table 2) are prepare by using KEEL software tool [50]. Datasets are gone through Min-max normalization to scale them in interval −1 to 1 before training and testing phase. Other parameters are set as follows.

- **Harmony Search Parameter Setting**:
  Harmony Memory Size (HMS) = 50; Harmony Memory Consideration Rate (HMCR) = 0.9; Pitch Adjustment Rate (PAR) = 0.45; Bandwidth (BW) = 0.0001.
- **FLANN Parameter Setting**:
  Number of functional expansion term for a single input value of a input pattern: 11 (By setting n = 5 in 2n + 1 input terms to be generated for functional expansion in FLANN); Initialization of weight vector: Values between −1 and 1; Number of epochs: 1,000.
- **Gradient Descent Learning Parameter Setting**:
  Learning rate: 0.13 (Tested values in the range 0–2 with an interval of 0.01 and found best for the value 0.13.)

Table 1 Datasets information

Dataset	Number of pattern	Number of attributes	Number of classes	Number of pattern in class-1	Number of pattern in class-2	Number of pattern in class-3
Monk 2	256	07	02	121	135	–
New thyroid	215	06	03	150	35	30
Heart	256	14	02	142	114	–
Hayesroth	160	05	03	65	64	31

Table 2 Comparison of classification accuracy in cross-validated datasets

Datasets		Task	Accuracy of classification in %			
Data	5-folds		FLANN	GA-FLANN	PSO-FLANN	HS-GDL-FLANN
Monk-2	monk-2-5-1tra dat	Training	92.03125	94.14065	96.36718	98.08593
	monk-2-5-1tst.dat	Testing	91.95403	93.10344	96.23697	97.7011
	monk-2-5-2tra.dat	Training	90.62501	94.60938	96.13281	97.96874
	monk-2-5-2tst.dat	Testing	89.54023	91.26433	93.56322	96.55172
	monk-2-5-3tra.dat	Training	93.39845	96.4844	97.65626	96.3672
	monk-2-5-3tst.dat	Testing	89.18602	90.93026	94.186	95.81396
	monk-2-5-4tra.dat	Training	95.23436	98.8281	97.89062	98.2422
	monk-2-5-4tst.dat	Testing	95.34881	94.186	95.81396	95.3488
	monk-2-5-5tra.dat	Training	97.8516	98.67192	99.2188	98.90628
	monk-2-5-5tst.dat	Testing	94.18605	96.51163	97.5291	97.26748

5 Simulation Results

In this section, we present the comparative study on the efficiency of our proposed method under 5 fold cross validation and the results of proposed HS-GDL-FLANN model is compared with FLANN, GA-FLANN based on Genetic Algorithm [51] and PSO-FLANN based on Particle Swarm Optimization [52, 53]. Classification accuracy (Eq. 4) of classification models are calculated in terms of confusion matrix and listed in Table 3 and average performance of datasets are demonstrated in Table 3. The root mean square error (RMSE) generated up to 1,000 epochs are obtained from all four models in different datasets are compared and demonstrated in Figs. 2, 3, 4 and 5.

If 'cm' is confusion matrix of order m x n then, accuracy of classification is computed as:

$$Classification\ Accuracy = \frac{\sum_{i=1}^{n}\sum_{\substack{j=1 \\ i==j}}^{m} cm_{i,j}}{\sum_{i=1}^{n}\sum_{j=1}^{m} cm_{i,j}} \times 100 \tag{4}$$

6 Statistical Analysis

The simulated results of proposed method are analysed under ANOVA by using SPSS-16.0 statistical tool to prove the result statistically significant. The test has been carried out using one-way ANOVA in Duncan multiple test range with 95 %

Table 3 Performance comparisons in terms of accuracy

Dataset	Accuracy of classification in average (%)							
	FLANN		GA-FLANN		PSO-FLANN		HS-GDL-FLANN	
	Train	Test	Train	Test	Train	Test	Train	Test
Monk 2	93.82813	92.04303	96.54689	93.19913	97.453134	95.46585	97.91407	96.53661
Thyroid	93.918596	76.55813	94.197648	77.53487	94.3023	78.79069	94.406952	79.2558
Heart	88.962966	78.48149	89.407408	79.07408	89.777762	79.85185	89.91665	80.22221
Hayesroth	90.35938	82.3125	91.062504	83.5625	91.265628	83.9375	91.546876	85.0625

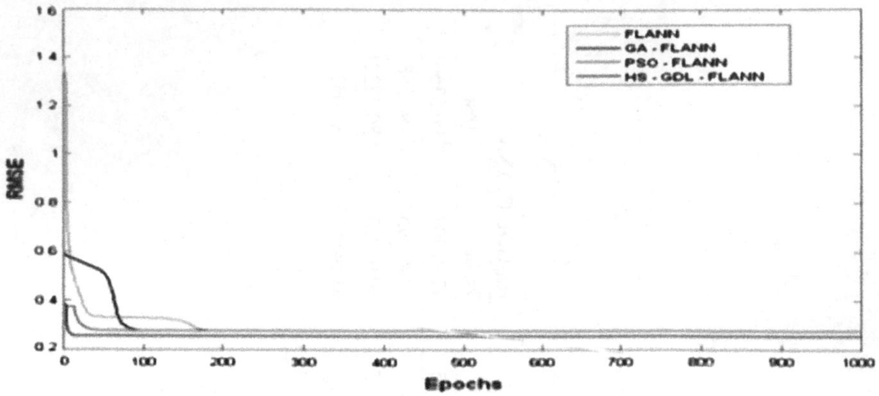

Fig. 2 Monk-2 dataset

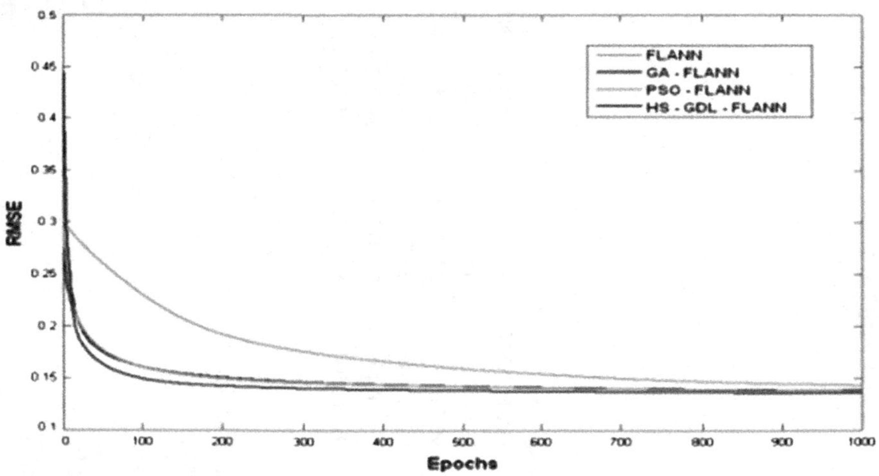

Fig. 3 New thyroid dataset

confidence interval, 0.05 significant levels and linear polynomial contrast. The snap shot of the test is listed below (Figs. 6 and 7). The result of one-way ANOVA test shows that, the proposed method is statistically significant and has minimum standard error.

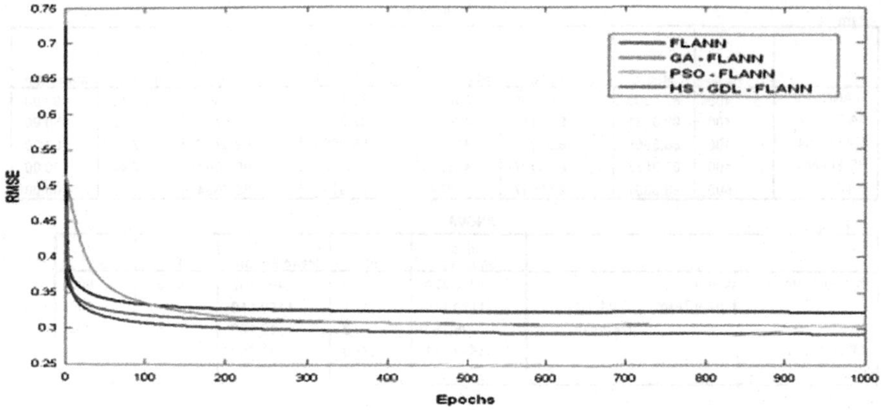

Fig. 4 Heart dataset

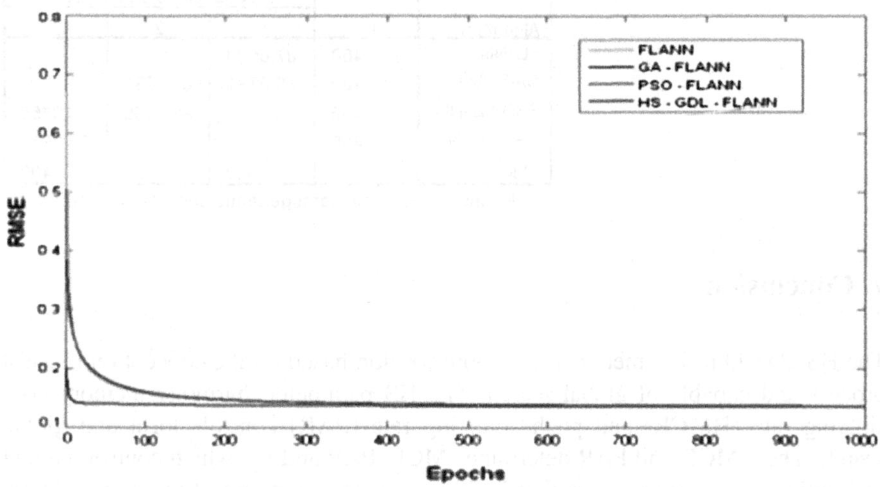

Fig. 5 Hayesroth dataset

Descriptives

Sample

	N	Mean	Std. Deviation	Std. Error	95% Confidence Interval for Mean		Minimum	Maximum
					Lower Bound	Upper Bound		
FLANN	400	87.0580	8.96739	.44837	86.1766	87.9395	65.12	100.00
GA-FLANN	400	88.0731	8.52111	.42606	87.2355	88.9107	65.12	100.00
PSO-FLANN	400	88.8556	8.31227	.41561	88.0385	89.6727	67.44	100.00
HS-FLANN	400	89.3577	8.06437	.40322	88.5650	90.1504	67.44	100.00
Total	1600	88.3361	8.50927	.21273	87.9189	88.7534	65.12	100.00

ANOVA

Sample

			Sum of Squares	df	Mean Square	F	Sig.
Between Groups	(Combined)		1206.470	3	402.157	5.602	.001
	Linear Term	Contrast	1180.110	1	1180.110	16.439	.000
		Deviation	26.360	2	13.180	.184	.832
Within Groups			114573.387	1596	71.788		
Total			115779.857	1599			

Fig. 6 One-way test and ANOVA

Fig. 7 Post-hoc

Sample

Duncan

Algorithm	N	Subset for alpha = 0.05		
		1	2	3
FLANN	400	87.0580		
GA-FLANN	400	88.0731	88.0731	
PSO-FLANN	400		88.8556	88.8556
HS-FLANN	400			89.3577
Sig.		.090	.192	.402

Means for groups in homogeneous subsets are displayed.

7 Conclusion

The HS algorithm is a meta-heuristic optimization based on the concept on musical process and capable of global search. The HS parameter, harmony memory considering rate (HMCR) and pitch adjusting rate (PAR) controls local and global search. The HMCR and PAR determines MCP, PAP and RP which control amount of solution vector migrated to NHM without changes, amount of previous solution vector undergoes changes through pitch adjustment, and amount of solution vector undergoes random changes. HS is free from complicated operator like in GA (crossover and mutation) and requires less few mathematical requirements, which make this flexible algorithm and produce better solutions. In this study, HS is used to enhance the performance of traditional gradient descent learning for FLANN classifier (HS-GDL-FLANN). The proposed method (HS-GDL-FLANN) is compared with FLANN, GA based FLANN and PSO based FLANN and found better in all most all cases.

Future work includes the use of various HS variants algorithms with Gradient descent learning in FLANN classifier.

References

1. Geem, Z.W., et al.: A new heuristic optimization algorithm: harmony search. Simulation **76** (2), 60–68 (2001)
2. Geem, Z.W.: School bus routing using harmony search. In: Genetic and Evolutionary Computation Conference, Washington, DC (2005)
3. Geem, Z.W., et al.: Harmony search for generalized orienteering problem: best touring in China. Lect. Notes Comput. Sci. **3612**, 741–750 (2005)
4. Geem, Z.W.: Optimal cost design of water distribution networks using harmony search. Eng. Optim. **38**, 259–280 (2006)
5. Yazdi, E. et.al.: A new biped locomotion involving arms swing based on neural network with harmony search optimizer. In: IEEE International Conference on Automation and Logistics, pp. 18–23 (2011)
6. Xu, H., et al.: Harmony search optimization algorithm: application to a reconfigurable mobile robot prototype. Stud. Comput. Intell. **270**, 11–22 (2011)
7. Tangpattanakul, P., et al.: Optimal trajectory of robot manipulator using harmony search algorithms. Stud. Comput. Intell. **270**, 23–36 (2010)
8. Coelho, L.S., et al.: A harmony search approach using exponential probability distribution applied to fuzzy logic control optimization. Stud. Comput. Intell. **270**, 77–88 (2010)
9. Das Sharma, K., et al.: Design of a hybrid stable adaptive fuzzy controller employing lyapunov theory and harmony search algorithm. IEEE Trans. Contr. Syst. Tech. **18**, 1440–1447 (2010)
10. Javaheri, H., Goldoost-Soloot, R.: Locating and sizing of series facts devices using lineoutage sensitivity factors and harmony search algorithm. In: 2nd International Conference on Advances in Energy Engineering, vol. 14, pp. 1445–1450 (2012)
11. Sirjani, R., et al.: Optimal allocation of shuntvar compensators in power systems using a novel global harmony search algorithm. Int. J. Electr. Power Energy Syst. **43**(1), 562–572 (2012)
12. Afshari, S., et al.: Application of an improved harmony search algorithm in well placement optimization using streamline simulation. J. Petrol. Sci. Eng. **78**(3–4), 664–678 (2011)
13. Geem, Z.W.: Discussion on combined heat and power economic dispatch by harmony search algorithm. Int. J. Electr. Power Energy Syst. **33**(7), 1348 (2011)
14. Panchal, A.: Harmony search in the rapeutic medical physics. Stud. Comput. Intell. **191**, 189–203 (2009)
15. Gandhi, T.K., et al.: Discrete harmony search based expert model for epileptic seizure detection in electroencephalography. Expert Syst. Appl. **39**(4), 4055–4063 (2012)
16. Landa-Torres, I., et al.: A multi-objective grouping harmony search algorithm for the optimal distribution of 24-hour medical emergency units. Expert Syst. Appl. **40**(6), 2343–2349 (2012)
17. Shariatkhah, M.H., et al.: Duration based reconfiguration of electric distribution networks using dynamic programming and harmony search algorithm. Int. J. Electr. Power Energy Syst. **41**(1), 1–10 (2012)
18. Degertekin, S.O.: Improved harmony search algorithms for sizing optimization of truss structures. Comput. Struct. **92–93**, 229–241 (2012)
19. Askarzadeh, A., Rezazadeh, A.: An innovative global harmony search algorithm for parameter identification of a PEM fuel cell model. IEEE Trans. Ind. Electron. **59**(9), 3473–3480 (2012)
20. Manjarres, D., et al.: A survey on applications of the harmony search algorithm. Eng. Appl. Artif. Intell. **26**, 1818–1831 (2013)
21. Klaseen, M., Pao, Y.H.: The functional link net in structural pattern recognition. In: IEEE Region 10 Conference on Computer and Communication Systems, vol. 2, pp. 567–571 (1990)
22. Liu, L. M. et al.: Image classification in remote sensing using functional link neural networks in Image analysis and interpretation. In: Proceedings of the IEEE Southwest Symposium, pp. 54–58 (1994)
23. Raghu, P.P., et al.: A combined neural network approach for texture classification. Neural Networks **8**, 975–987 (1995)

24. Patra, J.C., Pal, N.R.: A functional link artificial neural network for adaptive channel equalization. Sig. Process. **43**, 181–195 (1995)
25. Teeter, J., Mo-Yuen, C.: Application of functional link neural network to HVAC thermal dynamic system identification. IEEE Trans. Industr. Electron. **45**, 170–176 (1998)
26. Park, G.H., Pao, Y.H.: Unconstrained word-based approach for off-line script recognition using density-based random-vector functional-link net. Neurocomputing **31**, 45–65 (2000)
27. Patra, J.C., Kot, A.C.: Nonlinear dynamic system identification using Chebyshev functional link artificial neural networks. IEEE Trans. Syst. Man Cybern. Part B: Cybern. **32**, 505–511 (2002)
28. Abu-Mahfouz, I.-A.: A comparative study of three artificial neural networks for the detection and classification of gear faults. Int. J. Gen Syst. **34**, 261–277 (2005)
29. Patra, J.C. et.al.: Financial prediction of major indices using computational efficient artificial neural networks. In: International Joint Conference on Neural Networks, Canada, pp. 2114–2120, 16–21 July 2006
30. Mishra, B.B., Dehuri, S.: Functional link artificial neural network for classification task in data mining. J. Comput. Sci. **3**(12), 948–955 (2007)
31. Dehuri, S., et al.: Genetic feature selection for optimal functional link artificial neural network in classification, pp. 156–163. Springer, Berlin (2008)
32. Patra, J.C. et.al.: Computationally efficient FLANN-based intelligent stock price prediction system. In: Proceedings of International Joint Conference on Neural Networks, IEEE, Atlanta, Georgia, USA, pp. 2431–2438, 14–19 June 2009
33. Abbas, H.M.: System identification using optimally designed functional link networks via a fast orthogonal search technique. J. Comput. **4**, 147 (2009)
34. Sun, J. et.al.: Functional link artificial neural network-based disease gene prediction. In: Proceedings of International Joint Conference on Neural Networks, IEEE, Atlanta, Georgia, USA, pp. 3003–3010, 14–19 June 2009
35. Nanda, S.J. et al.: Improved identification of nonlinear mimo plants using new hybrid FLANN-AIS model. In: International on Conference Advance Computing, pp. 141–146 (2009)
36. Chakravarty, S., Dash, P.K.: Forecasting stock market indices using hybrid network. In: World Congress on Nature & Biologically Inspired Computing, IEEE, pp. 1225–1230 (2009)
37. Majhi, R., et al.: Development and performance evaluation of FLANN based model for forecasting of stock markets. Expert Syst. Appl. **36**, 6800–6808 (2009)
38. Patra, J.C., Bornand, C.: Nonlinear dynamic system identification using Legendre neural network. Imn: International Joint Conference on Neural Networks, pp. 1–7 (2010)
39. Emrani, S. et al.: Individual particle optimized functional link neural network for real time identification of nonlinear dynamic systems. In: 5th IEEE Conference on Industrial Electronics and Applications, pp. 35–40 (2010)
40. Majhi, R. et.al.: Classification of consumer behavior using functional link artificial neural network. In: International Conference on Advances in Computer Engineering, IEEE, pp. 323–325 (2010)
41. Bebarta, D.K. et.al.: Forecasting and classification of indian stocks using different polynomial functional link artificial neural networks. In: 2012 Annual IEEE India Conference (INDICON), pp. 178–182 (2012)
42. Mili, F., Hamdi, H.: A comparative study of expansion functions for evolutionary hybrid functional link artificial neural networks for data mining and classification, pp. 1–8 (2013)
43. Mishra, S., et al.: A new meta-heuristic bat inspired classification approach for microarray data. Procedia Technol. **4**, 802–806 (2012). (C3IT)
44. Mahapatra, R. et.al.: Reduced feature based efficient cancer classification using single layer neural network. In: 2nd International Conference on Communication, Computing & Security. Procedia Technol. **6**, 180–187 (2012)
45. Mishra, S. et.al.: An enhanced classifier fusion model for classifying biomedical data. Int. J. Comput. Vis. Robot. **3**(½), 129–137 (2012)

46. Dehuri, S. et.al.: An improved swarm optimized functional link artificial neural network (ISO-FLANN) for classification. J. Syst, Softw. **85**, 1333–1345 (2012)
47. Naik, B., Nayak, J., Behera, H.S.: A novel FLANN with a hybrid PSO and GA based gradient descent learning for classification. In: Proceedings of the 3rd International Conference on Frontiers of Intelligent Computing (FICTA), vol. 1. Advances in Intelligent Systems and Computing 327, pp. 745–754 (2014). doi:10.1007/978-3-319-11933-5_84
48. Pao, Y.H., Takefuji, Y.: Functional-link net computing: theory, system architecture, and functionalities. Computer **25**, 76–79 (1992)
49. Bache, K., Lichman, M.: UCI machine learning repository. http://archive.ics.uci.edu/ml. University of California, School of Information and Computer Science, Irvine, CA (2013)
50. Alcalá-Fdez, J., et al.: KEEL data-mining software tool: data set repository. Integration of algorithms and experimental analysis framework. J. Multiple-Valued Logic Soft Comput. **17** (2–3), 255–287 (2011)
51. Holland, J.H.: Adaptation in natural and artificial systems. University of Michigan Press, Ann Arbor (1975). ISBN 9780262581110
52. Kennedy, J., Eberhart, R.: Particle swarm optimization. In: IEEE International Conference on Neural Network, vol.4, pp. 1942–1948 (1995)
53. Eberhart, R., Kennedy, J.: A new optimizer using particle swarm theory. In: Sixth International Symposium on Micro Machine and Human Science (MHS), pp. 39–43 (1995)

Improved AODV Performance in DOS and Black Hole Attack Environment

Anurag Gupta, Bhupendra Patel, Kamlesh Rana and Rahul Pradhan

Abstract A mobile Ad hoc network has been constructed as virtual backbone or spine of wireless medium. Due to nomadic habit of the node, MANET network are easily establish to a group of nodes with self configuring manner. MANET is simple to set and has changing Topology. Routing in a MANET is a particularly challenging task compared to a conventional network. For data transfer in MANET one of the most commonly used routing protocol is Ad hoc on demand distance vector (AODV), where routes provide on demand and node act as router. It's performance is based on the services of the network, however MANET are vulnerable to attacks due to it's dynamic, nomadic and open environment. However, network is vulnerable to the attacks by presence of malicious nodes. DOS and Black Hole attacks are two security threats in AODV.DOS attacks aims to disrupt the service provided by network or server by continuously sending unwanted packets using malicious node. In Black hole attack, the aim of this attack to capture the path and send fake route reply message and routing information. So to overcome from these consequences we propose an algorithm which introduce a mechanism of DOS and Black hole attacks prevention and check network performance in malicious environment on the analysis of throughput and end to end delay.

Keywords Denial of service (DOS) · Black hole attack · AODV · MANET

A. Gupta (✉) · K. Rana
CSE Department, Galgotias College of Engineering and Technology,
Greater Noida, India
e-mail: 03anuragupta@gmail.com

K. Rana
e-mail: kamlesh.rana@galgotiacollege.edu

B. Patel
CSE Department, Dr. Jivraj Mehta Institute of Technology, Mogar,
Anand, Gujarat, India
e-mail: Patelbhupendra01@gmail.com

R. Pradhan
CSE Department, GLA University, Mathura, India
e-mail: rahul.pradhan@gla.ac.in

L.C. Jain et al. (eds.), *Computational Intelligence in Data Mining - Volume 2*,
Smart Innovation, Systems and Technologies 32, DOI 10.1007/978-81-322-2208-8_49

1 Introduction

MANET network are easily establish to a group or collection of mobile nodes that connect dynamically. No centralized gateway device for connecting and monitoring node. On behalf of this we can say MANET has self configuring, self monitoring and self organizing behavior. Fundamentally two type of mobile Ad hoc network: first is infrastructure based and second are those networks with fixed and wired gateways. In terms of wireless networks bridges for this network are known as base station [1]. Since the network medium is dynamic, nomadic and open operational malicious node access it easily. Malicious node is the variant of mobile node but it's working is completely different compared to normal mobile nodes. Malicious nodes have ability to change or remove routing information, sends the fake route request to access user's data.

Routing performance is effected by malicious node due to carrying different type of attacks on the networks. There are two type of Network routing is defined in MANET for data transfer. One is Proactive and another is Reactive. Reactive Routing protocols are used on time when node wants to send packet or information to the destination [2] opposite the proactive routing protocols. In this type of routing protocols every node should have stored the routing information of its neighbors. Proactive routing protocols discover and maintain a complete set of routes for the lifetime of the network.

The structure of this paper is as follows: Sect. 1 Introduce about MANET, Malicious (Selfish) node and AODV Routing Protocol. Section 2 Present the Security issues of AODV. Section 3 Related Studies, Sect. 4, presents the brief introduction of AODV Routing Protocol. Section 5 presents the nature of DOS and Black Hole attacks Sect. 6 presents the research work done on DOS and Black Hole attack and its solution. In Sect. 7 shows how to implement the dos and Black Hole attack s on network and Experimental outcomes and Sect. 8 Conclusion.

2 Security Issues of AODV

Because in MANET network medium is dynamic, nomadic and open operational, due to this malicious node easily attack on physical link, as they can easily manipulated in ad hoc network. So Ad hoc network are vulnerable to security problems than the wired networks. In this section various security issues are explored.

1. **No centralized gateway Facility**: No centralized gateway device for connecting and monitoring node. Node work in the nomadic and open operational environment, by this intruder can easily attack and disrupt the network this lead to lack of security, we cannot identify the trusted or untrusted node.

2. **No predefine range of Network**: In Ad hoc network there is no predefine range of network. Node work in dynamic environment. Node can easily join and leave the wireless medium. Attacks include eaves dropping, Impersonation, tempering, replay and denial of service attack [3].
3. **End to End packet Delay**: Ad hoc wireless network is based on mobile node. Due to hidden terminals and path break, it increases the error rate and End to End packet delivery ratio in wireless medium.
4. **Lack of Energy Resources**: In Ad hoc network limited energy resource, No alternate power resource. Node operation is depend on the battery power. When attacker attack on the target node, Congestion in the network increase due to requesting fake connection request causing it's battery power lost.

3 Related Studies

Here we have analyzed some related works of avoid Denial of Service attacks and Black Hole attack:

1. The performance of AODV routing protocol with existing malicious nodes has been analyzed using NS2.34 simulator. To measure the performance evaluation different metrics like Throughput, Packet Delivery Ratio and End to end delay have been used. In all these scenarios the number of malicious node varies from 0 to 5 [4].
2. In this approach CORE mechanism that enhances watchdog for monitoring and isolating selfish nodes based on a subjective, indirect and functional reputation is presented. The reputation is calculated based on various types of information on each entity's rate of collaboration. Since there is no motivation for a node to maliciously spread harmful information or data about further nodes, denial of service attacks using the collaboration technique itself are prevented [5].
3. In this paper the algorithm to prevent the DoS and Black Hole attack is proposed. We summarized the node categorized as friends and strangers based on their relationships with their neighbor nodes. A dependence estimator is used in each node to evaluate the trust level of its nearest nodes. The trust level is a function of various parameters like Packet Delivery Ratio, End-to End Delay [6].

4 AODV Routing Protocol

One of the most precious routing protocol for routing is the Ad hoc on demand distance vector (AODV). AODV means it is a collection of Ad hoc, ON demand, Distance and Vector. Ad hoc means node movement or connection or disconnection with the networks at any time, On demand means when source wants to send data to

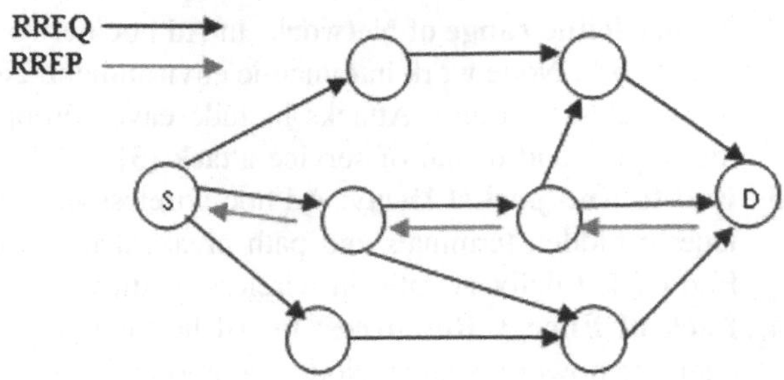

Fig. 1 AODV routing protocol with RREQ and RREP message

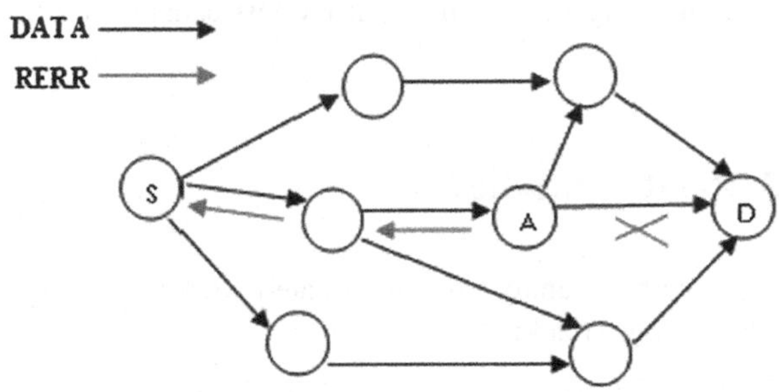

Fig. 2 AODV routing protocol with RERR message

the destination, Destination means find the distance between source to destination in term of number of hope counts, vector means list which store the node information. AODV work on the OSPF method/algorithm. open shortest path first (OSPF) is based on the DISKJETRA's.

Algorithm. In AODV, at every node routing information are store. AD hoc Network routing protocols handle discovering, establishing, recovering and maintaining routing paths, In [4, 7, 8], AODV use some approaches for path or route establishment. Route Request (RREQ) (Fig. 1) Source node broadcast the route request message to neighbor node, neighbor node pass the message to specific destination.

Route Reply(RREP) Destination Node unicast a route reply message to source. Neighbor node maintains the next entry for destination and forwards the reply. In the case of multiple replies source node choose the shortest path with minimum hope count.

Source sequence number and destination sequence number play a important role in AODV. Source node broadcast the packet with SSN and destination use DSN number which define the freshness of the path.

In the route maintenance when link are break, it broad casts a route error packet to its neighbor, when node receive route error message then source restart a route discovery process. In Fig. 2 link break between A and D, So Node A informs or RERR to previous node that this link is broken chose another shortest path.

5 Network Layer Attacks

Despite the efforts that were done to secure the on demand routing protocols [8], in the network layer AODV services are disrupting by several attacks. Through decreasing computation power, bandwidth, energy and resource consumption.

1. **Denial of Service attack**: DOS attack is one of the most dangerous attacks. In this attack, the Malicious node continuous broadcast or send the false control or data packets in the network due to this kind of sending data or false packet network bandwidth is wasted largely and the original packets are not able to reach their destinations. This attack is implemented on the reactive protocols by broadcasting the false data packets and RREQ messages [8].
2. **Black Hole attack**: A Black hole attack is another attack possible on MANET. The aim of this attack to drops the Routing packet or data packet during the performing the operation. It is claiming that it has shortest and fresh path with smaller number of hope count and large sequence number to destination. Due to this Claiming it attracts all the packets and absorbed them without forward to destination node. It increase the Congestion and traffic in the network, and therefore attacker can misuse the traffic.

Figure 3 Shows an example of Black Hole attack, where S is source node, D is Destination node and A is Attacker node. A drop the Routing packet or data packet and claim that I have a fresh and shortest path with minimum hope count.

6 Proposed Solution

The Work is mainly focused on to avoid the Denial of service and Black Hole attacks in mobile Ad hoc network. In this scheme first malicious node is detected and then without interrupting middle nodes and destination node. In this research

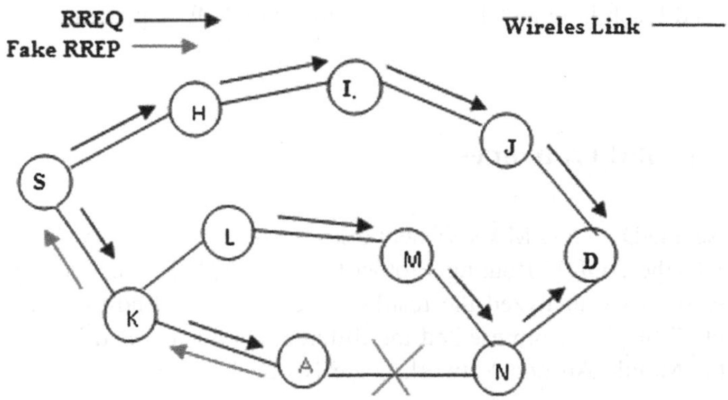

Fig. 3 Black hole attack on AODV

scheme we propose the combined solution for both DOS attack and Black Hole attack.

Step I: First we take all node in node list. Define that neighbor node is malicious or not. First we check whether it is nodes first request, node adds in RREQ table.

Step II: Here Node in RREQ table. We apply the different method for checking the behavior of the node. Node is malicious or not.

1. First process, we apply the counter on each node. Which count the node RREQ to the neighbor.
2. Second process, check node route request to the maximum route request, if route request is greater than the maximum route request then the node add in the malicious list (here we take maximum route request is 5).
3. Third process, It is basis on the current time and expire time. If node RREQ expire time is less than or equal to current time the continue default AODV process otherwise node add in malicious list.
4. Fourth process, apply on each node, It is basis on the Destination sequence number, Current Time and Execute receive reply time required. If Node get the route reply message With The destination sequence number (DSN) then source node compare the Neighbors DSN to each other and DSN to SSN and also compare the route reply time of the neighbors to the current time. If it's excided then Node add in malicious list. Than after source node will send the route request or send the data packet on another path. So this type of process doing till source node will not identify original destination node.

Step III: It is the Step of removing malicious node from the malicious node list after the session expires. May be possible after some time malicious node stop doing malicious things when malicious node expire time is less than current time than node remove from malicious node list. So like that we can catch the malicious node from network, and stop the DOS and black hole attack and continuous the AODV process.

7 Experimental Outcomes

In this phase, the Dynamic MANET network is created with 25, 50, 75, 100 mobile nodes. Apply the AODV Routing Protocol on this MANET network with some parameters. We have analyzed the results in terms of the End-to-End Delay and Throughput. Table 1 has represented the different parameters and its value use for the build the Mobile Ad hoc Network using NS 2.

Table 1 Parameter values used in simulation

Parameters	Values
Number of nodes	25, 50, 75, 100
Area size	1,000 × 1,000
MAC	802.11
Simulation time	100, 200, 300, 400
Traffic source	CBR
Packet size	1,000
Bandwidth	10 mb
Data rate	10 mb
Routing protocol	AODV scenario has been SIMULATED using NS2
Transmission	UDP
Number of malicious node	2, 4, 6, 8 (formula used for calculating malicious node: N * 8/100, where N represent number of nodes)

7.1 Performance Metrics and Results Analysis

Here we show implementation results performance of AODV on the basis of Throughput and End to End Delay.

7.1.1 When We Have Implemented the Dos Attack and It's Prevention Results & Analysis

Simulation result shows variation of the average throughput for 25 nodes with simulation time 100–400 s. It is observed that with attack scenario throughput is 0 and with prevention the throughput decreases as simulation time increases (refer Fig. 4).

Simulation result shows variation of the average End-to-End Delay for 25 nodes with various simulation time 100–400 s. It is observed that with attack scenario

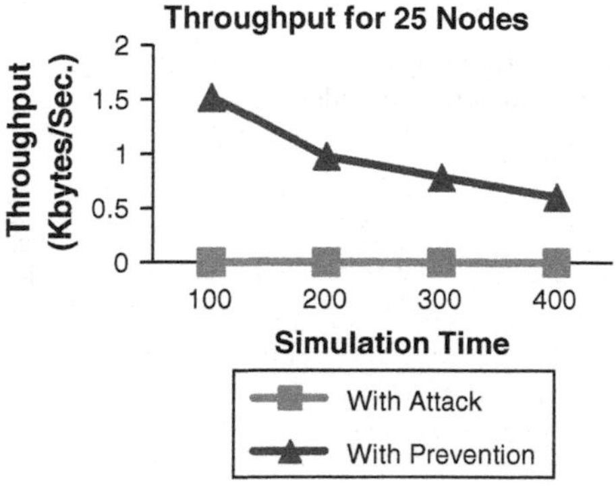

Fig. 4 Throughput versus simulation time (seconds)

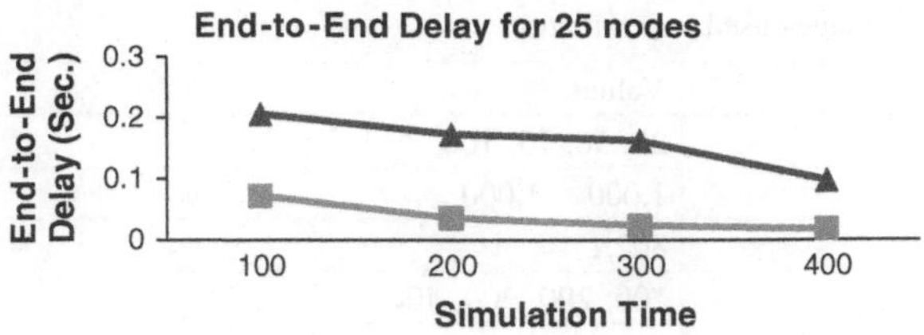

Fig. 5 Average end-to-end delay versus simulation time (seconds)

End-to-End delay decreases and with prevention End-to-End delay increases as simulation time increases.(refer Fig. 5)

7.1.2 When We Have Implemented the Black Hole Attack and Its Prevention Results & Analysis

Simulation result shows variation of the average throughput for 25 nodes with simulation time 100–400 s. It is observed that with attack scenario throughput is 0 and with prevention the throughput decreases as simulation time increases (refer Fig. 6).

Simulation result shows variation of the average End-to-End Delay for 25 nodes with various simulation time 100–400 s. It is observed that with attack scenario End-to-End delay decreases and with prevention End-to-End delay increases as simulation time increases.(refer Fig. 7)

8 Conclusions and Futures Work

Selfish node is the main security danger that special affects the performance of AODV routing protocol. In malicious environment this problem has found because primarily necessary the routing performance in malicious environments. In general

Fig. 6 Throughput versus simulation time (seconds)

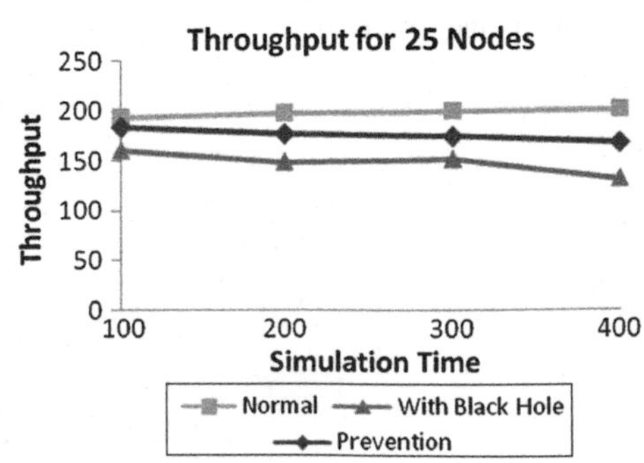

Fig. 7 Average end-to-end delay versus simulation time (seconds)

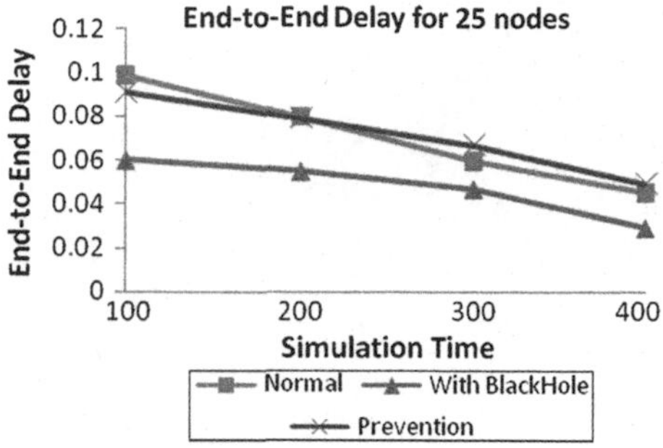

scenario many attacks happen in mobile ad hoc networks. Therefore this work is focus on mechanism to detect and prevent the DOS and Black Hole attack. Work will be focused on securing the network in malicious environment with less delay. Work is still to be done in completely dynamic scenario.

In future we will try to provide the solution of multiple attack on same routing protocols using combine approaches.

References

1. Ismail, I.I., Jaafar, M.H.F.: Mobile ad hoc network overview. In: Proceedings of Asia-Pacific Conference on Applied Electromagnetics (2014)
2. Yau, P.-W., Hu, S., Mitchell, C.J.: Malicious attacks on ad hoc network routing protocols
3. Mishra, A., Nadkarni, K.M.: Security in wireless ad hoc network. In: The Hand Book of Ad Hoc Wireless Network. CRC Press Inc., Boca Raton (2003)
4. Kumar, V., Sharma, R., Kush, A.: Effect of malicious nodes on AODV in mobile ad hoc networks. Int. J. Comput. Sci. Manage. Res. **1**(3), 395 (2012)
5. Michiardi, P., Molva, R.: CORE: A Collaborative reputation mechanism to enforce node cooperation in mobile ad hoc networks. In: Proceedings of IFIP TC6/TC11 Sixth Joint Working Conference of Communication and Multimedia Security (2002)
6. Al-Omari, S.A.K., Sumari, P.: An overview of mobile ad hoc networks for the existing protocols and application. J. Appl. Graph Theory Wireless Ad-hoc Netw. Sens. Netw **2**(1), 87 (2010)
7. Sheikhl, R., Chandee, M.S., Mishra, D.K.: Security issues in MANET: a review. In: 7th International Conference on Wireless and Optical Communications Networks (2010)
8. Issariyakul, T., Hossain, E.: Introduction to Network Simulator NS2 (2009)
9. Gambhir, S., Sharma, S.: PPN: prime product number based malicious node detection scheme for MANETS
10. Sheikhl, R., Chandee, M.S., Mishra ,D.K.: Security issues in MANET: a review. In: IEEE, pp. 1–4, Sept 2010. ISBN 978-1-4244-7203-1
11. Dr. Tamilarasan, S.: Securing and Preventing AODV routing protocol from black hole attack using counter algorithm. Int. J. Eng. Res. Technol. (IJERT) **1**(5) (2012)
12. Akanksha Saini, Harish Kumar : Effect of black hole attack on AODV routing protocol In MANET. IJCST **1**(2) (2010)

Airfoil Self Noise Prediction Using Linear Regression Approach

Shiju Sathyadevan and M.A. Chaitra

Abstract This project attempts to predict the scaled sound pressure levels in decibels, based on the aerodynamic and acoustic related attributes. Each attribute can be regarded as a potential feature. The problem is how to predict the sound pressure level accurately based on those features. This paper describes the approaches of using linear regression models and other optimization algorithms used for the better predictions. The comparative results and analysis are also provided in experiment and results section.

Keywords Linear regression · Airfoil · Gradient descent · Stochastic gradient descent · Linear least squares

1 Introduction

Regression comes under the prediction methods of data mining and under supervised learning in machine learning. Supervised learning is a machine learning task that takes in a known set of input data and known responses to the data, and seeks to build a model that generates reasonable predictions against a new data set. Supervised learning can be split into two broad categories:

Classification: for responses that can have just a few values such as "True" or "False". Regression: for responses that are real numbers such as miles per gallon for a particular car.

S. Sathyadevan (✉)
Amrita Center for Cyber Security, Amrita Vishwa Vidyapeetham,
Clappana, P.O., Kollam, India
e-mail: shiju.s@am.amrita.edu

M.A. Chaitra
Department of Computer Science, Amrita Vishwa Vidyapeetham, Mysore, India
e-mail: chaitrama1991@gmail.com

L.C. Jain et al. (eds.), *Computational Intelligence in Data Mining - Volume 2*,
Smart Innovation, Systems and Technologies 32, DOI 10.1007/978-81-322-2208-8_50

This paper considers Linear Regression based model which is used for noise level prediction of airfoils in wind turbines. This model assumes that the input variables (attributes) have a linear relationship with the output.

There are two ways to performing the Linear Regression. One is the direct way of computing the closed form expression, the other one is to compute numerically or iteratively.

Linear least square approach is being adopted for this research where an over determined system of linear equations is being solved approximately. The best approximation is defined as that which minimizes the sum of squared differences between the data values and their corresponding distance to the regression line. When it comes to optimization algorithms Gradient descent a first order optimization algorithm was used. Although many methods are available for predicting the sound levels of airfoil noise, Linear regression is a very basic and simple algorithm which tells how the target values vary with respect to different independent attribute or features [1, 2].

2 Related Works

Regression analysis is a statistical method for knowing the relationship among the variables. There are many regression techniques available to know the relationship among the variables in those types Linear regression is one of the very simple and basic algorithms for fitting the data it assumes the given data has a linear relationship between dependent variable and one or more independent variables. This algorithm is used in many disciplines to fit a given data. There are other types of linear regression such as like polynomial regression it is a form of linear regression where the relationship between independent variable and dependent variable is modelled as nth order polynomial. This paper talks about a how a linear regression approach is used to fit the airfoil self noise dataset. There are many advanced methods that predict the sound noise level of an airfoil. Regression analysis in combination with Data Binning technique are used, which is a more sophisticated method for finding the noise levels and to analyze the relationship between sound pressure level and the wind speed. The co-relation co-efficient or R^2 is greater than 0.8 the 4th order regression analysis is used or if it is less than that value than a linear regression with data binning techniques are used which gives a more appropriate results [3, 4].

There are many challenges in predicting the airfoil trailing edge noise; the most difficult one is modelling the turbulence generated at the boundary layer. Analytical solutions are available to predict noise generated by airfoil however, the solution requires an estimate of the turbulent velocity or surface pressure spectrum. Disturbance in air flow or more specifically turbulence is a random, complex and highly nonlinear process so it's difficult to get closed form solution. To resolve this problem many advance techniques have been developed example large eddy simulation and RANS modelling etc. But these methods are computationally costs very high.

3 Scope

Wind turbines are considered as renewable source of energy, which are used to produce energy from wind. It uses the kinetic energy from wind and turns into electric power and are used by many countries. But the noise from wind turbines is often perceived as disturbing and affects residents in the residential areas. The guidelines for which sound pressure levels from wind turbines that are allowed varies between countries, but is a main concern for manufacturer, owner and people living nearby the wind turbine. A guideline has been issued stating that the noise levels from wind turbine should be below 40 dB [5, 6].

4 Methods Used

The Linear Regression with multiple features model is used in the proposed method to understand that the regression can also be used to predict the noise generated by the wind turbines.

4.1 Problem Statement

The Objective: The objective is to adjust the model parameters for the linear function to best fit the dataset. A sample dataset consists of n data points $(x_i; y_i)$ where i = 1, 2…n. Where x_i is an independent variable and y_i is a dependent variable. The Model function has the form of $f(x; \theta)$ when the m adjustable parameters are held in the vector θ. The Least Square methods find optimum solution when the sums of squared residuals are minimum.

Residuals are defined as the difference between the actual value and the predicted value by the model. To build the model Linear Regression with Linear least squares and Gradient descent for calculating the model parameters or regression coefficient for the model are derived first.

5 Dataset

An extensive experimentation is carried out on the NASA dataset with attributes Frequency, Angle of attack, Chord length, Free stream velocity, Suction side displacement and Scaled sound pressure level. The proposed approach is critically analyzed and also evaluated. Table 1 provides a snapshot of a subset of the records contained in the dataset. Table 2 provides the attribute, no instances and class variable information and Table 3 provides the detailed meaning of each attribute in

Table 1 Sample dataset

Fq (Hz)	AOA	CL	FSV	SST	SSPL
800	0	0.30	7.13	0.0026	126.2
1,000	0	0.30	7.13	0.0026	125.2
1,250	0	0.30	7.13	0.0026	125.9
1,600	0	0.30	7.13	0.0026	127.5

Table 2 Dataset information

Dataset	Airfoil self noise
No of instances	1,503
No of attributes	6
No of class	1

Table 3 Attribute information of airfoil self noise dataset

Attribute	Information
Frequency	Frequency in Hertz (Hz)
Angle of attack	Angle of attack, in degrees
Chord length	Chord length, in meters
Free-stream velocity	Free-stream velocity, in meters per second
Suction side displacement	Suction side displacement thickness in meters
Scaled sound pressure level	Scaled sound pressure level, in decibels

NASA dataset, obtained from a series of aerodynamic and acoustic tests of two and three-dimensional airfoil blade sections conducted in an anechoic wind tunnel. It is also observed that in Table 1 the dataset have only numeric attributes [7].

5.1 Dataset Information

See Table 2.

5.2 Attribute Information

See Table 3.

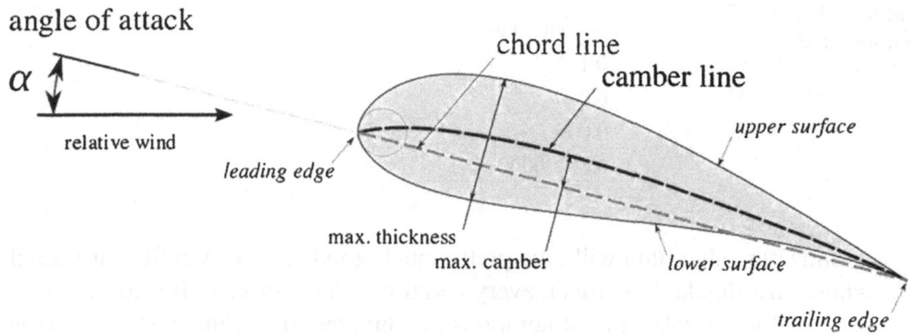

Fig. 1 Airfoil of a wind turbine

5.3 Sample Airfoil Figure

See Fig. 1.

5.4 Pre-processing

The dataset has to be pre processed before experimenting for the better performance. The Pre-processing in Linear Regression involves:

(a) Feature Scaling or parameter regularization (i.e., bringing each attribute to similar levels, e.g., between 0 and 1 (or) 1).
(b) Only numerical values for the attributes are allowed, no categorical or character values are allowed.
(c) The outcome variable is usually continuous.

5.5 Customization

In order to apply this algorithm across different domains, all factors that will vary are termed as customization factors and not hard coded within the algorithm. These parameters can be fed as inputs while executing the algorithm.

(a) Option of selecting the algorithm that is used to compute the model parameters can be selected, Batch Gradient Descent or any other algorithms that can converge faster can be used.
(b) The Learning Rate a customizable parameter used in the optimization algorithm plays an important role while calculating the model parameters. For different datasets the learning rate can be different. Finding or tuning proper alpha for

Table 4 Experimented learning rates

Learning rate
0.1
0.01
0.005
0.000000001

optimization algorithm will give pretty much good results. A sufficiently small whose cost should decrease on every iteration can be selected. If learning rate is too small it will take several iterations to converge, if learning rate is too large then the cost may not decrease on every iteration and may not even converge. Table 4 shows some of the learning rate values that have been experimented.

5.6 Explanation

The dataset is experimented on linear regression algorithm and different optimization algorithm to obtain the best model parameters. The algorithm is explained in detail below.

Step 1: Feature Scaling is applied initially i.e. a pre-processing step to make the values to fall in a range that makes optimization algorithm to converge faster.

Step 2: The hypothesis is set as

$$h_\theta(x) = \theta^T x = \theta_0 + \theta_1 x_1 + \cdots + \theta_n x_n \tag{1}$$

Step 3: To calculate the model parameters, optimization algorithm is used iteratively. Initial model parameters are set to 0 and cost is computed using the linear least square method.

Optimization algorithm calculates the model parameters iteratively and cost is calculated at every iteration by using the obtained model parameters. Checks are performed to ensure that the cost is less than a threshold value (example: $€ < 10^{-9}$) and cost is decreasing in every iteration. If the cost value is more than the threshold value, the process is repeated again and if cost diverges tuning of learning rate is necessary. The obtained parameters are than tested with the hypothesis for prediction as shown in Eq (1).

5.7 Optimization Algorithms

The Linear regression with multiple features is experimented with two optimization algorithm Batch Gradient Descent and Stochastic Gradient Descent.
Batch Gradient Descent: This is the first order minimization algorithm to find the local minimum of a function. Batch gradient descent needs to run over entire dataset before updating the model parameters, which means larger dataset takes more time to converge to optimum value. Learning rate is an important parameter to this algorithm.
Stochastic gradient descent: The stochastic gradient descent (SGD) algorithm is a drastic simplification. Stochastic gradient descent has a convergence rate which is independent of the size of the dataset [8].

The Convergence of Stochastic Gradient Descent usually requires decreasing learning rates. Algorithm updates model parameters by considering one training example at every iteration so algorithm may not converge but, good model parameters can be obtained. When the learning rate is decreased in every iteration we can end up with an optimum solution.
Learning rates: plays an important role in convergence. It will vary for different datasets even though it is possible to set a default value but it has to be experimented with different values for better results.

6 Experiment Results and Analysis

The result of the model is a set of regression coefficients also called as model parameters and associated cost. For the given problem the Hypothesis is framed as given in Table 5.

The hypothesis is framed on the assumption Linear regression model assumes that the input variables (attributes) have a linear relationship with the output. The entire dataset is assumed as set linear equations. The obtained model parameters or regression coefficients are substituted in place of θ_1–θ_6 and x_1–x_6 are the independent variables or the attributes values of the dataset. The value of $h_\theta(x)$ is computed as shown in Table 5 which is the predicted outcome. The minimized model parameters obtained for airfoil dataset is shown in Table 6.

In Table 7 few predicted values obtained by the model for airfoil self noise dataset is mentioned.

Residuals are helpful in evaluating how well a linear model fits a dataset. Figure 2 shows the residuals plot for airfoil self noise data. It is observed that the

Table 5 Hypothesis for airfoil self noise dataset

Hypothesis function
$h_\theta(x) = \theta_1 x_1 + \theta_2 x_2 + \theta_3 x_3 + \theta_4 x_4 + \theta_5 x_5 + \theta_6 x_6$

Table 6 Minimized model parameters of airfoil self noise data

Minimized parameters
0.556660
0.027508
−0.013729
0.409982
0.242270
−0.006914

Table 7 Predicted values by the model

Predicted output
39.9623
45.4640
52.3411
61.9691
72.9725
86.7268
104.6074

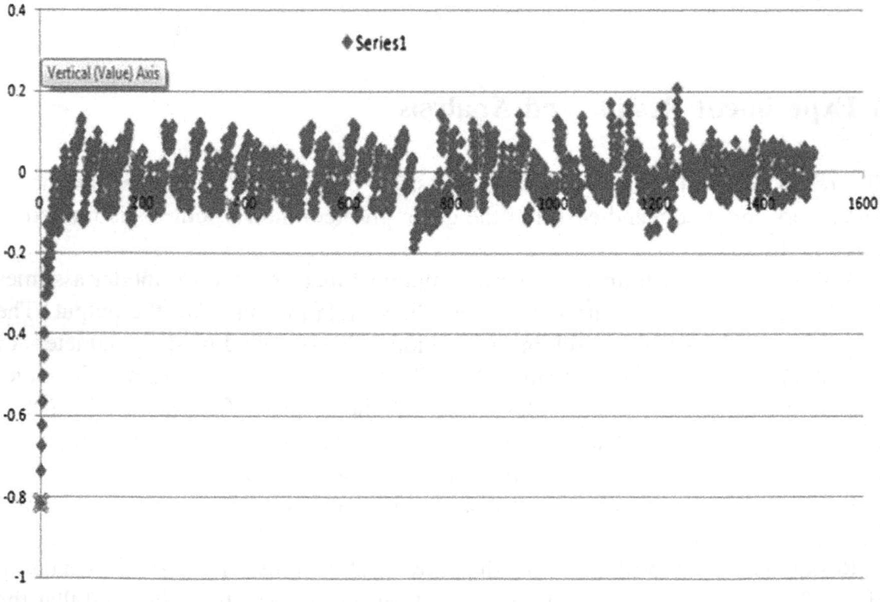

Fig. 2 Residuals plot for airfoil self noise dataset

residuals are randomly shuffled and no pattern is observed. If any pattern in observed in residual plot that indicates a higher order algorithms should be applied to model the data.

The cost values are plotted in a graph as shown in Fig. 3 x-axis represents the number of iterations and y-axis represents successive cost values. It is observed that cost is decreased in successive iterations which inferences that the algorithm is almost converging to the global minimum.

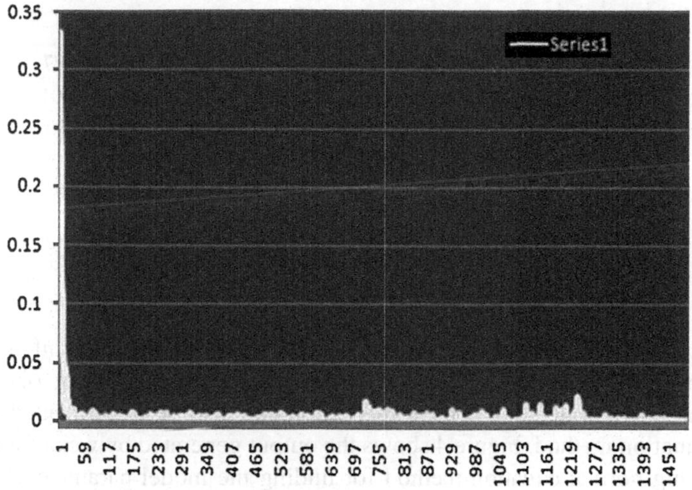

Fig. 3 Successive iteration's cost values plotted in a graph

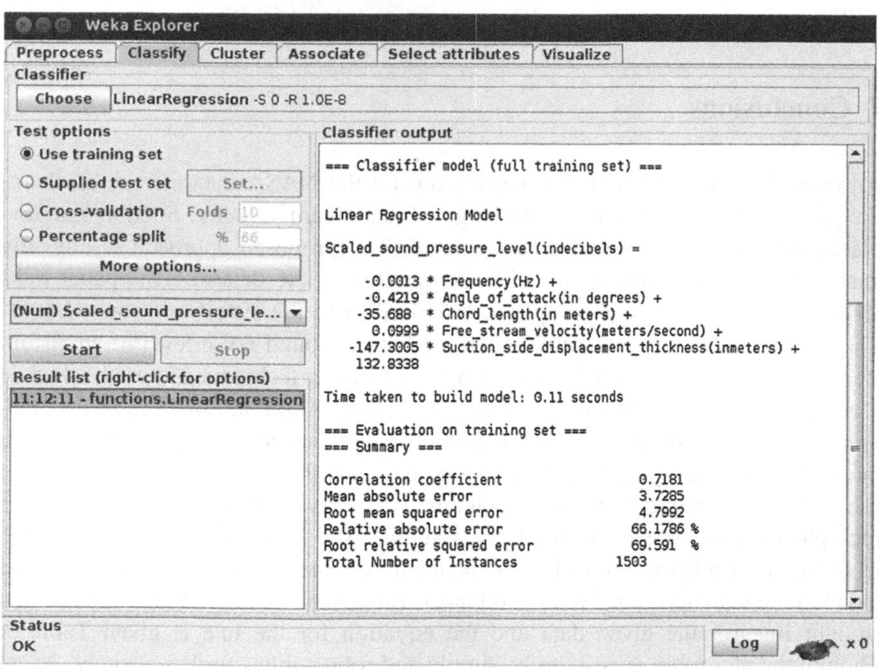

Fig. 4 Weka sample output for airfoil self noise dataset

Table 8 Equation of line

Equation of line
$0.556660x_1 + 0.027508x_2 - 0.013729x_3 + 0.409982x_4 + 0.242270x_5 - 0.006914x_6$

Table 9 Model statistics

Model statistics
Residual sum squares = 1.3815319611668326E7
Explained sum of squares = 1.3051506724666014E7
Sum of squared total = 2.686682633633434E7
R square (R^2) = 0.4857852044480344
Adjusted R square (R^2) = 0.4854426229719838
Mean squared error = 9,197.949142255877
Root mean squared error = 95.90593990353688

In Fig. 3 it is also observed a random moment in the graph which inferences that stochastic gradient descent calculates model parameters by taking one training example at a single iteration. Linear regression can also be implemented by using normal equation method. Figure 4 shows the output generated by weka for the same dataset using normal equation method for finding the model parameters.

Table 9 shows the Model Statistics or the Accuracy measures of the generated model by the Linear Regression algorithm and Table 8 provides the Equation of the Line.

7 Conclusions

An extensive experimentation is carried out on the NASA dataset with attributes Frequency, Angle of attack, Chord length, Free stream velocity, Suction side displacement and Scaled sound pressure level. The proposed approach is critically analyzed and also evaluates for different natures of the dataset. This paper talks about how a linear regression approach is used to fit the airfoil self noise dataset. There are many advanced methods that predict the sound noise level of an airfoil. Regression analysis in combination with Data Binning technique are used, which is a more sophisticated method for finding the noise levels and to analyze the relationship between sound pressure level and the wind speed. The Linear Regression model is used in the proposed method to understand that the regression can also be used. This paper has reviewed predicting noise level of a airfoil by using the multiple linear regression method, by assuming the set of independent attributes have the linear relationship with the dependent variable sound pressure level. To get the model parameters iterative method has is proposed. The experiment tried to fit a straight line for the given data and the equation for the line is given Table 9. Though the algorithm appears to be simple and robust it has further avenues. It can be experimented in combination with other sophisticated methods.

References

1. De Jesús Rubio Avila, J., Ramírez, A.F., Flores, G.D., Pereyra, M.S., Posada, F.B.S.: The wind turbine. In: Proceedings of the 12th WSEAS International Conference on Computers, ICCOMP'08, pp. 607–615. USA (2008)
2. Hastie, T., Tibshirani, R., Friedman, J.: The elements of statistical learning: data mining, inference and prediction, 2n edn. Springer, Heidelberg (2008)
3. John, A.: Ekaterinaris and Nikolaos Kampanis: a numerical prediction of acoustic fields generated by wind turbines. Syst. Anal. Model. Simul. **39**(1), 49–73 (2000)
4. Guarnaccia, C., Mastorakis, N.E., Quartieri, J.: A mathematical approach for wind turbine noise propagation. In: Proceedings of the 2011 American Conference on Applied Mathematics and the 5th WSEAS International Conference on Computer Engineering and Applications, pp. 187–194, USA (2011)
5. Coppi, R., D'Urso, P., Giordani, P., Santoro, A.: Least squares estimation of a linear regression model with LR fuzzy response. Comput. Stat. Data Anal. **51**(1), 267–286 (2006)
6. Chiang, H.W.D., Fleeter, S.: Prediction of loaded airfoil unsteady aerodynamic gust response by a locally analytical method. Math. Comput. Model. **10**(3), 193–206 (1988)
7. Bache, K., Lichman, M.: UCI Machine Learning Repository (2013)
8. Doel, K., Ascher, U.: The chaotic nature of faster gradient descent methods. J. Sci. Comput. **51** (3), 560–581 (2012)

Detection of Outliers in an Unsupervised Environment

M. Ashwini Kumari, M.S. Bhargavi and Sahana D. Gowda

Abstract Outliers are exceptions when compared with the rest of the data. Outliers do not have a clear distinction with respect to regular samples in the dataset. Analysis and knowledge extraction from data with outliers lead to ambiguity and confused conclusions. Therefore, there is a need for detection of outliers as a pre-processing stage for data mining. In a multidimensional perspective, outlier detection is a challenging issue as an object may deviate in one subspace and may appear perfectly regular in another subspace. In this paper, an ensemble meta-algorithm has been proposed to analyze and vote the samples for outlier identification in multidimensional subspaces. Cook's distance, a regression based model has been applied to detect the outliers voted by the ensemble meta-algorithm. Extensive experimentation on real datasets demonstrates the efficiency of the proposed system in detecting outliers.

Keywords Outlier detection · Outlier ensemble · Multidimensional subspace analysis · Cook's distance

1 Introduction

Data mining task aims to find the general patterns applicable to majority of the data, but due to occurrence of abnormal behaviour of the data, the valuable knowledge hidden behind may be suppressed and aid in ambiguous or confused conclusions.

M. Ashwini Kumari (✉) · M.S. Bhargavi · S.D. Gowda
Department of Computer Science and Engineering, BNM Institute of Technology, Bangalore, India
e-mail: mk_ashwini@yahoo.com

M.S. Bhargavi
e-mail: ms.bhargavi@gmail.com

S.D. Gowda
e-mail: sahanagowda@rediffmail.com

L.C. Jain et al. (eds.), *Computational Intelligence in Data Mining - Volume 2*,
Smart Innovation, Systems and Technologies 32, DOI 10.1007/978-81-322-2208-8_51

Abnormal behaviour of data are outliers. In practise, however there is not always a clear distinction between outlier and regular data as data have different roles with respect to different attribute sets in a multidimensional space. Therefore, outlier detection in a multidimensional subspace is a challenging issue.

Detection of outliers in an unsupervised environment makes the problem more complex as the basic knowledge about the number of clusters and the behaviour would not be available. Thus, to ensure robustness and generality of outlier detection across variety of data, an outlier ensemble with diverse outlier factor analysis is required.

In this paper, an outlier detection method has been proposed by devising an ensemble meta-algorithm. The ensemble meta-algorithm is based on three diverse factors. Each factor analyses the data in different perspective in a sub-space. Multidimensional subspace is a permutable combination of attribute set. The number of sub-spaces depends on the number of the attributes considered for analysis. The three factors considered are: Distribution based factor, depth based factor and proximity based factor. Distribution based factor uses the Mahalanobis distance measure to model the data distribution and portray the occurrence of the samples to the centre of the distribution. Depth based factor utilizes the Mahalanobis depth function which gives the outward ordering of the data points from the deepest point among the data samples. Proximity based factor utilizes the k-nearest neighbour distances that shows the proximity of the samples in the local neighbourhood.

Based on the analysis of the dataset by every factor, each sample is scored by 1 or 0. Score 1 indicates the rise in voting level where as 0 indicates no change. The cumulative scores indicate the number of votes gained by every sample. This may not suffice in declaring the samples as outliers as it is just the cumulative rise in number. To assimilate the voted samples of subspace to declare outliers into multidimensional level, a regression based Cook's distance has been adapted to analyze the slope deviation of the regression due to change in membership occurrence of voted samples as outliers. If the slope deviation is within the specified threshold computed dynamically, the sample is not an outlier else it is considered as outlier.

The rest of this paper is organized as follows. Section 2 signifies the state-of-art, in Sect. 3, a detailed description of the proposed methodology is given. Section 4 depicts experimental analysis and results. This paper is concluded in Sect. 5.

2 State-of-Art

Detecting and eliminating outliers is of great significance to knowledge extraction. Various techniques have been proposed for outlier detection. Available techniques vary from single conceptual model [1–10] to multiple conceptual models [11–17]. Single conceptual models build detection strategies based on single outlier factor for analysis whereas multiple conceptual models build the detection strategies with a set of outlier factors.

In literature, outlier detection methods with single outlier factors can be classified as statistical [1–3, 7, 10] and non-statistical [4–6, 8, 9] methods. Statistical methods depend on statistical reasoning of the data distribution and prominent categories include distribution-based [1, 10, 18] and depth-based [2, 19, 20] approaches. Distribution-based approaches indicate those observations that are located relatively far from the center of the data distribution as outliers. Depth-based approaches are non-parametric and do not assume any underlying distribution for the data. They indicate that an observation is an outlier based on its "center-outward" ordering in the data. Non-statistical or proximity-based methods focus on detecting outliers based on compactness of the samples in the local neighborhood. Prominent approaches for non-statistical outlier detection include distance-based [4, 5], density-based [14, 15] and clustering-based [6, 9] approaches.

Single conceptual models for outlier detection make specific assumptions about the data to define what constitutes an outlier. They may not be effective if those assumptions do not hold for the given data. Multiple conceptual models, also called outlier ensembles have been proposed to overcome this problem [11–13, 16, 17, 21–25]. Outlier ensembles can be classified into those that involve multiple detection factors [12, 13, 17], those with multiple executions of a single outlier factor using different parametric values each time [14, 15] and those which perform single outlier factors in multiple feature subspaces [11, 22–24]. The ensembles that explore single factors in multiple subspaces are also called data-centered ensembles [26].

Data-centered outlier ensembles analyze the outliers from various perspectives. These methods select random subspaces [11, 23] or use statistical methods for selection of relevant subspaces [22, 24, 25] for outlier analysis. OutRank [21] is one technique which performs outlier detection in all subspaces. Majority of the work that use feature bagging, use single factor for analysis. There is very limited work in literature that uses a combination of model centered and data centered models [12, 13].

In any form of outlier ensemble, an important aspect is the combination and interpretation of results from different runs. Prominent approaches for combining the scores are model averaging [22], best fit [14], aggregation [11] and product of scores [12]. For final interpretation of outlier scores, user specified information such as number of outliers L [12] or threshold [13] is the usually adopted method. There are certain methods proposed to overcome such static cut-off. Gao [17] use calibration approaches to fit outlier scores provided by different detectors into probability values. The probability estimates are then used to select the appropriate threshold for declaring outliers using a Bayesian risk model. Papadimitriou et al. [15] introduce a new definition of density-based outliers. The outlier score of each data sample is used to compare against the normalized deviation of its neighborhood's scores and standard-deviation is employed in the outlier detection. However, above techniques are limited to specific outlier ensembles.

It is evident from survey that for a robust detection of outliers, outlier ensembles have been widely used. The existing methods for outlier ensembles can be classified into data-centered methods and model-centered methods. Data-centered methods use the concept of feature bagging and model-centered methods use different detection strategies in full data space. The methods that use both the concepts, concentrate on the selection of relevant subspaces for outlier detection. But these are considered weak guesses and the true subsets of attributes relevant for outlier analysis in an unsupervised environment may not be accurately identified [26]. Therefore, there is a need for an ensemble meta-algorithm with outlier factors at multidimensional subspace level.

3 Proposed Methodology

Multidimensional subspace analysis explores the data in different subspaces. All permutable combinations of attributes are considered to analyze the abnormal behavior of outliers. The abnormality of a sample is due to its occurrence, neighborhood relationship and nearness to the deepest point in the data clutter. To measure all these factors, an ensemble meta-algorithm is devised with three factors such as distribution based outlier factor to measure the occurrence based on distribution, depth based outlier factor to measure the nearness and proximity based outlier factor to measure the nearest neighborhood relationship.

3.1 Distribution Based Outlier Factor

Distribution based factor uses Mahalanobis distance measure to model the data distribution and portray the occurrence of the samples to the center of the distribution. The samples that are highly deviating from the center of the data distribution are outliers. For multidimensional data, the distance of sample from the center or the mean of the data can be obtained by using multivariate distance measures. Mahalanobis distance measure is a multivariate distance measure which has been used for the distribution based factor in the proposed ensemble meta-algorithm. Mahalanobis distance calculates the distance of a sample from its mean considering the variance and the covariance of the features.

Mahalanobis distance [1] is defined as follows. Consider a d-dimensional multivariate dataset with n samples, $x1,\ldots,xn \in R^d$. The Mahalanobis distance for each multivariate sample $xi = (x1i,\ldots,xid)^T$ is given by

$$MDi = \sqrt{(xi - \mu)^T C^{-1} (xi - \mu)} \tag{1}$$

where, $\mu = (\mu 1,\ldots,\mu d)^T$ denotes the overall sample mean and C = variance-covariance matrix given by,

$$\mathrm{C} = \begin{bmatrix} c11\ldots c1d \\ \ldots \\ \ldots \\ cd1\ldots cdd \end{bmatrix} \tag{2}$$

and each term is a covariance between two features A and B given by

$$CAB = \sum_{i=1}^{n} (xiA - \mu A)(xiB - \mu B)/(n-1) \tag{3}$$

Distances obtained follow the Chi-squared distribution with d-degrees of freedom. Cut-off value for the distribution is obtained from the Chi-squared distribution table at 97.5th percentile. Distance values above the cutoff are selected as outliers.

3.2 Depth Based Outlier Factor

The depth factor in the ensemble provides an outward ordering of the data points or samples from the deepest point in the data. This ordering of the points does not depend on the distribution of the data. Mahalanobis depth function [2] is used and is defined as follows.

Consider a d-dimensional dataset with n samples $x1,\ldots,xn \in R^d$. The depth function for each multivariate sample $xi = (x1i,\ldots,xid)^T$ is given by

$$MDEi = 1/(1 + SMDi) \tag{4}$$

where, SMD_i = squared Mahalanobis distance of the sample given by

$$SMDi = (xi - \mu)^T C^{-1}(xi - \mu) \tag{5}$$

where, μ and C are the overall sample mean and the variance-covariance matrix respectively.

The depth values (distances) obtained shows bigger values for samples nearer to the deepest point in the data clutter and smaller values for the samples far away. Samples with smaller values are considered outliers. Through trial and learning process, a threshold of 0.15 is set to identify the outliers.

3.3 Proximity Based Outlier Factor

The proximity based outlier function is the distance between a sample and its k-nearest neighbors [5]. For every sample, the average distance is computed to the corresponding k-nearest neighbors considered.

For a d-dimensional multivariate dataset with n samples $x1, \ldots, xn \in R^d$, the distance function for a sample $xi = (x1i, \ldots, xid)^T$ is given by,

$$AvgDisti = \sum_{j=1}^{k} Euclidean\_Disti, k/k \quad (6)$$

where, k = number of neighbors,

Euclidean_Dist $_{i,k}$ = Euclidean distance between sample x_i and its *j*th neighbor is given by,

$$EDij = \sqrt{(xij - xj1)^2 + \cdots + (xid - xjd)^2} \quad (7)$$

The distance of the samples obtained is sorted and the samples with bigger values are considered outliers. The threshold to declare outliers is based on the average of the number of outlier samples obtained in the above two factors.

3.4 Interpretation of Outlier Scores

The outliers identified in each factor vary depending on the metric utilized. Due to the abnormal behavior, a sample may be voted as outlier by a factor in a subspace and may not be true in another factor in the same space. If there are n dimensional attributes, then the total number of subspaces considered are $2^n - n - 1$, where n is the number of single feature subspaces and 1 is an empty subspace. Single feature subspaces are eliminated which drastically reduces the computational complexity by n without eliminating any permutable combination of attribute sets.

To assimilate the analyses of every subspace of every factor, cumulative voting record is generated. In the cumulative record, there is a rise in value by 1 if any sample is voted as outlier. The cumulative record is the additive value of every sample in the dataset voted outlier in a subspace by a factor. There is a rise in value by 1 if the sample is voted as outlier else the numeric value remains unchanged for non-outliers. The samples which are never voted always have the value zero.

The comprehensive outlier scores obtained for the data samples do not provide a direct indication of whether the samples are outliers in the cumulative record. Outlier scores that not only have high values but also have a large deviation from the rest of the scores are to be identified dynamically. To analyze the voted samples on a single plot, the cumulative records of all the samples are assimilated by a regression based model called Cook's distance [27]. Based on the observation, when the outlier scores are plotted, a majority of the outlier scores (representing the weak outliers) form a linear band and can be represented by a linear model. Those samples with high scores affect the linearity of the data. To identify such scores, regression analysis using Cook's distance has been adapted.

For a regression Y on the outlier scores $(y1, y2, \ldots, yn)$, the Cook's distance for a point yi is given by,

$$Di = \sum_{j=1}^{n} (\hat{y}j - \hat{y}j(i))^2 / (k+1)s^2 \tag{8}$$

where,

$\hat{y}j$ is the predicted value of the jth observation.
$\hat{y}j(i)$ is the predicted value of the jth observation with the ith point removed and s^2 is the mean squared error or the variance from the fit based on all the observations (variance is the squared difference between the predicted value $\hat{y}$ and observed value y).

The predicted value of y from a regression i.e. $\hat{y}$ is given by

$$\hat{y} = Hy \tag{9}$$

where, H is called the Hat matrix [28] given by,

$$H = X(X'X)^{-1}X' \tag{10}$$

Here, X is the predictor variable for the response variable Y (vector of outlier scores).

The point i is treated as an outlier if

$$Di \geq 4/(n - (k+1)) \tag{11}$$

where, k is the number of predictor variables which is equal to 1 as the outlier score is a single vector.

4 Experimental Analysis and Results

In this section, an experimental study is presented to evaluate the efficacy of the proposed method in detecting true outliers. Experiments are conducted on four real world datasets from the UCI Machine Repository [29]. The real world datasets used for experimental analysis are Wisconsin breast cancer dataset, New Thyroid dataset 1 (with class 2 as rare class), New Thyroid dataset 2 (with class 3 as rare class) and Pima Indian diabetes dataset. The datasets selected are binary class sets with the larger class being the normal data and the smaller class being the rare class data. Datasets with multiple classes are converted to binary class sets.

In the following section, a brief overview of all the datasets used for experimental analysis is provided along with the results achieved using the proposed method. The details of the datasets are as follows:

Wisconsin breast cancer dataset: The original Wisconsin breast cancer data consists of 699 records with nine attributes each. The records are labeled either benign (458 records, 65.5 %) or malignant (241 records, 34.5 %). For experimental purpose, 444 benign records and 39 malignant records are chosen. This leads to 92 % normal records and 8 % abnormal or rare class records.

New Thyroid dataset 1: The New thyroid dataset gives information about the thyroid disease. The task is to detect whether a given patient is normal or suffers from hyperthyroidism or hypothyroidism. The original dataset consists of 215 records out of which 150 records are normal, 35 records denote hyperthyroidism and 30 records denote hypothyroidism. For experimental purpose, 150 normal records and 15 records from class 2 are considered to form 90 % normal records and 10 % abnormal or rare class records.

New Thyroid dataset 2: The New thyroid dataset is used with another combination of binary class set. 150 normal records and 15 records from class 3 are considered to form 90 % normal records and 10 % abnormal or rare class records.

Pima Indian diabetes dataset: The original Pima Indian Diabetes dataset consists of 500 'tested negative' records and 268 'tested positive' records. For experimental purpose, 150 normal records and 15 records from class 2 are considered to form 90 % normal records and 10 % abnormal or rare class records.

Precision and Recall are used for measuring the quality of outlier detection by the proposed method. Precision, also called the Positive predictive rate, is the percentage of the reported outliers, which turn to be true outliers. Recall, also called True Positive Rate (TPR), is the percentage of the true outliers that have been reported as outliers at a given threshold. False Positive Rate (FPR) is also shown, which is the percentage of falsely reported outliers out of the true inliers.

From Table 1, it is evident that the recall values for the majority of the datasets are high indicating that the proposed system can detect outliers effectively (high percentage of outliers are detected). High values of precision and low values of FPR indicate that true outliers are more likely to be detected as compared to false outliers. Hence the quality of outlier detection is high for the proposed system.

The overall expected performance of the proposed method is evaluated using the well-established Receiver Operating Characteristics (ROC) curve. ROC curve is obtained using the plot of TPR against the FPR rate. The ROC curve characterizes the trade-off between the TPR and FPR values. It is preferred that the outlier detection method has high TPR and low FPR values. This will have an ROC curve

Table 1 Precision, recall and FPR metrics for the proposed meta-algorithm

Dataset	Precision (%)	Recall (TPR) (%)	FPR (%)
Wisconsin breast cancer data	74	94.87	2.92
New thyroid data 1	91.67	73.3	0.67
New thyroid data 2	92.3	80	0.67
Pima Indian Diabetes data	27.3	24	6.4

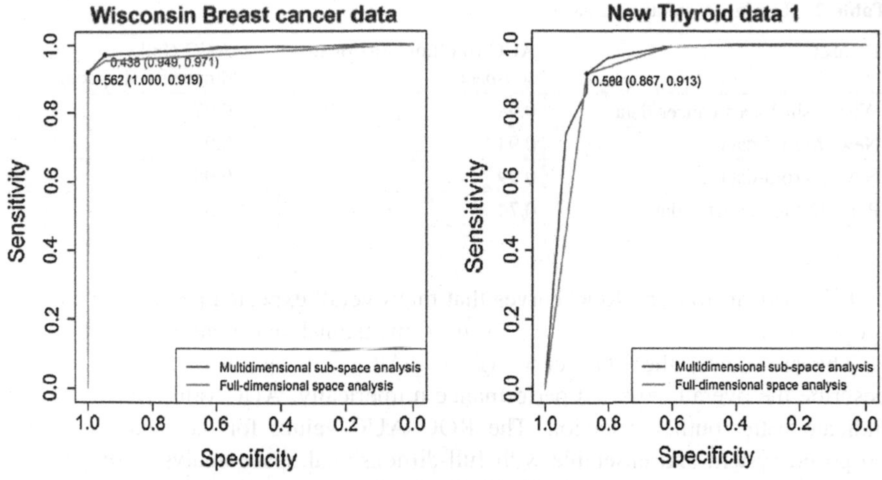

Fig. 1 ROC curves for wisconsin breast cancer and new thyroid 1 datasets using the proposed meta-algorithm as against full-space outlier detection

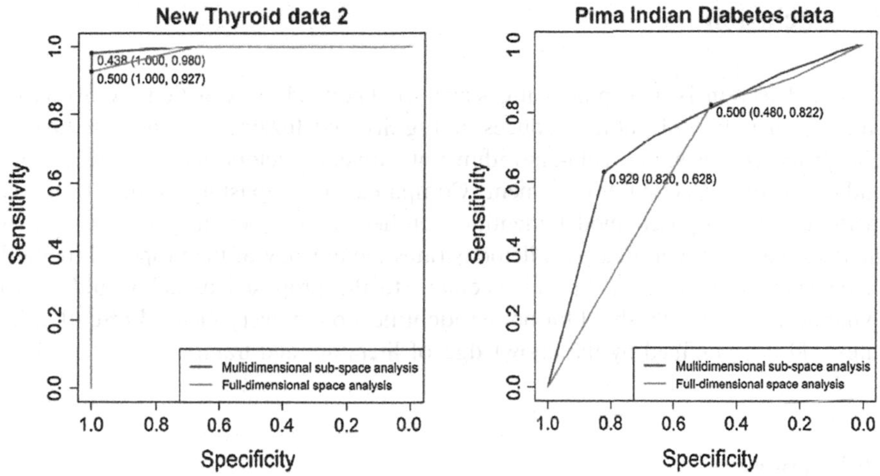

Fig. 2 ROC curves for new thyroid 2 and pima Indian diabetes datasets using the proposed meta-algorithm as against full-space outlier detection

that is closer to the upper left corner of the graph indicating high TPR and low FPR. The ROC curves for the datasets are shown in the Figs. 1 and 2. Results obtained for the proposed model with multidimensional subspaces are compared with full dimensional space with the same factors. The ROC curves in the plots demonstrate the improvement in performance due to multi-dimensional subspace analysis.

Table 2 Results on mentioned datasets

Dataset	AUC (multidimensional subspace)	AUC (full dimensional space)
Wisconsin breast cancer data	0.98	0.97
New thyroid data 1	0.94	0.91
New thyroid data 2	0.99	0.98
Pima Indian diabetes data	0.74	0.65

It is evident from the ROC curves that that overall expected performance of the proposed system is high compared to full-dimensional space analysis.

The area under the ROC curve (ROC AUC) is a summary statistic used to describe the overall expected performance numerically. AUC values closer to one indicate better outlier detection. The ROC AUC values for the datasets for the proposed system and ensemble with full-dimensional space analysis are given in Table 2. It can be seen that the ROC AUC values for the proposed multi-dimensional subspace analysis performs better than full-space analysis.

5 Conclusion

Outlier detection is a preprocessing stage for knowledge extraction. Analysis of data which is free of outliers reduces ambiguity and fuzziness in the conclusion. The proposed ensemble meta-algorithm with diverse factors, aims at identifying outliers in unsupervised environment. Compared to the existing outlier detection methods, the proposed model identifies outliers more accurately. Experimental analysis with real world datasets demonstrates the efficacy of the proposed method in detecting true outliers. An enhancement to the proposed model would be to dynamically set the threshold for outlier identification in every factor. Presently, the threshold is predefined by the knowledge of literature and training.

References

1. Mahalanobis, P.C.: On the generalized distance in statistics. Proc. Natl. Inst. Sci. India **12**, 49–55 (1936)
2. Liu, R.Y., Singh, K.: A quality index based on data depth and multivariate rank tests. J. Am. Stat. Assoc. **88**, 252–260 (1993)
3. Arning, A., Agrawal, R., Raghavan, P.: A linear method for deviation detection in large databases. In: Proceedings of Data Mining and Knowledge Discovery, pp. 164–169. Portland, Oregon (1996)
4. Knorr, E., Ng. R.: Algorithms for mining distance-based outliers in large datasets. In: Proceedings of 24th International Conference on Very Large Data Bases (VLDB), pp. 392–403, 24–27 (1998)

5. Ramaswamy S., Rastogi, R., Shim, K.: Efficient algorithms for mining outliers from large data sets. In: Proceedings of the ACM SIGMOD International Conference on Management of Data. Dallas, TX (2000)
6. Jiang, M.F., Tseng, S.S., Su, C.M.: Two-phase clustering process for outliers detection. Pattern Recogn. Lett. **22**(6), 691–700 (2001)
7. Filzmoser, P.: A multivariate outlier detection method (2004)
8. Hawkins, S., et al.: Outlier detection using replicator neural networks. Data Warehousing and Knowledge Discovery, pp. 170–180. Springer, Berlin (2002)
9. He, Z., Xu, X., Deng, S.: Discovering cluster-based local outliers. Pattern Recogn. Lett. **24**(9), 1641–1650 (2003)
10. Rousseeuw, P.J., Van Zomeren, B.C.: Unmasking multivariate outliers and leverage points. J. Am. Stat. Assoc. **85**(411), 633–639 (1990)
11. Lazarevic, A., Kumar, V.: Feature bagging for outlier detection. In: Proceedings of KDD. pp. 157–166 (2005)
12. He, Z., Deng, S., Xu, X.: A unified subspace outlier ensemble framework for outlier detection. In: Fan, W., Wu, Z., Yang, J. (eds.). LNCS, vol. 3739 pp. 632–637. Springer, Heidelberg, (2005)
13. Nguyen, H.V., Ang, H.H., Gopalkrishnan V: Mining outliers with ensemble of heterogeneous detectors on random subspaces. Database Systems for Advanced Applications. Springer Berlin Heidelberg (2010)
14. Breunig, M., Kriegel, H.P., Ng, R., Sander, J.: LOF: Identifying Density-based Local Outliers. ACM SIGMOD Conference (2000)
15. Papadimitriou, S., et al.: Loci: Fast outlier detection using the local correlation integral. In: Proceedings of 19th International Conference on Data Engineering. IEEE (2003)
16. Zimek, A., et al.: Subsampling for efficient and effective unsupervised outlier detection ensembles. In: Proceedings of the 19th ACM SIGKDD International Conference on Knowledge Discovery and Data Mining. ACM (2013)
17. Gao, J., Tan, P.N.: Converting output scores from outlier detection algorithms into probability estimates. In: Sixth International Conference on Data Mining. IEEE (2006)
18. Rousseeuw, P.J., Leroy, A.M.: Robust Regression and Outlier Detection. Wiley (1987)
19. Tukey, J.: Exploratory Data Analysis. Addison-Wesley (1977)
20. Ruts, I., Rousseeuw, P.J.: Computing depth contours of bivariate point clouds. Comput. Stat. Data Anal. **23**, 153–168 (1996)
21. Müller, E., et al.: Outlier Ranking via Subspace Analysis in Multiple Views of the Data. ICDM (2012)
22. Keller, F., Muller, E., Bohm, K.: HiCS: high contrast subspaces for density-based outlier ranking. In: IEEE 28th International Conference on. Data Engineering (ICDE) (2012)
23. Foss, A., Zaïane, O.R.: Class separation through variance: a new application of outlier detection. Knowl. Inf. Syst. **29**(3), 565–596 (2011)
24. Nguyen, H.V., et al.: CMI: an information-theoretic contrast measure for enhancing subspace cluster and outlier detection. SDM (2013)
25. Aggarwal, C.C., Yu, P.S.: Outlier detection for high dimensional data. ACM Sigmod Record, vol. 30. No. 2, ACM, New York (2001)
26. Aggarwal, C.C.: Outlier ensembles. Position paper. ACM SIGKDD Explorations Newsletter. pp. 49–58, (2013)
27. Cook, R.: Detection of influential observations in linear regression. Technometrics **19**, 15–18 (1977)
28. Hoaglin, D., Welsch, R.: The hat matrix in regression and anova. Am. Stat. **32**, 17–22 (1978)
29. Bache, K., Lichman, M.: UCI machine learning repository. http://archive.ics.uci.edu/ml

Discovering Latent Relationships Among Learning Style Dimensions Using Association Rule Mining

C. Beulah Christalin Latha, E. Kirubakaran and Ranjit Jeba Thangaiah

Abstract Learning is the process of acquiring knowledge or skill. It is the process of filtering, storing and organizing information in our brains. Learners differ by the ways in which perceive, process and receive information. Based on the means of processing the information, learners are considered to possess their own style of learning. The learning process tends to be more effective if the knowledge is disseminated to the learners in their own learning styles. In order to disseminate knowledge in different ways, learners should be categorized based on their styles and they should be trained in an appropriate manner. This paper proposes a novel method to detect the relationships between different dimensions in Felder-Silverman learning style model using a data mining technique known as association rules, thereby providing a simpler way to disseminate the knowledge in the context of technology enhanced learning.

Keywords Intelligent tutoring systems · Association rule mining · Pedagogical issues · Teaching and learning style

1 Introduction

Learners differ in the way they learn. They have different means of perceiving the information, receiving the information, processing the information and understanding it. In traditional classroom teaching, teachers can teach different learners in

C. Beulah Christalin Latha (✉) · R. Jeba Thangaiah
Department of Computer Applications, Karunya University, Coimbatore, Tamil Nadu, India
e-mail: beulsonline@gmail.com

R. Jeba Thangaiah
e-mail: ranjit@karunya.edu

E. Kirubakaran
Bharat Heavy Electricals Ltd, Tiruchirappalli, Tamil Nadu, India
e-mail: e_kiru@yahoo.com

L.C. Jain et al. (eds.), *Computational Intelligence in Data Mining - Volume 2*,
Smart Innovation, Systems and Technologies 32, DOI 10.1007/978-81-322-2208-8_52

a way that suits each type of learner. In technology enhanced learning scenarios, the presence of a teacher or a trainer is not possible always. Though there is a flexibility of pervasive learning in such learning systems, the absence of a teacher becomes a bottleneck in achieving the best results in the learning process. The physical presence of a teacher, however, is difficult to be compromised, but there are other means in which the learner can be assisted. One of the means in which the efficiency of learning in e-learning systems can be improved is by delivering the contents to the learners in their own style. Every learner has a mix of learning styles. A learning style model classifies learners based on their styles of acquiring and processing information. This paper focuses on Felder-Silverman Learning style model that has been widely used in technology enhanced learning [1]. Felder has pointed out that learners with a strong preference for a specific learning style may have difficulties in learning if the teaching style does not match their preferred learning styles [2].

Felder Silverman Learning Style Model (FSLSM) has been claimed as the most appropriate model for hypermedia courseware through various researches since it provides a detailed description about different dimensions of the learner [2, 3]. There are four dimensions in FSLSM as shown in Fig. 1. Each learner has a specific preference in each of these dimensions. Learners are divided into active and reflective learners based on the way they process information, sensing learners and intuitive learners based on their information perception, visual and verbal learners on the basis of information reception and sequential and global learners based on their understanding. This paper discusses a data mining approach for finding the relationships across the dimensions of FSLSM using a data mining approach.

The style in each dimension is determined using the Index of Learning Styles [ILS]. It is a psychometric assessment instrument designed to assess the preferences and degree of preferences for a learner on each of the four dimensions of FSLSM [4–7]. Each dimension is characterized by the answers to 11 questions in ILS. The index of preference of the learner in each dimension is represented with values between −11 and +11 with steps ±2. Each question is dichotomous with two responses a or b. Answer a corresponds to the first pole of each dimension and is awarded a score of +1. Answer b corresponds to the second pole of each dimension and is awarded a score of −1. If a learner scores a positive score, he is classified to be strong in the first pole and if he gets a negative score, he is strong in the second

Fig. 1 Felder-Silverman learning style model

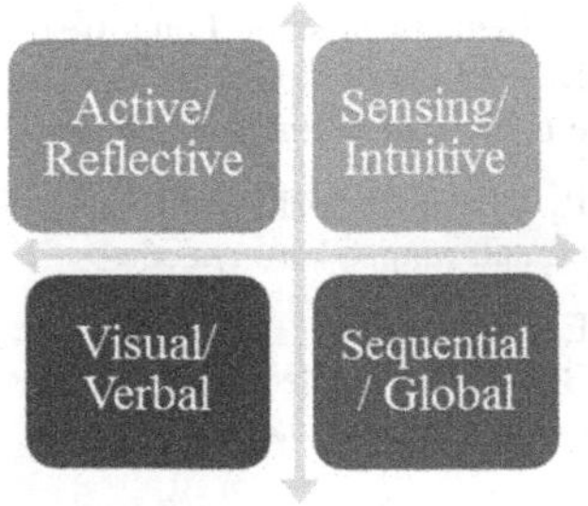

pole in each dimension. The Index of Learning Styles has been proved to be a reliable and valid tool for evaluating the learning styles of learners by many researchers [6].

2 Background

Several researches have been carried out in analyzing the Felder-Silverman model and also in analyzing the ILS questionnaire. The literatures show that there are relationships among various dimensions of the FSLSM and various techniques have been used for detecting those relationships. Silvia Rita Viola, Sabine Graf et al. discusses a data-driven and graphical exploratory analysis to discover unexpected latent knowledge and relationships inside educational data. Results show that two styles belonging to different ILS dimensions, namely active-visual and sequential-sensing are highly correlated; Pearson Correlation coefficients show many dependencies even between styles belonging to the same dimension like active/reflective, sequential/sensing, and sensing/intuitive. This may indicate that some ILS dimensions assess two relatively independent latent variables rather than one variable having two opposite expressions. The Euclidean distance measures indicate relationship between sensing-sequential and intuitive-global which is supported by a cognitive point of view also. The results on Cronbach's alpha coefficients also lead to conclude for the co-existence of various latent dimensions underlying each ILS dimension that could not be peculiar to one dimension only. It has been concluded that learning style identified by ILS presents consistent dependencies between some styles [7].

Sabine Graf et al. in their paper carry out an analysis on data about learning styles with respect to FSLSM. They have collected data from a mixed group of students and conclude that there are more number of learners who prefer a visual presentation of contents and their results are found to be in agreement with that of Felder and Spurlin [3, 6, 8]. In yet another paper, Sabine Graf et al. demonstrate relationships between the working memory capacity and learning style of learners. The results show dependency between learning styles and cognitive traits [9]. Patricio Garcia et al. evaluate the precision of Bayesian Networks (BN) at detecting the learning style of a learner in a web-based education system. The most popular processing style is found to be reflective by this model. It has also been observed that most of the learners in the study are sequential and no global learners are discovered [10].

Ali Alharbi et al. conclude that aspects of learners' preferred learning styles have a significant impact on academic performance. It also suggests the usage of self-regulated learning strategies. Correlation analysis indicates that only the perception dimension has a significant impact on the performance of the learners. Suitable learning materials have also been suggested [11]. Nor Bahiah Hj Ahmad and Siti Mariyam Shamsuddin compare the performance of several classifiers such as Bayes, Decision Tree etc. in classifying learners' learning styles. The study provides an approach to intelligently detect learner's learning style based on FSLSM [12]. Malgorzata S. Zywno provides an analysis of psychometric properties of the ILS and concludes that it is a suitable psychometric tool to assess learning styles of engineering students. The reliability of scales is found to be higher than that in Van Zwanenberg et al. study and the resulting coefficients meet acceptable limits as suggested by Tuckman [13]. The consistency of the scale is also found to be good and the results are in agreement with the results of Livesay et al. [14]. Factor analysis shows that an overlap occurs between Sensing-Intuitive and Sequential-Global dimensions [5]. Litzinger et al. [15] conducted a study to assess reliability, factor structure and construct validity of ILS and conclude that the ILS is reliable and valid and a change or revision is not needed.

3 Proposed Work

The proposed work aims at discovering some hidden relationships among the dimensions of the Felder-Silverman Learning Style model. A survey was conducted among 238 students from engineering colleges in Tamil Nadu, India, using the ILS questionnaire. The responses were collected and they were analyzed using data mining techniques. The results indicate that there are relationships among different dimensions of the learner. The framework for the proposed work is shown in Fig. 2. The learner is presented with an ILS questionnaire and the responses of the learner are considered as the input for constructing the learning model of the learner based on FSLSM. The score in each dimension and the strength of preference are also calculated. The ILS responses are segregated and the type of the learner in each dimension is identified. Then the responses of each type of learner are mined in each dimension using Apriori algorithm for Association rules.

Association rule is a data mining technique that is used for finding hidden patterns in the data. Association rules are in general, used for extracting interesting

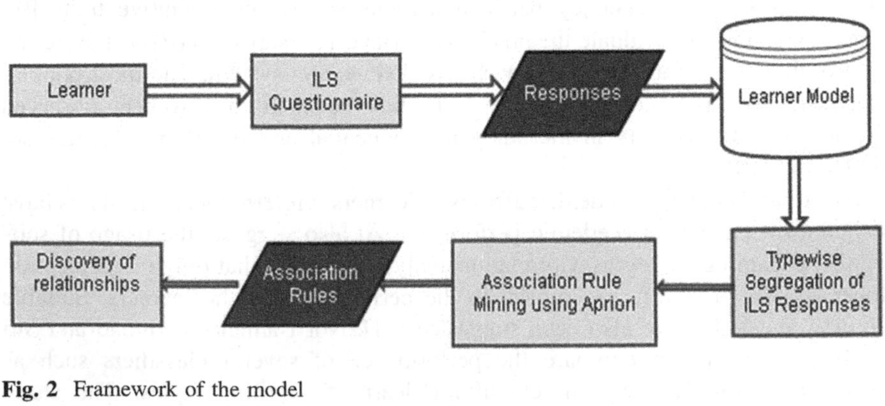

Fig. 2 Framework of the model

correlations, to discover frequent patterns and for finding associations or casual structures among sets of items in datasets.

The strength of the association rules is calculated by support and confidence. The support of a rule indicates how frequently the items in the rule occur together. Confidence is the conditional probability that the consequent will occur given the occurrence of the antecedent. The results of the association rules indicate that relationships exist between the dimensions of FSLSM which is discussed in detail in the results section.

4 Results and Discussion

The ILS responses from 238 students were collected. The scores in each dimension are calculated and the learners are divided into different types. Table 1 shows the percentage of learners in each type. The results are in agreement with the previous studies also [1].

The results show that there is a high preference for visual learning type which indicates that the effectiveness of learning can be increased by presenting the content in visual format like images, animations, charts etc. The strength of the preferences of the learners is calculated in each dimension. The chart in Fig. 3 shows the results. Dim 1,…, Dim 4 indicate the dimensions. Mod indicates moderate preference, Str indicates strong preference and Bal indicates balanced preference. The suffixes 1 and 2 indicate the first and second pole of each dimension. From the chart, it has also been inferred that, the learners who prefer the second pole of the dimension are less in all the dimensions and a majority of the learners have a balanced style. And there are no learners who have strong preference for verbal style and intuitive style. Next, the ILS responses for each style of the learner in each dimension are segregated and mined using association rules using Apriori algorithm. The following are the inferences made from the results.

A comparison of active learners with the sensing learners indicates that more than 90 % of the sensing learners have also chosen the same option. This leads to infer that most of the active learners are likely to be sensing learners in the second dimension. The rules generated for the active learners in the third dimension indicate that most of the learners have chosen the same option as the visual learners for most of the questions. From these results, it can be inferred that there is considerable interception between the active and the visual learners. This result is

Table 1 Typewise percentage of learners

Dimensions	Processing		Perception		Reception		Understanding	
Type of learners	Active	Reflective	Sensing	Intuitive	Visual	Verbal	Sequential	Global
% of learners	58.40	41.60	57.56	42.44	80.25	19.75	57.56	42.44

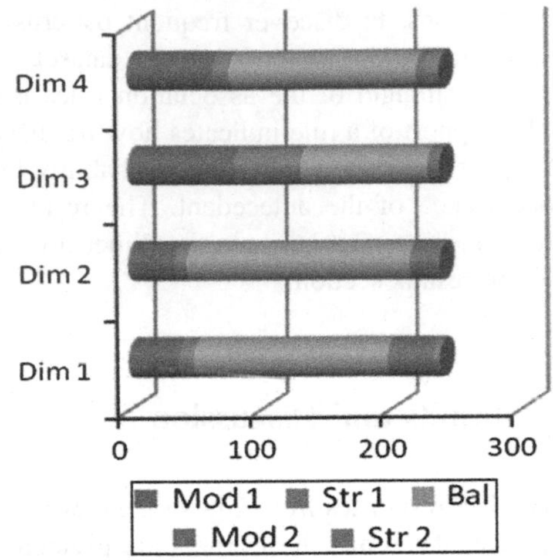

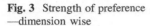

Fig. 3 Strength of preference —dimension wise

supported by previous researches also [1, 8] using data driven and statistical approaches. When the responses of active learners for the fourth dimension are mined, the results do not show much relationship between these dimensions.

The rules for the responses in the second dimension suggest that reflective learners have more inclination towards the sensing dimension. It is interesting to note that most of the active learners also have the same pattern in the second dimension but with a higher support. The responses of the reflective learners show that the reflective learners also prefer visual representations to a certain extent. Though the minimum support is only 0.4, the best rules do not indicate strong relationship with the verbal dimension. However, the patterns of the responses differ from the pattern selected by the active learners. The rules in the last dimension have a very less support of only 0.35. So the reflective learners may be considered to have a very moderate inclination towards the sequential style which is also supported by Viola et al. [8].

In the second dimension, the pattern of sensing learners matches very well with that of active learners. Based on the results, the sensing learners may be considered to have a balanced active/reflective style in the first dimension. The results also indicate that some sensing learners may prefer sequential style and others global style and sequential style is more preferred. The results are in agreement with [6]. The responses for the intuitive learners in dimension 1 show that they may be considered to have a balanced or moderate preference towards active or reflective scale in this dimension. The responses for intuitive learners in the third dimension indicate that the intuitive learners may also prefer visual instruction. Analysis of the responses of intuitive learners in the last dimension indicates that intuitors can favor either sequential learning or global learning which is also supported by [6].

When the responses of visual learners are combined with dimension 1 data and subject to mining, the results show that a majority of the learners may have a moderate preference in the reflective dimension. The patterns of the reflective learners also indicate that they prefer visual learning style and most of the visual learners tend to be sensing. The rules have a minimum support of 0.55 and a minimum confidence of 0.92. The results in the fourth dimension show that visual learners have a strong preference for sequential style. The verbal learners show an inclination towards reflective learners and they tend to have balanced preference in both sensing and intuitive poles. The fourth dimension results indicate that verbal learners are more sequential than global.

Sequential learners tend to be active learners in first dimension. The analysis of the responses of these learners in the second dimension indicates that most of them are sensing learners. These results are in contrast with the results in [6] which claim that sequential learners can be either sensors or intuitors. We don't have enough rules for supporting intuitors. The responses of the sequential learners in the third dimension indicate that most of them prefer the visual learning style in that dimension. This inference also varies from the results in [6] which say that sequential learners are more likely to be verbal processors based on brain hemisphere research and clinical observations. But it has also been specified that the linkage may not apply to the categories of the Felder-Silverman model since in this model visual style is based on more of internal processing than of sensory input [6]. Our rules have a minimum support of 0.55 and a minimum confidence of 0.91.

The responses of the global learners in the last dimension indicate that they may also prefer the sequential style in certain circumstances. The responses in the first dimension show that most of them are active. But the rules have minimum support 0.35 only and a minimum confidence of 0.91. The responses in the second dimension indicate that they can have a balanced style in sensing/intuitive dimension. The responses of the global learners in the third dimension show that they like visual style.

A graphical representation of the results is showed in the following diagrams. It shows the patterns of the various types of learners in each of the four dimensions of the Felder-Silverman model. The graph is plotted with responses against learner types for each dimension.

Figure 4 shows the graphical patterns generated dimension wise, considering all 'a' responses as '1' and all 'b' responses as '−1'. The graph shows that Active, Intuitive, Sequential and global learners have more positive scores. Sensing and intuitive learners may have balanced preference towards active or reflective style. Sequential learners also tend to be active learners in dimension 1. Similarly, reflective, visual and verbal learners have more negative scores. Therefore, visual and verbal learners tend to be reflective learners in dimension 1. In dimension 2, active, reflective, sensing, visual and sequential learners have more positive scores. Hence, active, reflective, visual and sequential learners tend to be sensing learners. Intuitive and verbal learners have more negative scores. So verbal learners may have a balanced preference in sensing/intuitive dimension. Global learners have equal scores and so they may also be balanced sensing/intuitive. In dimension 3, all types of learners tend to have more preference towards visual style. Only verbal learners have more preference towards the verbal style. Active learners seem to be moderate preference towards sequential/global; Global learners also tend to have a preference towards sequential dimension.

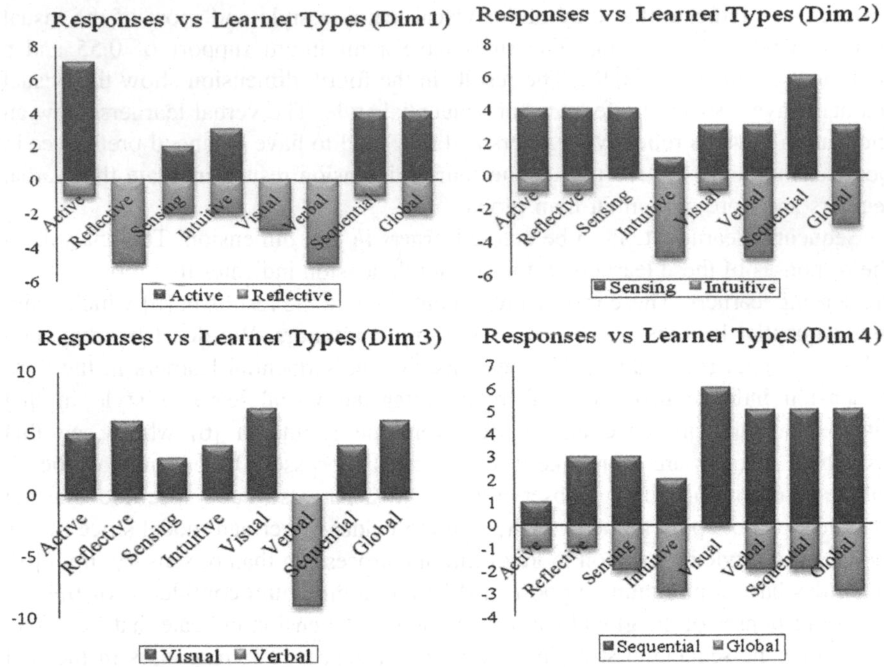

Fig. 4 Responses versus learner types (Dimensions 1–4)

5 Conclusions

In this study, a novel method of detecting the hidden relations between the dimensions of the Felder-Silverman learning style is carried out. Association rules are well known to detect relationships. The results show good accuracy and have discovered hidden relationships among the dimensions. The study might prove to be a useful step in authoring and delivering the learning objects in self-directed learning scenarios. For instances, almost all learners have a preference towards the visual dimension except a very few of them. Therefore, learning objects may be authored in such a way that they present the contents more visually. Inclusion of more pictures, animations, tables and charts can make the learning process more effective. Similarly more learners prefer a sequential approach of learning. Therefore presenting the contents in an orderly, step by step manner might improve the process of learning. The results generated show many hidden relationships in various dimensions of a single learner and also assist in determining the similarities among learners which might be considered as an effective method for designing the learning objects for different types of learners. Some of the previous researches also are in agreement with the above results. The results of this study have been chosen as the base for personalization of content delivery of the learners which is the second phase of this research. Based on the relationships revealed by the association rule mining, the learning objects are personalized and then recommended for the learner, the results of which show a considerable improvement in the efficiency of learning.

References

1. Graph, S., Viola, S.R., Kinshuk T.L.: In-Depth analysis of the felder-silverman learning style dimensions. J. Res. Technol Educ. **40**, 79–94 (2007)
2. Felder, R.M., Silverman, L.K.: Learning and teaching styles in engineering education. Eng. Educ. **78**, 674–681 (1988)
3. Graf, S., Viola, S.R., Kinshuk, T.L.: Representative characteristics of Felder-Silverman learning styles: an empirical model. In: Proceedings of the IADIS International Conference, pp. 235–242 (2006)
4. Ortigosa, A., Paredes, P., Rodriguez, P.: An adaptive hierarchical questionnaire for learning styles. Comput. Educ. pp. 999–1005 (2010)
5. Zywno, M.S.: A contribution to validation of score meaning for Felder-Soloman's index of learning styles. In: Proceedings of the American Society for Engineering Education, Annual Conference and Exposition (2003)
6. Felder, R.M., Spurlin, J.: Applications, reliability and validity of the index of learning styles. Int. J. Eng. Educ. **21**, 103–112 (2005)
7. Viola, S.R., Graph, S., Kinshuk, T.L.: Analysis of Felder-Silverman index of learning styles by a data-driven statistical approach. In: Eighth IEEE International Symposium on Multimedia, pp. 959–964 (2006)
8. Viola, S.R., Graf, S., Leo, T., Kinshuk, T.L.: Investigating relationships within the index of learning styles: a data-driven approach. Interac. Technol. Smart Educ. **4**, 7–18 (2007)
9. Graf, S., Lin, T., Kinshuk, T.L.: The relationship between learning styles and cognitive traits—getting additional information for improving student modelling. Comput. Hum. Behav. **24**, 122–137 (2008)
10. Garcia, P., Amandi, A., Schiaffino, S., Campo, M.: Evaluating Bayesian networks' precision for detecting students' learning styles. Comput Educ **49**, 794–808 (2007)
11. Alharbi, A., Paul, D., Henskens, F., Hannaford, M.: An investigation into the learning styles and self-regulated learning strategies for computer science students. In: Proceedings Ascilite, pp. 36–46 (2011)
12. Ahmad, N.B.H., Shamsuddin, S.M: A comparative analysis of mining techniques for automatic detection of student's learning. In: 10th International Conference on Intelligent Systems Design and Applications (ISDA), pp. 877–882 (2010)
13. Tuckman, B.W.: Conducting educational research, 5th edn., Harcourt Brace College Publishers (1999)
14. Livesay, G., Dee, K., Felder, R., Hites, L.: Statistical evaluation of the index of learning styles. ASEE Annual Conference and Exposition, pp. 101–109 (2002)
15. Litzinger, T.A.,, Lee, S.H., Wise J.C., Felder, R.M.: A psychometric study on the index of learning styles. J. Eng. Educ. **96**, 306–319 (2007)
16. Varde, A.S., Takakshi, M., Rundensteiner, E.A., Ward, M.O.: Apriori algorithm and game-of-life for predictive analysis in materials science. Int. J. Knowl. Based Intell. Eng. Syst. **8**, 213–228 (2004)
17. Handbook of teaching and learning in medicine, Peter Honey Publications (1992)
18. Honey, P., Mumford, A.: The manual of learning styles, 3rd edn., Peter Honey Publications (1992)
19. Jiawei, H., Micheline, K.: Data mining, concepts and techniques. Morgan Kaufmann publisher (2006)

20. Kolb, D.A.: Experiential learning: experience as the source of learning and development. Prentice-Hall, Englewood Cliffs, N.J (1984)
21. Zatarain Cabada, R., Barron Estrada, M.L., Angulo, V.P., Garcia ,A.J., Garcia, C.A.R.: Identification of Felder-Silverman learning styles with a supervised neural network. Advanced Intelligent Computing Theories and Applications, Lecture Notes in Computer Science, **6216**, 479–486 (2010)

Empirical Analysis of Low Power and High Performance Multiplier

K. Hemavathi and G. Manmadha Rao

Abstract In this paper, A Modified Booth Recoding with Zero Bypassing technique for Low Power and High Performance Multiplier is proposed. In CMOS digital circuit design, the power and area are the main concerns. The proposed multiplier reduces the area and power consumption compared to the existing multipliers. In this paper, modified booth recoding technique and Encoder circuit and Zero bypassing Techniques are used. In this proposed multiplier, the numbers of partial products are reduced by using booth recoding with zero bypassing technique. The proposed multiplier is simulated and synthesized by using Xilinx 10.1 ISE design tool. The Total power is calculated by using XPower Analyzer.

Keywords Switching activity reduction · Booth recoding · Full adder · Zero bypassing

1 Introduction

Multiplication is a fundamental component in digital signal processing, image processing and filtering. The multiplication is a two-folded process i.e. the first step is the generation of partial products and the second step is collecting the partial products and adds them. The need for portable devices places a restriction on area and power. The battery life and size are most attractive in recent years. The portable devices have more demand on low power consumption and high throughput [1]. The switching activities are more in CMOS design. The major sources of power consumption in CMOS circuits [2] are shown in Eq. (1).

K. Hemavathi (✉) · G. Manmadha Rao
Department of E.C.E, GMR Institute of Technology, Rajam, AP, India
e-mail: hemakorrai@gmail.com

G. Manmadha Rao
e-mail: manmadharao.g@gmrit.org

L.C. Jain et al. (eds.), *Computational Intelligence in Data Mining - Volume 2*,
Smart Innovation, Systems and Technologies 32, DOI 10.1007/978-81-322-2208-8_53

$$P = \alpha C\ V^2 f \quad (1)$$

where C is the load capacitance, α is switching activity, V is the operating voltage, and f is operating frequency. αC is an effective switching capacitance of the transistor nodes on discharging and charging. The less number of switching activities has low power dissipation. The general multipliers are combined of additions, subtractions if multiplier coefficients are constant. But this multiplication process is inflexible. To achieve flexible multiplication reconfigurable techniques [3] are developed to multiplier for efficient power. But this process is more complex. Compare to all arithmetic operations multiplication is power hungry and takes more time to execute. The critical path delays [4] effects on speed limit of the entire design.

2 Literature Survey

For pipelined multipliers, the main component is the carry save adder. The clock speed of the multiplier is determined by the delay time of a CSA cell to form and add the partial product, and generate the Carry [5]. The main disadvantage of this multiplier is circuit complexity is more.

The Baugh Wooley multiplier is an efficient technique to handle the signed bit multiplication [6]. The Baugh Wooley technique is designed to direct multiplication of 2's complement numbers. When 2's complement numbers are multiplied directly, each partial product to be added is a signed number. So each partial product is sign extended to the width of the final product to get the correct sum by carry save adder tree. In this technique an extra entries are used in the matrix. Due to this the negative bits in the partial product are avoided. But this technique requires an additional circuitry which results in more power consumption.

The Braun multiplier is a simple parallel multiplier. The Braun multipliers are performing multiplication on unsigned numbers. The architecture consists a network of full adders, AND gates arranged in iterative structure that does not need logic registers. It is also known as carry save array multiplier. The drawback of this multiplier is the no of components increases quadratically with the no of bits which will make multiplier to be inefficient.

The Booth algorithm multiplies two signed binary numbers in 2's complement notation. This process reduces the number of repeated steps to perform multiplication. In this algorithm the multiplier bits are grouped. Figure 1 shows the Booth Multiplier.

In Booth multiplier, the multiplier bits are recoded and each recoded format assigned to recoded values i.e. 0, 1, −1. Then this algorithm requires examination of the multiplier bits, and shifting of the partial product. Table 1 shows an Encoding Table of Radix-2 Booth multiplier. In Encoding Table when recoded bit is a 0 then shift operation will apply to multiplicand. When recoded bit is +1 then addition

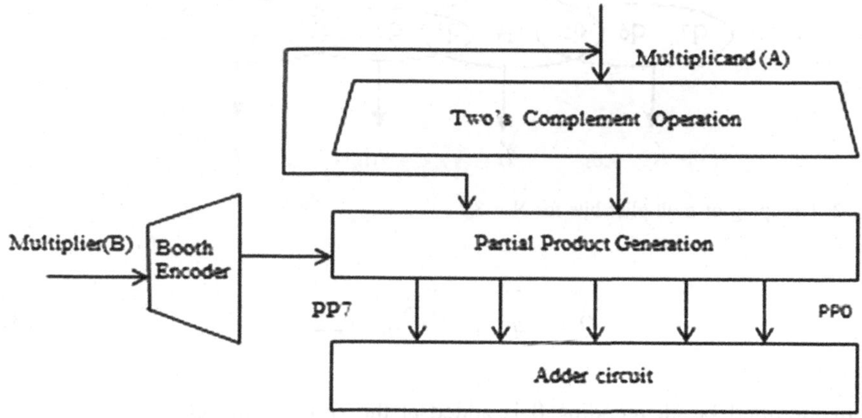

Fig. 1 Booth multiplier

Table 1 Encoding table of radix-2 booth multiplier

Q_n	Q_{n-1}	Recoded bits	Operation performed
0	0	0	Shift
0	1	+1	Addition
1	0	−1	Subtraction
1	1	0	Shift

operation can be performed and recoded bit is −1 then subtraction can be performed. The drawback of booth multiplier is the number of add/subtract operations and shift operations are becoming variable. In Bypass zero, feed A directly (BZ-FAD) [9] is a Low-Area and Low-power Multiplier based on Shift-and-Add Architecture. In this Multiplier Bypassing technique is used, but it takes into account each and every bit of the multiplier then it passes through the zero bypass technique. In this method a ring counter is used for bypassing.

3 Modified Booth Radix-4 Multiplier

In this paper, MBAR4 is used. Modified booth recoding is an important technique for reduction of the number of partial products [7]. Consider multiplication of two 2's complement N bit numbers P and Q are shown in Eqs. (2) and (3).

$$P = -p_{N-1} + \sum_{i=1}^{N-1} p_{N-1-i} 2^{-i} \tag{2}$$

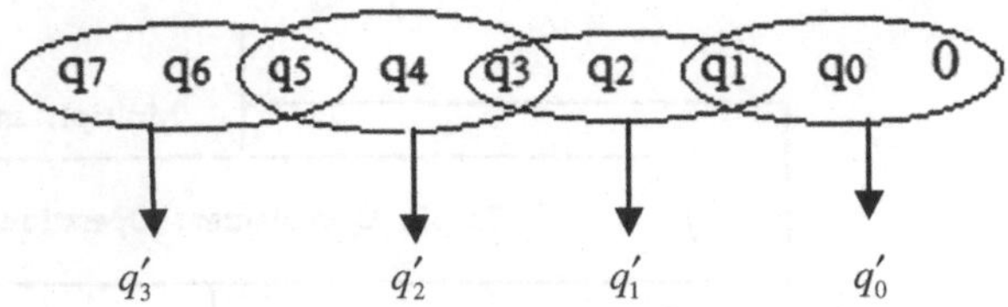

Fig. 2 Grouping of multiplier bits for N = 8

$$Q = -q_{N-1} + \sum_{i=1}^{N-1} q_{N-1-i} 2^{-i} \tag{3}$$

In modified booth recoding 0 is added at the right of the Q. Figure 2 shows grouping process.

In grouping process the multiplier is grouped into 4 groups of 8 bits. The right of the LSB bit of the multiplier is considered as always zero. The grouping process starts from LSB bit of multiplier and each group considers 3 bits of multiplier from next it considers 3 bits in which one bit from previous group.

After applying modifying booth recoding to the Q. The modified Q is shown in Eq. (4).

$$Q = \sum_{i=0}^{\frac{N}{2}-1} q'_{\frac{N}{2}-1-i} 2^{-(2i+1)} \tag{4}$$

where

$$q'_i = -2q_{2i+1} + q_{2i} + q_{2i-1} \tag{5}$$

3.1 Booth Recoding Process

The partial products are produced by multiplying multiplicand with recoded values i.e. 0, 1, −1, 2, −2. In which 1 indicates partial product is same as the multiplicand, and −1 intends partial product is 2's complement of the multiplicand, and 2 intends partial product is shift left one bit of the multiplicand, and −2 intends partial product is shift left one bit of 2's complement of the multiplicand. The Radix-4 Booth Recoding Table is given in Table 2.

3.2 Booth Encoder

Booth encoder architecture consists 2:1 mux, AND gate, NOR gate, XOR gate to produce encoded output for reducing partial products. Figure 3 shows encoder

Table 2 Radix-4 booth recoding table

q_{2i+1}	q_{2i}	q_{2i-1}	q'_i
0	0	0	0
0	0	1	1
0	1	0	1
0	1	1	2
1	0	0	−2
1	0	1	−1
1	1	0	−1
1	1	1	0

Fig. 3 Encoder circuit

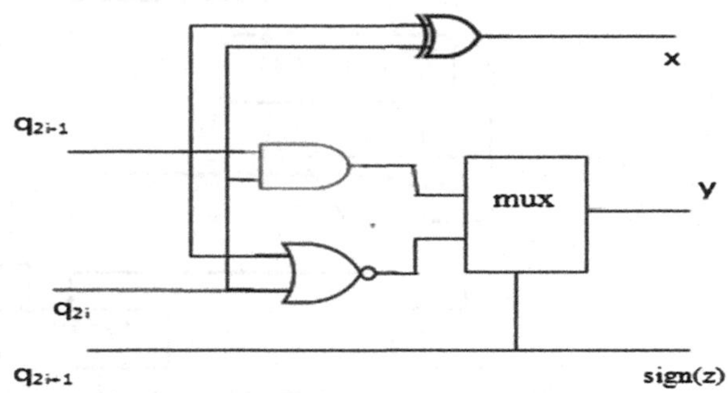

architecture. The encoder inputs are q_{2i-1}, q_{2i} and q_{2i+1}. The encoded output z is the sign bit is equal to q_{2i+1}. The Sign bit indicates signed multiplication, when q_{2i+1} is one, then y is AND operation between q_{2i-1} and q_{2i}, when q_{2i+1} is zero then y is NOR operation between q_{2i-1} and q_{2i}. The Encoded outputs are shown in Eq. (5).

$$\begin{aligned} &x = q_{2i-1} \text{XOR } q_{2i} \\ &a = q_{2i-1} \text{AND } q_{2i} \\ &b = q_{2i-1} \text{NOR } q_{2i} \\ &\text{If } q_{2i+1=1} \text{ then } y = a \text{ else } y = b \\ &z = \text{sign}(q_{2i+1}) \end{aligned} \tag{5}$$

In this process the numbers of multiplication steps are reduced.

4 Proposed Multiplier

The proposed multiplier is a combination of booth recoding and zero bypassing. For zero bypassing, the decoder circuit is used in this paper. After encoding (Fig. 3) the multiplier bits, the recoded outputs are directly fed to this multiplier architecture.

The recoded outputs are x, y, z and depends on these recoded values the partial products are found.

Figure 4 shows the multiplier architecture in this P is multiplicand, the 2:1 multiplexer selects the multiplicand P or 2's complement of multiplicand depending on sign bit (z). If z is zero, then multiplicand P is selected and stored in R register else the 2's complement of P is selected and stored in R register. M0 and M1 are counter signals required for no of addition operations can be performed. The recoded outputs

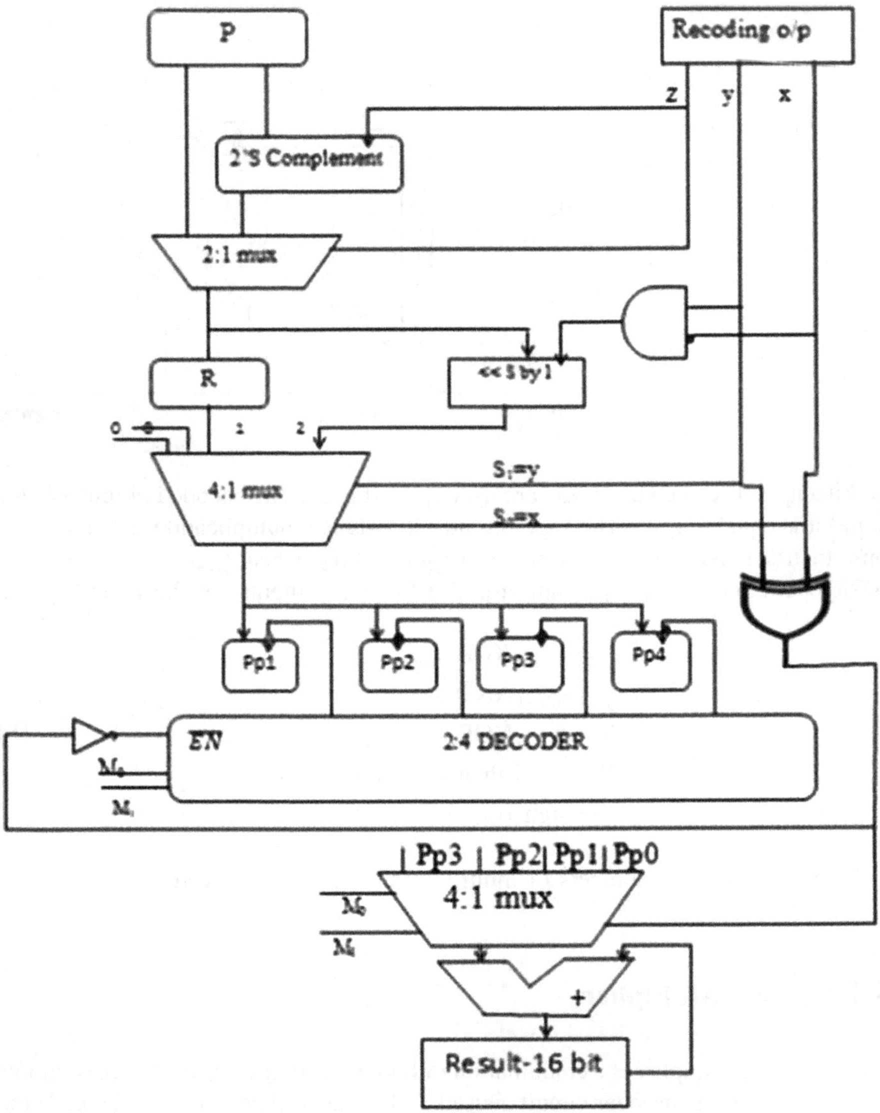

Fig. 4 Multiplier architecture

x, y act as select lines for 4:1 multiplexer. Then 1-bit shift register is used to perform multiplication of the negative or the positive multiplicand by decimal 2. While XOR output used to enable 4:1 multiplexer of the partial product register and for 2:4 decoder, and four partial product registers are used to store internal partial product values. The one input bubbled AND gate is used to enable shift operation. When XOR output is zero both decoder and adder circuits are disabled. When recoded outputs x, y, z are 0, 1, 0. The z value is zero, so it selects multiplicand P using 2:1 multiplexer, this value is stored in R register. In this case y, x outputs are 1, 0 respectively i.e. the recoded value is +2, then inverted x value of AND gate enables the shift register which shifts the value of R register by 1bit. In this case the XOR output is high, so the decoder and 4:1 multiplexer enabled, and the partial product has been selected and stored in register, the value is given through adder circuit. When recoded outputs x, y, z are 0, 0, 0. The z value is 0 so P is selected and stored in R register. In this case x, y are 0, 0 so the XOR gate output is low, the output is passed through the NOT gate so the output of the NOT gate is high, this output acts as enable input to the decoder, but the enable input of the decoder is active low so in this case decoder is disabled and 4:1 multiplexer also disabled, the partial product registers are disabled. Figure 5 shows partial product registers. In this case the partial product containing zeros is bypassed so the addition operation is saved due to this switching activities are minimized resulting in reduced power consumption.

The proposed architecture is implemented using Xilinx 10.1 ISE. The program is written in Verilog and the power is calculated using XPower Analyzer.

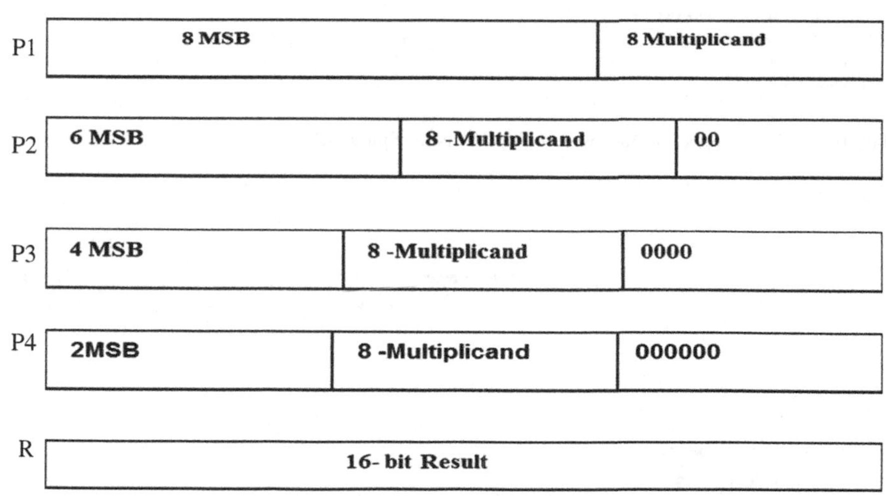

Fig. 5 Partial product registers

5 Experimental Results

In this section the simulation and synthesis results are obtained by using Xilinx 10.1 ISE. The results shown that the power required to proposed multiplier is reduced compared to existing booth multiplier architectures. The proposed multiplier requires less power, and less number of Slice Flip-flops, less Number of Slice LUTs and less Number of Bonded IOBs compared to radix-2 and radix-4 Booth multipliers.

Figure 6 shows the device utilization summary of the proposed multiplier for Spartan-2 family and number of slice registers and number of slice LUTs, number of bonded IOBs are reduced compared to Radix-2 and Radix-4 booth multiplier.

Figure 7 shows the device utilization summary of the proposed multiplier for Spartan-3E family. Here the Number of slices and Number of LUTs compared to Radix-2 and Radix-4 booth multiplier are reduced.

Device Utilization Summary				
Logic Utilization	**Used**	**Available**	**Utilization**	**Note(s)**
Number of Slice Flip Flops	28	864	3%	
Number of 4 input LUTs	53	864	6%	
Logic Distribution				
Number of occupied Slices	33	432	7%	
Number of Slices containing only related logic	33	33	100%	
Number of Slices containing unrelated logic	0	33	0%	
Total Number of 4 input LUTs	54	864	6%	
Number used as logic	53			
Number used as a route-thru	1			
Number of bonded IOBs	43	92	46%	
Number of GCLKs	1	4	25%	
Number of GCLKIOBs	1	4	25%	

Fig. 6 Synthesis report of proposed multiplier using Spartan-2

Device Utilization Summary				
Logic Utilization	**Used**	**Available**	**Utilization**	**Note(s)**
Number of Slice Flip Flops	28	9,312	1%	
Number of 4 input LUTs	47	9,312	1%	
Logic Distribution				
Number of occupied Slices	29	4,656	1%	
Number of Slices containing only related logic	29	29	100%	
Number of Slices containing unrelated logic	0	29	0%	
Total Number of 4 input LUTs	47	9,312	1%	
Number of bonded IOBs	44	66	66%	
Number of BUFGMUXs	1	24	4%	

Fig. 7 Synthesis report of proposed multiplier using Spartan-3E

Table 3 Comparison of different multipliers

Multipliers / Parameters	Radix-2 booth multiplier	Radix-4 booth multiplier	Proposed multiplier	Proposed multiplier
Device and family	Spartan-2	Spartan-2	Spartan-2	Spartan-3E
Number of slice flip flops	77	72	28	28
Number of LUTs	140	129	53	47
Number bonded IOBs	32	32	43	44
Delays (ns)	27.11	26.103	8.468	5.89
Total power (mW)	15	11	9	5.3

Table 4 Comparision of the proposed multiplier at different frequencies

Different frequencies / Devices and total power (mW)	25 MHz	50 MHz	75 MHz	100 MHz
Spartan-2	14.34	28.68	43.02	57.36
Spartan-3E	13.28	26.56	39.84	53.12

In Table 3 the power consumption and delay of the proposed multiplier is reduced compared to Radix-2 and Radix-4 booth algorithm.

The total power of the proposed multiplier at different frequencies for Spartan-2 and Spartan-3E family is calculated is shown in Table 4. From the results the total power required to proposed multiplier is reduced compared to low power design multipliers.

6 Conclusion

In this paper empirical analysis of low power and high performance multiplier was proposed. In this paper, Compare to conventional multiplier architectures i.e. Radix-2 and Radix-4 Booth multipliers and other low power design multipliers the dynamic power and delay and area are reduced. The proposed design of a multiplier successfully simulated and synthesized using XILINX 10.1 ISE design. The power is calculated using XPower Analyzer and performance of the proposed multiplier is improved and No. of LUTs and slice Flip flops are reduced which influences on power consumption and speed. The results are shown for power and delay reduction for different devices compare to other multiplier design.

References

1. Chandrakasan, A., Brodersen, R.: Low-power CMOS digital design. IEEE J. Solid- State Circ. **27**(4), 473–484 (1992)
2. Chen, O.T., Wang, S., Yi-Wen, Wu: Minimization of switching activities of partial products for designing low-power multipliers. IEEE Trans. VLSI Syst. **11**, 418–433 (2003)
3. Kuang, S.: Design of power-efficient configurable booth multiplier. IEEE Trans. circ. syst., **57**(3), 568–580 (2010)
4. He, Y., Chang, C.: A new redundant binary booth encoding for fast 2n-bit multiplier design. IEEE Trans. Circ. Syst., **56**(6), (2009)
5. Wu, A.: High performance adder cell for low power pipelined multiplier. IEEE Int. Symp. Circuits Syst. **4**, 57–60 (1996)
6. Juang, T.B., Hsiao, S.F.: Low-power carry-free fixed-width multipliers with low-cost compensation circuit. IEEE Trans. 5 Circuits Syst. II Analog Digital Signal Process **52**(6), 299–303 (2005)
7. Kim, J.M., Lee, J.S., Cho, J.D.: A low power booth multiplier based on operand swapping in instruction level. J. Korean Phys. Soc., **33**, S258–S261 (1998)
8. Devi V., Lokku, G.K., Natarajan, A.: Fixed width booth multiplier based on peb circuit. Int. J. Artificial Intell. Appl. (IJAIA), **3**(2), 255–259 (2012)
9. Mottaghi-Dastjerdi, M., Afzali-Kusha, A., Pedram, M.: BZ-FAD- A low-power low-area multiplier based on shift-and-add architecture. IEEE Transactions on very large scale integration (VLSI) systems, vol. 17, no. 2 pp. 302–306 (February 2009)

Facial Expression Recognition Using Modified Local Binary Pattern

Suparna Biswas and Jaya Sil

Abstract In this paper a low computation feature space has been proposed to recognize expressions of face images. The image is divided into number of blocks and binary pattern corresponding to each block is generated by modifying the Local Binary Pattern (LBP). The proposed method generates compressed binary pattern of images and therefore, reduced in size. Features are extracted from transformed image using block wise histograms with variable number of bins. For classification we use two techniques, template matching and Support Vector Machine (SVM). Experiments on face images with different resolutions show that the proposed approach performs well for low resolution images. Considering Cohn-Kanade database, the proposed method is compared with LBP feature based methods demonstrating better performance.

Keywords Facial expression · Template matching · Local binary pattern

1 Introduction

Human face communicates useful information about a person's emotional state or expressions. Accurate recognition of facial expression is difficult due to facial occlusions, pose variation and illumination differences. The existing expression recognition methods perform well for the high resolution images. However, in real world applications such as visual surveillance and smart meeting, the input face

S. Biswas (✉)
Department of Electronics and Communication Engineering,
Gurunanak Institute of Technology, Sodepur, Kolkata, India
e-mail: suparna_b80@yahoo.co.in

J. Sil
Department of Computer Science and Technology, Indian Institute of Engineering Science and Technology University, Shibpur, Howrah, India
e-mail: js@cs.becs.ac.in

L.C. Jain et al. (eds.), *Computational Intelligence in Data Mining - Volume 2*,
Smart Innovation, Systems and Technologies 32, DOI 10.1007/978-81-322-2208-8_54

images are often at low resolutions. In this paper, a method has been presented to recognize the expression of face images by creating a compressed image space. The main contribution of this paper is summarized as pre-processing to overcome the illumination differences, next transformation to a compressed size image by generating block-wise binary patterns, block based histogram calculation as features and finally classification of expression. By introducing modified LBP based transformation the image size is reduced and computation speed is improved. We comprehensively study facial expression recognition with different classifiers and compared to previous work based on Local Binary Pattern (LBP) and Gabor Wavelet [1–3]. Our technique provides better performance with respect to computation speed and recognition accuracy. The recognition accuracy for the low resolution images are specifically good and look promising for the real world applications.

Automatic facial expression recognition involves two important steps: facial feature representation and designing of classifier. A number of methods have been developed for extracting features from face images like Facial Action Coding System (FACS), Principal Component Analysis (PCA) [4–7], Local Binary Pattern (LBP) [1, 2, 8], Independent Component Analysis (ICA), Linear Discriminate Analysis (LDA) [9], Edge detection, Active Appearance Model (AAM), Gabor Wavelet [10, 11], Contourlet Transform [12–14].

The rest of the paper is organized as follows. We present a brief review of related work in this section. Section 2 presents the methodology followed, Sect. 3 presents the experimental results analysis and finally concluding remarks are summarized in Sect. 4.

2 Proposed Method

The block diagram of the proposed method for expression recognition is shown in Fig. 1, consisting of five main modules: face detection, pre-processing, modified LBP, feature extraction and classification.

2.1 Face Detection

The first step of expression recognition is face detection. In the paper face detection has been carried out using Viola Jones method [15], and satisfactory results are achieved with a correct detection rate of 99 %, tested on Caltech Image Database.

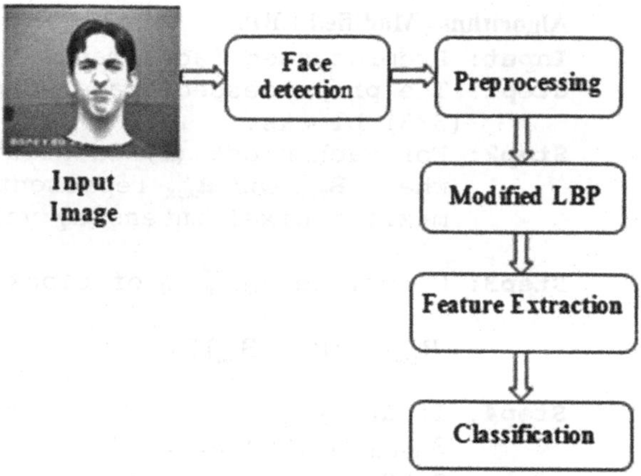

Fig. 1 Flow diagram of expression detection

2.2 *Preprocessing*

After face detection, the feature area is extracted by cropping the face image 15 % from right and left and 20 % from top of the image for removing ears and hairs. Then in the next step Gaussian smoothing filter is applied and image is resized. At the last step of preprocessing, histogram equalization is performed to overcome the illumination differences.

2.3 *Modified LBP*

The preprocessed image is transformed to an image of reduced dimension by introducing modified LBP. The preprocessed image is first divided into (3 × 3) blocks, and we calculate the threshold value for each block. The binary pattern for each block is obtained after thresholding and represented by equivalent decimal value. The threshold for generating the binary pattern is calculated using a new technique. The algorithm is described below:

Algorithm - Modified LBP.

Input: Preprocessed face Image.

Step1: The preprocessed image is divided into (3×3) blocks.

Step2: For each block (**B**) calculate B_{min} and B_{max}. where $\mathbf{B}_{min}$ and $\mathbf{B}_{max}$ represent minimum and maximum pixel intensity value of block **B**.

Step3: Calculate $B_{threshold}$ of block **B**

$$B_{threshold} = (B_{min} + B_{max})/2.$$

Step4: If any element in **B** is greater equal to $B_{threshold}$ then write '1' to that element, else '0'.

Step5: The eight bit pattern of **B** is converted to decimal number, representing transformed block **B**.

End.

A binary pattern is thus obtained for block **B**, as shown in Fig. 2. In this way for each block transformed intensity value is thus obtained considering decimal number of eight bit pattern. Therefore, the preprocessed image is transformed into compressed image as shown in Fig. 3.

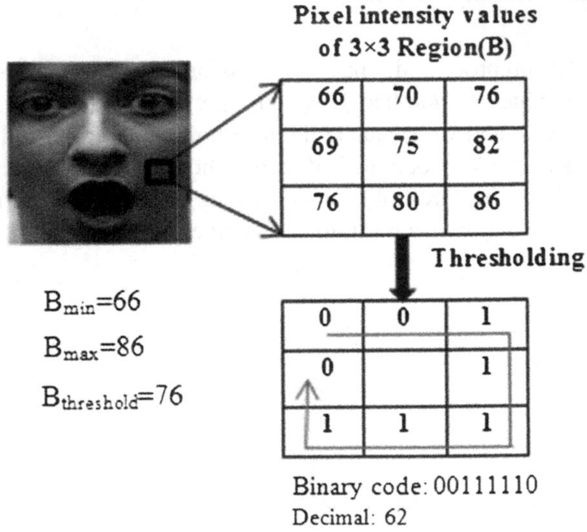

Fig. 2 Modified LBP for block **B**

Fig. 3 Compressed pattern image

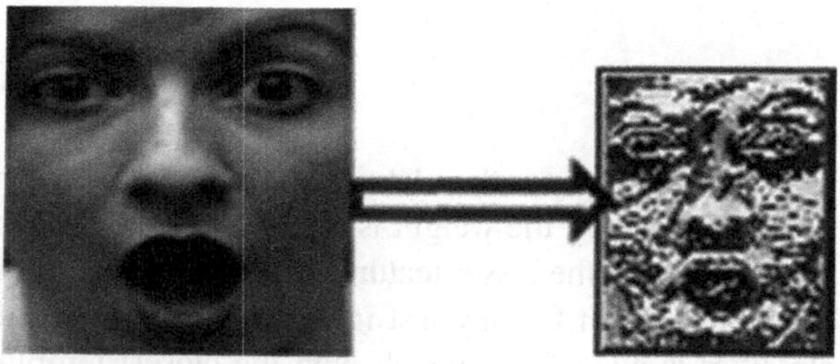

2.4 Feature Extraction

To extract features from the compressed modified LBP image, divide the compressed pattern image into (M × N) blocks, and calculate histogram for each block. Concatenating the histogram of all (M × N) blocks, features are obtained, as shown in Fig. 4. During experiment we have extracted features for different block size, and for each block size we have calculated the histogram by increasing no. of bin from 5 to 59.

2.5 Classification

Facial expressions are classified using template matching and Support Vector machine. Computation speed is an important issue in real time system development, which depends on feature extraction and classification procedures.

Template Matching. We first adopt template matching technique for expression classification. Average value of histograms of the transformed image for a particular class is obtained to generate template1. In template2, weighted average technique is applied. Consider n number of training images and extract the features say, $\{x_1, x_2, \ldots, x_n\}$ where x_i represents average feature value of histograms for image i. Generate non negative weight set randomly $\{w_1, w_2, \ldots, w_n\}$ and calculate average feature for template2 using Eq. (1).

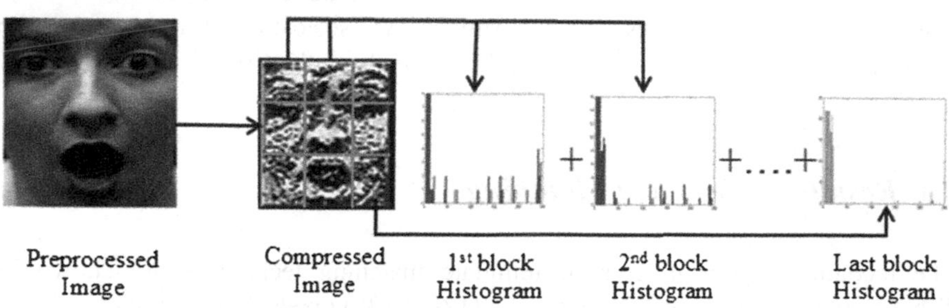

Fig. 4 Histogram generation procedure

$$\bar{x} = \frac{\sum_{i=1}^{n} w_i x_i}{\sum_{i=1}^{n} w_i} \tag{1}$$

Then by sorting the data set (features) in ascending order and weight in descending order, the weight is multiplied with corresponding feature, which imply more weight to the lesser feature value. To create a template, 50 images are used for each class. Then for any test image, after extracting the features, nearest-neighbor classifier is adopted to match with the closest template. We have used Euclidean distance to measure similarity.

Support Vector Machine (SVM). Support vector machine performs an implicit mapping of data into a higher dimensional feature space, and finds a linear separating hyper plane with the maximal margin to separate data in the higher dimensional space. Given a training set of labeled examples $F_{train} = \{(x_i, y_i), i = 1, \ldots, p\}$, where $x_i \in R^n$ and $y_i \in \{1, -1\}$, then the new test data set is classified by the function, described in Eq. (2).

$$f(x) = \mathrm{sgn}(\sum_{i=1}^{p} \alpha_i y_i K(x_i, x) + b) \tag{2}$$

where α_i is the Lagrange multiplier, of dual optimization problem, $K(x_i, x)$ is the kernel function and b is the threshold parameter of the hyperplane. Given a non-linear mapping Φ that embeds the input data into high dimensional space, kernels have the form of $K(x_i, x_j) = \langle \Phi(x_i) \cdot \Phi(x_j) \rangle$. The most frequently used kernels are polynomial kernels and radial basis functions.

3 Results and Discussions

The proposed algorithm was trained and tested on the Cohn Kanade facial expression database [16] consists of 100 university students aged between 18 and 30 years, among which 65 % female, 15 % African–American, and 3 % Asian or Latino.

Database contains anger, disgust, happy, neutral, sadness, surprise and some of feared face image sequences. For experiments, we selected 600 images from the database where Fig. 5 shows some sample images from the Cohn-Kanade database.

3.1 Results of Template Matching

The recognition performances of template matching techniques are shown in Table 1 considering facial images of size (128 × 128) pixels. For feature extraction, after image compression using modified LBP, image is divided into blocks of

Fig. 5 Sample face expression images from the Cohn Kanade database

Table 1 Recognition performance of template matching techniques

No. of bin	Template1 (%)	Template2 (%)
5	83	89
9	78	87
15	72	80
23	60	70

(8 × 8) pixels per region. The template matching technique achieves maximum accuracy of 89 % for the weighted average method (template2) and for simple average (template1) it is 83 %. We have tested the template matching techniques, for images of different resolutions and observed that images of (128 × 128) resolution give best result. We compared the results with [1, 2], where template matching technique has been used to classify the expression. Comparison in Table 2 illustrates that our template matching technique performs better.

Table 2 Comparison with other template matching for 6-class expression recognition

Method	Recognition (%)
LBP + template matching [2]	84.5
LBP + template matching [1]	79.1
Proposed template1	83
Proposed template2	89

3.2 Results of SVM

SVM can classify objects or training samples into two categories, so the multi classification can be performed using one-against-rest technique, which trains binary classifiers to differentiate one expression from all others. The performance has been achieved with different kernels are shown in Table 3.

For Table 3, the degree of the polynomial kernel is 1, and the standard deviation for the RBF kernel is 2^{15} is considered. Facial images of (128 × 128) pixels are compressed (nine times) by our proposed technique and divided into block of (8 × 8) pixels per region. Thus compressed images are divided into 36 regions and then features are extracted using 5 bin histograms with the length of 180 (36 × 5). From Table 3 we conclude that surprise, sad, happy, angry are recognized with high accuracy (95.67–100 %) while the recognition rate for disgust and neutral is greater than 90 %. For the SVM implementation we have used Matlab and 10-fold cross-validation technique. We compare the computation time and number of features, is shown in Table 4. It is observed that our technique is better than the LBP and Gabor wavelet based feature extraction technique.

To examine the performances of the proposed method for low resolution images we have studied 4 different resolutions of the face mages (110 × 150, 55 × 75, 36 × 48, 27 × 37) based on Cohn-Kanade database. Recognition performance for different resolution images is shown in Fig. 6.

For the images of resolution 110 × 150 is divided into 18 × 21 pixels per region and for 55 × 75, 36 × 48, 27 × 37 resolution images are divided into 10 × 10 pixels per region. For the recognition of different expression we have used SVM with polynomial.

Table 3 Classification accuracy for SVM classifier

Expression	Linear (%)	RBF (%)	Polynomial (%)
Surprise	96.33	96	96.67
Disgust	90.67	90.00	91.00
Sadness	100	100	100
Happy	95.67	93.67	96.00
Angry	95.67	96.00	96.00
Neutral	91.67	94.00	92.33

Table 4 Comparison of computation time and no of features using SVMs

Method	No. of features	Feature extraction time
LBP [2]	22,478	84.5 0.03 s
Gabor [2]	42,650	30 s
Gabor [3]	92,160	–
Proposed	180	0.027 s

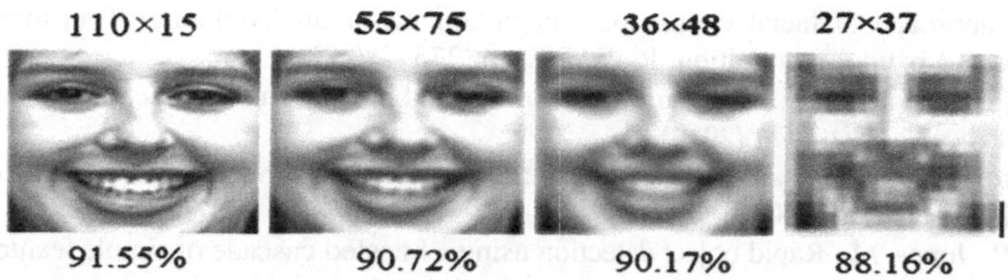

Fig. 6 Recognition performance of different resolution images for 59 bin

4 Conclusions

This paper presented a new method for facial expression recognition. Classification accuracy shows effectiveness of the proposed feature extraction method. Compare to Gabor wavelet and LBP features, the proposed technique save more computational time and resources. This feature extraction technique is robust and stable over a useful range of low resolution images. For the low resolution images when geometric features are not available, our technique can be applied for the expression recognition.

References

1. Shan, C., Gong, S., McOwan, P.W.: Robust facial expression recognition using local binary patterns. Image Process. ICIP 2 **2**, 370–373 (2005)
2. Shan, C., Gong, S., McOwan, P.W.: Facial expression recognition based on local binary patterns: a comprehensive study. Image Vis. Comput. **27**, 803–816 (2009)
3. Bartlett, M.S., Littlewort, G., Frank, M., Lainscsek, C., Fasel, I., Movellan, J.: Recognizing facial expression: machine learning and application to spontaneous behavior. In: IEEE conference on computer vision and pattern recognition (2005)
4. Yeasi, M., Bullot, B., Sharma, R.: Recognition of facial expressions and measurement of levels of interest from video. IEEE Trans. Multimedia **8**(3), 500–508 (2006)
5. Thai, L.H., Nguyen, N.D.T., Hai, T.S.: A facial expression classification system integrating canny, principal component analysis and artificial neural network. Int. J. Mach. Learn. Comput. **1**(4) (2011)
6. Sumathi1, C.P., Santhanam2, T., Mahadevi, M.: Automatic facial expression analysis a survey. Int. J. Comput. Sci. Eng. Surv. (IJCSES) **3**(6) (2012)
7. Kalita, J., Das, K.: Recognition of facial expression using eigenvector based distributed features and euclidean distance based decision making technique. (IJACSA) Int. J. Adv. Comput. Sci. Appl. **4**(2) (2013)
8. Sarode, N., Bhatia, S.: Facial expression recognition. Int. J. Comput. Sci. Eng. **2**(5), 1552–1557 (2010)
9. Shih, F.Y., Chuang, C.H., Wang, P.S.P.: Performance comparisons of facial expression recognition in jaffe database. Int. J. Pattern. Recogn. Artif. Intell. **22**(3), 445–459 (2008)
10. Deng, H.B., Jin, L.W., Zhen, L.X., Huang, J.C.: A new facial expression recognition method based on local Gabor filter bank and PCA plus LDA. Int. J. Inf. Technol. **11**(11), 86–96 (2005)
11. Liu, C., Wechsler, H.: Gabor feature based classification using the enhanced fisher linear discriminant model for face recognition. IEEE Trans. Image Process. **11**(4), 467–476 (2002)

12. A new approach of facial expression recognition based on Contourlet Transform, wavelet analysis and pattern recognition. ICWAPR, pp. 275–280 (2009)
13. Suresh, R., Audithan, S.: Contourlet transform based human emotion recognition system. Int. J. Sig. Process. Syst. **2**(1) (2014)
14. Lajevardi, S.M., Hussain, Z.M.: Contourlet structural similarity for facial expression recognition. ICASSP (2010)
15. Viola, P., Jones, M.: Rapid object detection using a boosted cascade of simple features. CVPR, Kauai (2001)
16. Kanade, T., Cohn, J.F., Tian, Y.: Comprehensive database for facial expression analysis. IEEE FG, pp. 46–53 (2000)

On Multigranular Approximate Rough Equivalence of Sets and Approximate Reasoning

B.K. Tripathy, Prateek Saraf and S.Ch. Parida

Abstract As the notion of equality in mathematics is too stringent and less applicable in real life situations, Novotny and Pawlak introduced approximate equalities through rough sets. Three more types of such equalities were introduced by Tripathy et al. as further generalisations of these equalities. As rough set introduced by Pawlak is unigranular from the granular computing point of view, two types of multigranulations rough sets called the optimistic and the pessimistic multigranular rough sets have been introduced. Three of the above approximate equalities were extended to the multigranular context by Tripathy et al. recently. In this paper, we extend the last but the most general of these approximate equalities to the multigranular context. We establish several direct and replacement properties of this type of approximate equalities. Also, we illustrate the properties as well as provide counter examples by taking a real life example.

Keywords Rough sets · Approximate equalities · Approximate equivalence · Optimistic multigranulation · Pessimistic multigranulation · Replacement properties

B.K. Tripathy (✉) · P. Saraf
School of Computer Science and Engineering, VIT University, Vellore 632014
Tamil Nadu, India
e-mail: tripathybk@vit.ac.in

P. Saraf
e-mail: prateeksaraf2010@gmail.com

S.Ch. Parida
Department of Mathematics, K.B.V. Mahavidyalaya, Kabisurya Nagar,
Ganjam, Odisha 761104, India
e-mail: sudamparida32@gmail.com

L.C. Jain et al. (eds.), *Computational Intelligence in Data Mining - Volume 2*,
Smart Innovation, Systems and Technologies 32, DOI 10.1007/978-81-322-2208-8_55

1 Introduction

The notion of equality of sets used in mathematics is not of much use in real life situations because of its stringent definition, which can only be used when the two sets have the same elements. Moreover, it does not use the knowledge of the observer regarding the domain while considering the equality. In real life situations we use user knowledge as a supporting tool, which also determines the equality or otherwise of sets under consideration. In an attempt to achieve this Novotny and Pawlak [1–3] introduced three types of equalities of sets through rough sets, where the equivalence relation plays the deciding role. Since human knowledgc according to Pawlak is determined through their classification capability, which is dependent upon classification of universes and in turn is equivalent to equivalence relations defined over the domain, this definition while being more general than the mathematical equality, takes care of the human knowledge, making it more natural. This early notion of approximate equality was not considered further until Tripathy et al. [4, 5] introduced another such notion called the rough equivalence, which was proved later to be the most general of this kind of equalities and also is free from the notion of mathematical equality. In fact, two more types of approximate equalities using rough sets were introduced by Tripathy [6] in 2011, by the way completing the four types of possible approximate equalities using rough sets. The notion of rough sets introduced by Pawlak [7, 8] is unigranular from the granular computing point of view in the sense that it considers only one equivalence relation at a time. Extending this notion the concept of optimistic multigranular rough sets was introduced by Qian and Liang [9] in 2006. Later on they defined another type of multigranular rough sets called the pessimistic multigranular rough sets in 2010 [10]. Three of the four types of approximate equalities were extended to the setting of multigranular rough sets by Tripathy and Mitra [5, 11, 12] very recently. In this paper we extend the last but the most important type of approximate equality that is we define multigranular rough equivalence of sets and establish their properties. Also, we prove the replacement properties. In this paper we provide two diagrams which provide a comparative analysis of the unigranular rough set notions of lower and upper approximation and the multigranular upper and lower approximations for both the optimistic and pessimistic multigranular rough sets. We use a real life database to illustrate the concepts of the paper and also provide counter examples wherever required using this example.

2 Definitions and Notations

In this section we provide some of the definitions and notations to be used. First we start with the basic rough sets in the next section.

2.1 Basic Rough Sets

Let *U* be a universe of discourse and *R* be an equivalence relation over *U*. By *U/R* we denote the family of all equivalence class of *R*, referred to as categories or concepts of *R* and the equivalence class of an element $x \in U$ is denoted by $[x]_R$. By a knowledge base, we understand a relational system $K = (U, R)$, where *U* is as above and *R* is a family of equivalence relations over *U*. For any subset $P(\neq \phi) \subseteq R$, the intersection of all equivalence relations in *P* is denoted by IND (P) and is called the indiscernibility relation over *P*. Given any $X \subseteq U$ and $R \in \text{IND}(K)$, we associate two subsets, $\underline{R}X = \cup\{Y \in U/R : Y \subseteq X\}$ and $\bar{R}X = \cup\{Y \in U/R : Y \cap X \neq \phi\}$, called the R-lower and R-upper approximations of X respectively.

The R-boundary of *X* is denoted by $BN_R(X)$ and is given by $BN_R(X) = \bar{R}X - \underline{R}X$. The elements of $\underline{R}X$ are those elements of *U*, which can certainly be classified as elements of X, and the elements of $\bar{R}X$ are those elements of U, which can possibly be classified as elements of X, employing knowledge of *R*. We say that *X* is rough with respect to *R* if and only if $\underline{R}X \neq \bar{R}X$, equivalently $BN_R(X) \neq \phi$. *X* is said to be R-definable otherwise.

2.2 Multigranular Rough Sets

We introduce the two types of multigranulations in this direction using the notations in recent papers by Tripathy et al. [13–15] followed by some properties of these multigranulations.

Definition 2.2.1 Let K = (U, **R**) be a knowledge base, **R** be a family of equivalence relations, $X \subseteq U$ and $R, S \in \mathbf{R}$. We define [9] the optimistic multi-granular lower approximation and optimistic multi-granular upper approximation of X with respect to R and S in U as

$$\underline{R+S}\,X = \{x \mid [x]_R \subseteq X \text{ or } [x]_S \subseteq X\} \tag{2.2.1}$$

$$\overline{R+S}\,X = \sim(\underline{R+S}(\sim X)). \tag{2.2.2}$$

Definition 2.2.2 Let K = (U, **R**) be a knowledge base, **R** be a family of equivalence relations, $X \subseteq U$ and $R, S \in \mathbf{R}$. We define [10] the pessimistic multi-granular lower approximation and pessimistic multi-granular upper approximation of X with respect to R and S in U as

$$\underline{R*S}\,X = \{x \mid [x]_R \subseteq X \text{ and } [x]_S \subseteq X\}, \tag{2.2.3}$$

$$\overline{R*S}\,X = \sim(\underline{R*S}(\sim X)). \tag{2.2.4}$$

2.2.1 Properties of Multigranular Approximations

We present some properties of multigranular rough sets, which shall be used in the proofs of the results of this paper [14, 15].

$$\underline{R+S}(X\cap Y)\subseteq \underline{R+S}(X)\cap \underline{R+S}(Y) \tag{2.2.5}$$

$$\underline{R+S}(X\cup Y)\supseteq \underline{R+S}(X)\cup \underline{R+S}(Y) \tag{2.2.6}$$

$$\overline{R+S}(X\cap Y)\subseteq \overline{R+S}(X)\cap \overline{R+S}(Y) \tag{2.2.7}$$

$$\overline{R+S}(X\cup Y)\supseteq \overline{R+S}(X)\cup \overline{R+S}(Y) \tag{2.2.8}$$

$$\underline{R*S}(X\cup Y)\supseteq \underline{R*S}(X)\cup \underline{R*S}(Y) \tag{2.2.9}$$

$$\overline{R*S}(X\cap Y)\subseteq \overline{R*S}(X)\cap \overline{R*S}(Y) \tag{2.2.10}$$

$$\underline{R*S}(X\cap Y)= \underline{R*S}(X)\cap \underline{R*S}(Y) \tag{2.2.11}$$

$$\overline{R*S}(X\cup Y)= \overline{R*S}(X)\cup \overline{R*S}(Y) \tag{2.2.12}$$

We would like to note some cases when equalities hold in (2.2.8) and (2.2.10). These results will be helpful to us in establishing some properties later.

Lemma 2.2.3.1 *If* $\overline{R+S}(X)=\overline{R+S}(Y)$ *then equality holds in* (2.2.8).

Proof It is easy to see from definition (2.2.2) that

$$\overline{R+S}(X)=\{x:[x]_R\cap X\neq \phi \ or\ [\mathrm{x}]_S\cap X\neq \phi\}$$

and

$$\overline{R+S}(\mathrm{Y})=\{x:[x]_R\cap Y\neq \phi \ or\ [\mathrm{x}]_S\cap Y\neq \phi\}.$$

Also,

$$\overline{R+S}(\mathrm{X}\cup\mathrm{Y})=\{x:[x]_R\cap (X\cup Y)\neq \phi \ or\ [\mathrm{x}]_S\cap (X\cup Y)\neq \phi\}.$$

$$So,\ x\in \overline{R+S}(X\cup Y)\Rightarrow\{[x]_R\cap X\neq\phi\ or\ [x]_R\cap Y\neq\phi\ or$$
$$[x]_S\cap X\neq\phi\ or [x]_S\cap Y\neq\phi\}$$

By our assumption this implies that

$$\{[x]_R\cap X\neq\phi\ or\ [x]_R\cap X\neq\phi\ or\ [x]_S\cap X\neq\phi\ or\ [x]_S\cap X\neq\phi\}$$

So, $x\in\overline{R+S}(X)$ and hence $x\in\overline{R+S}(\mathrm{Y})$. This completes the proof. □

Lemma 2.2.3.2 *If* $\overline{R*S}(X) = \overline{R*S}(Y)$ *then equality holds in* (2.2.10).

Proof It is easy to see from definition (2.2.4) that

$$\overline{R*S}(X) = \{x : [x]_R \cap X \neq \phi \text{ and } [x]_S \cap X \neq \phi\}$$

$$\overline{R*S}(Y) = \{x : [x]_R \cap Y \neq \phi \text{ and } [x]_S \cap Y \neq \phi\}.$$

Also,

$$\overline{R*S}(X \cap Y) = \{x : [x]_R \cap (X \cap Y) \neq \phi \text{ and } [x]_S \cap (X \cap Y) \neq \phi\}.$$

So, $x \in \overline{R*S}(X) \cap \overline{R*S}(Y) \Rightarrow x \in \{y : [y]_R \cap X \neq \phi \text{ and } [y]_S \cap X \neq \phi\}$ *and* $x \in \{y : [y]_R \cap Y \neq \phi \text{ and } [y]_S \cap Y \neq \phi\}$

By our assumption this implies that

$$x \in \{y : [y]_R \cap X \neq \phi \text{ and } [y]_R \cap Y \neq \phi\} \text{ and}$$
$$x \in \{y : [y]_R \cap X \neq \phi \text{ and } [y]_S \cap Y \neq \phi\}$$

So, $[x]_R \cap (X \cap Y) \neq \phi$ *and* $[x]_S \cap (X \cap Y) \neq \phi$. Hence, $x \in \overline{R*S}(X \cap Y)$. This completes the proof. □

3 Approximate Multigranular Rough Equivalences

In this section we introduce the notions of approximate rough equivalences and study their properties. First, we define the two types of approximate multigranular rough equivalence below.

Definition 3.1 Let R and S be two equivalence relations on U and X, $Y \subseteq U$. Then

(3.1) X and Y are pessimistic bottom multigranular approximate rough equivalent to each other with respect to R and S (X b_R*S_aeqv Y) if *and* only if $\underline{R*S}X$ and $\underline{R*S}Y$ are ϕ or not ϕ together.

(3.2) X and Y are pessimistic top multigranular approximate rough equivalent to each other with respect to R and S (X t_R*S_aeqv Y) if and only if $\overline{R*S}X = \overline{R*S}Y$.

(3.3) X and Y are pessimistic multigranular approximate rough equivalent to each other with respect to R and S (X R*S_aeqv Y) if and only if $\underline{R*S}X$ *and* $\underline{R*S}Y$ are ϕ or not ϕ together and $\overline{R*S}X = \overline{R*S}Y$.

Definition 3.2 Let R and S be two equivalence relations on U and X, $Y \subseteq U$. Then

(3.4) X and Y are optimistic bottom multigranular approximate rough equivalent with respect to R and S (X b_R + S_aeqv Y) if and only if $\underline{R+S}X$ and $\underline{R+S}Y$ are ϕ or not ϕ together .

(3.5) X and Y are optimistic top multigranular approximate rough equivalent with respect to R and S (X t_R + S_aeqv Y) if and only if $\overline{R+S}X = \overline{R+S}Y$

(3.6) X and Y are optimistic multigranular approximate rough equivalent to each other with respect to R and S (X R + S_aeqv Y) if and only if $\underline{R+S}X$ and $\underline{R+S}Y$ are ϕ *or not* ϕ together and $\overline{R+S}X = \overline{R+S}Y$.

It may be noted here that we do not specify the optimistic or pessimistic case specifically as it is clear from the context. We will use Table 1 to prove the properties in sections to follow.

Table 1 Faculty database

S.no.	Name	Division	Grade	Top degree
1.	Sam	Network	Assistant professor	MCA
2.	Ram	Information system	Professor	PhD
3.	Shyam	Software engineering	Assistant professor (junior)	M.Sc.
4.	Peter	Artificial intelligence	Associate professor	PhD
5.	Roger	Embedded system	Professor	PhD
6.	Albert	Artificial intelligence	Assistant professor (Junior)	M.Sc.
7.	Mishra	Embedded system	Assistant professor (junior)	M.Sc.
8.	Hari	Information systems	Senior professor	PhD
9.	John	Software engineering	Assistant professor	MCA
10.	Smith	Network	Associate professor	PhD
11.	Linz	Artificial intelligence	Senior professor	PhD
12.	Keny	Software engineering	Professor	PhD
13.	Williams	Embedded systems	Associate professor	PhD
14.	Martin	Information systems	Assistant professor (junior)	M.Sc.
15.	Jacob	Network	Assistant professor (junior)	M.Sc.
16.	Lakman	Software engineering	Associate professor	PhD
17.	Sita	Artificial intelligence	Assistant professor	PhD
18.	Fatima	Embedded systems	Assistant professor	M.Tech
19.	Biswas	Information systems	Senior professor	M.Tech
20.	Pretha	Software engineering	Senior professor	PhD

3.1 Properties of Optimistic Multigranular Approximate Rough Equalities

In this section, we shall deal with the properties of optimistic multigranular approximate equivalence of rough sets. First, we establish some basic properties in the next subsection. Taking Table 1 into consideration, we have

U = {Sam, Ram, Shyam … Pretha}. Also, the three attributes "Division", "Grade" and "Top Degree" induce three equivalences relations. The three equivalence classes are as given below.
U/Division = {{Sam, Smith, Jacob}, {Shyam, John, Keny, Lakman, Pretha}, {Peter, Albert, Linz, Sita}, {Roger, Mishra, Williams, Fatima}, {Ram, Hari, Martin, Biswas}}.
U/Grade = {{Shyam, Albert, Mishra, Martin, Jacob}, {Sam, John, Sita, Fatima}, {Peter, Smith, Williams, Lakman}, {Ram, Roger, Keny}, {Linz, Biswas, Pretha, Hari}}.
U/Top Degree = {{Shyam, Albert, Mishra, Martin, Jacob}, {Sam John}, {Sita, Fatima}, {Ram, Peter, Roger, Hari, Smith, Keny, Linz, Williams, Lakman, Biswas, Pretha}}.

3.1.1 Basic Properties

3.1.1.1. X b_R + S_aeqv Y if $X \cap Y$ b_R + S_aeqv X and $X \cap Y$ b_R + S_aeqv Y. The converse may not be true.

Proof The first part follows directly from the definition of optimistic multigranular approximate bottom rough equivalence. The converse follows from the following example. We refer to Table 1, Let X = {Jacob, John, Peter, Albert, Linz, Sita} and Y = {Jacob, John, Sam, Sita, Fatima}. Then $X \cap Y$ = {Jacob, John, Sita}. Hence, $\underline{R+S}$(X) = {Peter, Albert, Linz, Sita} $\neq \phi$, $\underline{R+S}$(Y) = {Sam, John, Sita, Fatima} $\neq \phi$ and $\underline{R+S}\,(X \cap Y) = \phi$. So, although X and Y are b_R + S_aeqv, none of them is b_R + S_aeqv to $X \cap Y$. □

3.1.1.2. X t_R + S_aeqv Y if $X \cup Y$ t_R + S_aeqv X and $X \cup Y$ t_R + S_aeqv Y. The converse is also true.

Proof If part follows directly from definition of optimistic multigranular approximate top rough equivalence. Conversely, X t_R + S_aeqv Y then $\overline{R+S}(\mathrm{X}) = \overline{R+S}(\mathrm{Y})$. Then by (2.2.8), $\overline{R+S}(X \cup Y) \supseteq \overline{R+S}(X) = \overline{R+S}(\mathrm{Y})$. The converse follows from Lemma 2.2.3.1. □

3.1.1.3. If X t_R + S_aeqv X' and Y t_R + S_aeqv Y' then we have $X \cup Y$ t_R + S_aeqv $X' \cup Y'$.

Proof The proof follows from Lemma 2.2.3.1. □

3.1.1.4. X b_R + S_aeqv X' and Y b_R + S_aeqv Y' may not imply that $X \cap Y$ b_R + S_aeqv $X' \cap Y'$.

Proof We provide one example to show this. Referring to Table 1, let us take X = {Ram, Hari, Martin, Biswas, Peter}, Y = {Peter, Smith, Williams, Lakman}, X' = {Peter, Albert, Linz, Sita, Biswas, Pretha, Hari} and Y' = {Linz, Biswas, Pretha, Hari}. Then $\underline{R+S}$ (X) = {Ram, Hari, Martin, Biswas} and $\underline{R+S}(X')$ = {Peter, Albert, Linz, Sita, Biswas, Pretha, Hari} are not ϕ together. So, X b_R + S_aeqv X'. Again, $\underline{R+S}$ (Y) = {Peter, Smith, Williams, Lakman} and $\underline{R+S}(Y')$ = {Linz, Biswas, Pretha, Hari} are not ϕ together. So, Y t_R + S_aeqvY', whereas $X \cap Y$ = {Peter}. So, $\underline{R+S}$ $(X \cap Y) = \phi$. Again, $X' \cap Y'$ = {Linz, Biswas, Pretha, Hari}. So, $\underline{R+S}$ $(X' \cap Y')$ = {Linz, Biswas, Pretha, Hari} $\neq \phi$. Hence $X \cap Y$ b_R + S_aeqv $X' \cap Y'$ is not true. □

3.1.1.5. If X t_R + S_aeqv Y then X ∪ ~Y t_R + S_aeqv U.

Proof We have by hypothesis, $\overline{R+S}(X) = \overline{R+S}(Y)$. So, $\overline{R+S}(X \cup \sim Y) \supseteq \overline{R+S}(X) \cup \overline{R+S}(\sim Y) \supseteq \overline{R+S}(Y) \cup \overline{R+S}(\sim Y) = \overline{R+S}(Y) \cup (\sim \underline{R+S}(Y)) \supseteq \overline{R+S}(Y) \cup \sim BN_{R+S}(Y) \cup \sim \overline{R+S}(Y) \supseteq \overline{R+S}(Y) \cup \sim \overline{R+S}(Y) = U$. So, $\overline{R+S}$ $(X \cup \sim Y) = U$. This completes the proof. □

3.1.1.6. if X b_R + S_aeqv Y then we may not have X ∩ ~Y b_R + S_aeqv ϕ.

Proof An example can be provided as in the case of 3.1.1.4.

The proofs of the following two properties are obvious and hence omitted. □

3.1.1.7. If $X \subseteq Y$ and X t_R + S_aeqv U then Y t_R + S_aeqv U.
3.1.1.8. If $X \subseteq Y$ and Y t_R + S_aeqv ϕ then X t_R + S_aeqv ϕ.
3.1.1.9. If X t_R + S_aeqv Y then it is true that ~X b_R + S_aeqv ~Y

Proof By hypothesis, we have $\overline{R+S}(X) = \overline{R+S}(Y)$. So, $\sim \underline{R+S}(\sim X) = \sim \underline{R+S}(\sim Y)$ and hence $\underline{R+S}(\sim X) = \underline{R+S}(\sim Y)$. This implies that $\underline{R+S}(\sim X)$ and $\underline{R+S}(\sim Y)$ are ϕ or not ϕ together. This completes the proof. □

3.1.1.10. If X b_R + S_aeqv ϕ or Y b_R + S_aeqv ϕ then $X \cap Y$ b_R + S_aeqv ϕ.

Proof By hypothesis, we have $\underline{R+S}(X) = \phi$ or $\underline{R+S}(Y) = \phi$. In any case, $\underline{R+S}$ $(X) \cap \underline{R+S}(Y) = \phi$. Thus by (2.2.5) $\underline{R+S}(X \cap Y) = \phi$. This completes the proof. □

3.1.1.11 If X t_R + S_aeqv U or Y t_R + S_aeqv U then $X \cup Y$ t_R + S_aeqv U.

Proof By hypothesis, $\overline{R+S}(X) = U$ or $\overline{R+S}(Y) = U$. In any case, $\overline{R+S}(X) \cup \overline{R+S}(Y) = U$. Now, by (2.2.8), $\overline{R+S}$ (X ∪ Y) ⊇ $\overline{R+S}$ (X) ∪ $\overline{R+S}$ (Y) = U. Hence, the proof follows. □

3.1.2 Replacement Properties

We would like to note that in the properties below, we have avoided providing examples due to scarcity of space. These examples can be constructed as in the earlier cases.

3.1.2.1. X t_R + S_aeqv Y if $X \cap Y$ t_R + S_aeqv X and $X \cap Y$ t_R + S_aeqv Y. The converse may not be true.

Proof The first part follows directly from the definition of optimistic multigranular approximate top rough equivalence. To establish the second part we can provide an example. □

3.1.2.2. X b_R + S_aeqv Y if $X \cup Y$ b_R + S_aeqv X and $X \cup Y$ b_R + S_aeqv Y. The converse may not be true.

Proof The proof of the first part follows from the definition of bottom optimistic almost equivalence. For the second part we can provide an example. □

3.1.2.3. X b_R + S_aeqv X′ and Y b_R + S_aeqv Y′ may not imply that $X \cup Y$ b_R + S_aeqv X′ ∪ Y′.

Proof We can provide an example to establish our claim. □

3.1.2.4. X t_R + S_aeqv X' and Y t_R + S_aeqv Y' may not imply that $X \cap Y$ t_R + S_aeqv $X' \cap Y'$.

Proof An example can be constructed to establish our claim. □

3.1.2.5. If X b_R + S_aeqv Y then X ∪ ∼ Y may not be b_R + S_aeqv U.

Proof An example can be provided to establish our claim. □

3.1.2.6. If X t_R + S_aeqv Y then it may not be true that X ∩ ∼Y t_R + S_aeqv ϕ.

Proof We can provide an example in favour of our claim.

The next two properties follow directly from definition. □

3.1.2.7. If $X \subseteq Y$ and X b_R + S_aeqv U then Y b_R + S_aeqv U

3.1.2.8. If $X \subseteq Y$ and Y b_R + S_aeqv ϕ then X b_R + S_aeqv ϕ.

3.1.2.9. If X b_R + S_aeqv Y then it may not be true that ∼X t_R + S_aeqv ∼ Y.

Proof From hypothesis $\underline{R+S}$ (X) and $\underline{R+S}$ (Y) are ϕ or not ϕ together. So, $\sim\overline{R+S}$ ($\sim$X) and $\sim\overline{R+S}$ ($\sim$Y) are ϕ or not ϕ together. This implies that $\overline{R+S}$ ($\sim$X) and $\overline{R+S}$ ($\sim$Y) are U or not U together. When both are equal to U, there is no problem. They are equal. So the conclusion is true. This can be shown through an example. □

3.1.2.10. If X t_R + S_aeqv ϕ or Y t_R + S_aeqv ϕ then $X \cap Y$ t_R + S_aeqv ϕ.

Proof From the hypothesis, we have by definition, $\overline{R+S}$ (X) = ϕ or $\overline{R+S}$ (Y) = ϕ. Hence, $\overline{R+S}$ (X ∩ Y) ⊆ $\overline{R+S}$ (X) ∩ $\overline{R+S}$ (Y) ⊆ ϕ ∩ $\overline{R+S}$ (Y) or $\overline{R+S}$ (X) ∩ ϕ ⊆ ϕ. So, $\overline{R+S}$ (X ∩ Y) = ϕ. □

3.1.2.11. If X b_R + S_aeqv U or Y b_R + S_aeqv U then $X \cup Y$ may not be b_R + S_aeqv U.

Proof As $\underline{R+S}$ (U) = U, it follows from the hypothesis that $\underline{R+S}$ (X) and $\underline{R+S}$ (U) are not ϕ together or $\underline{R+S}$ (Y) and $\underline{R+S}$ (U) are not ϕ together. Now, $\underline{R+S}(X \cup Y) \supseteq \underline{R+S}X \cup \underline{R+S}Y = \phi$. This completes the proof. □

3.2 Properties of Pessimistic Multigranular Approximate Rough Equivalences

In this section, we shall deal with the properties of pessimistic multigranular approximate equivalence of rough sets. Due to shortage of space, we only state the properties below.

3.2.1 Basic Properties

3.2.1.1. X b_R*S_aeqv Y if $X \cap Y$ b_R*S_aeqv X and $X \cap Y$ b_R*S_aeqv Y. The converse may not be true.

3.2.1.2. X t_R*S_aeqv Y iff $X \cup Y$ t_R*S_aeqv Y and $X \cup Y$ t_R*S_aeqv Y.

3.2.1.3. If X t_R*S_aeqv X' and Y t_R*S_aeqv Y' then $X \cup Y$ t_R*S_aeqv$X' \cup Y'$.

3.2.1.4. X b_R*S_aeqv X' and Y b_R*S_aeqv Y' may not imply that $X \cap Y$ b_R*S_aeqv $X' \cap Y'$.

3.2.1.5. If X t_R*S_aeqv Y then X ∪ $\sim$ Y t_R*S_aeqv U.

3.2.1.6. If X b_R*S_aeqv Y then X ∩ $\sim$Y may not be b_R*S_aeqv ϕ.

3.2.1.7. If $X \subseteq Y$ and X t_R*S_aeqv U then Y t_R*S_aeqv U.

3.2.1.8. If $X \subseteq Y$ and Y t_R*S_aeqv ϕ then X t_R*S_aeqvϕ.

3.2.1.9. If X t_R*S_aeqv Y then it is true that $\sim$ X b_R*S_aeqv $\sim$ Y.

3.2.1.10. If X b_R*S_aeqv ϕ or Y b_R*S_aeqv ϕ then $X \cap Y$ b_R*S_aeqvϕ.

3.2.1.11. If X t_R*S_aeqv U or Y t_R*S_aeqv U then $X \cup Y$ t_R*S_aeqv U.

3.2.2 Replacement Properties

3.2.2.1. X t_R*S_aeqv Y if $X \cap Y$ t_R*S_aeqv X and $X \cap Y$ t_R*S_aeqv Y. The converse may not be true.

3.2.2.2. X b_R*S_aeqv Y if $X \cup Y$ b_R*S_aeqv X and $X \cup Y$ b_R*S_aeqv Y. The converse is also true.

3.2.2.3. X b_R*S_aeqv X' and Y b_R*S_aeqv Y' may not imply that $X \cup Y$ b_R*S_aeqv$X' \cup Y'$.

3.2.2.4. X t_R*S_aeqv X' and Y t_R*S_aeqv Y' may not imply that $X \cap Y$ t_R*S_aeqv $X' \cap Y'$.

3.2.2.5. If X b_R*S_aeqv Y then X ∪ ~Y may not be b_R*S_aeqv U.

3.2.2.6. If X t_R*S_aeqv Y then X ∩ ~Y may not be t_R*S_aeqv ϕ.

3.2.2.7. If $X \subseteq Y$ and X b_R*S_aeqv U then Y b_R*S_aeqv U

3.2.2.8. If $X \subseteq Y$ and Y b_R*S_aeqv ϕ then X b_R*S_aeqv ϕ.

3.2.2.9. If X b_R*S_aeqv Y then it may not be true that ~X t_R*S_aeqv ~ Y.

3.2.2.10. If X t_R*S_aeqv ϕ or Y t_R*S_aeqv ϕ then $X \cap Y$ t_R*S_aeqvϕ.

3.2.2.11. If X b_R*S_aeqv U or Y b_R*S_aeqv U then $X \cup Y$ may not be b_R*S_aeqv U.

4 Rough Equivalence Based Approximate Reasoning

As mentioned by Zadeh, approximate reasoning is viewed as a process of approximate solution of a system of relational assignment equations. We can consider the approximate equalities in this sense providing approximate reasoning. The usual practice is to generalize the modus ponens used in discrete mathematics for generation of rules. But here, we have used it in the first sense when we mention approximate reasoning.

5 Conclusions

In this paper the notion of multigranular rough equivalence of sets for both the optimistic and pessimistic multigranular rough sets are introduced. Several of their direct as well as replacement properties have been established. Two diagrams showing the comparison of the lower and upper approximations for unigranular rough sets and two types of multigranular rough sets have been presented. We have taken a real life database for the description of the concepts of this paper and also provided counter examples using this real life database.

References

1. Novotny, M., Pawlak, Z.: Characterization of rough top equalities and rough bottom equalities. Bull. Polish Acad. Sci. Math. **33**, 91–97 (1985)
2. Novotny, M., Pawlak, Z.: On rough equalities. Bull. Polish Acad. Sci. Math. **33**, 99–103 (1985)
3. Novotny, M., Pawlak, Z.: Black box analysis and rough top equality. Bull. Polish Acad. Sci. Math. **33**, 105–113 (1985)
4. Tripathy, B.K.: On approximation of classifications, rough equalities and rough equivalences. In: Studies in computational intelligence. Rough set theory: a true landmark in data analysis, vol. 174, pp. 85–136. Springer, Berlin (2009)
5. Tripathy, B.K. and Mitra, A.: On the approximate equalities of multigranular rough sets and approximate reasoning. In: Proceedings of 4th IEEE International Conference on Computing, Communication and Networking Technologies (ICCCNT 2013), 4–6 July 2013
6. Tripathy, B.K.: An analysis of approximate equalities based on rough set theory. Int. J. Adv. Sci. Technol. **31**, 23–36 (2011)
7. Pawlak, Z.: Rough sets, Int. J. Inf. Comput. Sci. 341–346 (1982)
8. Pawlak, Z.: Rough Sets, Theoretical Aspects of Reasoning about Data. Kluwer Academic Publishers, Dordrecht (1991)
9. Qian, Y.H and Liang, J.Y.: Rough set method based on multi-granulations. In: Proceedings of the 5th IEEE Conference on Cognitive Informatics, vol. 1, pp. 297–303 (2006)
10. Qian, Y.H., Liang, J.Y., Dang, C.Y.: Pessimistic rough decision. In: Proceedings of RST 2010, pp. 440–449, Zhoushan, China (2010)
11. Tripathy, B.K., Mitra, A.: On approximate equivalences of multigranular rough sets and approximate reasoning. Int. J. Inf. Technol. Comput. Sci. **10**, 103–113 (2013)
12. Tripathy, B.K., Rawat, R., Divya, V Parida, S.Ch.: Approximate reasoning through multigranular approximate rough equalities. Int. J. Intell. Syst. Appl. **6**, 69–76 (2014)
13. Tripathy, B.K., Nagaraju, M.: A comparative analysis of multigranular approaches and on topological properties of incomplete pessimistic multigranular rough fuzzy sets. Int. Intell. Syst. Appl. **11**, 99–109 (2012)
14. Tripathy, B.K., Raghavan, R.: Some algebraic properties of multigranulations and an analysis of multigranular approximations of classifications. Int. J. Inf. Technol. Comput. Sci. **7**, 63–70 (2013)
15. Tripathy, B.K., Raghavan, R.: On some comparison properties of rough sets based on multigranulations and types of multigranular approximations of classifications. Int. J. Intell. Syst. Appl. **06**, 70–77 (2013)
16. Tripathy, B.K., Mitra, A., Ojha, J.: On Rough Equalities and Rough Equivalences of Sets, SCTC 2008-Akron, U.S.A., LNAI, vol. 5306, pp. 92–102. Springer, Berlin (2008)
17. Tripathy, B.K., Mitra, A., Ojha, J.: Rough equivalence and algebraic properties of rough sets. Int. J. Artif. Intell. Soft Comput. **1**(2/3/4), 271–289 (2009)

Spatial and Temporal Analysis of VOC Concentration in Capital City of Odisha in Indian Sub-continent

S. Rath, S.K. Pandey, D.P. Sandha, M. Mohapatra, B. Rath, T. Grahacharya and B.P.S. Sahoo

Abstract Since time immemorial, it has been a constant endeavour of human being to pursue various aspects of life with ease. This has generated requirement of building equipments and in turn instigate industrialization and urbanization, which is moving at very rapid pace so also the population. The over grown population, industrialisation and heavy transportation together have greatly impacted the normal human life with poor air quality, noise and water body pollution. The rate at which urban air pollution has grown across the country is alarming. This paper deliberates the air quality of Bhubaneswar, the capital of the Indian State of Odisha. Different air pollutants are being measured and analysed with respect to its temporal and geo-spatial behaviour. This paper focused on Volatile Organic Compound (VOC) with its temporal and spatial variation within the city area. The result shows a demand of high resolution of sensing data to understand the overall air quality profile.

Keywords Air quality · Spatial analysis · Temporal analysis · VOC

S. Rath (✉) · S.K. Pandey · D.P. Sandha · M. Mohapatra · B. Rath · T. Grahacharya · B.P.S. Sahoo
Computer Networking and e-Management Division, CSIR-Institute of Minerals and Materials Technology, Bhubaneswar, India
e-mail: satya@immt.res.in

S.K. Pandey
e-mail: skpandey@immt.res.in

D.P. Sandha
e-mail: dpsandha@immt.res.in

M. Mohapatra
e-mail: minakshi3333@gmail.com

B. Rath
e-mail: bimalrath2008@gmail.com

T. Grahacharya
e-mail: tarini106624@gmail.com

B.P.S. Sahoo
e-mail: bpratap.cse@gmail.com

L.C. Jain et al. (eds.), *Computational Intelligence in Data Mining - Volume 2*,
Smart Innovation, Systems and Technologies 32, DOI 10.1007/978-81-322-2208-8_56

1 Introduction

Worldwide, outdoor air pollution caused 3.2 million premature deaths and over 74 million years of healthy life cost in 2010 and has the fifth largest killer in India after high blood pressure, indoor air pollution, tobacco smoking, says a new set of findings of the Center for Science and Environment report [1]. Hence, in this context there is good reason to study and analyze the sensational aspects of air pollutants. There may have some general principles which creates this irregularity and results in a tragic human loss. Therefore it is necessary to gather information about the air pollutants and the environmental changes in a local scale as the pollution is entirely depends on its in and around areas with the major sources being fuel wood and biomass burning, fuel adulteration, vehicle emission and traffic congestion.

So far there are many groups worldwide those who are working in their respective city and monitor the air quality data. A 2014 study of World Health Organization [2] has found that 13 of the 20 most-polluted cities in the world are in India. Bhubaneswar, the capital of the Indian state of Odisha is one of the country's fastest developing cities. This necessitates an early step by the local authorities before any alarming situation. A study has been initiated to capture the air quality data in and around the city. The Odisha State Pollution control board currently runs only 3 monitoring stations across Bhubaneswar city which studies the atmospheric particle matter include suspended particulate matter (SPM), respirable suspended particle matter (RSPM) and SO2 and Nox [3]. Looking at the over growing population and so also vehicle emission and traffic congestion, these current system are not sufficient and capable enough to measure the air quality of the city. There are two fundamental limitations to the existing approach: First, the spatial and temporal resolution of the pollution sampling is very poor. Second, current observations of concentrations do not reflect actual exposures experienced by individuals, due to spatial heterogeneity of pollutant concentrations and individuals mobility patterns, such as time spent and activity levels at home, at work, and commuting. Developing a pollution monitoring system to study and analyze the ambient air quality requires its domain expertise, not just to build the instruments setup but to use them properly and interpret their output for rational purpose.

In this paper, we proposed a framework to develop a real-time monitoring, prediction and visualization system analyzing the causing parameters in and around Bhubaneswar city. This paper first articulates the real-time fine-grained air quality data observations to assess the current state of the city. The goal of this system is to acquire a global view of entire city rather than the location specific view to design a fully functional application. A pilot study has been made at the different site to identify the causing parameters and subsequently their benefaction which resulted in health hazards and economic losses.

2 Related Work

Worldwide, in the last couple of years the research in environmental monitoring is motivated towards the necessity of air quality monitoring due to its dangerousness which causes havoc life loss. In this paper we have studied some works which concentrate on the spatial and temporal analysis of various air pollutants.

In [4], the author proposed a model to estimate the emissions and spatial distribution of traffic pollutants and subsequently predict the emissions changes in specific traffic conditions. However the behaviour and chance of the emitted air pollutants are not clearly understood; it also does not consider the wind direction and terrain analysis which affects the spatial modelling for air pollution. In the similar line, the authors in [5], proposed a prediction map to show the concentration of air pollutants. They have also done the spatial interpolation analysis variability of some of the major air pollutants like SO_2, NO_2 and PM_{10} across the city Chengdu, China. One significant failure of this system is that they failed to show the spatial variation of major air pollutants in real-time basis.

In [6], the author compared various spatial methods for the estimation of air quality data. They evaluated the spatial interpolation method to assess the degree to which different methods influence the estimated air pollutant levels using the Kriging method to produce most realistic estimates for ozone (O_3) concentration. It is quite difficult from the study to conclude which method is suitable for which concentration due to insufficiency of records, as they have collected from static monitoring stations.

The authors in [7] used two well-known spatial methods namely Inverse Distance Weighting (IDW) (Deterministic) and Ordinary Kriging (OK) (Geostatistics) in order to generate map of the major source of ambient air concentration such as Nitrogen Oxide (NO_2), Benzene (C_6H_6). The performance of these two methods has been compared with cross-validation method. They simulate the OK method for C_6H_6 concentration but for NO_2, the IDW outperformed the OK method. Therefore for each individual concentration, it need to be analyse using different mechanism every time, which limiting the approach as it do not support the real-time analysis.

In many instances we may wish to model feature as continuous field, however we only have data values for limited number of points. So it became necessary to interpolate/estimate the values for intervening points. So far, all the system for spatial prediction map has been developed in Arc GIS software and also not in fully real time approach. Here we proposed a solution to find out the problem and build up an extended system to show the immediate map of each concentration in the capital city of Odisha in Indian sub-continent. With this objective a pilot study on VOCs variation was carried out to understand the temporal and spatial variation vis-à-vis land use patterns in Bhubaneswar city.

3 Pollution Data Aggregation and Analysis

3.1 Data Aggregation

In any natural calamities or any environmental changes, the limiting factor for acquiring large amounts of real-time data is often difficult and sensitive too. We have installed the monitoring station nearer to the high traffic density and industrial zone so as to get a better assessment and accurate data. The data has collected data from five different locations as shown in Fig. 1 at different time and date.

In initial phase of the work the real time monitoring system was designed keeping in mind to collect and gather a huge amount of fine-grain air pollution data and study the behavioral changes and the impact of different pollutants in the air.

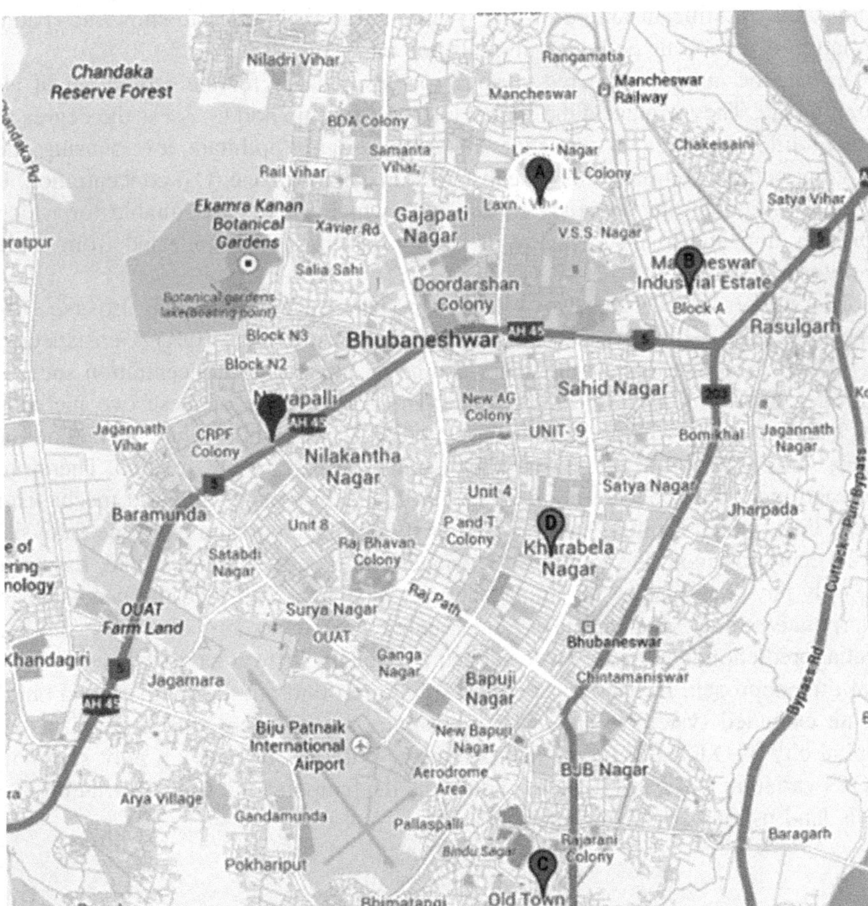

Fig. 1 Monitored stations around Bhubaneswar city

The pollution monitoring instrumentation system includes electrochemical sensors for various gaseous pollutant monitoring, namely Ozone, Volatile Organic Compound (VOC), Carbon dioxide along with other parameters such as temperature and humidity. In this current scope of the paper, the VOC data have been measured using a hand held device (aeroqual S 500). The daily sensor data is being stored in the embedded database in the monitoring instrumentation setup then the data is being transferred to the local monitoring system for a larger extent of study. However a system has been designed to capture the realtime sensor data of different air pollutants and transmit the same to central monitoring system. Due to the lesser number of instrumentation setup we collect the data from various location at different time and date. The detail analysis of data from numerous location, time period is presented in Sect. 4. It has been observed that, there is a substantial disparity in the reading non-identical pollutants in different locations and time duration. The analysis reveals that it is quite sensitive and difficult to major the impact of air pollutants in the city as a whole by collecting data from various static location. Hence it is suggested for a moving data collection system; where it is possible to collect data from vehicular emission in the city as well as industrial zone and residential pockets with geospatial and temporal variant.

It is proposed to developed vehicular sensor network [8] based air pollution data collection system for better assessment and accurate prediction of pollution. Our proposed overall system architecture is influenced by [9–11]. In the similar context an analysis on black carbon concentration in and around city area has also been carried out at CSIR-IMMT [10]. Due to page constraint we have not elaborate the proposed system in detail.

3.2 Data Analysis

In this section we categorize the collected sensor data according to the site specific environment. The data sample has been recorded in and around Bhubaneswar city from five different sites as shown in Fig. 1. The sites were carefully chosen based on the following criteria: accessibility to the locations, freedom from any obstacle to free flow of air in the vicinity and security of the sampler, road intersection in the city, industrial zone and residential pockets. The locations were chosen to reflect kind of activities in the area and its impact. The study is mostly concentrated on the measurement of the VOC level in air as it is the major factor deciding the impact of pollution rate [12].

The primary idea is to distinguish and analyze the variations in readings, and come up with a rational difference. The main contribution of this paper is to show the differences in readings in different scale with respect to area such as: city, residential and industrial zone. In [12], the readings of VOCs in industrial has been extensively studied and presented the detailed nature of air quality situation in the area. Their study reveals that there is an increase in the VOCs concentration from November to December due the dry season. With one step ahead, in this paper we have studied industrial site which is adjacent to a road intersection, as the vehicular traffic emission rate is high in these locations.

The air quality data collected from different sites records the measurement at one location at a time so it doesn't depict the overall scenario of the study area. Also it is not potential to setup a data collection instrument in every location of the city for air pollutants information. It is a tradeoff between the locations and running vehicular emission. The both system scenario has its own limitations but if we could collect the city air quality real-time data along with industrial pollution data, the overall scenario may be presented as a whole. As described in [12], the VOC concentration has huge discrepancy in time and date wise. So we propose a spatial distributed mechanism which will cover up the entire region and gives us a better resolution of data. The spatial analysis of the measured data has been elaborated in the next section.

The study suggests a spatial analysis mechanism for absolute prediction of fine-grain air quality data. As we studied in literature in Sect. 2, use of various types of spatial interpolation methods could be solution for better assessment. The IDW [4, 5] and OK [7] are among the well known methods for this purpose. IDW and OK methods are required huge amount of data to analyze and depict the city scenario as a whole.

4 Result and Analysis

In this section, we briefly outline our filed experiences and presented the VOCs study results with site wise analysis.

The temporal variation of VOC sensor data at two residential sites has been shown in Fig. 2. The Fig. 2a indicates the temporal variation is not much however the VOC reading in evening time is beat higher in comparison to day time. This site is nearer to the famous Lingaraj temple in Bhubaneswar, so this could be because of lighting of lamps in the temple and increase in traffic volume to the temple in the evening. In Fig. 2b, the VOC reading shows a normal distribution. The increase in VOC readings during the morning time may be due to the cooking at residential area.

In the similar context the VOC reading in Fig. 2c which is adjacent to National Highway (NH) and the site in Fig. 2d, which is in an industrial area, have been shown. Both the location shows a very high concentration of VOC, which indicates the influence of heavy vehicular traffic and industrialization on urban population.

The VOC readings at all the five sites are superimposed in Fig. 3, shows the geo-spatial variation in the readings. This indicates, the local influencing parameters like traffic volume, industry and other anthropogenic effects demands a high resolution sensing data to understand the regional VOC readings vis-e-vis the air quality. To understand the spatial variation of all the sites, the data has been mapped in Quantum GIS, open source geographic information system (GIS) software. Spatial interpolation has been carried out using IDW method in GIS environment.

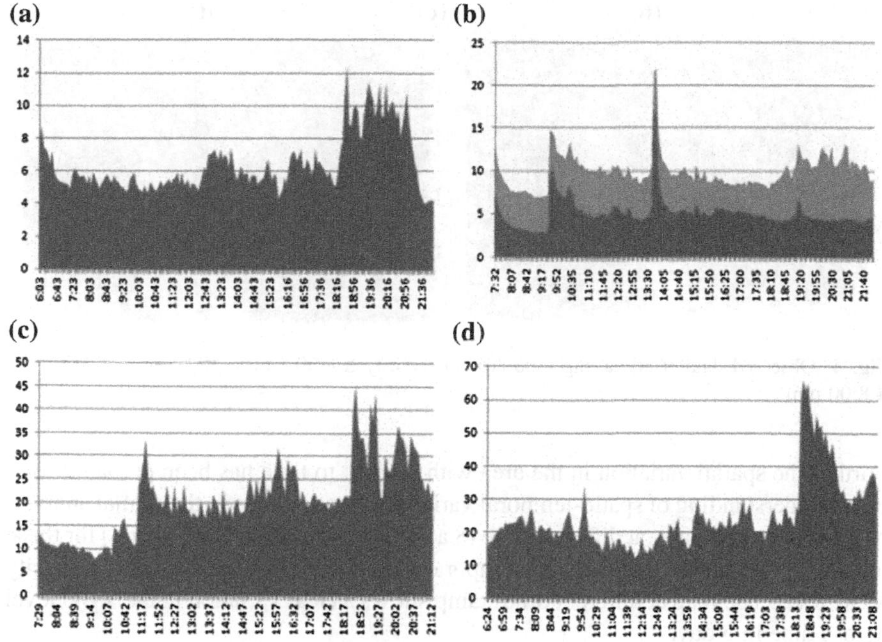

Fig. 2 VOCs reading in and around Bhubaneswar city **a** residential site near to NH **b** residential site inside city **c** NH with road intersection **d** industrial zone near to NH

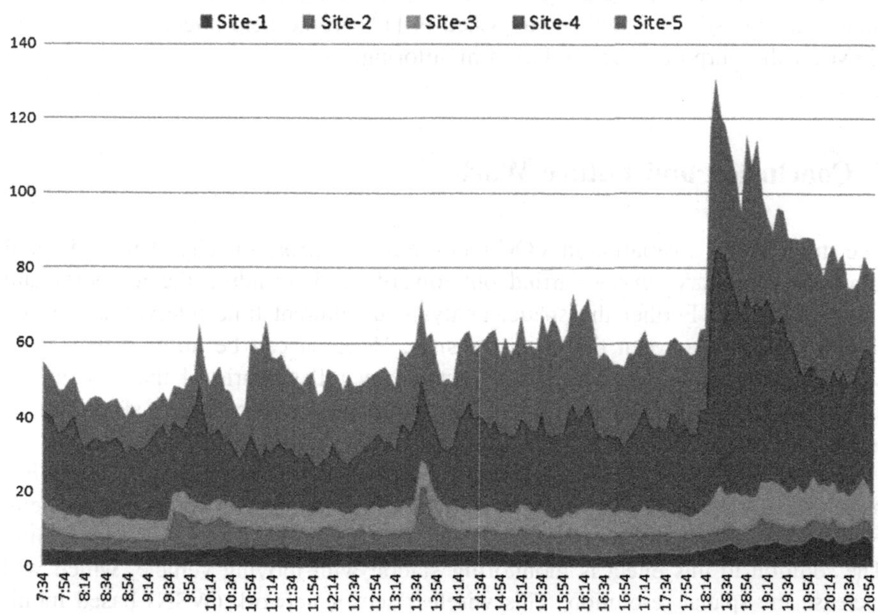

Fig. 3 Geo-spatial variation of VOCs around Bhubaneswar city

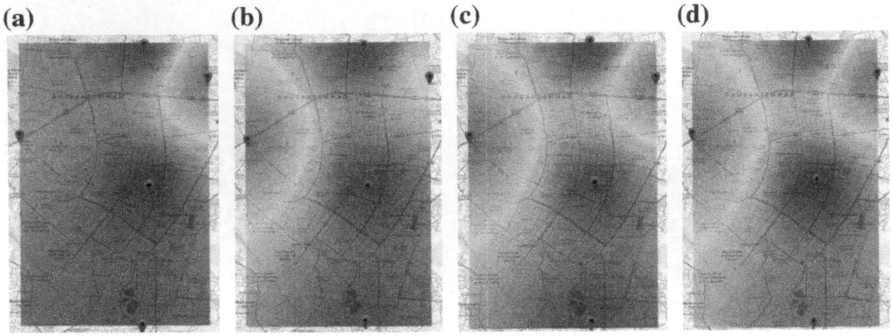

Fig. 4 Observed dataset at varying time around the city **a** 8:00 a.m. **b** 12:00 p.m. **c** 4:00 p.m. **d** 8:00 p.m

Further the spatial variation in the area with respect to time has been prepared for a better understanding of spatio-temporal variation. Figure 4 shows the spatial analysis of the observed dataset at different time (8 a.m., 12 noon, 4 p.m. and 8 p.m.) for these locations. This clearly indicates the temporal variation in the overall profiling the city VOC level. The map at different time stamp shows a clear variation in the VOC level across time in a day.

Establishing several monitoring stations across the city would not be feasible. This demands a location based vehicular sensor network for sensing pollution parameters for profiling the air quality in city area. A pollution monitoring system with moving node is in the design phase for high resolution sensing data with real-time data analysis. The method suggested in [11] would help integrating GPS and GSM for the purpose of air pollution monitoring.

5 Conclusion and Future Work

The study shows a variation in VOC measurements across the city. This is a result analysis of primary survey carried out towards understanding the temporal and spatial variation. Further the spatial analysis at different time interval shows the spatial variation between the observed sites. Hence it can be inferred that Measurement at one location does not depict the overall scenario of the city. In the similar context, it also can be inferred that a continuous observation of air pollutants is the necessity of the time.

Accuracy of spatial profiling depends on the density of measurement station. For more granular profiling more number of measuring stations will be the requirement. However the number of monitoring station would increase the cost of monitoring. The solution to this effect is monitoring station with moving vehicle, which will increase the spatial resolution. This wireless sensor network (WSN) based monitoring will help to understand the air pollution in a better way.

Acknowledgments This research is supported and funded by the Department of Electronics and Information Technology (DeitY), Govt. of India. The authors are thankful to Shri B.M. Baveja, Group Coordinator, CC and BT, DeitY, MCIT, New Delhi. Authors are also thankful to the Director of CSIR-IMMT and other members of this project, for their kind support towards carrying out this research pursuit.

References

1. A Briefing Note: Burden of Disease—Outdoor Air Pollution among Top Killers. Center for Science and Environment, Feb 2013
2. World Health Organization: Ambient (outdoor) air pollution in cities database 2014, Online available: http://www.who.int/countries/ind/en/
3. Orissa State Pollution Control Board: Available online http://www.orissapcb.nic.in/airquality.asp
4. Lin, M.D., Lin, Y.-C.: The application of GIS to air quality analysis in Taichung City, Taiwan, ROC. Environ. Model. Softw. **17**(1), 11–19, ISSN 1364-8152, http://dx.doi.org/10.1016/S1364-8152(01)00048-2
5. Li, S., Song, S., Fei, X.: Spatial characteristics of air pollution in the main city area of Chengdu, China. Geoinformatics, 2011 19th International Conference on, pp. 1–4 (24–26 June 2011). doi:10.1109/GeoInformatics.2011.5981082
6. Wong, D.W., Susan, L.Y., Perlin, A.: Comparison of spatial interpolation methods for the estimation of air quality data. J. Exp. Anal. Environ. Epidemiol. **14**, 404–415 (2004)
7. Fontes, T., Barros, N.: Interpolation of air quality monitoring data in an urban area sensitive the Oporto/Asprela case. J. Fac. Sci. Technol. Port. **7**, 6–18 (2010) ISSN 1646-0499
8. Sahoo, B.P.S., Rath, S., Puthal, D.: Energy efficient protocols for wireless sensor networks: a survey and approach. Int. J. Comput. Appl. **44**(18), 43–48 (2012)
9. Sivaraman, V., Carrapetta, J., Hu, K., Luxan, B.G.: HazeWatch: a participatory sensor system for monitoring air pollution in Sydney. Local computer networks workshops (LCN Workshops), 2013 IEEE 38th Conference on, pp. 56–64 (21–24 Oct 2013)
10. Mahapatra, P.S., Panda, S., Das, N., Rath, S., Das, T.: Variation in black carbon mass concentration over an urban site in the eastern coastal plains of the Indian sub-continent. Theor. Appl. Climatol. pp. 1–15 (2013). doi:10.1007/s00704-013-0984-z
11. Sahoo, B. P. S., Satyajit, R.: Integrating GPS, GSM and cellular phone for location tracking and monitoring. The 3rd International Conference on Geospatial Technology and Applications (Geomatrix'12). (2012) IIT Bombay, India
12. Ojiodu, C.C.: Spatial and temporal variation of volatile organic compounds (VOCS) pollution in Isolo industrial areas of Lagos State, Southwestern—Nigeria. Ethiop. J. Environ. Stud. Manage. **7**(1), 65–72 (2014)

Generating Prioritized Test Sequences Using Firefly Optimization Technique

Vikas Panthi and D.P. Mohapatra

Abstract The aim of this study is to propose an algorithm for generating minimal test sequences by applying Firefly optimization technique. In this study, we use state machine diagram for the behavioral specification of software. This paper generates the important test sequences for composite states in the state machine diagram under consideration. The generated test sequences are then prioritized based on a software coverage criterion. The use of firefly technique results in efficient prioritization of the generated test sequences.

Keywords Test sequence · Firefly optimization algorithm · State machine · Composite model

1 Introduction

Software testing is one of the most important parts of quality assurance in software development life cycle. Increase in complexity and size of software requires more manpower and time for testing the software systems. Manual testing is very much labor-intensive and error-prone. Therefore software testers need novel approaches, which are reliable and automatic for test case generation [2].

Test sequence generation contributes a large part to testing cost. It is generally agreed that manual testing is a construction and is a frequent cause of project delays, especially for large projects. Therefore, automatic test sequence generation has become important to ensure the quality of present day large software products and to contain the apace growing testing costs. As it is very difficult to generate test

V. Panthi (✉) · D.P. Mohapatra
Department of Computer Science and Engineering, National Institute of Technology, Rourkela 769008, Odisha, India
e-mail: 512CS103@nitrkl.ac.in

D.P. Mohapatra
e-mail: durga@nitrkl.ac.in

L.C. Jain et al. (eds.), *Computational Intelligence in Data Mining - Volume 2*, Smart Innovation, Systems and Technologies 32, DOI 10.1007/978-81-322-2208-8_57

sequences from source codes, a substitute approach is to generate test sequences from UML models constructed during the design process. UML is a widely accepted modeling language for object-oriented systems. It has various diagrams for depicting the behavior of objects in a system.

Test sequence generation from design models has several rewards. Model based test sequence generation technique can detect sealed categories of faults that are not easily detected by source code based testing. It also allows test cases to be available much earlier in the software development process, thereby making test planning more efficient, and reduces the rescue time and usage of resources [1, 4].

To generate optimal test sequences and to prioritize the test sequences is still a challenging task. So we use Firefly optimization algorithm for prioritization of test sequences. Firefly algorithm is becoming powerful in solving the modern numerical optimization problems, especially for the NP-hard problem (traveling salesman). So we can also apply this approach on CFG for test sequences generation which is also an optimization problem. As compared to Genetic Algorithm (GA) and Particle Swarm Optimization (PSO) techniques, Firefly Algorithm (FA) reduces the overall computational effort by 86 and 74 %, respectively [6].

We consider composite states in the state machine diagram to extract test sequences and to verify the behavior of state machines. This paper generates all possible valid test sequences for the software to be tested and employs FA to prioritize the generated test sequences.

The rest of the paper is structured as follows: Sect. 2 provides an overview of the Firefly optimization algorithm and vector space basis. Section 3 presents the proposed approach for prioritizing the test sequences generated from the state machine model using FA. Section 4 describes the state machine model of a Bank ATM system taken as a case study. This section also discusses the experimental studies and presents the corresponding result analysis. Section 5 concludes the paper and provides some insight into our future work.

2 Preliminaries

In this section we discuss some of the basic concepts required for understanding of our proposed approach. In Sect. 2.1, we discuss the basics of Firefly algorithm and Sect. 2.2 discusses the concept of Vector space basis.

2.1 Firefly Algorithm (FA)

Firefly is a bio-inspired meta-heuristic algorithm introduced by Yang [5] and is used to solve optimization problems. According to Yang [5] there are three important assumptions in firefly algorithm [3, 5]:

1. All fireflies are unisex, which means any firefly can be attracted to any other brighter one.
2. Attractiveness is proportional to their brightness, thus for any two flashing fireflies, the less bright one will move towards the brighter one. The attractiveness is proportional to the brightness and they both decrease as their distance increases.
3. The firefly will move towards the brighter one, and if there is no brighter one, it will move randomly.

Based on these three rules, the basic steps of a firefly algorithm (FA) can be summarized into two basic concepts: variation of light intensity and formulation of the attractiveness. For simplicity, this paper assumes that the attractiveness of a firefly is determined by its brightness. The intensity of light is inversely proportional to the square of the distance, say d, from the source. Thus, the intensity at I (d) varies according to the inverse square law, i.e. $I(d) = I_s/d^2$ where I_s is the intensity at source point. In the simplest format, the brightness on intensity I of a firefly are at a particular location x can be chosen as $I(x) \propto f(x)$.

When light passes through a medium with light absorption coefficient of λ, the light intensity I, varies with distance say d as given below [5]:

$$I(d) = I_0\, e^{-\lambda\, d^2} \tag{1}$$

where I_0 = Intensity at the origin point.

The approximation of the Gaussian form in Eq. 1 obtained by using combined effect of inverse square law and absorption is given below:

$$I(d) = \frac{I_0}{1 + \lambda d^2} \tag{2}$$

For a shorter distance Eqs. 1 and 2 are essentially the same. This is because the series expansions at d = 0 is given in Eq. 3.

$$e^{-\lambda d^2} \approx 1 - \lambda d^2 + \frac{1}{2}\,\lambda^2 d^4 + \cdots, \frac{1}{1 + \lambda d^2} \approx 1 - \lambda d^2 + \frac{1}{2}\,\lambda^2 d^4 + \cdots, \tag{3}$$

Similarly, the attractiveness of a firefly can be defined as follows:

$$A(d) = \frac{A_0}{1 + \lambda d^2} \tag{4}$$

where A_0 is the attractiveness at d = 0, and $A(d)$ is the attractiveness of the node at a distance d. The movement of a firefly i, that is attracted to another more attractive (brighter) firefly is determined by using Eq. 5 as follows:

$$x_i = x_i + A_0 e^{-\lambda d_{ij}^2} (x_i - x_j) + \alpha (rand + \frac{1}{2}) \quad (5)$$

where

$$d_{ij} = \|x_i - x_j\| = \sqrt{\sum_{k=1}^{n} (x_{i,k} - x_{j,k})^2} \quad (6)$$

Here, $x_{i,k}$ is the *k*th component of the spatial coordinate x_i of the *i*th firefly. In 2D form, we have

$$d_{ij} = \sqrt{(x_i - x_j)^2 + (y_i - y_j)^2} \quad (7)$$

In Eq. 5 the second term corresponds to the attraction and the third term is for randomization with α being the randomization parameter and $0 \leq \alpha \leq 1$. *rand* is a random number generated between [0, 1].

2.2 Vector Space Basis

Vector space is also called linear space. Let V be the set of n-dimensional vectors. If the set V is nonempty, and closed under finite vector addition and scalar multiplication, where the scalars are the members of a field F, we say V is a vector space over F. Furthermore, a vector space should satisfy some axioms. We shall not list all those assumptions. The details may be found in any linear algebra book. If a vector set $\{v_1, v_2, \ldots, v_k\}$ in a vector space V is linear independent and this set spans V, we say that elements $v_1, v_2, \ldots, v_k$ form a basis of V, which means that all v_n in V can be expressed uniquely as a linear combination.

3 Proposed Algorithm

In this section, we discuss our proposed approach for automatically generating and prioritization test sequences using FA. The schematic representation of our approach is given in Fig. 1. Our approach for generation and prioritization of the test sequences is explained below.

Step 1: Construct the state machine diagram for the given project or the program.
Step 2: Convert the State Machine Diagram (SMD) to CCFG (Composite Control Flow Graph).
Step 3: Convert CCFG to Feasible Control Flow Graph (defined in Sect. 4).

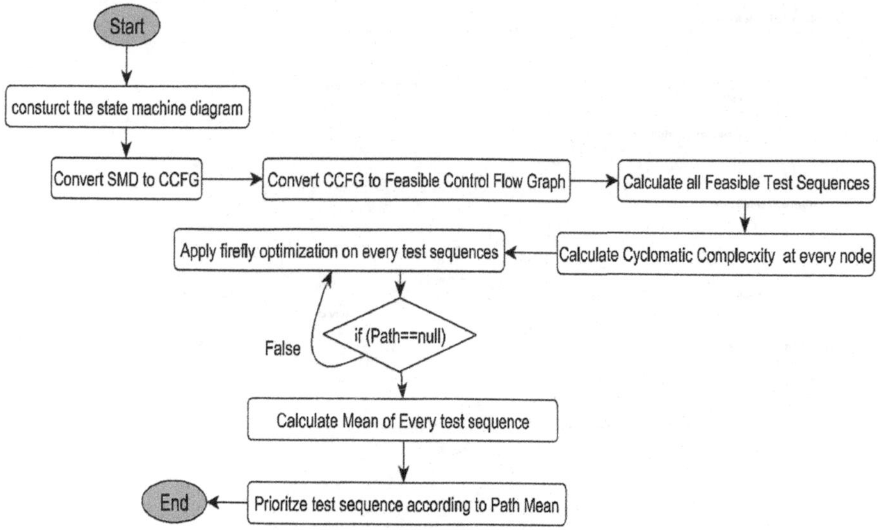

Fig. 1 Schematic procedure of the proposed approach

Step 4: Analyze and collect all the possible Feasible Test sequences using Feasible Control Flow Graph.
Step 5: Choose number of fireflies at each node.
Step 6: Generate the fireflies at each node by using $\mathbf{I}(\mathbf{d}) = \frac{I_0}{1+\lambda d^2}$
where $I_0 = 10$, λ = Cyclomatic complexity at node of graph, d = random distance for brightness.
Step 7: Calculate mean of brightness at every path.
Step 8: Calculate priority of the paths using the following formula.
Highest priority ∝ *Highest mean brightness*
Step 9: Exit.

4 Case Study

We consider a Bank ATM system for our case study. We construct the state machine diagram for the Bank ATM system as shown in Fig. 2 using IBM Rational Software Architecture (RSA) tool. The initial state of the ATM is shown to be *turned off*. After the power is *turned on*, the ATM performs *startup* action and enters *selfTest* state. If the test fails, then the ATM goes into *outofService* state, otherwise, there is a triggerless transition to the *Idle* state. In this state, ATM waits for customer interaction. The ATM state changes from *Idle* to *ServingCustomer* state as soon as the customer inserts banking or credit card details through the ATM card reader. On entering the *ServingCustomer* state, the entry action *readCard* is

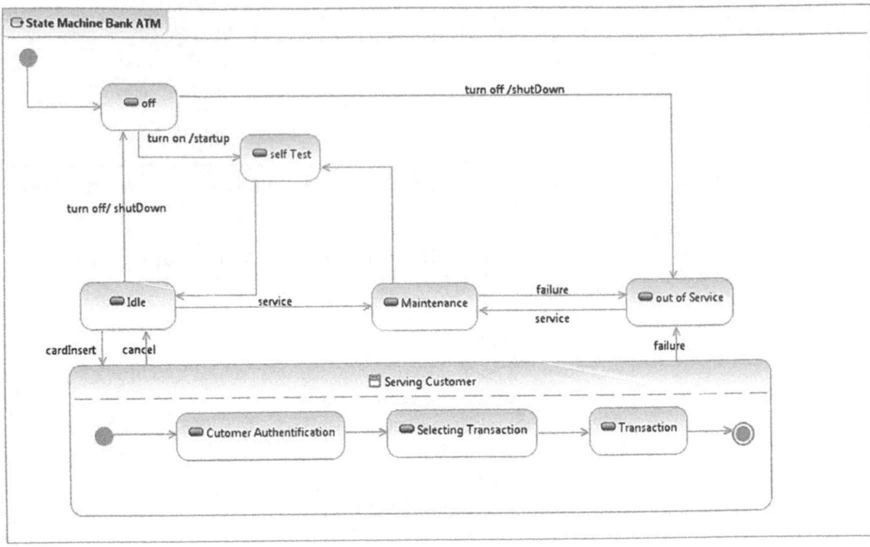

Fig. 2 State machine diagram of bank ATM system

performed. Note that the transition from *ServingCustomer* state back to the *Idle* state could be triggered by cancel event, as the customer can cancel the transaction at any time. The servingcustomer state is a composite state with the sequential substates such as *CustomerAuthentication*, *SelectingTransaction* and *Transaction*. *CustomerAuthentication* and *Transaction* are composite states by themselves, which are shown with hidden decomposition indicator icon.

ServingCustomer state has a triggerless transition back to the *Idle* state after the transaction is finished. The state also has an exit action, *ejectCard,* which ejects the customer card on exiting the state, no matter what caused the transition *outofService* state.

We generate a XML document from State Machine Diagram (SMD). Then, we parse the XML file and calculate the number of states, transitions and cyclomatic complexity for every node and construct the corresponding composite control flow graph (CCFG). The nodes of the CCFG correspond to the different states in SMD and the edges correspond to the transition events. The CCFG of the SMD in Fig. 2 is shown in Fig. 3a. After calculating all the required elements, we convert the CCFG into Feasible Control Flow Graph (FCFG) for generating the feasible test sequences of the state machine diagram. This conversion is carried out to remove the presence of loops and redundant states in CCFG due to the composite nature of the states (Table 1).

In FCFG, every node is associated with number of fireflies present in the test sequences. Through these test sequences, fireflies can transmigrate to the next state of the state sequence according to the attraction of fireflies by brightness. After completing the process of attraction, fireflies are able to reach the end node. Table 2 shows the cyclomatic complexity computed for every node by using Eq. 7.

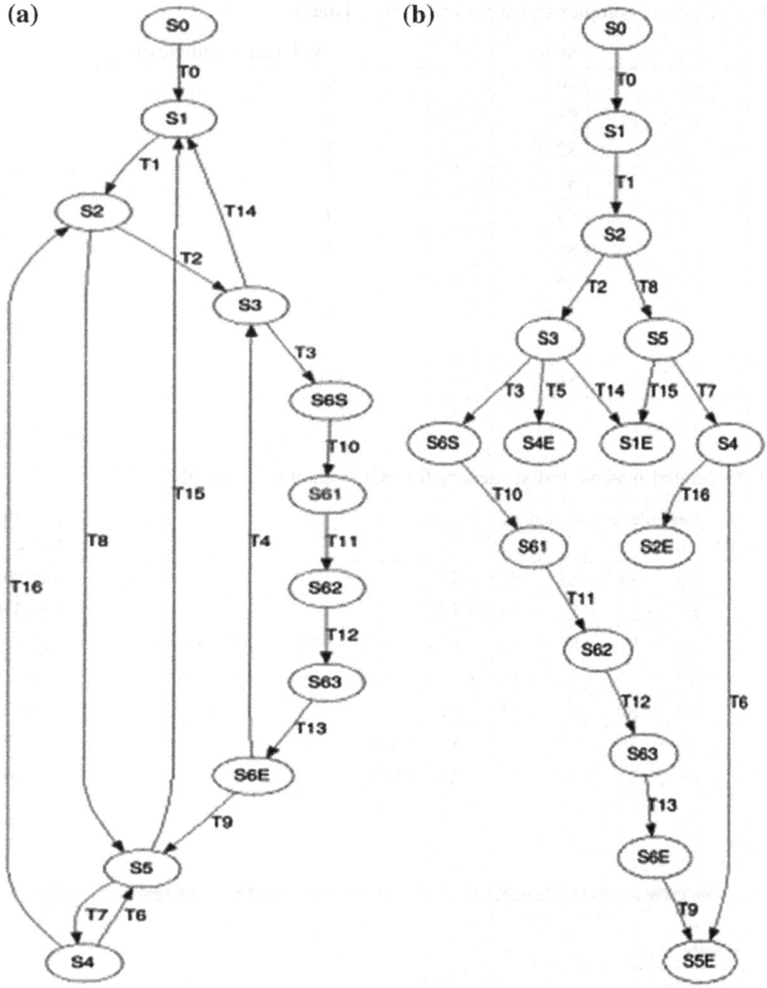

Fig. 3 Composite control flow graph (**a**) and feasible graph structure of composite control flow graph (**b**) of bank ATM system in Fig. 2

$$CC = E - N + 2 \quad (7)$$

Here, CC = cyclomatic complexity of CCFG, E = No. of edges in CCFG and N = No. of nodes in graph.

Test sequence prioritization includes scheduling of the test cases in sequential manner to improve the performance of the regression testing. In the proposed approach, hundred fireflies are at each node and they traverse each feasible test sequence by using the firefly optimization algorithm. The screenshot of the obtained prioritized test sequences is shown in Fig. 4. For all the test sequences, hundred fireflies are transmitted at every node as inputs in accordance to Eq. 4.

Table 1 Cyclomatic complexity list for every node (state)

S. no	State	Cyclomatic complexity
1	S0	8
2	S1	8
3	S2	7
4	S3	5
5	S4	4
6	S5	3
7	S6	3
8	S61	3
9	S62	3
10	S63	10

Table 2 Generated feasible test sequences for FCFG shown in Fig. 3b

Test Sequence ID	Feasible test sequences	Brightness value
TS1	S0 → S1 → S2 → S3 → S1E	40.8087
TS2	S0 → S1 → S2 → S3 → S4E	55.5104
TS3	S0 → S1 → S2 → S3 → S6S → S61 → S6 → S63 → S6E	227.928
TS4	S0 → S1 → S2 → S3 → S6S → S61 → S6 → S63 → S6E → S5E	224.315
TS5	S0 → S1 → S2 → S5 → S1E	62.5211
TS6	S0 → S1 → S2 → S5 → S4 → S2E	77.3287
TS7	S0 → S1 → S2 → S5 → S4 → S5E	100.8452

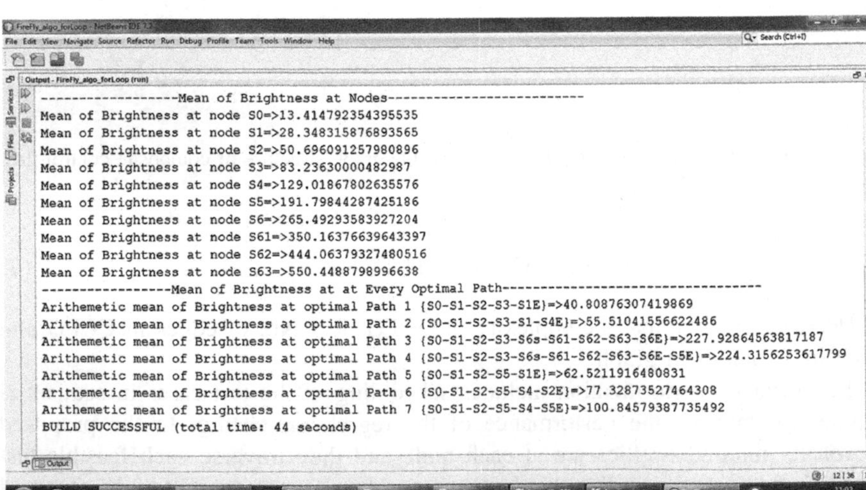

Fig. 4 Prioritized test sequences for stat machine diagram of ATM system

We assume,

$$A_0 = 10,\ \mathrm{d} = [0.9, \ldots, 1.0]\ \text{to}\ [0, \ldots, 0.1],$$

λ = Cyclomatic complexity of the node under consideration.

The prioritized optimal test sequences of state machine diagram for Bank ATM system is based on the brightness values. Prioritized test sequences:

TS3 → TS4 → TS7 → TS6 → TS5 → TS2 → TS1

5 Conclusion and Future Work

This paper presented a Firefly optimization based approach for test sequence generation and prioritization using composite state in state machine diagram. Using the proposed algorithm, a group of fireflies can effectively explore the UML state machine diagram and automatically generate test sequences to achieve the test adequacy requirement. To our knowledge, there exists no such technique that generates test sequences by considering the composite states. In such a scenario, we believe that the generated test sequences are both optimal and minimal. Redundant exploration of the state diagrams and the iteration over the state loops are avoided by the construction of the Feasible Control Flow Graph. The use of Firefly Algorithm, resulted in efficient prioritization of the generated test sequences. In future, we focus to prioritize the test sequences by adopting some heuristic algorithms and compare their respective results.

References

1. Kalaji, R., Hierons, M., Swift, S.: A search-based approach for automatic test generation from extended finite state machine (EFSM). In Testing: Academic and Industrial Conference-Practice and Research Techniques (IEEE) (2009) pp. 131–132
2. Shirole, M., Kumar, R.: UML behavioral model based test case generation: a survey. SIGSOFT Softw. Eng. Notes **38**, 1–13 (2013)
3. Surafel, L.T., Hong, C.: Modified firefly algorithm. J. Appl. Math. **39**, 01–12 (2012)
4. Swain, R., Panthi, V., Behera, P.K., Mohapatra, D.P.: Automatic test case generation from UML state chart diagram. Int. J. Comput. Appl. **42**, 26–40 (2012)
5. Yang, X.S.: Firefly algorithms for multi-modal optimization. Stoch. Algorithms Found. Appl. (SAGA) **2264**, 169–178 (2009)
6. Yang, X.S., Xin, S.H.: Firefly algorithm: recent advances and applications. Int. J. Swarm Intell. **1**, 36–50 (2013)

Directional Multi-scaled Fusion Based Median Filter for Removal of RVIN

Aparna Sarkar, Suvamoy Changder and J.K. Mandal

Abstract In this paper Directional Multi-scaled Fusion Based Median Filter for Removal of RVIN (DMFBMF) has been proposed to restore images corrupted with random valued impulse noises. This filter employs a technique of impulsive noise detection-suppression strategy from images, which is simple but efficient and uses a fusion technique to suppress the impulses from the corrupted images. At first, using two methodologies two different restored images are obtained from the same corrupted source image. Then using a simple fusion rule these two restored images are fused. Thus, a well impulse suppressed image has been obtained. Extensive simulations showed that this scheme provides good performances of suppressing impulse with low noise level as well as for highly corrupted benchmarked images.

Keywords DMFBMF · PSNR · SNR · Median · Fusion and gray-scale image

1 Introduction

The principal sources of noise in digital images arise during image acquisition (digitization) and/or transmission [1, 2]. The performance of imaging sensors is affected by variety of sensors, such as environmental conditions during image

A. Sarkar (✉)
Department of Computer Science, Ramananda College, Bishnupur, Bankura
West Bengal, India
e-mail: sarkaraparna09@gmail.com

S. Changder
Department of Computer Application, NIT Durgapur, Durgapur, West Bengal, India
e-mail: suvamoy.nitdgp@gmail.com

J.K. Mandal
Department of Computer Science and Engineering, University of Kalyani, Kalyani, Nadia, West Bengal, India
e-mail: jkm.cse@gmail.com

L.C. Jain et al. (eds.), *Computational Intelligence in Data Mining - Volume 2*,
Smart Innovation, Systems and Technologies 32, DOI 10.1007/978-81-322-2208-8_58

acquisition, and by the quality of the sensing elements themselves. Image denoising/noise reduction is the process of removing noise from an image [3]. Based on the noise values, the noise can be classified as the easier-to-restore salt-and-pepper noise and the more difficult random valued impulse noise [4]. Among all the methods for removal of impulse noise, the median filter [5, 6] is used widely because of its effective noise suppression capability and high computational efficiency [7]. However, it uniformly replaces the gray-level value of every pixel by the median of its neighbors. Consequently, some desirable details are also removed, especially when the window size is large. Multiscale filters [8] work efficiently preserving edges while denoising. Image Fusion is the process of combining two or more images into a single image while retaining the important features of each image [9]. Image fusion is used to integrate the complementary information from multi-sensor images to generate one composite image containing a more accurate description of the true scene than any one of the individual sensor images. The goal of image fusion is to decrease the uncertainty about the true scene. Standard deviation describes how tightly all the values are clustered around the mean in the set of pixels. In an image, distinct gray levels are to be like their neighbors [10]. So if a pixel value got corrupted, then considering its neighboring gray values we can restore the actual value [11, 12]. The organization of this paper is as follows. The working procedure of the proposed scheme is described in Sect. 2. Section 3 fabricates a number of experimental results and simulations to demonstrate the performance of the proposed DMFBMF. Conclusions are drawn in Sect. 4.

2 Working Procedures

The scheme proposed here is a technique of impulsive noise detection-suppression strategy from images, corrupted with random impulse noises. Using two methodologies two different restored images (image1, image2) are obtained from the same corrupted source image. Through a fusion rule both converted images are fused. Thus, a well impulse suppressed image has been obtained. Both methodologies work in two phases. The first phase detects contaminated pixels and the second phase filters only those pixels keeping others intact. The working procedure of the proposed scheme is depicted in Fig. 1.

2.1 Impulse Detector

Noise-free image consists of locally smoothly varying areas separated by edges [13–15]. Here, we only focused on the edges aligned with four main directions as shown in Fig. 2.

Usually the pixels located in the neighborhood of a test pixel are correlated to each other and they possess almost similar characteristics. The differences between

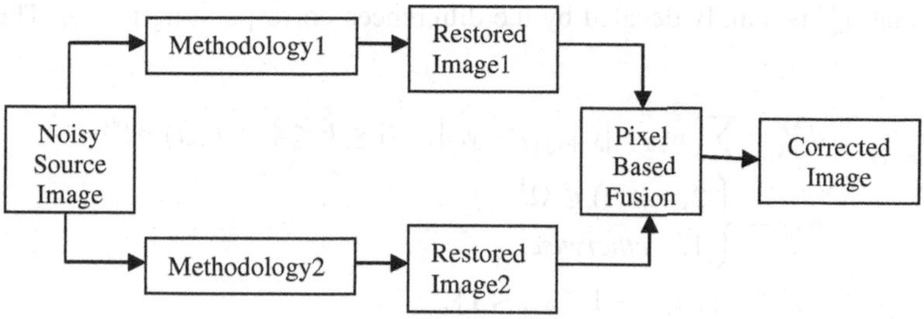

Fig. 1 Flow diagram of the proposed scheme

Fig. 2 Alignment of edges in four directions

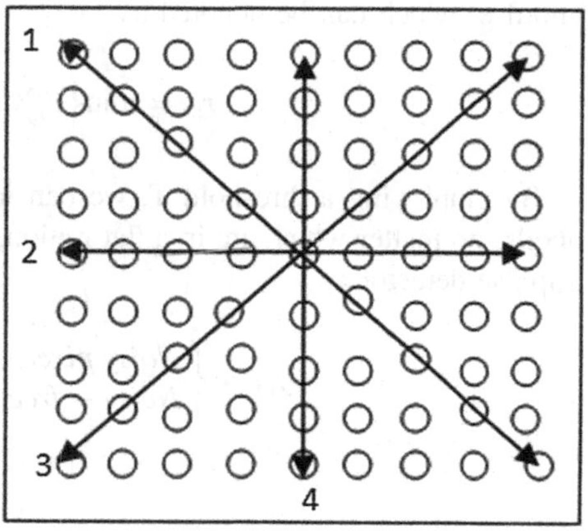

the test pixel and its neighbors aligned with four main directions are considered for impulse detection. A direction index is assigned to each edge aligned with a given direction. The minimum of these four direction indexes is used to detect the impulse in each mask window of size (9 × 9). Let $S_k^{(9)}$ (k = 1–4) denote a set of coordinates aligned with kth direction centered at (0,0), for a (9 × 9) window, i.e.,

$$\begin{aligned} S1 &= \{(-4,-4),(-3,-3),(-2,-2),(-1,-1),(0,0),(1,1),(2,2),(3,3),(4,4)\} \\ S2 &= \{(0,-4),(0,-3),(0,-2),(0,-1),(0,0),(0,1),(0,2),(0,3),(0,4)\} \\ S3 &= \{(4,-4),(3,-3),(2,-2),(1,-1),(0,0),(-1,1),(-2,2),(-3,3),(-4,4)\} \\ S4 &= \{(-4,0),(-3,0),(-2,0),(-1,0),(0,0),(1,0),(2,0),(3,0),(4,0)\}. \end{aligned} \tag{1}$$

For 9 × 9 window centered at (i, j), in each direction, define $d_{i,j}^{(k)}$ as the sum of all absolute differences of gray-level values between $y_{i+s,\ j+t}$ and $y_{i,j}$ with (s,t) ϵ S_k^0, where $S_k^0 = S_k \backslash (0,0)$ for all k from 1 to 4. Considering that two pixels whose spatial distance is small, their grey-level values should be close, we will weight the absolute differences between the two closest pixel with a larger value $w_{s,t}$, it will cause that $d_{i,j}^{(k)}$ is mainly decided by the differences corresponding to $w_{s,t}$. Thus we get,

$$\begin{aligned} d_{i,j}^{(k)} &= \sum w_{s,t} * \left|y_{i+s,j+t} - y_{i,j}\right|, \quad 1 \le k \le 4, \quad (s,t) \in S_k^0 \\ w_{s,t} &= \begin{cases} 2, & (s,t) \in \Omega^3 \\ 1, & otherwise \end{cases} \\ \Omega^3 &= \{(s,t): -1 \le s,t \le 1\}. \end{aligned} \tag{2}$$

$d_{i,j}^{(k)}$ is the direction index. Each direction index is sensitive to the edge aligned with a given direction. The minimum of these four direction indexes is used to detect the impulse, which can be denoted as

$$r_{i,j} = \min\left\{d_{i,j}^{(k)} : 1 \le k \le 4\right\}. \tag{3}$$

By employing a threshold T, we can identify the impulse from the noise-free pixels, no matter which are in a flat region, edge or thin line. Hence, we define the impulse detector as

$$y_{i,j} = \begin{cases} Noisy\ pixel, & if\ r_{i,j} > T \\ Noise - free\ pixel, & if\ r_{i,j} \le T \end{cases}$$

2.2 Methodology1

This technique is used to obtain a version of restored image (image1) from the corrupted source image.

Input: corrupted source image
Output: first restored image, i.e., image1

Using the following steps the output image is obtained from the input image.

Step 1: take 9 × 9 masking window
Step 2: s_k are four 1 × 9 matrices containing all the pixel values in kth direction, where k = 1, 2, 3, 4 respectively as in (1).
Step 3: Find the value of $r_{i,j}$, using Eq. (2) followed by Eq. (3).
Step 4: find the threshold value T as below

$$\begin{aligned} &T = \max(t1, t2, t3, t4) \\ &t1 = median\ of\ \{p|p \in s1\}, \quad t2 = median\ of\ \{p|p \in s2\} \\ &t3 = median\ of\ \{p|p \in s3\}, \quad t4 = median\ of\ \{p|p \in s4\} \end{aligned}$$

Step 5: divide the 9×9 masking window into nine equal square 3×3 pixels sub-windows.
Step 6: compute the median of the pixels of each of the sub-windows to obtain nine median values. Let the medians are $y^{(i)}$, $i = 1(1)9$ for nine 3×3 sub-windows.
Step 7: compute the median y_2 of the nine median values. That is, y_2 = median of $\{y^{(1)}, y^{(2)}, y^{(3)}, y^{(4)}, y^{(5)}, y^{(6)}, y^{(7)}, y^{(8)}, y^{(9)}\}$
Step 8: compute $y_m^{(2)}$ as $y_m^{(2)}$ = median of $\{y^{(i)} - y_2\}$, $i = 1(1)9$
Step 9: compute $x_2 = y_m^{(2)} + y_2$
Step 10: compute y1 as the median of the 9 pixels of the central 3×3 sub-window (x^i, $i = 1(1)9$) of the 9×9 masking window. That is, y1 = median of $\{x^i, i = 1(1)9\}$.
Step 11: compute $y_m^{(1)}$ as $y_m^{(1)}$ = median of $\{x^i - y_1\}$, $i = 1(1)9$
Step 12: compute x_1 as $x_1 = y_1 + y_m^{(1)}$
Step 13: compute x as $x = (x_1 - y_1) + x_2$
Step 14: compute the value of u as shown in (4) and replace the central pixel $y_{i,j}$ of the 9×9 window by the value of u.

$$u = b * y_{i,j} + (1 - b) * x.$$
$$b = \begin{cases} 0, & \text{if } r_{i,j} > T \\ 1, & \text{if } r_{i,j} \le T \end{cases} \tag{4}$$

Step 15: repeat all the above steps on each 9×9 window of the input image in row major order to obtain the restored image.

2.3 Methodology2

This technique is used to obtain a version of restored image (image2) from the corrupted source image.

Input: corrupted source image
Output: second restored image, i.e., image2

Using the following steps the output image is obtained from the input image.

Step 1: take 9×9 masking window
Step 2: $s_{(k)}$ are four 1×9 matrices containing all the pixel values in kth direction, where k = 1, 2, 3, 4 respectively as in (1).
Step 3: Find the values of $d_{i,j}^{(1)}, d_{i,j}^{(2)}, d_{i,j}^{(3)}, d_{i,j}^{(4)}$, using (2)
Step 4: Find the value of $r_{i,j}$, using (3).
Step 5: assign the value of t as below

$$t = \begin{cases} \text{median}\{p|p \in s_1\}, & \text{if } r_{i,j} = d_{i,j}^{(1)} \\ \text{median}\{p|p \in s_2\}, & \text{if } r_{i,j} = d_{i,j}^{(2)} \\ \text{median}\{p|p \in s_3\}, & \text{if } r_{i,j} = d_{i,j}^{(3)} \\ \text{median}\{p|p \in s_4\}, & \text{if } r_{i,j} = d_{i,j}^{(4)} \end{cases}$$

Step 6: divide the 9 × 9 masking window into nine equal square sub-windows. Each of which contains 3 × 3 pixels, as shown in Fig. 3.

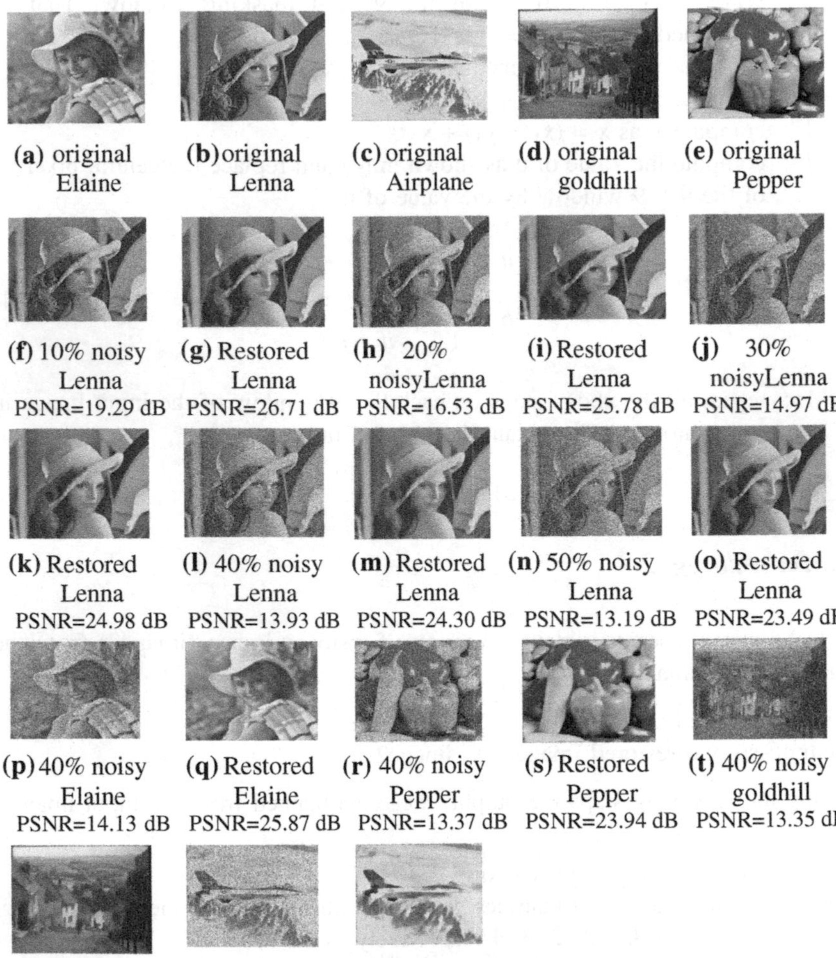

(a) original Elaine **(b)** original Lenna **(c)** original Airplane **(d)** original goldhill **(e)** original Pepper

(f) 10% noisy Lenna PSNR=19.29 dB **(g)** Restored Lenna PSNR=26.71 dB **(h)** 20% noisyLenna PSNR=16.53 dB **(i)** Restored Lenna PSNR=25.78 dB **(j)** 30% noisyLenna PSNR=14.97 dB

(k) Restored Lenna PSNR=24.98 dB **(l)** 40% noisy Lenna PSNR=13.93 dB **(m)** Restored Lenna PSNR=24.30 dB **(n)** 50% noisy Lenna PSNR=13.19 dB **(o)** Restored Lenna PSNR=23.49 dB

(p) 40% noisy Elaine PSNR=14.13 dB **(q)** Restored Elaine PSNR=25.87 dB **(r)** 40% noisy Pepper PSNR=13.37 dB **(s)** Restored Pepper PSNR=23.94 dB **(t)** 40% noisy goldhill PSNR=13.35 dB

(u) Restored goldhill PSNR=24.67 dB **(v)** 40% noisy Airplane PSNR=12.76 DB **(w)** Restored Airplane PSNR=22.18 dB

Fig. 3 Visual effect of results using DMFBMF on *gray* images

Step 7: compute the median of the pixels of each of the sub-windows to obtain nine median values. Let the medians are $y^{(i)}$, i = 1(1)9 for nine 3 × 3 sub-windows.
Step 8: compute the median y_2 of the nine median values. That is, y_2 = median $\{y^{(1)}, y^{(2)}, y^{(3)}, y^{(4)}, y^{(5)}, y^{(6)}, y^{(7)}, y^{(8)}, y^{(9)}\}$
Step 9: compute $y_m^{(2)}$ as $y_m^{(2)}$ = median of $\{y^{(i)} - y_2\}$, i = 1(1)9
Step 10: compute $x_2 = y_m^{(2)} + y_2$
Step 11: initially assign threshold T as T = 510 and while T ≥ t repeat from step 12 to step 17
Step 12: compute y_1 as the median of the 9 pixels of the central 3 × 3 sub-window (x^i, i = 1(1)9) of the 9 × 9 masking window. That is, y_1 = median $\{x^i$, i = 1(1)9$\}$.
Step 13: compute $y_m^{(1)}$ as $y_m^{(1)}$ = median$\{x^i - y_1\}$, i = 1(1)9
Step 14: compute x_1 as $x_1 = y_1 + y_m^{(1)}$
Step 15: compute x as $x = (x_1 - y_1) + x_2$
Step 16: compute the value of u as shown in (4) and replace the central pixel $y_{i,j}$ of the 9 × 9 window by the value of u.
Step 17: decrease the threshold T as T = T * 0.8
Step 18: repeat all the above steps on each 9 × 9 window of the input image in row major order to obtain the restored image.

2.4 Pixel Wise Fusion of Multiple Corrected Images

Two restored images obtained using two techniques are fused using a simple averaging fusion rule to obtain a single fused image. Two versions of images are the two restored images in identical dimensions. Let, M × N be the dimension of both the images. Then, the simple averaging rule is as in (5).

$$\text{Fused\_image}(i,j) = (\text{image1}(i,j) + \text{image2}(i,j))/2, \tag{5}$$

where i = 1, …, M; j = 1, …, N.

3 Simulation and Results

The phrase peak signal-to-noise ratio, often abbreviated as PSNR, is most commonly used as a measure of quality of reconstructed image. This can be defined using Eq. (6).

$$PSNR(\text{dB}) = 10\log\frac{255^2}{\frac{1}{M*N}\sum[I1(m,n) - I2(m,n)]^2}. \tag{6}$$

The results are simulated using PSNR.

A Number of gray benchmark images [16] have been taken for the experimental purpose. Noises have been injected randomly into the original images to produce noisy images. The filter is applied on these noisy images to obtain restored images. Figure 3a–e are original benchmarks Elaine and Lenna, Airplane, gold hill and Pepper gray images respectively. Figure 3f, h are the noisy images, with 10 and 20 % noise density, of Lenna where PSNR is 19.29 and 16.53 dB respectively that of Fig. 3g, i are the restored images using DMFBMF where the PSNR is 26.71 and 25.78 dB respectively. Figure 3j shows 30 % corrupted Lenna benchmark image whose PSNR value is 14.97 dB but when DMFBMF filter has been applied on it, the PSNR obtained is 24.98 dB (Fig. 3k). Figure 3l, n are the noisy images, with 40 and 50 % noise density, of Lenna where PSNR is 13.93 and 13,16 dB respectively that of Fig. 3m, o are the filtered images using DMFBMF where the PSNR is 24.30 and 23.49 dB respectively. Figure 3p, r, t, v are the noisy images, with 40 % noise density, of Elaine, Pepper, gold hill and Airplane where PSNR are 14.13, 13.38, 13.36 and 12.76 dB respectively that of Fig. 3q, s, u, w are the filtered images using DMFBMF where the PSNR are 25.87, 23.94, 24.67 and 22.18 dB respectively. So, for each corrupted image a corresponding restored version is generated where PSNR values are considerably increased.

Table 1 shows the different PSNR using various filters PWMAD [7], ACWM filter [13, 17], AMF [18] including proposed DMFBMF applied on Lenna gray image corrupted by various percentages of noise density. Table 2 shows the effect of applying DMFBM filter on various images corrupted by 40 % noise. In each case the enhancement factor is better than existing techniques. So from the analysis of results it can be inferred that the proposed filter will be an efficient filter which will remove the RVIN better than the existing technique compared in this literature.

Table 1 Comparitive results in PSNR of different algorithms applied to "Lenna" gray image corrupted by various rates of random-valued impulse noise

Filters	PSNR of restored image in dB					
	10 % noise	20 % noise	30 % noise	40 % noise	50 % noise	60 % noise
PWMAD	34.86	30.58	25.94	22.41	19.42	17.08
ACWM filter	–	36.07	32.59	28.79	25.19	21.19
AMF	28.06	26.79	24.03	23.17	21.99	–

Table 2 Results in PSNR of different algorithms applied to various kinds of gray images corrupted with 40 % of random-valued impulse noise

Filters	PSNR of restored image in dB			
	Elaine	Goldhill	Pepper	Airplane
PWMAD [19]	24.66	24.16	24.63	24.37
Trilateral [19]	19.38	19.14	19.53	19.54
TSM [4, 19]	20.26	20.02	20.14	19.37
Proposed	25.87	24.67	23.94	22.18

4 Conclusion

In this paper, a directional multi-scaled fusion based median filter has been proposed, for removing random-valued impulse noise from gray image. It makes full use of the characteristics of impulses and edges to detect and restore noise. Simulation results showed that this filter performs well in suppressing impulses in both subjective and objective (PSNR) evaluations. Especially, on some specific corrupted images, the proposed filter (DMFBMF) obtains better restored images than many other existing median filters.

References

1. Sa, P.K., Dash, R., Majhi B.: Second order difference based detection and directional weighted median filter for removal of random valued impulse noise. 4th IEEE International Conference on Industrial and Information Science (2009)
2. Abreu, E., Lightstone, M., Mitra, S.K., Arakawa, K.: Anew efficient approach for the removal of impulse noise from highly corrupted images. IEEE Trans. Image Process **5**(6), 1012–1025 (1996)
3. Image denoising. http://wikipedia.org. 15 June 2012
4. Chen, T., Wu, H.R.: Space variant median filters for the restoration of impulse noise corrupted images. IEEE Trans. Circ. Syst. II **48**(8), 784–789 (2001)
5. Pratt, W.K.: Median filtering. Image Process. Inst. University Southern California Los Angeles. Technical Report (1975)
6. Zhang, S., Karim, M.A.: A new impulse detector for switching median filters. IEEE Signal Process. Lett. **9**(11), 360–363 (2002)
7. Crnojević V, V., Trpovski, Z.: Advanced impulse detection based on pixel-wise MAD. IEEE Signal Process. Lett. **11**, 589–592 (2004)
8. Wang, X.: Multiscale median filter for image denoising. IEEE ICSP Proceedings, pp. 4244–5900 (2010)
9. Harikiran, J., Saichandana, B., Divakar, B.: Impulse noise removal in digital images. Int. J. Comput. Appl. **10**(8), 0975–8887 (2010)
10. Pakhira, M.K.: Digital Image Processing and Pattern Recognition. Book PHI Learning Private Limited (2011)
11. Chanda, B., Majumder, D.D.: Digital Image Processing and Analysis, 1st edn. Prentice-Hall of India (2002)
12. Gonzalez, Rafeal C., Woods, Richard E.: Digital Image Processing, 3rd edn. PHI Learning Private Limited (2008)
13. Dong, Y., Xu, S.: A new directional weighted median filter for removal of random-valued impulse noise. IEEE Signal Process. Lett. **14**(3) (2007)
14. Mandal, J.K., Sarkar, A.: A novel modified directional weighted median based filter for removal of random valued impulse noise. International Symposium on Electronic System Design, pp. 230–234 (2010)
15. Mandal, J.K., Sarkar, A.: A modified weighted based filter for removal of random impulse noise. Second International Conference on Emerging Applications of Information Technology, pp. 173–176 (2011)
16. The USC-SIPI Image Database. Signal and Image Processing Institute. USC University of California. http://sipi.usc.edu/database. 5 June 2012

17. Chen, T., Wu, H.R.: Adaptive impulse detection using center-weighted median filters. IEEE Signal Process. Lett. **8**(1), 1–3 (2001)
18. Chen, T., Ma, K.-K., Chen, L.-H.: Tri-State median filter for image denoising. IEEE Trans. Image Process. **8**(8), 1–3 (1999)
19. Nawaz, W., Jaffar, A., Hussain, A.: Fast directional weighted median filter for removal of random-valued impulse noise. IEEE Conference Paper, pp. 4244–8003 (2010)

Design of Area Optimized Sobel Edge Detection

Sunil Kumar Kuppili and P.M.K. Prasad

Abstract Edge attribute Extraction is an indispensable and central subject in computer vision. In topical years edge detection performance has progressively been widely used because it filter out hopeless data from the image. Sobel edge detection is one of the classic edge detection operator, used to distinguish the edge pixels in an image and property of less weakening in high level of noise. This method exploits modify in intensity with respect to neighboring pixels in the image. The gradient edge detection is preferred in order to optimize the area which is crucial feature. For the purpose of reducing the FPGA area resource Sobel edge detection is implemented in VDHL and simulation and synthesis will be done using Xilinx.

Keywords Red green blue (RGB) · Very high speed integrated circuit (VHSIC) · Hardware description language (VHDL) · Field programmable gate array (FPGA)

1 Introduction

Edge detection refers to the course of action of identifying and locating of sharp discontinuities in an image. Edges classify the boundaries stuck between regions in an image, which helps with segmentation and entity recognition. They generally show where shadows fall in an image or any other diverse change in the intensity of an image. Edge detection is an elementary of low-level image processing and good edges are obligatory for higher level processing. The discontinuities are hasty changes in pixel intensity which portray restrictions of objects in an outstandingly landscape. Since the edges of a picture are thought-about to be most important

S.K. Kuppili (✉) · P.M.K. Prasad
Department of ECE, GMRIT, Rajam, Andhra Pradesh, India
e-mail: sunilkumarkuppili@gmail.com

P.M.K. Prasad
e-mail: mkprasad.p@gmrit.org

L.C. Jain et al. (eds.), *Computational Intelligence in Data Mining - Volume 2*,
Smart Innovation, Systems and Technologies 32, DOI 10.1007/978-81-322-2208-8_59

image attributes that offer valuable facts to user, the tingle detection is one in every of the key in stages in image/video processing, object recognition and tracking. The ambition of an edge detection algorithm is to establish the sharp changes within the image brightness. There are many ways to perform edge detection. However, the preponderance of various ways might be grouped into two, Gradient primarily based edge detection that detects the sides by trying for the utmost and minimum in the first derivative of the image and Laplacian chiefly based edge detection that detects edges with zero crossings in the second order derivative of the image [1, 2]. The second order derivative is very responsive to noise gift in the image and hence second order derivative operators are not habitually used for edge detection operation [3]. For the purpose detecting the edge sharply so the implementation is done in VHDL, simulation and synthesis is done using Xilinx [4].

Consequently, over the history of Digital image Processing a variety of edge detectors have been devised which differ in their purpose (i.e., the photometrical and geometrical properties of edges which they can able to extract) and their mathematical and algorithmic properties.

This paper is organized as follows; Sect. 2 explains the existing model of the Sobel Edge Detection and Sect. 3 explains the proposed model of Optimization in terms of Area of Sobel Edge Detection. Simulation and Results are analyzed in the Sects. 4 and 5 with the conclusion.

2 Sobel Edge Detection

The Sobel operator is far and wide or broadly used for edge detection in image processing. It has advantage of simple gradient operator over the remaining gradient operator in image processing because of its property to counteract the noise sensitivity. The operator is mainly based on computing a ballpark figure of the gradient of the image intensity utility. It principally uses two 3 × 3 spatial masks (Gx and Gy), which are convolved with the original image to gauge the ballpark Figure of the gradient utility. The Sobel operator mainly uses two filter masks they are has shown in Fig. 1.

These masks compute the middling gradient components across the neighbouring lines or columns respectively. The local edge potency is defined as the gradient magnitude prearranged by:

-1	0	+1
-2	0	+2
-1	0	+1

=

+1	+2	+1
0	0	0
-1	-2	-1

Fig. 1 Convolution kernels

$$G = \sqrt{G_x^2 + G_y^2} \quad (1)$$

where 'G' represents the rate of change in direction of gradient vector value.

The precision of the Sobel operator used for edge detection is fairly low because it uses two masks which detect the edges nearby in horizontal and vertical directions only. The basic block echelon data flow diagram for working out of the Sobel edge detector. It consists of essentially four stages. In the first stage, the homeward bound pixel data is stored in buffer memory. Four gradients along poles apart directions are computed in the second stage. The utmost gradient is selected, and final edge map is computed by comparing the maximum gradient value with a threshold in the third and fourth stages, respectively.

The Sobel Edge Detector consists of chiefly three blocks are:

1. Memory module
2. Gradient calculation
3. Binary segmentation

MEMORY MODULE: In the Sobel compass operator [3 × 3], the masks are used to work out gradient values along poles apart directions over an input image. Consequently, it is obligatory to store at slightest two rows of input image data in FPGA on-chip memory before the handing out begins. To achieve this, we have used elegant buffer-based memory structural design which utilizes a set of registers in order to shift the image data into computing window. The length of the shift register depends on input image width [5].

GRADIENT CALCULATION: By using these convolution kernels we can reckon the gradient values. From these masks we can effortlessly use the subsequent equations which are obtained from modified gradient calculations from Fig. 2

$$G_x = (P3 - P1) + 2 * (P6 - P4) + (P9 - P7) \quad (2)$$

$$G_y = (P7 - P1) + 2 * (P8 - P2) + (P9 - P3) \quad (3)$$

$$G_{dr} = (P8 - P2) + 2 * (P9 - P1) + (P6 - P4) \quad (4)$$

$$G_{dl} = (P2 - P8) + 2 * (P3 - P7) + (P6 - P8) \quad (5)$$

$$\mathbf{G} = |\mathbf{G_x}| + |\mathbf{G_y}| + |\mathbf{G_{dr}}| + |\mathbf{G_{dl}}| \quad (6)$$

By using the gradient the complete combinational logic is split keen on quite a few multiple levels [6].

BINARY SEGMENTATION: The representation of binary segmentation module is shown in Fig. 3 input is the gradient value and then it compares with the given threshold value [6] and either of two values one will correspond edges can be detected in this section.

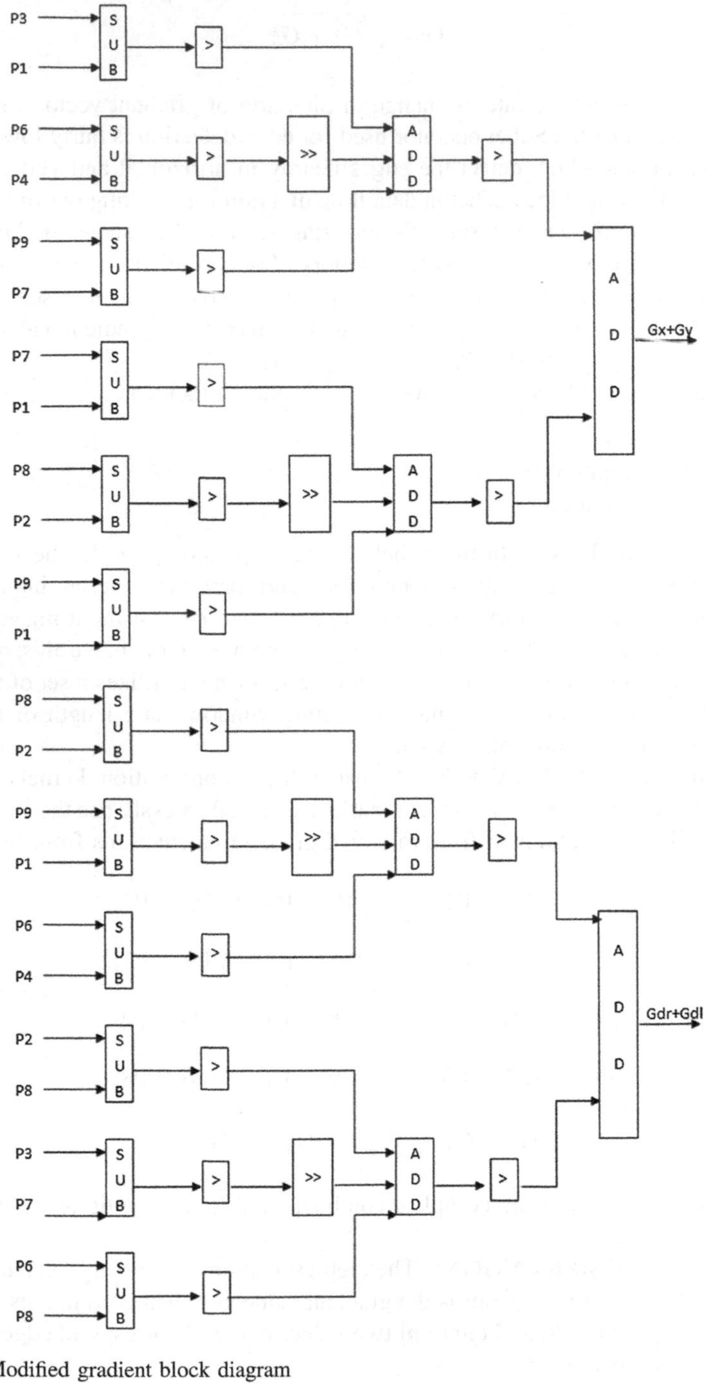

Fig. 2 Modified gradient block diagram

Fig. 3 Binary segmentation

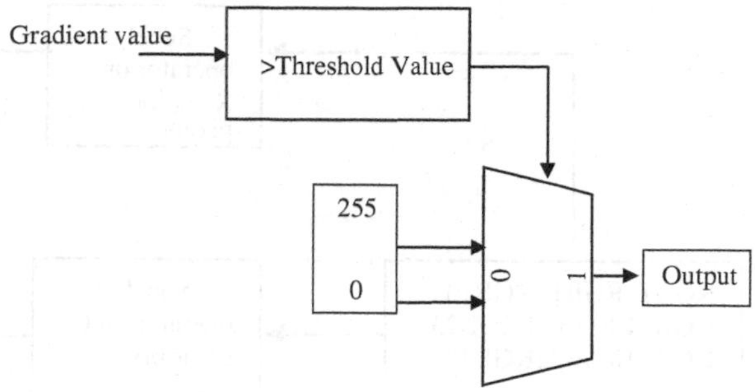

3 Proposed Architecture

The Proposed architecture consists of input pixels from the image and that pixels are given as the input. These pixels are stored in buffer memory and for that pixels values Sobel edge detection have been used to find out edge as well as filter out useless information. Gradient operation have been used. From that gradient values the maximum value will comes as the output. And from that we can tell that whether the edge present in the given input pixels values of the image. The implement of the proposed architecture is shown in Fig. 4. So, in order to reduce the power the memory module has been design with RS flip flop with two separate clock pluses and for that we proposed to use the ring counter. For the buffer memory the color edge detection separates the basic components which is shown in Fig. 5 and that separated components are applied to the Sobel edge operator in which it generally detects the edges [7]. And now the color edge map is calculated by merging the edges of each basic color channel.

3.1 From Buffer Memory

The Sobel edge detection architecture used for three basic color channel for gradient calculation. It determines horizontal and vertical gradients for each channel.

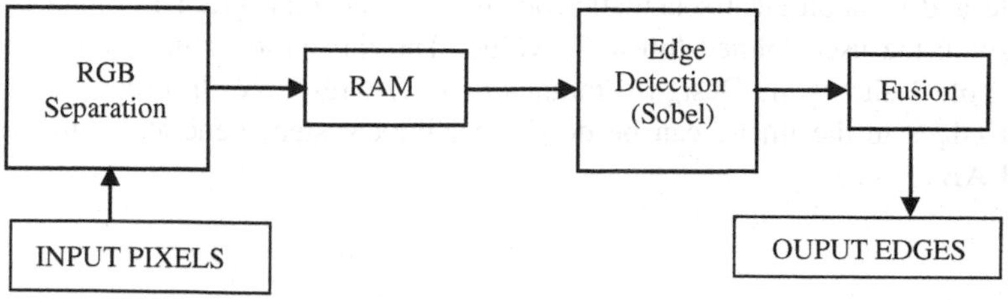

Fig. 4 Proposed architecture

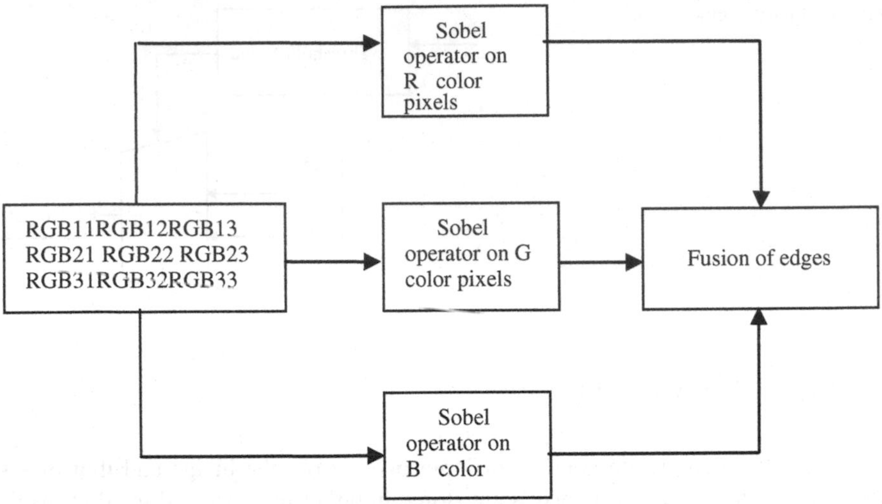

Fig. 5 Implementation of proposed architecture

The combined gradient for each channel is computed by adding the absolute values of both gradients values to find out the edges. The edge map is calculated by comparing the gradient value with threshold of our edges requirements. And from the Figure the Sobel operator module is same on RBG components. The only change is in the input pixels applied to each Sobel operator module. The final stage outputs are obtained and in which we can see the edges and corresponding FPGA resources are reduced and summarized in the simulation results [8].

4 Simulation Results

The proposed of Sobel edge detection algorithm is described in VHDL. Design and testing of individual module has been carried out. The final output consists of only basic components of the image, Fig. 6 depicts the simulation results when there is no edge in the image and Fig. 7 depicts simulation results when the edges present in the image. Figures 8 and 9 depicts the simulation results of RTL schematic top module and technological schematic top module and final gradient value is compared with the user defined threshold value. The threshold value is taken here is 200. Table 1 shows the Design summary of FPGA resource. In order to view the output edges in the image can be done by Xilinx system generator with help of MATLAB.

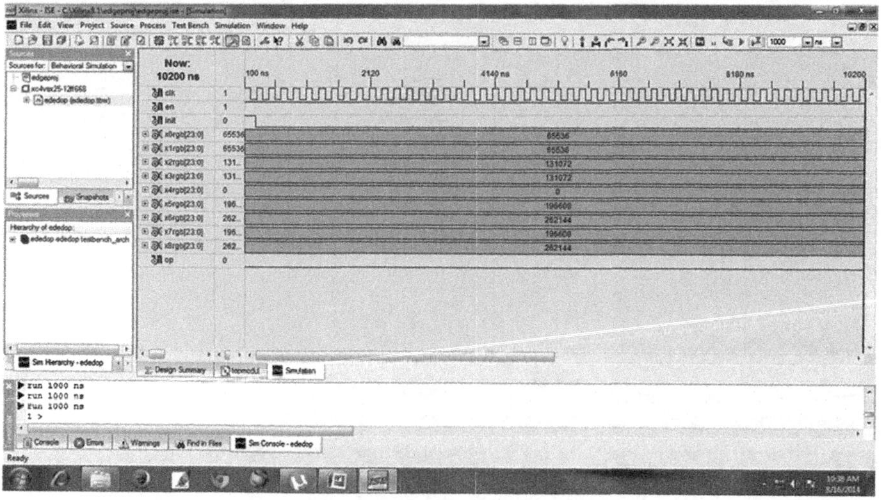

Fig. 6 Simulation result for Sobel edge detection when edges absent

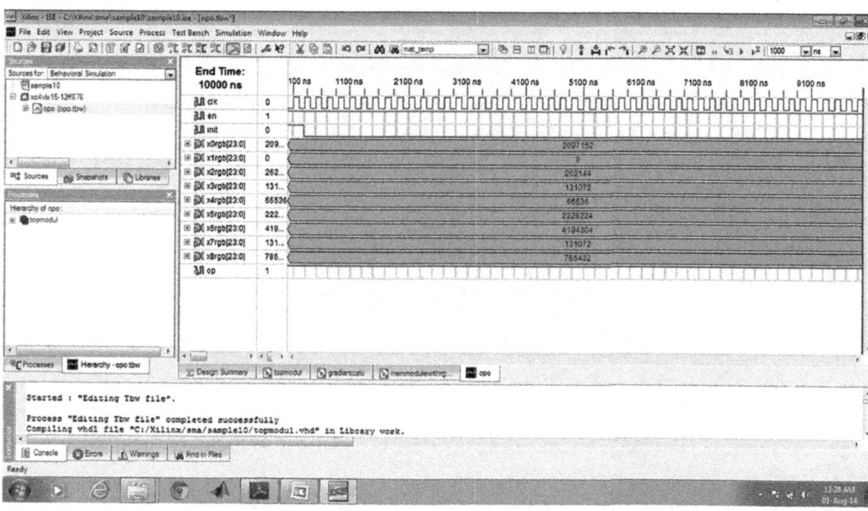

Fig. 7 Simulation result for Sobel edge detection when edges present

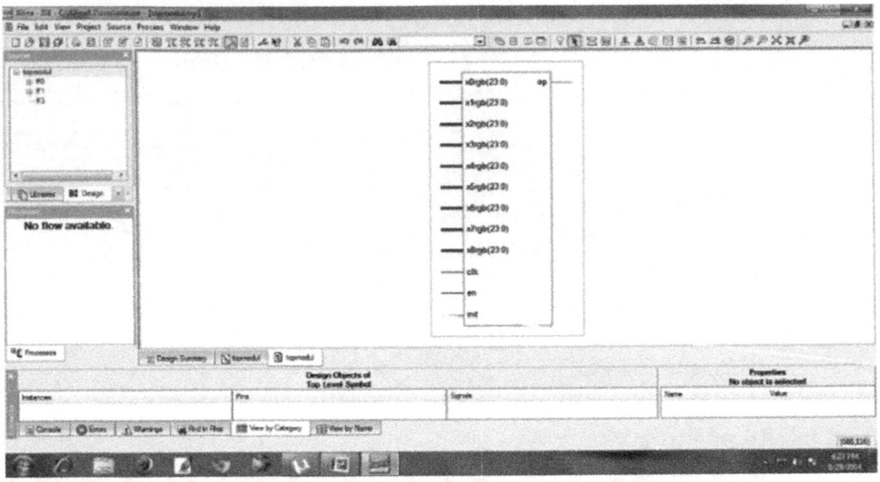

Fig. 8 Top module of RTL schematic

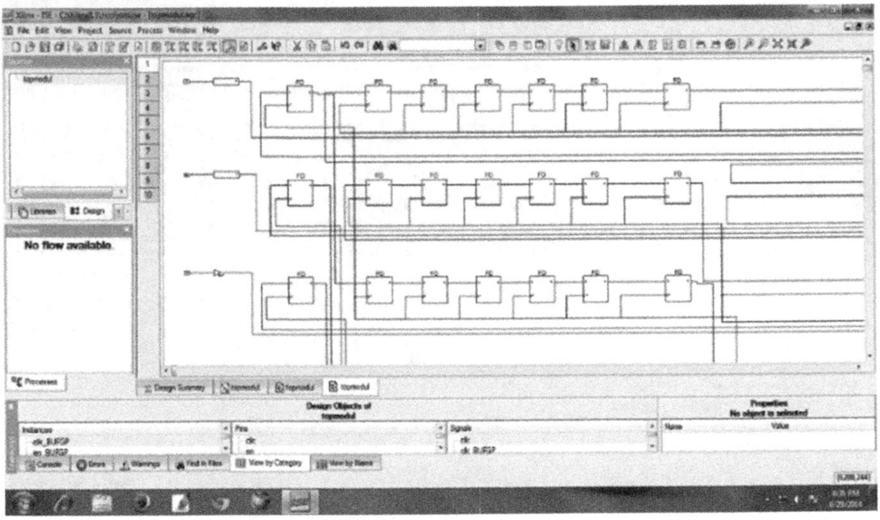

Fig. 9 Technological schematic view of the top module

Table 1 Design summary of top module

Device utilization summary			
Logic utilization	Used	Available	Utilization (%)
Number of slices	11,289	10,240	110
Number of slice flip flops	13,636	20,480	66
Number of 4 input LUTs	9,466	20,480	46
Number of bonded IOBs	68	320	21
Number of GCLKs	2	32	6

5 Conclusion

Sobel edge detection operator is insensitive to noise and the masks of Sobel operator is relatively small as compare to the other edge detection operator (theoretically studied) that's why Sobel edge detection operator is used. From that the area is optimized with gradient calculation and the memory module. And the resource of FPGA area has been reduced. The process of finding edges can also be implemented in Xilinx system generator with help of MATLAB. For more accurate results we can use the four convolutions to detect the edges of the images.

References

1. Singh, S., Sekhar, C.: Area optimized FPGA implementation of color edge detection. 2013 International Conference on Advanced Electronics Systems (ICAES)
2. Ziou, D., Tabbone, S.: Edge detection techniques—an overview. Int. J. Pattern Recog. Image Anal. **8**, 537–559 (1998)
3. Maini, R., Aggarwal, H.: Study and comparison of various image edge detection techniques. Int. J. Image Process. (IJIP) **3**
4. Nosrat, A., Kavian, Y.S.: Hardware description of multi-directional fast sobel edge detection processor by VHDL for implementing on FPGA. Int. J. Comput. Appl. **47**(25), 1–7 (2012)
5. Moore, C., Devos, H., Stroobandt, D.: Optimizing the FPGA memory design for a sobel edge detector. In: Proceedings of the 20th Annual Workshop on Circuits, Systems and Signal Processing (2009)
6. Ahmad, M.B., Choi, T.S.: Local threshold and boolean function based edge detection. IEEE Trans. Consum. Electron. **45**(3), 674–679 (1999)
7. Koschan, A., Abidi, M.: Detection and classification of edges in color images. IEEE Signal Process. Mag. **22**(1), 64–73 (2005)
8. Sudeep, K.C., Majumdar, J.: A novel architecture for real time implementation of edge detectors on FPGA. Int. J. Comput. Sci. Issues **8**(1), 193–202 (2011)

Accelerated FFT Computation for GNU Radio Using GPU of Raspberry Pi

S. Sabarinath, R. Shyam, C. Aneesh, R. Gandhiraj and K.P. Soman

Abstract This paper presents the effective exploitation of Graphical Processing Unit (GPU) in Raspberry Pi for fast Fourier transform (FFT) computation. Very fast computation of FFT is found useful in computer vision based navigation system, Global Positioning System (GPS), HAM radio and on Raspberry Pi. A comparison is performed over the speed of FFT computation on BCM2835 GPU with that of 700 MHz ARM processor available in Raspberry Pi and also with intel-COREi5 processors. The FFT is computed for any one dimensional input signal and its analysis is done on different processors with varying signal lengths. The GNU radio is installed on Raspberry Pi, and the FFT computation done on GNU radio is accelerated using GPU of Raspberry Pi. Even though the Raspberry Pi GPU is primarily built for video enhancement, the parallel computational ability of GPU is utilized in this paper for accelerated FFT computation.

Keywords GPU · FFT · GNU radio · Raspberry Pi

S. Sabarinath (✉) · R. Shyam · C. Aneesh · K.P. Soman
Centre for Excellence in Computational Engineering and Networking,
Amrita School of Engineering, Amrita Vishwa Vidyapeetham, Coimbatore
641112, Tamil Nadu, India
e-mail: sabarin8@gmail.com

R. Shyam
e-mail: shyam.neezhoor@gmail.com

C. Aneesh
e-mail: a4aneeshc@gmail.com

R. Gandhiraj
Department of Electronics and Communication Engineering, Amrita School
of Engineering, Amrita Vishwa Vidyapeetham, Coimbatore
641112, Tamil Nadu, India
e-mail: r_gandhiraj@cb.amrita.edu

L.C. Jain et al. (eds.), *Computational Intelligence in Data Mining - Volume 2*,
Smart Innovation, Systems and Technologies 32, DOI 10.1007/978-81-322-2208-8_60

1 Introduction

Fast Fourier Transform enables fast computation of Discrete Fourier Transform [1]. The pronunciation of data can be very well executed using Fourier transform. But still FFT computation is time consuming. So, faster method of computation is introduced using GPU of Raspberry Pi [2] using GNU radio software.

1.1 Raspberry Pi

Raspberry Pi is a cost effective hardware developed in UK, with the aim of teaching computer science in schools. Mainly two versions of Raspberry Pi are available, model A and model B. Model A has 256 MB RAM with single USB port and a power rating of 300 mA (1.5 W). Model B has 512 MB RAM with two USB ports and a power rating of 700 mA (3.5 W). System on chip RAM (Broadcom BCM2835) of Raspberry Pi contains CPU, GPU, DSP and SDRAM. The 256 MB RAM in model A is equally split in the ratio 50:50 in the default case. If graphical intensive work load is less, then the split ratio can be changed by the user to give the CPU some more RAM. The CPU and GPU performance are independent to each other. Communication between them is done through mail boxes. The value of flag is set when communication stops. The parameter-data is left in mail boxes. GPU even have the ability to generate its own interrupt events that stops the CPU's current execution. Code for the split is of the form,

sudo cp/boot/arm n_start.elf/boot/start.elf

where *n* specifies the memory allocation for CPU and the remaining 256-*n* for GPU.

1.2 GNU Radio

GNU Radio is software that is purely open-source [3]. It is extremely useful in signal processing applications because it contain most of the signal processing tools as blocks, which can be used for simulation purpose. The FFT is one such block using which any signal can be given as the input of the block, that will return Fourier transform of the given input. The GNU radio is installed on Raspberry Pi. The software has an FFT block that takes any signal as its input and gives FFT of that signal as its output and also a plot for both input and output signals [4]. The FFT computation is slower process as this is done using the CPU of Raspberry Pi initially. The paper presents a method of using the GPU of Raspberry Pi for FFT computation instead of CPU and a comparison between the times of computation.

1.3 Continuous, Discrete, Fast Fourier Transforms

Any signal can be represented in time domain, the values of the physical process as some quantity *h*, e.g., *h*(*t*), or in Fourier domain, where *H* is the amplitude of the process which is specified, that is *H*(*f*), where *f* is the frequency represented in cycles per second. A transformation from *h*(*t*) to *H*(*f*) or vice versa is given by the Fourier transform equation as,

$$H(f) = \int_{-\infty}^{\infty} h(t)e^{-j\omega t}dt \tag{1}$$

$$h(t) = \int_{-\infty}^{\infty} H(f)e^{j\omega t}df \tag{2}$$

ω and f are related as, $\omega = 2\pi f$.

The spectrum will be continuous in nature. Fourier transform uses complex bases and it's conjugate to represent the signal. Complex values of information has both phase and magnitude values. This kind of representation is found helpful for the analysis of multi-tone, non-periodic signals which are time invariant in nature.

Fourier transform of a signal from finite number of sample points can be estimated, called discrete Fourier transform [5]. The function has non zero values at these discrete levels of time. Levels are chosen based on Nyquist criteria. The number of independent outputs obtained will evidently be the same as that of number of input given that is *N*. Fourier transform need to be only estimated at these discrete values. The discrete Fourier transform H_n,

$$H_n = \sum_{k=0}^{N-1} h_k e^{2\pi ikn/N} \tag{3}$$

The discrete Fourier transform [6] also exhibits symmetry property exactly like continuous Fourier transform. Recovering back h_k s from H_n s called discrete inverse Fourier transform. Transformation relation is given by,

$$h_n = \frac{1}{N}\sum_{n=0}^{N-1} H_n e^{-2\pi ikn/N} \tag{4}$$

Computational speed of fast Fourier transform is much more than that of discrete Fourier transform. It defines a complex number *W*,

$$W = e^{2\pi i/N}$$

Each value of h_k is multiplied by matrix which has $W^{(n \times k)}$ as its components, *n* and *k* are the number of rows and columns of the matrix containing complex values.

Discrete Fourier transform is $O(N^2)$ process, that is, the matrix multiplication need N^2 complex multiplications, and additional small operation to calculate the required power values of W. Using fast Fourier transform the discrete Fourier transform can be computed in $O(N \log_2 N)$ operations. Fourier transform having length N can be represented as sum of two $N = 2$ length discrete Fourier transform. One is even numbered point and other is odd numbered point of the original N. That is,

$$F_k = \sum_{j=0}^{N-1} e^{2\pi ik/N} f_j \tag{5}$$

$$F_k = F_k^e + W^K F_k^o \tag{6}$$

where, W is complex constant, kth component of the Fourier transform is denoted by F_k^e and F_k^o having length $N = 2$ generated from even and odd components. Both the components are periodic in k having lengths $N = 2$. So repetitions of two cycles are necessary to retrieve F_k.

2 FFT in Raspberry Pi and Proposed Methodology

FFT calculation need to be executed in an accelerated way because most of the real time application of signal processing need FFT computation, which consumes much of the execution time. FFT calculation is done initially over the normal Raspberry Pi processor or the CPU of Raspberry Pi using GNU radio software. Then done over conventional MATLAB software and both results are compared with the speed of FFT computed from GNU radio using GPU of the same Raspberry Pi. Videocore IV the graphical processing unit (GPU) in BCM2835, which is the microchip available in Raspberry Pi, is effectively made use for writing general purpose code for accelerated fast Fourier transform library. Main application of GPU in Raspberry Pi was for video streaming. The FFT block used in GNU radio is forced to work using FFT library that access GPU rather than Fastest Fourier Transform in the West (FFTW) library, which is commonly used in most of the operating systems. The time of computation is accurately calculated and studied.

The architecture of GPU (BCM2835 Videocore IV) has 2 processors that are distinct in nature. One is Programmable Graphic Core (PGC) and other is its co-processor. The programmable portion has many slices that are organized. Each slice has four single input-multiple data (SIMD) units, each is capable of executing different instructions per cycle, thus making the processing parallel.

2.1 FFT in GPU of Raspberry Pi

The use of processing power of GPU for calculating FFT can reduce the computational limitation of normal CPU. Andrew Holme has designed such library which uses the GPU for calculating the FFT in Raspberry Pi [1]. This library is purely open source and can be installed on to Raspberry Pi platform by running its command in command prompt of Raspberry Pi. We implemented code for FFT computation of an input signal of given length on this library. The code has three main functions *GPU_FFT_prepare* that creates structure for storing the input. The memory is effectively utilized by using a *maloc* function, *GPU_FFT_execute* is the core function that is used to execute the FFT and also it computes the time of computation, and finally *GPU_FFT_release* that will release all the previously stored memory locations for further reuse. The program is written in such a way that it take input of size as power of 2, and also takes the number of batches containing the number of arrays of input. Each row of the batch corresponds to each input and result is calculated row wise. This will return the FFT of each row along with the time it took to calculate the same. Any signal that has been generated randomly is given as its input to compute its FFT. This is compared with the previously generated time of execution obtained from Raspberry Pi CPU and MATLAB software.

2.2 FFT in CPU of Raspberry Pi

The CPU in Raspberry Pi belongs to 700 MHz ARM1176JZF-S core. It is more similar in configuration to that of old Pentium II processor, which is comparatively slower as compared with the newer versions available and hence it cannot provide much of execution speed. The execution power in the CPU cannot be up to the expectation while computing a high mathematical computation incorporated FFTs. FFT is ideally used to make DFT having computation of $O(N^2)$ faster. FFT has number of computation of $O(N\log_2 N)$. This is incorporated into CPU of Raspberry Pi using executables like *FFT_processor*, *GNU plot_driver*, *Signal_Source* and *Sound_Source*. The time taken for execution of a real time signal is computed for various lengths of the signal and is tabulated for comparison. The real time input signal is shown in Fig. 1. The obtained FFT plot is depicted in Fig. 2.

2.3 FFT in Intel-COREi5

FFT computed on intel-COREi5 is done using MATLAB software version R2013a (64 bit). The processor is having a clock speed of 3.3 GHz and maximum turbo frequency of 3.7 GHz. MATLAB has got capability of reading any one dimensional signal which can be passed as the input for computing FFT. We generated a signal

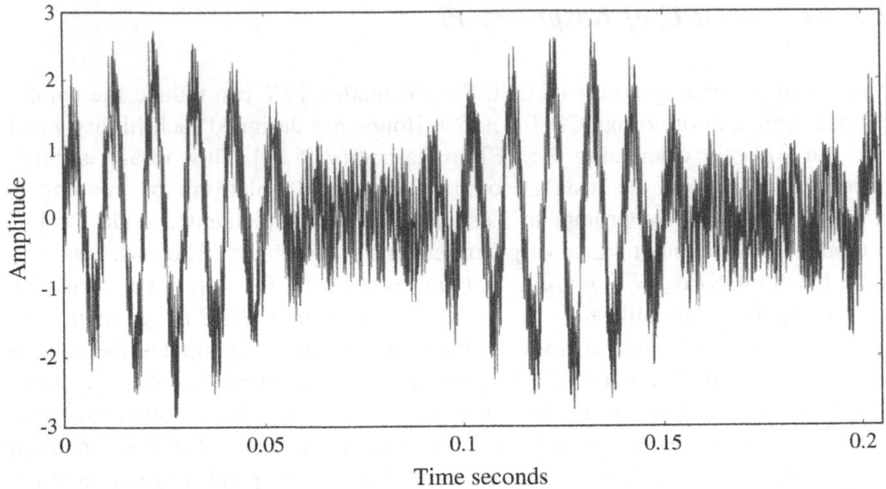

Fig. 1 Noisy input signal in time domain

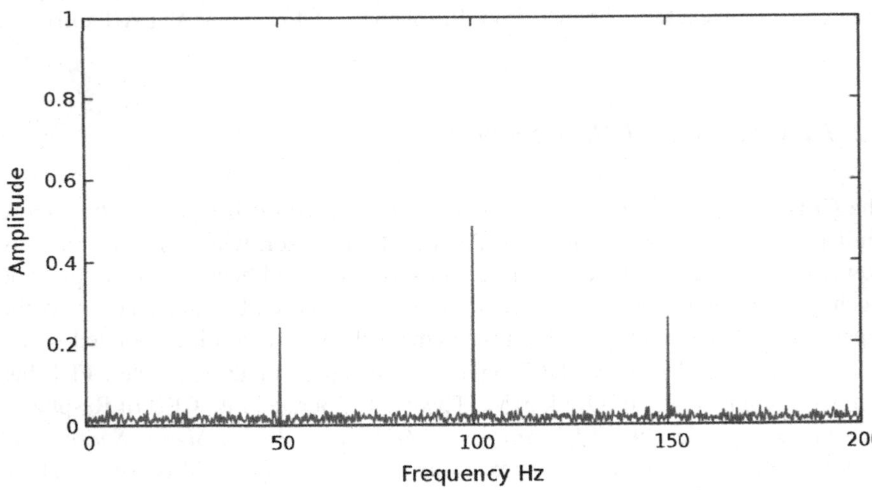

Fig. 2 Frequency spectrum

using this software and passed as the input of FFT function. MATLAB software already has inbuilt FFT command which is used to calculate the FFT. Time of computation for each input is calculated within MATLAB itself. Batches of inputs are given each batch having a particular number of inputs. So a batch 2 will contain two inputs. The computation time of both these inputs are calculated together in a batch.

3 Results and Discussion

A comparison of time taken for FFT computation by CPU, GPU and MATLAB software is done here. Signal having various lengths are considered. Here we consider length 256, 512, 1,024 and 2,048. Each signal contains four batches, each batch having array of input signal whose FFT need to be computed. Results obtained are tabulated for easy comparison. Table 1 shows the execution time for FFT in CPU of Raspberry Pi, Table 3 shows the execution time for FFT in intel-COREi5. The computation of FFT using GNU radio is found to be time consuming, as it is done using GPU in Raspberry Pi. This paper presents an effective solution to the above mentioned problem. From detailed analysis of sequences having various lengths (for different batches), it is found that CPU of Raspberry Pi consumes much more time as compared with MATLAB software which runs in conventional CPU as shown in Tables 1 and 3. But at the same time GPU of Raspberry Pi performs faster FFT computation than both of the above methods. Execution time of GPU is found to come in the order of micro seconds, which is shown in Table 2.

Table 1 Execution time of FFT in CPU of Raspberry Pi given in seconds

Batches	Length of FFT			
	256	512	1,024	2,048
1	0.2235	0.3134	0.534	1.0335
2	0.4229	0.7378	1.817	2.1096
3	0.6602	1.08	1.20	3.1788
4	0.9298	2	2.448	4.185

Table 2 Execution time of FFT in GPU of Raspberry Pi given in seconds

Batches	Length of FFT			
	256	512	1,024	2,048
1	0.23×10^{-6}	36×10^{-6}	57×10^{-6}	117×10^{-6}
2	46×10^{-6}	76×10^{-6}	112×10^{-6}	228×10^{-6}
3	72×10^{-6}	113×10^{-6}	174×10^{-6}	346×10^{-6}
4	114×10^{-6}	154×10^{-6}	238×10^{-6}	455×10^{-6}

Table 3 Execution time of FFT in intel-COREi5 given in seconds

Batches	Length of FFT			
	256	512	1,024	2,048
1	0.014089	0.052547	0.038113	0.086689
2	0.039796	0.043302	0.095895	0.188929
3	0.036306	0.075456	0.167499	0.278827
4	0.098299	0.081684	0.149205	0.320916

4 Conclusion and Future Work

In this paper, a comparison is done on the speed of FFT computation in GPU with that of CPU of Raspberry Pi and computation performed in intel-COREi5 processor (using MATLAB software). It was found that GPU allows us to compute the FFT much faster than both CPU of Raspberry Pi and intel-COREi5 processor. In future this GPU can be synchronized with normal processor in such a way that whenever faster computational power is needed, processor automatically shifts to GPU. The motivation behind doing this work was the slower computation speed of FFT in GNU Radio Companion. The slow computational speed is compensated by the use of GPU in Raspberry Pi. This computation can even be performed on environments where there is absence of computer, since Raspberry Pi itself act as a mini computer. As a future work this FFT computation could be utilized in computer vision based navigation system using Raspberry Pi for path tracking applications.

References

1. Duhamel, P., Vetterli, M.: Fast Fourier transforms: a tutorial review and a state of the art. Sig. Process. **19**, 259–299 (1990)
2. Accelerating Fourier transforms using the GPU | Raspberry. http://www.raspberrypi.org/accelerating-fourier-transforms-using-the-gpu/
3. WhatIsGR—GNU Radio—gnuradio.org. http://gnuradio.org/redmine/projects/gnuradio/wiki/WhatIsGR/
4. Gandhiraj, R., Soman, K.P.: Modern analog and digital communication systems development using GNU radio with USRP. Telecommun. Syst. 1-15 (2013)
5. Brigham, E.O., Yuen, C.: The fast Fourier transform. Syst. Man Cybern. IEEE Trans. **8**, 146–146 (1978)
6. Sorensen, H.V., Jones, D.L., Heideman, M., Burrus, C.S.: Real-valued fast Fourier transform algorithms. Acoust. Speech Sig. Process. IEEE Trans. **35**, 849–863 (1987)
7. Gandhiraj, R., Ram, R., Soman, K.P.: Analog and digital modulation toolkit for software defined radio. Proc. Eng **30**, 1155–1162 (2012)

Feature Extraction and Performance Analysis of EEG Signal Using S-Transform

Monorama Swain, Rutuparna Panda, Himansu Mahapatra and Sneha Tibrewal

Abstract Feature can be described as a functional component observed from a data set. The extracted features give the information related to a signal, thus it requires to calculate cost of information processing and complexity of analyzing a huge data set. This paper presents a feature extraction method using S transform. Five data sets are taken and feature extraction has been performed by implementing two methods: first by applying S-transform and other without S-transform. The performance of the neural model is evaluated on the basis of training performance and classification accuracies and the results confirmed that the proposed scheme has potential in classifying the EEG signals.

Keywords Feature abstraction EEG signal · Neural network · S-transform

1 Introduction

The German psychiatrist, Hans Berger performed the first electroencephalographic (EEG) recording in humans (Berger 1929). EEG signals are electrical signals obtained due to brain activity. These signals are then processed by a computer. Feature extraction is required for construction of an efficient brain computer interface. EEG signal processing follows a three step procedure. The first step

M. Swain (✉) · H. Mahapatra · S. Tibrewal
Silicon Institute of Technology, Bhubaneswar, India
e-mail: mswain@silicon.ac.in

H. Mahapatra
e-mail: rintusilicon@gmail.com

S. Tibrewal
e-mail: tibrewalsneha24@gmail.com

R. Panda
Department of ECE, Veer Surendra Sai University of Technology, Sambalpur, India
e-mail: r_ppanda@yahoo.co.in

L.C. Jain et al. (eds.), *Computational Intelligence in Data Mining - Volume 2*,
Smart Innovation, Systems and Technologies 32, DOI 10.1007/978-81-322-2208-8_61

acquisition of brain signals and subsequent processing to remove the unwanted noise components. Secondly a distinct feature is identified and extracted. In the final step machine learning algorithms are used for pattern recognition.

A number of signal processing techniques have been widely used to study various EEG signals after the first EEG record [1, 2] was discovered. In the past few years, researchers in applied mathematics and signal processing have developed many methods like wavelet transform [3, 4] and S transform for the extraction of features of brain signals. Feature extraction techniques are required to distinguish the requisite signals from background.

Numerous signal processing techniques have been applied to obtain representations and extract the features of a signal. Feature extraction simplifies the analysis of complex data by reducing data dimensions. Careful selection of features is an important criterion to obtain discriminative information for classification. Recent advances in the field of neural networks have made them attractive for analyzing signals. Artificial neural networks [5] provide an insight into the functioning of real network of neurons. The application of neural networks has opened a new area of solving problems not resolvable by other signal processing techniques. ANNSs not only model the signal, but also make a decision as to the classify the signal.

Many methods of feature extraction have been applied to extract the relevant characteristics from a given EEG data. Here our objective is to analyze the EEG signals and classify the EEG data into different classes. Our main target is to improve the accuracy of EEG signals. Analysis of EEG signals provides a crucial tool for diagnosis of neurobiological diseases [6]. The problem of EEG signal classification [7] into healthy and pathological cases is primarily a pattern recognition problem using extracted features.

The rest of the paper is organized as follows: Sect. 2 describes the database and the techniques used in our work, Discrete S transform and artificial neural network, Sect. 3 gives the feature extraction and classification results.

2 Data Acquisition and Methods

2.1 Data Selection and Recording

We have used the publicly available data with Bonn University. The complete dataset consists of five classes viz. A, B, C, D, E each of which contains 100 single channel EEG segments of 23.6 s duration and having 4,097 data points. Sets A and B consists of signals taken from surface EEG recordings that were carried out on five healthy volunteers using a standardized electrode placement scheme given in Fig. 1.

Data in set A belongs to volunteers relaxed in an awakened state with eyes open while data in set B belongs to the same group of individuals having eyes closed. Set C–E contains data from five patients all of whom have achieved complete seizure control. Set C contains data from the hippocampal formation of the opposite hemisphere of the brain. In Set D recordings were taken from the epileptic zone but

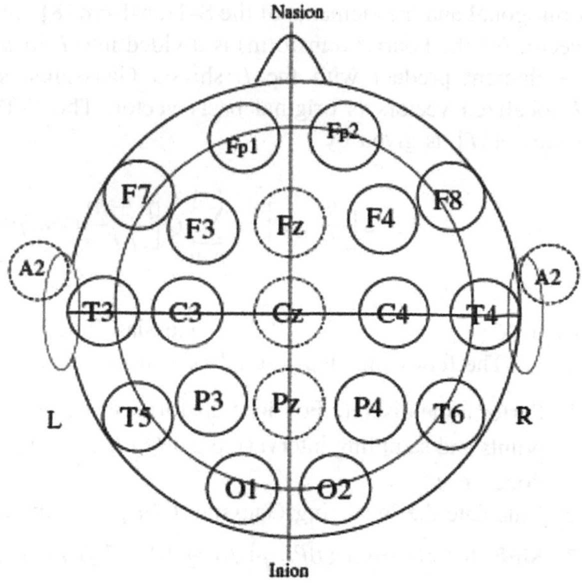

Fig. 1 The 10–20 system of electrode placement

during seizure free interval while Set E contained only seizure activity. From the data available, a rectangular window of length 256 discrete data was selected to form a single EEG segment.

2.2 Analysis Using S Transform

The discrete S-transform is defined as follows:

Let, $q[kT]$, $k = 0, 1, \ldots, L-1$, denote a discrete time series corresponding to a signal $q(t)$ with a time sampling interval of T [8]. The discrete Fourier transform of the signal can be obtained as follows:

$$Q\left[\frac{n}{LT}\right] = \frac{1}{L}\sum_{k=0}^{L-1} q[kT]\mathrm{e}^{-(i2\Pi nk/L)} \tag{1}$$

where $n = 0, 1, \ldots, L-1$ and the inverse discrete Fourier transform is

$$q[kT] = \sum_{n=0}^{L-1} Q\left[\frac{n}{LT}\right]\mathrm{e}^{i2\Pi nk/L} \tag{2}$$

In the discrete case, the S-Transform is the projection of the vector defined by the time series $q[kT]$, onto a spanning set of vectors. The spanning vectors are not orthogonal and the elements of the S-Transform [8] are not independent. Each basis vector (of the Fourier transform) is divided into L localized vectors by an element-by-element product with the L shifted Gaussians, such that the sum of these L localized vectors is original basis vector. The S-Transform of a discrete time series $q[kT]$, is given by

$$S\left[\frac{n}{NT}, jT\right] = \sum_{p=0}^{L-1} Q\left[\frac{p+n}{LT}\right] G(n,p) e^{i2\Pi pj/L} \tag{3}$$

where $G(n,p) = e^{-\left(2\Pi^2 p^2/n^2\right)}$ = Gaussian function and where, $j, p, n = 0, 1, \ldots,$ $L-1$. The following steps are adapted for computing the discrete S-Transform.

1. Perform the discrete Fourier transform of the original time series $q[kT]$ (with N points and sampling interval) to get $Q[^p/_{LT}]$ using the FFT routine. This is only done once.
2. Calculate the localizing Gaussian $G(n,p)$ for the required frequency $^n/_{LT}$.
3. Shift the spectrum $Q[^p/_{LT}]$ to $Q\left[^{(p+n)}/_{LT}\right]$ to for the frequency $^n/_{LT}$ (one pointer addition).
4. Multiply $Q\left[^{(p+n)}/_{LT}\right]$ by $G[n,p]$ to get $A\left[^n/_{LT}, ^p/_{LT}\right]$ (L multiplications).
5. Inverse Fourier transform of $A\left[^n/_{LT}, ^p/_{LT}\right] ^p/_{LT}$ to j to give the row of $S\left[^n/_{LT}, jT\right]$ corresponding to the frequency $^n/_{LT}$.
6. Repeat steps 3, 4, and 5 until all the rows of $S\left[^n/_{LT}, jT\right]$ corresponding to all discrete frequencies $^n/_{LT}$, have been defined. From (3), it is seen that the output from the S-Transform is an $L \times P$ matrix called the S-matrix whose rows pertain to frequency and columns to time. Each element of the S-matrix is complex valued. The choice of windowing function is not limited to the Gaussian function; other windowing functions were also implemented successfully.

The plots of the EEG signals from the first electrode and their corresponding S-transforms given in Fig. 2.

2.3 Artificial Neural Networks

Artificial neural networks (ANNs) are computing systems made up of large number of firmly interconnected adaptive processing elements (neurons) that are able to perform massively parallel computations for data processing and knowledge representation. Learning in ANNs is accomplished through special training algorithms developed based on learning rules presumed to mimic the learning mechanisms of

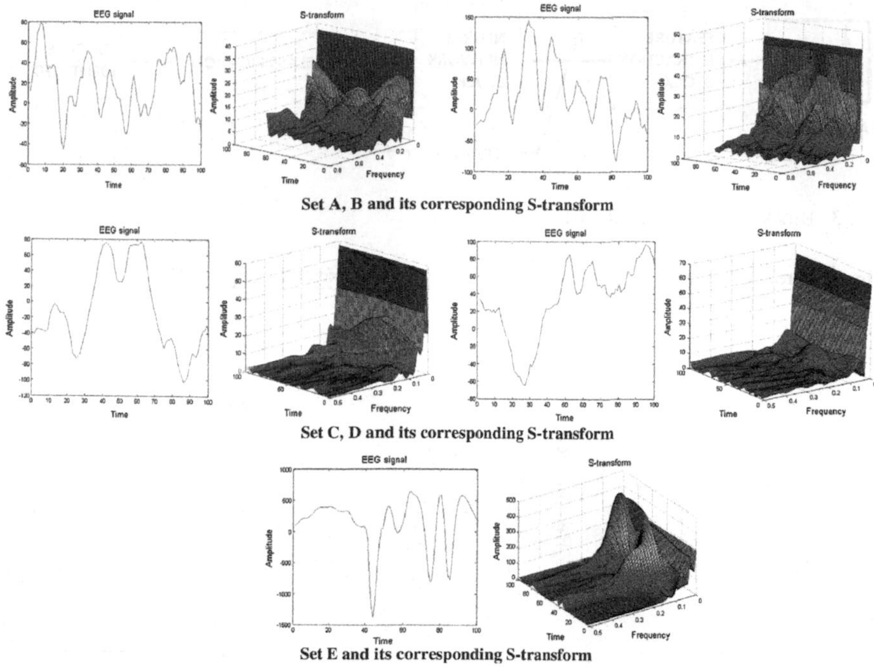

Fig. 2 EEG signals and their corresponding S-transforms

biological systems [5, 9]. ANNs can be trained to recognize patterns and the nonlinear models developed during training allow neural networks to generalize their conclusions and to make application to patterns not previously encountered.

3 Feature Extraction and Classification

The feature extraction of the above five sets of data are done by 2 different methods followed by classification.

3.1 Method-I

The features i.e. mean, standard deviation, skewness and kurtosis are directly calculated from each electrode data signal of each class of the data sets available. Then, some of the data are fed to a neural network for training it to be able to classify those types of EEG signals and the rest data samples are used for testing the accuracy of the classification process given Figs. 3, 4 (Simulation 1) and 5 (Simulation 2). Table 1 shows simulation result.

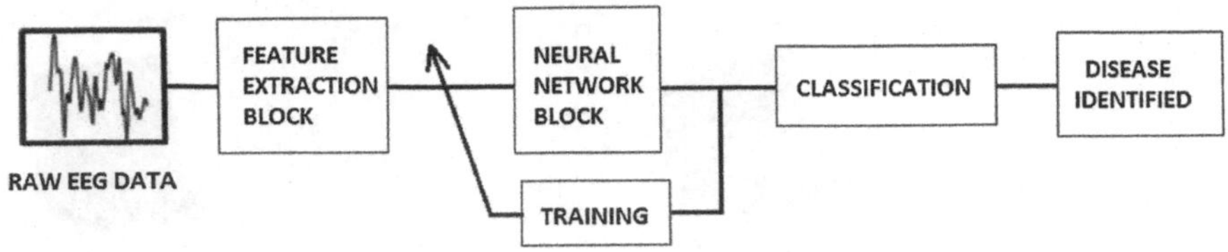

Fig. 3 Block diagram of method-1

Fig. 4 Error minimization plot [classification accuracy (in %): 32]

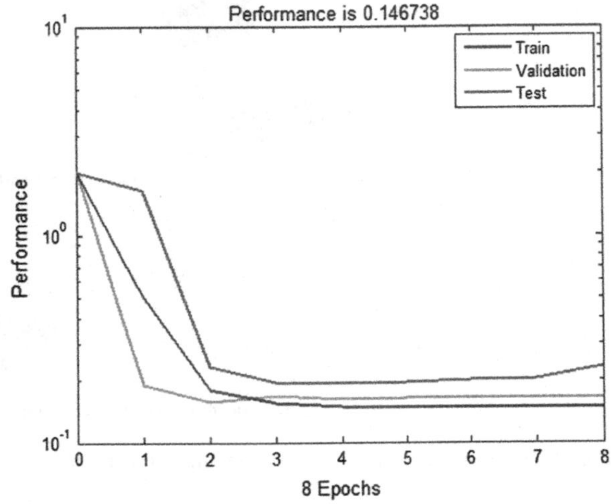

Fig. 5 Error minimization plot [classification accuracy (in %): 56]

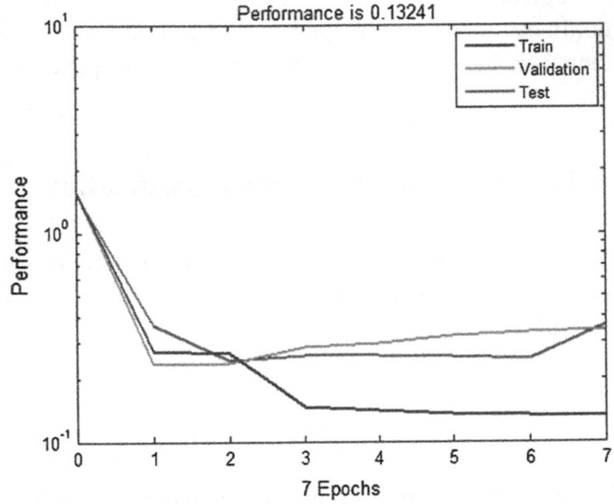

3.2 Method-II

The features i.e. mean, standard deviation, skewness and kurtosis are calculated after obtaining the S-transform [8] of each electrode data signal of each class of the

Table 1 Simulation and results

Classes used	Total number of electrodes	Number of data points used for each electrode	Total number of samples	No. of samples used for training	No. of hidden layer nodes in neural network	Training percentage	Validation percentage	Testing percentage	Performance analysis
A, B, C, D, E	100	256	500	450	20	70	5	25	0.146738
A, E	100	256	200	100	20	70	5	25	0.13241

data sets available. Then, some of the data is fed to a neural network for training so as to be able to classify those types of EEG signals and the rest data samples are used for testing the accuracy of the classification process given in Figs. 6, 7 (Simulation 1) and 8 (Simulation 2). Table 2 shows simulation result.

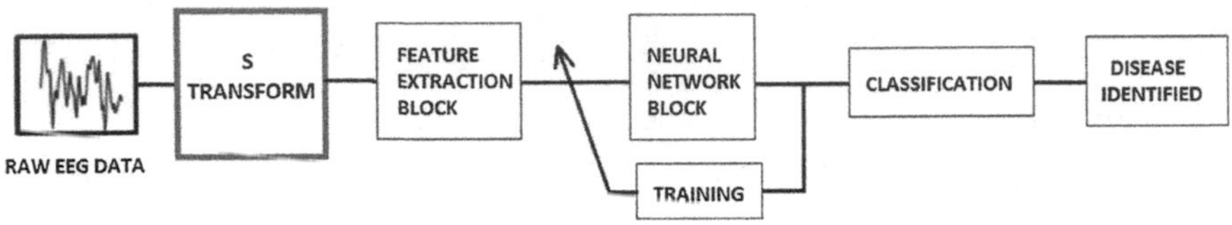

Fig. 6 Block diagram of method-2

Fig. 7 Error minimization plot [classification accuracy (in %): 60]

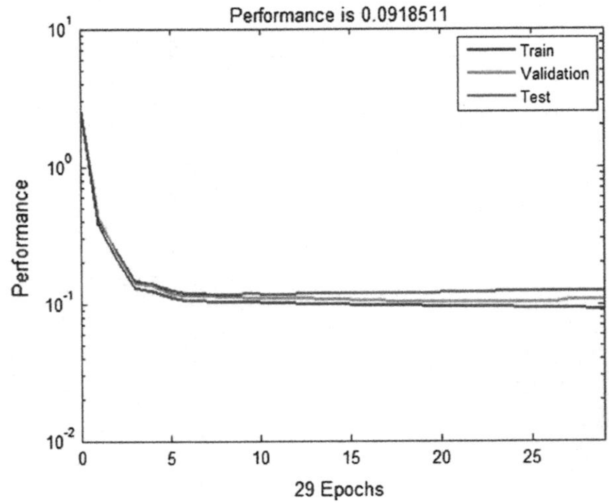

Fig. 8 Error minimization plot [classification accuracy (in %): 94]

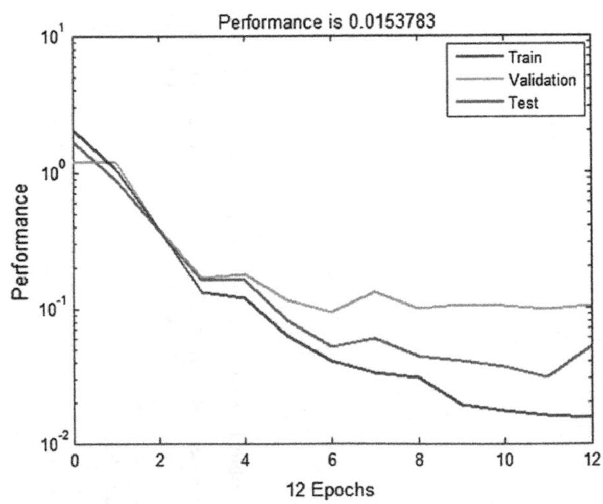

Table 2 Simulation and results

Classes used	Total number of electrodes	Number of data points used for each electrode	Total number of samples	No. of samples used for training	No. of hidden layer nodes in neural network	Training percentage	Validation percentage	Testing percentage	Performance analysis
A, B, C, D, E	100	256	500	450	20	70	5	25	0.146738
A, E	100	256	200	100	20	70	5	25	0.13241

4 Conclusion

Feature extraction is being performed using two methods: S transform and without S transform. Performance analysis is evaluated using pattern recognition toolbox. In this paper we have extracted features directly from the five data sets which give a poor percentage of accuracy. Hence, we have used S-transform before feature extraction which gives a very good classification accuracy score of 94 %.

References

1. Subasi, A.: EEG signal classification using wavelet feature extraction and a mixture of expert model. Expert Syst. Appl. **32**, 1084–1093 (2007)
2. Adeli, H., Zhou, Z., Dadmehr, N.: Analysis of EEG records in an epileptic patient using wavelet transform. J. Neurosci. Methods. **123**, 69–87 (2003)
3. Jahankhani, P., Kodogiannis, V., Revett, K.: EEG signal classification using wavelet feature extraction and neural networks. International Symposium on Modern Computing, pp. 52–57 (2006)
4. Subasi, A.: Automatic recognition of alertness level from EEG by using neural network and wavelet coefficients. Expert Syst. Appl. **28**, 701–711 (2005)
5. Subasi, A., Ercelebi, E.: Classification of EEG signals using neural network and logistic regression. Comput. Methods Programs Biomed. **78**, 87–99 (2005)
6. Petrosian, A., Prokhorov, D., Homan, R., Dashei, R., Wunsch, D.: Recurrent neural network based prediction of epileptic seizures intra and extra cranial EEG. Neurocomputing, **30**, 201–218 (2000)
7. Jian-feng, H.U.: Multifeature analysis in motor imagery EEG classification. In: Proceedings of 3rd International Symposium on Electronic Commerce and Security of IEEE, pp. 114–117 (2010)
8. Mishra, S., Bhende, C.N., Panigrahi, B.K.: Detection and classification of power quality disturbances using S-transform and probabilistic neural network. IEEE Trans. Power Deliv. **23** (1), 280–287 (2008)
9. Krusienski, D.J., McFarland, D.J., Wolpaw, J.R.: An evaluation of autoregressive spectral estimation model order for brain-computer interface applications. In: EMBS Annual International Conference of IEEE, New York, pp. 1323–1326 (2006)

Image Super Resolution Reconstruction Using Iterative Adaptive Regularization Method and Genetic Algorithm

S.S. Panda, G. Jena and S.K. Sahu

Abstract Super resolution is a technique to obtain high resolution images from several degraded low-resolution images. This has got attention in the research society because of its wide use in many fields of science and technology. Even though many methods exist for super resolution, adaptive regularization method is preferred because of its simplicity and the constraints used to get better image restoration result. In this paper first adaptive algorithm is considered to restore better edge and texture of image. Further Genetic algorithm is used to smooth the noise and better frequency addition into the image to get an optimum super resolution image.

Keywords Peak signal to noise ratio (PSNR) · Regularization · Low/high resolution (LR:HR) · Genetic algorithm (GA)

1 Introduction

Image Resolution is an important term in image processing which deals with the quality of various image acquisitions and processing devices. Resolution can be defined as the smallest measurable detail in an image. In digital image processing we have three different types of image resolution parameters, such as: spatial resolution, brightness resolution and temporal resolution. This work considered spatial resolution instead of spectral or temporal resolution for super resolution process, as it says about the spacing of pixels in an image and is measured in pixels per inch (ppi). The higher the spatial resolution of an image, greater the number of pixels in the image accordingly, smaller the size of individual pixels will be. The spatial

S.S. Panda (✉)
AMET University, Chennai, Tamil Nadu, India
e-mail: sudamshekhar@gmail.com

G. Jena · S.K. Sahu
Roland Institute of Technology, Berhampur, Odisha, India
e-mail: g_jena@rediffmail.com

L.C. Jain et al. (eds.), *Computational Intelligence in Data Mining - Volume 2*,
Smart Innovation, Systems and Technologies 32, DOI 10.1007/978-81-322-2208-8_62

resolution of a display device is often expressed in terms of dots per inch (dpi) and it refers to the size of the individual spots created by the device [1].

The goal of super-resolution is to increase the resolution of an image. Resolution is a measure of frequency content in an image: high-resolution (HR) images are band limited to a larger frequency range than low-resolution (LR) images. The resolution of digital photographs is limited by the optics of the imaging devices [2].

In conventional cameras the resolution depends on CCD sensor density, which may not be sufficiently high. Infrared and X-ray devices have their own limitations. Super-resolution is an approach that attempts to resolve this problem with software rather than hardware as hardware based are costly. The concept behind this is time-frequency resolution [3].

In the context of super-resolution for images, it is assumed that several LR images (e.g. from a video sequence) can be combined into a single HR image: we are decreasing the time resolution, and increasing the spatial frequency content. The LR images cannot all be identical, of course. Rather, there must be some variation between them, such as translational, rotation, moving away or toward the camera, or viewing angles. In theory, the information contained about the object in multiple frames, and the knowledge of transformations between the frames, can enable us to obtain a much better image of the object. In practice, there are certain limitations such as it is sometimes difficult to deduce the transformation [4].

For example, the image of a cube viewed from an angle will appear distorted or deformed in shape from the original one, because the camera is projecting a 3-D object onto a plane, and without a priori knowledge of the transformation, it is impossible to tell whether the object was actually deformed [5].

In general, however, super-resolution can be broken down into two broad parts:

1. Registration of the changes between the LR images, and
2. Restoration, or synthesis, of the LR images into a HR image;

this is a conceptual classification only, as sometimes the two steps are performed simultaneously. A huge number of papers has been published on super resolution and related topics since Tsai and Huang's [6] first work in 1984, and it would be impossible to mention every one of them. Here, we try briefly presenting some the main developments on the topic. Belge et al. [7] used the wavelet domain regularization method and edge preserving techniques to get a super-resolution image.

2 Image Observation Model

High Resolution (HR) and low-resolution images (LR) frames between the forward models can be expressed as

$$y_k = DBM_kX + n_k = H_kx + n_k, \quad 1 \le k \le p \tag{1}$$

where y_k is the k frame image degradation, X For the high-resolution images to be restored. n_k is the line for the additive noise row vector, and can be considered an additional Gaussian white noise (AWGN). Matrix D, B, M_k were under-sampled imaging system operator, fuzzy operators and k is the frame displacement operator, H_k is the Degradation of the k-frame matrix, p is the number of measured low-resolution images.

Suppose X is an image of size $L_1N_1 \times L_2N_2$ can be written as a vector sequence $X = [x_1, x_2, \ldots, x_N]$, Where $N = L_1N_1 \times L_2N_2$. If L_1 and L_2 are horizontal and vertical sampling factor, then the low-resolution images of each test y_k is of size $N_1 \times N_2$. Then the kth Low-resolution image frame sequence can be expressed as a symbol $y_k = [y_{k1}, y_{k2}, \ldots, y_{kM}]$, $k = 1, 2, \ldots p$, and $M = N_1 \times N_2$.

Here X is the value function and is used to determine the reconstruction algorithm. The super-resolution image reconstruction is to use all technical means from the observed degraded image y to recover as much as possible to reconstruct the original high-resolution images.

3 Adaptive Regularization

Panda [8] used global regularization method and the major shortcoming of this is smooth noise, at the same time smoothing the edges. Image restoration results cannot be maintained between the edges and smooth noise to achieve good balance.

In order to overcome the global regularization method deficiencies, Adaptive regularization method is used in image restoration process. Further to properly recover the details of the edge of the image, it is necessary to recover the image edges in the smooth and flat areas to suppress noise amplification.

Adaptive regularization method can be divided into the following categories; distribution method based on noise, image restoration based on a priori characteristics of the method and image information. Among them, the characteristics of image restoration method based on the first right, right down to partition the image quality to determine the edge of the image area, the texture area and flat area, then the marginal zone and the flat area before and after the partial recovery of the variance of the change in binding items to restore and build the edge and artifacts to eliminate constraints items; based on a priori image information is through the use of standard image as a priori information to improve the image quality of image restoration, such as facial image restoration method is based on a priori information on the typical image application.

Adaptive algorithm used to achieve still iterative compared to the adaptive regularization image restoration method, iterative form as (2):

$$x_{k+1} = x_k + H'y - (H'H + \alpha(x_k)C'W'WC)x_k \tag{2}$$

where W is the MN × MN the diagonal matrix whose diagonal elements are the image corresponding to the local regularization parameters, weights, weight coefficient values corresponding to the image by the local characteristics of the decision.

In this paper, the weight coefficient matrix is:

$$\mathrm{W(i,j)} = 1 - \frac{\sigma_g^2(\mathrm{i,j})/(y - Hx)_{\mathrm{i,j}}^2}{\max[\sigma_g^2(\mathrm{i,j})/(y - Hx)_{\mathrm{i,j}}^2]} \tag{3}$$

where $(y - Hx)_{\mathrm{i,j}}^2$, can prevent noise-point edge for the miscarriage of justice, avoid the amplification of strong noise. W(i, j) value is small (close to 0), can reduce smooth, edge detail to achieve enhanced; in flat areas and, W(i, j) value is large (close to 1), can effectively suppress parasitic ripple generation. And, it satisfies $0 < \mathrm{W(i, j)} < 1$.

4 Genetic Algorithm

Genetic Algorithm (GA) is a global optimization algorithm derived from evolution and natural selection. Although genetic algorithm cannot always provide optimal solution, It solves many complex problems [9]. It always produces original population randomly and generates next generation by the process of crossover and mutation. Forming a new population of N individual is also an important task of it by satisfying the conditions determined by the fitness function.

In this work genetic algorithm is used to get an optimum solution from a set of solution, the algorithm is designed is as follows:

Step 1. Determine the number of chromosomes, generation, and mutation rate and crossover rate value
Step 2. Generate number of the population, and the initialization value of the genes with a random value
Step 3. Process steps 4–7 until the number of generations is met
Step 4. Evaluation of fitness value of chromosomes by calculating objective function
Step 5. Chromosomes selection
Step 6. Crossover
Step 7. Mutation
Step 8. New Chromosomes (Offspring)
Step 9. Solution (Best Chromosomes)

The block diagram of the proposed algorithm is also shown in Fig. 1.

Fig. 1 Block diagram for the proposed algorithm

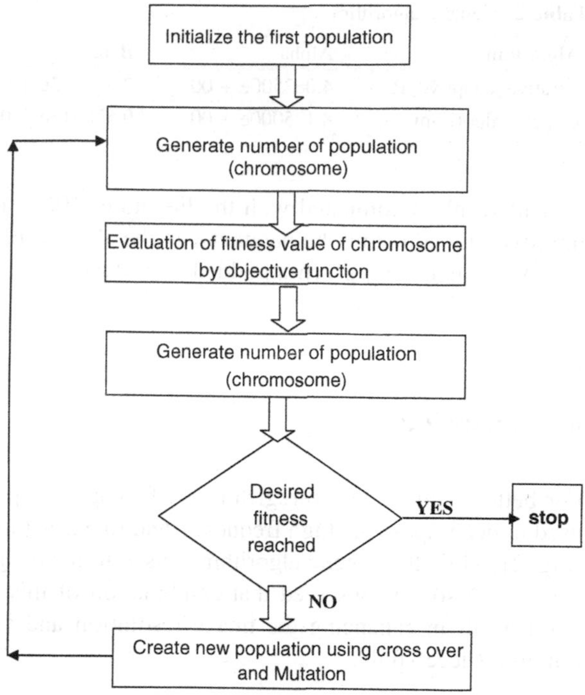

5 Experimental Results

The regularization parameter alpha and beta can be obtained by the experiment. L1 (horizontal sampling factor) and L2 (Vertical sampling factor) with the difference Dx with corresponding PSNR value is given in Table 1.

To optimize the result and to get a better restoration of the image genetic algorithm is implemented and it is given bellow. By iterative adaptive and genetic algorithm the following result of alpha and beta are obtained.

It shows that (Table 2) the values of alpha Beta and the time cost between them is significantly more in Genetic algorithm as compare to Iterative Adaptive algorithm. This is the reason we got a good optimize super resolution image.

Table 1 Iterative_adaptive_R

L1	L2	Dx	PSNR
200	204	1.3346e + 07	52.9223
424	636	2.0463e + 09	21.4815
424	636	1.9877e + 09	21.7723

Where *L1* Horizontal sampling *L2* vertical sampling factor

Table 2 Genetic algorithm

Algorithm	Alpha	Beta	Alpha–Beta time cost
iterative_adaptive_R	4.062500e + 00	7.539063e + 00	110.6429
Genetic algorithm	8.125000e + 00	9.570313e + 00	127.9496

This result is compared with the literature [10], it is observed that sharpness is improved by adjusting the parameter in the algorithm, further here implementation of genetic algorithm given an added advantage to increase the resolution with the expectation.

6 Conclusion

For better estimate of the registration of image, adaptive regularization method is used to decompose the high frequency and to extract the distribution of information (Fig. 2), while the genetic algorithm is used to have a good image restoration result (Fig. 3). Also it is observed that combination of this two algorithms show a very good result in enhancing the image resolution and taken care of high frequency removal successfully.

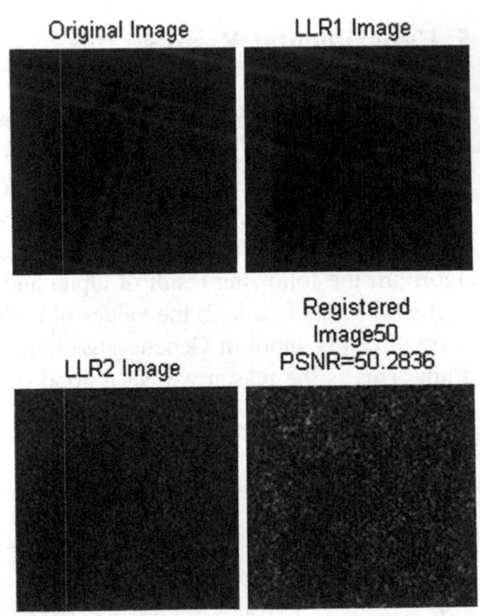

Fig. 2 Shows the images got from the adaptive method to decompose high frequency implementation

Fig. 3 Shows the optimize image of super resolution image

References

1. Bing, T., Qing, X., Xun, G., Shuai, X.: Super-resolution image reconstruction technology development status of the information engineering university **4**(4) (2003)
2. Borman, S., Stevenson, R.: Spatial Resolution Enhancement of Low-resolution Image Sequences a Comprehensive Review with Directions for Future Research [online]. http://citeseer.nj.nec.com
3. Gold, W.W.: Adaptive regularized image restoration (Ph.D. thesis). National Defense University, Washington (2006)
4. Geman, D., Yang, C.: Nonlinear image recovery with half-quadratic regularization. IEEE Trans. Image Process. **4**(7), 932–946 (1995)
5. Kang, M.G., Katsaggelos, A.K., Schafer, R.W.: A regularized iterative image restoration algorithm. IEEE Trans. Signal Process. **39**(4) (1991)
6. Tsai, R.Y., Huang, T.S.: Multiframe image restoration and registration. Adv. Comput. Vis. Image Process. Greenwich **1**(2), 317–339 (1984)
7. Belge, M., Kilmer, M.E., Miller, E.L.: Wavelet domain image restoration with adaptive edge-preserving regularization. IEEE Trans. Image Process. **9**(4), 597–608 (2000)
8. Panda, S.S. : (IJAEST) International Journal of Advance Engineering Science and Technologies, **11**(Issue No. 1), pp. 008–014
9. Yugeng, X., Tianyou, C., Weimin, Y. : Summarization of genetic algorithm. Control Theory Appl. 697–708 (1996)
10. Efrat, N., et al.: Accurate Blur Models versus image priors in single image super-resolution. In: IEEE International Conference on Computer Vision (ICCV). IEEE (2013)
11. Dai, S.S., et al.: Super-resolution reconstruction of images based on uncontrollable microscanning and genetic algorithm. Optoelectron. Lett. **10**, 313–316 (2014)
12. Ling, F., et al.: Post-processing of interpolation-based super-resolution mapping with morphological filtering and fraction refilling. Int. J. Remote Sens. **35**(13), 5251–5262 (2014)

Semi-markov Process Based Cooperation Enforcement Mechanism for MANETs

J. Sengathir and R. Manoharan

Abstract Network Survivability is defined as the potential of the network to maintain its connectivity even during the event of failures and attacks. This network survivability is highly affected by the presence of selfish nodes in the ad hoc network. In this paper, we focus on developing a Semi-Markov Process based Cooperation Enforcement Model (SMPCEM) for isolating selfish nodes and further analyze the factors that affect network survivability through transition probability matrix derived from stochastic events. Furthermore, we examine the dynamic change in behavior of mobile nodes based on parameters like residual energy and packet delivery rate. Finally, we evaluate the proposed SMPCEM approach through simulation and numerical analysis. The simulation results make it evident that SMPCEM efficiently isolates selfish nodes and upholds the network survivability by increasing the packet delivery ratio and decreasing the end to end delay, drop rate, energy consumption when compared to Correlated Node Behavior Model (CNBM).

Keywords Semi-markov decision process · Cooperation · Selfish nodes · Stochastic properties · Network survivability · Transition probability matrix

1 Introduction

In MANETs, the dynamic change in node behavior plays a key role for analyzing the performance of the network in terms of network survivability. Further, the change in behavior of mobile nodes form cooperativeness to selfishness directly affects the survivability of the network [1]. Furthermore, mobile ad hoc networks

J. Sengathir (✉) · R. Manoharan
Department of Computer Science and Engineering, Pondicherry Engineering College, Pillaichavady, Puducherry 605014, India
e-mail: j.sengathir@gmail.com

R. Manoharan
e-mail: rmanoharan@pec.edu

L.C. Jain et al. (eds.), *Computational Intelligence in Data Mining - Volume 2*,
Smart Innovation, Systems and Technologies 32, DOI 10.1007/978-81-322-2208-8_63

are highly vulnerable to random failures and selfishness due their characteristics features like dynamic network topology, limited availability of energy and error prone communication link [2]. Thus, the primary goal for a highly survivable ad hoc network is to establish and maintain a cooperative environment between the mobile nodes of the network during reliable dissemination of data. In addition, the selfish behavior of mobile nodes not only affects an individual node, but it may also affect multiple nodes participating in the routing process.

We observe that network connectivity has been considered as an important measure in assessing the survivability of the ad hoc network by a number of works available in the literature. Further, these studies clearly portrays that a node is said to be connected in the network until or unless it has a cooperative neighbor node. But, this connectivity may be affected by the existence of selfish nodes. Since, selfish nodes do not forward either data or control packets of their neighbor nodes for conserving its own energy. We also notice that adequate numbers of works are available in the literature to characterize node misbehaviors and to evaluate their impact and performance of the network [3–5]. However, only the little amount of effort has been made for analyzing and quantifying the extent of impact posed by the node misbehaviors towards the survivability of the networks. Hence the challenges imposed by the presence of selfish nodes in the network motivate us to identify their impacts of network survivability.

In this paper, we study and analyze the impacts of selfish nodes towards network connectivity through quantitative process. The contributions of the proposed work are:

(a) A Semi-Markov decision process (SMDP) is proposed to identify and isolate selfish nodes based on stochastic properties determined through possible node behaviors.
(b) The probability of a mobile node to get transited into selfish node is also examined.
(c) The upper and lower bounds of network survivability are also derived statistically for quantifying the impact of nodes' selfishness.

The rest of the paper is organized as follows. Section 2 presents some of the related works on Semi-Markov models proposed for mitigating malicious nodes. Section 3 elaborates the Semi-Markov Process based Cooperation Enforcement Model for detecting selfish nodes in MANETs. The detailed analytical and simulation investigation conducted are discussed in Sect. 4. Section 5 concludes the paper with scope of future works.

2 Related Work

A number of reputation mechanisms have been proposed for enforcing cooperation in mobile ad hoc networks. These reputation based approaches are classified into (a) History aware approach (b) Conditional probabilistic approach and (c) Markov process reputation approach. Some of the Semi-Markov based reputation mechanisms proposed for selfish node detection available in the literature are enumerated below.

Xing and Wang [6] contributed a survivability model based on k-connectivity factor for quantifying the effects of misbehaving mobile nodes towards the survivability of the network. This model incorporated a Semi-Markov process that classifies the probability of node behavior through stochastic parameters. They also formulated a loose upper bound and tight lower bound based on factors like network density and transmission range for estimating the extent of network survivability. Hernandez-Orallo et al. [7] proposed a watch dog mechanism based on Markov process for quantifying the time and cost incurred for detecting selfish nodes. Authors classified mobile into selfish and cooperative using two states viz., POSITIVE and NOINFO derived through Poisson distribution. Authors also proved that minimization of periodic diffusion of packets could greatly decrease the detection time of selfish nodes. Their proposed approach incorporated two factors viz., total overhead and detection time for identifying nodes' misbehavior.

Further, Tang and Cheng in [8] contributed a Markov decision process for malicious node detection. They utilized a reward function for detecting non-cooperating nodes. The authors also evaluated the decision process based on CUSUM test conducted with respect to average decision delay, misdetection ratio and average false positive rate. Xing in [9] proposed a mechanism for clustering mobile nodes based on stochastic properties obtained through transient and limited probability. They also derived a parameter called k-connectivity for the manipulation of coordination time required for both cooperative and malicious transition.

Furthermore, Cardenas et al. [10] proposed an analytical model called DOMINO, an adapted version of CUSUM test. They compared DOMINO with sequential probability test for detecting worst case of malicious attacks. Vallam et al. [11] contributed a discrete time Markov chain process for assessing the impact of selfish behavior of nodes manipulated based on steady state probabilities. Authors also used a non linear optimization model for classifying the mobile nodes into colluding attackers and detectors. Komathy and Narayanasamy in [12], proposed a Semi-Markov process based node behavior model for enforcing cooperation between the mobile nodes participating in active communication. In addition, this model also aids in estimating the mean cooperation level of mobile nodes present in the correlated cluster formed by similarity in node behavior. Wang and Park [13] also proposed a mechanism for modeling and analyzing multi-type failures for formulating a Markov process in order to present its occurrence in the wireless ad hoc network.

Since, Most of the Semi-Markov based reputation mechanism proposed for detecting and isolating selfish nodes considered either discrete or continuous time parameter into account. Hence, we propose a Semi-Markov Process based Cooperation Enforcement Model to handle selfishness by considering both discrete and continuous time parameter for detecting and isolating selfish nodes.

3 Markov Process Based Cooperation Enforcement Model (SMPCEM)

3.1 Network Model

In this selfish node detection model 'M' number of mobile nodes are randomly distributed over the two dimensional square area 'A' with 'r' as the uniform radius for node transmission. Hence, this wireless ad hoc environment can be modeled by a geometric random graph represented as G (M, r).

3.2 Node Classification

We extend this selfish node detection model by assuming that mobile node present in the ad hoc environment could exhibit any one of three states as proposed in [14] are listed below:

(a) Cooperative State (C)
(b) Selfish State (S)
(c) Failed State (F)

A mobile node is said to be in cooperative state when they actively participate in routing activity. In contrast, a node is said to be in selfish mode when it refuses to forward packet for their neighbors in order to conserve its own energy.

Based on the assumptions of the network model and node behavior, our proposed Semi-Markov Process based Cooperation Enforcement Model is incorporated as a distributed mechanism which makes its decision of isolating selfish nodes based on information obtained from routing packets. This SMPCEM approach detects and isolates selfish nodes through the following steps.

(a) Determination of Stochastic properties from the mobile nodes.
(b) Modeling Node Behavior based on Transition probability matrix.
(c) Computation of Futuristic Probability of selfishness.
(d) Isolation of selfish nodes based on the threshold of selfishness detection.

3.3 Determination of Stochastic Properties from the Mobile Nodes

The estimation of stochastic properties of a mobile necessitates adequate elicitation of input factors from the packets that are routed between the source and the destination. This stochastic property depicts the exhaustive set of behaviors that a mobile node could exhibit during communication. The input factors that are extracted from the routing packets are information related to nodes' residual energy, number of packets forwarded and the number of bytes of packets relayed or received. The stochastic probabilities computed by this model are categorized into four types viz.,

(i) P_{ns}—Transition probability of a cooperative node to exhibit selfish
(ii) P_{sn}—Transition probability of a selfish node to rehabilitate into normal node
(iii) $P_{ns\text{-}f}$—Transition probability of a node in cooperative or selfish state to attain failure.
(iv) P_{fn}—Transition probability of failed mobile node to rehabilitee into a cooperative node.

3.4 Modeling of Stochastic Probabilities

We consider the stochastic probability 'P_{ns}' as the likelihood measurement for a cooperative mobile node to get transited into a selfish node. The stochastic probability (P_{ns}) obtained through (1) depicts that this probability depends on residual energy level possessed by a mobile node, Energy required for transmitting single packet and the number of packets forwarded by that mobile node.

$$P_{ns} = \frac{R_p}{R_p - N_p P_{sp}} \quad (1)$$

where
R_p Residual Energy of the mobile node
P_{sp} Energy required for transmitting single packet
N_p Maximum number of packet forwarded by the mobile node.

In this scenario, a mobile node is identified as selfish when the value of 'P_{ns}' is found to be less than threshold value of selfishness as proposed in [15]. Hence, it is transparent that a mobile node with value 'P_{ns}' exhibit selfishness. In contrast, the same mobile node is said to be cooperative with probability $(1 - P_{ns})$.

Similarly, the stochastic probability 'P_{sn}' represents the likelihood of a mobile node to get transited from selfish mode to cooperative mode. This stochastic probability 'P_{sn}' is defined as the ratio of sum of packets forwarded by a mobile node for their neighbor nodes (SP_f) to the sum of packets received from their neighbor node (SP_r) derived through (2)

$$P_{sn} = \frac{SP_f}{SP_r} \quad (2)$$

In this case, a mobile node in selfish mode gets converted into cooperative nodes when the value of 'P_{sn}' reaches as formulated in [16].

At the same time, a cooperative node or selfish node may become failed when the path link loss between mobile nodes derived through (3) reaches below a threshold reduces of communication as defined in [15]

$$L_{loss} = \frac{R_p}{\mathrm{d}^2\mathrm{k}} \tag{3}$$

With 'd' is the distance between mobile nodes and k is the mobility variable.

Further, the mobile node is re-confirmed as failure based on stochastic probability 'P_{ns-f}' derived through the ratio of sum of packets dropped (SP_d) to the sum of packet received from their neighbor (SP_r) obtained through (4). When this stochastic probability reaches below a threshold of packet dissemination as proposed in [17]

$$P_{ns-f} = \frac{SP_d}{SP_r} \tag{4}$$

Finally, the transition probability for a failure node to rehabilitate back into cooperative node (P_{fn}) highly depends on the mean rehabilitation time (M_{RT}) defined by (5)

$$P_{fn} = \frac{1}{MRT} \tag{5}$$

Furthermore, it is also evident that a failed node rapidly rehabilitates into cooperative node when the mean rehabilitation time of the one network is minimum.

3.5 Modeling Node Behavior Based on Transition Probability Matrix

In this model, four modeling parameters $(P_{ns}, P_{sn}, P_{ns-f},\ and\, P_{fn})$ are computed based on P_{mn} (stochastic probabilities) derived for a mobile node at time t given by 'N_t' as defined in (6)

$$\begin{aligned} P_{ns} &= P[N_{t-1}^{(i)} = C/N_{(t)}^{(i)} = S] \\ P_{sn} &= P[N_{t-1}^{(i)} = S/N_{(t)}^{(i)} = N] \\ P_{ns-f} &= P[N_{t-1}^{(i)} = N/N_{(t)}^{(i)} = F] \\ &\text{or} \\ P_{ns-f} &= P[N_{t-1}^{(i)} = S/N_{(t)}^{(i)} = F] \end{aligned} \tag{6}$$

Based on the above defined exhaustive set of possibilities in node behavior, the semi Markov transition probability matrix is formulated with three states viz.,

Cooperative (C)	when	P_{ns} > threshold and P_{sn} < threshold
Selfish (S)	when	P_{ns} < threshold and P_{sn} < threshold
Failed (F)	when	P_{ns-f} > threshold and P_{ns} < threshold

Hence, the transition probability matrix can be represented as

$$P_{u,v} = \begin{bmatrix} 1-(P_{ns}+P_{ns-f}) & P_{ns} & P_{ns-f} \\ P_{sn} & 1-(P_{sn}+P_{ns-f}) & P_{ns-f} \\ P_{fn} & 0 & 1-P_{fn} \end{bmatrix} \quad (7)$$

where m and n are two successive states of a mobile node.

3.6 Computation of Futuristic Probability of Selfishness

The Futuristic Probability vector of selfish estimates the degree of cooperativeness, selfishness and failureness that could be exhibited by the mobile node at any time instant during lifetime. Initially, Futuristic Probability is considered as (1,0,0) by assuming that the node is initially cooperative in the routing process. Then, The Futuristic Probability vector exhibiting the likelihood of possible node behavior obtained through (8) and (9).

$$V_{fp} = V_{fp} \cdot P_{m,n} \quad (8)$$

and

$$V_{fp}(i) = \frac{v_{fp}(i)E(T_i)}{\Sigma v_{fp}E(T_i)} \quad (9)$$

where $\sum_{i=1}^{N} V_{fp_i} = 1, V_{fp} \geq 0$.

Based on determined V_{fp}, the node can be identified as cooperative or selfish or failed.

3.7 Isolation of Selfish Nodes Based on the Threshold of Selfishness Detection

Based on estimated futuristic probability vector of selfishness, $V_{fp}(P_c, P_s, P_t)$, a mobile node is isolated from the routing path. This is achieved by analyzing the three values of probability viz.,

(i) Probability of cooperation (P_c)
(ii) Probability of selfishness (P_s)
(iii) Probability of failure (P_t)

Hence, the decision on detection and isolation of selfish node is incorporated when the value of 'P_s' reaches above 0.6 as proposed in [16].

4 Experimental Results and Analysis

In this paper, SMPCEM approach is studied based on both analytical results and simulation analysis. The analytical results are evaluated based on the upper and lower bound of network survivability as proposed in [18].

The analytical results are studied based on cooperative probability and selfish probability. The decrease in the Cooperative probability affects the degree of coordination maintained between the mobile nodes during communication. The following Fig. 1a shows the analytical results obtained under varying level of cooperative probability with degree of cooperation 'k' kept as constant.

Similarly, the network survivability is also highly affected by selfish probability, Since, the connectivity gets affected when the energy level of mobile nodes reaches below 40 % of the estimated energy required for routing. Figure 1b makes it evident that the degree of selfishness is inversely proportional to degree of cooperation maintained between the mobile nodes.

Further, the simulation of the proposed SMPCEM approach is implemented using network simulator ns-v2.32. The simulated topology utilizes AODV routing with CBR traffic flows. The performance of SMPCEM is compared with CNBM [19] based on evaluation metrics like packet delivery ratio, average end-to-end delay, packet drop rate and Energy consumption. The simulation environment contains 50 mobile nodes distributed in a terrain size of 1,000 × 1,000 in a random

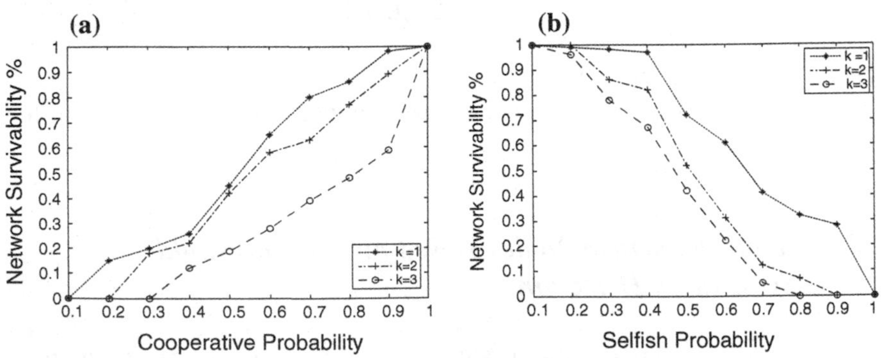

Fig. 1 Analytical results of SMPCEM based on **a** cooperative probability **b** selfish probability

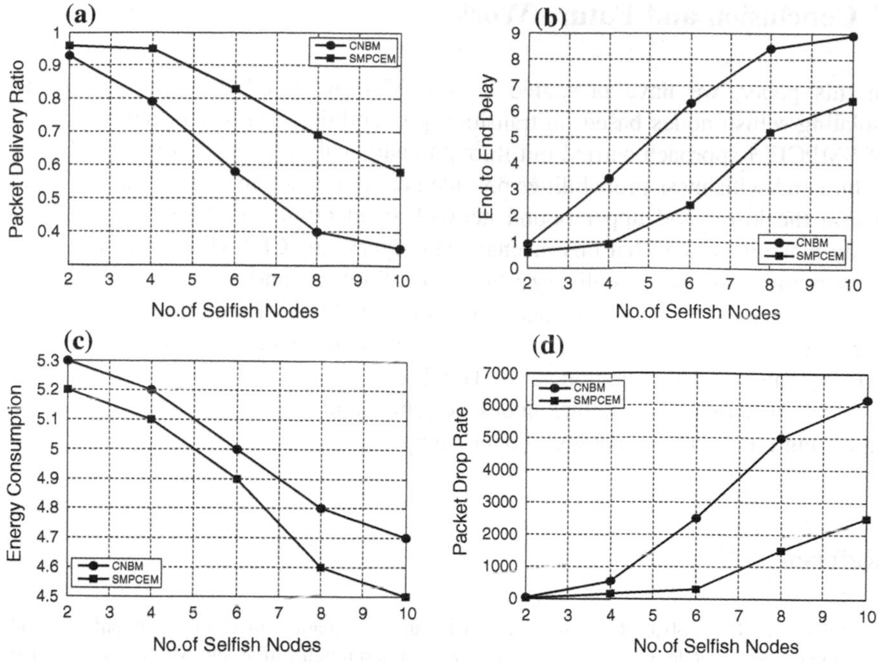

Fig. 2 Performance analysis of SMPCEM approach based on **a** packet delivery ratio **b** end to end delay **c** energy consumption **d** packet drop rate

way point model with node's initial energy set as 8.5 j and the simulation is run for 4,800 s with maximum packet size used per connection as 100,000.

The impact of the proposed SMPCEM towards selfish node isolation in terms of packet delivery ratio, end-to-end delay, energy consumption and packet drop rate are evaluated under different initial energy conditions and presented by Fig. 2a–d. When the number of selfishness nodes are high in an ad hoc environment, the average end-to-end delay decreases. Since, the amount of packet loss is comparatively high when the traffic load crosses the medium level of threshold.

The Fig. 2a shows that the packet delivery ratio of proposed Semi-Markov process model increases to 30 % when compared to the existing CNBM model Likewise, Fig. 2b demonstrates that the end-to-end delay increases when the number of selfish nodes grows gradually from 2 to 10 in increments of 2. Further, the proposed Semi-Markov process model also decreases the delay probability by 30–35 %.

Similarly, Fig. 2c presents the energy consumption rates measured when the number of selfish nodes increases in an ad hoc environment. The proposed Semi-Markov process model increases the energy conservation rate by 30–40 %. Finally, from Fig. 2d, it is obvious that the drop rates get increased when the number of attackers increases from 2 to 10 in increments of 2. The proposed SMPCEM decreases the drop probability by 50–55 %, when compared to the existing correlated node behavior model (CNBM).

5 Conclusion and Future Work

In this paper, we have presented a Semi-Markov based decision process for isolating selfish nodes based on transition probabilities. The performance analysis of SMPCEM approach carried out through analytical investigation proves that the failure and selfishness probabilities provide greater impact on network survivability within the lower and upper bound network connectivity. Further the simulation study make it evident that this mechanism outperforms CNBM in terms of energy consumption rate, packet delivery ratio, packet drop rate and end to end delay, since this model isolated selfish nodes at the rate of 24 %. Furthermore, as part of our future work, this SMPCEM model can be extensively evaluated by varying (a) The number of mobile nodes, (b) The threshold set for selfish nodes detection, (c) The threshold set of failure nodes detection and (d) the bounds proposed for analyzing the degree of network connectivity.

References

1. Michiardi, P., Molva, R.: CORE: a collaborative reputation mechanism to enforce node coperationin mobile adhoc networks. Presented at Communication and Multimedia Security. Protoroz, Solvenia (2002)
2. Buchegger, S., Boudec, J.Y.: Nodes bearing grudges: towards routing security, fairness and robustness in mobile ad-hoc network. Presented at Tenth Eurominicro Workshop on Parallel, Distributed and Network Based Processing. Canary Islands, Spain (2002)
3. Marti, S., Giuli, T.J., Lai, K., Baker, M.: Mitigating routing misbehavior in mobile ad hoc networks. Mobile Comput. Networking **1**, 255–265 (2000)
4. Hwang, S.K., Kim, D.S.: Markov model of link connectivity in mobile ad hoc networks. Telecommun. Syst. **34**(1–2), 51–58 (2006)
5. Guang, L., Chadi, M., Benslimane, A.: Enhancing IEEE 802.11 random backoff in selfish environments. IEEE Trans. Veh. Technol. **57**(3), 1806–1822 (2008)
6. Xing, F., Wang, W.: Modeling and analysis of connectivity in mobile ad hoc networks with misbehaving nodes. IEEE (2006)
7. Orallo, E., Serraty, M.D., Cano, J.C., Calafate, T., Manzoni, P.: Improving selfish node detection in MANETs using a collaborative watchdog. IEEE Lett. **16**(5), 642–645 (2012)
8. Tang, J., Cheng, Y.: Selfish misbehavior detection in 802.11 based wireless networks. In: An Adaptive Approach Based on Markov Decision Process IEEE (2013)
9. Xing, F.: Modeling, Design, and Analysis on the Resilience of Large-scale Wireless Multi-hop Networks. University of North Carolina, North Carolina (2009)
10. Cárdenas, A.A., Radosavac, S., Baras, J.S.: Evaluation of detection algorithms for MAC layer misbehavior: theory and experiments. IEEE (2009)
11. Vallam, R.D., Franklin, A.A., Murthy, C.S.: Modelling co-operative MAC layer misbehaviour in IEEE 802.11 ad hoc networks with heterogeneous loads. IEEE (2008)
12. Komathy, K., Narayanasamy, P.: A probabilistic behavioral model for selfish neighbors in a wireless ad hoc network. IJCSNS **7**(7), 77 (2007)
13. Wang, S., Park, J.T.: Modeling and analysis of multi-type failures in wireless body area networks with Semi-Markov model. Commun. Lett. IEEE. **14**(1), 6–8 (2010)
14. Patil, A.P., Rajanikanth, K., BatheySharanya, M.P., Kumar, D., Malavika, J.: Design of energy effiecient routing protocol for MANETs based on AODV. IJCSI **8**(1), 215–220 (2011)

15. Akhtar, A.K. Md., Sahoo, G: Mathematical model for the detection of selfish nodes in MANETs. Int. J. Comput. Sci. Inform. **1**(3), 25–28 (2008)
16. Sundarajan, T., Shanmugam, A.: Modeling the behavior of selfish forwarding nodes to simulate cooperation in MANET. Int. J. **2**(2), 147–160 (2010)
17. Corradi, G., Janssen, J., Manca, R.: Numerical treatment of homogenous Semi-Markov rocesses in transient case—a straightforward approach. Methodol. Comput. Appl. Probab. **6**, 233–246 (2004)
18. Sundarajan, T., Shanmugam, A.: Modeling the behavior of selfish forwarding nodes to simulate cooperation in MANET. Int. J. **2**(2), 147–160 (2010)
19. Aznil, A.H., Ahmad, R., Zul, A. Md., Samad, A., Basari, H., Hussin, B.: Correlated node behavior model based on Semi Markov process for MANETS. IEEE (2013)

Reduction Combination Determination for Efficient Microarray Data Classification with Three Stage Dimensionality Reduction Approach

Rasmita Dash and B.B. Misra

Abstract Classification of microarray data with high dimension and small sample size is a complex task. This work explores the optimal search space appropriate for classification. Here the crush of dimensionality is handled with a three stages dimension reduction technique. At the first stage, statistical measures are used to remove genes that do not contribute for classification. In the second stage, more noisy genes are removed by considering signal to noise ratio (SNR). In the third stage, principal component analysis (PCA) method is used to further reduce the dimension. Further, how much to reduce at each stage is crucial to develop an efficient classifier. Combination of different proportion of reduction at each stage is considered in this study to find appropriate combination for each dataset which maximizes the classifier performance. Help of naïve Bayes classifier is taken here to find appropriate combination of reduction.

Keywords Microarray data · Data classification · Feature selection

1 Introduction

There are different types of microarray data analysis technique such as classification, clustering, informative gene identification and many more. In classification a classifier finds a rule that allows to assign new sample to one class [1]. To design a

R. Dash (✉)
Department of Computer Science and Information Technology, Institute of Technical Education and Research, Siksha 'O' Anusandhan University, Khandagiri Square, Bhubaneswar, India
e-mail: rasmitadash@soauniversity.ac.in

B.B. Misra
Department of Computer Science and Engineering, Silicon Institute of Technology, Bhubaneswar 751024, Odisha, India
e-mail: misrabijan@gmail.com

L.C. Jain et al. (eds.), *Computational Intelligence in Data Mining - Volume 2*,
Smart Innovation, Systems and Technologies 32, DOI 10.1007/978-81-322-2208-8_64

classifier model, sufficient number of sample is required to train the model. But in microarray data the number of samples are much less (within hundred) in comparison to the number of attributes (in the range of tens of thousands) [2]. It is therefore highly essential to reduce the dimension or number of genes in the dataset in such a way that the remaining genes can contribute significantly in designing the classifier. In fact prior to developing any type of model for any category of task of microarray data analysis, dimensionality reduction is almost mandatory [3].

Several feature selection techniques are developed to address the problem of reducing genes but identifying an appropriate feature reduction technique is challenging in microarray data analysis. This is due to the presence of enormous number of genes compared to the number of samples. Gene expression data may contain thousands of genes that are highly correlated, one feature out of the several correlated feature is good enough to represent the data. The dependent variables provide no extra information about the classes and thus serve as noise for the predictor. This means that the total information content can be obtained from fewer unique features which contain maximum discrimination information about the classes. Hence by eliminating the dependent variables, the amount of data can be reduced. If the original data is experimented with the classifier, then its performance degrades drastically [2, 4]. So before classification identification of relevant or significant genes is important. Therefore an ensemble feature reduction and classification model can be used for efficient processing of microarray data.

In this paper a three stage dimensionality reduction technique for microarray data classification is used [5] with an objective to determine suitable combination of reduction at each stage of reduction that maximizes the efficiency of the classifier.

Rest of the paper is organized as follows. Three stage dimension reduction explained at Sect. 2. Simulation results and analysis is presented in Sect. 3. The paper is concluded with Sect. 4.

2 Three Stages Dimensionality Reduction Scheme Considered

In three dimension reduction technique, the number of genes in microarray data is reduced in three different stages (Fig. 1).

DRMD1: In the 1st stage on analysis of the datasets, it is observed that the values of the genes in some of the attribute are highly similar irrespective of the class labels. The dissimilarity in the feature value belonging to different class contributes to design any classifier model. Therefore less dissimilar genes can be considered less important for the task of classifier design and may be excluded. In this stage the standard deviation of the attribute is evaluated. The attributes with high dissimilarity are allowed to remain in the database and other attributes are removed. Comparing the standard deviation values of different attributes, a threshold value δ is fixed. All the genes having standard deviation value less than

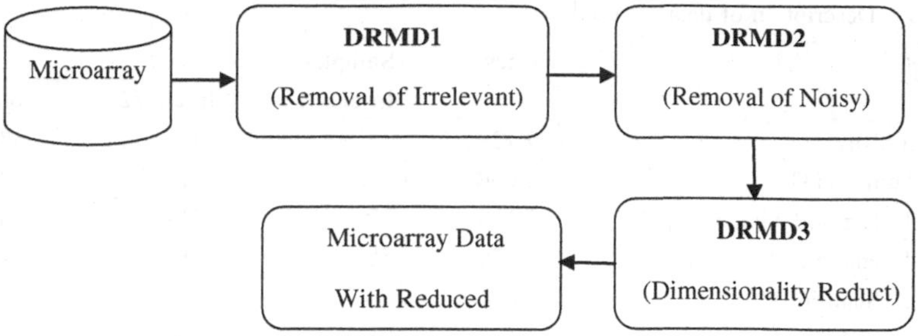

Fig. 1 Three stage feature reduction model

that of threshold value are eliminated from the microarray dataset. The remaining database is referred as dimensionally reduced microarray database1 (DRMD1).

DRMD2: After removal of irrelevant genes at stage 1, it is observed that DRMD1 contains significant number of attributes which can be considered as noise for developing a robust classifier model. So in the 2nd stage signal to noise ratio (SNR) technique is used to measure the level of desired signal to the level of background noise. A maximal difference in mean expression between two groups and minimal variation of expression within each group [6] of an expression pattern is represented with SNR score. According to this method genes using SNR test statistic first genes are ranked. The SNR is defined as follows:

$$SNR(j) = \frac{(\mu_-^j - \mu_+^j)}{(\sigma_-^j - \sigma_+^j)} \qquad (1)$$

where SNR(j) is the signal to noise ratio of jth attribute, μ_-^j and μ_+^j is the mean value of jth attribute that belongs to negative class and positive class respectively. σ_-^j and σ_+^j are the standard deviations of jth attribute for the respective classes. A predefined number of attribute with higher SNR value are selected and the reduced data is termed as dimensionally reduced microarray database2 (DRMD2).

DRMD3: Dimensionally reduced microarray database3 represents another level of feature reduction takes place at the 3rd stage. After noise reduction, principal component analysis (PCA) is used for further reduction of the dimensionality.

3 Experimental Results

To obtain the best set of features of microarray data different proportion of features are selected in each stage of the three stage dimensionality reduction. In each stage 50–90 % of reduction is considered with a step size of 10 %. To find the combination of reduction, 10 gene expression databases are considered. The dimension of these microarray databases are presented in Table 1.

Table 1 Description of datasets used

Dataset	Genes	Samples		
		In class1	In class2	Total
AllAml [10]	7,129	27	11	38
ColonTumor [11]	2,000	40	22	62
DLBCL_Harvard_Outcome [14]	7,129	32	26	58
DLBCL_Harvard_Tumor [14]	7,129	58	19	77
DLBCL_Stanford [13]	4,026	24	23	47
LungCancerMichigan [8]	7,129	86	10	96
LungCancerOntario [12]	2,880	24	15	39
ProstateTumor [14]	12,600	52	50	102
Adcalung [7]	12,533	15	134	149
CNS [9]	7,129	39	21	60

In microarray data to remove the noisy and insignificant features, standard deviation of each feature is calculated and maximum value of standard deviation is determined. Reduction of 50 % at stage one, does not mean reduction of 50 % features but if standard deviation of a feature is less than or equal to 50 % of the maximum standard deviation value then it is removed. In the second stage, signal to noise ratio is evaluated, with an objective to remove more noisy data. Here reduction of 50 % represents, reduction of 50 % of features left after stage one that contains higher noise. The features left after stage two are dimensionality reduced by PCA and specified % of features is rejected.

Figure 2 shows the confusion matrix, where, true positive is represented with TP, false negative with FN false positive with FP, true negative with TN, P is total actual positive, N is total actual negative, P' is total predicted positive, N' is total predicted negative and SM is the total number of records.

The classifier performance measures considered here for analysis of result are presented below.

Fig. 2 Confusion matrix

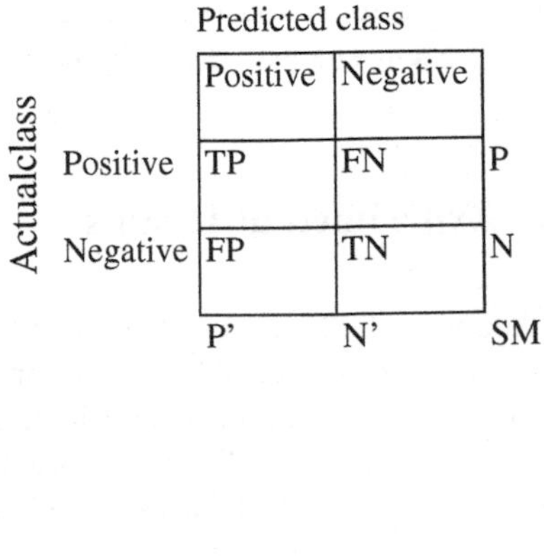

P1. Accuracy: $\frac{\mathrm{TP}+\mathrm{TN}}{\mathrm{SM}}$

P2. Sensitivity: $\frac{\mathrm{TP}}{\mathrm{TP}+\mathrm{FN}}$

P3. Specificity: $\frac{\mathrm{TN}}{\mathrm{TN}+\mathrm{FP}}$

P4. Jaccard: $\frac{\mathrm{TP}}{\mathrm{TP}+\mathrm{FP}+\mathrm{FN}}$

P5. M−estimate: $\frac{\mathrm{TP}+22.4\left(\frac{\mathrm{TP}}{\mathrm{SM}}\right)}{\mathrm{TP}+\mathrm{FP}+22.4}$

P6. Klosgen: $\left(\frac{\mathrm{TP}+\mathrm{FP}}{\mathrm{SM}}\right)^{0.43}\left(\frac{\mathrm{TP}}{\mathrm{TP}+\mathrm{FP}}-\frac{\mathrm{TP}+\mathrm{FN}}{\mathrm{SM}}\right)$

P7. F−measure: $\frac{2\mathrm{TP}}{(\mathrm{TP}+\mathrm{FN})+(\mathrm{TP}+\mathrm{FP})}$

P8. Kappa Index: $\frac{\mathrm{SM}(\mathrm{TP}+\mathrm{TN})-((\mathrm{TP}+\mathrm{FP})(\mathrm{TP}+\mathrm{FN})+(\mathrm{FN}+\mathrm{TN})(\mathrm{FP}+\mathrm{TN}))}{\mathrm{SM}^2-((\mathrm{TP}+\mathrm{FP})(\mathrm{TP}+\mathrm{FN})+(\mathrm{FN}+\mathrm{TN})(\mathrm{FP}+\mathrm{TN}))}$

This experiment is conducted, taking help of naïve Bayesian classifier. The performance measures P1–P8 are measured for all the combinations of the reduction. Here the experiment is performed for minimum reduction of 50 % and maximum reduction of 90 % with a step size of 10 % for each stage of the reduction. Starting with a reduction combination of [0.5, 0.5, 0.5] and ending with a reduction combination of [0.9, 0.9, 0.9], total 125 simulations are made for each dataset. On analysis of the results it is observed that there is a specific pattern for obtaining the best performance. The best results and the pattern of reduction leading to these best results are presented in Table 2.

In Table 2, 'Red1', 'Red2' and 'Red3' refers to the percentage of reductions considered at each stage of the three stage reduction respectively. Though maximum performance is obtained at different combinations, but here the maximum reduction at which the best result is obtained for the database is presented. In case of Adcalung database, best result of 100 % with maximum reduction is obtained for the performance measures such as accuracy, sensitivity, specificity, Jaccard, F-measure and Kappa Index. Further, performance measures M-estimate and Klosgen are dependent on the number of samples, also produced the best result possible for the sample in the database. Similar result is obtained for Allaml with max reduction at stage 1 and stage 2 but with 70 % reduction at stage 3. Here further reduction at stage 3 degrades the performance. Equivalent performance is obtained from DLBCL Stanford with 60 % reduction at stage 1, 90 % at stage 2 and 50 % reduction at stage 3.

After obtaining the best performing combination of reductions, it is considered to find the worst performing combinations of reductions. Table 3 presents the maximum reduction combination at which the lowest performance is noticed.

The above analysis reveals that, without a proper reduction strategy, the designing of a classifier model may significantly affect its performance. DBCL

Table 2 Best performing combination of reduction

Dataset	Red1	Red2	Red3	P1	P2	P3	P4	P5	P6	P7	P8
Allaml	0.9	0.9	0.7	1	1	1	1	0.868	0.249	1	1
Colon tumor	0.7	0.9	0.8	0.730	0.632	0.909	0.602	0.692	0.197	0.751	0.479
DLBCL Harvard outcome	0.8	0.6	0.7	0.863	0.887	0.834	0.785	0.715	0.248	0.878	0.723
DLBCL Harvard tumor	0.9	0.7	0.8	0.874	0.832	1	0.832	0.881	0.201	0.908	0.711
DLBCL stanford	0.6	0.9	0.5	1	1	1	1	0.763	0.366	1	1
Lung cancer Michigan	0.7	0.9	0.9	0.992	0.991	1	0.991	0.976	0.099	0.995	0.963
Lung cancer Ontario	0.6	0.7	0.9	0.753	0.85	0.6	0.679	0.657	0.133	0.809	0.463
Prostate tumor	0.9	0.9	0.5	0.660	0.903	0.408	0.576	0.579	0.091	0.730	0.314
Adcalung	0.9	0.9	0.9	1	1	1	1	0.461	0.335	1	1
CNS	0.8	0.9	0.7	0.703	0.651	0.8	0.588	0.670	0.153	0.740	0.409

Table 3 Worst performing combination of reduction

Dataset	Red1	Red2	Red3	P1	P2	P3	P4	P5	P6	P7	P8
AllAml	0.5	0.5	0.5	0.289	0.00	1.00	0.00	0.00	0.00	0.00	0.00
Colon tumor	0.8	0.6	0.9	0.659	0.547	0.863	0.509	0.630	0.158	0.674	0.355
DLBCL Harvard outcome	0.6	0.5	0.5	0.448	0	1	0	0	0	0	0
DLBCL Harvard tumor	0.5	0.5	0.5	0.246	0	1	0	0	0	0	0
DLBCL Stanford	0.9	0.5	0.8	0.948	0.9	1	0.9	0.724	0.350	0.947	0.898
Lung cancer Michigan	0.6	0.5	0.5	0.104	0	1	0	0	0	0	0
Lung cancer Ontario	0.5	0.5	0.7	0.676	0.829	0.433	0.613	0.616	0.077	0.759	0.274
Prostate tumor	0.7	0.5	0.5	0.490	0	1	0	0	0	0	0
Adcalung	0.9	0.6	0.5	0.899	0	1	0	0	0	0	0
CNS	0.7	0.5	0.5	0.35	0	1	0	0	0	0	0

Stanford, Lung cancer Ontarion and colon tumor the minimum performance obtained is not very discouraging. But in almost all other cases, zero performance is noticed for performances Sensitivity, Jacquard, M-estimate, Klosgen, F-measure and Kappa index. The designer of classifier should be very careful about such combinations which may disrupt his efforts.

4 Conclusion

This paper has experimented on the dimensionality reduction strategies and has tried to find out the search space in each problem that favors classifier designing and also finds out the search space for which the designer should remain careful. For this experimentation, a three stage reduction approach is considered. In the first stage, the less relevant genes are dropped. In the second stage, attributes with higher noise are discarded. PCA is applied in stage 3, to reduce the dimension and then a subset of attributes resulting from PCA are selected. Help of naïve Bayesian classifier is taken to find out suitable combination of reduction which favors classifier design. Different combinations of reductions are obtained by considering reduction at each stage from 50 to 90 % with a step size of 10 %. From the analysis of the results it is observed that a specific pattern exists for each data set which maximizes the performance of classifier and also there exists specific pattern which minimizes the performance.

References

1. Quackenbush, J.: Computational analysis of microarray data. Nat. Rev. Genet. **2**(6), 418–427 (2001)
2. Zhou, X., Tuck, D.P.: MSVM-RFE extensions of SVM-REF for multiclass gene selection on DNA microarray data. Bioinformatics **23**(9), 1106–1114 (2007)
3. Mutch, D.M., Berger, A., Mansourian, R., Rytz, A., Roberts, M.A.: Microarray data analysis. a practical approach for selecting differentially expressed genes. Genome Biol. 2(12) (2001)
4. Chien-Pang, L., Yungho, L.: A novel hybrid feature selection method for microarray data analysis. Appl. Soft Comput. **11**, 208–213 (2011)
5. Dash, R., Misra, B.B., Dehuri, S., Cho, S.B.: Efficient microarray data classification with three stage dimensionality reduction. International Conference on Intelligent Computing, Communication and Devices ICCD (2014)
6. Yan, J., Zhang, B., Ning, L., Shuicheng, Y., Qiansheng, C., Weiguo, F., Qiang, Y., Wensi, X., Zheng, C.: Effective and efficient dimensionality reduction for large-scale and streaming data preprocessing. IEEE Trans. Knowl. Data Eng. **18**(3), 320–333 (2006)
7. Gordon, G.J.: Translation of microarray data into clinically relevant cancer diagnostic tests using gege expression ratios in lung cancer and mesothelioma. Cancer Res. **62**, 4963–4967 (2002)
8. Beer, D.G.: Gene-expression profiles predict survival of patients with lung adenocarcinoma. Nat. Med. **8**, 816–823 (2002)

9. Pomeroy, S.L.: Prediction of central nervous system embryonal tumour outcome based on gene expression. Nature **415**, 436–442 (2002)
10. Golub, T.R., Slonim, D.K., Tamayo, P., Huard, C., Gaasenbeek, M., Mesirov, J.P., Coller, H., Loh, M.L., Downing, J.R., Caligiuri, M.A., Bloomfield, C.D., Lander, E.S.: Molecular classification of cancer: class discovery and class prediction by gene expression monitoring. Science **286**(5439), 531–537 (1999)
11. Alon, U., Barkai, N., Notterman, D.A., Gish, K., Ybarra, S., Mack, D., Levine, A.J.: Broad patterns of gene expression revealed by clustering analysis of tumor and normalcolon tissues probed by oligonucleotide arrays. Proc. Natl. Acad. Sci. USA **96**(12), 6745–6750 (1999)
12. Wigle, D.A.: Molecular profiling of non-small cell lung cancer and correlation with disease-free survival. Cancer Res. **62**, 3005–3008 (2002)
13. Alizadeh, A.A.: Distinct types of diffuse large B-cell lymphoma identified by gene expression profiling. Nature **403**, 503–511 (2000)
14. Kent Ridge Bio-medical Dataset. Retrieved 25 Aug 2013, from http://datam.i2r.a-star.edu.sg/datasets/krbd/

Author Index

L.C. Jain et al. (eds.), *Computational Intelligence in Data Mining - Volume 2*,
Smart Innovation, Systems and Technologies 32, DOI 10.1007/978-81-322-2208-8

Printed in the United States
By Bookmasters